PLEASE STAMP DATE DUE, BOTH BELOW AND ON C

| DATE DUE | DATE DUE | DATE DUE |

Aspects of the Tectonic Evolution
of China

Special Publication reviewing procedures

The Society makes every effort to ensure that the scientific and production quality of its books matches that of its journals. Since 1997, all book proposals have been refereed by specialist reviewers as well as by the Society's Books Editorial Committee. If the referees identify weaknesses in the proposal, these must be addressed before the proposal is accepted.

Once the book is accepted, the Society has a team of Book Editors (listed above) who ensure that the volume editors follow strict guidelines on refereeing and quality control. We insist that individual papers can only be accepted after satisfactory review by two independent referees. The questions on the review forms are similar to those for *Journal of the Geological Society*. The referees' forms and comments must be available to the Society's Book Editors on request.

Although many of the books result from meetings, the editors are expected to commission papers that were not presented at the meeting to ensure that the book provides a balanced coverage of the subject. Being accepted for presentation at the meeting does not guarantee inclusion in the book.

Geological Society Special Publications are included in the ISI Index of Scientific Book Contents, but they do not have an impact factor, the latter being applicable only to journals.

More information about submitting a proposal and producing a Special Publication can be found on the Society's web site: www.geolsoc.org.uk.

It is recommended that reference to all or part of this book should be made in one of the following ways:

MALPAS, J., FLETCHER, C. J. N., AITCHISON, J. C. & ALI, J. (eds) 2004. *Aspects of the Tectonic Evolution of China*. Geological Society, London, Special Publications, **226**.

HACKER, B., RATSCHBACHER, L. & LIOU, J. G. 2004. Subduction, collision and exhumation in the ultrahigh-pressure Qinling-Dabie orogen. *In*: MALPAS, J., FLETCHER, C. J. N., ALI, J. & AITCHISON, J. C. (eds) *Aspects of the Tectonic Evolution of China*. Geological Society, London, Special Publications, **226**, 157–176.

GEOLOGICAL SOCIETY SPECIAL PUBLICATION NO. 226

Aspects of the Tectonic Evolution of China

EDITED BY

J. MALPAS, C. J. N. FLETCHER, J. R. ALI & J. C. AITCHISON
The University of Hong Kong, Hong Kong

2004
Published by
The Geological Society
London

THE GEOLOGICAL SOCIETY

The Geological Society of London (GSL) was founded in 1807. It is the oldest national geological society in the world and the largest in Europe. It was incorporated under Royal Charter in 1825 and is Registered Charity 210161.

The Society is the UK national learned and professional society for geology with a worldwide Fellowship (FGS) of 9000. The Society has the power to confer Chartered status on suitably qualified Fellows, and about 2000 of the Fellowship carry the title (CGeol). Chartered Geologists may also obtain the equivalent European title, European Geologist (EurGeol). One fifth of the Society's fellowship resides outside the UK. To find out more about the Society, log on to www.geolsoc.org.uk.

The Geological Society Publishing House (Bath, UK) produces the Society's international journals and books, and acts as European distributor for selected publications of the American Association of Petroleum Geologists (AAPG), the American Geological Institute (AGI), the Indonesian Petroleum Association (IPA), the Geological Society of America (GSA), the Society for Sedimentary Geology (SEPM) and the Geologists' Association (GA). Joint marketing agreements ensure that GSL Fellows may purchase these societies' publications at a discount. The Society's online bookshop (accessible from www.geolsoc.org.uk) offers secure book purchasing with your credit or debit card.

To find out about joining the Society and benefiting from substantial discounts on publications of GSL and other societies worldwide, consult www.geolsoc.org.uk, or contact the Fellowship Department at: The Geological Society, Burlington House, Piccadilly, London W1J 0BG: Tel. +44 (0)20 7434 9944; Fax +44 (0)20 7439 8975; E-mail: enquiries@geolsoc.org.uk.

For information about the Society's meetings, consult *Events* on www.geolsoc.org.uk. To find out more about the Society's Corporate Affiliates Scheme, write to enquiries@geolsoc.org.uk.

Published by The Geological Society from:
The Geological Society Publishing House
Unit 7, Brassmill Enterprise Centre
Brassmill Lane
Bath BA1 3JN, UK

(*Orders*: Tel. +44 (0)1225 445046
          Fax +44 (0)1225 442836)
Online bookshop: http://bookshop.geolsoc.org.uk

British Library Cataloguing in Publication Data

A catalogue record for this book is available from the British Library.

ISBN 1-86239-156-4

Typeset by Techset Composition, Salisbury, UK
Printed by The Alden Press, Oxford, UK

Distributors

USA
  AAPG Bookstore
  PO Box 979
  Tulsa
  OK 74101-0979
  USA
  Orders: Tel. +1 918 584-2555
           Fax +1 918 560-2652
           E-mail bookstore@aapg.org

India
  Affiliated East–West Press PVT Ltd
  G-1/16 Ansari Road, Daryaganj,
  New Delhi 110 002
  India
  Orders: Tel. +91 11 2327-9113
           Fax +91 11 2326-0538
           E-mail affiliat@nda.vsnl.net.in

Japan
  Kanda Book Trading Company
  Cityhouse Tama 204
  Tsurumaki 1-3-10
  Tama-shi
  Tokyo 206-0034
  Japan
  Orders: Tel. +81 (0)423 57-7650
           Fax +81 (0)423 57-7651
           E-mail geokanda@ma.kcom.ne.jp

Contents

Preface

The subject of this Special Publication is one of the most interesting in global geoscience: the evolution of the tectonic collage that forms China. Many of the processes that shape the Earth's lithosphere are best exemplified by the geology of this part of Asia, but, since geology has no national boundaries, the geological evolution of China cannot be appropriately discussed without reference to key features in adjacent areas. The geology of China provides outstanding opportunities to elucidate global processes, but there are some features that are unique to the region. Many have been the focus of recent attention and have attracted international research teams because of their world-wide significance.

Since 1980 it has become clear that the geology of China is a collection of crustal terranes juxtaposed along belts of major tectonic activity representing orogenic events of both Tethyan and circum-Pacific affinity (Fig. 1). This first became abundantly obvious through the 1985 Preliminary Tectonostratigraphic Terrane Map of the Circum Pacific Region constructed by Howell *et al.* This compilation showed that the addition of fragments of Gondwana on to the southern margin of the Eurasian continent went hand-in-hand with the accretion of island-arc terranes and microcontinental blocks derived primarily from Tethys. In terms of terrane analysis, the tectonic collage is composed of fragments, including the Siberian Block; the Mongolian island-arc assemblages forming the Central Asian Orogenic Belt; the Lhasa Block; the Qiangtang Block; the Songpan–Ganzi Fold Belt; the Qiadam Block; the Tarim Block; and the North China (Sino-Korea) and South China (Yangtze and Cathaysia) blocks, all separated by suture zones that have been considerably modified by successive tectonic events. Not least of these events has been the continuing collision of the Indian subcontinent to the south and accompanying extrusion tectonics. Surrounding areas of great interest include the Indochina and Sibumasu Blocks in the south, and terranes of the Pacific Active Margin to the east.

The geology of China therefore represents a fascinating natural laboratory in which to study plate tectonic processes in general, as well as those plate interactions that have resulted in the assembly of the constituent terranes over the last 400 Ma. In addition, it is host to features that are unique in the world, including examples of rapid crustal uplift and the exhumation and the preservation of ultra-high-pressure metamorphic assemblages.

In an effort to bring together leading experts in the field, the Department of Earth Sciences at The University of Hong Kong hosted a workshop on 'Tectonic processes in the evolution of China' in April 2002. This workshop was attended by some 60 leading Earth scientists from China and overseas, and was intended to evaluate present knowledge of the geological development of China and the immediately surrounding areas, and to discuss the most recent geological discoveries made by both Chinese and international teams. The workshop was sponsored by The Croucher Foundation of Hong Kong under the Advanced Studies Institute (ASI) programme. The Croucher Foundation was set up in 1979 to promote excellence in natural science, technology and medicine through education and research in Hong Kong. An ASI is a high-level activity where a carefully defined subject, systematically presented, is treated in depth by scientists of international standing. ASIs are essentially short courses aimed at established scientists who wish to learn of and discuss recent developments in the subject area.

The assembly of the geological terranes that form China has not been straightforward. The classic concept of terranes docking with the Siberian Craton successively from north to south, has been challenged on numerous occasions. The terranes are themselves internally complex. The scientific contributions to this volume are therefore arranged not simply in terms of successive age of docking of the blocks, which remains disputed, but also in terms of the age of those processes that best define the construction of the blocks.

The Precambrian section includes contributions by **Wilde *et al.*** (1) **Zhao *et al.*** (2) and **Zhai** (3), and deals with events in the central portion of the North China Block. **Wilde *et al.*** re-interpret the evolution of the central part of the North China Craton on the basis of precise, high-resolution, single-zircon geochronology, and suggest that the Precambrian succession is a lithotectonic package assembled during a previously unrecognized Palaeoproterozoic

From: MALPAS, J., FLETCHER, C. J. N., ALI, J. R. & AITCHISON, J. C. (eds) 2004. *Aspects of the Tectonic Evolution of China*. Geological Society, London, Special Publications, **226**, 1–4.
0305-8719/04/$15 © The Geological Society of London 2004.

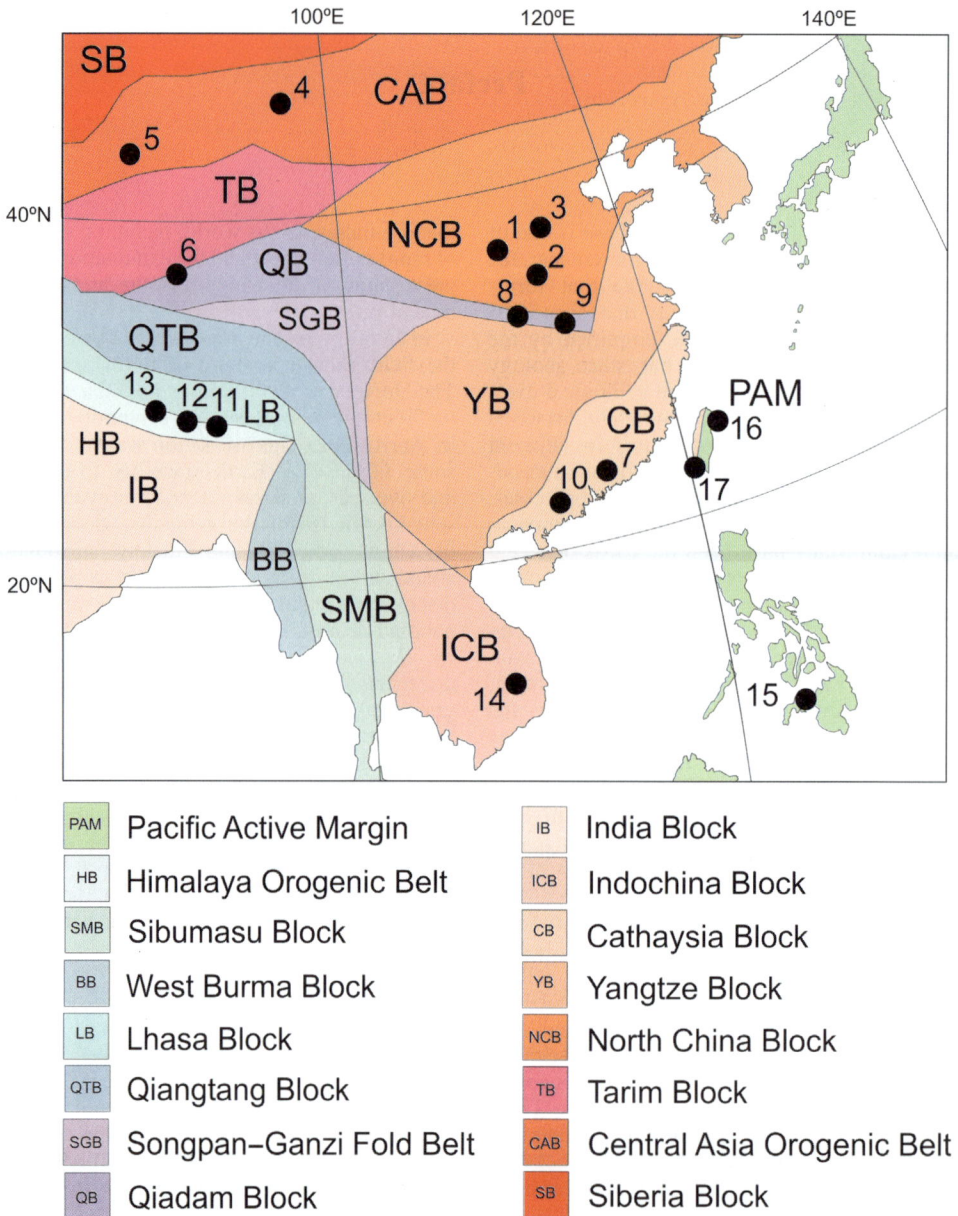

Fig. 1. Location of the study areas presented in this volume with respect, to the main tectonic elements of China and neighbouring regions. 1, Wilde *et al.*; 2, Zhao *et al.*; 3, Zhai; 4, Jahn; 5, Buckman & Aitchison; 6, Yuan *et al.*; 7, Fletcher *et al.*; 8, Hacker *et al.*; 9, Wang & Cong; 10, Li *et al.*; 11, Aitchison & Davis; 12, Davis *et al.*; 13, Robinson *et al.*; 14, Cung & Dorobek; 15, Yumul *et al.*; 16, Teng & Lin; 17, Wu *et al.*

collisional event. This concept is taken up by **Zhao *et al.*** in a study of the Hengshen–Wutai–Fuping orogenic belt, which was previously considered to result from the interaction of separate microcontinental blocks, but which they show

to be parts of a single Late Archaean/Early Palaeoproterozoic magmatic arc, subsequently incorporated into the Trans-North China orogen that separates the eastern and western parts of the North China Block. **Zhai**, in describing the tectonic evolution of the North China Block, suggests that it had a more complicated early tectonic history than the other cratons, with high-grade metamorphic events recording amalgamation at 2.5 Ga and 'cratonization' at 1.8 Ga followed swiftly by rifting, and little recognizable effect of incorporation into Rodinia.

The Palaeozoic section includes contributions on four key areas, including the Central Asian Orogenic Belt by **Jahn** (4), and **Buckman & Aitchison** (5), West Kunlun by **Yuan et al.** (6), and Cathaysia by **Fletcher et al.** (7). Jahn describes the overall accretionary tectonic style and generation of juvenile crust in the form of granites and associated volcanic rocks, characteristic of the Central Asian Orogenic Belt, concluding that crustal accretion was achieved through both lateral and vertical processes, i.e. the juxtapositioning of arc complexes and magmatic underplating. **Buckman & Aitchison** describe in some detail the amalgamation of nine separate Cambrian to Carboniferous terranes in West Jungaar as a result of the southward migration of subduction, which eventually resulted in the widespread intrusion of I-type granites carrying epithermal and porphyry gold mineralization. **Yuan et al.** have investigated the Akaz metavolcanic rocks of the West Kunlun Mountains, interpreting them as rift-related basalts formed as the Tarim Block split from the Gondwana margin in the Sinian to Cambrian, contemporaneous with rifting in East Kunlun and North Qilian further to the east. **Fletcher et al.** show that the Cathaysia Block of the South China craton is made up of a series of Late Archaean to Mid-Proterozoic crustal slices, intruded and overlain by Mesozoic plutonic and volcanic rocks. Using a variety of geophysical techniques, the discontinuities between these basement terranes, which have significantly influenced the nature and evolution of the younger Phanerozoic sequences, can be identified.

Hacker et al. (8) and **Wang & Cong** (9) describe the occurrence of ultra-high-pressure (UHP) metamorphic rocks in the Dabieshan Orogenic Belt: the Mesozoic collision zone between the North and South China blocks. The complex accretion of terranes and associated subduction tectonics along the southern margin of the North China Block can be recorded from 1.0 Ga (Grenvillian) to Late Permian–Early Triassic, at which time the northern edge of the South China Block was subducted to >150 km, creating the diamond- and coesite-bearing eclogitic rocks of Dabie-Sulu. These are rocks that have lately become the centre of considerable international interest, mainly because they provide almost unique examples resulting from processes of extremely rapid exhumation from mantle depths. **Li et al.** (10) describe aspects of the Mesozoic geology of South China, outlining three episodes of Jurassic magmatism, including alkaline basalts at 175 Ma, shoshonitic intrusions at 160 Ma and transitional basalts at 150 Ma. They consider these all to be related to lithosphere extension and thinning that eventually led to rifting and formation of the South China Sea.

Contributions on Cenozoic and Recent tectonic events range from Tibet and Indochina to the presently active areas of the Philippines and Taiwan. **Aitchison & Davis** (11), **Davis et al.** (12) and **Robinson et al.** (13), summarize the results of recent investigations of The University of Hong Kong Tibet Research Group along the Yarlung–Tsangpo suture zone in southern Tibet. These results confirm that existing models of a single India–Asia collisional event at 55 Ma are somewhat naïve, and that multiple collisional events can be recognized; involving the northern Indian margin and an intra-oceanic island arc at that time, with final India–Eurasia collision some 20 million years later. Such interpretations are afforded by the detailed examination of individual conglomerate units of clearly different age and provenance, and investigation of the various ophiolite massifs preserved in the suture zone. Amongst these, the Luobusa ophiolite displays a complex history of formation, involving activity at a mid-ocean ridge, followed by suprasubduction zone magmatism. The preservation of an interesting suite of UHP minerals in the mantle rocks suggests that the mid-ocean ridge magmatism was associated with plume upwelling from sources below the 450 km transition zone. On the basis of palaeomagnetic data, **Cung & Dorobek** (14) show that the Thai–Vietnam–Sundaland region can be divided into a series of tectonic domains that underwent different rotational histories during the Late Cretaceous and Early Tertiary, and that the Indochina Block has not acted as a rigid body, having undergone significant internal deformation during the Cenozoic. Such observations allow for refinement of the models of indentation tectonics associated with India–Eurasia collision.

The last section of this volume contains contributions on the tectonics of the Pacific Active Margin, from **Yumul et al.** (15) concerning a possible fragment of continental lithosphere in the southern Philippines, and **Teng & Lin** (16) and **Wu et al.** (17) on active orogenic events in Taiwan. **Yumul et al.** put forward the hypothesis that the Zamboanga Peninsula is part of the Palawan microcontinental block which separated from southern China during the opening of the South China Sea in Oligocene/Miocene times, and thereby represents the southernmost part of the southeast China margin. **Teng & Lin** describe the recent geological history of Taiwan, relating it to the overall evolution of the China continental margin, noting that, regardless of the presence of subduction zones, this margin has been dominated by extensional tectonics since the end of the Cretaceous. This extension finally culminated in the formation of the South China Sea and later, to the north, the Okinawa trough. **Wu *et al.*** describe recent seismic activity in Central Taiwan, and associate seismicity with focal mechanisms to delineate many of the key structures within the active orogen.

Thus, this volume contains a complete spectrum of contributions both in terms of time and space, from the Late Archaean to the present, and from Siberia and Mongolia, Tibet and Indochina to the Pacific Active Margin. Many of the papers resulted in vigorous discussion at the workshop, emphasizing the fact that there is no simple model for the tectonic evolution of this vast subcontinent. That debate will surely continue, especially as more collaborative research evolves between Chinese and international geologists. These papers present some of our most up-to-date understanding of this fascinating tectonic collage.

We would like to thank the Croucher Foundation, The University of Hong Kong and its Department of Earth Sciences for their sponsorship of the workshop. Also the many participants, including those not represented here as authors. In particular, Sun Shu of Academia Sinica, Beijing, who acted as co-director of the meeting. Grateful thanks also go to the referees of the manuscripts: A. Collins, A. Carroll, B. Collins, Cheng-Hong Lin, C. Blake, D. Jones, H. Williams, J. Evans, J. Monger, J. Milsom, K. Condie, M. de Wit, M. Pubellier, M. Allen, M. Menzies, M. Fuller, N. Arndt, N. Culshaw, P. O'Brien, P. Robinson, P. Black, P. Leloup, R. Flood, R. Coleman, R. Sewell, S. Suzuki, T. Rivers, T. Barber, U. Knittel and Zheng-Xing Li.

John Malpas
Hong Kong
November 2003

Reference

HOWELL, D. G., SCHERMER, E. R., JONES, D. L., BEN-AVRAHAM, Z. & SCHEIBNER, E. 1985. *Preliminary Tectonostratigraphic Terrane Map of the Circum-Pacific Region.* Scale 1:17 000 000. American Association of Petroleum Geologists, Tulsa, Oklahoma.

Determining Precambrian crustal evolution in China: a case-study from Wutaishan, Shanxi Province, demonstrating the application of precise SHRIMP U–Pb geochronology

SIMON A. WILDE[1], PETER A. CAWOOD[1], KAIYI WANG[2],
ALEXANDER NEMCHIN[1] & GUOCHUN ZHAO[3]

[1]*Tectonics Special Research Centre, Department of Applied Geology,
Curtin University of Technology, PO Box U1987, Perth,
Western Australia, 6845 (e-mail: s.wilde@curtin.edu.au)*
[2]*Institute of Geology and Geophysics, Chinese Academy of Sciences,
Beijing, 100029, China*
[3]*Department of Earth Sciences, The University of Hong Kong,
Pokfulam Road, Hong Kong, China*

Abstract: SHRIMP U–Pb zircon analyses from eight samples of metamorphosed intermediate to felsic volcanic rocks from the lower, middle and upper 'subgroups' of the Wutai sequence in the North China Craton define a weighted mean ^{207}Pb/^{206}Pb age of 2523 ± 3 Ma. Although individual rock ages range from 2533 ± 8 Ma to 2513 ± 8 Ma, all overlap within the error of the mean and do not support a stratigraphic interpretation for the sequence, since variations within individual previously assigned 'formations' in the sequence match the total age range. Contrary to previous interpretations, there is no correlation in age with metamorphic grade. These features highlight the need to reformulate stratigraphic schemes when defining the Precambrian geology of the North China Craton. The similarity in age between volcanic rocks of the Wutai Complex and higher-grade gneisses of the adjacent Fuping and Hengshan complexes supports the view that all three complexes represent portions of a Late Archaean arc complex that was tectonically dismembered and then re-assembled. There is no Fuping or Wutai orogeny in this, its type area: all three complexes were deformed and metamorphosed during collision of the eastern and western blocks of the North China Craton in the Lüliang orogeny *c.*1.8 Ga ago.

As elsewhere, Chinese geology grew up on the tradition of stratigraphy. However, these techniques have their limitations, as evidenced by major advances in both structural geology and tectonics in the 1960s. These techniques are only now being applied in China, and many older geological interpretations need to be revised. As examples, we can note the application of stratigraphic terms such as 'group' and 'formation' to non-sedimentary rocks – including high-grade gneisses – and the previous widely-held view that the higher the metamorphic grade or degree of structural complexity, the older the rock. Similarly, the late recognition of the nature and significance of ductile shear zones means that many stratigraphic relationships now need to be reinvestigated, in particular, key 'unconformities'. When this is coupled with a local terminology related to tectonic events such as 'movements', 'uplifts' and 'tectonic stages/cycles' – which predate the acceptance of plate-tectonic

theory – we can see some of the reasons for confusion and possible misinterpretation.

We demonstrate here, through the use of precise, high-resolution, single-zircon geochronology, that the traditionally accepted interpretation of the geology in the Wutaishan area, located approximately 200 km west-southwest of Beijing in the Central Zone (Fig. 1) of the North China Craton (Zhao *et al.* 2001*b*), requires significant modification. The study area was selected because the traditional Chinese view (Bai 1986; Tian 1991; Tian *et al.* 1996) is that the Precambrian succession here is essentially a stratigraphic sequence that can be subdivided on the basis of orogenic 'movements'; and the type areas of two of these – the Fuping and Wutai 'movements' – are located in this region (Fig. 2). Instead, our results will establish that we are dealing with a lithotectonic package that has been assembled during a major collisional event in the Palaeoproterozoic.

From: MALPAS, J., FLETCHER, C. J. N., ALI, J. R. & AITCHISON, J. C. (eds) 2004. *Aspects of the Tectonic Evolution of China*. Geological Society, London, Special Publications, **226**, 5–25.
0305-8719/04/$15 © The Geological Society of London 2004.

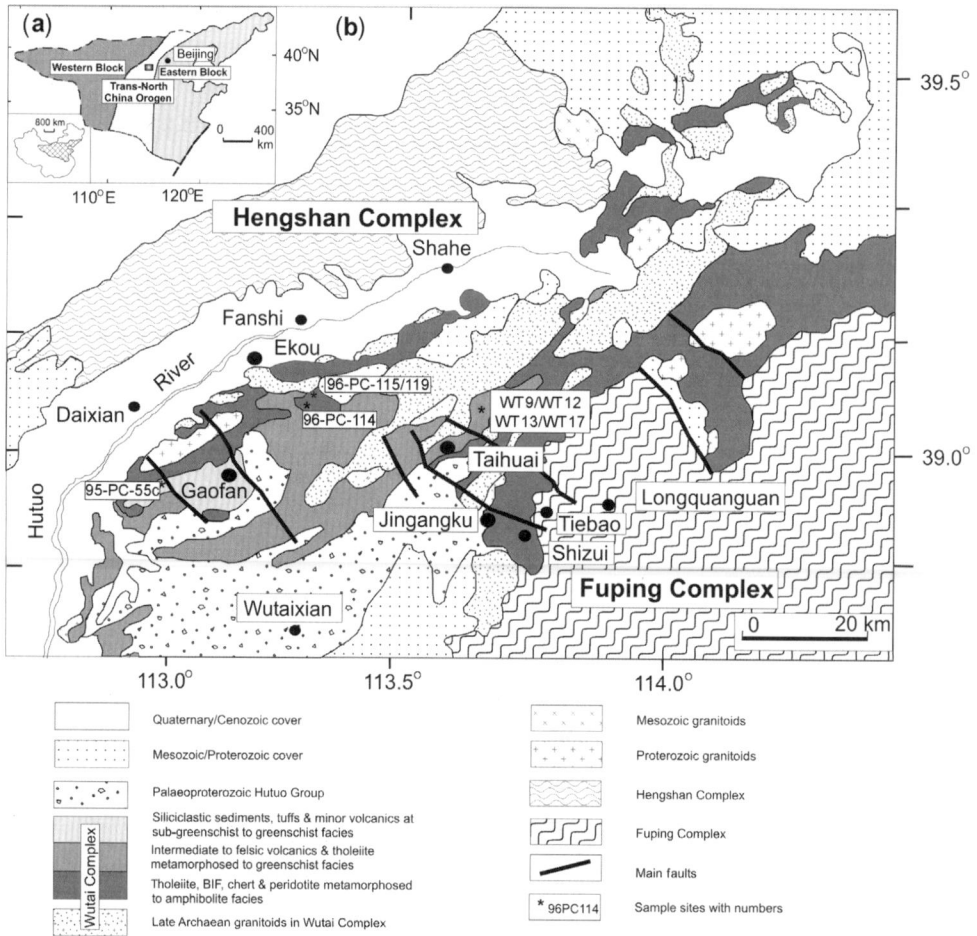

Fig. 1. (a) Map showing the three-fold subdivision of the North China Craton into the Trans-North China Orogen and Eastern and Western blocks (study area shown as a small rectangle southwest of Beijing); inset shows the location of North China Craton within China. (b) Simplified geological map of the Wutai–Hengshan–Fuping area, showing sample locations (modified from Sun *et al.* 1992 and Zhao *et al.* 2001*b*).

Previous work

The Precambrian rocks in the Wutaishan area (Fig. 2) are considered in the Chinese literature to represent three main 'cycles' (see Bai 1986; Tian *et al.* 1996), characterized by discrete events of sedimentation, granitoid intrusion, metamorphism and deformation, and separated by two major orogenic episodes or 'movements'. The older Fuping 'cycle' extended from 2.9 to 2.5 Ga, and consisted of volcanogenic sediments (Hengshan and Fuping 'groups'), which were metamorphosed to upper amphibolite–granulite facies prior to intrusion of extensive bodies of tonalite–trondhjemite–granodiorite (TTG) and mafic dykes. The entire package was then deformed and metamorphosed

to upper amphibolite–granulite facies at *c.*2.5 Ga and underwent local partial melting. This event is referred to as the Fuping 'movement' and considered to represent the culmination of a tectonic 'stage'. The succeeding Wutai 'cycle' of volcanic rocks and volcanogenic sediments was deposited on the deformed Fuping 'Group' between 2.5 and 2.4 Ga, and variably metamorphosed between 2.3 and 2.2 Ga, with the lower rocks attaining amphibolite facies and the upper rocks only sub-greenschist facies metamorphism; this likewise is considered to be the termination of a tectonic 'stage', and the event is referred to as the Wutai 'movement'. Rocks of the Wutai 'Group' have been considered to

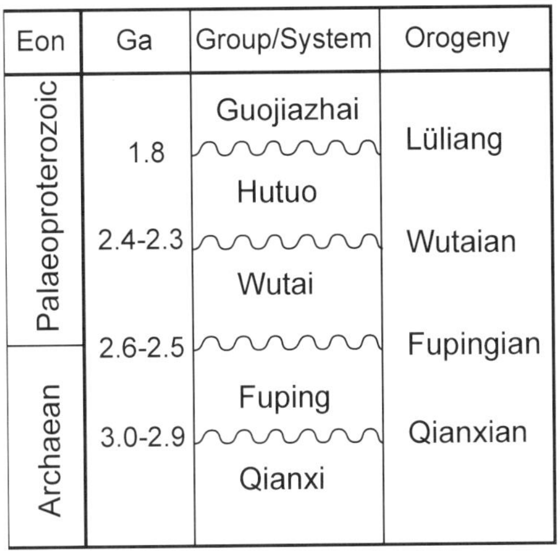

Eon	Ga	Group/System	Orogeny
Palaeoproterozoic	1.8	Guojiazhai 〰〰〰 Hutuo	Lüliang
Palaeoproterozoic	2.4-2.3	Hutuo 〰〰〰 Wutai	Wutaian
Archaean	2.6-2.5	Wutai 〰〰〰 Fuping	Fupingian
Archaean	3.0-2.9	Fuping 〰〰〰 Qianxi	Qianxian

Fig. 2. Major subdivisions of the Precambrian of China, using the traditional breakdown into 'groups' ('cycles') and orogenies (or 'movements') (based on Ma & Bai 1998).

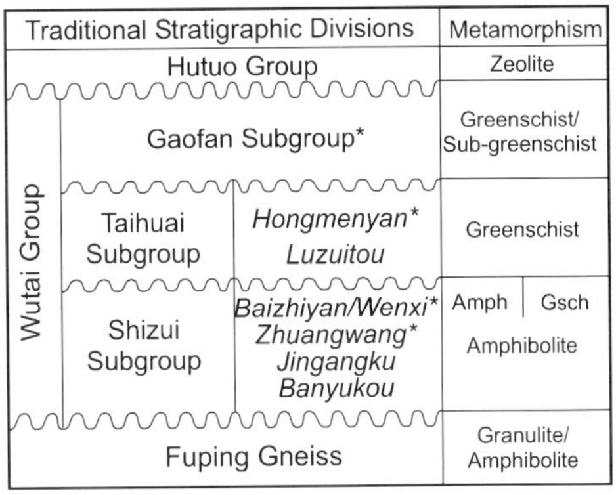

Traditional Stratigraphic Divisions		Metamorphism
Hutuo Group		Zeolite
Wutai Group — Gaofan Subgroup*		Greenschist/ Sub-greenschist
Taihuai Subgroup	Hongmenyan* Luzuitou	Greenschist
Shizui Subgroup	Baizhiyan/Wenxi* Zhuangwang* Jingangku Banyukou	Amph | Gsch — Amphibolite
Fuping Gneiss		Granulite/ Amphibolite

Fig. 3. Traditional subdivision of the Wutai 'Group' of Shanxi province, based on Tian (1991) and Ma & Bai (1998). The asterisks refer to 'formations' dated in this study.

have typical Archaean greenstone belt affinities (Bai 1986). Finally, the Hutuo 'Group', a sequence of shallow-water carbonates and siliciclastic sedimentary rocks, was deposited over the deformed and metamorphosed Wutai and Fuping 'groups', between 2.3 and 2.0 Ga, and metamorphosed at 2.0 to 1.8 Ga. This constitutes the Lüliang 'cycle', and the peak metamorphic event was considered to be between 1.85 and 1.7 Ga, and is referred to as the Lüliang 'movement'.

This sequence of events is considered to be typical for China and is applied to areas not just within the North China Craton, but to the South China and Tarim cratons as well (see Ma & Bai 1998).

The rocks at Wutaishan have traditionally been considered to form a complete stratigraphic succession – known as the Wutai 'Group' – composed of a lower sequence of amphibolite, banded iron formation, paragneiss and minor carbonate, that was metamorphosed to amphibolite facies. These are overlain by middle and upper sequences composed of clastic sedimentary rocks and intermediate to felsic volcanic rocks, that have been metamorphosed to greenschist or sub-greenschist facies. Thus the older rocks are considered to be of higher metamorphic grade. The rocks have been further subdivided into a number of 'formations' (Fig. 3), locally separated by 'unconformities' (Tian et al. 1996), and represent one of the classical stratigraphic sequences in the Chinese Precambrian (Bai 1986). Although a modern plate-tectonic interpretation has recently been applied to the rock association

(Ma & Bai 1998), it is still considered to maintain a coherent stratigraphy.

Based on preliminary results, we have shown that this interpretation is inadequate on several grounds, including:

(1) the apparent similarity in age of volcanic rocks throughout the Wutai sequence (Wilde & Wang 2000);
(2) many of the so-called 'unconformities' being tectonic contacts (Cawood et al. 1998); and,
(3) the provenance age of zircons within the metasedimentary units indicate that the basal sequence of the Wutai Complex is younger than the overlying volcanic rocks (Cawood et al. 1998).

In view of these findings, we refer here to the rocks as belonging to the Wutai Complex (Wilde et al. 1997). All previous stratigraphic terms will be marked within parentheses.

Prior to commencement of our work in the area in 1995, the traditional view, as outlined above (Figs 2 and 3), was being questioned, and controversy surrounded the nature and age of the rocks at Wutaishan. They were variously considered to be Archaean in age (Wang & Bai 1986), or else marked the base of the Proterozoic in China (Yang et al. 1986). Liu et al. (1985) provided a broad geochronological framework for the area, using conventional multigrain U–Pb zircon techniques. They dated what was referred to as a low-grade keratophyre lava in the central part of the Wutai sequence (Hongmenyan Formation of Li S. et al. 1990; see Table 1) just north of Taihuai (Fig. 1), obtaining an age of 2520 ± 17 Ma. However, Wang & Bai (1986) considered these rocks to be part of a granitic intrusion, implying that it may be younger than the volcanic succession. Subsequent Rb–Sr, Pb/Pb and Sm–Nd isotopic analyses yielded significantly younger ages for this sequence. A combined Sm–Nd isochron age of *c.*2250 Ma from samples of the so-called spilite–keratophyre sequence near Taihuai was interpreted as an 'errorchron', resulting from mobilization of REE during low-grade metamorphism (Li S. *et al.* 1990). Separate isochrons from these spilite and quartz–keratophyre samples yielded ages of around 1980 Ma, within the error of the Rb–Sr isochron age (Li S. *et al.* 1990). A later Rb–Sr, Pb/Pb and Sm–Nd study by Sun *et al.* (1992) on mafic rocks from Wutaishan yielded similar dates. These post-Archaean ages have tended to be interpreted as primary ages (see Tian *et al.* 1996), rather than the more robust U–Pb zircon dates of Liu *et al.* (1985).

A SHRIMP ^{207}Pb/^{206}Pb date of 2524 ± 8 Ma for a rhyodacitic lava from the Hongmenyan 'Formation' within the 'middle' Wutai clearly confirms its Archaean age (Wilde *et al.* 1997).

The minimum age of the Wutai Complex is constrained by the age of the overlying Hutuo 'Group' and that of the intrusive, essentially post-tectonic Dawaliang Granite. Conventional multigrain U–Pb zircon analyses from a Hutuo 'Group' metabasalt gave a date of 2366 +103/−94 Ma (Wu *et al.* 1986), which was considered to be the age of igneous crystallization. Sun *et al.* (1992) later argued that it could be a metamorphic age. The Dawaliang Granite has a SHRIMP ^{207}Pb/^{206}Pb date of 2176 ± 12 Ma (Wilde *et al.* 1997), interpreted as the igneous crystallization age.

In this paper, we will present new data on the ages of felsic volcanic rocks within the regional 'stratigraphy' (confirmed by Yongqing Tian, pers. comm.). The results are discussed in relation to recent evidence on the nature of adjacent terrains and the implications that these have for the Late Archaean to Palaeoproterozoic evolution of the North China Craton.

Timing of felsic volcanism within the Wutai Complex

We collected a suite of eight metamorphosed intermediate to felsic volcanic rocks from four units previously recognized as 'formations', including samples from what were considered to be the lower, middle and upper 'subgroups' (Fig. 3). These are described below in their previously assumed stratigraphic order, using the traditional formational names (e.g. Bai 1986; Tian 1991). The Hongmenyan 'Formation' is considered to have the best-developed sequence of intermediate to felsic volcanic rocks, and we selected four samples from this unit for analysis from the well-exposed roadside section on Wutai Mountain, approximately 15 km by road north-northeast of Taihuai (Fig. 1). All coordinates quoted in this paper were collected with a hand-held Trimble or Garmin GPS.

Sample descriptions

All rocks in the Wutai Complex are strongly deformed and have undergone either greenschist- or amphibolite-facies metamorphism. The samples selected for analysis are fine-grained, meso- to leucocratic schistose rocks, composed of variable amounts of quartz, feldspar, biotite, muscovite and hornblende, with minor epidote, chlorite, magnetite, pyrite and calcite. Geochemical data for these rocks (Wilde, unpublished data; Wang *et al.* in press) and the local preservation of igneous textures establish their volcanic parentage and, in the following sections, we use their volcanic names for brevity.

Zhuangwang 'Formation'. Sample *96-PC-114* was obtained from an exposure along the side road leading to the Ekou Iron Mine, *c.*20 m on the uphill side of a short tunnel (Fig. 1). The rock is a fine-grained mesocratic andesite. It has a well-developed metamorphic fabric defined by strongly aligned biotite laths, with local development of poikiloblastic blue-green amphibole. The remainder is composed of an aggregate of weakly aligned quartz and untwinned feldspar, with granular epidote and aligned laths of chlorite. Pyrite is a common accessory mineral. There are a few calcite veins, and the whole rock has a crenulated fabric. It has been metamorphosed to the epidote amphibolite facies.

The zircons are colourless to pale pink and mostly euhedral with well-formed prismatic and pyramidal faces. The length to width ratio ranges from 2:1 to 3.5:1. Some crystals contain rounded titanite or elongate apatite inclusions. A number of grains are fractured and metamict.

Sample 96-PC-119 was obtained from a roadside exposure on the Ekou–Yantou road, *c.*100 m north of the turn-off to the Ekou Iron Mine. The rock is a buff-grey mesocratic andesite with a micaceous foliation that is interbanded with garnet–mica schist. It consists of a granoblastic aggregate of quartz and untwinned feldspar (mostly plagioclase), interspersed with strongly aligned biotite and muscovite. The mica laths locally show tight to isoclinal folding, with

Table 1. *SHRIMP U–Pb–Th zircon data for Wutai Volcanic samples 96-PC-114, 96-PC-119 and 96-PC-115*

Spot	U (ppm)	Th (ppm)	Th/U	Pb (ppm)	$^{204}Pb/^{206}Pb$	$\% f^{206a}$	$^{208}Pb^*/^{206}Pb^*$	$^{208}Pb^*/^{232}Th$	$^{206}Pb^*/^{238}U$	$^{207}Pb^*/^{235}U$	$^{207}Pb^*/^{206}Pb^*$	% conc.	Age $^{207}Pb^*/^{206}Pb^*$
Sample 96-PC-114													
pc114-1	46	24	0.51	25	0.00069	1.108	0.1408 ± 69	0.1303 ± 75	0.4721 ± 129	10.86 ± 39	0.16691 ± 345	99	2527 ± 35
pc114-2	113	81	0.72	61	0.00020	0.323	0.2018 ± 34	0.1271 ± 40	0.4545 ± 114	10.45 ± 30	0.16674 ± 177	96	2525 ± 18
pc114-3	73	69	0.95	43	0.00075	1.205	0.2540 ± 62	0.1248 ± 46	0.4652 ± 121	10.90 ± 36	0.16989 ± 296	96	2557 ± 29
pc114-4	52	31	0.59	28	0.00063	1.011	0.1701 ± 67	0.1309 ± 64	0.4518 ± 122	10.53 ± 37	0.16907 ± 330	94	2548 ± 33
pc114-5	34	42	1.23	23	0.00107	1.707	0.3377 ± 98	0.1349 ± 58	0.4930 ± 142	12.32 ± 49	0.18126 ± 448	97	2664 ± 41
pc114-6	64	43	0.66	37	0.00082	1.318	0.1815 ± 60	0.1292 ± 56	0.4727 ± 125	10.96 ± 37	0.16817 ± 297	98	2539 ± 30
pc114-7	50	43	0.87	28	0.00059	0.945	0.2447 ± 69	0.1288 ± 52	0.4578 ± 125	10.25 ± 37	0.16238 ± 326	98	2481 ± 34
pc114-8	186	163	0.88	90	0.00028	0.455	0.2656 ± 27	0.1159 ± 31	0.3828 ± 94	10.07 ± 27	0.19084 ± 143	76	2749 ± 12
pc114-9	136	130	0.96	66	0.00186	2.978	0.2771 ± 67	0.1037 ± 37	0.3581 ± 90	8.40 ± 27	0.17015 ± 305	77	2559 ± 30
pc114-10	62	31	0.49	35	0.00067	1.065	0.1331 ± 52	0.1311 ± 62	0.4874 ± 127	11.05 ± 36	0.16449 ± 263	102	2502 ± 27
pc114-11	118	76	0.65	67	0.00033	0.522	0.1792 ± 30	0.1338 ± 41	0.4833 ± 120	11.16 ± 31	0.16749 ± 158	100	2533 ± 16
pc114-12	157	127	0.81	90	0.00026	0.413	0.2229 ± 24	0.1303 ± 36	0.4742 ± 116	11.11 ± 29	0.16986 ± 128	98	2556 ± 13
pc114-13	79	35	0.44	44	0.00053	0.842	0.1241 ± 41	0.1356 ± 58	0.4841 ± 123	11.17 ± 33	0.16728 ± 216	101	2531 ± 22
pc114-14	95	66	0.70	55	0.00035	0.557	0.1969 ± 34	0.1356 ± 43	0.4809 ± 121	11.10 ± 32	0.16734 ± 178	100	2531 ± 18
pc114-15	76	44	0.57	43	0.00032	0.511	0.1586 ± 39	0.1339 ± 49	0.4848 ± 124	11.23 ± 33	0.16799 ± 204	100	2538 ± 20
pc114-16	67	52	0.78	40	0.00060	0.957	0.2126 ± 45	0.1348 ± 46	0.4951 ± 127	11.42 ± 35	0.16735 ± 224	102	2531 ± 22
pc114-17	41	22	0.55	24	0.00113	1.810	0.1422 ± 75	0.1268 ± 76	0.4874 ± 132	11.02 ± 41	0.16399 ± 364	102	2497 ± 37
pc114-18	146	148	1.02	87	0.00024	0.389	0.2820 ± 26	0.1308 ± 35	0.4727 ± 116	10.80 ± 29	0.16576 ± 130	99	2515 ± 13
pc114-19	47	29	0.61	27	0.00049	0.778	0.1740 ± 61	0.1394 ± 64	0.4912 ± 132	11.37 ± 39	0.16789 ± 306	102	2537 ± 31
pc114-20	75	78	1.04	46	0.00041	0.650	0.2898 ± 44	0.1343 ± 41	0.4822 ± 123	11.10 ± 33	0.16689 ± 211	100	2527 ± 21
pc114-21	86	89	1.04	53	0.00045	0.724	0.2896 ± 48	0.1351 ± 42	0.4835 ± 123	11.28 ± 34	0.16919 ± 225	101	2550 ± 22
pc114-22	63	52	0.83	38	0.00044	0.706	0.2318 ± 45	0.1357 ± 46	0.4877 ± 126	11.23 ± 34	0.16694 ± 226	101	2527 ± 23
pc114-23	92	27	0.30	52	0.00055	0.887	0.0770 ± 36	0.1316 ± 71	0.5068 ± 128	12.80 ± 37	0.18322 ± 202	99	2682 ± 18
pc114-24	88	62	0.70	50	0.00071	1.142	0.1876 ± 47	0.1261 ± 46	0.4708 ± 120	10.48 ± 32	0.16149 ± 229	101	2471 ± 24
pc114-25	35	19	0.53	21	0.00113	1.811	0.1440 ± 87	0.1326 ± 90	0.4907 ± 140	11.00 ± 45	0.16265 ± 420	104	2483 ± 44
Sample 96-PC-119													
96pc119-1	240	113	0.47	108	0.00011	0.182	0.1010 ± 10	0.0878 ± 11	0.4077 ± 23	9.36 ± 7	0.16657 ± 66	87	2523 ± 7
96pc119-2	238	166	0.70	127	0.00011	0.177	0.1980 ± 12	0.1283 ± 12	0.4521 ± 26	10.35 ± 7	0.16607 ± 64	95	2518 ± 7
96pc119-3	258	401	1.56	168	0.00008	0.125	0.4340 ± 15	0.1312 ± 10	0.4704 ± 26	10.78 ± 7	0.16618 ± 59	99	2519 ± 6
96pc119-4	47	16	0.34	26	0.00027	0.431	0.0924 ± 34	0.1339 ± 52	0.4938 ± 55	11.28 ± 19	0.16571 ± 181	103	2515 ± 18
96pc119-5	205	199	0.97	119	0.00004	0.062	0.2693 ± 13	0.1295 ± 11	0.4652 ± 28	10.77 ± 8	0.16798 ± 63	97	2538 ± 6
96pc119-6	331	164	0.50	162	0.00007	0.104	0.1400 ± 9	0.1233 ± 11	0.4362 ± 23	9.91 ± 6	0.16473 ± 53	93	2505 ± 5
96pc119-7	683	289	0.42	275	0.00005	0.074	0.1143 ± 6	0.0994 ± 7	0.3673 ± 17	7.85 ± 4	0.15503 ± 37	84	2402 ± 4
96pc119-8	225	159	0.71	115	0.00005	0.074	0.1961 ± 11	0.1205 ± 11	0.4343 ± 24	9.73 ± 7	0.16243 ± 60	94	2481 ± 6
96pc119-9	336	172	0.51	131	0.00013	0.207	0.1480 ± 11	0.0994 ± 10	0.3444 ± 18	7.42 ± 5	0.15617 ± 61	79	2415 ± 7
96pc119-10	581	308	0.53	252	0.00003	0.047	0.1419 ± 6	0.1035 ± 7	0.3871 ± 18	8.53 ± 5	0.15989 ± 39	86	2454 ± 4
96pc119-11	175	120	0.69	98	0.00005	0.085	0.1926 ± 13	0.1334 ± 13	0.4755 ± 30	11.02 ± 9	0.16804 ± 70	99	2538 ± 7
96pc119-12	775	889	1.15	315	0.00047	0.753	0.3092 ± 12	0.0858 ± 5	0.3184 ± 14	6.20 ± 4	0.14135 ± 52	79	2244 ± 6
96pc119-13	272	123	0.45	107	0.00011	0.180	0.1730 ± 13	0.1292 ± 13	0.3378 ± 19	7.74 ± 6	0.16617 ± 74	74	2519 ± 7
96pc119-14	410	173	0.42	202	0.00006	0.093	0.1173 ± 7	0.1234 ± 11	0.4451 ± 22	10.16 ± 6	0.16549 ± 49	94	2513 ± 5

(continued)

Table 1. *Continued*

Spot	U (ppm)	Th (ppm)	Th/U	Pb (ppm)	$^{204}Pb/^{206}Pb$	$\%f^{206}a$	$^{208}Pb/^{206}Pb*$	$^{208}Pb*/^{232}Th$	$^{206}Pb*/^{238}U$	$^{207}Pb*/^{235}U$	$^{207}Pb*/^{206}Pb*$	% conc.	Age $^{207}Pb*/^{206}Pb*$
96pc119-15	481	248	0.52	217	0.00006	0.090	0.1403 ± 8	0.1091 ± 9	0.4009 ± 20	9.07 ± 5	0.16407 ± 48	87	2498 ± 5
96pc119-16	1712	787	0.46	331	0.00040	0.637	0.1374 ± 10	0.0529 ± 5	0.1767 ± 7	2.60 ± 2	0.10684 ± 49	60	1746 ± 8
96pc119-17	202	99	0.49	89	0.00013	0.210	0.1567 ± 14	0.1228 ± 14	0.3851 ± 23	8.56 ± 7	0.16116 ± 76	85	2468 ± 8
96pc119-18	159	109	0.68	78	0.00026	0.413	0.2147 ± 19	0.1280 ± 15	0.4081 ± 27	9.26 ± 9	0.16447 ± 95	88	2502 ± 10
96pc119-19	171	174	1.02	99	0.00016	0.260	0.2899 ± 18	0.1311 ± 13	0.4606 ± 29	10.53 ± 9	0.16585 ± 81	97	2516 ± 8
96pc119-20	250	193	0.77	137	0.00006	0.099	0.2094 ± 11	0.1255 ± 11	0.4617 ± 25	10.59 ± 7	0.16635 ± 60	97	2521 ± 6
96pc119-21	275	154	0.56	138	0.00004	0.070	0.1555 ± 9	0.1227 ± 11	0.4410 ± 24	10.16 ± 7	0.16713 ± 57	93	2529 ± 6
96pc119-22	1710	3177	1.86	433	0.00013	0.204	0.4669 ± 11	0.0462 ± 2	0.1837 ± 9	3.17 ± 2	0.12504 ± 37	54	2029 ± 5
96pc119-23	433	511	1.18	230	0.00005	0.087	0.3130 ± 10	0.1107 ± 7	0.4172 ± 21	9.28 ± 6	0.16138 ± 44	91	2470 ± 5
96pc119-24	220	210	0.95	102	0.00014	0.228	0.1714 ± 14	0.0721 ± 8	0.4007 ± 24	8.98 ± 7	0.16256 ± 73	88	2482 ± 8
96pc119-25	457	313	0.68	211	0.00006	1.010	0.1856 ± 9	0.1076 ± 8	0.3968 ± 19	9.01 ± 5	0.16461 ± 46	86	2504 ± 5
96pc119-26	455	212	0.47	160	0.00011	0.181	0.1262 ± 9	0.0859 ± 8	0.3174 ± 16	6.69 ± 4	0.15291 ± 51	75	2379 ± 6
96pc119-27	229	169	0.74	125	0.00023	0.374	0.1934 ± 14	0.1213 ± 12	0.4631 ± 26	10.55 ± 8	0.16519 ± 72	98	2510 ± 7
96pc119-28	167	97	0.58	85	0.00013	0.203	0.1454 ± 13	0.1126 ± 13	0.4468 ± 28	10.13 ± 8	0.16441 ± 74	95	2502 ± 8
96pc119-29	177	98	0.56	91	0.00011	0.173	0.1537 ± 13	0.1240 ± 14	0.4493 ± 28	10.19 ± 8	0.16445 ± 74	96	2502 ± 8
96pc119-30	253	98	0.39	122	0.00006	0.103	0.1088 ± 9	0.1231 ± 13	0.4390 ± 24	9.87 ± 7	0.16307 ± 60	94	2488 ± 6
96pc119-31	483	306	0.63	202	0.00002	0.035	0.1639 ± 7	0.0949 ± 7	0.3667 ± 18	8.03 ± 5	0.15878 ± 43	82	2443 ± 5
96pc119-32	443	297	0.67	212	0.00004	0.069	0.1890 ± 7	0.1161 ± 8	0.4109 ± 19	9.21 ± 5	0.16262 ± 40	89	2483 ± 4
96pc119-33	424	225	0.53	149	0.00030	0.474	0.1835 ± 13	0.1036 ± 9	0.2997 ± 14	6.69 ± 4	0.16182 ± 64	68	2475 ± 7
96pc119-34	745	552	0.74	256	0.00004	0.056	0.1983 ± 7	0.0791 ± 5	0.2957 ± 13	6.03 ± 3	0.14798 ± 36	72	2323 ± 4
96pc119-35	691	380	0.55	245	0.00009	0.144	0.1373 ± 6	0.0794 ± 5	0.3177 ± 14	6.74 ± 4	0.15395 ± 38	74	2390 ± 4
96pc119-36	288	211	0.73	157	0.00007	0.114	0.2048 ± 10	0.1287 ± 10	0.4609 ± 25	10.55 ± 7	0.16597 ± 53	97	2517 ± 5
96pc119-37	351	256	0.73	160	0.00009	0.141	0.1813 ± 9	0.0976 ± 8	0.3925 ± 19	8.81 ± 5	0.16279 ± 52	86	2485 ± 5
96pc119-38	616	291	0.47	239	0.00008	0.130	0.1295 ± 7	0.0959 ± 7	0.3500 ± 16	7.42 ± 4	0.15378 ± 40	81	2388 ± 4
Sample 96-PC-115													
pc115-1	70	24	0.34	37	0.00096	1.537	0.0976 ± 60	0.1332 ± 91	0.4644 ± 123	10.76 ± 37	0.16806 ± 309	97	2538 ± 31
pc115-2	113	42	0.38	55	0.00071	1.139	0.1098 ± 40	0.1249 ± 56	0.4270 ± 107	9.76 ± 29	0.16585 ± 210	91	2516 ± 21
pc115-3	56	27	0.48	32	0.00070	1.123	0.1324 ± 70	0.1363 ± 82	0.4895 ± 131	11.26 ± 40	0.16685 ± 346	102	2526 ± 35
pc115-4	64	38	0.59	35	0.00056	0.895	0.1561 ± 55	0.1238 ± 56	0.4699 ± 125	10.72 ± 36	0.16547 ± 201	99	2512 ± 29
pc115-5	362	376	1.04	80	0.00086	1.379	0.1629 ± 41	0.0292 ± 10	0.1859 ± 45	4.23 ± 12	0.16514 ± 199	44	2509 ± 20
pc115-6	190	195	1.03	95	0.00055	0.873	0.1938 ± 29	0.0786 ± 23	0.4162 ± 101	9.50 ± 25	0.16562 ± 146	89	2514 ± 15
pc115-7	83	34	0.41	47	0.00058	0.921	0.1195 ± 45	0.1447 ± 67	0.4993 ± 128	11.51 ± 35	0.16714 ± 230	103	2529 ± 23
pc115-8	74	25	0.34	37	0.00084	1.338	0.1057 ± 53	0.1366 ± 78	0.4354 ± 123	10.17 ± 33	0.16939 ± 270	91	2552 ± 27
pc115-9	87	31	0.35	47	0.00067	1.069	0.0994 ± 46	0.1341 ± 72	0.4789 ± 124	11.26 ± 35	0.17056 ± 241	98	2563 ± 24
pc115-10	260	190	0.73	143	0.00025	0.395	0.2001 ± 20	0.1266 ± 34	0.4620 ± 112	10.62 ± 27	0.16673 ± 106	97	2525 ± 11
pc115-11	124	89	0.72	71	0.00056	0.902	0.2038 ± 37	0.1338 ± 42	0.4713 ± 118	10.79 ± 31	0.16606 ± 186	99	2518 ± 19
pc115-12	113	96	0.85	52	0.00099	1.585	0.1113 ± 54	0.0525 ± 29	0.3993 ± 101	8.97 ± 28	0.16290 ± 266	87	2486 ± 27
pc115-13	84	41	0.49	47	0.00092	1.472	0.1262 ± 50	0.1228 ± 59	0.4759 ± 122	10.94 ± 34	0.16678 ± 252	99	2526 ± 25
pc115-14	86	31	0.36	44	0.00040	0.640	0.1023 ± 39	0.1311 ± 62	0.4587 ± 117	10.57 ± 32	0.16705 ± 213	96	2528 ± 21
pc115-15	253	361	1.43	74	0.00160	2.560	0.1611 ± 52	0.0268 ± 11	0.2375 ± 58	5.31 ± 16	0.16206 ± 248	55	2477 ± 26

$a f^{206}Pb$ is the percentage of common ^{206}Pb in the total measured Pb.

*Radiogenic lead corrected using ^{204}Pb.

% conc. = % concordance defined as $[(^{206}Pb/^{238}U$ age$)/(^{207}Pb/^{206}Pb$ age$)] \times 100$.

alteration of biotite to chlorite at fold hinges, commonly accompanied by calcite. The rock is partially retrogressed from the amphibolite facies.

The zircons are stubby, subhedral, pale-pink to red grains with well-formed prism faces, but with somewhat rounded terminations, with an average length to width ratio of 1.5:1. The darker grains are commonly more metamict. Inclusions of titanite and apatite are common.

Baizhiyan 'Formation'. Sample 96-PC-115 was obtained from within the Ekou Iron Mine, where it is overlain by amphibolite and BIF. The rock is a fine-grained, dark-grey, mesocratic andesite of fairly massive appearance, with numerous carbonate veins that transgress the fabric. It is composed of equidimensional quartz and untwinned feldspar, with abundant laths of chlorite and muscovite showing strong alignment. Later calcite also forms isolated crystals associated with the felsic minerals. There are some irregular opaque mineral grains. Based on assemblages from other rock-types in the mine, the rock is considered to be retrogressed from the amphibolite facies.

The zircons are colourless to pale pink, elongate crystals with well-developed prism and pyramid faces, although some surface pitting is evident. The length to width ratio varies from 2:1 to 5:1. Some crystals show distinct oscillatory zoning. There are a few inclusions of apatite in certain grains.

Hongmenyan 'Formation'. Samples were obtained between the original 32 and 35 km road markers on the Taihuai–Shahe road across Wutai Mountain. The rocks range from basaltic andesite to rhyodacite in composition and are associated with units of basalt and volcanoclastic sediments. Some of the more felsic units are locally transgressive, and may include subvolcanic intrusions. Petrographically, they all show evidence of strong deformation, with local mylonitic fabrics, and have been metamorphosed to greenschist facies. Geochemical data indicate that the felsic volcanics are calc-alkaline in nature, with basaltic andesites and some andesite samples showing slight LREE enrichment and weak to negligible Eu anomalies, whereas the more dacitic and rhyolitic samples show greater LREE enrichment and more pronounced Eu anomalies (Wilde & Wang 1995; Wang *et al.* in press).

Sample WT 9 is from a 2–2.5-metre-wide unit of fine-grained, grey, leucocratic dacite, with strong layering defined by chlorite. The rock has a pronounced deformation fabric, with aligned chlorite and muscovite grouped into weakly defined layers. The felsic components have straight to curved boundaries and consist of equidimensional quartz and untwinned feldspar. There are also scattered relict microphenocrysts of feldspar and quartz, partially wrapped around by chlorite and white mica. Some epidote veins occur oblique to the foliation.

The zircons are pale pink, euhedral, stubby to elongate crystals with length to width ratios of 1.5:1 to 3:1. Most crystals are clear and devoid of inclusions, but rare grains have rounded inclusions of titanite and needle-like apatite.

Sample WT 12 is from an 8-metre-wide unit of massive, fine-grained, white, leucocratic rhyodacite that is slightly transgressive to the regional foliation. The rock may have been intrusive. It preserves a microporphyritic igneous texture, consisting of abundant, albite-twinned plagioclase and rarer quartz phenocrysts, set in a groundmass of equidimensional to weakly aligned quartz and untwinned feldspar and strongly aligned muscovite. The latter is evenly distributed, although it does form local clusters, especially where it partially wraps around the relict microphenocrysts. There are a few late veins of coarser-grained muscovite and calcite.

The zircons are clear and pale pink, with euhedral prismatic faces and pyramidal terminations. Most are elongate, with length to width ratios of 1.5:1 to 5:1. There is weak oscillatory zoning, and some faces show surface frosting.

Sample WT 13 is from a 4-metre-wide unit of fine-grained, creamy-white, leucocratic rhyolite. Its overall fabric is similar to Sample WT 9, with chlorite concentrated into layers that define the main foliation. Muscovite is also strongly aligned and is associated with equidimensional quartz and weakly aligned untwinned feldspar. A few relict microphenocrysts of plagioclase and quartz occur scattered throughout the rock.

The zircons are well-formed, colourless to pale-pink, stubby grains with length to width ratios of 1.5:1 to 2:1. There are rare rod-like inclusions of apatite.

Sample WT 17 is a rhyodacitic lava also collected from this road section. It is a pale-pink to cream, leucocratic rock, composed of somewhat larger crystals of plagioclase and quartz set in a polygonal matrix of quartz and feldspar with minor chlorite. Muscovite is abundant and strongly aligned.

The zircons are pale pink, with well-formed prismatic faces and pyramidal terminations. Although the age and concordia diagram have previously been published (Wilde *et al.* 1997), the full data-set has not, and we include it in this paper for completeness.

Gaofan 'Subgroup'. Sample 95-PC-55c was collected from a riverbank near Xiazhuan village. It is a pale greenish-grey, leucocratic, felsic schist, interleaved with fine-grained chlorite-rich mafic volcanics, some of which contain garnet, and associated with strongly deformed granitoids which are locally mylonitic. It contains original porphyroblasts of garnet, now totally replaced by chlorite, and is composed of an aggregate of quartz and untwinned feldspar (mostly zoned plagioclase), between schistose layers dominated by chlorite and muscovite. Granular epidote is common within the felsic areas, but is independent of the feldspar. There is evidence of local recrystallization of the initial straight quartz and feldspar grain boundaries to abundant sub-grains. Certain layers are coarser grained and composed of quartz, muscovite, chlorite and calcite. Although now at greenschist facies, the rock was originally at a higher grade, possibly reaching the amphibolite facies, and is interpreted as a tuff.

The zircons are pink, rather stubby grains, mostly with length to width ratios of 1.5:1, but reaching a maximum of 2:1. Crystal faces tend to be slightly irregular, and there is extensive surface pitting. Many grains are slightly metamict and show internal fractures.

Analytical procedures

Following crushing, zircon crystals were extracted from the samples using a combination of heavy liquids and magnetic separation techniques. Individual crystals were hand picked and mounted, along with pieces of the Curtin University Sri Lankan zircon standard (CZ3), on to double-sided adhesive tape and then enclosed in epoxy resin discs. The discs were ground and polished, so as to effectively cut all zircon grains in half, and the samples were then gold coated.

The U–Th–Pb analyses of the zircons were performed using the SHRIMP II ion microprobe at Curtin University, Perth, Western Australia, following standard operating techniques (Nelson 1997; Williams 1998). An average mass resolution of 4800 was recorded during measurement of the Pb/Pb and Pb–U isotopic ratios and Pb–U ratios were normalized to those measured on the standard zircon [CZ3 – (^{206}Pb–^{238}U $= 0.0914$)], to compensate for elemental discrimination that occurs during sputter ionization (Kinny *et al.* 1993). The conventionally measured age of the standard is 564 Ma (Pidgeon *et al.* 1993) and the error associated with the measurement of Pb–U isotopic ratios for the standard, at one standard deviation, averaged *c.*1.64%. The measured

^{204}Pb values in the unknowns were similar to those recorded for the standard zircon, and so common lead corrections were made, assuming an isotopic composition of Broken Hill lead, since the common lead is considered to be mainly from surface contamination in the gold coating (Nelson 1997). The analytical spot size averaged *c.*30 μm during each analytical run, and each spot was rastered over 100 μm for three to five minutes prior to analysis to remove common Pb on the surface or contamination from the gold coating. Data reduction was performed using the Krill 007 program of P. D. Kinny at Curtin University and applying the ^{204}Pb correction. Errors on individual analyses are at the 1σ level whereas errors on pooled analyses are quoted at the 2σ level.

Results

Zhuangwang 'Formation'. The data for samples 96-PC-114 and 96-PC-119 are presented in Table 1 and shown on concordia plots in Figs 4a and 4b. For sample 96-PC-114, 25 analyses on 25 zircons were made, along with 12 analyses of the standard (CZ3) that gave an error of 2.75% over the analytical period. The data can be broadly grouped into two sets, those with low U and Th, and those with slightly higher values (Table 1). The former have U values ranging from 34 to 95 ppm, and Th from 19 to 89 ppm, whereas the latter have U contents of 113 to 186 ppm, and Th contents of 127 to 148 ppm. However, both sets show similar Th–U ratios, which range from 0.30 to 1.23 (average 0.74) (Table 1). With two exceptions, the data are concordant in the range of 94–104% (Table 1). The majority of the analyses, 21 zircons, define a ^{207}Pb/^{206}Pb age of 2529 $\pm$ 10 Ma (Fig. 4a). Three analyses are considerably older, with the two most concordant data points giving a ^{207}Pb/^{206}Pb age of 2679 $\pm$ 16 Ma (Fig. 4a).

A total of 38 zircons, along with 11 analyses of the standard that gave an error of 0.52%, were analysed from sample 96-PC-119. This sample is quite distinct in zircon chemistry from the above, revealing considerably higher U and Th values and having much higher total Pb contents (Table 1). The U ranges from 47 to 1712 ppm, and the Th from 16 to 3177 ppm, although the range in Th/U ratios is similar, varying from 0.34 to 1.86 (average 0.69) (Table 1). On the concordia diagram (Fig. 4b), the data define a pronounced discordia line trending to zero million years – suggestive of recent lead loss. However, in terms of (^{206}Pb–^{238}U)/ (^{207}Pb/^{206}Pb) systematics (see Table 1), the data are fairly concordant (Table 1). If the 16

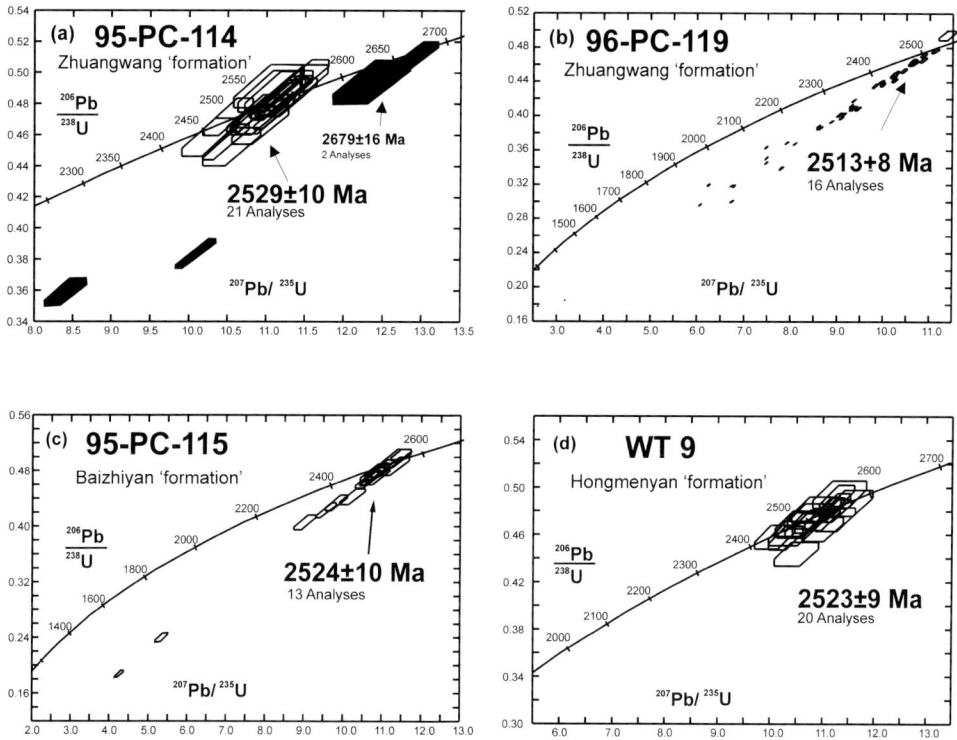

Fig. 4. (**a**) Concordia diagram of sample 96-PC-114 from the Zhuangwang 'Formation'; (**b**) Concordia diagram of sample 96-PC-119 from the Zhuangwang 'Formation'; (**c**) Concordia diagram of sample 96-PC-115 from the Baizhiyan 'Formation'; (**d**) Concordia diagram of sample WT 9 from the Hongmenyan 'Formation'.

most concordant data points are taken, the weighted mean $^{207}Pb/^{206}Pb$ date is 2513 ± 8 Ma (Fig. 4b), and this is taken to be the time of igneous crystallization. This is within the error of the main population of zircons in sample 96-PC-114 and indicates a possible range of crystallization age for the Zhuangwang 'Formation' of c.2530–2515 Ma.

Baizhiyan 'Formation'. The data for sample 96-PC-115 are presented in Table 1 and on a concordia diagram in Figure 4c. A total of 15 analyses were made on 15 zircons, along with 12 of the standard which recorded an error of 2.75%. There is some variation in both the U and Th data, with the former ranging from 56–362 ppm and the latter from 24–376 ppm. However, variation in the Th–U ratio is similar to samples from the Zhuangwang 'Formation', ranging from 0.34 to 1.43 (average 0.64). With two exceptions (analyses 115-PC-5 and 115-PC-15 in Table 1) the data are concordant with respect to the $(^{206}Pb/^{238}U)/(^{207}Pb/^{206}Pb)$ system, although they do define a distinct discordia line trending to zero million years in the concordia plot (Fig. 4c). The 13 most concordant analyses

from 13 individual zircon crystals give a weighted mean $^{207}Pb/^{206}Pb$ date of 2524 ± 10 Ma. This is interpreted as the crystallization age of the Baizhiyan 'Formation' and is within the range exhibited by the two samples from the Zhuangwang 'Formation'.

Hongmenyan 'Formation'. The data from the four samples analysed in this study are presented in Table 2 and on concordia diagrams in Figure 4d and Figure 5a, b and c. The most striking features of the data are their close similarity in age, their concordance and the lack of any inheritance.

For sample WT 9, a total of 20 analyses were made on 20 zircons, along with seven analyses of the standard, which recorded an error of 1.00% during the analytical session. The zircons are remarkably uniform in their chemistry, with U ranging from 32 to 73 ppm, and Th from 16 to 52 ppm; the Th–U ratio shows a small range from 0.47 to 0.71 (average 0.58), (Table 2). The zircons are concordant (92–103%), and define a single population with a weighted mean $^{207}Pb/^{206}Pb$ age of 2523 ± 9 Ma (Fig. 4d).

Table 2. *SHRIMP U–Pb–Th zircon data for Wutai Volcanic samples WT 9, WT 12 and WT 13*

Spot	U (ppm)	Th (ppm)	Th/U	Pb (ppm)	$^{204}Pb/^{206}Pb$	$\%f^{206a}$	$^{208}Pb^*/^{206}Pb^*$	$^{208}Pb^*/^{232}Th$	$^{206}Pb^*/^{238}U$	$^{207}Pb^*/^{235}U$	$^{207}Pb^*/^{206}Pb^*$	% conc.	Age $^{207}Pb^*/^{206}Pb^*$
Sample WT 9													
WT9-1	37	22	0.60	30	0.00158	2.623	0.1723 ± 105	0.1371 ± 91	0.4778 ± 107	10.76 ± 41	0.16334 ± 469	101	2491 ± 48
WT9-2	43	20	0.47	25	0.00155	2.491	0.1323 ± 90	0.1341 ± 98	0.4748 ± 104	11.27 ± 39	0.17210 ± 413	97	2578 ± 40
WT9-3	73	52	0.71	43	0.00054	0.859	0.2007 ± 56	0.1369 ± 49	0.4824 ± 100	11.26 ± 30	0.16932 ± 253	99	2551 ± 25
WT9-4	33	19	0.57	19	0.00242	3.307	0.1191 ± 134	0.0948 ± 110	0.4515 ± 105	9.48 ± 45	0.15224 ± 597	101	2371 ± 67
WT9-5	46	32	0.70	27	0.00079	1.276	0.1954 ± 71	0.1332 ± 59	0.4760 ± 103	11.02 ± 34	0.16788 ± 324	99	2537 ± 32
WT9-6	34	18	0.54	20	0.00162	2.596	0.1537 ± 114	0.1389 ± 110	0.4888 ± 113	11.53 ± 46	0.17115 ± 514	100	2569 ± 50
WT9-7	33	18	0.57	19	0.00167	2.666	0.1441 ± 123	0.1228 ± 111	0.4830 ± 113	10.89 ± 47	0.16355 ± 554	102	2493 ± 57
WT9-8	69	37	0.54	39	0.00090	1.444	0.1424 ± 60	0.1246 ± 61	0.4732 ± 99	10.75 ± 31	0.16476 ± 278	100	2505 ± 28
WT9-9	41	27	0.65	24	0.00118	1.885	0.1716 ± 88	0.1240 ± 72	0.4693 ± 104	10.67 ± 37	0.16495 ± 399	99	2507 ± 41
WT9-10	39	17	0.43	21	0.00156	2.499	0.1017 ± 99	0.1081 ± 109	0.4597 ± 103	10.1 ± 39	0.15941 ± 451	100	2449 ± 48
WT9-11	35	16	0.47	19	0.00190	3.034	0.1164 ± 123	0.1150 ± 126	0.4629 ± 106	10.43 ± 45	0.16347 ± 556	98	2492 ± 57
WT9-12	35	22	0.63	20	0.00180	2.879	0.1773 ± 116	0.1252 ± 89	0.4450 ± 102	10.52 ± 42	0.17139 ± 522	92	2571 ± 51
WT9-13	39	24	0.63	22	0.00095	1.526	0.1813 ± 93	0.1346 ± 78	0.4663 ± 104	11.02 ± 39	0.17138 ± 419	96	2571 ± 41
WT9-14	52	28	0.54	29	0.00134	2.145	0.1313 ± 77	0.1122 ± 72	0.4590 ± 99	10.32 ± 33	0.16299 ± 352	98	2487 ± 36
WT9-15	47	33	0.70	28	0.00057	0.914	0.1954 ± 58	0.1354 ± 53	0.4829 ± 104	11.13 ± 32	0.16710 ± 266	100	2529 ± 27
WT9-16	56	33	0.59	33	0.00097	1.549	0.1543 ± 65	0.1301 ± 64	0.4937 ± 105	11.29 ± 34	0.16588 ± 301	103	2516 ± 31
WT9-17	36	20	0.56	20	0.00126	2.016	0.1362 ± 106	0.1153 ± 95	0.4697 ± 107	10.55 ± 42	0.16293 ± 480	100	2486 ± 50
WT9-18	59	41	0.68	35	0.00052	0.832	0.1945 ± 56	0.1385 ± 52	0.4870 ± 103	11.25 ± 31	0.16751 ± 258	101	2533 ± 26
WT9-19	35	19	0.56	20	0.00172	2.755	0.1385 ± 118	0.1173 ± 105	0.4711 ± 109	10.56 ± 45	0.16263 ± 530	100	2483 ± 55
WT9-20	32	18	0.54	20	0.00205	3.285	0.1414 ± 134	0.1290 ± 128	0.4961 ± 117	11.36 ± 52	0.16604 ± 602	103	2518 ± 61
Sample WT 12													
WT12-1	119	76	0.64	66	0.00047	0.746	0.1721 ± 30	0.1268 ± 38	0.4731 ± 108	10.89 ± 28	0.16697 ± 145	99	2528 ± 15
WT12-2	40	27	0.67	25	0.00049	0.792	0.2081 ± 68	0.1570 ± 67	0.5026 ± 123	12.10 ± 39	0.17459 ± 311	101	2602 ± 30
WT12-3	57	24	0.42	32	0.00132	2.119	0.1088 ± 75	0.1235 ± 91	0.4816 ± 115	10.92 ± 37	0.16451 ± 344	101	2503 ± 35
WT12-4	89	49	0.55	52	0.00111	1.777	0.1329 ± 53	0.1187 ± 56	0.4894 ± 114	11.20 ± 33	0.16591 ± 246	102	2517 ± 25
WT12-5	54	40	0.73	33	0.00132	2.106	0.1915 ± 69	0.1263 ± 57	0.4822 ± 116	10.94 ± 35	0.16462 ± 311	101	2504 ± 32
WT12-6	48	23	0.47	28	0.00175	2.802	0.1122 ± 88	0.1145 ± 95	0.4833 ± 118	10.78 ± 40	0.16184 ± 399	103	2475 ± 42
WT12-7	38	23	0.60	24	0.00273	4.369	0.1524 ± 120	0.1209 ± 101	0.4794 ± 120	10.70 ± 47	0.16195 ± 534	102	2476 ± 56
WT12-8	39	22	0.57	19	0.00193	3.095	0.1553 ± 104	0.1314 ± 96	0.4815 ± 119	11.06 ± 44	0.16656 ± 469	100	2523 ± 47
WT12-9	32	17	0.52	19	0.00286	4.569	0.1185 ± 134	0.1057 ± 124	0.4669 ± 119	10.17 ± 49	0.15804 ± 600	101	2435 ± 64
WT12-10	59	43	0.74	36	0.00150	2.402	0.1788 ± 71	0.1201 ± 57	0.4955 ± 118	11.08 ± 36	0.16216 ± 320	105	2478 ± 33
WT12-11	94	63	0.67	55	0.00094	1.501	0.1754 ± 49	0.1267 ± 47	0.4830 ± 111	11.30 ± 32	0.16961 ± 222	99	2554 ± 22
WT12-12	85	42	0.49	49	0.00140	2.242	0.1166 ± 57	0.1136 ± 63	0.4817 ± 112	10.76 ± 32	0.16200 ± 263	102	2477 ± 27
WT12-13	33	18	0.55	20	0.00206	3.301	0.1481 ± 116	0.1316 ± 110	0.4875 ± 124	11.05 ± 47	0.16448 ± 519	102	2502 ± 53
WT12-14	93	73	0.79	57	0.00059	0.947	0.2165 ± 42	0.1374 ± 43	0.4986 ± 115	11.56 ± 31	0.16817 ± 191	103	2539 ± 19
WT12-15	32	18	0.56	19	0.00163	2.606	0.1513 ± 118	0.1326 ± 111	0.4949 ± 128	11.41 ± 50	0.16727 ± 532	102	2531 ± 53
WT12-16	64	45	0.71	40	0.00077	1.226	0.1948 ± 59	0.1417 ± 57	0.5129 ± 124	11.76 ± 36	0.16635 ± 268	106	2521 ± 27
WT12-17	30	18	0.60	19	0.00129	2.063	0.1655 ± 99	0.1372 ± 93	0.4991 ± 132	11.45 ± 46	0.16636 ± 451	104	2521 ± 46
WT12-18	52	40	0.77	34	0.00099	1.579	0.2023 ± 70	0.1374 ± 62	0.5260 ± 131	11.88 ± 40	0.16389 ± 318	109	2496 ± 33
WT12-19	73	48	0.65	41	0.00070	1.116	0.1805 ± 56	0.1306 ± 53	0.4711 ± 111	10.74 ± 32	0.16538 ± 258	99	2511 ± 26

WT12-20	195	205	1.05	98	0.00045	0.720	0.2729 ± 30	0.1029 ± 26	0.3966 ± 89	9.02 ± 22	0.16497 ± 130	86	2507 ± 13
WT12-21	70	51	0.73	40	0.00100	1.602	0.1933 ± 60	0.1226 ± 50	0.4641 ± 110	10.79 ± 33	0.16863 ± 272	97	2544 ± 27
WT12-22	66	37	0.57	38	0.00114	1.819	0.1570 ± 58	0.1312 ± 59	0.4742 ± 112	10.96 ± 33	0.16757 ± 266	99	2533 ± 27
WT12-23	80	55	0.69	47	0.00094	1.507	0.1795 ± 49	0.1252 ± 46	0.4827 ± 112	10.80 ± 31	0.16228 ± 221	102	2480 ± 23
WT12-24	59	30	0.51	34	0.00096	1.539	0.1308 ± 61	0.1242 ± 67	0.4852 ± 116	11.18 ± 35	0.16717 ± 282	101	2530 ± 28
WT12-25a	23	9	0.40	13	0.00178	2.843	0.1039 ± 135	0.1236 ± 166	0.4770 ± 128	10.89 ± 53	0.16560 ± 615	100	2514 ± 62
WT12-25b	24	9	0.39	14	0.00181	2.904	0.0981 ± 147	0.1208 ± 186	0.4813 ± 129	10.73 ± 56	0.16173 ± 666	102	2474 ± 70
WT12-25c	32	10	0.30	18	0.00247	3.953	0.0380 ± 141	0.0592 ± 220	0.4741 ± 124	10.01 ± 52	0.15317 ± 640	105	2382 ± 71
WT12-26	33	20	0.61	21	0.00166	2.661	0.1552 ± 115	0.1281 ± 103	0.5033 ± 129	11.39 ± 49	0.16417 ± 518	105	2499 ± 53
Sample WT 13													
WT13-1	86	39	0.46	50	0.00077	1.221	0.1296 ± 41	0.1418 ± 52	0.4989 ± 79	11.86 ± 24	0.17247 ± 195	101	2582 ± 19
WT13-2	44	28	0.63	25	0.00024	0.388	0.1696 ± 40	0.1343 ± 42	0.4953 ± 83	11.37 ± 25	0.16648 ± 194	103	2523 ± 20
WT13-3	35	18	0.52	26	0.00022	0.357	0.1542 ± 70	0.1465 ± 75	0.4917 ± 87	11.52 ± 32	0.16995 ± 328	101	2557 ± 32
WT13-4	39	23	0.59	22	0.00035	0.991	0.1602 ± 44	0.1406 ± 49	0.5163 ± 89	11.98 ± 27	0.16824 ± 218	106	2540 ± 22
WT13-5	80	54	0.68	46	0.00033	0.523	0.1839 ± 36	0.1327 ± 36	0.4890 ± 79	11.20 ± 22	0.16617 ± 169	102	2519 ± 17
WT13-6	65	34	0.52	29	0.00021	0.340	0.1496 ± 35	0.1467 ± 45	0.5148 ± 84	11.86 ± 24	0.16706 ± 176	106	2528 ± 18
WT13-7	34	18	0.54	19	0.00067	1.075	0.1422 ± 65	0.1242 ± 65	0.4717 ± 87	10.59 ± 30	0.16285 ± 311	100	2485 ± 32
WT13-8	35	20	0.58	22	0.00082	1.102	0.1472 ± 74	0.1347 ± 74	0.5279 ± 96	11.99 ± 35	0.16477 ± 341	109	2505 ± 35
WT13-9	42	28	0.67	24	0.00053	0.852	0.1831 ± 65	0.1300 ± 54	0.4733 ± 82	10.83 ± 28	0.16596 ± 296	99	2517 ± 30
WT13-10	44	29	0.66	20	0.00015	0.234	0.1910 ± 64	0.1414 ± 56	0.4858 ± 83	11.64 ± 30	0.17373 ± 292	98	2594 ± 28
WT13-11/3	38	18	0.48	20	0.00031	0.507	0.1230 ± 59	0.1224 ± 64	0.4791 ± 81	10.88 ± 28	0.16473 ± 283	101	2505 ± 29
WT13-12	27	15	0.56	15	0.00076	1.223	0.1395 ± 77	0.1164 ± 70	0.4702 ± 86	10.71 ± 32	0.16524 ± 361	99	2510 ± 37
WT13-13	114	78	0.68	63	0.00004	0.058	0.1880 ± 17	0.1320 ± 25	0.4788 ± 72	11.01 ± 18	0.16670 ± 89	100	2525 ± 9
WT13-14	44	31	0.71	26	0.00024	0.380	0.1936 ± 44	0.1344 ± 40	0.4896 ± 80	11.17 ± 24	0.16549 ± 205	102	2513 ± 21
WT13-15	40	20	0.49	22	0.00030	0.479	0.1302 ± 50	0.1314 ± 57	0.4978 ± 82	11.26 ± 26	0.16410 ± 236	104	2498 ± 24
WT13-16	43	26	0.59	25	0.00021	0.341	0.1617 ± 37	0.1345 ± 41	0.4943 ± 81	11.38 ± 24	0.16695 ± 185	102	2527 ± 19
WT13-17a	39	19	0.50	21	0.00027	0.430	0.1354 ± 50	0.1303 ± 55	0.4786 ± 79	10.99 ± 26	0.16661 ± 240	100	2524 ± 24
WT13-17b	40	22	0.55	22	0.00021	0.332	0.1486 ± 38	0.1317 ± 43	0.4857 ± 81	11.23 ± 24	0.16770 ± 194	101	2535 ± 19
WT13-18	60	32	0.53	33	0.00012	0.191	0.1477 ± 28	0.1345 ± 36	0.4867 ± 77	11.25 ± 21	0.16766 ± 145	101	2534 ± 15
WT13-19	34	13	0.53	19	0.00010	0.168	0.1478 ± 44	0.1398 ± 51	0.4980 ± 84	11.73 ± 26	0.17080 ± 221	102	2565 ± 22
WT13-20	34	19	0.56	26	0.00041	0.650	0.1495 ± 54	0.1301 ± 55	0.4914 ± 83	11.31 ± 28	0.16687 ± 258	102	2526 ± 26
WT13-21	32	20	0.62	18	0.00004	0.318	0.1773 ± 31	0.1407 ± 39	0.4893 ± 84	11.47 ± 24	0.17002 ± 165	100	2558 ± 16
WT13-22	41	20	0.48	22	0.00018	0.293	0.1311 ± 40	0.1310 ± 48	0.4826 ± 79	11.18 ± 24	0.16802 ± 199	100	2538 ± 19
WT13-23	29	16	0.56	16	0.00033	0.522	0.1580 ± 50	0.1356 ± 52	0.4785 ± 82	11.05 ± 26	0.16744 ± 244	100	2532 ± 24
WT13-24	48	27	0.56	27	0.00015	0.233	0.1573 ± 35	0.1365 ± 40	0.4839 ± 78	11.30 ± 23	0.16930 ± 174	100	2551 ± 17

[a] $f^{206}Pb$ is the percentage of common ^{206}Pb in the total measured Pb.

* Radiogenic lead corrected using ^{204}Pb.

% conc. = % concordance defined as $[(^{206}Pb-^{238}U \ age)/(^{207}Pb-^{206}Pb \ age)] \times 100$.

a, b and c represent multiple analyses within a single zircon crystal.

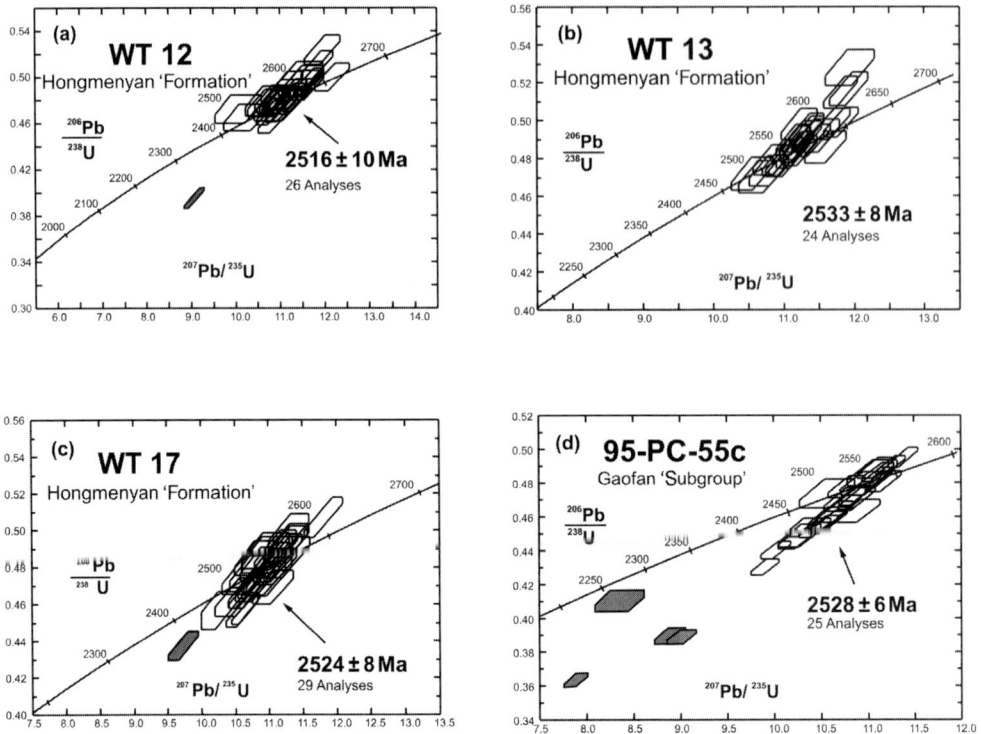

Fig. 5. (a) Concordia diagram of sample WT 12 from the Hongmenyan 'Formation'; (b) Concordia diagram of sample WT 13 from the Hongmenyan 'Formation'; (c) Concordia diagram of sample WT 17 from the Hongmenyan 'Formation'; (d) Concordia diagram of sample 95-PC-55c from the Gaofan 'Subgroup'.

A total of 27 analyses were made on 26 zircons from sample WT 12, along with seven analyses of the standard, which recorded an error of 1.61% during the analytical session. With two exceptions (WT 12-1 and WT 12-20 in Table 2), the zircons show little variation in their U and Th contents – these ranging from 23–94 ppm and 9–73 ppm, respectively. The Th–U ratio, with the exception of WT 12-20 which is higher at 1.05, shows little variation, ranging from 0.30 to 0.79 (average 0.61). The data are reasonably concordant (86–109% in Table 2), and the 26 most concordant analyses define a single population with a weighted mean $^{207}Pb/^{206}Pb$ age of 2516 ± 10 Ma (Fig. 5a).

For sample WT 13, a total of 24 analyses on 23 zircons were determined, along with 10 analyses on the standard zircon, which gave an error of 1.45%. With one exception (WT 13-13 in Table 2), the data reveal little variation in U and Th contents, ranging from 27–86 ppm and 15–54 ppm, respectively, with Th–U ratios varying from 0.46–0.71 (average 0.57). The data are reasonably concordant (98–109%) and define a weighted mean $^{207}Pb/^{206}Pb$ age of 2533 ± 8 Ma (Fig. 5b).

For completeness, the data-set for sample WT 17 is also included in this paper (Table 3 and on a concordia plot in Fig. 5d). Thirty analyses were made, along with 10 measurements of the standard, which gave an error of 2.06% during the analytical session. A total of 29 zircon analyses (the exception being WT 17-11) reveal a small range in U and Th contents, varying from 24 to 74 ppm and 9 to 58 ppm, respectively; the Th–U ratio of the total population varies from 0.33 to 0.75 (average 0.53) (Table 3). The data are tightly concordant (95–104%), and define a weighted mean $^{207}Pb/^{206}Pb$ age of 2524 ± 8 Ma.

In summary, all four samples from the Hongmenyan 'Formation' have $^{207}Pb/^{206}Pb$ dates within the range c.2533–2516 Ma, that is, indistinguishable from the data from the Zhuangwang 'Formation'.

Gaofan 'Subgroup'. The data for sample 95-PC-55c are presented in Table 3 and on a concordia plot in Figure 5d. A total of 29 zircons were analysed, along with seven analyses of the standard, which gave an error of 1.00% during the analytical session. The U and Th contents of the zircons range

Table 3. *SHRIMP U–Pb–Th zircon data for Wutai Volcanic samples WT 17 and 95-PC-55c*

Spot	U (ppm)	Th (ppm)	Th/U	Pb (ppm)	$^{204}Pb/^{206}Pb$	% f^{206a}	$^{207}Pb^*/^{206}Pb^*$	$^{208}Pb^*/^{206}Pb^*$	$^{206}Pb^*/^{238}U$	$^{207}Pb^*/^{235}U$	$^{208}Pb^*/^{232}Th$	% conc.	Age $^{207}Pb^*/^{206}Pb^*$
Sample WT 17													
WT-17-1	34	16	0.48	19	0.00009	0.150	0.16887 ± 307	0.1363 ± 64	0.4853 ± 110	11.30 ± 35	0.1372 ± 75	100	2546 ± 30
WT-17-2	37	21	0.58	20	0.00091	1.414	0.16457 ± 320	0.1446 ± 69	0.4564 ± 102	10.36 ± 32	0.1135 ± 62	97	2503 ± 33
WT-17-3	37	22	0.60	21	0.00085	1.156	0.16294 ± 371	0.1542 ± 82	0.4760 ± 107	10.69 ± 36	0.1222 ± 72	101	2486 ± 38
WT-17-4	71	52	0.74	43	0.00033	0.514	0.16961 ± 160	0.2031 ± 34	0.5069 ± 107	11.85 ± 29	0.1397 ± 40	104	2554 ± 16
WT-17-5	36	19	0.54	20	0.00077	1.227	0.16854 ± 322	0.1496 ± 70	0.4752 ± 105	11.04 ± 34	0.1324 ± 71	99	2543 ± 32
WT-17-6	39	19	0.47	23	0.00104	2.010	0.16645 ± 291	0.1360 ± 62	0.4927 ± 108	11.31 ± 34	0.1420 ± 74	102	2522 ± 29
WT-17-7	42	19	0.45	23	0.00025	0.397	0.16712 ± 233	0.1274 ± 48	0.4873 ± 105	11.23 ± 31	0.1385 ± 63	101	2529 ± 23
WT-17-8	38	18	0.48	21	0.00093	1.503	0.16796 ± 312	0.1313 ± 67	0.4739 ± 91	10.97 ± 31	0.1297 ± 73	99	2537 ± 31
WT-17-9	24	11	0.45	13	0.00058	1.358	0.16808 ± 303	0.1368 ± 62	0.4805 ± 97	11.13 ± 32	0.1455 ± 76	100	2539 ± 30
WT-17-10	33	16	0.50	18	0.00068	1.094	0.16214 ± 320	0.1198 ± 69	0.4848 ± 93	10.84 ± 32	0.171 ± 72	103	2478 ± 33
WT-17-11	105	79	0.75	59	0.00031	0.490	0.16580 ± 124	0.2017 ± 26	0.4659 ± 83	10.65 ± 21	0.1253 ± 29	98	2516 ± 13
WT-17-12	61	26	0.43	33	0.00026	0.423	0.16832 ± 176	0.1187 ± 35	0.4807 ± 88	11.16 ± 25	0.1331 ± 48	100	2541 ± 17
WT-17-13	28	14	0.52	16	0.00167	2.671	0.16414 ± 418	0.1320 ± 92	0.4866 ± 97	11.01 ± 38	0.1236 ± 91	102	2499 ± 43
WT-17-14	55	23	0.41	30	0.00014	0.222	0.16657 ± 170	0.1163 ± 33	0.4891 ± 90	11.23 ± 25	0.1378 ± 49	102	2523 ± 17
WT-17-15	26	9	0.33	15	0.00089	1.426	0.16310 ± 372	0.0828 ± 79	0.4889 ± 97	10.99 ± 35	0.1241 ± 123	103	2488 ± 38
WT-17-16	34	22	0.63	21	0.00079	1.269	0.16473 ± 269	0.1688 ± 58	0.4995 ± 96	11.34 ± 30	0.1335 ± 55	104	2505 ± 27
WT-17-17	41	22	0.55	23	0.00044	0.707	0.16605 ± 218	0.1437 ± 45	0.4843 ± 91	11.09 ± 27	0.1275 ± 49	101	2518 ± 22
WT-17-18	68	40	0.60	36	0.00011	0.305	0.16961 ± 115	0.1661 ± 21	0.4573 ± 83	10.69 ± 30	0.1270 ± 30	95	2554 ± 11
WT-17-19	49	22	0.46	27	0.00123	3.622	0.16355 ± 205	0.1153 ± 42	0.4818 ± 89	10.87 ± 26	0.1206 ± 51	102	2493 ± 21
WT-17-20	51	22	0.43	27	0.00073	1.134	0.16277 ± 231	0.1090 ± 49	0.4776 ± 88	10.72 ± 26	0.1222 ± 61	101	2485 ± 24
WT-17-21	74	58	0.78	42	0.00024	0.381	0.16607 ± 138	0.2182 ± 30	0.4715 ± 85	10.80 ± 22	0.1314 ± 31	99	2518 ± 14
WT-17-22	54	2±	0.43	27	0.00028	0.450	0.16182 ± 175	0.1263 ± 35	0.4374 ± 81	9.76 ± 22	0.1279 ± 45	95	2475 ± 18
WT-17-23	38	24	0.64	21	0.00045	0.728	0.16623 ± 245	0.1734 ± 52	0.4669 ± 89	10.70 ± 27	0.1260 ± 47	98	2520 ± 25
WY-17-24	59	25	0.41	30	0.00025	0.396	0.16919 ± 180	0.1130 ± 36	0.4560 ± 84	10.64 ± 24	0.1245 ± 47	95	2550 ± 18
WT-17-25	27	15	0.55	15	0.00030	0.478	0.16692 ± 263	0.1507 ± 54	0.4707 ± 95	10.83 ± 29	0.1290 ± 56	98	2527 ± 26
WT-17-26	34	17	0.51	19	0.00051	0.815	0.17160 ± 295	0.1443 ± 63	0.4690 ± 92	11.10 ± 31	0.1324 ± 66	96	2573 ± 29
WT-17-27	39	21	0.55	22	0.00057	0.921	0.16517 ± 241	0.1504 ± 51	0.4893 ± 107	11.14 ± 31	0.1329 ± 56	102	2509 ± 25
WT-17-28	41	28	0.68	24	0.00056	0.896	0.16667 ± 266	0.1816 ± 58	0.4914 ± 107	11.29 ± 32	0.1319 ± 53	102	2524 ± 27
WT-17-29	30	19	0.63	17	0.00073	1.174	0.16525 ± 330	0.1729 ± 72	0.4637 ± 105	10.57 ± 34	0.1274 ± 63	98	2510 ± 34
WT-17-30	35	15	0.43	19	0.00062	0.986	0.16656 ± 284	0.1137 ± 59	0.4768 ± 106	10.95 ± 32	0.1264 ± 74	100	2523 ± 29

(continued)

Table 3. *Continued*

Spot	U (ppm)	Th (ppm)	Th/U	Pb (ppm)	$^{204}Pb/^{206}Pb$	$\%f^{206a}$	$^{207}Pb*/^{206}Pb*$	$^{208}Pb*/^{206}Pb*$	$^{206}Pb*/^{238}U$	$^{207}Pb*/^{235}U$	$^{208}Pb*/^{232}Th$	% conc.	Age $^{207}Pb*/^{206}Pb*$
Sample 95-PC-55c													
95pc55c-1	113	83	0.74	66	0.00002	0.032	0.1999 ± 19	0.1342 ± 22	0.4945 ± 59	11.40 ± 16	0.16724 ± 111	102	2530 ± 11
95pc55c-2	66	26	0.40	33	0.00000	0.000	0.1155 ± 14	0.1331 ± 27	0.4575 ± 61	10.63 ± 17	0.16852 ± 133	95	2543 ± 13
95pc55c-3	68	33	0.48	36	0.00015	0.244	0.1297 ± 29	0.1285 ± 35	0.4772 ± 65	10.79 ± 19	0.16400 ± 169	101	2497 ± 17
95pc55c-4	81	38	0.46	44	0.00019	0.299	0.1235 ± 24	0.1299 ± 37	0.4883 ± 65	11.10 ± 20	0.16487 ± 166	102	2506 ± 17
95pc55c-5	121	55	0.45	60	0.00012	0.186	0.1211 ± 39	0.1181 ± 28	0.4399 ± 53	10.03 ± 16	0.16542 ± 141	94	2512 ± 14
95pc55c-6	63	26	0.42	32	0.00001	0.020	0.1191 ± 39	0.1314 ± 49	0.4628 ± 68	10.92 ± 23	0.17120 ± 218	95	2569 ± 21
95pc55c-7	72	37	0.52	37	0.00011	0.170	0.1391 ± 38	0.1237 ± 40	0.4585 ± 65	10.58 ± 21	0.16742 ± 205	96	2532 ± 21
95pc55c-8	108	53	0.49	55	0.00008	0.132	0.1355 ± 26	0.1246 ± 30	0.4528 ± 58	10.45 ± 17	0.16739 ± 152	95	2532 ± 15
95pc55c-9	45	18	0.40	24	0.00031	0.503	0.1017 ± 51	0.1228 ± 66	0.4852 ± 79	11.07 ± 27	0.16551 ± 267	101	2513 ± 27
95pc55c-10	160	85	0.53	89	0.00008	0.133	0.1431 ± 18	0.1311 ± 24	0.4900 ± 57	11.26 ± 16	0.16672 ± 109	102	2525 ± 11
95pc55c-11	171	179	1.05	95	0.00150	2.396	0.1824 ± 41	0.0776 ± 20	0.4456 ± 51	10.24 ± 18	0.16661 ± 194	94	2524 ± 20
95pc55c-12	207	188	0.91	106	0.00087	1.395	0.1645 ± 30	0.0777 ± 17	0.4297 ± 47	9.91 ± 15	0.16734 ± 151	91	2531 ± 15
95pc55c-13	63	25	0.39	31	0.00026	0.422	0.0985 ± 43	0.1131 ± 53	0.4481 ± 67	10.29 ± 22	0.16660 ± 231	95	2524 ± 23
95pc55c-14	241	498	2.06	126	0.00206	3.290	0.2509 ± 45	0.0472 ± 10	0.3884 ± 43	9.02 ± 16	0.16841 ± 205	83	2542 ± 20
95pc55c-15	190	244	1.28	86	0.00068	1.088	0.2447 ± 34	0.0694 ± 13	0.3629 ± 41	7.89 ± 13	0.15761 ± 163	82	2430 ± 18
95pc55c-16	51	17	0.34	24	0.00127	2.027	0.1002 ± 82	0.1217 ± 10	0.4097 ± 62	8.36 ± 26	0.14792 ± 386	95	2322 ± 45
95pc55c-17	40	18	0.45	21	0.00044	0.698	0.1125 ± 63	0.1194 ± 72	0.4730 ± 83	10.59 ± 30	0.16229 ± 322	101	2480 ± 33
95pc55c-18	128	156	1.22	59	0.00139	2.220	0.1230 ± 46	0.0393 ± 16	0.3889 ± 47	8.89 ± 17	0.16585 ± 222	84	2516 ± 22
95pc55c-19	148	106	0.72	77	0.00032	0.505	0.1394 ± 25	0.0881 ± 20	0.4546 ± 53	10.54 ± 16	0.16817 ± 135	95	2539 ± 13
95pc55c-20	146	96	0.66	80	0.00015	0.235	0.1769 ± 22	0.1264 ± 23	0.4717 ± 55	10.87 ± 16	0.16715 ± 125	98	2529 ± 13
95pc55c-21	185	92	0.50	100	0.00004	0.060	0.1358 ± 16	0.1310 ± 22	0.4813 ± 54	11.14 ± 15	0.16786 ± 99	100	2536 ± 10
95pc55c-22	138	97	0.70	71	0.00038	0.606	0.1365 ± 26	0.0880 ± 21	0.4530 ± 54	10.30 ± 16	0.16498 ± 142	96	2507 ± 14
95pc55c-23	98	62	0.63	55	0.00010	0.159	0.1696 ± 29	0.1297 ± 29	0.4804 ± 62	11.20 ± 19	0.16904 ± 158	99	2548 ± 16
95pc55c-24	62	24	0.38	32	0.00017	0.270	0.1025 ± 38	0.1253 ± 51	0.4695 ± 69	10.77 ± 22	0.16640 ± 211	98	2522 ± 21
95pc55c-25	86	45	0.53	45	0.00004	0.068	0.1523 ± 30	0.1338 ± 34	0.4620 ± 62	10.67 ± 19	0.16756 ± 171	97	2533 ± 17
95pc55c-26	126	57	0.45	68	0.00010	0.156	0.1231 ± 21	0.1314 ± 29	0.4826 ± 58	11.11 ± 17	0.16699 ± 127	100	2528 ± 13
95pc55c-27	55	25	0.45	28	0.00000	0.000	0.1177 ± 16	0.1228 ± 28	0.4702 ± 71	10.90 ± 20	0.16813 ± 147	98	2539 ± 15
95pc55c-28	121	63	0.52	66	0.00005	0.078	0.1437 ± 23	0.1337 ± 28	0.4858 ± 58	11.13 ± 17	0.16619 ± 132	101	2520 ± 13
95pc55c-30	55	23	0.41	30	0.00037	0.596	0.1149 ± 50	0.1332 ± 62	0.4789 ± 71	10.98 ± 25	0.16623 ± 256	100	2520 ± 26

$^a f^{206}Pb$ is the percentage of common ^{206}Pb in the total measured Pb.

* Radiogenic lead corrected using ^{204}Pb.

% conc. = % concordance defined as $[(^{206}Pb/^{238}U\ age)/(^{207}Pb/^{206}Pb\ age)] \times 100$.

from 40 to 241 ppm and 17 to 498 ppm, respectively, with a range in the Th–U ratio from 0.38 to 2.06 (average 0.64). All bar four of the analyses are reasonably concordant (91–102%, Table 3) and these remaining 25 analyses define a weighted mean ^{207}Pb/^{206}Pb age of 2528 ± 6 Ma.

Interpretation

Zircons from all specimens show strong oscillatory zoning in cathodoluminesence imagery, indicative of magmatic growth. Furthermore, the metamorphic grade of greenschist to lower amphibolite facies is too low to allow metamorphic zircon growth and all grains, including the inherited population in sample 96-PC-114, are considered to reflect igneous ages.

The dating results from the seven new specimens analysed, together with sample WT 17 from Wilde *et al.* (1997), are summarized in Table 4. It is evident that:

(1) all the dates are essentially similar within error;
(2) there is no consistent trend of upward younging from what has been interpreted as the base of the Wutai succession (Tian 1991); and,
(3) there is more variability within what were considered to be single 'formations' than there is between units supposedly near the top and bottom of the sequence.

Although the weighted mean ages of the eight individual samples do not overlap within two sigma error (Table 4), collectively, the data sets do overlap without a time break, extending from the youngest date of 2513 ± 8 Ma (96-PC-119) to the oldest date of 2533 ± 8 Ma (WT 13). If the 149 most concordant zircon analyses are treated as a single population from one magmatic episode, the calculated weighted mean ^{207}Pb/^{206}Pb age is 2523 ± 3 Ma; we consider this to be the best estimate of the general age of felsic volcanism in the Wutai Complex. More than one age is possibly represented in the suite, but the fairly large error recorded for individual populations, coupled with the slightly discordant nature of several analyses, precludes a thorough evaluation of this aspect. However, the overall similarity in both age and zircon chemistry of the volcanic rocks, together with their whole-rock geochemistry (Wang *et al.* in press), suggests that they represent essentially the products of one magmatic event.

The age of the Zhuangwang 'Formation' andesite (96-PC-114) is 2529 ± 10 Ma and this is considered to be within the lower Shizui 'Subgroup' (Fig. 3). It is virtually identical to the age of the Gaofan 'Subgroup' tuff (95-PC-55c) at 2528 ± 6 Ma, considered to be at the top of the sequence (Tian *et al.* 1996; Ma & Bai 1998). This indicates that: (1) the difference in metamorphic grade between the Zhuangwang 'Formation' (amphibolite facies) and the Gaofan 'Subgroup' (low greenschist facies) has no apparent age significance and cannot be used as a discriminator to map-out stratigraphy (contrast with Bai 1986; Tian 1991; Ma & Bai 1998); (2) the rocks evolved rapidly, so that there is no recognizable stratigraphy to the felsic volcanics at the resolution

Table 4. *Sample numbers, locations, ^{207}Pb/^{206}Pb zircon ages and average Th/U ratios of the eight felsic volcanic samples discussed in this study. The 'formations' are listed from top to bottom in the stratigraphic order identified in Bai (1986) and Tian (1991)*

Formation name	Sample no.	Location	Age (Ma)	Th/U Ratio (av.)
Gaofan	95-PC-55c	Lat. 38°56′13″; Long. 113°00′40″	2528 ± 6	0.64
Hongmenyan	WT 17	Lat. 39°02′52″; Long. 113°36′45″	2524 ± 8	0.53
	WT 13	Lat. 39°02′53″; Long. 113°37′13″	2533 ± 8	0.57
	WT 12	Lat. 39°02′49″; Long. 113°36′54″	2516 ± 5	0.61
	WT 9	Lat. 39°02′52″; Long. 113°36′54″	2523 ± 9	0.58
Baizhiyan	96-PC-115	Lat. 39°03′42″; Long. 113°15′60″	2524 ± 10	0.64
Zhuangwang	96-PC-119	Lat. 39°05′19″; Long. 113°16′50″	2513 ± 8	0.69
	96-PC-114	Lat. 39°04′01″; Long. 113°16′38″	2529 ± 10	0.74

of our data and, by implication, to the mafic volcanics and metasediments interleaved with them.

In the two 'formations' where we have analysed more than one sample, there is considerable variation between the ages. In the Zhuangwang 'Formation', sample 96-PC-119 (the most discordant of the samples analysed) has a date of 2513 ± 8 Ma, whereas sample 96-PC-114 has a date of 2529 ± 10 Ma. These just overlap within error, but are at the two extremes of the dates that we have obtained; 2513 ± 8 Ma being the youngest date recorded and 2529 ± 10 Ma being the second oldest (Table 4). Similarly, the spread of dates recorded from the so-called Hongmenyan 'Formation' range from 2516 ± 5 Ma for sample WT 12 (second youngest) to 2533 ± 8 Ma for sample WT 13 (the oldest age obtained from the suite). One possibility is that this may be attributable to an error in the original mapping of the 'formations'. However, samples from the Hongmenyan 'Formation' were all collected from a single, well-exposed 3-km-long traverse and considered to represent a simple stratigraphic sequence on all published maps and by Y. Tian (pers. comm., 1996). As with the data for the units considered to represent the basal and upper 'subgroups' of the sequence (Zhuangwang and Gaofan, respectively), the validity of the individual 'formations' is also severely compromised. The slightly younger age of 2516 ± 5 Ma for sample WT 12 may be real, since the field evidence suggested that it was slightly transgressive and might therefore be a somewhat younger intrusive.

Significance of the results

The new data presented here require revision of the 'stratigraphic model' for the Wutai 'Group' (see also Cawood *et al.* 1998), since the previously accepted stratigraphic divisions are no longer valid. Because the Wutai area has been used to define type-locations for greenstone belt stratigraphy, orogenies and tectonic events (Fig. 2) throughout China (Ma & Bai 1998), this has important ramifications.

Relationship to associated granitoids

A number of granitic intrusions are associated with the Wutai Complex, and field evidence indicates a range of relationships. Many granitoids are strongly deformed and in tectonic contact, so that original contact relations cannot be unequivocally interpreted. Wilde *et al.* (1997) identified that some granitoids were older than the felsic volcanics of the Wutai 'Group' – the Lanzhishan granite new Longquanguan and Ekou granite at Ekou (Fig. 1) having SHRIMP ^{207}Pb/^{206}Pb ages of *c.*2545 Ma

and 2555 ± 6 Ma, respectively. They also established the younger age of the Dawaliang granite near Gaofan at 2176 ± 12 Ma. Two distinct suites of granitoid have been identified (Wang *et al.* 2000; Liu *et al.* 2002): those older than the Wutai volcanics, including some bodies coeval with the volcanics with ages of 2530–2525 Ma, and a younger set of intrusions (including both the Dawaliang Granite and the pink phase of the Wangjiahui granite southeast of Daixian – Fig. 1) with ages of 2170–2120 Ma.

The Lanzhishan Granite was originally considered to be unconformably overlain by the Wutai Complex (Liu *et al.* 1985). However, we have questioned this interpretation (Wilde *et al.* 1997), considering the granite to be overlain by the Hutuo Group at several localities. Furthermore, its contacts are sheared, like those between the Ekou Granite and the Wutai Complex. Details on the age and relationship of the granitoids to the Wutai Complex will be discussed in a subsequent paper.

Relationship to adjacent terranes

Three main lithotectonic components are recognized in the area: the Wutai, Fuping and Hengshan complexes (Fig. 1). Traditionally, in the Chinese literature, the Wutai Complex has been considered to unconformably overlie both the Fuping and Hengshan complexes (Tian *et al.* 1996). Both the Fuping and Hengshan complexes consist of a variety of grey granitoid gneisses, amphibolites and metasediments at amphibolite- to granulite-facies metamorphic grade (Bai 1986; Tian 1991). The Hutuo Group, which is mostly composed of low-grade metasediments with minor metavolcanics, is considered to be the youngest unit in the region, and to unconformably overlie the Wutai and Fuping Complexes (Tian 1991).

The Wutai Complex was generally considered to have developed within a continental rift environment, due to fracturing of pre-existing continental crust, now represented by the Fuping and Hengshan complexes (Tian 1991). Alternatively, Li, J. *et al.* (1990), Li & Wang (1992) and Sun *et al.* (1992) suggested that plate-tectonic-styled terrane accretion may have occurred, resulting in the collision of a typical low-grade Archaean greenstone sequence (the Wutai Complex), with island-arc affinity, and older continental fragments (the Fuping and Hengshan complexes) during the Late Archaean. The Fuping Complex was considered to be an ancient continental nucleus Liu *et al.* (1985). Sun *et al.* (1992), on the other hand, proposed that it was the basal portion of an island-arc. Li & Wang (1992) further developed the plate-tectonic model, suggesting that the Fuping Complex formed part of a *c.*2.9 Ga microcontinent which collided with both the Hengshan Complex (another microcontinent) and the Wutai Complex (containing both island-arc and fore-arc components) in the Early Proterozoic (*c.*2.0 Ga). A major shear zone (the Longquanguan Shear Zone) was considered to be the surface expression of a décollement that controlled the thin-skin tectonics of the foreland.

SHRIMP U–Pb zircon geochronology (Guan *et al.* 2002; Zhao *et al.* 2002), has shown that the Fuping Complex is composed of four lithotectonic units: the Fuping gneisses, Longquanguan augen gneisses, Wanzi supracrustals and Nanying granitic gneisses (Zhao *et al.* 2002). Cathodoluminescence (CL) and backscattered electron (BSE) images reveal the coexistence of magmatic and metamorphic zircons in nearly all rock types of the Fuping Complex (Guan *et al.* 2002; Zhao *et al.* 2002). SHRIMP U–Pb analyses on magmatic zircons reveal that the Fuping TTG gneisses were emplaced between 2523 ± 14 Ma and 2486 ± 8 Ma ago (Guan *et al.* 2002; Zhao *et al.* 2002), whereas the protoliths of the Longquanguan augen gneisses were intruded between 2543 ± 7 Ma and 2507 ± 11 Ma (Wilde *et al.* 1997; Zhao *et al.* 2002). SHRIMP data also reveal that the Nanying granitic gneisses formed between 2077 ± 13 Ma and 2024 ± 21 Ma (Guan *et al.* 2002; Zhao *et al.* 2002). Zircon grains and new overgrowth zircon rims from all components of the Fuping Complex yielded similar concordant $^{207}Pb/^{206}Pb$ dates in the range 1875–1802 Ma, interpreted as approximating the age of regional metamorphism of the Fuping Complex (Guan *et al.* 2002; Zhao *et al.* 2002). In addition, Guan *et al.* (2002) obtained a SHRIMP U–Pb zircon age of 2708 ± 8 Ma for a hornblende gneiss, which occurs as enclaves in the Fuping gneisses. Work on the Hengshan Complex (Kroner *et al.* 2001; Zhao *et al.* 2001a) has shown a similar range of ages to those recorded from the Fuping Complex. In both areas, the preferred interpretation is that they represent lower components of a complex arc system, of which the Wutai Complex represents the upper portion (Fig. 6), and which was assembled into its current position at *c.*1.8 Ga (see also Zhao *et al.* 2001b; Wilde *et al.* 2002). No components older than *c.*2.7 Ga have so far been identified in either the Fuping or Hengshan Complexes. Indeed, the majority of rocks are similar in age to the Wutai Complex; this implies that neither the rift model (Tian 1991) nor the plate-tectonic model of Li & Wang (1992), involving older cratonic blocks, is viable.

Tectonic setting

Zhao *et al.* (2000 & 2001b) have proposed that Wutaishan and the adjacent Fuping and Hengshan complexes (Fig. 1), lie within the Central Zone of the North China Craton. This zone, also referred to as the Trans-North China Orogen (Zhao *et al.* 2001a), is a linear zone of crustal thickening characterized by clockwise $P-T$ paths, which separates two major crustal blocks – referred to as the eastern and western blocks (Zhao *et al.* 1998), that are considered to be separate continental nuclei which collided at *c.*1.8 Ga ago (Zhao *et al.* 2001b; Wilde *et al.* 2002). The western block consists of Archaean tonalite–trondhjemite–granodiorite (TTG) gneisses and mafic igneous rocks, unconformably overlain by Palaeoproterozoic high-grade metasedimentary rocks referred to locally as 'khondalites' (Zhao *et al.* 1999). The Archaean rocks underwent greenschist- to granulite-facies metamorphism at 2.6–2.5 Ga and are characterized by anticlockwise $P-T$ paths. The eastern block also consists of TTG gneisses, accompanied by syntectonic granitoids and interleaved supracrustal rocks, including ultramafic to felsic volcanic rocks and metasediments, metamorphosed from greenschist to granulite facies; also exhibiting anticlockwise $P-T$ paths (Zhao *et al.* 2001b). In contrast to the western block, some basement rocks have considerably older protolith ages, up to *c.*3.8 Ga (Song *et al.* 1996).

Our new results are important on the craton-scale, for the following reasons. The age data from felsic volcanic rocks in the Wutai Complex suggest a single magmatic source that extended over a period of approximately 20 Ma, from *c.*2533 to *c.*2515 Ma, with a mean age of 2523 ± 3 Ma. Because there is no difference in the age of the felsic volcanic rocks within the Wutai Complex, these findings may also apply equally to other associated components of the complex, with the proviso that some younger rocks of the Hutou Group might be interleaved (Cawood *et al.* 1998). The complex had previously been considered an unconformity-bounded, layer-cake stratigraphic succession up to 5 km thick (Tian *et al.* 1996) that was either formed in a rift environment (Tian 1991) or island-arc setting (Li J. *et al.* 1990; Li & Wang 1992). Both models imply that the Wutai rocks were younger than the adjacent Fuping and Hengshan complexes, which is not supported by our new data.

Investigations of the Wutai Complex (Wilde & Wang 1995; Wilde *et al.* 1997, 1998 and 2001; Wang *et al.* in press) have established that this complex represents relics of an island and/or

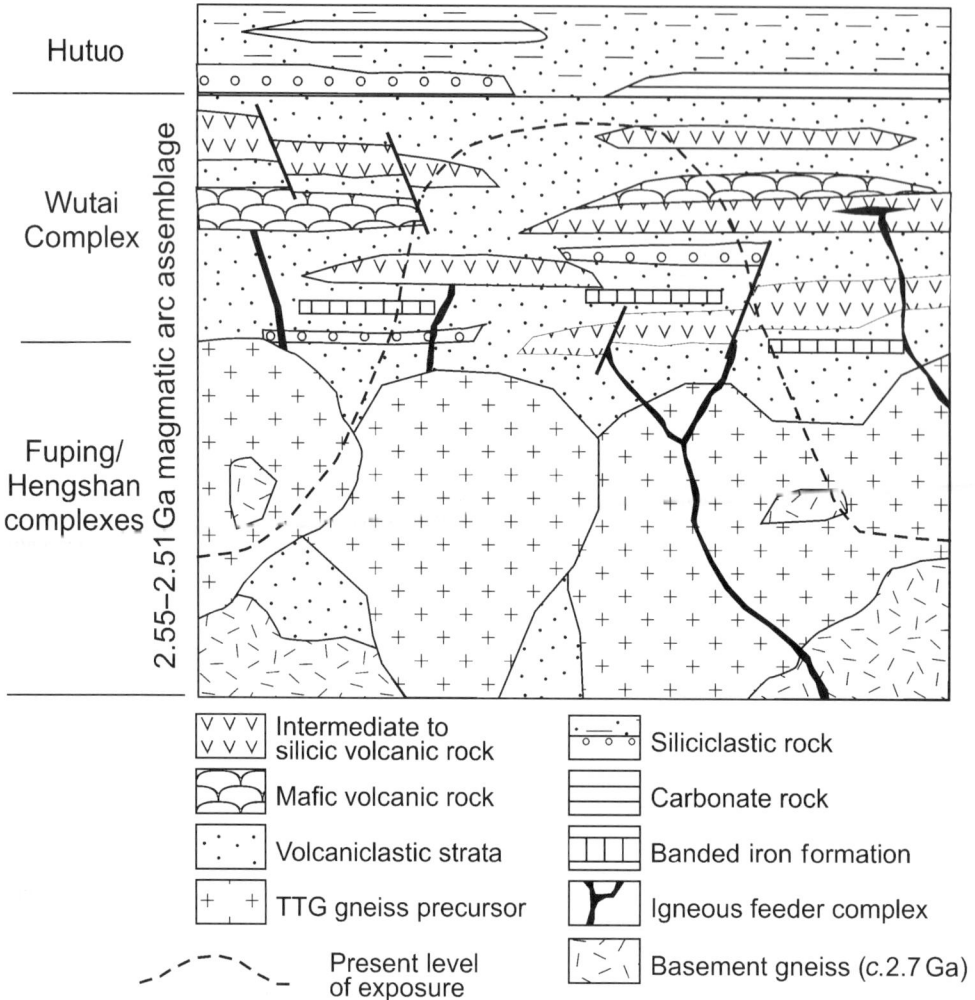

Fig. 6. Schematic diagram showing the relationship of the major lithological components within the Wutai arc system prior to dismemberment at 1.8 Ga. Approximate width of diagram is 75 km.

magmatic arc composed of mafic, intermediate and felsic volcanic rocks, with associated volcaniclastic and chemical sediments (Fig. 6). We consider that the components of the arc have been tectonically disrupted during a major collisional event and interleaved with granitoids, some of which predate the volcanic rocks and others which evolved coevally with them (Wilde *et al.* 1997; Wang *et al.* 2000). Our present results further substantiate this interpretation, with the implication of repetition of the sequence rather than a true stratigraphy.

With respect to the timing of deformation and metamorphism, geochronological evidence shows that metamorphic ages of *c.*1800 Ma are present throughout the area, especially in the higher-grade rocks. This age was originally considered to represent a local thermal event (Tian 1991; Bai *et al.* 1992). However, since all three crustal complexes show evidence of major tecto-nothermal activity at this time, the event is of regional extent. Evidence includes K–Ar data on hornblendes from the low-grade Wutai Complex of 1782 ± 20 Ma (Wang *et al.* 1997) and metamorphic overgrowths on existing zircons and generation of new zircons in rocks of the higher-grade Hengshan and Fuping Complexes at *c.*1.8 Ga (Kroner *et al.* 2000; Guan *et al.* 2002; Zhao *et al.* 2001a). It now appears that the *c.*1.8 Ga event (the so-called Lüliang orogeny) records the timing of collision resulting from amalgamation of the east and west blocks of the North China Craton and its formation as a crustal entity (see also Zhao *et al.* 2001b; Wilde *et al.* 2002).

Our new data indicate that, in one sample (96-PC-115), a $^{207}Pb/^{206}Pb$ date of 2679 ± 16 Ma (Fig. 4c) was recorded from two of three older, inherited zircons. This is the only volcanic sample to contain inheritance, and this in turn suggests that at least some of the components of the Wutai Complex were extruded through pre-existing crust (Fig. 6). It has previously been noted (Wilde *et al.* 1997) that some of the earlier granitoids at Wutaishan also contain inherited zircons as old as *c.*2.7 Ga, considered to most likely represent components of the eastern block of the North China Craton (Wilde *et al.* 2002); a similar origin is proposed for the older zircons contained in the Wutai volcanics, and this suggests that part of the Wutai Complex might also have evolved at the western continental margin of the eastern block.

Locally, tonalitic gneisses as old as *c.*2.7 Ga are also present in the Fuping (Guan *et al.* 2002) and Hengshan (Kroner *et al.* 2001) complexes. Therefore, not only are the dominant rock-types in the Wutai, Fuping and Hengshan Complexes 2.52 Ga old, but all three areas contain evidence of earlier *c.*2.7-Ga crustal components.

Geochemical data (Sun *et al.* 1992; Liu *et al.* in press) support the view that the high-grade granitic gneisses of the Fuping and Hengshan Complexes are arc-related. Because of their similarity in age (Guan *et al.* 2002; Zhao *et al.* 2001b) and chemistry (Liu *et al.* in press; Wang *et al.* in press) to the Wutai Complex, they may represent part of the same arc system (Kroner *et al.* 2001; Wilde *et al.* 2002). We therefore interpret both the Hengshan and Fuping Complexes as the lower part of a Late Archaean arc complex (Fig. 6), with the intervening Wutai Complex representing the upper portion of the arc (Kroner *et al.* 2001; Zhao *et al.* 2001b; Guan *et al.* 2002; Wilde *et al.* 2002; Zhao *et al.* 2002). It appears likely that the arc may have been complex, since it contains island-arc and back-arc components (Wang *et al.* in press), as well as segments that erupted through pre-existing continental crust.

At the present time, there are insufficient data to adequately model the convergent margin components of the arc, or to elucidate the full sequence of events that occurred between arc formation at 2.53 Ga – possibly along the western margin of the eastern block of the North China Craton – and collision with the western block at *c.*1.8 Ga (Zhao *et al.* 2001b). One possibility is that the

island-arc/back-arc components amalgamated with the western margin of the eastern block during the late Archaean/earliest Proterozoic and subsequently evolved as a passive margin. The age of the overlying Hutuo Group at 2366 +103/−94 Ma (Wu *et al.* 1986) is poorly constrained, although it may be a correlative of the *c.*2100 Ma Wanzi Supracrustal Suite that is tectonically interleaved with the Fuping TTG gneisses (Guan *et al.* 2002). There is also a suite of younger granitoids with Palaeoproterozoic ages of *c.*2150 Ma in the Wutai, Fuping and Hengshan Complexes (Wilde *et al.* 1997; Guan *et al.* 2002). These data suggest that the complexes had a common evolution from *c.*2.2 Ga.

Our favoured interpretation, therefore, is that the Wutai, Fuping and Hengshan Complexes evolved at a convergent margin along the western continental margin of the Eastern block of the North China Craton (Zhao *et al.* 2001b; Wilde *et al.* 2002). They were subsequently tectonically dismembered, interleaved with Palaeoproterozoic shelf sediments and granitoids and metamorphosed at *c.*1.8 Ga during a major collision that brought together the eastern and western blocks of the North China Craton (Zhao *et al.* 2001b; Wilde *et al.* 2002).

Conclusions

New SHRIMP U–Pb data from felsic volcanic rocks collected throughout the Wutai sequence in the North China Craton establish that:

(1) The rocks have similar dates within error, indicating that there is only one period of volcanism recorded in the Wutai Complex. This shows a range in age from 2533 ± 8 Ma to 2513 ± 8 Ma, with a mean of 2523 ± 3 Ma.

(2) There is no layer-cake stratigraphy to the Wutai Complex and, coupled with evidence of structural complexity, this indicates that the rocks are tectonically juxtaposed. Remapping of this classic area of Chinese geology will probably reveal a complex tectonostratigraphy.

(3) There is no correlation between metamorphic grade and age of volcanic rocks in the Wutai Complex, contrary to previous views (Bai 1986; Tian 1991; Tian *et al.* 1996).

(4) There is a similarity in age between the igneous rocks of the low-grade Wutai Complex and those in the high-grade Fuping and Hengshan complexes (Guan *et al.* 2002; Kroner *et al.* 2001; Zhao *et al.* 2001b). There is also a consistency in the geochemical data (Sun *et al.* 1992; Wang *et al.* in press), tending to support a common origin within a complex arc system.

(5) There is no evidence for rocks older than *c.*2.7 Ga in any of the three complexes and there

is thus no apparent difference between the so-called Fuping and Wutai 'movements' or orogenic events. Furthermore, there is no direct evidence that the Wutai or Fuping/Hengshan complexes were deformed prior to *c.*1.8 Ga ago (Kroner *et al.* 2001), and so the very existence of the Fuping and Wutai orogenies is in need of re-evaluation.

(6) The marked similarity in age between the main metamorphic events in the Wutai, Fuping and Hengshan Complexes at *c.*1800 Ma indicates that the Lüliang 'movement' (orogeny) is the major tectonic event in the region. It resulted in formation of the North China Craton by amalgamation during collision of the eastern and western blocks (Zhao *et al.* 2000).

We express thanks to the late Xingyuan Ma for emphasizing to the first author the reasons why the Wutaishan area should be studied in more detail; Yonquing Tian for kindly accompanying us in the field on one of our visits; and Allen Kennedy for initial advice on SHRIMP data collection. We also thank Kent Condie and Maarten de Wit for their valuable reviews. This work was supported by an Australian Research Council Grant (A39532446) to S. A. Wilde and P. A. Cawood, and National Natural Science Foundation of China awards to K. Y. Wang. The SHRIMP II consortium in Perth is operated jointly by Curtin University of Technology, the University of Western Australia and the Geological Survey of Western Australia. This is Tectonics Research Centre Publication No. 226.

References

Bai, J. 1986. *The Early Precambrian Geology of Wutaishan.* Tianjin Science and Technology Press, Tianjin, 190 pp.

Bai, J., Wang, R. Z. & Guo, J. J. 1992. *The Major Geologic Events of Early Precambrian and their Dating in Wutaishan Region.* Geological Publishing House, Beijing, 1–55 (in Chinese with English abstract).

Cawood, P. A., Wilde, S. A., Wang, K. Y. & Nemchin, A. A. 1998. Integrated geochronology and field constraints on subdivision of the Precambrian of China: data from the Wutaishan. Abstracts of ICOG-9, 1998 Beijing. *Chinese Science Bulletin*, **43**, 17.

Guan, H., Sun, M., Wilde, S. A., Zhou, X. H. & Zhai, M. G. 2002. SHRIMP U–Pb zircon geochronology of the Fuping Complex: implications for formation and assembly of the North China craton. *Precambrian Research*, **113**, 1–18.

Kinny, P. D., Black, L. P. & Sheraton, J. W. 1993. Zircon ages and the distribution of Archaean and Proterozoic rocks in the Rauer Islands. *Antarctic Science*, **5**, 193–206.

Kroner, A., Wilde, S., O'Brien, P. J. & Li, J. H. 2001. The Hengshan and Wutai complexes of northern China: lower and upper crustal domains of a late Archaean to Palaeoproterozoic magmatic arc and significance for the evolution of the North

China Craton, 4th International Archaean Symposium, Perth, Australia. *AGSO Record*, **2001/37**, 327.

Li, J. L. & Wang, K. Y. 1992. Accretion tectonics of Early Precambrian in North China, *Scientia Geologia Sinica*, **1**, 15–29.

Li, J. L., Wang, K. Y., Wang, Q., Liu, X. & Zhao, Z. 1990. Early Proterozoic collision orogenic belt in Wutaishan Area, China. *Scientia Geologica Sinica*, **1**, 1–11.

Li, S., Hart, S. R. & Wu, T. 1990. Rb–Sr and Sm–Nd isotopic dating of an early Precambrian spilite–keratophyre sequence in the Wutaishan area, North China: preliminary evidence for Nd-isotopic homogenization in the mafic and felsic lavas during low-grade metamorphism. *Precambrian Research*, **47**, 191–203.

Liu, D. Y., Page, R. W., Compston, W. & Wu, J. 1985. U–Pb zircon geochronology of Late Archaean metamorphic rocks in the Taihangshan–Wutaishan area, North China. *Precambrian Research*, **27**, 85–109.

Liu, S. W., Pan, Y. M., Zhang, J. & Li, Q. G. 2002. Archean geodynamics in the Central Zone, North China Craton: constraints from geochemistry of two contrasting series of granitoids in the Fuping, Hengshan and Wutaishan complexes. *Precambrian Research*.

Ma, X. Y. & Bai, J. 1998. *Precambrian Crustal Evolution of China.* Springer, 331 pp.

Nelson, D. R. 1997. Compilation of SHRIMP U–Pb zircon geochronology data, 1996. *Geological Survey of Western Australia*, **Record 1997/2**, 189 pp.

Pidgeon, R. T., Furfaro, D., Kennedy, A. K., Nemchin, A. A., van Bronswijk, W. & Todt, W. A. 1994. Calibration of zircon standards for the Curtin SHRIMP II, *Eighth International Conference on Geochronology, Cosmochronology and Isotope Geology*, Berkeley, June 5–11, Abstracts, 251.

Song, B., Nutman, A. P., Liu, D. Y., & Wu, J. S. 1996. 3800 to 2500 Ma crustal evolution in Anshan area of Liaoning Province, Northeastern China. *Precambrian Research*, **78**, 79–94.

Sun, M., Armstrong, R. L. & Lambert, R. St. J. 1992. Petrochemistry and Sr, Pb, and Nd isotopic geochemistry of Early Precambrian rocks, Wutaishan and Taihangshan areas, China. *Precambrian Research*, **56**, 1–31.

Tian, Y. 1991. *Geology and Gold Mineralization of Wutai–Hengshan Greenstone Belt.* Shanxi Science and Technology Press (in Chinese with English abstract), 235–240.

Tian, Y., Ma, Z., Yu, K., Liu, Z. & Peng, Q. 1996. The Early Precambrian geology of Wutai–Hengshan Mt., Shanxi, China – T315 Field Trip Guide, *30th International Geological Congress.* Geological Publishing House, Beijing, 55 pp.

Wang, K. Y., Hao, J., Cawood, P. A. & Wilde, S. A. 1997. High-pressure metamorphism in kyanite-bearing schists from the original Jingangku Formation of the Wutaishan. *Proceedings of the 30th International Geological Congress: Precambrian Geology and Metamorphic Petrology*, **17**, VSP

International Science Publishers, Amsterdam, 213–220.

WANG, K. Y., CHAI, Y. C. & LI, J. L. 1992. On the late Archaean granitoids and their tectonic environments of the Wutai area. *Memoir of Lithospheric Tectonic Evolution Research*, **1**, 107–112, Seismological Press, Beijing.

WANG, K. Y., LI, J. L., HAO, J., LI, J. H. & ZHOU, S. P. 1996. The Wutaishan orogenic belt within the Shanxi Province, northern China: a record of late Archaean collision tectonics, *Precambrian Research*, **78**, 95–103.

WANG, K. Y., HAO, J., WILDE, S. A., CAWOOD, P. A. 2000. Reconsideration of key geological problems of late Archaean–early Proterozoic in the Wutaishan–Hengshan area: constraints from SHRIMP U–Pb zircon data. *Scientia Geologica Sinica*, **35**, 175–184.

WANG, R. & BAI, J. 1986. Ages of the Wutai Group. *In:* BAI, J. (ed), *The Early Precambrian Geology of Wutaishan*, Tianjin Science and Technology Press, Tianjin 364-370 (in Chinese).

WANG, Z., WILDE, S. A., WANG, K. & YU, L. in press. A MORB–arc basalt–adakite association in the 2.5 Ga Wutai greenstone belt: late Archaean magmatism and crustal growth in the North China Craton. *Precambrian Research*.

WILDE, S. A. & WANG, K. 1995. The nature and age of felsic volcanism within the Late Archaean Wutai Complex, Sino-Korean Craton, China. *Precambrian '95, Tectonics and Metallogeny of Early/Mid Precambrian Orogenic Belts*, Montreal, Canada, Abstracts, 61.

WILDE, S. A. & WANG, K. Y. 2000. SHRIMP U–Pb zircon dating of felsic volcanism in the Wutai Complex, North China Craton: overturning the stratigraphic model. *Abstract Vol. CD, 31st IGC*, Brazil, 2000.

WILDE, S. A., CAWOOD, P. A. & WANG, K. Y. 1997. The relationship and timing of granitoid evolution with respect to felsic volcanism in the Wutai Complex, North China Craton. *Proceedings of the 30th International Geological Congress, Beijing*, **17**, VSP International Science Publishers, Amsterdam, 75–87.

WILDE, S. A., CAWOOD, P. A., WANG, K. Y. & NEMCHIN, A. A. 1998. SHRIMP U–Pb zircon dating of granites and gneisses in the Taihangshan–Wutaishan area: implications for the timing of crustal growth in the North China Craton. Abstracts of

ICOG-9, 1998 Beijing, *Chinese Science Bulletin*, **43**, 144.

WILDE, S. A., ZHAO, G. C. and SUN, M. 2002. Development of the North China Craton during the Late Archaean and its amalgamation along a major 1.8 Ga collision zone; including speculations on its position within a global Palaeoproterozoic Supercontinent. *Gondwana Research*, **5**, 85–94.

WILLIAMS, I. S. 1998. U–Th–Pb geochronology by ion microprobe. *In:* MCKIBBEN, M. A., SHANKS III, W. C. and RIDLEY, W. I. (eds) *Applications of Microanalytical Techniques to Understanding mineralizing processes*, Reviews in Economic Geology, **7**, 1–35.

WU, J., LIU, D. & JIN, L. 1986. The zircon U–Pb age of metabasic volcanic lavas from the Hutuo Group in the Wutai mountain area, Shanxi Province. *Geology Reviews*, **32**, 178–184 (in Chinese).

YANG, Z., CHENG, Y. & WANG, H. 1986. *The Geology of China*. Clarendon Press, Oxford.

ZHAO, G. C., CAWOOD, P. A., WILDE, S. A. & LU, L. Z. 2001a. High-pressure granulites (retrograded eclogites) from the Hengshan Complex, North China Craton: petrology and tectonic implications. *Journal of Petrology*, **42**, 1141–1170.

ZHAO, G. C., WILDE, S. A., CAWOOD, P. A. & LU, L. Z. 1998. Thermal evolution of Archean basement rocks from the eastern part of the North China craton and its bearing on tectonic setting. *International Geology Review*, **40**, 706–721.

ZHAO, G. C., WILDE, S. A., CAWOOD, P. A. & LU, L. Z. 1999. Tectonothermal history of the basement rocks in the western zone of the North China Craton and its tectonic implications. *Tectonophysics*, **310**, 37–53.

ZHAO, G. C., WILDE, S. A., CAWOOD, P. A. & LU, L. Z. 2000. Petrology and P–T path of the Fuping mafic granulites: implications for tectonic evolution of the central zone of the North China Craton. *Journal of Metamorphic Geology*, **18**, 375–391.

ZHAO, G. C., WILDE, S. A., CAWOOD, P. A. & SUN, M. 2001b. Archean blocks and their boundaries in the North China Craton: lithological, geochemical, structural and P–T path constraints and tectonic evolution. *Precambrian Research*, **107**, 45–73.

ZHAO, G.C., WILDE, S. A., CAWOOD, P. A. & SUN, M. 2002. SHRIMP U–Pb zircon ages of the Fuping Complex: implications for Late Archean to Paleoproterozoic accretion and assembly of the North China Craton. *American Journal of Science*, **302**, 191–226.

Late Archaean to Palaeoproterozoic evolution of the Trans-North China Orogen: insights from synthesis of existing data from the Hengshan–Wutai–Fuping belt

GUOCHUN ZHAO[1], MIN SUN[1], SIMON A. WILDE[2] and JINGHUI GUO[3]

[1]*Department of Earth Sciences, The University of Hong Kong,*
Pokfulam Road, Hong Kong. (E-mail: gzhao@hkucc.hku.hk)
[2]*Department of Applied Geology, Curtin University of Technology,*
GPO Box U1987, Perth 6845, Australia
[3]*Institute of Geology and Geophysics, Chinese Academy of Sciences,*
Beijing, 100029, China

Abstract: The Hengshan–Wutai–Fuping mountain belt constitutes the middle segment of the Trans-North China Orogen, which separates the North China Craton into the Eastern and Western Blocks. The belt consists of the high-grade Hengshan and Fuping complexes, and the intervening low- to medium-grade Wutai Complex. Previous tectonic models assumed that the high-grade complexes were an older basement (Archaean to Palaeoproterozoic) to the low-grade Wutai Complex. However, new geochronological data show that the emplacement of granitoid rocks and eruption of volcanic rocks in the Wutai Complex occurred essentially coeval with or slightly earlier than intrusion of the tonalitic–trondhjemitic–granodioritic (TTG) suites in the Hengshan and Fuping complexes. New isotopic data also reveal the widespread presence of Palaeoproterozoic granitoid rocks in these complexes. Structural and metamorphic data demonstrate similar tectonothermal histories for the three complexes, which are characterized by peak medium- to high-pressure metamorphism accompanied by the development of thrusting, isoclinal folding (F_2) and penetrative foliations, followed by near-isothermal decompression and cooling and retrogression associated with the formation of large-scale ductile shear zones and asymmetrical folds (F_3) with nearly vertical axial planes. These geochronological, structural and metamorphic data suggest that the tectonic evolution of the Hengshan–Wutai–Fuping mountain belt may not be related to local interaction of the three complexes, as suggested in earlier models, either through closure of a Wutai rift or collision between a Wutai arc and the Hengshan and Fuping microcontinental blocks. Instead, they may represent elements of a single Late Archaean to Early Palaeoproterozoic magmatic arc that was subsequently incorporated into the Trans-North China Orogen along which the Eastern and Western blocks amalgamated to form the North China Craton at around 1.85 Ga.

Recent progress in understanding the basement architecture of the North China Craton (Zhao *et al.* 1998, 1999*a*, 2000*a*, 2001*a*; Wilde *et al.* 2002) has resulted in the recognition of the Trans-North China Orogen, which separates the craton into two blocks – named the Eastern and Western blocks. The Trans-North China Orogen is characterized by the presence of fragments of ancient oceanic crust, mélanges, high-pressure (HP) granulites and retrograded eclogites, strike-slip ductile shear zones, large-scale thrusts and folds, mineral stretching lineations and minor sheath folds (Bai 1986; Li *et al.* 1990; Zhai *et al.* 1992, 1995; Guo *et al.* 1993, 2001*a*, 2001*b*, 2002; Wang *et al.* 1996, 1997*a*; Bai & Dai 1998; Guo & Zhai 2001; Zhao *et al.* 2001*b*, 2001*c*; Guan *et al.* 2002; Liu *et al.* 2002; Wang *et al.* 2003*a*). These lithotectonic elements contrast with the dominant Late Archaean tonalitic–trondhjemitic–granodioritic (TTG) gneiss domes surrounded by minor supracrustal rocks of the Eastern and Western blocks (Ma & Wu 1981; Jahn & Zhang 1984; He & Ye 1998; Zhao *et al.* 1998). In addition, petrographic and thermobarometric data have revealed that basement rocks in the Trans-North China Orogen differ in metamorphic $P–T$ evolution from those in the Eastern and Western blocks (Zhao *et al.* 1998, 1999*a*, 1999*b*). The former underwent metamorphism characterized by clockwise $P–T$ paths involving isothermal decompression (Zhao *et al.* 2000*a*, 2000*b*),

From: MALPAS, J., FLETCHER, C. J. N., ALI, J. R. & AITCHISON, J. C. (eds) 2004. *Aspects of the Tectonic Evolution of China*. Geological Society, London, Special Publications, **226**, 27–55.
0305-8719/04/$15 © The Geological Society of London 2004.

whereas the latter experienced metamorphism at *c.*2.5 Ga, with anticlockwise *P*–*T* paths involving isobaric cooling (Zhao *et al.* 1998). These differences led Zhao *et al.* (2001*a*) to propose that the Trans-North China Orogen represents a collisional zone along which the Eastern and Western blocks amalgamated to form the North China Craton. Although there is now considerable knowledge of the pre-amalgamation history of the Eastern and Western Blocks that were subsequently incorporated into the North China Craton (Wu & Zhong 1998; Zhao *et al.* 1998, 1999*b*, 2000*b*, Wu *et al.* 2000; Zhao 2001), details of the magmatic, structural and metamorphic history of the Trans-North China Orogen are still poorly constrained.

As the largest and most lithologically representative basement exposure across the Trans-North China Orogen (Fig. 1), the Hengshan–Wutai–Fuping mountain belt is probably the most promising area for investigating the detailed magmatic, structural and metamorphic history of the orogen. Of particular significance is the presence of a low- to medium-grade granite–greenstone terrane (Wutai Complex) located between two high-grade gneiss complexes (Hengshan and Fuping complexes). Two contrasting tectonic models were originally proposed for the evolution of the Hengshan–Wutai–Fuping mountain belt. One suggested that the Fuping and Hengshan complexes developed as a single continental block that underwent Late Archaean rifting associated with formation of the Wutai greenstones and followed by closure in the Palaeoproterozoic (Tian 1991; Yuan & Zhang 1993), whereas the other proposed that the belt is a Late Archaean continent–arc–continent collision system, in which the Fuping and Hengshan complexes represent two exotic Archaean continental blocks, and the Wutai granite–greenstone represents an intervening island arc (Bai 1986; Li *et al.* 1990; Bai *et al.* 1992; Wang *et al.* 1996). In order to further test and/or refine these models, geologists from China, Australia, and Germany have carried out extensive magmatic, structural, metamorphic, geochemical and geochronological investigations on the Hengshan, Wutai and Fuping complexes in the last few years. However, the data obtained from these investigations do not support either of the above models, and have thus led to a re-interpretation of the geological history of the Hengshan–Wutai–Fuping mountain belt (Zhao *et al.* 1999*a*, 2000*a*, 2001*a*; Wu *et al.* 2000; Guo & Zhai 2001; Guo *et al.* 2002; Kröner *et al.* 2004; Wilde 2002; O'Brien *et al.* 2004). It has become increasingly clear that the tectonic evolution of the Hengshan, Wutai and Fuping complexes was not related to a

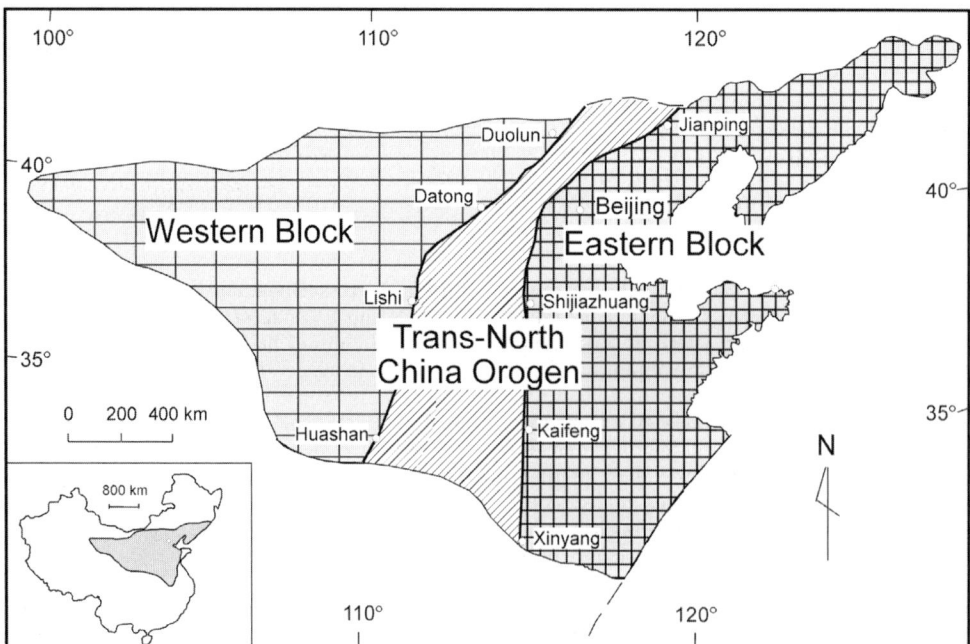

Fig. 1. Three-fold tectonic subdivision of the North China Craton by Zhao *et al.* (1998, 2001*a*).

local interaction, either through closure of a rift or collision between two micro-continental blocks; it appears more likely that the complexes represent an imbricated section of the upper to lower crust, which was deformed, metamorphosed and exhumed during the amalgamation of the North China Craton. In this contribution, we present a compilation, summary and re-assessment of these new data from the Hengshan, Wutai and Fuping complexes, which, in combination with previous studies, provide insights into the Late Archaean to Palaeoproterozoic evolution of the Trans-North China Orogen.

Regional setting

As mentioned above, the North China Craton can be tectonically divided into the Archaean to Palaeoproterozoic Eastern and Western blocks, separated by the Palaeoproterozoic Trans-North China Orogen (Fig. 1; Zhao et al. 2001a).

The Eastern Block consists of early Archaean to Palaeoproterozoic basement, partially overlain by Palaeoproterozoic to Cenozoic platform cover. The early Archaean (>3.4 Ga) basement is extremely limited, represented by c.3.8 Ga trondhjemitic gneisses in the Anshan area and 3.6–3.4 Ga supracrustal rocks in Eastern Hebei (Liu et al. 1992; Song et al. 1996; Wu et al. 1998); their original extent and tectonic history are unknown, owing to extensive reworking by the c.2.5 Ga tectonothermal event. The Middle Archaean (3.4–2.9 Ga) basement is also limited, represented by the Lower Anshan, Qianxi and Longgang groups and associated granite plutons (e.g. Tiejiashan granite). The Late Archaean (2.9–2.5 Ga) basement is more widespread, making up 80% of the total exposure of Archaean basement in the Eastern Block. It consists of 2.6–2.5 Ga tonalitic–trondhjemitic–granodioritic (TTG) gneisses, with minor supracrustal rocks ranging in age from 2.8 to 2.5 Ga. These rocks were intruded by syntectonic granite/charnockite and experienced greenschist- to granulite-facies metamorphism at c.2.5 Ga (Pidgeon 1980; Compston et al. 1983; Jahn & Zhang 1984; Kröner et al. 1998; Zhao et al. 1998). The Palaeoproterozoic (>1.8 Ga) basement comprises I-type and A-type granites, and bimodal volcanic and sedimentary rocks, most of which are interpreted to have developed in intracontinental rift basins (Li et al. 1997, 2004).

Much of the basement in the Western Block underlies deeply subsided Mesozoic to Cenozoic basins, with exposures largely restricted to the northern portion of the block. The exposed basement rocks are poorly constrained in terms of modern geochronology, and are generally considered to be composed of Late Archaean rocks in the north, flanked to the south by a belt of Palaeoproterozoic high-grade supracrustal rocks. Like those in the Eastern Block, the Late Archaean basement rocks in the Western block are also composed predominantly of TTG gneisses with minor supracrustal rocks, metamorphosed from greenschist to granulite facies at c.2.5 Ga (Zhao et al. 1999a). The Palaeoproterozoic supracrustal rocks consist of graphite-bearing, sillimanite–garnet gneisses, garnet quartzites, calc-silicate rocks and marbles, representing stable continental margin deposits (Lu 1991; Condie et al. 1992; Lu & Jin 1993; Lu et al. 1996), with a maximum depositional age of c.2.3 Ga and a metamorphic age of c.1.85 Ga (Zhao et al. 1999a).

Between the eastern and Western blocks is the Trans-North China Orogen, a nearly south–north-trending zone, c.1200 km long and 100–300 km wide, which is separated from the Eastern and Western blocks by the Xingyang–Kaifeng–Shijiazhuang–Jiianping Fault and the Huashan–Lishi–Datong–Duolun Fault, respectively. Both faults strike north–south in the central and southern parts and E–N in the north (Fig. 2). The presence of voluminous mantle-derived basalts exposed along the two faults suggests that they are deep structures, possibly reaching into the lower crust or upper mantle (Ren 1980). The main lithotectonic features of the Trans-North China Orogen include:

(1) dominant Late Archaean to Palaeoproterozoic juvenile crust with minor reworked basement rocks (Zhao et al. 2000a);
(2) linear structural belts defined by strike-slip ductile shear zones, large-scale thrusting and folding, and transcurrent tectonics (Li & Qian 1991; Zhao et al. 1999a);
(3) sheath folds and mineral lineations (Wu & Zhong 1998);
(4) high-pressure granulites and retrograde eclogites (Zhai et al. 1992; Guo et al. 1993, 2002; Zhai 1997; Guo & Zhai 2001; Zhao et al. 2001b, 2001c);
(5) clockwise metamorphic P–T paths involving near-isothermal decompression (Zhao et al. 2000a);
(6) ancient oceanic fragments and ophiolitic mélange (Li et al. 1990; Bai et al. 1992; Wang et al. 1996, 1997a; Wu & Zhong 1998);
(7) syn- or post-tectonic granites (Liu et al. 2000); and
(8) post-collisional mafic dyke swarms (Halls et al. 2000; Zhao et al. 2001a). Lithotectonic elements 1–7 are classical indicators of collision tectonics.

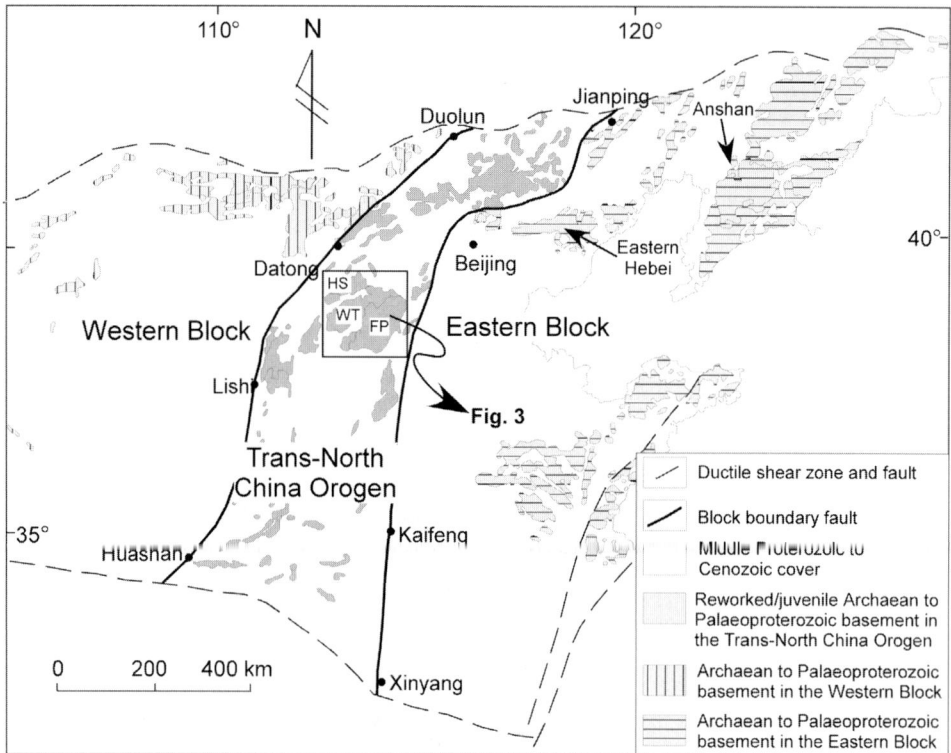

Fig. 2. Spatial distribution of the Trans-North China Orogen in the North China Craton, and the locality of the Hengshan–Wutai–Fuping mountain belt.

The Hengshan Complex is separated from the Wutai Complex by a broad valley of the Hutuo River, and bounded in the NW by the valley of the Sanggan River. The Fuping Complex was previously considered to be unconformably overlain by the Wutai Complex along the 'Tiebao unconformity' (Bai 1986; Wu *et al.* 1989), but recent research has revealed that the so-called 'Tiebao unconformity' is a regional-scale ductile shear zone, named the Longquanguan Ductile Shear Zone by Li & Qian (1991); thus, the nature of the boundary between the Fuping and Wutai Complexes is a tectonic feature.

Lithologies of the Hengshan–Wutai–Fuping mountain belt

Hengshan complex

The Hengshan Complex is composed of five main lithotectonic units (Fig. 3):

(1) the Hengshan TTG gneisses;
(2) the Hengshan mafic granulites;

(3) the Hengshan supracrustal assemblage;
(4) the Zhujiafang supracrustal assemblage; and
(5) the Yixingzhai gneisses (Tian 1991; Li & Qian 1994).

The Hengshan TTG gneisses are strongly deformed layered orthogneisses of tonalitic–trondhjemitic–granodioritic–granitic composition. These rocks are extensively migmatized, and some of the migmatized zones show evidence of *in situ* melting and advanced anatexis, generating reddish granites. The compositional layering in the TTG gneisses ranges from dark, hornblende-rich dioritic to tonalitic compositions to K-feldspar-dominated leucocratic granitoid varieties. Li & Qian (1994) reported major- and trace-element data for the granitoid gneisses, and concluded that they belong to a high-alumina calc-alkaline suite. The tonalites and trondhjemites show no negative Eu-anomalies, which is typical for Archaean TTG gneisses worldwide (Taylor & McLennan 1985).

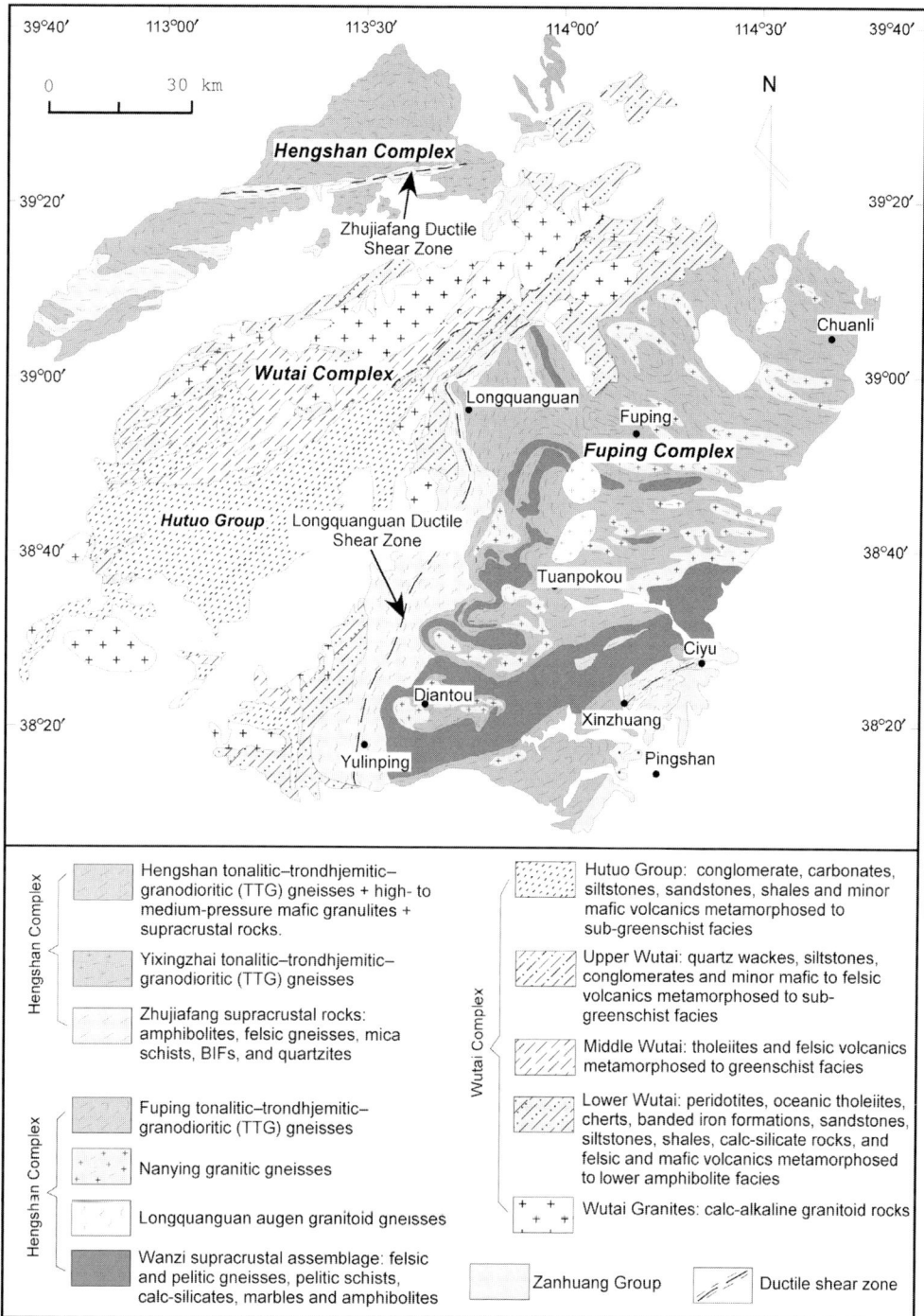

Fig. 3. Geological map of the Hengshan–Wutai–Fuping mountain belt.

Within the TTG gneisses there are discontinuous lenses or layers of amphibolite and high-and medium-pressure mafic granulite, ranging from 0.1 to 5 m in width and 0.1 to 50 m in length and interpreted as boudinaged gabbroic dykes (Kröner *et al.* 2004). In the field, the high-pressure granulite lenses can be distinguished from the surrounding medium-pressure granulites by the presence of coarse-grained textures and lack of brown orthopyroxene in the former. The long axes of the granulite lenses are parallel to the regional foliation of TTG gneisses. In some places, the interiors of amphibolites and mafic granulites preserve a relict igneous gabbroic/doleritic texture. There can be no doubt that these amphibolites and mafic granulites are remnants of mafic dykes that originally intruded into the granitoid rocks, as can still be observed at a few localities in low-strain zones (Kröner *et al.* 2004). It remains to be seen, however, whether all amphibolites and mafic granulites in the Hengshan Complex were derived from metamorphosed mafic dykes.

The Hengshan supracrustal assemblage is interdigitated with the Hengshan TTG gneisses, and consists of lenses and sheets of felsic paragneiss, quartzite, magnetite quartzite, garnet quartzite and minor Al-rich gneiss. These rocks, combined with amphibolite and mafic granulite, were previously called the Dongzhuang or Qianzhuangwang supracrustal rocks, and interpreted as a metagreywacke association (e.g. Li & Qian 1994; Tian 1991) but, on the basis of zircon morphology and age, Kröner *et al.* (2004) interpreted fine-grained biotite gneisses to be derived from felsic volcanic rocks such as dacites and rhyodacites. However, other rocks, including Al-rich gneisses, quartzite and thin layers of BIF, are undoubtedly of sedimentary origin.

The Zhujiafang assemblage consists predominantly of amphibolite, felsic gneiss, mica schist, BIF and quartzite, which occur largely along two nearly east–west-trending belts that cut across the regional layering/foliation of the Hengshan TTG gneisses and Yixingzhai gneisses (Fig. 3). These rocks are characterized by strong ductile deformation and ubiquitous mylonite fabrics, and are distinctively lower in metamorphic grade than the other supracrustal rocks within the Hengshan TTG gneisses. Most Chinese workers have considered the Zhujiafang assemblage to be the equivalent of the Wutai greenstones in the Hengshan area (Tian 1991; Li & Qian 1994), whereas Kröner *et al.* (2004) interpreted them as mylonitized orthogneisses and gabbroic dykes defining major ductile shear zones.

The Yixingzhai gneisses occur in the southern part of the Hengshan Complex. They are chemically similar to the Hengshan TTG gneisses, but are only metamorphosed to greenschist to lower-amphibolite facies and are less deformed, locally preserving igneous textures.

Wutai complex

The Wutai Complex consists of Late Archaean to Palaeoproterozoic granitoids and metamorphosed volcanic and sedimentary rocks, named the Wutai Group in the Chinese literature. As shown in Figure 3, the Wutai Group can be subdivided into three lithotectonic units. The first unit, previously named the Shizui Subgroup that includes the Banyukou, Jingangku, Zhuangwang and Wenxi Formations, consists of peridotites, oceanic tholeiites, dacites, rhyolite, cherts, banded iron formations, sandstones, siltstones, shales, calc-silicate rocks and minor limestones metamorphosed to lower amphibolite facies. Of these, the peridotites, oceanic tholeiites and cherts are considered to represent relict oceanic crust, whereas sandstones, siltstones, shales, calc-silicate rocks and minor limestones are interpreted as continental margin sediments (Li *et al.* 1990; Bai & Dai 1998; Wu & Zhong 1998). The second unit, previously named the Taihuai Subgroup that includes the Baizhiyan and Hongmenyan Formations, comprises felsic volcanic rocks and tholeiites of volcanic-arc affinity, intruded by calc-alkaline granitoid plutons and metamorphosed to greenschist facies, representing a Late Archaean to Palaeoproterozoic accretionary arc formation (Bai & Dai 1998; Wu & Zhong 1998). The third unit (called the Gaofan Subgroup and Hutuo Group) contains Palaeoproterozoic conglomerates, quartz wackes, siltstones, limestones and minor mafic to felsic volcanics, which are interpreted as developing in an intra-arc basin and/or a retro-arc foreland basin (Zhao *et al.* 2001a). The three imbricated lithotectonic units were structurally disrupted and juxtaposed along a series of NE–SW-trending ductile shear zones.

The granitoid plutons in the Wutai Complex can be largely divided into pre-, syn- and post-tectonic intrusions. The pre- to syn-tectonic intrusions, including the Chechang–Beitai, Dazhaikou, Duyu, Ekou, Shifu, Lanzhishan, Guangmingshi and Wangjiahui bodies, are composed predominantly of diorite–tonalite–trondhjemite–granodiorite suites, with minor quartz monzonite and monzogranite, and are variably deformed. The post-tectonic intrusions, including the Dawaliang, Fengkuangshan, Lianhuashan and Pingxingguan bodies, are composed of biotite granites and porphyritic granites, with a massive structure.

Fuping complex

The Fuping Complex comprises four distinct lithotectonic units, herein called the Fuping TTG gneisses, Longquanguan augen gneisses, Wanzi supracrustal assemblage, and Nanying granitic gneisses (Fig. 3; Liu 1997; Liu et al. 2000; Zhao et al. 2000a).

The Fuping gneisses make up about 60% of the complex, and consist of medium-grained tonalitic, trondhjemitic and granodioritic (TTG) gneisses enclosing mafic granulite, amphibolite and hornblende gneiss that have undergone a complex history of upper amphibolite- to granulite-facies metamorphism and polyphase deformation. Petrological and geochemical data suggest that the Fuping gneisses were derived from partial melting of mantle-derived basaltic rocks (Wang et al. 1991). They are interpreted to be in tectonic contact with the Wanzi supracrustal assemblage.

The Longquanguan augen gneisses, previously called the Longquanguan Group (Wu et al. 1989), are mainly exposed along the Longquanguan–Yulinping and Ciyu–Xinzhuang ductile shear zones (Fig. 3) and are composed predominantly of coarse-grained to porphyritic granodioritic and monzogranitic gneisses and mylonitized granitic pegmatites containing K-feldspar phenocrysts – most of which have been intensely deformed to form augen. Enclosed in the Longquanguan augen gneisses are amphibolite and hornblende gneiss enclaves similar to those in the Fuping gneisses. Along the Ciyu–Xinzhuang ductile shear zone, a gradual transition from weakly mylonitized to intensely mylonitized granodioritic and monzogranitic gneisses has been observed. The Longquanguan augen gneisses display a tectonic contact with the Wanzi supracrustal assemblage, but have a transitional relationship with the Fuping gneisses. They may represent reworking of the Fuping gneisses (Wu et al. 1989; Li & Qian 1991).

The Wanzi supracrustal assemblage forms a 100-km-long, northeast–southwest-trending belt in the southern part of the complex (15 km wide) that swings northward to the central part of the complex, where it is extensively folded (Fig. 3). The supracrustal rocks are metamorphosed to amphibolite facies and comprise felsic and pelitic gneisses, pelitic schists, calc-silicates, pure and impure marbles and amphibolites (Liu & Liang 1997; Zhao et al. 2000a). Also associated with the supracrustal rocks are some small

sillimanite-bearing granites, which are considered to represent S-type granites derived from partial melting of pelitic gneisses and felsic paragneisses, respectively (Zhao et al. 2000a, 2002a).

The Nanying granitic gneisses only occur within the Fuping gneisses, and are dominated by medium- to fine-grained, weakly foliated, magnetite-bearing monzogranitic gneiss, with minor granodioritic gneiss (Fig. 3; Zhao et al. 2000a, 2002a). In addition to compositional differences, the Nanying gneisses are more massive in structure, and more homogeneous in composition than the Fuping gneisses. In the field, where contact relations are preserved, the Nanying granitic gneisses are clearly intrusive into the Fuping gneisses, but their relatively weak foliation is consistently parallel to the strong penetrative foliation of the Fuping gneisses, suggesting that they were most likely to have undergone the same deformational event that resulted in the development of the regional foliation of the Fuping Complex.

Structures of the Hengshan–Wutai–Fuping mountain belt

Although different in lithologies and metamorphic grade, the Hengshan, Wutai and Fuping Complexes display a similar deformational history that can be related to three major phases of deformation, as summarized in Table 1. The earliest (D_1) structural elements in all three complexes have nearly been completely obliterated by later structural events, except for a relict foliation (S_1). In the high-grade Hengshan and Fuping Complexes, the earliest foliation (S_1) is only found in mafic granulites and amphibolites enclosed in the TTG gneisses. Thin felsic veins within these mafic granulite/amphibolite enclaves show a planar fabric (S_1) that is truncated by the major foliation (S_2) in the TTG gneisses. In addition, clinopyroxene and hornblende crystals in the mafic granulites and amphibolites enclaves are also oriented to form a foliation S_1, parallel to the felsic veins. In the low- and medium-grade Wutai Complex, the oldest foliation (S_1) is represented by a preferred planar orientation of actinolite + chlorite + epidote + biotite or muscovite + biotite mineral inclusions within garnet porphyroblasts.

The main phase of deformation in all the three complexes is D_2, leading to the development of thrust slices, tight to isoclinal folds (F_2), a penetrative foliation (S_2) and a mineral lineation (L_2) (Table 1). In most cases in the field where S_1 is absent, a penetrative foliation in the three complexes is the earliest recognizable fabric

Table 1. *Deformation phases and their structural elements in the Hengshan, Wutai and Fuping Complexes (after Zhang* et al. *1983; Tian 1991)*

Complex	D_1	D_2	D_3
Hengshan Complex	The earliest deformational structures are only found in high- and medium-pressure mafic granulites enclosed in the TTG gneisses. Some thin felsic veins in these mafic granulites show a planar fabric (S_1) that is truncated by the major foliation (S_2) in the TTG gneisses. Some clinopyroxene and hornblende crystals in these enclaves are also oriented to form a foliation S_1.	Major deformation phase, including large-scale thrust tectonics, ubiquitous recumbent folds (F_2), a regional foliation (S_2), boudinage of high-pressure mafic granulites, and a mineral lineation (L_2). Most L_2 lineations are nearly NE or E-W trending. D_2 was coeval with the peak metamorphism.	Asymmetrical folds (F_3) with nearly vertical axial planes, locally forming strain-slip cleavage (S_3). Also associated with D_3 is development of several-kilometre-wide E–W transcurrent steep shear zones that extend the full length of the Hengshan Complex, e.g. the Xiheqiao–Zhujiafang–Shuangqianshu shear zone, Zhoujiazhuang–Nianzigou shear zone, etc.
Wutai Complex	The oldest structure is a mineral foliation (S_1), defined by a preferred orientation of actinolite + chlorite + epidote + biotite or muscovite + biotite within garnet porphyroblasts in amphibolites and mica-schists.	Main deformation phase, leading to large-scale thrust tectonics, isoclinal folds (F_2), penetrative foliation (S_2), a mineral lineation (L_2) and minor sheath folds, indicating a roughly eastward movement. Most L_2 lineations are nearly NE or E-W trending. D_2 was coeval with the peak metamorphism.	Asymmetrical folds (F_3) with nearly vertical axial planes, local strain-slip cleavage (S_3) and a mullion or boudin lineation (L_3). Also associated with D_3 is the development of a large number of transcurrent steep shear zones along the boundaries of the formations in the Wutai Group (see Fig. 3).
Fuping Complex	The earliest structure is only found in medium-pressure mafic granulites and amphibolites enclosed in the Fuping TTG gneisses. Thin felsic veins in these enclaves show a planar fabric (S_1) that is truncated by the major foliation (S_2) in the TTG gneisses.	Major deformation phase, represented by large-scale thrust tectonics, isoclinal folds (F_2), a regional foliation (S_2), boudinage of medium-pressure mafic granulites and a mineral lineation (L_2). Most L_2 lineations are nearly NE or E-W trending. D_2 was coeval with the peak metamorphism.	Asymmetrical folds (F_3) with nearly vertical axial planes, local strain-slip cleavage (S_3). Also associated with D_3 is development of transcurrent steep shear zones, e.g. the NE–SW-trending Ciyu–Xinzhuang ductile shear zone and the N–S-trending Longquanguan shear zones that extend the full length of the Fuping Complex.

and is assigned to S_2. In the Hengshan Complex, D_2 was roughly coeval with the peak high-pressure metamorphism, and erased all older structures except for relics of S_1 foliation in mafic granulite/amphibolite enclaves or garnet porphyroblasts (Passchier & Walte 2002). Intensely foliated D_2-high strain zones alternate with lower strain zones, showing a foliation strike from N–S and NE–SW in the north to E–W in the central and southern part of the complex (Zhujiafang supracrustal unit). Also associated with D_2 are small-scale shear zones, whose range of orientation is due to subsequent folding (F_3). The D_2 deformation also led to boudinage

of the high-pressure mafic granulite layers and a mineral lineation (L_2), defined by sillimanite and hornblende crystals on the foliation surface. In the low-grade Wutai Complex, D_2-related thrust slices and isoclinal recumbent folds (F_2) of variable scale are ubiquitous, especially in the Erkou, Yantou and Gaofan areas, where most F_2 axial planes have been refolded during later deformational phases (Fig. 4). Associated with these thrust slices and folds are a roughly NE–SW penetrative axial-planar schistosity (S_2) and a mineral stretching lineation (L_2) that generally plunges to the east or ENE (Fig. 4). Similarly, thrust slices, tight to isoclinal folds,

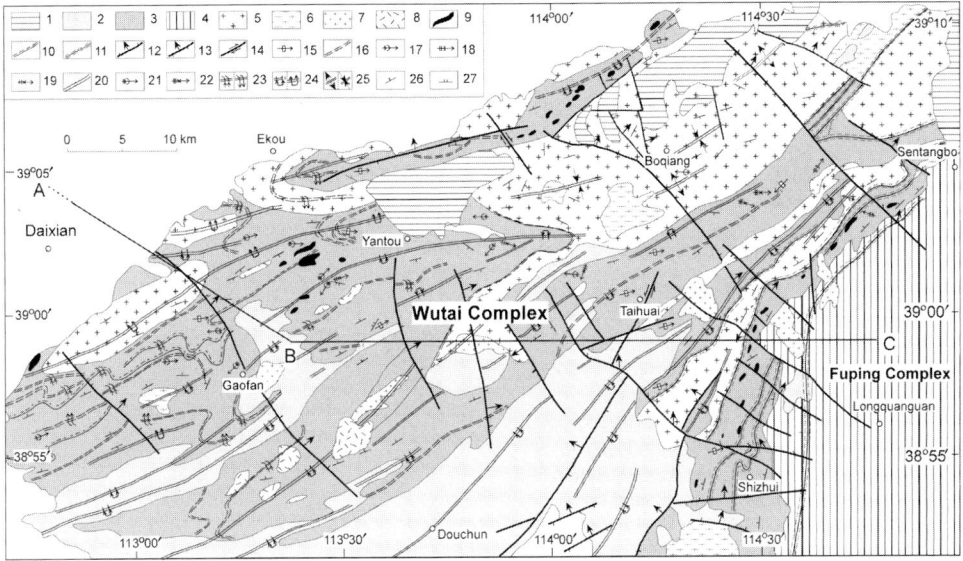

Fig. 4. Structural outline of the Wutai Complex (after Tian 1991). 1, Unmetamorphosed cover; 2, Hutuo Group; 3, Wutai greenstone assemblage; 4, Fuping Complex; 5, Wutai granitoid rocks; 6, Precambrian granitoid rocks of unknown age; 7, Mesozoic granites; 8, mafic intrusions; 9, ultramafic rocks (remnants of oceanic crust?); 10, décollement or thrusting fault; 11, ductile shear zone; 12, normal fault; 13, reverse fault; 14, strike-slip fault; 15, mineral stretching lineation; 16, axial line of F_2 fold; 17, F_2 hinge; 18, lineation defined by F_2 hinge; 19, intersection lineation defined by F_2 axial plane; 20, axial line of F_3 fold; 21, F_3 hinge; 22, intersection lineation defined by F_3 axial plane; 23, isoclinal antiform or synform; 24, Isoclinal antiform or synform; 25, upright antiform or synform; 26, foliation S_2; 27, fold axial plane.

penetrative foliations, boudins and mineral stretching lineations that formed simultaneously with the peak granulite-facies metamorphism also characterize the second phase of deformation (D_2) of the Fuping Complex (Zhao *et al.* 2000b; Fig. 5).

The third phase of deformation (D_3) in all the three complexes resulted in the development of large-scale ductile shear zones and associated open to tight asymmetrical folds (F_3) with nearly vertical axial planes, local strain-slip cleavages (S_3) and mullion or boudin lineations (L_3) (Table 1). In the Hengshan Complex, a number of several-kilometre-wide East–West transcurrent steep shear zones extend the full length of the complex (Fig. 3). These shear zones show a dextral sense of shear as recorded by asymmetrical feldspar clasts, deformation of syn-D_3 intruded pegmatite dykes, and C' shear bands found in thin sections of ultramylonite (Passchier & Walte 2002). The change in strike of S_2 in the northern part of the Hengshan Complex is probably an effect of dextral shear on these major shear zones. In the gneisses near the shear zones, F_3 is locally recorded as open to tight asymmetrical folds of the D_2 foliation. In the Wutai Complex, D_3 formed numerous regional-scale

NE-trending, axial plane-vertical, asymmetrical folds (F_3) that locally refolded F_2 isoclinal recumbent folds (Fig. 4). The structural outline of the Wutai Complex is characterized by a large composite 'synform' that was mainly shaped by the superposition of F_2 and F_3 (Fig. 6). S_3 in the Wutai Complex commonly occurs as local strain-slip cleavages or mylonitic foliations within regional-scale brittle and semi-brittle strike-slip shear zones, especially in the eastern part of the complex (Fig. 4). In the Fuping Complex, D_3 deformation also resulted in the development of numerous regional-scale east–west-trending, axial-plane-vertical, asymmetrical folds (F_3) that refolded F_2 recumbent folds (Fig. 5). Along the western margin of the Fuping Complex is a roughly south–north trending ductile shear zone, named the Longquanguan shear zone, which extends from Yulinping Village in the south to Longquanguan Village in the north, and is up to 70 km long (Fig. 5). The shear zone consists of highly strained granitoid rocks and associated amphibolite enclaves. Most of these rocks can be classified as augen gneisses (referred as to the Longquanguan augen gneisses in Fig. 3), with minor protomylonites and mylonites present in the central part of the shear zone. Generally,

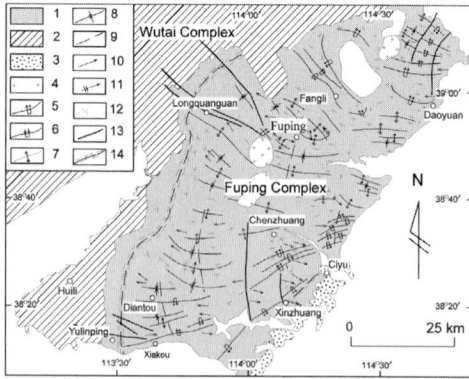

Fig. 5. Structural outline of the Fuping Complex, revised after Zhang *et al.* (1983) and Wu *et al.* (1989). 1, Fuping Complex; 2, Wutai Complex; 3; Palaeoproterozoic Zanhuang Group; 4, Mesozoic granites; 5, isoclinal folds or tight antiforms (F_2); 6, isoclinal folds or closed synforms (F_2); 7, upright antiforms (F_3 or F_2 refolded by F_3); 8, upright synform (F_3 or F_2 refolded by F_3); 9, F_3 axial trace; 10, mineral lineation (L_2); 11, mullion or boudin lineation (L_3); 12; penetrative foliation (S_2); 13, major faults; 14, ductile shear zone.

they show a NW-dipping ($5-30°$) mylonitic foliation defined by quartz ribbons, flattened plagioclase and hornblende grains and a stretching lineation defined by elongated hornblende and quartz ribbons. The plunge of the lineation varies slightly from WNW–ESE to east–west.

The shear zone and its mylonitic foliation cut the regional gneissosity (S_2) of the Fuping grey gneisses (Fig. 3), suggesting that the shear zone formed after the D_2 deformation of the Fuping complex, most probably coevally with D_3. This shear zone and those in the Hengshan and Wutai complexes may represent important tectonic boundaries along which high-grade metamorphic rocks were exhumed to the shallow levels of the crust. This conclusion is supported by the fact that these shear zone separate thrust slices with different metamorphic grades. For example, the Longquanguan ductile shear zone separates the high-grade Fuping Complex from the low- to medium-grade Wutai Complex (Fig. 3); the Zhujiafang ductile shear zone separates the northern high-pressure granulite-facies terrane from the southern medium-pressure granulite- and amphibolite-facies terrane (Fig. 3; Passchier & Walte 2002; Kröner 2002; Kröner *et al.* 2004).

Metamorphic evolution of the Hengshan–Wutai–Fuping mountain belt

The available petrographic and geothermobarometric data show that the Hengshan, Wutai and Fuping complexes all underwent a metamorphic history that is characterized by nearly isothermal decompression and then cooling following peak metamorphism.

In the Hengshan Complex, the high-pressure granulites preserve four distinct metamorphic

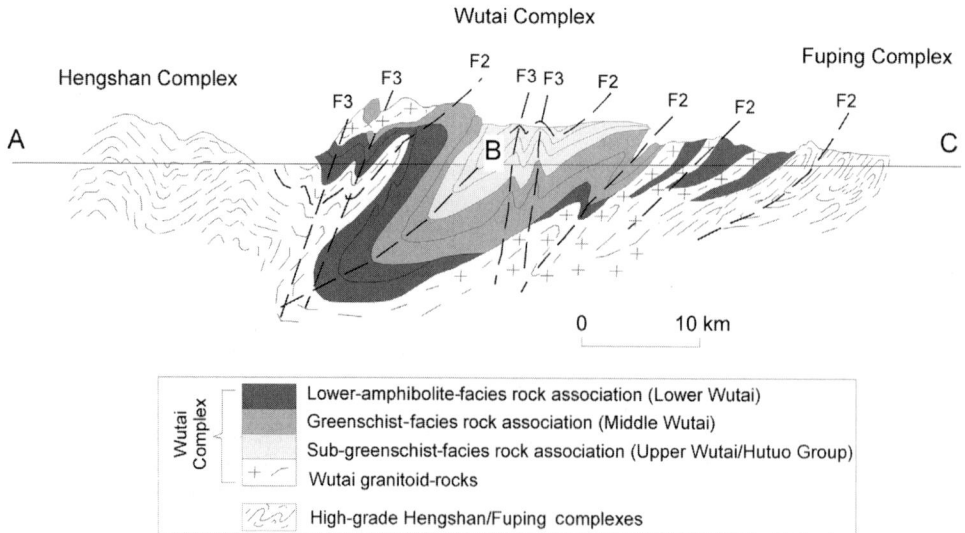

Fig. 6. Schematic cross-section of the Wutai Complex, showing the superposition of F_2 and F_3.

assemblages (Zhao *et al.* 2001*b*): prograde, peak, post-peak decompression and later retrogression cooling. The early prograde assemblage is represented by quartz and rutile inclusions within the cores of garnet porphyroblasts, and omphacite pseudomorphs that are represented by clinopyroxene + sodic plagioclase (An_{10-20}) symplectic intergrowths (Fig. 7a). The peak assemblage consists of clinopyroxene + garnet + sodic plagioclase + quartz ± hornblende. The post-peak decompression assemblage represents the development of orthopyroxene + clinopyroxene + plagioclase symplectites (Fig. 7b) and coronas (Fig. 7c) surrounding embayed garnet grains. The later retrogression cooling is represented by hornblende + plagioclase symplectites which occur as worm-like intergrowths adjacent to garnet (Fig. 7d). The $P-T$ conditions of the early prograde assemblage cannot be quantitatively estimated because of the absence of representative high-pressure minerals (e.g. omphacite). The $P-T$ conditions were estimated at 13–16 kbar and 770–840 °C for the peak mineral assemblage, 6.5–8.0 kbar and 750–830 °C for the decompression assemblage and 4.5–6.0 kbar and 680–790 °C for the later retrogression cooling assemblage (Zhao *et al.* 2001*b*). The combination of petrographic textures and thermobarometric data defines a near-isothermal decompressional clockwise $P-T$ path for the Hengshan high-pressure granulites (Fig. 8a). Medium-pressure granulites in the Hengshan Complex preserve peak, post-peak decompression and later retrogression cooling assemblages, represented by orthopyroxene + clinopyroxene + garnet + plagioclase + quartz in the matrix, orthopyroxene/clinopyroxene + plagioclase symplectites and coronas surrounding embayed garnet grains, and hornblende + plagioclase symplectites on garnet, respectively (Zhao *et al.* 2001*b*). The THERMOCALC program yielded $P-T$ conditions of 9–11 kbar and 820–870 °C for the peak assemblage, 6.5–8.0 kbar and 750–830 °C for the pyroxene + plagioclase symplectite and corona, and 4.5–6.0 kbar and 680–790 °C for the hornblende + plagioclase symplectite. These mineral assemblages and their $P-T$ estimates also defined a clockwise $P-T$ path involving near-isothermal decompression (Fig. 8b; Zhao *et al.* 2001b).

In the Wutai Complex, the amphibolites from the Jingangku Formation also preserve four metamorphic stages: prograde, peak, post-peak decompression and later retrogression cooling. The prograde assemblage is composed of plagioclase + quartz + actinolite + chlorite + epidote + biotite + rutile, preserved as inclusions within garnet porphyroblasts. This is a greenschist-facies assemblage that is stable

between 400 and 500 °C. The peak assemblage represents the growth of coarse garnet porphyroblasts and matrix minerals of amphibole + plagioclase + quartz + biotite ± clinopyroxene ± rutile ± ilmenite. The average $P-T$ conditions for this assemblage were estimated at 10–12 kbar and 600–650 °C (Zhao *et al.* 1999*c*). The post-peak decompression assemblage is represented by amphibole + Ca-rich plagioclase symplectites around the embayed garnet grains. The $P-T$ conditions for this decompression assemblage were estimated at pressure of 6.0–7.0 kbar and temperature of 610–650 °C (Zhao *et al.* 1999*c*). The later retrogression cooling assemblage represented by chlorite and epidote replacing garnet, chlorite replacing amphibole and epidote replacing plagioclase occurs under greenschist-facies conditions. Therefore, these mineral assemblages and their $P-T$ estimates from the amphibolites define an isothermal decompressional, clockwise, $P-T$ path for the Wutai Complex (Fig. 8c; Zhao *et al.* 1999*c*). In the Wutai pelitic gneisses/schists, three metamorphic mineral assemblages are recognized: the early prograde, peak and peak-post decompression assemblages (Zhao *et al.* 2000*a*). The early prograde assemblage is plagioclase + quartz + biotite + muscovite + staurolite, occurring as inclusions within garnet porphyroblasts. The peak assemblage is plagioclase + quartz + kyanite/sillimanite + biotite + garnet ± muscovite, representing the growth of porphyroblasts and matrix minerals. The decompressional assemblage is represented by cordierite + plagioclase symplectites surrounding garnet porphyroblasts (Liu *et al.* 1997; Wang *et al.* 1997*b*). The $P-T$ conditions were estimated at 7–8 kbar and 550–600 °C for the early prograde assemblage; 10–11 kbar and 600–650 °C for the peak assemblage; and 5.5–6.5 kbar and 600–700 °C for the peak-post-decompressional assemblage (Zhao *et al.* 2000*a*). These $P-T$ estimates also define an isothermal decompressional, clockwise, $P-T$ path (Fig. 8d; Liu *et al.* 1997; Wang *et al.* 1997*b*).

In the Fuping Complex, the mafic granulites preserve peak, peak-post decompression and later retrogression cooling assemblages. The peak is represented by the growth of garnet porphyroblasts and matrix quartz + plagioclase + orthopyroxene + clinopyroxene; the peak-post assemblage is documented by worm-like orthopyroxene + clinopyroxene + plagioclase ± magnetite symplectites and coronas around embayed garnet grains; and the retrogression cooling assemblage is represented by hornblende + plagioclase symplectites surrounding garnet grains. The peak $P-T$ conditions were estimated at 8.5–9.5 kbar and

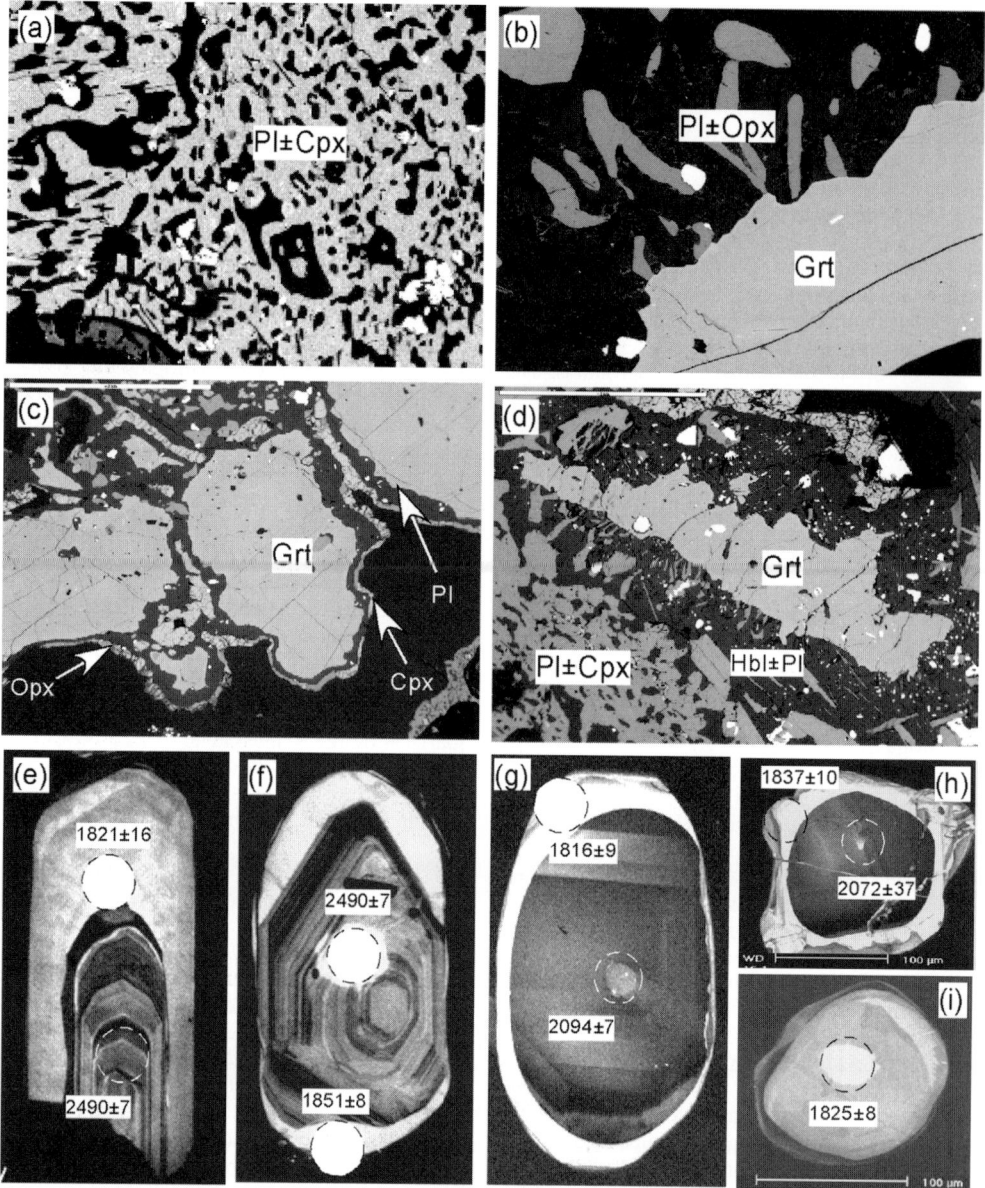

Fig. 7. Backscattered electron images showing representative metamorphic textures and cathodoluminescence (CL) images showing internal structures of zircons from the Fuping Complex. (**a**) Na-rich plagioclase + clinopyroxene symplectite forming by the breakdown of omphacites; (**b**) pyroxene + plagioclase symplectites surrounding the embayed garnet grains; (**c**) pyroxene + plagioclase coronas surrounding embayed garnet grains in the Hengshan mafic granulites; (**d**) hornblende + plagioclase symplectite on embayed garnet grains in the Hengshan mafic granulites; (**e**)–(**f**) metamorphic zircons occurring as overgrowth rims surrounding the magmatic zircon cores in the Archaean Fuping tonalitic gneisses; (**g**)–(**h**) Metamorphic zircons occurring as overgrowth rims surrounding the magmatic zircon cores in the Palaeoproterozoic Fuping granitic gneisses; (**i**) Metamorphic zircons occurring as single grains in the Archaean Fuping tonalitic gneisses. Mineral symbols are after Kretz (1983). Age unit in (**e**)–(**i**) is million years (Ma).

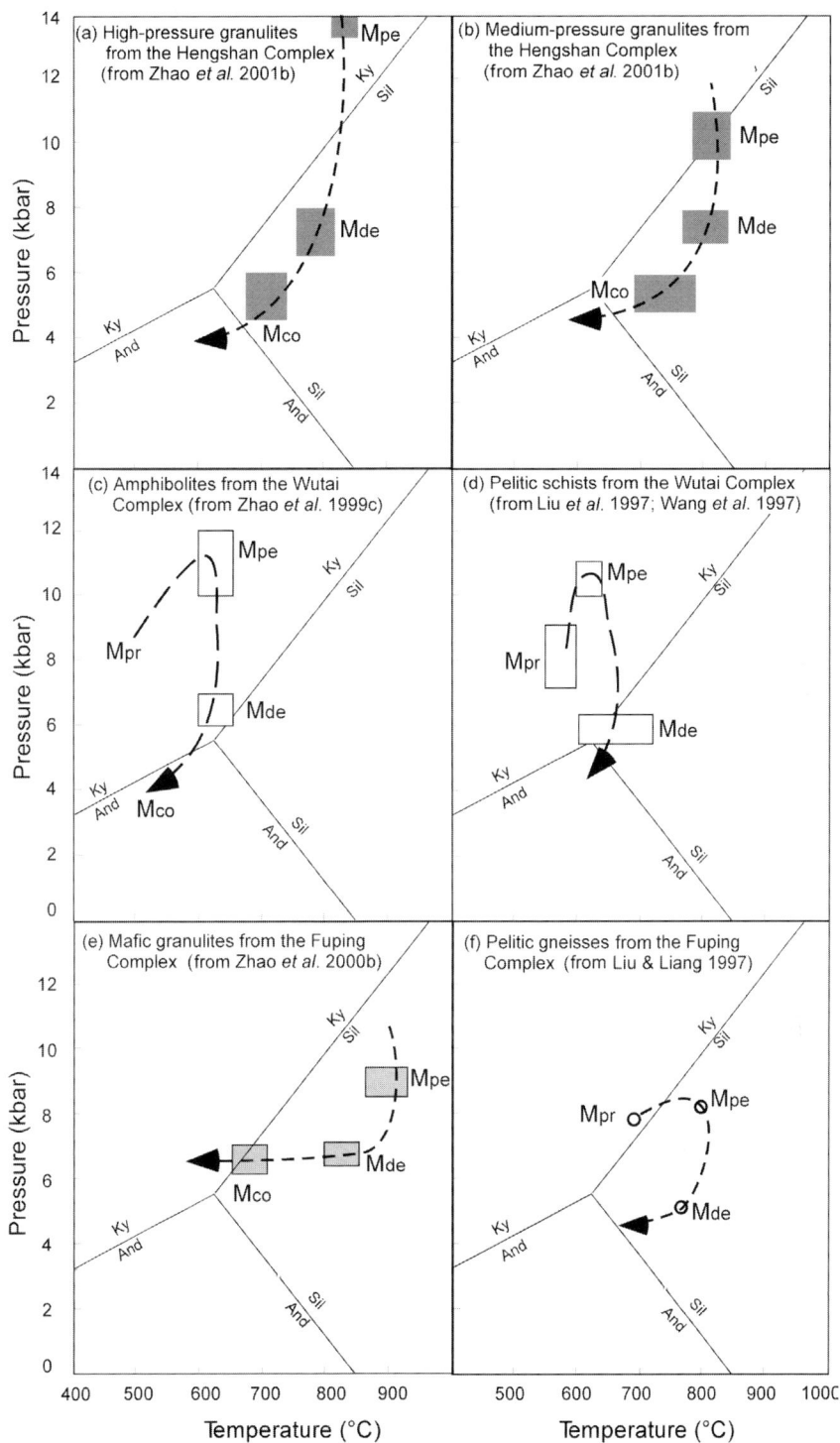

Fig. 8. Metamorphic $P-T$ paths of the Hengshan, Wutai and Fuping Complexes. Mineral symbols after Kretz (1983).

870–930 °C, based on the core compositions of garnet, matrix orthopyroxene, clinopyroxene and plagioclase. The peak-post $P-T$ conditions were estimated at 6.5–7.0 kbar and 800–850 °C, based on the garnet rim compositions and symplectic or coronitic orthopyroxene, clinopyroxene and plagioclase compositions. The $P-T$ conditions of the retrogression cooling assemblage were calculated at 6.0–7.0 kbar and 650–700 °C, based on garnet rim compositions and symplectic hornblende and plagioclase compositions. These $P-T$ estimates define a near-isothermal decompressional, clockwise $P-T$ path for the Fuping Complex (Fig. 8e; Zhao *et al.* 2000b). Metapelitic gneisses in the Fuping Complex preserve the pre-peak, peak and post-peak assemblages. The early prograde assemblage is represented by mineral inclusions of staurolite + plagioclase + muscovite + biotite + rutile within garnet porphyroblasts; the peak assemblage are matrix minerals of plagioclase + biotite + sillimanite + K-feldspar + ilmenite + gedrite and garnet porphyroblasts; and the peak-post decompressional assemblage is represented by spinel + corundum/sillimanite ± plagioclase ± biotite (Liu & Liang 1997; Zhao *et al.* 2000a). The $P-T$ conditions were determined at *c.* 8.0 kbar and *c.* 680 °C for the pre-peak assemblage, *c.* 8.0 kbar and *c.* 800 °C for the peak assemblage, and <6.5 kbar and *c.* 750 °C for the post-peak assemblage. These mineral assemblages and their $P-T$ estimates also define a clockwise $P-T$ path involving near-isothermal decompression (Fig. 8f; Liu & Liang 1997).

Therefore, the metamorphic evolution of the basement rocks in the Hengshan–Wutai–Fuping mountain belt, regardless of their metamorphic grade and composition, are all characterized by a clockwise $P-T$ path involving near-isothermal decompression following the peak metamorphism. Near-isothermal decompression paths require that unroofing of deep-seated metamorphic rocks is rapid relative to the rate of thermal relaxation and cooling. This can be typically be accompanied by rapid erosional exhumation or uplift following continent–continent collision (England & Richardson 1977; England & Thompson 1984; Thompson & England 1984; Oxburgh 1989; Brown 1993, 2001).

Geochronology of the Hengshan–Wutai–Fuping mountain belt

Most recently, the sensitive high-mass resolution ion micro probe (SHRIMP) technique combined with the single-grain zircon evaporation and mineral Sm–Nd and Ar/Ar methods has been applied to the dating of the Hengshan, Wutai and Fuping complexes, leading to a large amount of high-precision age data for these complexes (Wilde *et al.* 1997, 1998; Liu *et al.* 2000; Guan *et al.* 2002; Kröner *et al.* 2001, 2003; Kröner 2002; Wilde 2002; Zhao *et al.* 2002a). These new isotopic age data (Tables 2–4) enable improved resolution of magmatic and metamorphic events that contribute to a better understanding of the Late Archaean to Palaeoproterozoic history of the Hengshan–Wutai–Fuping mountain belt.

The oldest basement rocks

Recent isotopic data reveal that the known oldest basement components in the Hengshan–Wutai–Fuping mountain belt are medium-grained grey gneisses that occur interlayered with the younger TTG gneisses, but are difficult to recognize as separate units in the field (Kröner 2002; Kröner *et al.* 2004). A medium-grained hornblende gneiss sample collected from the Fuping Complex was dated by Guan *et al.* (2002) using SHRIMP at 2708 ± 8 Ma (Table 2), which was interpreted as the crystallization age of the tonalitic protolith. Two medium-grained biotite gneiss samples collected from the Hengshan Complex were dated by Kröner *et al.* (2004) at 2701 ± 5.5 Ma and 2697.1 ± 0.3 Ma (Table 3), also interpreted as the crystallization ages of granitoid plutons. Although rocks of similar age have not been found from the Wutai Complex, some Wutai granitoid rocks contain a small population of *c.* 2.7 Ga zircons (Wilde *et al.* 1998; Wilde 2002). These data indicate the existence of *c.* 2.7 Ga crustal material in the Hengshan–Wutai–Fuping region, which are considered to be remnants of an older basement to these complexes (Guan *et al.* 2002; Zhao *et al.* 2002a; Kröner *et al.* 2004).

Late Archaean to Early Palaeoproterozoic (2566–2450 Ga) granitoid magmatism

Many isotopic data published in the last few years indicate that TTG gneisses in the Hengshan and Fuping Complexes and gneissic granites in the Wutai Complex, which make up the bulk of these complexes (Fig. 3), were emplaced in the Late Archaean and Early Palaeoproterozoic (Tables 2–4). Moreover, an important conclusion from these new data is that the Hengshan and Fuping TTG gneisses are slightly younger than the Wutai gneissic granites; the latter were emplaced in the Late

Table 2. *SHRIMP U–Pb zircon data for the main lithologies of the Fuping Complex*

Rock assemblages	Lithologies	Sample no.	Magmatic crystallization age (a) or detrital age (b) (Ma)	Metamorphic age (Ma)	Sources
Old gneiss	Hornblende gneiss	FP50	2708 ± 8 (a)		Guan et al. (2002)
Longquanguan augen granite	Augen granite gneiss	WL12	2543 ± 7 (a)		Wilde et al. (1997)
	Augen tonalitic gneiss	WN11	2541 ± 14 (a)		Wilde et al. (1997)
	Augen granite gneiss	WL9	2540 ± 18 (a)		Wilde et al. (1997)
Fuping TTG gneisses	Tonalitic gneiss	FG1	2523 ± 14 (a)	1802 ± 43	Zhao et al. (2002a)
	Trondhjemitic gneiss	FP54	2513 ± 12 (a)		Guan et al. (2002)
		FP217	2499 ± 9.5 (a)		Zhao et al. (2002a)
	Granodioritic gneiss	FP216	2486 ± 8 (a)	1875 ± 43	Zhao et al. (2002a)
		FP08	2475 ± 8 (a)	1825 ± 12	Guan et al. (2002)
	Monzogranitic gneiss	FP236	2510 ± 22 (a)	1817 ± 26	Zhao et al. (2002a)
	Deformed pegmatite	FP224	2507 ± 11 (a)		Zhao et al. (2002a)
Wanzi supracrustals	Sillimanite leptynite	FP260	2502 ± 14 (b)		Zhao et al. (2002a)
		FP249	2502 ± 7 (b)		Zhao et al. (2002a)
			2109 ± 5 (b)		
Nanying gneisses	Monzogranitic gneiss	FP188-2	2077 ± 13 (a)	1826 ± 12	Zhao et al. (2002a)
		FP30	2045 ± 64 (a)		Guan et al. (2002)
	Granodioritic gneiss	FP204	2024 ± 21 (a)	1850 ± 9.6	Zhao et al. (2002a)
Pegmatites	Granitic pegmatite dyke	FG2	1790 ± 8 (a)	(1790 ± 8)	Wilde et al. (1998)

Table 3. *U–Pb zircon data for the main lithologies of the Hengshan Complex*

Rock assemblages	Lithologies	Sample no.	Magmatic zircon age (a) or detrital age (b) (Ma)	Metamorphic zircon age (Ma)	Methods	Sources
Old grey gneiss	Granodioritic gneiss	980811	2697 ± 0.3 (a)		SGEZ	Kröner (2002)
	Biotite gneiss	990838	2701 ± 5.5 (a)		SHRIMP	Kröner (2002)
Hengshan TTG gneisses	Dioritic gneiss	980814	2479 ± 3 (a)		SHRIMP	Kröner (2002)
		980814	2478.2 ± 0.3 (a)		SGEZ	Kröner (2002)
		990803	2455 ± 2 (a)	1881 ± 8	SHRIMP	Kröner (2002)
		990859	2506 ± 5 (a)		SHRIMP	Kröner (2002)
	Tonalitic gneiss	980802	2504.6 ± 0.3 (a)		SGEZ	Kröner (2002)
		990845	2500.5 ± 0.3 (a)		SGEZ	Kröner (2002)
		990871	2504.4 ± 0.4 (a)		SGEZ	Kröner (2002)
		HG1	2502.3 ± 0.3 (a)		SGEZ	Kröner (2002)
			2520 ± 15 (a)	1872 ± 17	SHRIMP	Wilde (2002)
	Trondhjemitic gneiss	980838	2503.0 ± 0.3 (a)		SGEZ	Kröner (2002)
		990843	2499.0 (a)		SGEZ	Kröner (2002)
		990847	2501.4 (a)		SGEZ	Kröner (2002)
		990854	2504.6 ± 0.3 (a)		SGEZ	Kröner (2002)
	Granodioritic/granitic gneiss	980809	2501 ± 3 (a)		SHRIMP	Kröner (2002)
			2503.5 ± 0.3 (a)		SGEZ	Kröner (2002)
		980825	2496.5 (a)		SGEZ	Kröner (2002)
		980833	2492.4 ± 0.3 (a)		SGEZ	Kröner (2002)
		990873	2499 ± 6 (a)		SHRIMP	Kröner (2002)
			2497.6 (a)		SGEZ	Kröner (2002)
		980803	2502.2 (a)		SGEZ	Kröner (2002)
Hengshan supracrustals	Felsic metavolcanics (?)	980824	2670.6 ± 0.4 (a)		SGEZ	Kröner (2002)
		990821	2526 ± 4.7 (a)		SHRIMP	Kröner (2002)
	Mafic granulite	HG2		1827 ± 10	SHRIMP	Wilde (2002)
	Garnet quartzite	HG5	2527 ± 10 (b)	1881 ± 86	SHRIMP	Wilde (2002)
Yixingzhai gneisses	Tonalitic gneiss	96PC153	2513 ± 5 (a)		SHRIMP	Wilde (2002)
		96PC154	2499 ± 4 (a)			Wilde (2002)
Zhujiafan supracrustals	Metagreywacke	HG7	2501 ± 15 (b)		SHRIMP	Wilde (2002)
Hengshan foliated granite	Red (anatectic?) gneissic granite	990844	2113 ± 8 (a)		SHRIMP	Kröner (2002)
Pegmatitic granite-gneiss	Pegmatite	990844	2112.3 ± 0.6 (a)		SGEZ	Kröner (2002)
		990881	2248.5 ± 0.5 (a)		SGEZ	Kröner (2002)
Mafic dyke swarms	Pegmatitic granite-gneiss	HG4	2331 ± 36 (a)		SHRIMP	Wilde (2002)
	Mafic dyke	GU12	1769.1 ± 2.5 (a)		SGDE	Halls et al. (2000)

Abbreviations: SGDZ, single-grain dissolution zircon U–Pb age; SGEZ, single-grain evaporation zircon U–Pb age; SHRIMP, Sensitive High-Resolution Ion Microprobe zircon U–Pb age.

Table 4. *SHRIMP U–Pb zircon ages of the Wutai Complex*

Lithotectonic unit	Rock	Sample no.	Location	Age (Ma)	Sources
Erkou	Deformed granitoid	95PC34	Erkou	2566 ± 13	Wilde *et al.* (1997)
Granitoid	Deformed granitoid	95-19	Erkou	2555 ± 6	
Lanzhishan	Sheared granitoid	95PC94	Changjiangtan	2553 ± 8	Wilde *et al.* (1997)
Granitoids	Sheared granitoid	95PC96	Hongyachun	2537 ± 10	
Chechang–Beitai	Hornblende-rich granitoid	95PC6b	Taihuai–Shahe	2552 ± 5	Wilde (2002)
	Fine-grained granitoid	WC6	Taipinggou	2546 ± 3	Wilde (2002)
Granitoids	Coarse-grained, granitoid	WC7	Shahe	2542 ± 7	Wilde (2002)
	Coarse-grained granitoid	WC5	Taipinggou	2538 ± 6	Wilde (2002)
Guangmingsi Granite	Medium-grained granitoid	95PC76	Guangmingsi	2531 ± 5	Wilde (2002)
Shifou Granite	Medium-grained granite	95PC98	Nanliang	2531 ± 4	Wilde (2002)
Wangjiahui	Deformed granite	95PC62	Shigang	2520 ± 9	Wilde (2002)
Granite (Phase I)	Deformed granite	95PC63	Shigang	2517 ± 12	Wilde (2002)
Wangjiahui	Medium-grained pink granite	95PC50	Daixian	2117 ± 18	Wilde (2002)
Granite (Phase II)	Medium-grained pink granite	95PC51	Daixian	2116 ± 16	Wilde (2002)
	Medium-grained pink granite	95PC60	Chungxie	2103 ± 20	Wilde (2002)
Dawaliang Granite	Porphyritic granite	D2	Dawaliang	2176 ± 12	Wilde *et al.* (1997)
	Porphyritic granite	D1	Dawaliang	2107 ± 15	
Gaofan Subgroup	Felsic tuff	95PC55c	Xiazhuang	2528 ± 6	Wilde (2002)
Hongmenyan Formation	Rhyolite lava	WT13	Taihuai–Shahe	2533 ± 8	Wilde (2002)
	Andesite lava	WT17	Taihuai–Shahe	2524 ± 8	Wilde (2002)
	Rhyodacite	WT9	Taihuai–Shahe	2523 ± 18	Wilde (2002)
	Subvolcanics	WT12	Taihuai–Shahe	2516 ± 8	Wilde (2002)
Baizhiyan Formation	Felsic tuff	95PC115	Erkou Iron Mine	2524 ± 10	Wilde (2002)
Zhuangwang Formation	Felsic tuff	95PC114	Ekou	2529 ± 10	Wilde (2002)

Table 5. *Geological events in the Hengshan–Wutai–Fuping mountain belt*

Geological events	Age (Ma)	Hengshan Complex	Wutai Complex	Fuping Complex
Remnants of an older basement	>2700	Biotite gneisses	?	Hornblende gneisses
Granitoid intrusion	2566–2517		Wutai gneissic granites	Longquangguan augen gneisses
Mafic to felsic volcanism	2533–2516		Wutai greenstone sequences	
TTG suite intrusion	2520–2455	Hengshan and Yixingzhai TTG gneisses		Fuping TTG gneisses
Supracrustal deposition	<2527 >1850	Hengshan and Zhuajiafang supracrustal rocks	Hutuo Group	Wanzi supracrustal rocks
Palaeoproterozoic granitoid intrusion	2360–2330	Pegmatitic granite-gneisses	?	?
	*c.*2250 ?176–2000	Pegmatite Red gneissic granite	? Wangjiahui and Dawaliang granites	? Nanying gneissic granitoids
Collisional orogeny	1900–1800	Medium- to high-pressure granulite-facies metamorphism and three phases of deformation	Medium- to high-pressure greenschist to lower amphibolite-facies metamorphism and three phases of deformation	Medium-pressure amphibolite- to granulite-facies metamorphism and three phases of deformation
Post-orogenic extension	1800–1750	Mafic dyke swarms/post-orogenic granites	Mafic dyke swarms/post-orogenic granites	Mafic dyke swarms/post-orogenic granites

Archaean, whereas the former were emplaced from the Late Archaean to the Early Palaeoproterozoic (Tables 2–4). This conclusion is in complete contrast to all previous models that regarded the high-grade Fuping and Hengshan Complexes as an older basement to the low-grade Wutai Complex. As shown in Tables 2–4, the Hengshan and Fuping TTG gneisses were formed at 2526–2455 Ma, and are thus younger than the Wutai gneissic granites that were emplaced at 2566–2517 Ma.

The Longquanguan augen gneisses were previously treated as the youngest lithological unit in the Fuping Complex. However, the recent SHRIMP U–Pb zircon data reveal that the Longquanguan augen gneisses were emplaced at *c.*2540 Ma, similar to the age of the Wutai gneissic granites, but older than the 2520–2480 Ma Fuping TTG gneisses. Therefore, the Longquanguan augen gneisses most likely represent the equivalents of the Wutai granites that were reworked in the Longquanguan shear zone. The similar fabric and petrographic features between the Longquanguan augen gneisses and some Wutai granites (e.g. Lanzhishan granite) support this conclusion.

In the Hengshan Complex, the weakly deformed, amphibolite-facies, Yingxingzhai TTG gneisses in the southern Hengshan have long been considered to be younger than granulite-facies, strongly deformed TTG gneiss in the northern Hengshan. However, recent isotopic data do not support this conclusion. Two Yingxingzhai TTG gneisses were dated by Wilde (2002) using the SHRIMP technique at 2513 ± 5 Ma and 2499 ± 4 Ma (Table 3); these ages are similar to 2526–2455 Ma age range of the Hengshan TTG gneisses (Kröner *et al.* 2004). Therefore, the Yingxingzhai TTG gneisses most likely represent the weakly deformed equivalents of the Hengshan TTG gneisses (Kröner *et al.* 2001, 2004).

Wutai greenstone formation

In the traditional Chinese literature, the Wutai greenstone belt was divided into the lower amphibolite-facies Lower Wutai (Banyukou, Jingangku, Zhuangwang and Wenxi formations), greenschist-facies Middle Wutai (Baizhiyan and Hongmenyan formations) and sub-greenschist-facies Upper Wutai sequences (Gaofan Subgroup) (Bai 1986; Tian 1991; Bai *et al.* 1992). However, recent SHRIMP zircon U–Pb data reveal no difference in age between volcanic rocks from the Lower, Middle and Upper Wutai greenstone sequences (Table 4). For example, zircons from felsic tuff of the Upper Wutai (Gaofan Subgroup) gave a weighted mean $^{207}Pb/^{206}Pb$ age of 2528 ± 6 Ma, whereas felsic tuffs from the Zhuangwang and Baizhiyan Formations of the Lower Wutai were dated at 2529 ± 10 Ma and 2524 ± 10 Ma, respectively (Table 4; Wilde 2002). Four intermediate to felsic volcanic rocks from the Hongmenyan Formation (Middle Wutai) have been analysed and their ages range from 2533 ± 8 Ma to 2516 ± 8 Ma, i.e. they are essentially coeval with the volcanic rocks of the Lower and Upper Wutai sequences. These results establish that:

(1) the age of the Wutai greenstones was Late Archaean, with all samples defining ages of 2516–2533 Ma – slightly younger than some of the Wutai gneissic granites;

(2) there is no difference in age between volcanic rocks previously considered to occupy different stratigraphic levels within the greenstone sequence. These results, together with our field observations, confirm that the Wutai Complex is not a simple stratigraphic succession, but is a tectonic assemblage of mafic, intermediate and felsic volcanic rocks, with associated metasediments.

High-grade metasedimentary rocks

The high-grade metasedimentary rocks include the Wanzi supracrustal assemblage in the Fuping Complex and the Hengshan and Zhujiafang supracrustal assemblages in the Hengshan Complex. These supracrustal rocks have long been considered to be the oldest lithologies in the Hengshan–Wutai–Fuping mountain belt (Liu *et al.* 1985; Wu *et al.* 1989; Tian 1991). However, new SHRIMP U–Pb zircon data do not support this interpretation. In the Wanzi supracrustal assemblage, the morphologies and internal structures of zircons from Al-rich gneisses confirm a detrital origin from an igneous source, and the cores of zircons from two samples yielded mean $^{207}Pb/^{206}Pb$ ages of 2502 ± 7 Ma and 2507 ± 14 Ma (Table 2), interpreted as the crystallization ages of igneous source rocks (Zhao *et al.* 2002a). Therefore, these rocks must have been deposited after c.2507 Ma ago. In the Hengshan supracrustal assemblage, a garnet quartzite yielded a SHRIMP U–Pb detrital zircon age of 2527 ± 10 Ma (Wilde 2002); a metadacite (biotite gneiss) was dated at 2526 ± 4.7 Ma, interpreted as the magmatic crystallization age (Kröner *et al.* 2004) (Table 3). A meta-greywacke from the Zhujiafang supracrustal assemblage was dated at 2501 ± 15 Ma, interpreted as the crystallization ages of the igneous source rocks (Wilde 2002). Thus, the Hengshan and Zhujiafang supracrustal rocks were most probably deposited after 2527 Ma and 2500 Ma, respectively.

Palaeoproterozoic granitoid magmatism

Recent geochronological data indicate the presence of Palaeoproterozoic granitoids in the Hengshan, Wutai and Fuping Complexes (Tables 2–4). In the Fuping Complex, the Palaeoproterozoic granitoids, named the Nanying gneissic granites, consist predominantly of monzogranitic and granodioritic gneisses that display similar fabrics and intrusive relationships with the Fuping TTG gneisses. SHRIMP results show that the monzogranitic component of the Nanying gneissic granite was emplaced at 2077–2045 Ma ago (Guan *et al.* 2002; Zhao *et al.* 2002a), somewhat earlier than the emplacement of the granodioritic component, which was dated at c.2024 Ma (Zhao *et al.* 2002a). In the Wutai Complex, the Palaeoproterozoic granitoids are represented by the Wangjiahui granite (phase II) and Dawaliang granite, which have been dated at c.2.1 Ga by Wilde *et al.* (1997) and Wilde (2002), also using the SHRIMP technique (Table 4). In the Hengshan Complex, gneissic granites of similar age have also been dated by Kröner *et al.* (2004) using the SHRIMP and single-grain evaporation zircon U–Pb techniques (Table 3). In addition, Kröner *et al.* (2004) also dated some Palaeoproterozoic gneissic granitoids at c.2360–2330 Ma and 2250 Ma (Table 3). It is particularly important to note that these Palaeoproterozoic granitoids in the Hengshan and Fuping complexes contain the same deformational features as the older TTG gneisses, and thus unambiguously demonstrate that the main deformational event is not Archaean but Palaeoproterozoic in age.

Timing of regional metamorphism

In the Chinese literature, the high-grade Hengshan and Fuping complexes and the low-grade Wutai Complex have long been assigned to two different metamorphic events, named the Fuping (*c.*2.5 Ga) and Wutai (2.4–2.3 Ga) 'movements', respectively (Wu *et al.* 1989; Tian 1991; Bai & Dai 1998). This conclusion was built up on the basis of a few 'unconformities', conventional multigrain TIMS U–Pb zircon geochronology, and a misconception that high-grade metamorphic rocks were older than low-grade ones. However, new geochronological data do not support the existence of either the Fuping or Wutai 'movements'. In the Fuping Complex, SHRIMP U–Pb zircon studies, when combined with cathodoluminescence images and U–Th chemistry, confirm the existence of metamorphic zircons in most lithologies of the complex (Guan *et al.* 2002; Zhao *et al.* 2002*a*). Minor lithologies (e.g. Wanzi supracrustal rocks) in the complex do not contain any metamorphic zircons, probably due to a Zr content insufficient for a metamorphic zircon-forming reaction to occur in these rocks. Metamorphic zircons in the Fuping Complex occur as either overgrowth rims surrounding older magmatic zircon cores (Fig. 7e–h) or single grains (Fig. 7i), and are structureless, highly luminescent and very low in Th and U contents. These features make them distinctly different from the magmatic zircons that are generally characterized by oscillatory zoning (Fig. 7e–h), low luminescence and comparatively high Th and U contents. The metamorphic zircons from different Late Archaean to Palaeoproterozoic rocks in the Fuping and Hengshan complexes yielded similar concordant $^{207}Pb/^{206}Pb$ ages in the range 1870 to 1800 Ma (Tables 2 and 3), which are 700 Ma to 150 Ma younger than their magmatic zircon cores. In the low-grade Wutai Complex, although metamorphic zircons have not been observed, a garnet–whole rock Sm–Nd isochron age of 1851 ± 9 Ma and a hornblende Ar/Ar age of 1781 ± 20 Ma have been obtained (Table 4; Wang *et al.* 1997*b*, 2001). Therefore, we conclude that the main regional metamorphism of the Hengshan, Wutai and Fuping Complexes occurred neither at the *c.*2.5 Ga Fuping 'Movement' nor the 2.4 Ga Wutai 'Movement', but at *c.*1.85 Ga.

Timing of intrusion of post-orogenic mafic dyke swarms

Post-orogenic mafic dyke swarms are widespread in the Hengshan–Wutai–Fuping mountain belt, with a predominant NW–SE to NNW–SSE trend (Fig. 9). Generally, they dip steeply, cutting both the Archaean and Palaeoproterozoic basement rocks, and are covered by Neoproterozoic and Cambrian strata. Individual dykes range in width from 10 to 50 m to a maximum of about 100 m, and in length from 10 to 40 km, to a maximum of about 100 km. Most dyke swarms are unmetamorphosed and undeformed, with chilled contacts. However, a few dykes were metamorphosed to upper amphibolite facies, as indicated by garnet and amphibole reaction rims surrounding igneous plagioclase and clinopyroxene grains. Using the single-grain evaporation zircon U–Pb method, Halls *et al.* (2000) dated an unmetamorphosed diabase dyke in the Hengshan Complex at 1769.1 ± 2.5 Ma, which suggests that a post-collisional extensional event led to the widespread intrusion of the mafic dyke swarms in the Hengshan–Wutai–Fuping mountain belt shortly after the *c.*1.8 Ga collisional event (Zhao *et al.* 2001*a*).

Discussion and conclusions

Table 5 lists the main geological events in the Hengshan–Wutai–Fuping mountain belt, based on recent lithological, structural, metamorphic and geochronological data as summarized in the present paper. These new data exclude the possibility of the high-grade Fuping and Hengshan Complexes being an older crystalline basement to the low-grade Wutai Complex, as suggested in previous tectonic models (Bai 1986; Li *et al.* 1990; Tian 1991; Bai *et al.* 1992; Yuan & Zhang 1993; Wang *et al.* 1996). Thus, the tectonic evolution of the Hengshan–Wutai–Fuping mountain belt may not be related to local interaction of the three complexes, either through closure of a Wutai rift (Tian 1991; Yuan & Zhang 1993) or collision between a Wutai arc and the putative Hengshan and Fuping microcontinental blocks (Bai 1986; Li *et al.* 1990; Bai *et al.* 1992; Wang *et al.* 1996). In contrast, we consider that the three complexes may represent elements of a single Late Archaean to Palaeoproterozoic magmatic arc that was subsequently incorporated into the Trans-North China Orogen, which, along with the Eastern and Western blocks, were amalgamated to form the North China Craton at around 1.85 Ga.

Geochemical data from the Hengshan, Wutai and Fuping Complexes support an arc derivation (Geng & Wu 1990; Bai *et al.* 1992; Sun *et al.* 1992; Wang *et al.* 1996; Zhao *et al.* 2000*a*; Kröner *et al.* 2004), although minor ultramafic to mafic rocks in the Lower Wutai may represent remnants of ancient oceanic crust (Bai *et al.*

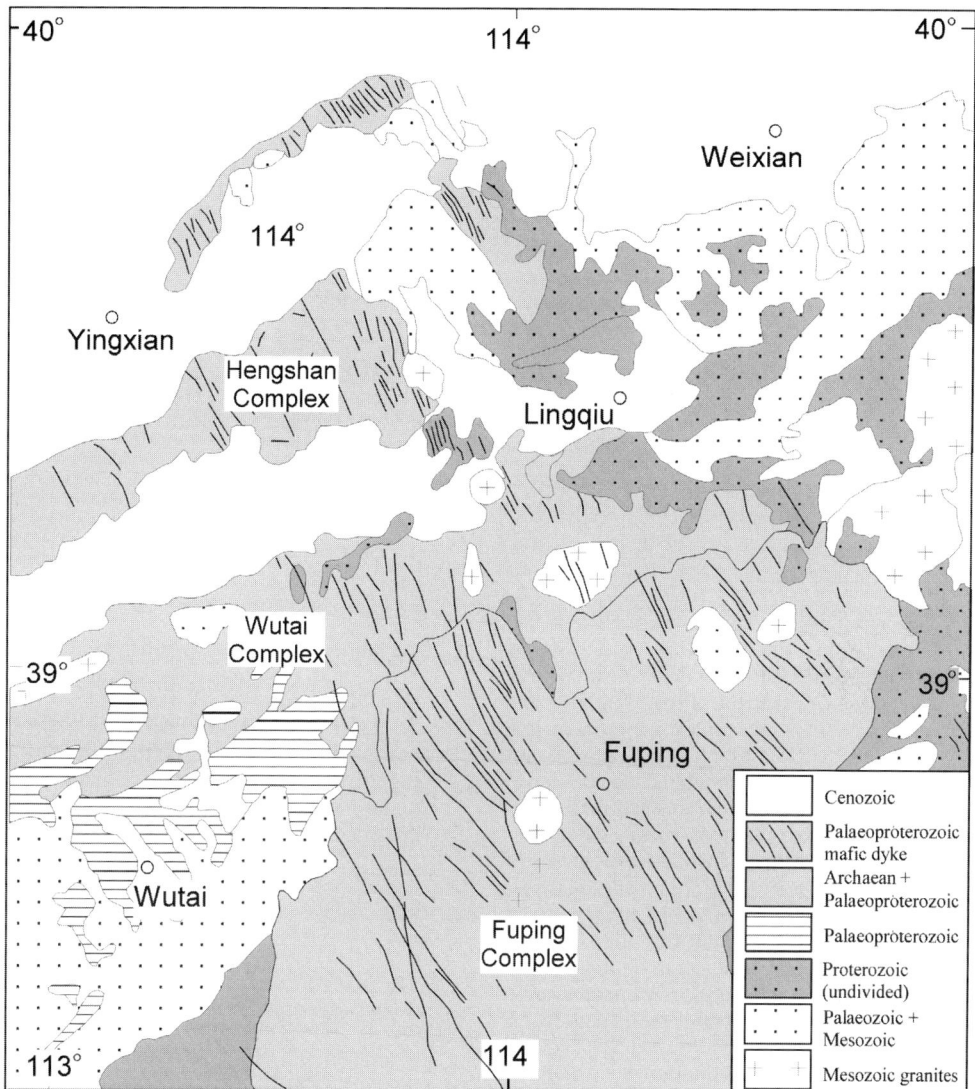

Fig. 9. Spatial distribution of mafic dykes in the Hengshan–Wutai–Fuping mountain belt (after Zhao *et al.* 2001*a*).

1992). For example, Bai *et al.* (1992) showed that the majority of amphibolites from the Lower Wutai and greenschists from the Middle Wutai have an affinity with calc-alkaline basalts or island-arc tholeiites (Fig. 10). Based on the features of major and trace elements and Nd isotopes, Kröner *et al.* (2004) showed that the Hengshan and Fuping TTG gneisses are characterized by a wide range in SiO_2 content, high Na_2O, Ba, Sr and low Y and HREE, and their selective enrichment in LIL-elements and depletion in Nb, Ta and Ti can be derived from

magmatic precursors with a strong mantle signature, modified by a subduction component and variable contributions from older crust. A particularly diagnostic discriminant for the tectonic setting is the Rb–Hf–Ta triangular plot for felsic to intermediate magmatic rocks, in which arc-derived suites occupy a field distinct from those generated in within-plate and ocean-floor settings (Harris *et al.* 1986). Figure 11 shows such a plot for the Hengshan TTG gneisses (Kröner *et al.* 2004; Liu *et al.* 2004), Wutai volcanic and granitoid rocks (Wang *et al.* 2004*b*; Liu *et al.* 2004)

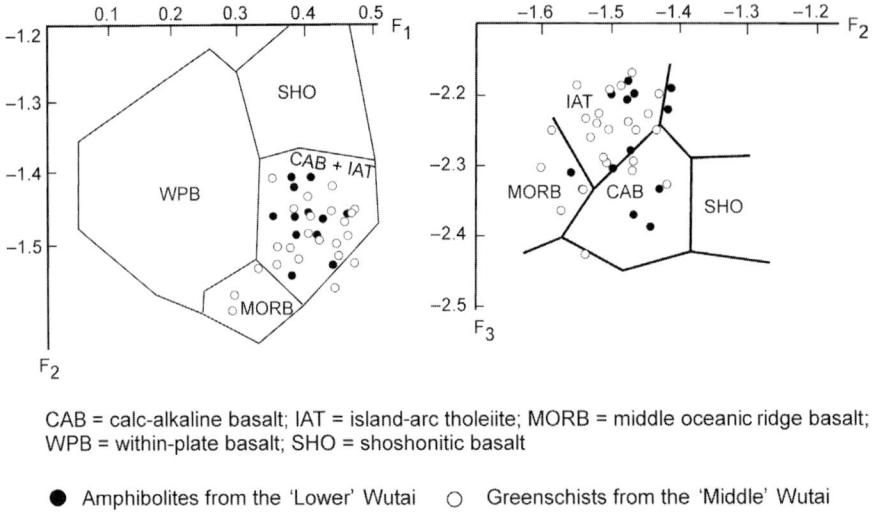

CAB = calc-alkaline basalt; IAT = island-arc tholeiite; MORB = middle oceanic ridge basalt;
WPB = within-plate basalt; SHO = shoshonitic basalt

● Amphibolites from the 'Lower' Wutai   ○ Greenschists from the 'Middle' Wutai

Fig. 10. Major element discrimination diagram (after Pearce 1976) for the Wutai amphibolites (Lower Wutai) and greenschists (Middle Wutai), based upon functions F1, F2 and F3 (Bai *et al.* 1992).

and Fuping TTG gneisses (Liu *et al.* 2004), clearly supporting their arc derivation (Kröner *et al.* 2004). Considering these features and those of the Wutai greenstones, we propose that the TTG suites in the Hengshan Fuping Complexes may have been generated in the lower levels of a continental-margin magmatic arc, whereas the Wutai granitoid rocks and greenstones may have formed in the upper levels of the arc and associated back-arc basins. The presence of retrograded eclogites and high-pressure granulites in the Hengshan Complex implies that this arc may have been partly subducted to the depth of the lower crust or upper mantle during collision to generate the Trans-North China Orogen.

On the basis of the lithological, structural, metamorphic, geochemical and geochronological data summarized in this paper, we propose the following scenario for the evolution of the Hengshan–Wutai–Fuping mountain belt.

(1) In the late Archaean to Palaeoproterozoic, the Hengshan–Wutai–Fuping region was part of a continental–margin arc along the western margin of the Eastern Block, which was separated from the Western Block by an old ocean, with subduction of the oceanic lithosphere beneath the western margin of the Eastern Block (Fig. 12a). At 2550–2520 Ma, the deep subduction caused partial melting of the medium–lower crust, producing large amounts of granitoid magma that was intruded into

the upper levels of the crust to form the Wutai and Longquanguan granitoid rocks (Fig. 12a).

(2) At 2530–2520 Ma, the subduction of the oceanic lithosphere caused the partial melting of the mantle wedge, which led to underplating of mafic magma in the lower crust and widespread mafic (and minor

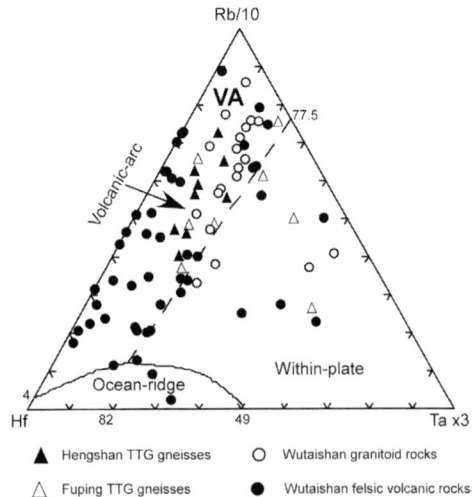

▲ Hengshan TTG gneisses   ○ Wutaishan granitoid rocks
△ Fuping TTG gneisses   ● Wutaishan felsic volcanic rocks

Fig. 11. The Hf–Rb/10–Ta × 3 discrimination diagram (after Harris *et al.* 1986) for the Hengshan TTG gneisses (Kröner *et al.* 2003; Liu *et al.* 2003), Wutai volcanic and granitoid rocks (Liu *et al.* 2003; Wang *et al.* 2003b) and Fuping TTG gneisses (Liu *et al.* 2003).

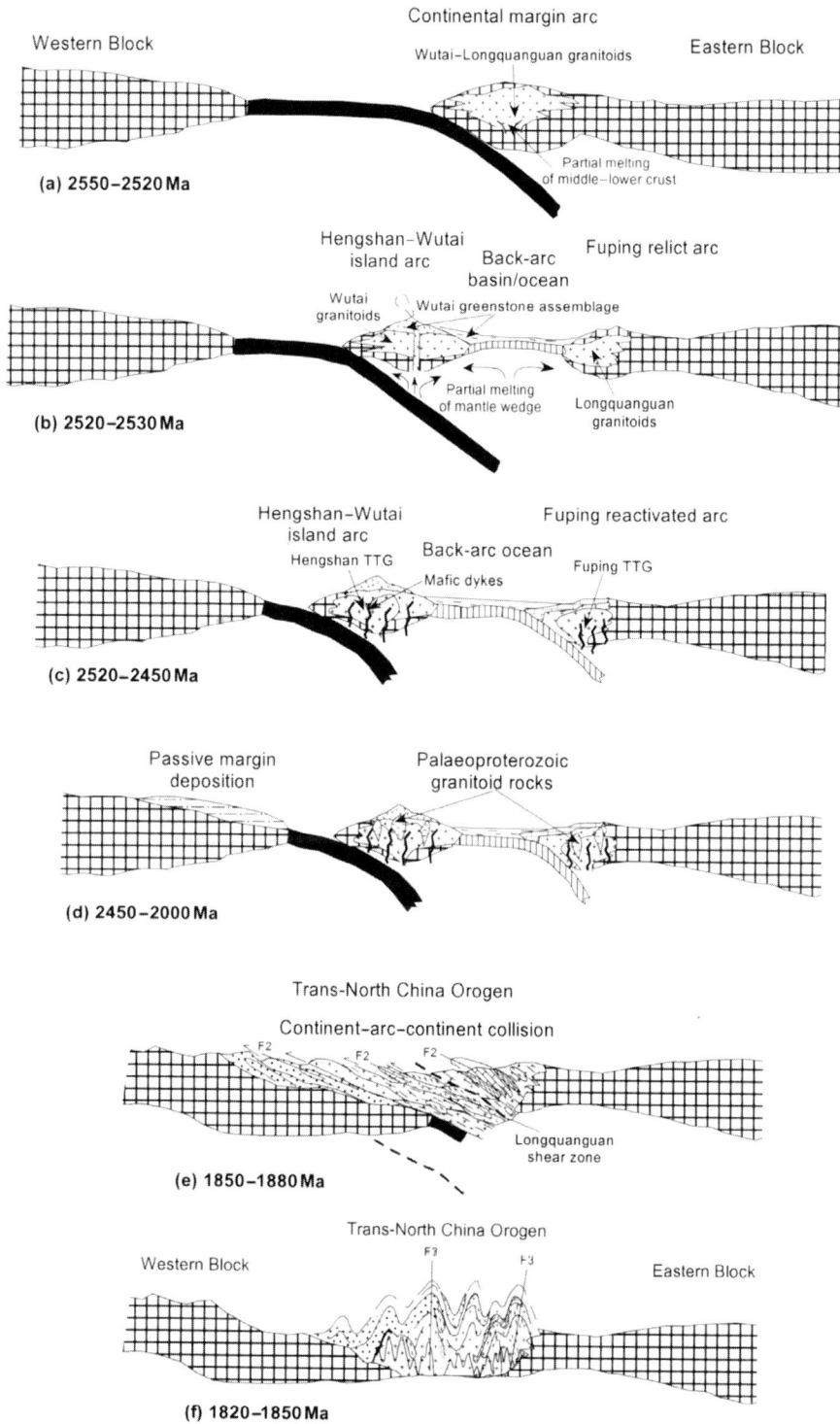

Fig. 12. A series of schematic sections showing the geological evolution of the Hengshan–Wutai–Fuping mountain belt and the amalgamation of the North China Craton.

felsic) volcanism in the continental margin arc, forming part of the Wutai greenstone assemblage (Fig. 12b). Extension driven by the widespread mafic to felsic volcanism led to the development of a back-arc basin or small ocean (represented by the ultramafic rocks in the Lower Wutai) between the Hengshan–Wutai island arc and the Fuping relict arc (Fig. 12b).

(3) At 2520–2450 Ma, the subduction beneath the Hengshan–Wutai island arc caused the further partial melting of the lower crust to generate large amounts of TTG magmatism that created the Hengshan TTG suite (Fig. 12c). Meanwhile, the eastward subduction of the back-arc ocean led to the reactivation of the Fuping relict arc, where the Fuping TTG suite was emplaced (Fig. 12c). Following the emplacement of the Hengshan and Fuping TTG suites is the emplacement of mafic dykes (Fig. 12c), now preserved as medium- to high-pressure granulites or retrograded eclogites within the Hengshan and Fuping TTG gneisses.

(4) In the Palaeoproterozoic, there were a number of episodes of granitoid magmatism in both the Hengshan–Wutai island arc and the Fuping reactivated arc (Fig. 12d), due to the partial melting of the Hengshan and Fuping TTG suites, resulting in the emplacement of 2360 Ma, *c.*2250 Ma and 2000–2100 Ma granites in the Hengshan Complex (Kröner *et al.* 2004), *c.*2100 Ma Wangjiahui and Dawaliang granites in the Wutai Complex (Wilde 2002), and 2100–2000 Ma Nanying gneissic granitoids in the Fuping Complex (Zhao *et al.* 2002*a*; Liu *et al.* 2002, 2004).

(5) At 1880–1820 Ma, the oceans between the Eastern and Western blocks disappeared by complete subduction, and the closing of these oceans led to the continent–arc–continent collision (Fig. 12e). Collision led to development of large-scale thrusting tectonics and isoclinal (F_2) folds (Fig. 12e) and transported part of the Hengshan and Fuping lithologies into lower crustal levels or upper mantle where granulite-facies or eclogite-facies metamorphism occurred, whereas the Wutai Complex underwent greenschist- to lower amphibolite-facies metamorphism in the upper crustal levels. Following the peak metamorphism, the thickened crust underwent exhumation, which resulted in the development of widespread asymmetrical F_3 folds (Fig. 12f) and symplectic textures in the rocks.

Further exhumation and inflow of fluids along the ductile shear zones led to widespread cooling and retrogression.

(6) The last magmatic event in the Hengshan–Wutai–Fuping mountain belt was the widespread emplacement of *c.*1789 Ma mafic dyke swarms, probably as a result of orogenic collapse or post-orogenic extension.

The above model implies that the Hengshan–Wutai–Fuping belt existed as a long-lived (2.55–1.85 Ga) convergent continental-margin arc during the Palaeoproterozoic. Similar long-lived continental-margin arcs are also thought to have prevailed in southeastern Laurentia, southern Baltica, central Australia and western Amazon during the Palaeo-Mesoproterozoic (Karlstrom *et al.* 2001; Bingen *et al.* 2002; Brewer *et al.* 2002; Rogers & Santosh 2002). In southeastern Laurentia and southern Baltica, a 1.8 to 1.2 Ga magmatic arc zone extends from Arizona through Colorado, Michigan, southern Greenland, Scotland, Sweden and Finland to western Russia, bordering the present southern margin of North America, Greenland and Baltica (Gower *et al.* 1990; Park 1992; Karlstrom *et al.* 2001; Rogers & Santosh 2002). It consists of the 1.8–1.7 Ga Yavapai and Central Plains belts, the 1.7–1.6 Ga Mazatzal Belt, the 1.5–1.3 Ga St Francois and Spavinaw Granite–Rhyolite belts and the 1.3–1.2 Ga Elzevirian Belt in southwestern North America; the 1.8–1.7 Ga Makkovikian Belt and the 1.7–1.6 Ga Labradorian Belt in northeastern North America; the 1.8–1.7 Ga Malin Belt in the British Isles; the 1.8–1.7 Ga Ketilidian Belt in Greenland; and the 1.8–1.7 Transscandinavian Igneous Belt, 1.7–1.6 Ga Kongsberggian–Gothian Belt, 1.6–1.5 Ga Southwest Sweden Granitoid Belt and the 1.3–1.2 Ga early Sveconorwegian Belt in Baltica (Gower *et al.* 1990; Park 1992; Karlstrom *et al.* 2001). Petrological and geochemical studies indicate that this large magmatic arc zone includes dominantly juvenile volcanogenic sequences and granitoid suites resembling those of present-day island arcs and active continental margins (Nelson & DePaolo 1985; Bennet & DePaolo 1987), representing subduction-related episodic outgrowth along the continental margin of a Palaeo-Mesoproterozoic supercontinent (Karlstrom *et al.* 2001; Rogers & Santosh 2002; Zhao *et al.* 2002*b*). A present-day example of long-lived convergent continental-margin arcs is the Andes, where the Pacific plate has been subducting under the west coast of South America for *c.*500 million years since the Cambrian (Dalziel 1997; Rivers & Corrigan 2000). These examples demonstrate that such a long-lived continental-margin arc is not unique to the Late Archaean to Palaeoproterozoic Hengshan–Wutai–Fuping belt in the North China Craton.

The geological record from the Hengshan–Wutai–Fuping mountain belt also indicates that there are apparently two periods of magmatic quiescence. The first occurred between c.2450 Ma and 2360 Ma, and the second occurred between approximately 2000 Ma and 1880 Ma. Similar long-lasting magmatic quiescence has also been recorded in other long-lived convergent continental-margin arcs, e.g. the Mesoproterozoic Laurentia (Rivers & Corrigan 2000) and the Palaeozoic–Mesozoic Andes (Soler & Bonhomme 1990; Kay et al. 1991). One of the possible interpretations for such apparent hiatuses of magmatism in a long-lived magmatic arc is the change in the dip the subducting plate, which generally varies from ca. c.10–70° (Rivers & Corrigan 2000). The geological record from the Andes indicates that periods of magmatic quiescence were associated with low-angle subduction and a compressional arc (e.g. Soler & Bonhomme 1990; Kay et al. 1991). This may explain the apparent gaps in the geological record in the Palaeoproterozoic Hengshan–Wutai–Fuping arc.

An integrated petrological, structural, metamorphic and geochronological overview of other metamorphic complexes in the Trans-North China Orogen shows that they share a late Archaean to Palaeoproterozoic history essentially similar to that of the Hengshan–Wutai–Fuping mountain belt. For example, the Huaian, Xuanhua, Chengde and Taihua Complexes can be compared with the Hengshan and Fuping Complexes, whereas the Lüliang, Zhongtiao, Dengfeng, Zanhuang and Northern Hebei Complexes are similar to the Wutai Complex (Zhao et al. 2000a). Therefore, the magmatic, structural and metamorphic history of the Hengshan–Wutai–Fuping mountain belt, reconstructed in this paper, provides important insights for understanding the tectonic evolution of the Trans-North China Orogen.

This research was financially supported by a NSFC Young Research Award (49929301) and Hong Kong RGC grants (HKU7090/01P, HKU7048/03P and HKU7055/03P). We would like to acknowledge A. Kröner, M. G. Zhai, C. H. Wu, S. W. Liu and K. Y. Wang for their many (controversial) discussions that influenced the content of this contribution. We thank Toby Rivers and Nicholas Culshaw for their critical but helpful comments on this paper.

References

BAI, J. 1986. The Precambrian crustal evolution of the Wutaishan area. In: BAI, J. (ed.) The Early Precambrian Geology of Wutaishan. Tianjin Science and Technology Press, Tianjin, 376–383 (in Chinese).

BAI, J. & DAI, F. Y. 1998. Archean crust of China. In: MA, X. Y. & BAI, J. (eds) Precambrian Crust Evolution of China. Springer–Geological Publishing House, Beijing, 15–86.

BAI, J., WANG, R. Z. & GUO J. J. 1992. The Major Geologic Events of Early Precambrian and their Dating in Wutaishan Region. Geological Publishing House, Beijing.

BENNET, V. C. & DePAOLO, D. J. 1987. Proterozoic crustal history of the western United States as determined by neodymium isotopic mapping. Geological Society America Bulletin, 99, 674–685.

BINGEN, B., MANSFELD, J., SIGMOND, E. M. O. & STEIN, H. 2002. Baltica–Laurentia link during the Mesoproterozoic: 1.27 Ga development of continental basins in the Sveconorwegian Orogen, southern Norway. Canadian Journal of Earth Sciences, 39, 1425–1440.

BREWER, T. S., AHALL, K. I., DARBYSHIRE, D. P. F. & MENUGE, J. F. 2002. Geochemistry of late Mesoproterozoic volcanism in southwestern Scandinavia: implications for Sveconorwegian/Grenvillian plate tectonic models. Journal of the Geological Society, 159, 129–144.

BROWN, M. 1993. P–T–t evolution of orogenic belts and the causes of regional metamorphism. Journal of the Geological Society of London, 150, 227–241.

BROWN, M. 2001. From microscope to mountain belt: 150 years of petrology and its contribution to understanding geodynamics, particularly the tectonics of orogens. Journal of Geodynamics, 32, 115–164.

COMPSTON, W., ZHANG, F. P., FOSTER, J. J., COLLERSON, K. D., BAI, J. & SUN, D. Z. 1983. Rubidium–strontium geochronology of Precambrian rocks from the Yanshan Region, North China. Precambrian Research, 22, 175–202.

CONDIE, K. C., BORYTA, M. D., LIU, J. Z. & QIAN, X. L. 1992. The origin of khondalites – geochemical evidence from the Archean to early Proterozoic granulite belt in the North China Craton. Precambrian Research, 59, 207–223.

DALZIEL, I. W. D. 1997. Neoproterozoic–Paleozoic geography and tectonics: review, hypothesis, environmental speculation. Geological Society of America Bulletin, 108, 16–42.

ENGLAND, P. C. & RICHARDSON, S. W. 1977. The influence of erosion upon the mineral facies of rocks from different metamorphic environments. Journal of the Geological Society, 134, 3201–3213.

ENGLAND, P. C. & THOMPSON, A. B. 1984. Pressure temperature–time paths of regional metamorphism, I. Heat transfer during the evolution of regions of thickened continental crust. Journal of Petrology, 25, 894–928.

GENG, Y. S. & WU, J. S. 1990. Geochemistry and evolution of the early Precambrian mafic rocks in Wutai–Taihangshan area. Precambrian Geology, 4, 167–174 (in Chinese).

GOWER, C. F., RYAN, A. B. & RIVERS, T. 1990. Mid-Proterozoic Laurentia-Baltic: an overview of its geological evolution and summary of the contributions by this volume. *In*: GOWER, C. F., RIVERS, T. RYAN, B. (eds), *Mid-Proterozoic Laurentia–Baltica*. Geological Association of Canada, Special Paper, **38**, 1–20.

GUAN, H., SUN, M., WILDE, S. A., ZHOU, X. H. & ZHAI M. G. 2002. SHRIMP U–Pb zircon geochronology of the Fuping Complex: implications for formation and assembly of the North China craton. *Precambrian Research*, **113**, 1–18.

GUO, J. H. & ZHAI, M. G. 2001. Sm–Nd age dating of high-pressure granulites and amphibolites from the Sanggan area, North China craton. *Chinese Science Bulletin*, **46**, 106–110.

GUO, J. H., O'BRIEN, P. J. & ZHAI, M. G. 2002. High-pressure granulites in the Sangan area, North China Craton: metamorphic evolution, *P–T* paths and geotectonic significance. *Journal of Metamorphic Geology*, **20**, 741–756.

GUO, J. H., WANG, S. S., SANG, H. Q. & ZHAI, M. G. 2001*b*. ^{40}Ar ^{39}Ar age spectra of garnet porphyroblasts: implications for metamorphic age of high-pressure granulites in the North China Craton. *Acta Petrologica Sinica*, **17**, 436–442.

GUO, J. H., ZHAI, M. G. & ZHANG, Y. G. 1993. Early Precambrian Manjinggou high-pressure granulites melange belt on the southern edge of the Huaian Complex, North China Craton: geological features, petrology and isotopic geochronology. *Acta Petrologica Sinica*, **9**, 329–341.

GUO, J. H., ZHAI, M. G. & XU, R. H. 2001*b*. Timing of the granulite facies metamorphism in the Sanggan area, North China craton: zircon U–Pb geochronology. *Science in China Series D*, **44**, 1010–1018.

HALLS, H. C., LI, J. H., DAVIS, D., HOU, T. G. & ZHANG, B. X. 2000. A precisely dated Proterozoic paleomagnetic pole from the North China Craton, and its relevance to paleocontinental reconstruction. *Geophysical Journal International*, **143**, 185–203.

HARRIS, N. B. W., PEARCE, J. A. & TINDLE, A. G. 1986. Geochemical characteristics of collision-zone magmatism. *In*: COWARD, M. P. & REIS, A. C. (eds) *Collision Tectonics*. Geological Society, London, Special Publications, **19**, 67–81.

HE, G. P. & YE, H. W. 1998. Two types of metamorphism of Paleoproterozoic basement in Eastern Liaoning and Southern Jilin area. *Acta Petrologica Sinica*, **14**, 152–162.

JAHN, B. M. & ZHANG, Z. Q. 1984. Archean granulite gneisses from eastern Hebei Province, China: rare earth geochemistry and tectonic implications. *Contributions to Mineralogy and Petrology*, **85**, 224–243.

KARLSTROM, K. E., HARLAN, S. S., ÅHÄLL, K. I., WILLIAMS, M. L., MCLELLAND, J. & GEISSMAN, J. W. 2001. Long-lived (1.8–1.0 Ga) convergent orogen in southern Laurentia, its extensions to Australia and Baltica, and implications for refining Rodinia. *Precambrian Research*, **111**, 5–30.

KAY, S. M., MPODOZIS, C., RAMOS, V. A. & MUNIZAGA, M. 1991. Magma source variations for mid–late Tertiary magmatic rocks associated with a shallow subduction zone and thickening crust in the central Andes (28–33°S). *In*: HARMON, R. S. & RAPELA, C. W. (eds) Andean Magmatism and its Tectonic Setting. Geological Society of America, Special Paper, **265**, 113–137.

KRETZ, R. 1983. Symbols for rock-forming minerals. *American Mineralogist*, **68**, 277–279.

KRÖNER, A. 2002. Zircon ages of the Hengshan Complex. *In*: KRÖNER, A., ZHAO, G. C. *et al.* (eds) *A Late Archaean to Palaeoproterozoic Lower to Upper Crustal Section in the Hengshan–Wutaishan Area of North China*. Guidebook for Penrose Conference Field Trip, September 2002, Chinese Academy of Sciences, Beijing, 28–32.

KRÖNER, A., CUI, W. Y., WANG, S. Q., WANG, C. Q. & NEMCHIN, A. A. 1998. Single zircon ages from high-grade rocks of the Jianping Complex, Liaoning Province, NE China. *Journal of Asian Earth Sciences*, **16**, 519–532.

KRÖNER, A., WILDE, S., O'BRIEN, P. J. & LI, J. H. 2001. The Hengshan and Wutai complexes of northern China: lower and upper crustal domains of a late Archaean to Palaeoproterozoic magmatic arc and significance for the evolution of the North China Craton. *In*: CASSIDY, K. F. (ed.) *4th International Archaean Symposium 2001, Extended Abstracts*. AGSO–Geoscience Australia, Record, **37**, p. 327.

KRÖNER, A., WILDE, S. A., LI, J. H. & WANG, K. Y. 2004. Age and evolution of a late Archaean to early Palaeozoic upper to lower crustal section in the Wutaishan/Hengshan/Fuping terrain of northern China. *Journal of Asian Earth Sciences* (in press).

LI, J. H. & QIAN, Q. L. 1991. A study on Longquanguan shear zone in northern part of Taihang Mountain. *Shanxi Geology*, **6**, 17–29 (in Chinese).

LI, J. H. & QIAN, X. L. 1994. *The early Precambrian Crustal Evolution of Hengshan Metamorphic Terrain, North China Craton*. Shanxi Science and Technology Press, Taiyuan.

LI, J. L., WANG, K. Y., WANG, C. Q., LIU, X. H. & Zhao Z. Y. 1990. Early Proterozoic collision orogenic belt in Wutaishan area, China. *Scientia Geologica Sinica*, **25**, 1–11.

LI, S. Z., YANG, Z. S., LIU, Y. J. & LIU, J. L. 1997. A model for the emplacement of Paleoproterozoic granites in Jiao–Liao–Ji area and its bearing on the uplift-delamination structure. *Acta Petrologica Sinica*, **13**, 190–202.

LI, S. Z., ZHAO, G. C., SUN, M., HAN, Z. Z., LUO, Y., HAO, D. F. & XIA, X. P. 2004. Deformational history of the Plaeoproterozoic Liaohe Group in the Eastern Block of the North China Craton. *Journal of Asian Earth Sciences* (in press).

LIU, D. Y., NUTMAN, A. P., COMPSTON, W., WU, J. S. & SHEN, Q. H. 1992. Remnants of 3800 Ma crust in the Chinese part of the Sino-Korean craton. *Geology*, **20**, 339–342.

LIU, D. Y., PAGE, R. W., COMPSTON, W. & WU, J. S. 1985. U–Pb zircon geochronology of Late Archean metamorphic rocks in the Taihangshan–Wutaishan area, North China. *Precambrian Research*, **27**, 85–109.

LIU, S. W. 1997. Study on fluid–rock equilibrium system of Fuping gneiss complex, Taihang Mountains. *Science in China Series D*, **40**, 239–245.

LIU, S. W. & LIANG, H. H. 1997. Metamorphism of Al-rich gneisses from the Fuping Complex, Taihang Mountain, China. *Acta Petrologica Sinica*, **13**, 303–312 (in Chinese with English abstract).

LIU, S. W., LIANG, H. H., ZHAO, G. C., HUA, Y. G. & JIAN, A. H. 2000. Isotopic chronology and geological events of Precambrian complex in the Taihangshan region. *Science in China Series D*, **43**, 386–393.

LIU, S. W., PAN, P. M., LI, J. H., LI, Q. G. & ZHANG, J. 2002. Geological and isotopic geochemical constraints on the evolution of the Fuping Complex, North China Craton. *Precambrian Research*, **117**, 41–56.

LIU, S. W., PAN, Y. M., XIE, Q. L., ZHANG, J. & LI, Q. G. 2004. Archean geodynamics in the Central Zone, North China Craton: constraints from geochemistry of two contrasting series of granitoids in the Fuping and Wutai Complexes. *Precambrian Research* (in press).

LIU, Z. H., WANG, A. J. & LI, X. F. 1997. Metamorphic evolution of Late Archaean Wutai collisional belt. *Geological Journal of China Universities*, **3**, 162–170.

LU, L. Z. 1991. Metamorphic P–T–t path of the Archean granulite-facies terrains in Jining area, Inner Mongolia and its tectonic implications. *Acta Petrologica Sinica*, **8**, 1–12.

LU, L. Z. & JIN, S. Q. 1993. P–T–t paths and tectonic history of an early Precambrian granulite facies terrane, Jining district, south-eastern Inner Mongolia, China. *Journal of Metamorphic Geology*, **11**, 483–498.

LU, L. Z., XU, X. C. & LIU, F. L. 1996. *Precambrian Khondalite Series in North China*. Changchun Press, Changchun.

MA, X. Y. & WU, Z. W. 1981. Early tectonic evolution of China. *Precambrian Research*, **14**, 185–202.

NELSON, B. K. & DEPAOLO, D. J. 1985. Rapid production of continental crust 1.7–1.9 b.y. ago: Nd and Sr isotopic evidence from the basement of the North American mid-continent. *American Geological Society Bulletin*, **96**, 746–754.

O'BRIEN, P. J., WALTE, N. & LI, J. H. 2004. The petrology of two distinct granulite types in the Hengshan Mts, China, and tectonic implications. *Journal of Asian Earth Sciences* (in press).

OXBURGH, F. R. 1989, Some thermal aspects of granulite history. *In*: VIELZEUF, D. & VIDAL, P. H. (eds) *Granulites and Crustal Evolution*. Kluwer Academic Publishers, Dordrecht, 569–580.

PARK, R. G. 1992. Plate kinematic history of Baltic during the Middle to Late Proterozoic: a model. *Geology*, **20**, 725–728.

PASSCHIER, C. W. & WALTE, N. 2002. Deformation of the Hengshan Complex. *In*: KRÖNER, A., ZHAO, G. C. *et al.* (eds) *A Late Archaean to Palaeoproterozoic Lower to Upper Crustal Section in the Hengshan–Wutaishan Area of North China*. Guidebook

for Penrose Conference Field Trip, September 2002, Chinese Academy of Sciences, Beijing, 11–12.

PEARCE, J. A. 1976. Statistical analysis of major element patterns in basalts. *Journal of Petrology*, **17**, 15–43.

PIDGEON, R.T. 1980. Isotopic ages of the zircons from the Archean granulite facies rocks, Eastern Hebei, China. *Geological Review*, **26**, 198–207.

REN, J. S. 1980. *Tectonics and Evolution of China*. Science Press, Beijing.

RIVERS, T. & CORRIGAN, D. 2000. Convergent margin on southeastern Laurentia during the Mesoproterozoic: tectonic implications. *Canadian Journal of Earth Sciences*, **37**, 359–383.

ROGERS, J. & SANTOSH, M. 2002. Configuration of Columbia, a Mesoproterozoic supercontinent. *Gondwana Research*, **5**, 5–22.

SOLER, P. & BONHOMME, M. G. 1990. Relation of magmatic activity to plate dynamics in central Peru from Cretaceous to present. *In*: KAY, S. M. & RAPELA, C. W. (eds) Plutonism from Antarctica to Alaska. *Geological Society of America, Special Paper*, **241**, 173–192.

SONG, B., NUTMAN, A. P., LIU, D. Y. & WU, J. S. 1996. 3800 to 2500 Ma crustal evolution in Anshan area of Liaoning Province, Northeastern China. *Precambrian Research*, **78**, 79–94.

SUN, M., ARMSTRONG, R. L. & LAMBERT, R. ST. J. 1992. Petrochemistry and Sr, Pb and Nd isotopic geochemistry of Early Precambrian rocks in the Wutaishan and Taihangshan areas, China. *Precambrian Research*, **56**, 1–31.

TAYLOR, S. R. & McLENNAN S. M. 1985. *The Continental Crust: its Composition and Evolution*. Blackwell Science Publishers, Oxford.

THOMPSON, A. B. & ENGLAND, P. C. 1984. Pressure–temperature–time paths of regional metamorphism, II. Their influences and interpretation using mineral assemblages in metamorphic rocks. *Journal of Petrology*, **25**, 929–955.

TIAN, Y. Q. 1991. *Geology and Mineralisation of the Wutai–Hengshan Greenstone Belt*. Shanxi Science and Technology Press, Taiyuan.

WANG, K. Y., HAO, J., CAWOOD, P. & WILDE, S. A. 1997b. High-pressure metamorphism in kyanite-bearing schists from the original Jinganku Formation of the Wutaishan. *Precambrian Geology and Metamorphic Petrology*, **17**, 213–220.

WANG, K. Y., LI, J. L. & LIU, L. Q. 1991. Petrogenesis of the Fuping grey gneisses. *Scientia Geologica Sinica*, **11**, 254–267 (in Chinese).

WANG, K. Y., LI, J. L., HAO, J., LI, J. H. & ZHOU, S. P. 1996. The Wutaishan mountain belt within the Shanxi Province, Northern China: a record of late Archean collision tectonics. *Precambrian Research*, **78**, 95–103.

WANG, K. Y., LI, J. L., HAO, J., LI, J. H. & ZHOU, S. P. 1997a. Late Archean mafic-ultramafic rocks from the Wuatishan, Shanxi Province: a possible ophiolite melange. *Acta Petrologica Sinica*, **13**, 139–151 (in Chinese).

WANG, K. Y., WANG, Z., YU, L., FAN, H., WILDE, S. A. & CAWOOD, P. A. 2001. Evolution of Archaean greenstone belt in the Wutaishan region, North China: constraints from SHRIMP zircon U–Pb and other geochronological and isotope information. *In*: CASSIDY, K. F. (ed.) *4th International Archaean Symposium 2001, Extended Abstracts.* AGSO–Geoscience Australia, Record, **37**, 104–105.

WANG, Y. J., FAN, W. M., ZHANG, Y. & GUO, F. 2003a. Structural evolution and ^{40}Ar/^{39}Ar dating of the Zanhuang metamorphic domain in the North China Craton: constraints on Paleoproterozoic tectonothermal overprinting. *Precambrian Research*, **122**, 159–182.

WANG, Z. H., WILDE, S. A., WANG, K. Y. & YU, L. J. 2004b. A MORB–arc–adakite association in the 2.5 Ga Wutai greenstone belt: late Archean magmatism and crustal growth in the north China Craton. *Precambrian Research* (in press).

WILDE, S. A. 2002. SHRIMP U–Pb zircon ages of the Wutai Complex. *In*: KRÖNER, A., ZHAO, G. C. *et al.* (eds) *A Late Archaean to Palaeoproterozoic Lower to Upper Crustal section in the Hengshan–Wutaishan Area of North China.* Guidebook for Penrose Conference Field Trip, September 2002, Chinese Academy of Sciences, Beijing, 32–34.

WILDE, S. A., CAWOOD, P., WANG, K. Y. & NEMCHIN, A. 1997. The relationship and timing of granitoid evolution with respect to felsic volcanism in the Wutai Complex, North China craton. *Precambrian Geology and Metamorphic Petrology*, **17**, 75–88.

WILDE, S. A., CAWOOD, P., WANG, K. Y. & NEMCHIN, A. 1998. SHRIMP U–Pb zircon dating of granites and gneisses in the Taihangshan–Wutaishan area: implications for the timing of crustal growth in the North China craton. *Chinese Science Bulletin*, **43**, 144.

WILDE, S. A., ZHAO, G. C. & SUN, M. 2002. Development of the North China Craton during the late Archean and its final amalgamation at 1.8 Ga: some speculations on its position within a global Paleoproterozoic supercontinent. *Gondwana Research*, **5**, 85–94.

WU, C. H. & ZHONG, C. T. 1998. The Paleoproterozoic SW–NE collision model for the central North China Craton. *Progress in Precambrian Research*, **21**, 28–50 (in Chinese).

WU, C. H., LI, H. M. & ZHONG, C. T. 2000. Single zircon U–Pb ages of the Fuping and Wanzi gneisses: geochronological evidence for non-uniform Archean basement of the Fuping Complex. *Progress in Precambrian Research*, **23**, 129–139 (in Chinese).

WU, J. S., GENG, Y. S., XU, H. F., JIN, L. G., HE, S. Y. & SUN, S. W. 1989. Metamorphic geology of Fuping Group. *Journal of Chinese Institute of Geology*, **19**, 1–213 (in Chinese).

WU, J. S., GENG, Y. S., SHEN, Q. H., WAN, Y. S., LIU, D. Y. & SONG, B. 1998. *Archean Geological Characteristics and Tectonic Evolution of China–Korea paleo-continent.* Geological Publishing House, Beijing.

YUAN, G. P. & ZHANG, R. Y. 1993. The structural environment of the paleorift in Wutai greenstone belt. *Shanxi Geology*, **8**, 21–28 (in Chinese).

ZHAI, M. G., GUO, J. H. & YAN, Y. H. 1992. Discovery and preliminary study of the Archean high-pressure granulites in the North China. *Science in China*, **12B**, 1325–1330.

ZHAI, M. G., GUO, J. H., LI, H. H., YAN, Y. H. & LI, Y. G. 1995. Discovery of retrograded eclogites in the Archaean North China Craton. *Chinese Science Bulletin*, **40**, 1590–1594.

ZHAI, M. G. 1997. Recent advance in the study of granulites from the North China Craton. *International Geological Review*, **39**, 325–341.

ZHANG, S. G., JIN, L. G. & XIAO, Q. H. 1983. Structural styles and deformational history of the Archean composite folds in the Fuping Complex. *Regional Geology of China*, **6**, 336–347.

ZHAO, G. C. 2001. Palaeoproterozoic assembly of the North China Craton. *Geological Magazine*, **138**, 87–91.

ZHAO, G. C., CAWOOD, P. A. & LU, L. Z. 1999c. Petrology and P–T history of the Wutai amphibolites: implications for tectonic evolution of the Wutai Complex, China. *Precambrian Research*, **93**, 181–199.

ZHAO, G. C., CAWOOD, P. A., WILDE, S. A. & LU, L. Z. 2000a. Metamorphism of basement rocks in the Central Zone of the North China Craton: implications for Paleoproterozoic tectonic evolution. *Precambrian Research*, **103**, 55–88.

ZHAO, G. C., CAWOOD, P. A. & WILDE, S. A. 2001b. High-pressure granulite (retrograded eclogites) from the Hengshan Complex, North China Craton: petrology and tectonic implications. *Journal of Petrology*, **42**, 1141–1170.

ZHAO, G. C., CAWOOD, P. A., WILDE, S. A. & SUN, M. 2001c. Polymetamorphism of granulites in the North China Craton: textural and thermobarometric evidence and tectonic implications. *In*: MILLER, J. A., HOLDSWORTH, R. E., BUICK, I. S., HAND, M. (eds) *Continental Reactivation and Reworking.* Geological Society, London, Special Publications, **184**, 323–342.

ZHAO, G. C., CAWOOD, P. A., WILDE, S. A. & SUN, M. 2002b. Review of global 2.1–1.8 Ga orogens: implications for a pre-Rodinia supercontinent. *Earth-Science Reviews*, **59**, 125–162.

ZHAO, G. C., WILDE, S. A., CAWOOD, P. A., LU, L. Z. 1998. Thermal evolution of the Archaean basement rocks from the eastern part of the North China Craton and its bearing on tectonic setting. *International Geological Review*, **40**, 706–721.

ZHAO, G. C., WILDE, S. A., CAWOOD, P. A., LU, L. Z. 1999a. Tectonothermal history of the basement rocks in the western zone of the North China Craton and its tectonic implications. *Tectonophysics*, **310**, 37–53.

ZHAO, G. C., WILDE, S. A., CAWOOD, P. A. & LU, L. Z. 1999b. Thermal evolution of two types of mafic granulites from the North China craton: implications for both mantle plume and collisional tectonics. *Geological Magazine*, **136**, 223–240.

ZHAO, G. C., WILDE, S. A., CAWOOD, P. A. & LU, L. Z. 2000b. Petrology and P–T path of the Fuping mafic granulites: implications for tectonic evolution of the central zone of the North China Craton. *Journal of Metamorphic Geology*, **18**, 375–391.

ZHAO, G. C., WILDE, S. A., CAWOOD, P. A. & SUN, M. 2001a. Archean blocks and their boundaries in the North China Craton: lithological, geochemical, structural and P–T path constraints and tectonic evolution. *Precambrian Research*, **107**, 45–73.

ZHAO, G. C., WILDE, S. A., CAWOOD, P. A. & SUN, M. 2002a. SHRIMP U–Pb zircon ages of the Fuping Complex: implications for accretion and assembly of the North China Craton. *American Journal of Science*, **302**, 191–226.

Precambrian tectonic evolution of the North China Craton

MINGGUO ZHAI

Key Laboratory of Mineral Resources, Institute of Geology & Geophysics, Chinese Academy of Sciences, Beijing, 100029, China (e-mail: mgzhai@mail.igcas.ac.cn)

Abstract: The North China Craton (NCC) is a major Archaean craton, covering an area of $c.300\,000\ km^2$ in north and northeast China. Almost all Archaean rocks on the craton experienced high-grade metamorphism and strong migmatization, so that the preserved greenstone belts underwent granulite–amphibolite-facies metamorphism, anatectic melting and strong deformation. This suggests that the NCC may have a more complicated early tectonic history than most other cratonic nuclei. The oldest NCC rocks are 3.8 Ga granitic gneisses in NE China and supracrustal rocks in eastern Hebei. Major continental growth occurred at 2.9–2.7 Ga. Two subsequent high-grade metamorphic events occurred at 2.6–2.45 Ga ('2.5 Ga event') and 1.9–1.75 Ga ('1.8 Ga event'). The older episode is considered to mark an amalgamation event, whereas the 1.8 Ga event represents the final cratonization of the NCC. Some researchers have divided the 1.8 Ga event into a 1.9–1.8 Ga metamorphic event (interpreted as a continent–continent collision) followed by a 1.8–1.65 Ga rifting episode. Other workers have suggested that the metamorphism and rifting could be parts of a single tectonic event related to Palaeo–Mesoproterozoic mantle upwelling. The general consensus on the NCC for the period 2.5–1.8 Ga is that the craton was then in an inactive stage. However, in this paper it is proposed that several Palaeoproterozoic mobile belts existed (showing many of the characteristics of Phanerozoic orogens). During the Mesoproterozoic–Neoproterozoic, a set of sedimentary sequences (the Changcheng–Jixian–Qingbaikou systems) constituted a disconformable–pseudoconformable succession within an intra-cratonic aulacogen. The signature of a 1.4–0.9 Ga orogen and the Rodinia breakup is very weak, indicating that the NCC did not experience major deformation as it was amalgamated into the Rodinia supercontinent.

The Archaean cratons of the world are often considered to contain two types of terrane:

(1) gneiss-dominated high metamorphic grade belts, and
(2) well-preserved, low-grade, volcanic-dominated greenstone belts (e.g. Windley 1995).

It is generally accepted that modern style plate tectonic processes were active by the early–middle Proterozoic when large stable cratons had formed, against which trailing margins (Dann 1991; Loukola-Ruskeeniemi *et al.* 1991), Andean-type margins, and collisional belts could develop (Windley 1993). Summarizing different models of Archaean continental evolution, Windley (1986) suggested that the most viable craton model combines the back-arc marginal basin idea for greenstone belt formation and the main arc (plutonic batholith) interpretation of the high-grade gneissic complex. Based on the similarities in structural deformation between ancient and young orogenic belts, Marshak (1999) suggested that plate tectonics was operational in the Early Precambrian.

The North China craton (NCC) is a major Archaean craton, covering >300 000 km². What is unusual about the block is that it shows many differences from the standard craton model (Zhai 1997). In particular, many of the Archaean rocks were subjected to granulite–upper amphibolite facies metamorphism and strong migmatization. This indicates that the North China Craton could have a more complicated early tectonic history than many other cratons, for example, the Pilbara (Krapez 1993) and Superior (Card 1990) cratons.

Cheng & Zhang (1982) suggested that two important metamorphic episodes took place in the NCC: a 2.6–2.5 Ga granulite facies event and a *c.*1.8 Ga amphibolite facies event. Shen *et al.* (1992) suggested that most protoliths of granulite facies had formed in the Palaeo–Mesoarchaean, their 2.6–2.45 Ga Sm–Nd and U–Pb isotopic data representing metamorphic ages. Considering rock associations and structures, Zhai & Windley (1990) divided the NCC Precambrian high-grade rocks into high-grade gneiss regions and metamorphosed greenstone belts with different ages (Table 1), indicating a

From: MALPAS, J., FLETCHER, C. J. N., ALI, J. R. & AITCHISON, J. C. (eds) 2004. *Aspects of the Tectonic Evolution of China*. Geological Society, London, Special Publications, **226**, 57–72.
0305-8719/04/$15 © The Geological Society of London 2004.

Table 1. *Correlation table of Early Precambrian rocks in the NCC (after Zhai & Windley 1990)*

District	Archaean					Early Proterozoic (Lower part)	
	Early	Middle		Late			
	HR1	HR2	GB2	HR3	GB3	P1	P2
E. Hebei	Caozhuang (>3.0)	Qianxi (2.8–3.0)		Zunhua (2.7–2.6) Dantazi (2.5)	Qinglong (2.5)	Zhuzhangzi (2.4) Shuangshanzi (2.4)	
Anshan		Baishanzheng Longgang (2.9–3.1)	Lower Qingyuan (2.9)	Anshan (2.7–2.6) Sandaogou (2.6)	Qingyuan (2.6–2.5)	Kuandian (2.4–2.2)	
						Lower Liaohe	Upper Liaohe (2.4–1.9) Hutuo (1.9) Lanhe
Wutaishan–Taihangshan (Shanxi Prov.)		Zabhuang Sushui		Fuping (2.7) Jiehekou (2.5)	Wutai (2.5)		
Yinshan (Inner Mongolia–north Shanxi Prov.)			Jiangouhe (>2.7)		Dongwufenzi (2.4–2.1?) Erdaogou	Fengzhen Wulashan	Majiadian
Taishan (Shandong Prov.)		Jiaodong (2.9–2.7)		Taishan (2.7–2.5)	Yarlingguan (2.7–2.6)	Fenzishan (2.4–2.2)	
Henan and Anhui Prov.				Taihua (2.7–2.5) (Huoqiu) (2.7–2.6)	Dengfeng (2.6–2.5)	Angou	Dongshan

HR – high-grade region; GB – greenstone belt; P – Proterozoic supracrustal belt. Numerals – isotopic ages in Ga.

multi-stage evolution of the NCC. Zhao (1993) proposed a tectonic model of three cratonization stages. Firstly, an old sialic crust underwent a Palaeoarchaean sodium–cratonization, then a Mesoarchaean sodium–potassium cratonization and a Neoarchaean potassium cratonization, associated with mainly vertical accretion with minor lateral accretion. In contrast, Geng *et al.* (2002) emphasized a crustal growth model of mantle-plume and magma underplating. In recent years, more tectonic models have been suggested to account for the presence of 2.5 Ga and/or 1.8 Ga Precambrian high-pressure granulites and retrograded eclogites (Zhai *et al.* 1992, 1995; Guo *et al.* 1993; Zhao 2001). The models include continent–continent collision

(Zhao *et al.* 1999), continent–arc collision (Wu *et al.* 1998) and the assembly of micro-blocks (Zhai *et al.* 2000). However, some aspects of Late Archaean or Mesoproterozoic tectonism are not well explained in these models. This review focuses on the oldest rock remnants, the main crustal growth period, and the important tectonic–metamorphic events and their chronological records. Finally, a new model for the early tectonic evolution of the NCC is presented.

Oldest continental nucleus

Liu *et al.* (1992) reported ion microprobe zircon U–Pb data of ≥3.8 Ga from the NCC, indicating the presence of old continental crust at two

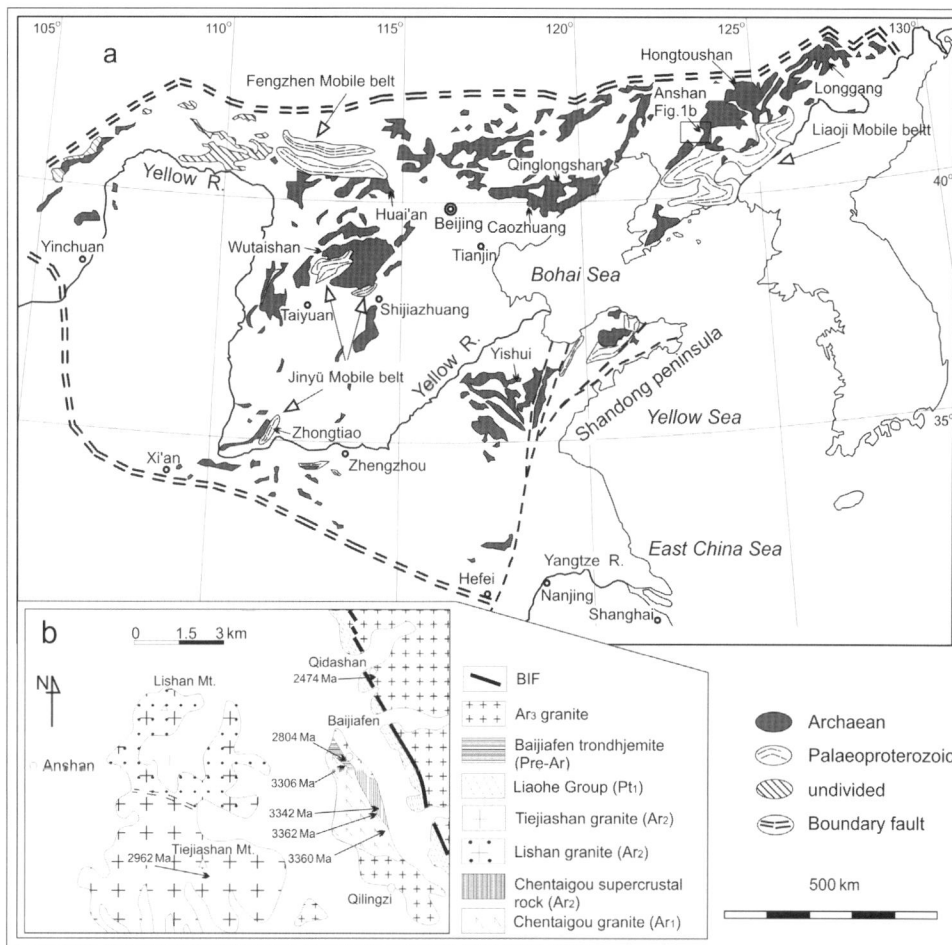

Fig. 1. Sketch geological map of the Early Precambrian NCC (**a**, after Zhai 1997) and the Tiejiashan complex (**b**, after Song *et al.* 1996).

localities: near Anshan, northeast China, and near Caozhuang, eastern Hebei (Fig. 1a).

The Neoarchaean Anshan Complex is divided into three parts: Tiejiashan gneiss, Anshan supracrustal series and Anshan gneiss (Zhai *et al.* 1990). The Anshan supracrustal series contains abundant banded iron formation rocks with Sm–Nd isochron ages of 2.7–2.66 Ga (Qiao *et al.* 1990). The granitoid protoliths of the Anshan gneiss intruded extensively into the Anshan supracrustal series at 2.55–2.47 Ga, based on their zircon U–Pb ages (Zhai *et al.* 1990). The Tiejiashan gneiss constitutes the basement of the Anshan supracrustal series, and includes a series of granitic and trondhjemitic orthogneisses and deformed felsic veins (Fig. 1b). Zircon U–Pb ages obtained from the unit range from *c.*3306 Ma in the northern part to *c.*2960 Ma in the central part (Liu *et al.* 1992). Two samples of sheared granitic gneiss were collected from Chentaigou village on the northwestern margin of the Tiejiashan gneisses. Liu *et al.* (1992) reported that ion microprobe U–Pb analyses show that the samples contain two generations of zircons with ages of 3805 ± 5 Ma and 3300 Ma, respectively. A presumed detrital zircon in a layered siliceous supracrustal rock near Chengtaigou yielded a SHRIMP age of 3362 ± 5 Ma (Song *et al.* 1996), but other supracrustal rocks contain detrital zircons not older than 3.0 Ga.

The oldest supracrustal remnant in the NCC is the Caozhuang complex in eastern Hebei Province (Fig. 1a). The Caozhuang complex can be divided into two parts: the Caozhuang Group and the Huangboyü grey banded gneiss. The Caozhaung Group is a slab 1.9 km long and 400–500 m wide that forms a complicated synformal fold structure. The Huangboyü grey gneiss intruded into the supracrustal rocks of the Caozhuang Group prior to intense multistage deformation. Zircons from the Huangboyü gneiss have U–Pb ages of 3.0–3.3 Ga (Zhao 1993). The main supracrustal rocks include amphibolites, serpentinized marbles, fuchsitic quartzites, metamorphosed calc-silicate rocks, banded iron formations, biotite gneisses, and sillimanite plagioclase gneisses. These rocks were metamorphosed to high-amphibolite/granulite facies (Yan *et al.* 1991). The amphibolites yielded Sm–Nb isochron ages of *c.*3.5 Ga (Huang *et al.* 1983; Jahn & Zhang 1984). Zircons from the fuchsitic quartzites are colourless to lilac and the rounded shape of the grains is attributed to abrasion during sedimentary transport. Eighty-two U–Pb SHRIMP analyses on 61 zircons yielded four age populations: 3.83–3.82 Ga, 3.8–3.78 Ga, 3.72–3.7 Ga and 3.68–3.6 Ga.

Higher-U zircon rims were also distinguished, which imply high-grade metamorphism at 2.5 Ga and 1.89 Ga (Liu *et al.* 1992).

Besides these two oldest crust remnants described above, several additional old continental nuclei in the NCC have also been proposed, for example, the Huani'an complex in the western–central NCC, the Yishui complex in the eastern NCC and the Longgang complex in the northeastern NCC (Fig. 1a). They have 3.5–3.3 Ga (Guo *et al.* 1991; Kröner *et al.* 1987), 3.1–2.97 Ga (Wu *et al.* 1998), and 3.1–3.0 Ga (Zhai & Windley 1990) isotopic ages, respectively.

Crustal growth

McLennan & Taylor (1982, 1983) envisaging episodic early growth, proposed that 65–75% of the continental crust formed during the period 3.2–2.5 Ga. Brown (1979) and Dewey & Windley (1981) also suggested that rapid growth was up to 2.5 Ga, by which time some 80% of the present continent had formed, with much slower growth after that. Zhai *et al.* (2001) reported that estimated volumetric crustal growth of the NCC was about 90% by 2.5–2.45 Ga, based on new geological maps, and geophysical and geochronological data. Neodymium isotope and trace-element characteristics of Early Precambrian rocks from the NCC, and their implication for the crustal growth, were discussed by Zhang (1998) and Jahn (1990). They considered that Sm–Nd T_{DM} ages can roughly infer the formation age of the crust. Figure 2a is an histogram of Nd T_{DM} ages from mafic igneous rocks. The samples older than 3.0 Ga account for *c.*15% of the total, whereas samples younger than 2.5 Ga account for only *c.*7%. Samples with T_{DM} ages between 3.0 and 2.5 Ga account for 78% of the total. Two discernible peaks occur at *c.*2.9 Ga and *c.*2.7 Ga (Fig. 2a). The diagram of $\varepsilon_{Nd}(T)$ versus t/Ga (Fig. 2b) shows two characteristics: all values of $\varepsilon_{Nd}(T)$ are positive, and there is an obvious change of $\varepsilon_{Nd}(T)$ with the change of T. The values of $\varepsilon_{Nd}(T)$ deviate from the depleted mantle evolution curve at about 3.0 Ga, which is attributable to contamination of crustal materials, and indicates that a thick continental crust existed in the NCC during the Neoarchaean. Rare-earth elements (REE) also demonstrate the same tendency, for example, the higher La/Nb ratios of pre-3.0 Ga mafic rocks indicate the presence of a considerable volume of continental crust at this time (Jahn 1990). Most mafic granulites and amphibolites from the NCC display REE patterns similar to those of basalts from island arc, continental margin and within-plate

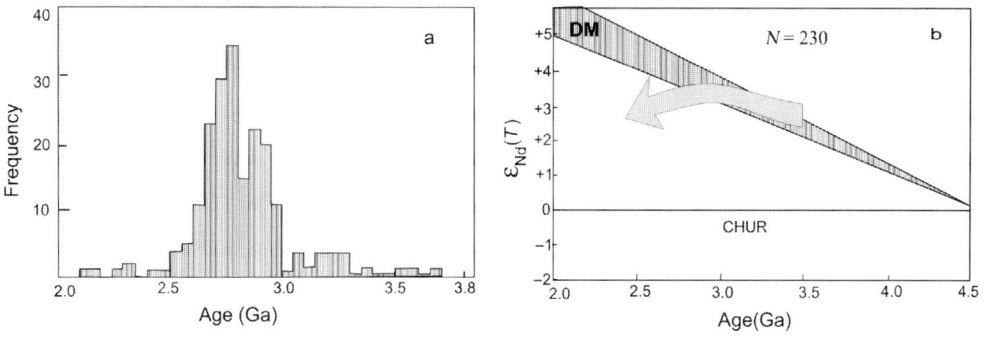

Fig. 2. Sm−Nd TDM age histogram (**a**) and diagram of $\varepsilon_{Nd}(T)-T/Ga$ (**b**) of mafic igneous rocks from the NCC (after Zhang 1998 and Zhai & Liu 2003).

settings, indicating that these rocks formed in different tectonic settings. However, a few samples have mid-ocean ridge basalt (MORB) characteristics.

The general rock distribution is of >3.0 Ga-old Archaean rocks in the centre of the NCC surrounded by Late Archaean and Proterozoic supracrustal rocks (Fig. 1a), − similar to other cratons (Bai *et al.* 1993). However, this simple picture is in reality more complicated, because there are a significant number of small Archaean-rock outcrops on the NCC that are too small to be represented on this map. Within the Archaean craton <3.0 Ga-old cores occur surrounded by 3.0−2.5 Ga rocks that formed as Mesoarchaean mobile belts. Deng *et al.* (1999) suggested that ten old continental nuclei existed in the NCC, in contrast to Wu *et al.* (1998) and Zhai *et al.* (2000), who proposed that only five

or six continental nuclei existed. The greenstone belts are interpreted as arc−back basin suites, while the 3.0−2.5 Ga high-grade regions are thought to represent continental margin, or island arc associations. For example (Fig. 3a), the Anshan Complex is composed of the Tiejiashan nucleus with a surrounding belt of the Anshan BIF-bearing supracrustal rocks that underwent high-grade metamorphism and were intruded by voluminous tonalites, granodiorites and granites which are chemically similar to modern Andean arc intrusive rocks (Fig. 3b; Zhai *et al.* 1990).

Late Neoarchaean craton and amalgamation event

A number of workers believe that the NCC comprises several micro-blocks which amalgamated

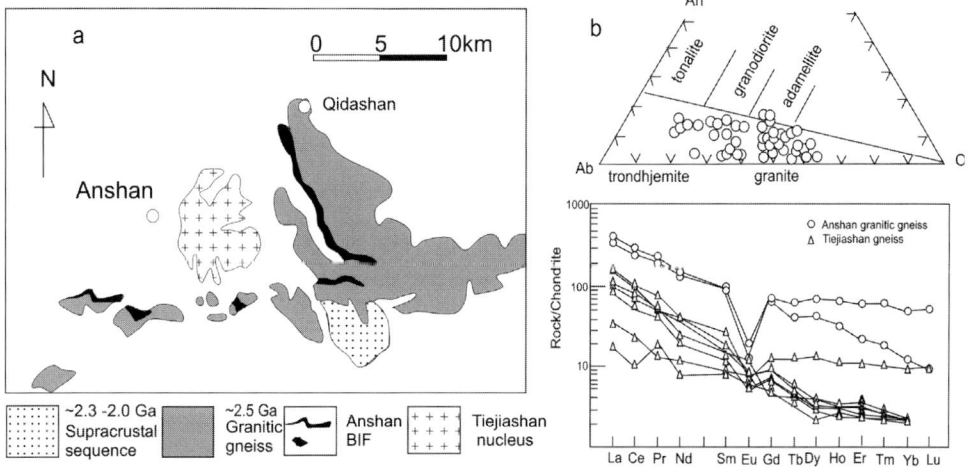

Fig. 3. Sketch map of Archaean rocks in Anshan area (**a**) and diagrams of Ab−An−Or and chondrite-normalized REE patterns for orthogneisses (**b**) (after Song *et al.* 1996).

to form a craton at or before 2.5 Ga (Geng 1998; Zhang 2000). Bai *et al.* (1993) and Wu *et al.* (1998) divided the NCC into six micro-blocks: Jiaoliao (JL), Qianhuai (QH), Fuping (FP), Ji'ning (JN), Xuchang (XCH), and the Alashan (ALS) blocks (Fig. 4). Exposed rock types and their distribution may vary considerably between blocks. For example, abundant Mesoarchaean

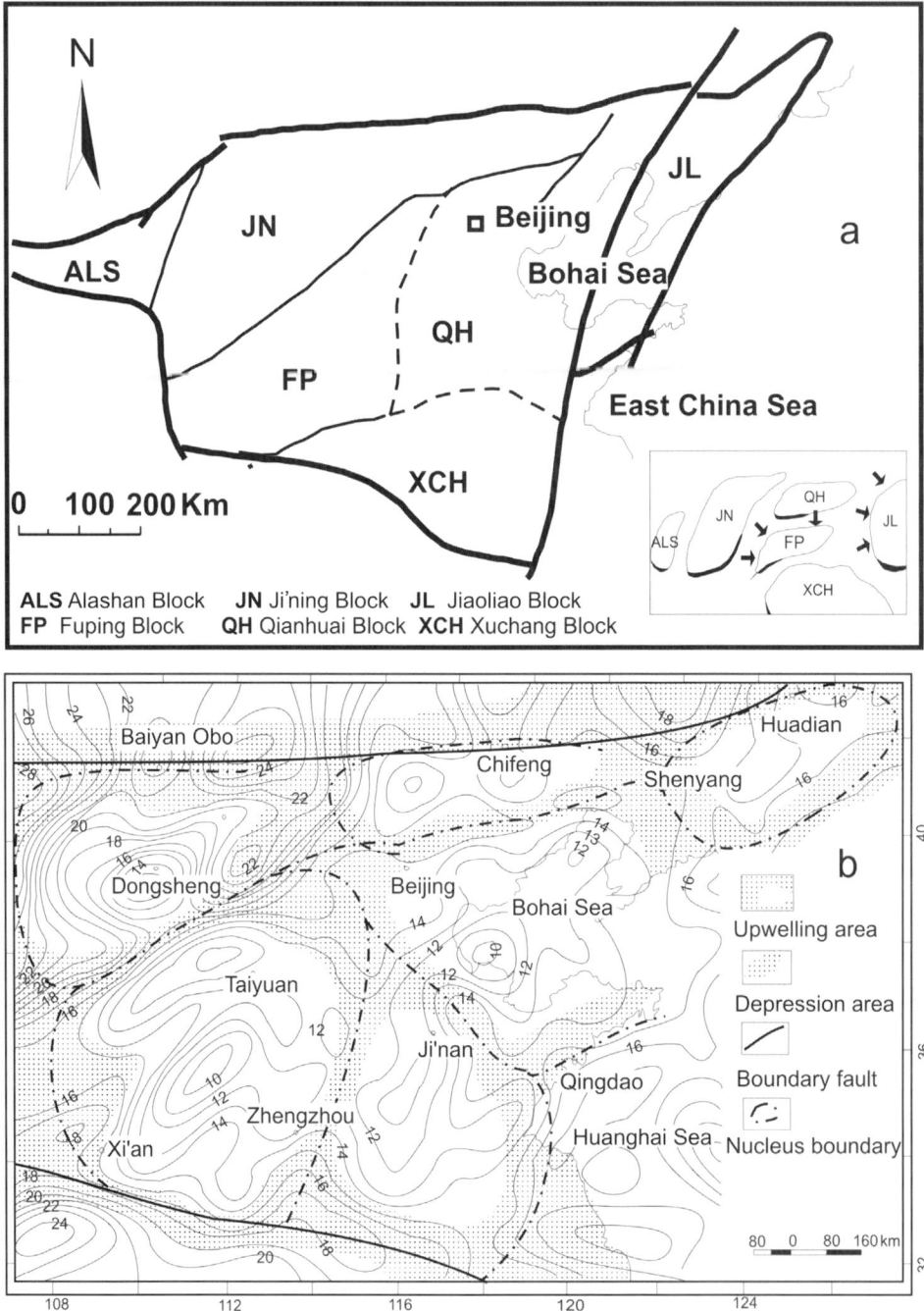

Fig. 4. Neoarchaean micro-blocks of North China Craton (**a**) and magnetic isobath map (**b**) (after Wu *et al.* 1998).

BIF-bearing supracrustal rocks are only developed in the Qianhuai Block. No rock more than 2.9 Ga is exposed on either the Ji'ning or the Fuping blocks, although geophysical data do not preclude the possibility that they exist in the deep crust (Wu et al. 1998). Neoarchean volcanism and magmatism in these blocks took place at 2.9–2.7 Ga and c.2.5 Ga, but their composition in different blocks varies greatly. The earlier phase of volcanism was, in general, voluminous in all blocks, especially in the Jiaoliao, Qianhuai and Xuchang blocks, and is associated with abundant BIFs. However, BIFs are not developed in the Alashan block. Volcanism at c.2.5 Ga was intense in the Jiaoliao and Fuping blocks. Basic-andesitic–acid volcanic rocks in the Fuping block are related closely to BIFs (i.e. Wutai Group). The Jiaoliao block contains massive sulfide Cu–Zn ores hosted in volcanic rocks (i.e. the Hongtoushan Group). It is worth noting that in all six blocks the Archaean rocks underwent metamorphism at 2.5 Ga, and were intruded by 2.5–2.45 Ga granitic sills and bodies. Emplacement of mafic dyke swarms at 2.5–2.45 Ga has also been recognized throughout the NCC (Liu 1989; J. H. Li et al. 1996; T. S. Li 1999). These dykes were metamorphosed to upper amphibolite–granulite facies and underwent retrograde metamorphic overprinting of amphibolite facies during pervasive deformation in the Proterozoic. The common post-2.5 Ga history of all six blocks suggests that the micro-blocks were assembled as the NCC by the end of the Archaean (J. H. Li et al. 1996).

There are a number of different tectonic models for the Late Archaean NCC. These vary from classical vertical accretion models that include multistage cratonization (e.g. Zhao 1993) and marginal accretion–reworking models (Jin & Li 1994). Some models proposed arc–continent collision or continent–continent collision similar to Phanerozoic tectonic processes (Li et al. 1997). Wu et al. (1998) suggested that an old arc volcanic–magmatic zone is located in the northern NCC from Hongtoushan, NE China, via Qinglongshan, eastern Hebei to western Shandong (Fig. 5a), sandwiched between two continental blocks. The meta-volcanic–sedimentary rocks with isotopic ages of 2.56–2.53 Ga in the Hongtoushan are chemically similar to modern arc rock sequences (Zhai et al. 1985). It is possible that the Qinglongshan rocks may have formed in an island arc setting (Wu et al. 1998), although the rocks are different from those in Hongtoushan, as no andesites are present. Many granitoid intrusive bodies occur within the arc. Tonalite, trondhjemite and granodiorite associations mainly occur along the western side, and calc-alkaline granitic rocks

on the eastern side; the former were dated at 2.55–2.5 Ga, whilst the latter were dated at 2.5–2.45 Ga (Zhao 1993; Li, T. S. 1999). Wu et al. (1998) suggested that this compositional polarity and diachronous intrusion pattern formed because of an ancient sea basin that lay to the west. Subduction of this ocean crust eastward, beneath the continent block, formed an island arc, which evolved into an arc–continent collisional zone. Zhang (2000) suggested that the greenstone belts that lie between the old continental blocks represent orogenic zones in the NCC (Fig. 5b). Greenstone belts of c.2.5 Ga mainly occur between the Qianhuai, Xuchang and Fuping blocks, and represent zones of continental amalgamation. Earlier 2.6–2.7 Ga greenstone belts suffered strong reworking at c.2.5 Ga. Zhai et al. (1992, 1995) proposed continent–continent collision between the Qianhuai and Fuping blocks, and the Qianhuai and Ji'ning blocks at 2.5–2.6 Ga, according to the distribution of high-pressure granulites. Zhai et al. (2000) also proposed that during the period 2.6–2.45 Ga the six micro-blocks in the NCC amalgamated as a result of continent–continent collision. Shortly after 2.5 Ga, the amalgamated NCC was intruded by 2.5–2.35 Ga granites and mafic dykes.

Palaeoproterozoic (2.35–1.97 Ga) mobile belts

Palaeoproterozoic rocks in the NCC were not well described in previous studies. Zhao (1993) pointed out that a Palaeoproterozoic volcanic and granitic event with an age of 2350–1970 Ma is stronger in the NCC than that in many other cratons in the world (e.g. Condie 1988; Hoffman 1988).

The representative rock sequences with ages of 2.35–1.97 Ga are the Liaohe Group in the Liaoji Mobile Belt, the Lüliang Group and the Hutuo Group in the Jinyü Mobile Belt, and the Fengzhen Group in the Fengzhen Mobile Belt (Fig. 1a). The Liaohe Group comprises basic–acid volcanic rocks and a thick boron-bearing sedimentary formation that has been metamorphosed to greenschist–amphibolite facies. Zhang (1984) suggested that these rocks formed in a rift environment. However, Bai et al. (1993) emphasized that basic volcanic rocks in the Liaohe Group have the characteristics of marine basalts and that sedimentary rocks were turbidites. Therefore they suggested that a small old ocean basin was located between the Longgang massif in the NE China and the Nangrim massif in the northern Korean peninsula. With a switch in tectonic regime from extension to

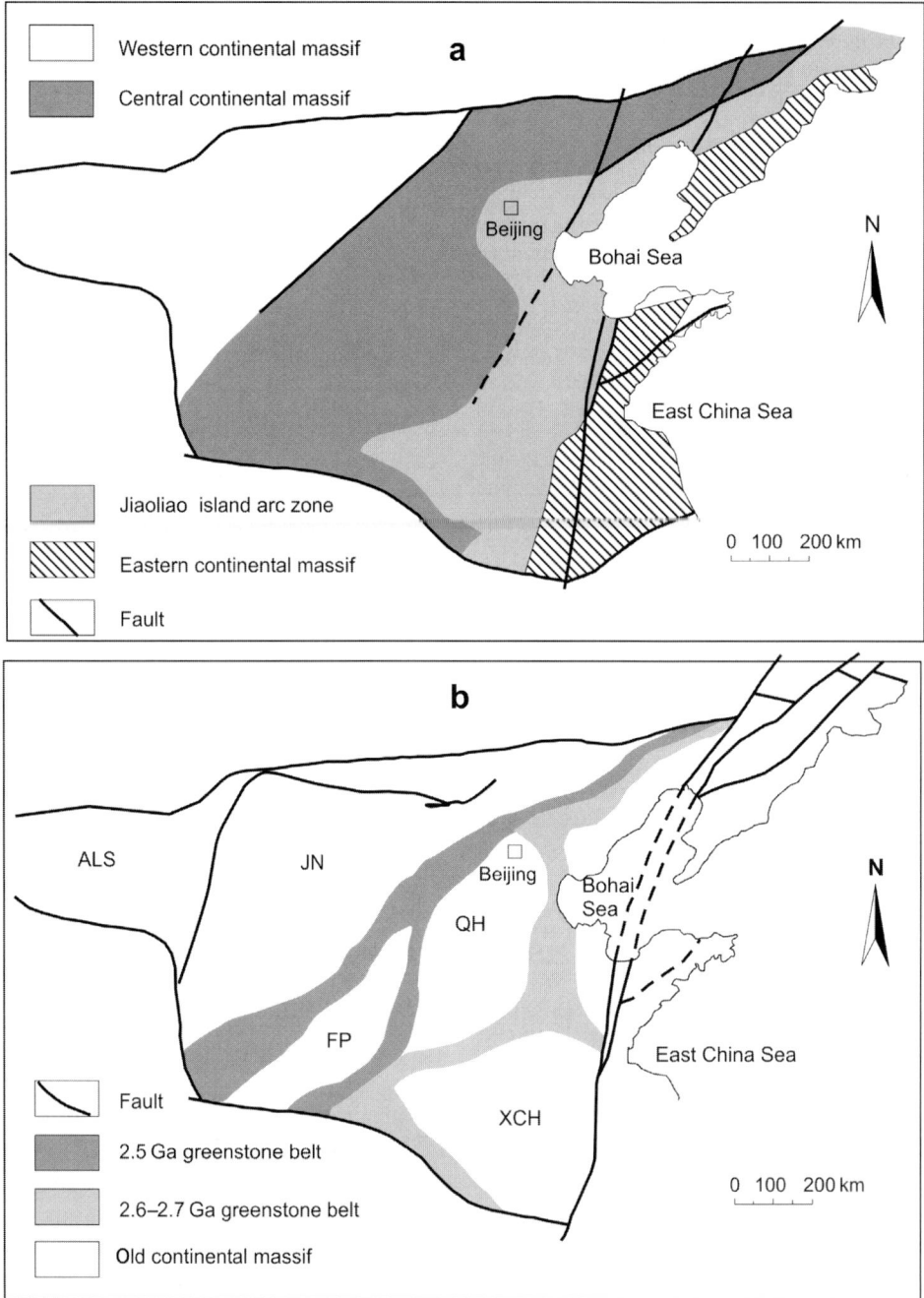

Fig. 5. Neoarchaean amalgamation of North China Craton by arc–continent collision (**a**, after Wu *et al.* 1998) and continent–continent collision (**b**, after T. S. Li 1999).

compression, the formation of the Liaohe Mobile Belt is related to a change in geological process from subduction to collision (Bai *et al.* 1993).

The Fengzhen Group is a thick sequence of metasedimentary rocks rich in aluminium, containing little volcanic material that underwent granulite-facies metamorphism (Qian *et al.* 1985; Shen *et al.* 1992; Wu *et al.* 1994). Although some geologists believe that the geological features of the Fengzhen Group indicate an Archaean age

(Zhao 1993; Lu *et al.* 1995; Qian 1996), zircon U–Pb ages of the khondalites support an interpretation of them being Palaeoproterozoic (Wu 1999). The protoliths of these rocks were deposited either in a cratonic basin (Condie *et al.* 1992; Qian 1996) or in a marginal sea (Lu *et al.* 1995). Another noteworthy idea is that the clockwise *P–T* path of the khondalites, from moderate–low pressure granulite facies to amphibolite facies, represents a continental collision (Wu *et al.* 1994).

The Jinyü Mobile Belt includes several Palaeoproterozoic formations, which occur in an area between 34°15′–39°N and 110°45′–114°50′E (Fig. 1). The rocks have similar protoliths, tectonic styles and metamorphic histories. These rocks have been interpreted to have been deposited in a continental rift called the Jinyü rift zone (Zhao 1993). However, Sun & Hu (1993) reported that the geochemical characteristics of the basic and intermediate–acid volcanic rocks are more similar to those associated with island arcs. Sedimentary rocks are mainly coarse- to medium-grained clastics, pelites and carbonates. Volcanic rocks are basalts, basaltic andesites and minor dacites. Their metamorphic grade is greenschist facies or amphibolite facies. Sun & Hu (1993) presented a geochronological study from the Zhongtiao Mountain area. The geochronological data for metamorphosed volcanic rocks, granites and related Cu-ores are consistent, and can be divided into three rock-forming periods: 2.36–2.32 Ga, 2.16–2.01 Ga, and 2.09–2.06 Ga, followed by a metamorphic event at 1.9–1.83 Ga.

In short, the Palaeoproterozoic mobile belts in the NCC formed and evolved within a craton or continental margin (epicontinental geosyncline) around 2400–2000 Ma. Various 2.35–1.95 Ma rift-margin and passive continental margin deposits (St-Onge & Lucas 1990; Windley 1995), ophiolites (Kontinen 1987; Helmstaedt & Scott 1992), orogens (Hoffman 1988) and BIF-bearing foreland basins (Hoffman 1987) associated with other cratons seem to have Late Phanerozoic analogues (Windley 1995), and this appears to be the case with the NCC Palaeoproterozoic mobile belts.

1.9–1.8 Ga metamorphism and 1.8–1.65 Ga rift event

The *c*.1.8 Ga event was traditionally named the Lüliang Movement or the Zhongtiao Movement (Wang 1955; Huang 1977). Many of the Early Precambrian rocks in the NCC have metamorphic and/or chronological imprints of the *c*.1.8 Ga event, which marked the final cratonization of

the NCC (Zhao 1993). Recently, different models have been proposed to interpret this event, e.g. continental extension (Li *et al.* 1997), continental uplift (Zhang *et al.* 1994) and continent–continent collision (Zhao 2001). However, the *c*.1.8 Ga event may in fact comprise a 1.9–1.8 Ga metamorphic episode followed by a rifting phase at 1.8–1.65 Ga, and thus marks two independent events. The major indications are:

(1) All the granulite facies and amphibolite facies rocks, no matter where they occur, be it on the margins of the NCC or within the core regions, underwent intense amphibolite facies retrograde metamorphism.

(2) The metamorphic rocks in the NCC were extensively migmatized at this time, as a result of which some meta-supracrustal rocks were converted into banded gneisses and some grey gneisses underwent strong potassium metasomatism (Zhao 1993).

(3) High-temperature ductile shear zones are common and form a complex network system, which is made up of alternating high and low strain zones. Extensional collapse and uplift structures with 1.85–1.8 Ga metamorphic ages are well developed in the NCC (Zhang *et al.* 1994; Zhang 1997; Xu *et al.* 1995).

(4) Continental rifts were developed in the NCC (Fig. 6a). Unmetamorphosed volcanic–sedimentary strata of the Xiong'er Group and the Changcheng system varying between 3–10 km in thickness and unconformably overlying Archaean and Palaeoproterozoic metamorphic basement (Fig. 6b).

(5) The anorogenic magmatic rock association (gabbro–anorthosite–rapakivi) at Miaoxiangshan on the northern Korean peninsula (Paek & Jon 1996) and Chengde–Jixian–Miyun in Hebei Province formed 1.78–1.63 Ga ago (Lu *et al.* 1995; Rämö *et al.* 1995; Zhao *et al.* 2002).

(6) Mafic dyke swarms of 1.8–1.7 Ga in age have been recognized in the NCC. They include low greenschist facies to greenschist facies diabases in Shanxi and Hebei Provinces, and granulite facies diabases in eastern and northwestern Hebei Province (Qian 1992; Li *et al.* 1997; T. S. Li 1999).

Three main possible interpretations for the tectonic evolution of the NCC between 1.9 and 1.65 Ga have been suggested:

(1) The *c*.1.8 Ga event represents a continent–continent collision event. Zhao (2001) suggests that the basement of the NCC consists of the Eastern Block and Western

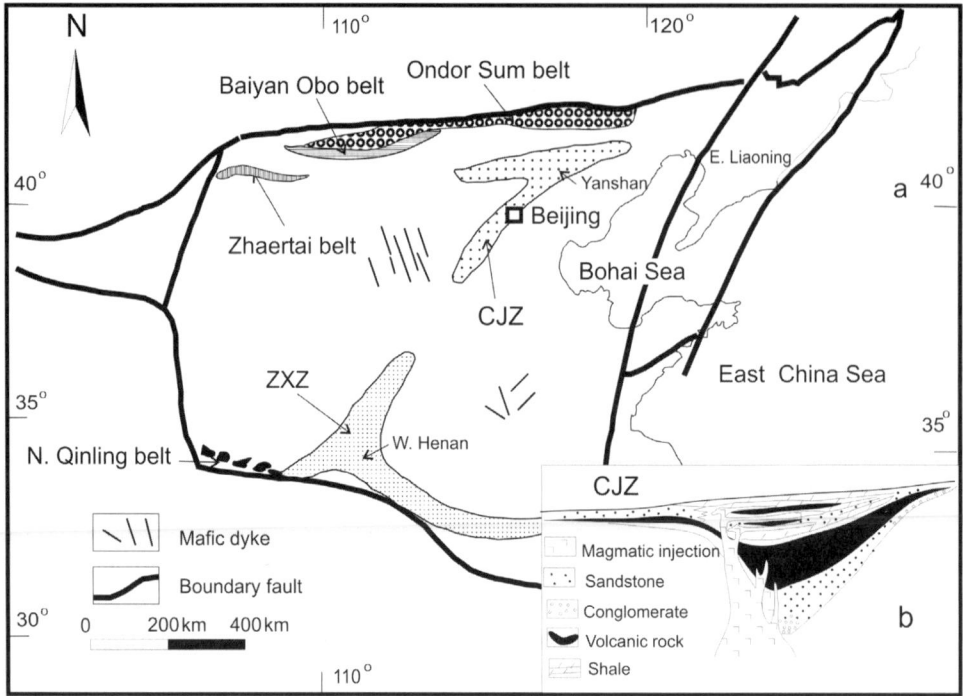

Fig. 6. Late Palaeoproterozoic–Mesoproterozoic rifts (**a**) and sketch section illustrating rift basin architecture in the Yanshan area (**b**) (after Zhai *et al.* 2000). ZCZ, Zhongtiao–Xiong'er Rift Zone; CJZ, Chengde–Jixian Rift Zone.

Block, separated by a central collision zone marked by the Hengshan–Wutai-Fuping mountain belt. The Central Zone extends as a roughly north–south trending belt and is composed of reworked Archaean basement and Late Archaean–Palaeoproterozoic sedimentary and igneous rocks. A tectonic evolution model include a >1.85 Ga ocean between two blocks, subduction of the Western Block under the Eastern Block, and amalgamation of the NCC along the Central Zone at *c.*1.8 Ga.

(2) The *c.*1.8 Ga event can be divided into two different events: a 1.9–1.8 Ga orogenic event and a 1.8–1.65 Ga rifting event (Zhai & Liu, 2003). The former is represented by 1890–1830 Ma high-pressure granulites, regional metamorphism with clockwise *P–T* paths involving near-isothermal decompression, and the presence of linear high-strain belts marked by an intense mineral stretching lineation and the presence of sheath folds. In contrast, the latter includes a rift-type volcanic–sedimentary cover lying unconformably on metamorphic basement. The 1800–1650 Ma rift event is independent, and

most of the strata and related intrusive rocks did not undergo metamorphism.

(3) A contrary view is that the 1.9–1.8 Ga event is an extensional-basement uplift episode. First, the strong regional metamorphism in the NCC is widespread and not restricted to one or several zones (Wang *et al.* 1987; Yan *et al.* 1991; Bai *et al.* 1993; Zhao 1993; Zhang 2000). Secondly, the typical structural style at 1900–1800 Ma is that of shallow-angle detachments associated with diapiric structures (Xu *et al.* 1995; Zhang 1997). Thirdly, no volcanic–sedimentary rocks formed during this period. From this, Zhai & Liu (2003) considered that the lower crust of the NCC was uplifted as a whole and was then exhumed at *c.*1800 Ma by a series of detachment structures. However, the causal tectonic mechanism is not so clear with this model. Instead, Zhai (1999) suggested a possible *c.*1.8 Ga upwelling mantle plume model, which induced uplifting of lower crust and was followed by rifting. Zhai (1999) also suggested that this event was linked with the breakup of the pre-Rodinia supercontinent, Columbia.

North China and Rodinia

From $c.1650$ Ma to the Middle Mesozoic, the NCC was tectonically rather inactive, and for this period it is termed the North China Platform/Paraplatform (Cheng *et al.* 1986; Zhao 1993). However, recent studies reveal a record of metamorphic–magmatic events of $c.1300$–1000 Ma and $c.800$–650 Ma in the NCC (Zhai 2001). Although these events were weak, Zhai (2001) and Shao *et al.* (2001) considered that they were related to the assembly and breakup of the supercontinent Rodinia. The main geological features of the Meso–Neoproterozoic NCC are summarized as follows:

(1) Precambrian metamorphic basement rocks within the NCC and along its eastern margins show no record of any Meso–Neoproterozoic tectonic–magmatic event, which indicates that the eastern NCC was tectonically inactive from 1.9–1.7 Ga until the Mesozoic. There is no evidence to support a Late Mesoproterozoic–Neoproterozoic convergent orogenic belt along the northern margin of the NCC (Fig. 6a). However, metamorphic rocks in the Zhaertai belt, Baiyan Obo belt, and Ondor Sum belt in northern-central Inner Mongolia yield Sm–Nd isochron ages of 1.0–1.5 Ga (Nie *et al.* 1993, 1995; Tao *et al.* 1998), which possibly reflects continental margin processes (Zhai *et al.* 2003).

(2) A late Mesoproterozoic to early Neoproterozoic convergent margin setting is represented by the North Qinling orogenic belt along the southern margin of the NCC. Here the main rocks consist of a metamorphosed volcano-sedimentary sequence (the Kuanping Group) and related gabbros and serpentinized peridotites that have been interpreted as the remains of an ophiolite (Zhang *et al.* 1997). The Kuanping Group is composed of turbidites and greenschists that have been suggested to be an oceanic flysch sequence (G. W. Zhang *et al.* 2000; Zhang *et al.* 2001). The greenschists demonstrate the chemical characteristics of MORB, with flat chondrite-normalized REE distribution patterns and $\varepsilon_{Nd}(T)$ values of $+2$ to $+6.5$. Their Sm–Nd isochrons range from 1.2 to 0.92 Ga (Zhou *et al.* 1995; Zhang *et al.* 1997; Zhang *et al.* 2001). A series of granitic intrusive rocks with ages of 1.0–0.8 Ga have been recognized along the North Qinling belt, and these show similarities with collisional and post-collisional granites.

(3) Above the basement of the NCC, Mesoproterozoic–Neoproterozoic sedimentary sequences (the Changcheng–Jixian–Qingbaikou systems) constitute a disconformable–pseudoconformable succession of shallow-marine siliciclastics, argillites and carbonates up to 10 km thick (Fig. 7), deposited in intra-cratonic aulacogen. Figure 6b is a sketch section illustrating rift basin architecture in the Yanshan area. Deposition marking the base of the Changcheng System started at $c.1800$ Ma. Zircon U–Pb and ^{40}Ar/^{39}Ar ages indicate that the volcanic activity of the Jixian System mainly took place at $c.1650$–1600 Ma (Ren 1986; Wang *et al.* 1987; Bai *et al.* 1993). The upper formation of the Jixian System represents an extensive marine transgression deposit that ended at $c.1.1$ Ga with a period of erosion between the Jixian System and the Qingbaikou System (Bai *et al.* 1993). This disconformity is called the Qinyü uplift, which corresponds to the widespread Late Mesoproterozoic orogenesis seen elsewhere. Therefore, Hao & Zhai (2004) suggested that the Qinyü unconformity is a signature marking the assembly of the NCC into the Rodinia supercontinent. The Qingbaikou System consists of sandstone–shale–carbonate rock associations that indicate a stable sedimentary environment. It was deposited between about 0.95 and 0.85 Ga ago. The Sinian-age sequence (0.85–0.65 Ga) was only locally formed on the southern margin of the NCC and in eastern Liaoning, NE China. Deposits of this age are termed the Huanglianduo Group and the Luoquan Group in Shanxi–W. Henan provinces, the Sanchakou Group in the North Qinling belt, the Huainan Group in Jiangsu–Anhui provinces, and the Jianchang Group in NE China. They are younger or equivalent in age to the Doushantuo Group (the South China Sinian sequences) in the South China Craton (SCC) in age (Hao & Zhai 2004), and differ from the Doushantuo Group in their sedimentary facies and rock association. The NCC Sinian-age sequences are shallow-marine carbonate and siliciclastic deposits with or without basal conglomerates, and have a top layer of shale and carbonate rocks. Glacial sediments (the Luoquan Group) have been only found in western Henai and the northern Qinling. The Luoquan Group may correspond with the Sinian System of the SCC (Hao & Zhai 2004). The Neoprotero-

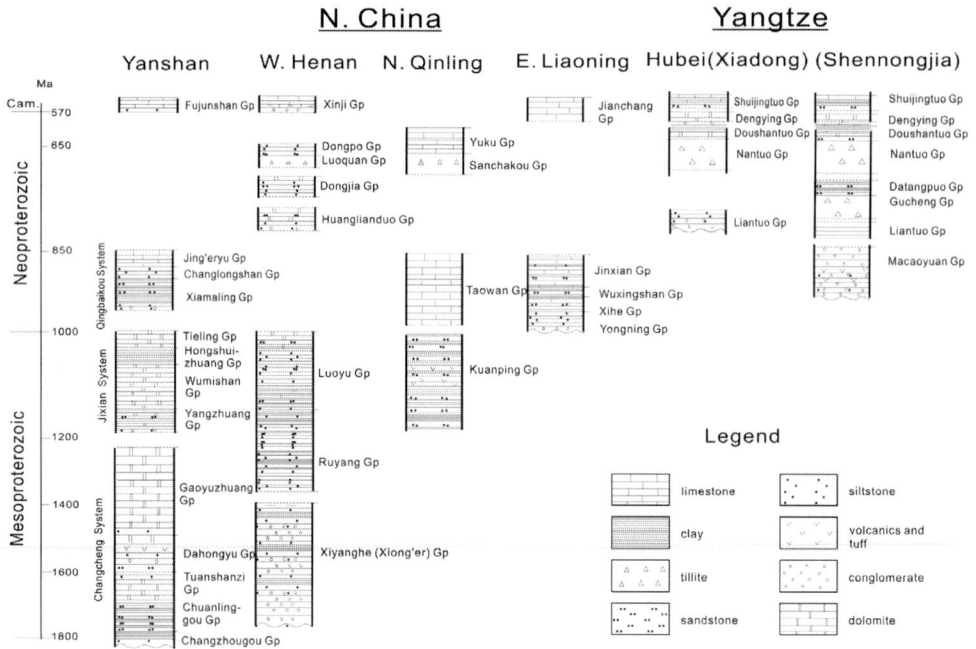

Fig. 7. Tectonostratigraphic correlations between the North China and Yangtze Cratons (after Hao 2003).

zoic fossils in the NCC are part of the 850–600 Ma Huainan biota. The Vermes fossils, which are represented by *Pararenicola–Palealina*, have been found on both the Siberian and the Indian Cratons, but not on the SCC (Xing & Liu 1982; Chen 1989; Niu & Sun 2000). This suggests that during the Neoproterozoic the NCC–Siberia–India Cratons were in close proximity, and that they were also quite separate from the SCC.

(4) Much palaeomagnetic work has been done on the Mesoproterozoic–Neoproterozoic sequences of the South China and North China Cratons. Z. X. Li *et al.* (1996) suggested that the SCC was located between east Australia, east Antarctica and Laurentia. In the same but in this reconstruction the NCC was positioned at a quite separate location on the northeastern margin of the Rodinia near the Siberian Craton. However, Guo *et al.* (1999) suggested that both the NCC and SCC were, as old continental micro-massifs, located in the centre of the Rodinian supercontinent, surrounded by Siberia, Australia and Laurentia. Based on new palaeomagnetic data, Zhang *et al.* (2000) suggest that the North China Craton was at low latitudinal positions for most of the

Neoproterozoic and could have been close to Siberia for much of this time.

(5) Anorogenic granites and mafic dyke swarms sourced from enriched mantle in the SCC have been interpreted as a result of an upwelling superplume that broke up the supercontinent Rodinia (Li X. H. 1999). However, no anorogenic magmatism of *c.*0.8–0.6 Ga age in the NCC has been reported. The North China Craton was therefore not significantly affected by the proposed Rodinian superplume (Zhai *et al.* 2003).

On the basis of geology and tectonostratigraphy, we suggest that the NCC was a part of the supercontinent Rodinia, and was located on the edge of the Rodinian supercontinent, in a similar position to that suggested by Z. X. Li *et al.* (1996) and Zhang & Li (2001). This location may partially explain why little evidence for the formation or breakup of Rodinia exists in the NCC.

Conclusions

(1) Old continental nuclei were recognized within the NCC – the oldest remnants of granitic gneiss and supracrustal rock are

3.8 billion years old. Main crustal growth in the NCC took place at 2.9–2.7 Ga.

(2) By 2.5 Ga, the NCC had basically amalgamated through a series of continent–continent, arc–continent and arc–arc collisions.

(3) The 2.35–1.97 Ga mobile belts represent intracratonic orogens, and share features that are common to Phanerozoic systems.

(4) The 1.8 Ga event marked an intense tectonic–metamorphic episode. Some have suggested that the metamorphosed basement of the NCC was brought to upper crustal levels by a series of detachment structures. However, I prefer to the rising mantle plume model, which uplifted the lower crust, and a process that was followed by rifting.

(5) The NCC shows little record of Neoproterozoic tectonic event; therefore it was suggested that the NCC was a part of the supercontinent Rodinia, and was probably located on one of its edges.

This study represents the research results of a project (Grant No. 40072061) supported by the National Nature Science Foundation of China, and a project (Grant No. KZCX1-07) supported by the Chinese Academy of Sciences. We especially thank J. Malpas, J. R. Ali and two reviewers for their help and discussions.

References

BAI, J., HUANG, X. G., DAI, F. Y. & WU, C. H. 1993. *The Precambrian evolution of China*. Geological Publishing House, Beijing 199–203.

BROWN, G. C. 1979. The changing pattern of batholith emplacement during earth history. *In*: ATHERTON, M. P. & TARNEY, J. (eds) *Origin of Granite Batholiths*. Shiva, Nantwich, 106–115.

CARD, K. D. 1990. A review of the Superior Province of the Canadian Shield: a product of Archaean accretion. *Precambrian Research*, **48**, 99–156.

CHEN, M. E. 1989. General description of Late Precambrian mega-fossils in China. *Acta Geologica Sinica*, **3**, 244–255.

CHENG, Y. Q. & ZHANG, S. G. 1982. Notes on the metamorphic series and metamorphic belts of various metamorphic epochs of China and related problems. *Regional Geology of China*, **2**, 1–14.

CHENG, Y. Q., SUN, D. Z. & WU, J. S. 1986. Certain geological and evolutional characteristics of the early Precambrian of the Proto-North China platform. *In*: *Proceedings of International Symposium on Precambrian Crustal Evolution (No. 2)*. Geological Publishing House, Beijing, 1–17.

CONDIE, K. C. 1988. Origin and early growth rate of continents. *Precambrian Research*, **32**, 261–278.

CONDIE, K. C., BORYTA, M. D., LIU, J. Z. & QIAN, X. L. 1992. The origin of khondalites: geochemical evidence from the Archaean to early Proterozoic granulite belt in the North China craton. *Precambrian Research*, **59**, 207–223.

DANN, J. C. 1991. Early Proterozoic ophiolite, Central Arizona. *Geology* **19**, 594–597.

DENG, J. F., WU, Z. X., ZHAO, G. C., ZHAO, H. L., LUO, S. H. & MO, X. X. 1999. Precambrian granitic rocks, continental crustal evolution and crust formation of the north China platform. *Acta Petrologica Sinica*, **15**, 190–198.

DEWEY, J. F. & WINDLEY, B. F. 1981. Growth and differentiation of the continental crust. *Philosophical Transactions of the Royal Society of London*, **A301**, 189–206.

GENG, Y. S. 1998. Archean granite pluton events of Qianan area, East Hebei province and its evolution. *In*: CHEN, Y. Q. (ed.), *Corpus on Early Precambrian Research of the North China Craton*. Geological Publishing House, Beijing, 105–121.

GENG, Y. S., WAN, Y. S. & SHEN, Q. H. 2002. Early Precambrian basic volcanism and crustal growth in the North China Craton. *Acta Geologica Sinica*, **76**, 199–206.

GUO, J. H., LIU, Y. G. & XIA, Y. L. 1991. Old continental nuclei–Early Archean Huai'an complex in the NCC: their Pb isotopic characteristics. *In*: INSTITUTE OF GEOLOGY, ACADEMIA SINICA (ed.), *Annual Bulletin (89–90) of Open Laboratory of Lithosphere Tectonic Evolution, Academia Sinica*. Chinese Scientific and Technological Press, 115–118.

GUO, J. H., ZHAI, M. G. & ZHANG, Y. G. 1993. Early Precambrian Manjinggou high-pressure granulites mélange belt on the southern edge of the Huai'an complex, North China Craton: geological features, petrology and isotope geochronology. *Acta Petrologica Sinica*, **9**, 329–341.

GUO, J. J., ZHANG, G. W., LU, S. N. & ZHAO, F. Q. 1999. Neoproterozoic continental assembly in China and Rodinia supercontinent. *Geological Journal, China Universities*, **5**, 148–156.

HAO, J. & ZHAI, M. G. 2004. Jinning movement and Sinian System in China: their relationship with Rodinia supercontinent. *Chinese Journal of Geology*, **39**, 139–152.

HELMSTAEDT, H. H. & SCOTT, D. J. 1992. The Proterozoic ophiolite problem. *In*: CONDIE, K. C. (ed), *Proterozoic Crustal Evolution*, Elsevier, Amsterdam, 55–95.

HOFFMAN, P. F. 1987. Early Proterozoic foredeeps, foredeep magmatism, and Superior-type iron-formations on the Canadian Shield. *In*: KRÖNER, A. (ed.), *Proterozoic Lithospheric Evolution*. Washington, D.C. American Geophysical Union, Geodynamics Series, pp. 85–98.

HOFFMAN, P. F. 1988. United plates of America, the birth of a craton: early Proterozoic assembly and growth of Laurentia. *Annual Review of Earth and Planetary Sciences*, **16**, 543–603.

HUANG, J. Q. 1977. The fundamental framework of the geotectonics of China. *Acta Geologica Sinica*, **2**, 117–134.

HUANG, X., BAI, Y. L. & DePAOLO, D. J. 1983. Sm–Nd isotope study of early Archaean rocks, Qian'an, Hebei province, China. *Geochimica et Cosmochimica Acta*, **50**, 625–631.

JAHN, B. M. 1990. Early Precambrian basic rocks of China, In: HALL, R. P. & HUGHES, D. J. (eds), Early Precambrian Basic Magmatism. Blackie, Glasgow, 294–316.

JAHN, B. M. & ZHANG, Z. Q. 1984. Radiometric ages (Rb–Sr, Sm–Nd, U–Pb) and REE geochemistry of Archaean granulite gneisses from eastern Hebei province, China, In: KRÖNER, A., HANSON, G. N. & GOODWIN, A. M. (eds), Archaean Geochemistry, Springer-Verlag, Berlin/Heidelburg, 183–204.

JIN, W. & LI, S. X. 1994. The lithological association and geological features of early Proterozoic orogenic belt in Daqingshan, Nei Mongol. In: QIAN, X. L. & WANG, R. M. (eds), Geological Evolution of the Granulite Terrain in North Part of the North China Craton. Seismological Press, Beijing, 32–42.

KONTINEN, A. 1987. An early Proterozoic ophiolite– the Jormua mafic–ultramafic complex, northeastern Finland. Precambrian Research, 35, 313–341.

KRAPEZ, B. 1993. Sequence stratigraphy of the Archaean supracrustal belts of the Pilbara block, Western Australia. Precambrian Research, 60, 1–45.

KRÖNER, A., COMPSTON, W., ZHANG, G. W., GUO, A. L. & CUI, W. Y. 1987. Single zircon ages for Archaean rocks from Henan, Hebei and Inner Mongolia, China and tectonic implication. In: Proceedings of the International Symposium on Tectonic Evolution and Dynamics of Continental Lithosphere. Beijing, 43.

LI, J. H., HE, W. Y. & QIAN, X. L. 1997. Genetic mechanism and tectonic setting of a Proterozoic mafic dyke swarm: its implication for palaeo-plate reconstruction. Geological Journal of China University, 3, 2–8.

LI, J. H., QIAN, X. L., ZHAI, M. G. & GUO, J. H. 1996. Tectonic division of high-grade metamorphic terrain and late Archaean tectonic evolution in north-central part of North China Craton. Acta Petrologica Sinica, 12, 179–192.

LI, T. S. 1999. Taipingzhai–Zunhua Neoarchaean island arc terrain and continental growth in Eastern Hebei, North China. PhD thesis, Institute of Geology, Chinese Academy of Sciences, Beijing, 10–129.

LI, X. H. 1999. U–Pb zircon ages of granites from the southern margin of the Yangtze block: timing of the Neoproterozoic Jining orogeny in SE China and implications for Rodinia. Precambrian Research, 97, 43–57.

LI, Z. X., ZHANG, L. & POWELL, C. McA. 1996. Position of the East Asian craton in the Neoproterozoic supercontinent Rodinia. Aust. J. Earth Sci, 43, 593–604.

LIU, D. Y., NUTMAN, A. P., COMPSTON, W., WU, J. S. & SHEN, Q. H. 1992. Remnants of ≥3800 Ma crust in the Chinese part of the Sino-Korea craton. Geology, 20, 339–342.

LIU, Y. G. 1989. Precambrian geology, petrology and geochemistry of NW Hebei and adjacent areas, China. PhD thesis, Institute of Geology, Academia Sinica, Beijing, 23–78.

LOUKOLA-RUSKEENIEMI, K., HEINO, T., TALVITIE, J. & VANNE, J. 1991, Base metal-rich metamorphosed black shales, associated with Proterozoic ophiolites in the Kainuu, Finland: a genetic link with the Outokumpu rock assemblage. Mineralium Deposita, 26, 143–151.

LU, L. Z., XU, X. C. & LIU, F. L. 1995. The Precambrian Khondalite Series in Northern China. Changchun Publishing House, Changchun, 1–99.

McLENNAN, S. M. & TAYLOR, S. R. 1982. Geochemical constraints on the growth of the continental crust. Journal of Geology, 90, 347–361.

McLENNAN, S. M. & TAYLOR, S. R. 1983. Continental freeboard sedimentation rates and growth of continental crust. Nature, 306, 169–172.

MARSHAK, S. 1999. Deformation style way back when: thoughts on the contrast between Archaean/Palaeoproterozoic and contemporary orogens. Journal of Structural Geology, 21, 1175–1182.

NIE, F. J., PEI, R. F. & WU, L. S. 1995. Nd–Sr isotope study on greenschist and granodiorite of the Bainaimiao area, Inner Mongolia, China. Acta Geoscientia Sinica, 1, 36–44.

NIE, F. J., PEI, R. F., WU, L. S. & ZHANG, H. T. 1993, Igneous Activity and Metallogeny in the Bainaimiao–Wenduermiao Area, South-Central Inner Mongolia, People's Republic of China. Beijing Science and Technology Press, Beijing, 1–239.

NIU, S. H. & SUN, S. F. 2000. The tentative establishment of Palaeopacific Huainan–Little Dal biogeographic demarcation and its significance. Progress in Precambrian Research, 23, 11–19.

PAEK, R.-J. & JON, G.-P. 1996. Stratigraphy. In: Geology of Korea [Institute of Geology, State Academy of Sciences, DPR of Korea], Foreign Languages Books Publishing House, Pyongyang, 31–79.

QIAN, X. L. 1992. Archaean cratonization in the northern China, collage of terrains and model of plate tectonic movement. In: Proceedings of 70th Anniversary of Founding of the Geological Society of China, Science and Technology Press, Beijing, 93–97.

QIAN, X. L. 1996. The nature of the early Precambrian continental crust and its tectonic evolution model. Acta Petrologica Sinica, 12, 169–178.

QIAN, X. L., CUI, W. Y. & WANG, S. Q. 1985. Evolution of the Inner Mongolia–eastern Hebei Archaean granulite belt in the North China craton. In: The Records of Geological Research. Beijing University Press, Beijing, 20–29.

QIAO, G. S., ZHAI, M. G. & YAN, Y. H. 1990. Isotopic chronology of Archaean Anshan complex. Acta Geologica Sinica, 2, 364–371.

RÄMÖ, O. T., HAAPALA, I., VAASJOKI, M., YU, J. H. & FU, H. Q. 1995. 1700 Ma Shachang complex, northeast China: Proterozoic rapakivi granite not associated with Palaeoproterozoic orogenic crust. Geology, 23, 815–818.

REN, F. G. 1986. Basic characteristics of the Dahongyu Group volcanic–sedimentary sequence in Jixian. Bulletin of the Tianjin Institute of Geology, 16, 91–106.

SHAO, J., ZHANG, L. Q., GUO, J. H. & LI, D. M. 2001. Response of the north margin of North China craton to breakup events of the global supercontinent in the Proterozoic. Gondwana Research, 4, 784–785.

SHEN, Q. H., XU, H. F., ZHANG, Z. Q., GAO, J. F., WU, J. S. & JI, C. L. 1992. *Precambrian Granulites in China*. Geological Publishing House, Beijing, 16–31, 214–223.

SONG, B., ALLEN, P. N., LIU, D. Y. & WU, J. S. 1996. 3800 to 2500 Ma crustal evolution in the Anshan area of Liaoning Province, northeastern China. *Precambrian Research*, **78**, 79–94.

ST-ONGE, M. R. & LUCAS, S. B. 1990. Evolution of the Cape Smith Belt: early Proterozoic continental underthrusting, ophiolite obduction, and thick-skinned folding. *In*: LEWRY, J. F. & STAUFFER, M. B. (eds), *The Early Proterozoic Trans-Hudson Orogen of North America*. Geological Association Canada, Waterloo, 313–351.

SUN, D. Z. & HU, W. X. 1993. *The Tectonic Framework of Precambrian in Zhongtiaoshan*. Geological Publishing House, Beijing, 108–117.

TAO, K. J., YANG, Z. M., ZHANG, P. S., & WANG, W. Z. 1998. Systematic geological investigation on carbonatite dykes in Bayan Obo, Inner Mongolia, China. *Scientia Geologica Sinica*, **33**, 73–83.

WANG, S. S., HU, S. L., ZHAI, M. G. & SANG, H. Q. 1987. An application of the $^{40}Ar/^{39}Ar$ dating technique to the formation time of Qingyuan granite–greenstone terrain in NE China. *Acta Petrologica Sinica*, **4**, 55–62.

WANG, Y. L. 1955. The division of Precambrian rocks in China. *Acta Geologica Sinica*, **35**, 327–360.

WINDLEY, B. F. 1986. *The Evolving Continents* (2nd edition). John Wiley & Sons, Ltd, Chichester, 1–65.

WINDLEY, B. F. 1993. Uniformitarianism today: plate tectonics is the key to the past. *Journal of the Geological Society of London*, **150**, 7–19.

WINDLEY, B. F. 1995. *The Evolving Continents* (3rd edition). John Wiley & Sons, Ltd, Chichester, 526.

WU, C. H. 1999. Timing of the khondalites and constrain to basement structure of the North China craton. *In*: LI, J. H., LIU, S. W. & ZHANG, C. (eds), *Precambrian Geology and Proterozoic Supercontinental Reconstruction of North China and Adjoining Area* (symposium abstract). Beijing University, Beijing, 36–40.

WU, C. H., GAO, Y. D., MEI, H. L. & ZHONG, C. T. 1994. Structural features and unconformity arguments between the khondalite suite and the granulite complex in the Huangtuyao area, Nei Mongol. *In*: QIAN, X. L. & WANG, R. M. (eds), *Geological Evolution of the Granulite Terrain in North Part of the North China Craton*. Seismological Press, Beijing, 145–154.

WU, J. S., GENG, Y. S. & SHEN, Q. H. 1998. *Archaean Geology Characteristics and Tectonic Evolution of Sino-Korea Paleo-continent*. Geological Publishing House, Beijing, 192–211.

XING, Y. S. & LIU, G. Z. 1982. Late Precambrian micro-palaeocoenosis in China and its stratigraphical significance. *Bulletin of the Academy of Chinese Geological Sciences*, **4**, 55–62.

XU, R. H., ZHU, M., CHEN, F. K. & GUO, J. H. 1995. A geochronological study of the Longquanguan ductile shear zone. *Quaternary Sciences*, **4**, 332–342.

YAN, Y. H., ZHAI, M. G. & GUO, J. H. 1991. Cordierite–silimanite mineral assemblage of granulite facies in North China Craton: a indicator to low-pressure granulite facies metamorphism. *Acta Petrologica Sinica*, **2**, 19–27.

ZHAI, M.G. 1997. Recent advances in the study of granulites from the North China Craton. *International Geology Review*, **39**, 325–341.

ZHAI, M. G. 1999. The 1.8 Ga thermal event in the North China Craton: a record of early upwelling mantle plumes. *In*: LEE, B.-J., LEE, S.-K., KIM, J. & LEE, H.-Y. (eds), *Crustal Evolution in Northeast Asia*. KIGMM, Taejon, 2–3.

ZHAI, M. G. 2001. Signature of North China Block in supercontinent Rodinia. *Gondwana Research*, **4**, 838–839.

ZHAI, M. G. & LIU, W. J. 2003. Proterozoic tectonic history of the North China Block: a review. *Precambrian Research*, **122**, 183–199.

ZHAI, M. G. & WINDLEY, B. F. 1990. The Archaean and Early Proterozoic banded iron formations of North China: their characteristics, geotectonic relations, chemistry and implications for crustal growth. *Precambrian Research*, **48**, 267–286.

ZHAI, M. G., BIAN, A. G. & ZHAO, T. P. 2000. The amalgamation of the supercontinent of North China craton at the end of the Neoarchaean, and its break-up during the late Palaeoproterozoic and Mesoproterozoic. *Science in China, Series D*, **43** Supp., 219–232.

ZHAI, M. G., GUO, J. H., YAN, Y. H., LI, Y. G. & ZHANG, W. H. 1992. The preliminary study and discovery of high pressure granulites in North China. *China Science*, **12B**, 1325–1330.

ZHAI, M. G., GUO, J. H., LI, J. H., YAN, Y. H., LI, Y. G. & ZHANG, W. H. 1995. The discoveries of retrograde eclogites in North China craton in Archaean. *Chinese Science Bulletin*, **40**, 1590–1594.

ZHAI, M. G., GUO JINGHUI & ZHAO, T.P. 2001. Study advances of Neoarchaean–Palaeoproterozoic tectonic evolution in the North China Craton. *Progress in Precambrian Research*, **24**, 17–27.

ZHAI, M. G., WINDLEY, B. F. & SILLS, J. D. 1990. Archaean gneisses, amphibolites, banded iron formation from Anshan area of Liaoning, NE China: their geochemistry, metamorphism and petrogenesis. *Precambrian Research*, **46**, 195–216.

ZHAI, M. G., YANG, R. Y., LU, W. J. & ZHOU, J. W. 1985. Geochemistry and evolution of the Qingyuan Archaean granite–greenstone terrain, NE, China. *Precambrian Research*, **27**, 37–62.

ZHAI, M.G., SHAO, J. A., HAO, J. & PEN, P. 2003. Signature and position of North China Block in the Supercontinent Rodinia. *Gondwana Research*, **6**, 171–183.

ZHANG, F. Q. 2000. Discussion of early Precambrian crustal accretion, reworking and metamorphic basement in North China Craton. *Acta Geophysica Sinica*, **41** Supp., 88–97.

ZHANG, G. W., YU, Z. P., DONG, Y. P. & YAO, A. P. 2000. On the Precambrian framework and evolution of the Qinling Belt. *Acta Petrologica Sinica*, **16**, 11–21.

ZHANG, G. W., ZHANG, B. R., YUAN, X. C. & XIAO, Q. H. 2001. *Qinling Orogenic Belt and Continental Dynamics*. Science Press, Beijing, 655–682.

ZHANG, J. S. 1997. Extension and uplift of the Datong-Huai'an granulite terrain. *Geological Reviews*, **43**, 503–514.

ZHANG, J. S., DRIKS, P. H. G. M. & PASSCHIER, C. W. 1994. Extensional collapse and uplift of a polyme-tamorphic granulite terrain in the Archaean of North China. *Precambrian Research*, **67**, 37–57.

ZHANG, Q. S. 1984. *Geology and Metallogeny of the Early Precambrian in China*. Jilin People's Press, Changchun, 196–230.

ZHANG, S. & LI, Z. X. 2001. Positions of the North China Block in Neoproterozoic Rodinia: a palaeo-magnetic constraint. *In*: SIRCOMBE, K. N. & LI, Z. X. (eds), *From Basin to Mountains: Rodinia at the Turn of the Century* (Abstract), Chris Powell Memorial Symposium, Fineline Print and Copy Service, Perth, Australia, 117–119.

ZHANG, S., LI, Z. X., WU, H. & WANG, H. 2000. New paleomagnetic results from the Neoproterozoic successions in southern North China Block and paleogeographic implications. *Science in China (Series D)*, **43** Supp., 233–244.

ZHANG, Z. Q. 1998. On main growth epoch of early Precambrian crust of the North China craton based on the Sm–Nd isotopic characteristics. *In*: Cheng, Y. Q. (ed.), *Corpus on Early Precambrian Research of the North China Craton*. Geological Publishing House, Beijing, 133–136.

ZHANG, Z. Q., TANG, S. H., SONG, B. & ZHANG, G. W. 1997. Strong Jinning geological event in Qinling orogenic belt and their geotectonic background. *Acta Geoscientica Sinica*, **18**, 43–45.

ZHAO, G.C. 2001. Palaeoproterozoic assembly of the North China craton. *Geological Magazine*, **138**, 87–91.

ZHAO, G. C., CAWOOD, P. A., WILDE, S. A. & SUN, M. 2002. Review of 2.1–1.8 Ga orogens: impli-cations for pre-Rodinia supercontinent. *Earth Science Reviews*, **59**, 125–162.

ZHAO, G. C., CAWOOD, P. A., WILDE, S. A., SUN, M. & LU, L. Z. 1999. Thermal evolution of two textural types of mafic granulites in the North China craton: evidence for both mantle plume and collisional tectonics. *Geological Magazine*, **136**, 223–240.

ZHAO, Z. P. 1993. *Precambrian Crustal Evolution of Sino-Korean Paraplatform*. Science Press, Beijing, 3–79, 366–384.

ZHOU, D. W., ZHANG, Z. J. & DONG, Y. P. 1995. Geological and geochemical characteristics of Pro-terozoic Songshugou ophiolite in Shangnan, eastern Qinling. *Acta Petrologica Sinica*, **11**, 154–164.

The Central Asian Orogenic Belt and growth of the continental crust in the Phanerozoic

BOR-MING JAHN

*Department of Geosciences, National Taiwan University, P.O. Box 13-318,
Taipei 106, Taiwan. (e-mail: jahn@ntu.edu.tw)*

Abstract: Asia is the world's largest composite continent, comprising numerous old cratonic blocks and young mobile belts. During the Phanerozoic it was enlarged by successive accretion of dispersed Gondwana-derived terranes. The opening and closing of palaeo-oceans would have inevitably produced a certain amount of fresh mantle-derived juvenile crust. The Central Asian Orogenic Belt (CAOB), otherwise known as the Altaid tectonic collage, is now celebrated for its accretionary tectonics and massive juvenile crustal production in the Phanerozoic. It is composed of a variety of tectonic units, including Precambrian microcontinental blocks, ancient island arcs, ocean island, accretionary complexes, ophiolites and passive continental margins. Yet, the most outstanding feature is the vast expanse of granitic intrusions and their volcanic equivalents. Since granitoids are generated in lower-to-middle crustal conditions, they are used to probe the nature of their crustal sources, and to evaluate the relative contribution of juvenile v. recycled crust in the orogenic belts. Using the Nd–Sr isotope tracer technique, the majority of granitoids from the CAOB can be shown to contain high proportions (60 to 100%) of the mantle component in their generation. This implies an important crustal growth in continental scale during the period of 500–100 Ma. The evolution of the CAOB undoubtedly involved both lateral and vertical accretion of juvenile material. The lateral accretion implies stacking of arc complexes, accompanied by amalgamation of old microcontinental blocks. Parts of the accreted arc assemblages were later converted into granitoids via underplating of basaltic magmas. The emplacement of large volumes of post-accretionary alkaline and peralkaline granites was most likely achieved by vertical accretion through a series of processes, including underplating of basaltic magma, mixing of basaltic liquid with lower-crustal rocks, partial melting of the mixed lithologies leading to generation of granitic liquids, and followed by fractional crystallization. The recognition of vast juvenile terranes in the Canadian Cordillera, the western US, the Appalachians and the Central Asian Orogenic Belt has considerably changed our view on the growth rate of the continental crust in the Phanerozoic.

In the last decade the Central Asian Orogenic Belt (CAOB) became celebrated for its orogenic style and the world's largest site of juvenile crustal formation in the Phanerozoic eon. Central Asia provides a prime example for study of accretionary tectonics and growth of the continental crust in the late part of the Earth's history. Asia has indeed grown in size, but a distinction must be made between the two terms commonly used in the discussion on the making of Asia:

(1) *amalgamation* of dispersed microcontinental fragments from the breakup of Gondwana. This process enlarged the size of Asia, but did not necessarily add a substantial amount of new (juvenile) crust to the continent and

(2) *growth* of the continental crust. This implies addition of juvenile crust and a net transfer of mantle-derived material to the continental crust. It is the latter process that is the focus of this article.

The CAOB, bounded by the Siberian and North China cratons (Fig. 1), represents a complex evolution of Phanerozoic orogenic belts (e.g. Tang 1990; Dobretsov *et al.* 1995). It has also been termed the Altaid tectonic collage by Sengör and his associates (Sengör *et al.* 1993; Sengör and Natal'in 1996). According to these authors, the CAOB was formed by successive accretion of arc complexes, accompanied by emplacement of immense volumes of granitic magmas (Fig. 2). In addition, they underlined the general absence of nappe complexes imbri-

From: MALPAS, J., FLETCHER, C. J. N., ALI, J. R. & AITCHISON, J. C. (eds) 2004. *Aspects of the Tectonic Evolution of China*. Geological Society, London, Special Publications, **226**, 73–100.
0305-8719/04/$15 © The Geological Society of London 2004.

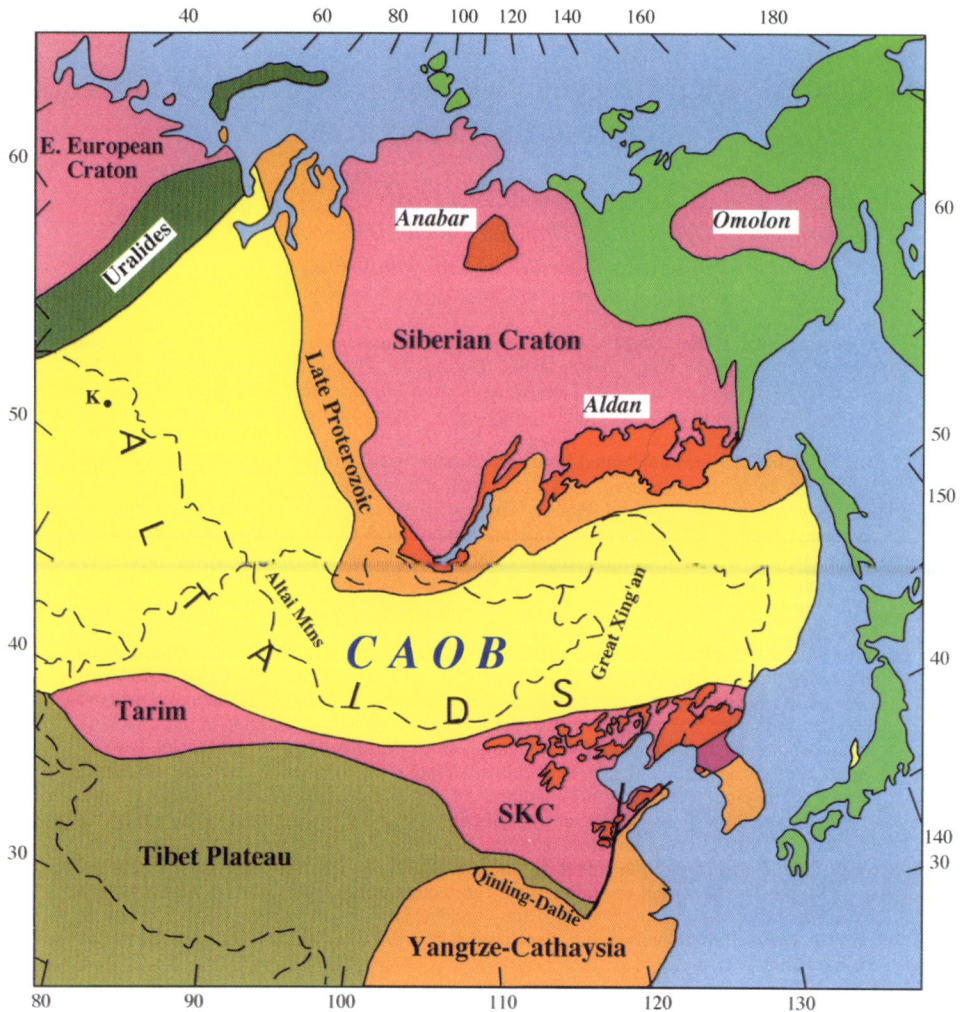

Fig. 1. Simplified tectonic divisions of Asia. The Central Asian Orogenic Belt (CAOB), also known as the Altaid tectonic collage (Sengör *et al.* 1993), is situated between two major Precambrian cratons: Siberian in the north and North China–Tarim in the south. Red areas are exposed Archaean to Early Proterozoic rocks. The light green pattern on the right-hand side, including the Japanese islands, represents Pacific fold-belts. The Hida Belt of Japan may belong tectonically to the CAOB. Abbreviation: K, Kokchetav (in northern Kazakhstan).

cating older continental crust as characteristic of the classic collisional orogenic belts. In general, the CAOB comprises a variety of tectonic units, including Precambrian cratonic blocks (=microcontinents), ancient island arcs, fragments of ocean island and seamount, accretionary complexes, ophiolites and passive continental margins (Sengör *et al.* 1993; Badarch *et al.* 2002; Dobretsov *et al.* 2004). However, it is the voluminous granitic intrusions, mostly of juvenile character, that distinguish the CAOB from other classic Phanerozoic orogenic belts, such

as the Caledonides and Hercynides in Europe (e.g. Kovalenko *et al.* 1996, 2003; Jahn *et al.* 2000*a, b*, 2003; Wu *et al.* 2000, 2001, 2003*a, b*). Because granitoids are derived mainly by melting of the lower to middle crust, they can be used as an invaluable tool to probe the nature of their sources at deep crustal levels.

In the following, the contrasting Nd–Sr isotopic characteristics of granitic rocks from the classic orogens (e.g. Caledonian and Hercynian in Europe, Cathaysia in SE China, South Korea, and eastern Australia) and from the CAOB will

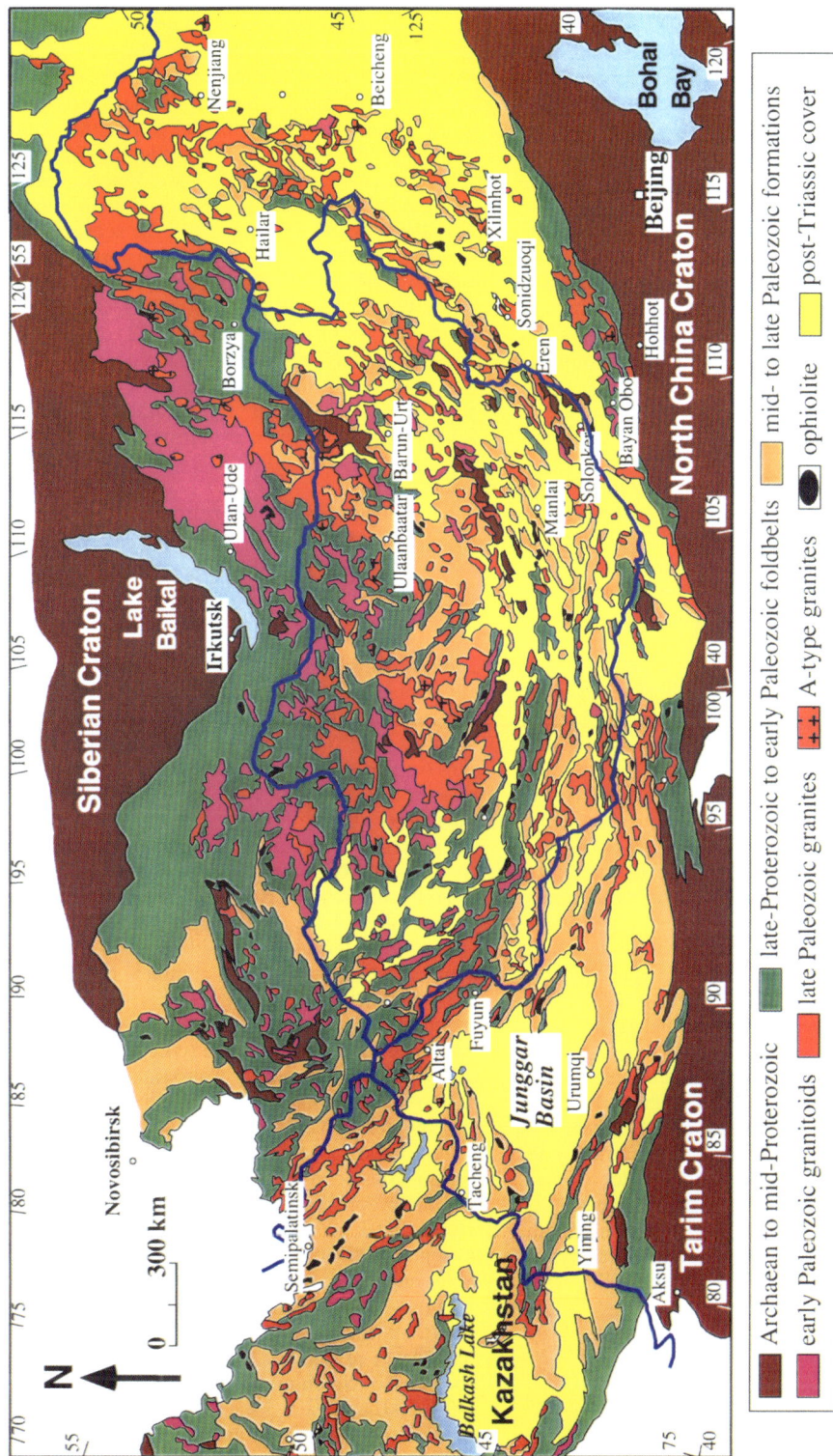

Fig. 2. Areal distribution of granitoids in the CAOB. The national boundaries between Russia, Mongolia, China and Kazakhstan are shown by heavy blue lines. Mongolia is situated in the heartland of Central Asia. The northern belt of CAOB represents the area from central–northern Mongolia to Transbaikalia (east of Lake Baikal), and the southern belt from east–central Kazakhstan (region to the north of Lake Balkash), Northern Xinjiang (regions surrounding the Junggar Basin), Inner Mongolia to NE China (Manchuria), of which the Great Xing'ar Mountains are shown. In Northern Xinjiang, the Altai terrane is situated to the north of the Junggar Basin, and the eastern and western Tianshan terranes to the south. The eastern and western Junggar terranes are found in the east and west of the Junggar Basin.

be summarized and compared. These data will then be used to discuss processes of the generation of the voluminous granitoids in Central Asia. At the end, some implications for the global Phanerozoic crustal growth will be addressed.

Lithological characters and emplacement periods of granites

According to Kovalenko *et al.* (1995, 2003), igneous activity in Central Asia continued throughout the entire Phanerozoic without significant interruption. Since the Early Palaeozoic, numerous granitic rocks have been emplaced. They include:

(1) the calc-alkaline series (tonalite–granodiorite–granite) of 'Caledonian' ages in northern Mongolia and Transbaikalia (e.g. Angara–Vitim batholith; Litvinovsky *et al.* 1992, 1994);

(2) the 'late Caledonian' calc-alkaline series in western Mongolia and the alkaline series in Tuva, Sayan, eastern Mongolian Altai, and vast areas in northern Mongolia and Transbaikalia;

(3) the 'Hercynian' (Late Carboniferous to Permian) alkaline series in southern Mongolia and in northern Mongolia to Transbaikalia; Permian granitoids of the calc-alkaline series, represented by the vast Hangay batholith (*c.*100 000 km^2) in west-central Mongolia; and

(4) the Early Mesozoic (*c.*200 Ma) granites of the calc-alkaline series and S-type granites in the Mongol–Okhotsk Belt, plus the alkaline to peralkaline series in Transbaikalia, of which the lithological types comprise alkaline and peralkaline granites, syenogranites, syenites and minor granodiorites.

It must be added that granitoid emplacement became very significant in NE China during the Cretaceous (Wu *et al.* 2000, 2002). Whether the Cretaceous event was related to the Central Asian or Pacific tectonic regime is still a matter of debate.

The ages of granites roughly decrease from north to south within the CAOB. In Transbaikalia, five main stages of K-rich magmatic activity have been distinguished (Zanvilevich *et al.* 1995; Wickham *et al.* 1995, 1996):

(1) Ordovician–Silurian (*c.*450 Ma);
(2) Devonian (*c.*375 Ma);
(3) Early Permian (*c.*280 Ma);
(4) Late Permian (*c.*250 Ma); and
(5) Triassic (*c.*220 Ma).

Litvinovsky and Zanvilevich (1998) later recognized one more stage in the Late Cambrian. It must be noted, however, that only a small minority of the plutons in Central Asia have been properly dated. Many of the ages reported in the literature were estimated from litho- and biostratigraphic correlations. During the last five years, systematic geochemical and geochronological studies on the granitoids from Central Asia have significantly changed the scenario of the thermal events and orogenic history of that region (Vladimirov *et al.* 1997; Yarmolyuk *et al.* 1997; Wilde *et al.* 1997, 2000; Chen and Jahn, 2002, 2003; Wu *et al.* 2000, 2002, 2003*a*, *b*; Jahn *et al.* 2001, 2004). Consequently, the use of the 'magmatic front' concept by Sengör *et al.* (1993) as a structural marker for delineating the tectonic evolution of the Altaid Collage must have involved a very large degree of uncertainty.

In east central Kazakhstan, important mineralizations (Au, Cu, Mo–W, Sn, REE, Nb–Ta) are associated with a variety of igneous rocks, including gabbros, diorites, granodiorites and granites (Heinhorst *et al.* 2000). These rocks were intruded in several episodes from 450 to 250 Ma. Heinhorst *et al.* (2000) considered that these rocks were formed in an active continental margin which developed from back-arc oceanic settings (for volcanic-hosted massive sulphide Cu–Au ore deposits) to subduction zone calc-alkaline magmatism (for Cu porphyries), with subsequent stages of differentiation (Mo porphyries) and finally to continental rifting magmatism (peralkaline REE–Zr–Nb deposits).

To the north in the Russian Gorny Altai, or the western part of the Altai–Sayan Folded Region (ASFR), the terrane is considered as a Caledonian accretion–collisional complex containing large fragments of Vendian to Early Cambrian island arcs (Dobretsov *et al.* 2004). A variety of tectonic units and lithological types can be identified: Vendian to Early Cambrian ophiolites, palaeo-oceanic islands or seamounts, and island arc complexes, Mid-Palaeozoic volcano-plutonic complexes of an active continental margin, and Permo-Triassic dyke swarms of alkali basalts and lamprophyres which are synchronous with the Kuznetsk and Siberian Traps. Moreover, massive post-collisional alkaline granite intrusions were emplaced in Permian to Jurassic times, roughly contemporaneously with, or slightly later than, the Siberian Traps. These rocks have been dated at 250 to 200 Ma using zircon U–Pb and Rb–Sr chronometry (Vladimirov *et al.* 1997, 2001).

In northern Xinjiang, granites of calc-alkaline to alkaline series appear to be dominant; most of these were emplaced in the period of 400–200 Ma but culminated around 300 Ma. A-type granites of the Ulungur River area were intruded at about 300 Ma (Rb–Sr ages, Wang et al. 1994; Han et al. 1997) and those in Inner Mongolia were emplaced slightly later at c.280 Ma (whole-rock Rb–Sr, Hong et al. 1995). In Inner Mongolia, arc-type calc-alkaline granitoids were intruded at c.310 Ma, but they were overwhelmed by much more widespread late orogenic granites of c.230 Ma (Chen et al. 2000). Further east to NE China the existing age data indicate four episodes of granitic intrusion (Fang 1992; Jahn et al. 2000a, b; Wu et al. 2000, 2002, 2003a):

(1)   Late Permian (270–250 Ma);
(2)   Late Triassic–Early Jurassic (220–180 Ma);
(3)   Middle Jurassic (170–150 Ma); and
(4)   Cretaceous (c.120 Ma).

These data support a younging trend of granitic emplacement from the west to the east within the southern belt of the CAOB.

In addition to the apparent regional age variation, two other trends are present:

(1)   a regular decrease in size for younger plutons, and
(2)   an increase in the proportion of syenite and alkaline granite to granite (s.s.), as well as in the ratio of K-feldspar to plagioclase, in the younger plutons. That is, the younger plutons tend to be more alkaline in nature.

This is particularly well demonstrated in the granitoids from the Mongolian–Transbaikalian belt (Litvinovsky and Zanvilevich 1998). In Transbaikalia, peralkaline granites and syenites containing aegirine and arfvedsonite only occur in the younger Permian and Triassic suites (Kuzmin and Antipin 1993; Wickham et al. 1995, 1996). However, we note that in the southern belt (Xinjiang–Inner Mongolia–NE China) such an increase of alkalinity in granitoids with the decrease of intrusive ages is not as clearly documented as in Transbaikalia (Hong et al. 1996).

In short, according to Sengör et al. (1993), Central Asia grew by successive accretion of subduction complexes along a single but migratory magmatic arc now found contorted between Siberia and Baltica. They recognized the main difference between the Altaids (=CAOB) and other classic collisional orogens such as the Alps and the Himalayas, in that the Altaids show the paucity of extensive ancient gneiss terrains or Precambrian microcontinents.

Besides, they underlined that no Alpine- or Himalayan-type crystalline nappe complexes inbricating pre-existing continental crust can be recognized within the Altaid collage, and that high-K granites were considered to be produced by anatexis, and only became abundant in the Permian. The above hypothesis on tectonic evolution and structural analyses has been a point of controversy in the last few years, and the model on the high-K granite genesis is not supported by the isotope data to be presented below. Nevertheless, the overall scheme for the growth of Central Asia by accretion of subduction complexes is probably correct, apart from the role of ancient microcontinents being greatly underestimated. The voluminous post-orogenic A-type granites could not be easily explained using a subduction model. It is argued that vertical accretion via basaltic underplating might be even more important than the horizontal accretion of subduction complexes for the growth of the Asian continent.

Nd–Sr isotopic data for Phanerozoic granitoids – a summary

The Nd–Sr isotope characteristics of granitoids from the world's classic Phanerozoic orogenic belts are summarized below using three types of diagrams: (1) initial Nd isotope composition $\varepsilon_{Nd}(T)$ v. intrusive ages; (2) $\varepsilon_{Nd}(T)$ v. depleted-mantle-based model age T_{DM}; and (3) $\varepsilon_{Nd}(T)$ v. initial Sr isotope composition I_{Sr} or $(^{87}Sr/^{86}Sr)_o$. For model ages, a linear Nd isotope evolution is assumed for the depleted mantle from $\varepsilon_{Nd} = 0$ at 4.56 Ga to +10 at the present, but the choice of a one- or two-stage model (DePaolo et al. 1991) is difficult, as each model has its own uncertainty and inconvenience. In the single-stage model, the main uncertainties are due to:

(1)   Sm/Nd fractionation between granitic melts and their sources during partial melting;
(2)   Sm/Nd fractionation during magma differentiation; and
(3)   mixing of melts or sources in petrogenetic processes (Arndt and Goldstein 1987; Jahn et al. 1990).

Many peralkaline granitoids of Central Asia show highly fractionated REE patterns, sometimes with the tetrad effect leading to enhanced or greater than chondritic Sm/Nd ratios and negative model ages (Masuda et al. 1987; Masuda and Akagi 1990; Bau 1996; Jahn et al. 2001,

2004). For this type of granite, single-stage model ages are clearly not appropriate. On the other hand, the two-stage model assumes that all the sources for granites follow the same isotope evolution as the average continental crust, regardless of their true lithological characteristics. This cannot always be realistic. Adoption of this model will result in most granitoid data forming a quasi-linear array in the $\varepsilon_{Nd}(T)$ v. T_{DM} plots (e.g. Wu *et al.* 2000).

Concerning initial Sr isotope ratios, it is important to know the uncertainty derived from the correction of *in situ* radiogenic growth. This problem is most severe for granitoids of very high Rb–Sr ratios, such as A-type or peralkaline granites, which are very abundant in Central Asia. Figure 3 shows a plot of initial Sr isotopic ratios (I_{Sr}) as a function of $^{87}Rb/^{86}Sr$ ratios for some granitoids from the southern belt of the CAOB. High $^{87}Rb/^{86}Sr$ ratios (up to 400) often occur with A-type and highly differentiated *I*-type granitoids. Note that as the I_{Sr} values were individually calculated by subtracting

the radiogenic components from the measured $^{87}Sr/^{86}Sr$ and Rb/Sr ratios, they may bear large uncertainties for high Rb–Sr rocks, and not uncommonly yield unreasonably low ratios (≤ 0.700; Fig. 3). The Rb–Sr induced errors (ξ) for calculated initial $^{87}Sr/^{86}Sr$ ratios are related to three factors: Rb–Sr ratio, assigned uncertainty for the Rb/Sr ratio, and age (true or assumed), all related by the equation:

$$\xi = {}^{87}Rb/{}^{86}Sr \times (\% \text{ error assigned}) \times (e^{\lambda t} - 1)$$

The error propagation envelope assuming $t = 200$ Ma (Fig. 3) shows that the Rb–Sr induced errors for $(^{87}Sr/^{86}Sr)_o$ are too large to have any petrogenetic significance for rocks with $^{87}Rb/^{86}Sr \geq 10$ or 20. Peralkaline rocks with ratios ≥ 100 do not provide a useful constraint to their genetic processes. Nevertheless, most I_{Sr} values for low Rb/Sr rocks seem to show a restricted range of $(^{87}Sr/^{86}Sr)_o$ of 0.705 ± 0.002, which is rather low for granitic rocks formed in Phanerozoic orogenic belts. Relative to Sr, the

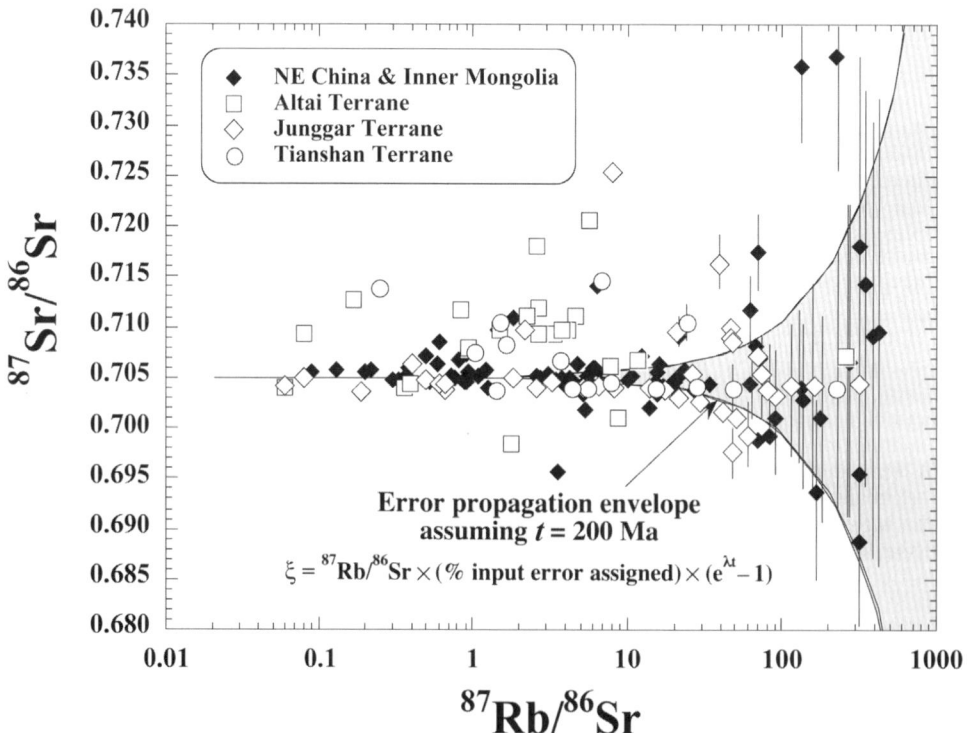

Fig. 3. Initial $^{87}Sr/^{86}Sr$ v. $^{87}Rb/^{86}Sr$ plot, only for young granitoids from the CAOB, showing the magnitude of uncertainty in $(^{87}Sr/^{86}Sr)_o$ induced by error in Rb/Sr ratio. More reliable $(^{87}Sr/^{86}Sr)_o$ for rocks with Rb/Sr ratios ≤ 20 range from 0.703 to 0.707. The $(^{87}Sr/^{86}Sr)_o$ calculated from rocks with high to very high Rb/Sr ratios, are too imprecise to have petrogenetic meanings. The grey area indicates the propagation of error in $(^{87}Sr/^{86}Sr)_o$ induced by 2% Rb/Sr uncertainty.

isotopic data of Nd are known to be more robust, and Sm/Nd ratios can be more accurately measured; thus they provide a much clearer and less ambiguous constraint to the origin of granitic rocks.

Classic Phanerozoic orogenic belts

European Caledonides and Hercynides and eastern Australia. Figures 4a and 4b show the available Sm–Nd isotope data for granitoids from the European Hercynides and Caledonides, and from the Lachlan Fold Belt (420–390 Ma) and New England Batholith (310–250 Ma) in eastern Australia. The data from young Himalayan leucogranites (*c.*20 Ma) are also shown for comparison, but their $\varepsilon_{Nd}(T)$ values are not adjusted to a Palaeozoic age and their T_{DM} were calculated using a two-stage model because most of them have f_{Sm-Nd} values higher than -0.2 (Fig. 4b). Note that almost all the Hercynian (450 data points) and Caledonian granitoids (80) and all Himalayan leucogranites (29) have negative $\varepsilon_{Nd}(T)$ values (Fig. 4a). This suggests that the granitoids were mainly generated from recycled sources containing large proportions of Precambrian crust. Most of the Hercynian granitoids with near-zero $\varepsilon_{Nd}(T)$ values represent the post-tectonic A-type granites from Corsica (Poitrasson *et al.* 1995). These rocks also have high Sm–Nd ratios, leading to very high T_{DM}, up to 3800 Ma (Fig. 4b), but their mantle component is significantly higher than the rest, as argued from the Nd isotope data. Figure 4b shows that if $f_{Sm/Nd}$ values are limited to -0.4 ± 0.2, then the majority of T_{DM} for the Hercynian and Caledonian granitoids would fall between 1000 and 2000 Ma. Note also that the Hercynian and Caledonian data-sets cannot be distinguished as a whole.

A significant proportion of Australian granitoids (black diamonds on Fig. 4) possess positive $\varepsilon_{Nd}(T)$ values, and their model ages (T_{DM}) are highly variable from *c.*500 Ma to >2000 Ma (Fig. 4a, b). The Australian granitoids differ from the European granites in their restricted range of f_{Sm-Nd} values for the same range of T_{DM}, which indicates their generation from sources of different mixing proportions between the mantle and crustal components. It has been estimated that in the Lachlan belt the added mantle component for the most primitive Moruya Suite ($\varepsilon_{Nd} = c.+4$) is *c.*40% (Keay *et al.* 1997), about 10% for the S-type Bullenbalong Suite, and between 10 and 40% for all other I-type granitoids (Collins, 1996, 1998). These estimates were based on a young crustal component

with relatively high $\varepsilon_{Nd}(T)$ values, which is not the case for Central Asia.

SE China and South Korea. Cathaysia is a tectonic unit of the South China Block. It is a major Phanerozoic orogenic belt in East Asia. Like the CAOB, it is also characterized by voluminous Phanerozoic granitoids with rich mineralizations. Thus, a brief comparison of their isotopic signatures with those of the CAOB appears instructive for the understanding of their respective crustal development. Cathaysia has been considered as the easternmost part of the Tethyside orogen (Hsü *et al.* 1990; Sengör *et al.* 1993). Cathaysia and the CAOB are situated to the south and north of the North China craton, respectively, and they exhibit very contrasting styles of tectonic and crustal evolution. Their principal characteristics and differences are summarized in Table 1. A-type granites also occur in Cathaysia, but their Nd isotopic signatures are generally 'crustal' (Charoy and Raimbault 1994; Martin *et al.* 1994; Darbyshire and Sewell 1997). Most granitic rocks in Cathaysia were produced by remelting of Proterozoic crustal sources; only very few Cretaceous granitic bodies in coastal Fujian and Taiwan have a significant depleted mantle component in their magma genesis (Jahn *et al.* 1976, 1986, 1990; Huang *et al.* 1986; Lan *et al.* 1995b; Gilder *et al.* 1996; Chen and Jahn, 1998). The principal heat source is thought to come from basaltic underplating (Zhou and Li 2000).

The Phanerozoic granitoids of SE China (Yangtze craton, Cathaysia and Taiwan) show negative $\varepsilon_{Nd}(T)$ values, except a few cases (Fig. 5a, b, c). Some Cretaceous granites from Dabieshan possess the lowest $\varepsilon_{Nd}(T)$ values from -15 to -25, suggesting their derivation from a protolith of Archean to Early Proterozoic age (Fig. 5a). In the $\varepsilon_{Nd}(T)$ v. initial $^{87}Sr/^{86}Sr$ diagram, the data indicate a dominance of both upper and lower crust in the generation of granitic rocks. The mantle component is subordinate (Fig. 5b). Single-stage model ages range from 1000 to 2500 Ma for the majority of the granitoids (Fig. 5c). Besides, there is an apparent oceanward younging of T_{DM} and an increase of $\varepsilon_{Nd}(T)$ within the whole of SE China (Chen and Jahn 1998; Zhou and Li 2000).

With respect to the granitoids of SE China, the Late Palaeozoic to Cretaceous granitoids of South Korea are characterized by even lower $\varepsilon_{Nd}(T)$ values (Figs 5d, e, f) but comparable T_{DM} model ages (Fig. 5f). The basement gneisses and metasediments have very radiogenic initial Sr isotope ratios (up to 0.775) and old to very old Nd model ages (1500–3800 Ma). The Sr isotope

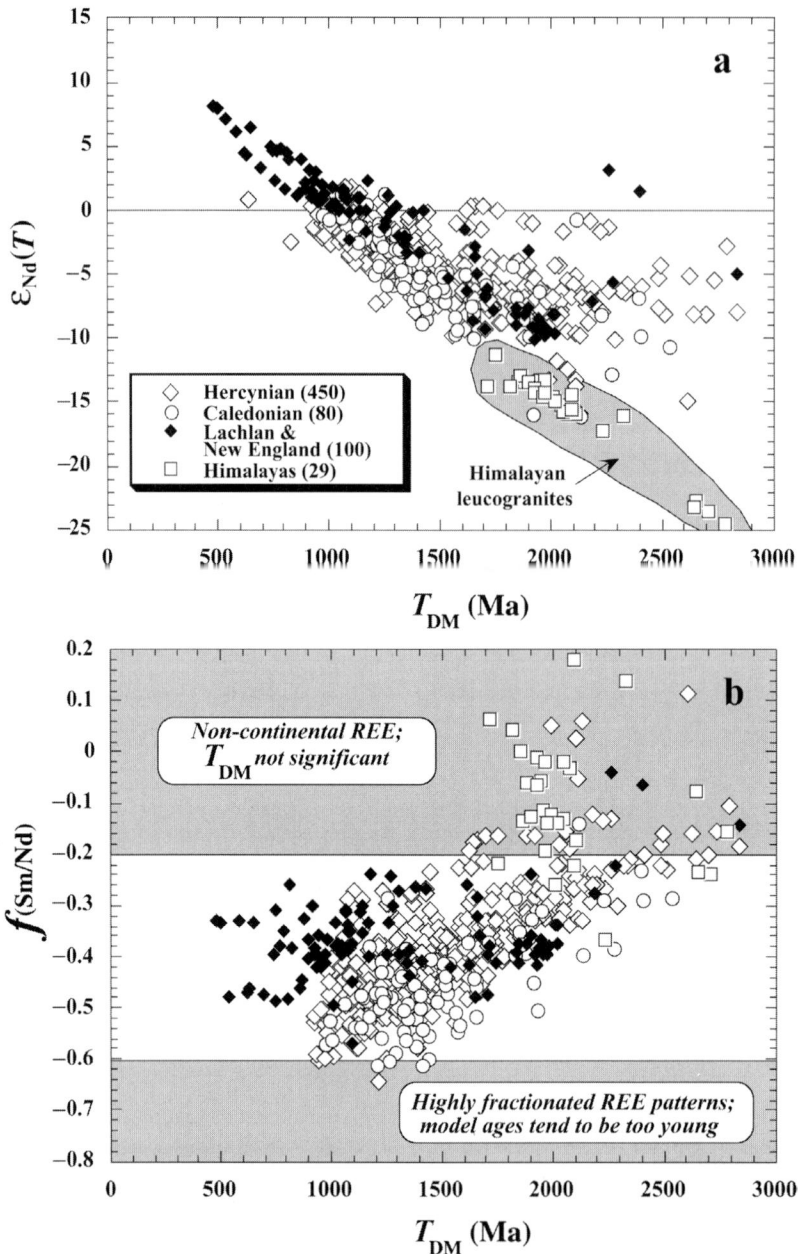

Fig. 4. Isotope diagrams for granitoids from European Hercynides and Caledonides, Australian Lachlan and New England fold-belts, and the Himalayas. Note that T_{DM} for Himalayan leucogranites shown in (**a**) are calculated using a two-stage model, because a large number of them have $f_{Sm-Nd} > -0.2$. Data sources: (**a**) **Caledonian Belt:** Hamilton *et al.* (1980), Halliday (1984), Frost and O'Nions (1985), Dempsey *et al.* (1990), Skjerlie (1992); (**b**) **Hercynian Belt:** Ben Othman *et al.* (1984), Bernard-Griffiths *et al.* (1985), Downes and Duthou (1988), Liew and Hofmann (1988), Liew *et al.* (1989), Pin and Duthou (1990), Turpin *et al.* (1990), Williamson *et al.* (1992), Cocherie *et al.* (1994), Darbyshire and Shepherd (1994), Dias and Leterrier (1994), Poitrasson *et al.* (1994, 1995), Moreno-Ventas *et al.* (1995), Siebel *et al.* (1995), Tommasini *et al.* (1995), Downes *et al.* (1997), Ajaji *et al.* (1998), Azevedo and Nolan (1998), Forster *et al.* (1999); (**c**) **Lachlan and New England Belts:** McCulloch and Chappell (1982), Hensel *et al.* (1985), Eberz *et al.* (1990), King *et al.* (1997), Keay *et al.* (1997), Mass *et al.* (1997); (**d**) **Himalaya:** Vidal *et al.* (1984), Deniel *et al.* (1987), Inger and Harris (1993), Gazis *et al.* (1998), Harrison *et al.* (1999).

Table 1. *Comparison of crustal evolution between the CAOB and Cathaysia of SE China*

	Central Asian Orogenic Belt	SE China (Cathaysia)
Type of orogen	Accretionary* (Altaid[†])	Collisional* (Tethyside)
Characteristics	immense Phanerozoic granitic intrusions	immense Phanerozoic granitic intrusions
Period of intrusion	550 to 120 Ma ΔT $c.$400 Ma	400 (?) to 80 Ma ΔT $c.$300 Ma (mainly 180–90 Ma)
Total volume (area)	$c.$5.3 M km^{2}[†] ($c.$11% of Asian total)	$c.$0.4 M km^{2}[‡]
Granitic type	Mainly I- and A-types Calc-alkaline, alkaline and peralkaline granites	I- > S- ≫ A-type Calc-alkaline granites dominate
Crustal type	Mainly juvenile	Mainly reworked
$\varepsilon_{Nd}(T)$	Mostly positive (+8 to 0)	Mostly negative (−2 to −17)
Tectonics	Assembly of numerous arc complexes; intruded by vast granitic plutons and covered in places by their volcanic equivalent	Assembly of ancient continental blocks; vast granitic plutons and rhyolite formed by remelting of old basement rocks
Granitoid generation	Melting of wet mantle wedge + differentiation; Basaltic underplating (lithosphere delamination or plume activities?) and melting of lower crust	Melting of the lower crust via basaltic underplating (Zhou & Li 2000)
Structure	Nappe complexes rare or absent[†]; suture zones broad	Nappe complexes common; suture zones narrow and elongate
Basement rocks	Precambrian basement rocks comparatively rare[†]	Proterozoic basement dominates

*Terminology of Windley (1993, 1995)
[†]According to Sengör *et al.* (1993)
[‡]Late Mesozoic granites + rhyolites = 240 000 km^{2} (Zhou and Li 2000)

data indicate that the granitoids have no direct genetic relationship with the metasediments or gneisses in the Ogcheon belt (Fig. 5e). On the other hand, the Early Tertiary (50 Ma) Namsan alkaline granite from the Kyongsang Basin in SE Korea is an exception. The chemical compositions indicate their A-type affinity, suggesting emplacement in a post-orogenic environment (Kim and Kim 1997). These rocks are characterized by low initial $^{87}Sr/^{86}Sr$ ratios of 0.704 to 0.705, and positive $\varepsilon_{Nd}(T)$ values of +3 to 0. Kim and Kim (1997) suggest that they were derived from a 'juvenile' source, presumably a young lower crust of underplated basalt with a small amount of old crustal material. As in the European Caledonides and Hercynides, the granitoids of SE China and South Korea, as a whole, are overwhelmed by the recycled continental crust (Jahn *et al.* 1990; Lan *et al.* 1995b; Chen and Jahn 1998).

Granitoids from the CAOB

NE China & Inner Mongolia. In NE China, ≥350 granitic bodies were intruded (mainly during the Mesozoic) in the Great Xing'an (or Khinggan), Lesser Xing'an and Zhangguangcai Ranges. Some of them were emplaced within

the domain of the Jiamusi Massif, a Proterozoic microcontinental block whose metamorphic age has been precisely dated at 500 Ma by SHRIMP zircon analyses (Wilde *et al.* 1997, 2000). The granites are composed mainly of I-type and subordinate A-type granites (Wu *et al.* 2000, 2002, 2003a, b). They are accompanied by extensive Mesozoic and Tertiary acid volcanic rocks. Isotope tracer analysis and age determination of deep-drilled cores revealed that the Songliao Basin in central NE China is underlain by granitic rocks and deformed granitic gneisses of Phanerozoic ages; no Precambrian zircons have been identified (Wu *et al.* 2001). This suggests that the true volume of granitic rocks is much greater than shown in the present geological map. The tectonic setting for the emplacement of such an immense distribution of granitic rocks in NE China has not been resolved. It appears to have a connection with continental rifting and no relation with subduction zone processes.

In Inner Mongolia, several periods of granitic intrusion took place from Devonian to Jurassic times. The samples used in this study came from a Palaeozoic anorogenic A-type suite (280 Ma; Hong *et al.* 1995, 1996), an arc-related calc-alkaline magmatic belt composed of gabbroic diorite, quartz diorite, tonalite and granodiorite

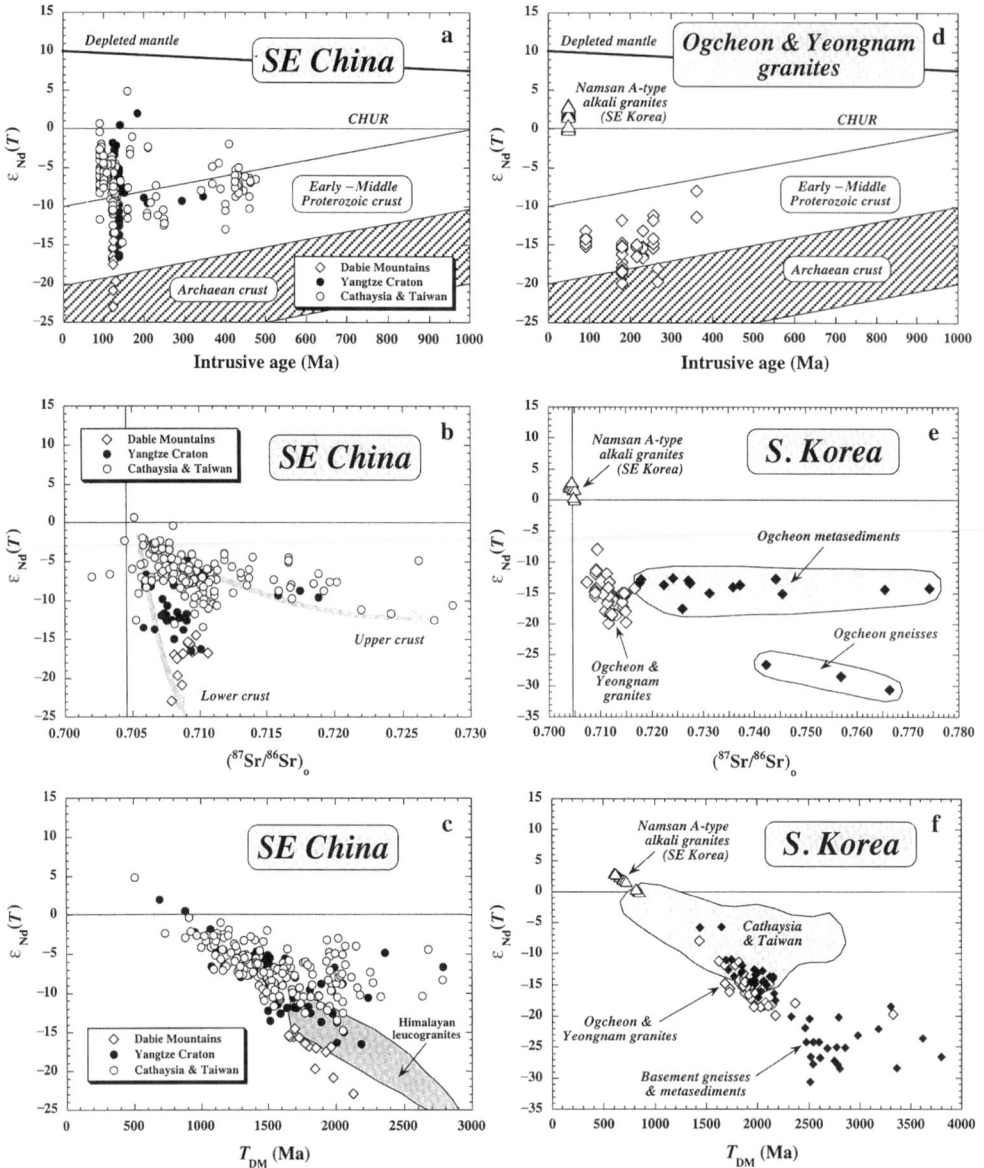

Fig. 5. Isotope diagrams for granitoids from SE China (**a**, **b** & **c**) and South Korea (**d**, **e** & **f**). Data sources: **SE China**: see references cited in Chen and Jahn (1998); **South Korea**: Lan *et al.* (1995a), Cheong and Chang (1997), Kim and Kim (1997), Lee *et al.* (1999).

(SHRIMP zircon age of 309 ± 8 Ma, Chen *et al.* 2000) and a Mesozoic collision-type granitic suite comprising adamellite, granodiorite and leucogranite (Rb–Sr isochron age of 230 ± 20 Ma; Chen *et al.* 2000).

The Nd–Sr isotope data, including all derivative parameters (intrusive and model ages, f_{Sm-Nd}) for the Phanerozoic granitic rocks from NE China, Inner Mongolia and the Hida Belt of Japan are presented in Figure 6. Figure 6a shows that the majority (75%) of the analysed samples have positive $\varepsilon_{Nd}(T)$ values. Most of the samples with negative $\varepsilon_{Nd}(T)$ values came from within the domain of the Precambrian Jiamusi Massif

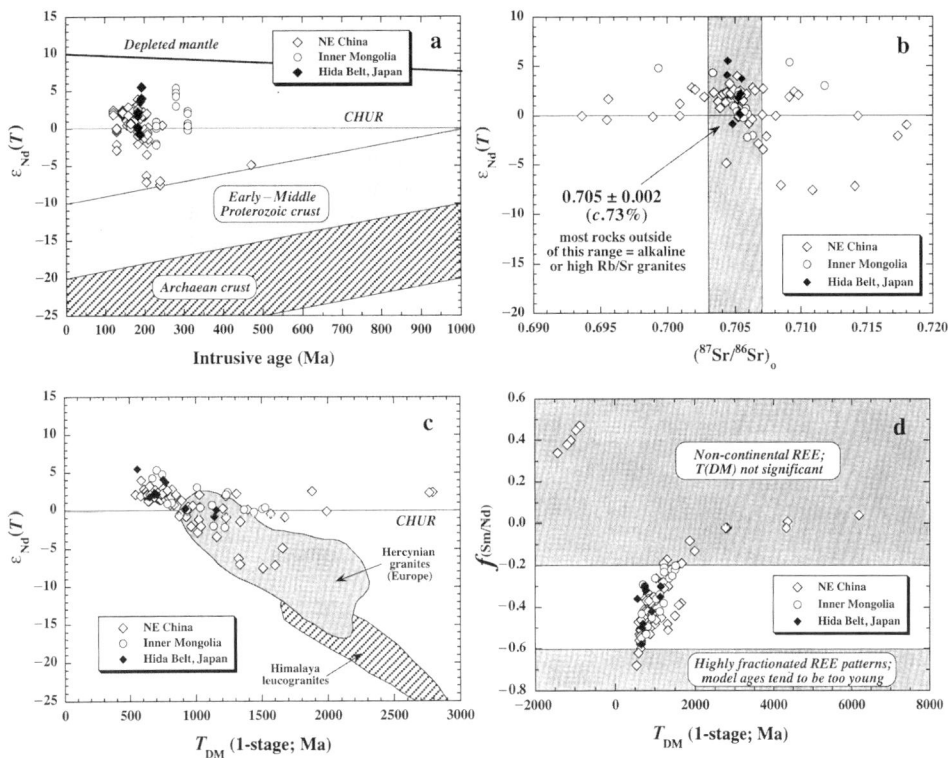

Fig. 6. Isotope diagrams for granitoids from NE China–Inner Mongolia–Hida Belt of Japan. (**a**) $\varepsilon_{Nd}(T)$ v. intrusive ages; (**b**) $\varepsilon_{Nd}(T)$ v. initial $^{87}Sr/^{86}Sr$ isotopic ratios; (**c**) $\varepsilon_{Nd}(T)$ v. T_{DM} (1-stage); and (**d**) f_{Sm-Nd} v. T_{DM} (1-stage). Data sources: **NE China**: Wu *et al.* (2000, 2002) and Jahn *et al.* (2001); **Inner Mongolia**: Chen *et al.* (2000), Jahn, unpubl.; **Hida Belt**: Arakawa and Shimura (1995), Arakawa *et al.* (2000).

($\varepsilon_{Nd}(200\,Ma) = -7$ to -12). Such a close relationship between the isotopic compositions of granitoids and the ages and nature of their intruded 'basement' rocks is also demonstrated by the granitoids from northern Xinjiang (Hu *et al.* 2000) and Mongolia (Kovalenko *et al.* 1996, 2004; Jahn *et al.* 2004). The lowering of the $\varepsilon_{Nd}(T)$ values was effected by the participation of old crustal rocks in their magma genesis.

Figure 6b shows a plot of initial Nd and Sr isotope ratios for the same suites of rocks. The Sr data are widely scattered, with I_{Sr} from 0.693 to 0.718. The scatter is due to the large correction of *in situ* radiogenic growth, as explained in the preceding section. However, the majority of the I_{Sr} appear to fall within a small range of 0.705 ± 0.002. This value is probably the most commonly found for granitoids of Central Asia. Figure 6c presents a $\varepsilon_{Nd}(T)$ v. model age plot with two reference fields (Hercynian and Himalayan granites). Much of the scatter in this plot results from the highly fractionated Sm–Nd ratios, whose effect would be reduced if a two-stage

model were used (figure not shown). Figure 6d illustrates the effect of f_{Sm-Nd} values on the single-stage model age calculation. Aberrant model ages (negative or ≥ 4000 Ma) are produced due to strong Sm–Nd fractionation through crystallization and magma–hydrothermal interaction which led to a tetrad REE distribution pattern (Masuda *et al.* 1987; Masuda and Akagi 1990; Bau, 1996; Irber 1999; Jahn *et al.* 2001, 2004). The model ages are interpretable only when f_{Sm-Nd} values fall in the range of -0.4 ± 0.2. Consequently, from Figures 6c and 6d, the granites from NE China and Inner Mongolia have young model ages ranging from 500 to 1200 Ma, except for a few plutons emplaced within the domain of the Jiamusi Massif. This is clearly distinguished from most of the European Caledonian and Hercynian granites, and even more clearly from the leucogranites of the Himalayas.

Junggar Basin. The geology of northern Xinjiang in NW China may be conveniently divided into five 'terranes' (from north to south): Altai, East and West Junggar, and East and West Tianshan (Fig. 2). A summary of the geological and isotopic characteristics of these terranes was given by Hu *et al.* (2000). The Junggar Basin is covered by Cenozoic desert sands and thick continental basin sediments (≥ 10 km) as old as the Permian. Drilling records indicate little deformation within the basin, suggesting stable configuration of the basement at least since the Permian (Coleman 1989). The nature of the Junggar basement has been much debated; some consider that the basin represents a microcontinent with Precambrian basement (Wu 1987), whereas others regard it as trapped Palaeozoic oceanic crust of various origins (Feng *et al.* 1989; Hsü 1989; Coleman, 1989; Carroll *et al.* 1990). Surrounding the Junggar Basin, numerous ophiolites are exposed in the East and West Junggar terranes, as well as in its southern margin. These terranes can be appropriately referred to as 'island-arc assemblages', and no rocks of Precambrian age have been documented. Coleman (1989) considered these terranes as oceanic arc assemblages, and compared them with those in the present western Pacific.

A variety of Phanerozoic granitoids occur throughout northern Xinjiang. The results of isotopic investigations are summarized in Figure 7. As for the case of NE China, the majority of granitoids have positive $\varepsilon_{Nd}(T)$ values which suggest the dominance of the mantle component in the generation of these rocks. This is particularly true for the granitoids from the E and W Junggar terranes (Fig. 7a; Zhao 1993; Han *et al.* 1997; Chen & Jahn 2004). Based on trace-element and Sr–Nd isotopic study, Chen and Jahn (2004) concluded that the basement is mostly likely underlain by Early to Middle Palaeozoic arc and oceanic crust assemblage that was trapped during the Late Palaeozoic tectonic consolidation of Central Asia. This is consistent with the very young model ages ranging from 400 to 1000 Ma (400 to 600 Ma in a two-stage model) for the Junggar granites (Fig. 7b).

Composite terranes of the Altai Mountains. The western Sayan and Gorny Altai in southern Siberia are considered to have formed during the complex evolution of the Palaeo-Asian Ocean (Sengör *et al.* 1993; Buslov *et al.* 2001; Dobretsov *et al.* 2004). The region is an Early Palaeozoic accretionary complex and island-arc system. It has been demonstrated to have large-scale strike-slip faults (up to several thousand kilometres) caused by subduction and collision of seamounts and island arcs (Buslov *et al.* 2001). Vendian to Lower Cambrian ophiolite,

HP/LT schists, Cambro-Ordovician turbidites, etc. are intruded by Ordovician and Devonian granitoids, and unconformably overlain by Silurian sediments, Devonian volcanics and non-marine clastics. The region was further intruded by post-orogenic Permian to Jurassic A-type granites (Vladimirov *et al.* 1997). Kruk *et al.* (2001) analysed 44 granitoid samples of Cambrian to Jurassic ages and the results are shown in Figure 7. The majority of the granitoids (35 out of 44) possess positive $\varepsilon_{Nd}(T)$ values, hence juvenile nature. Six samples with negative $\varepsilon_{Nd}(T)$ values (from -1.8 to -3.9) come from the 'Altai-Mongolian terrane', a contiguous part of the Chinese Altai. This obviously reflects the effect of Precambrian crust in the granitoid petrogenesis.

Similarly, the granitoids emplaced in the Chinese Altai composite terrane show a wide range of isotopic compositions and model ages (Fig. 7a, b). A tight relationship between the isotopic compositions of granitoids and the nature of their basement rocks can be established. An extensive Sm–Nd isotope study by Hu *et al.* (2000) reveals that the basement rocks of Altai and Tianshan were largely produced in the Proterozoic, but that of Junggar seem to represent very young accreted terrane with little Precambrian history (Chen & Jahn 2004). The parallel manifestation of isotopic compositions and model ages between basement rocks and intrusive granites argue for the significant role of crustal 'contamination' in the genesis of the Phanerozoic granitoids. An implication is that the presence of old Precambrian microcontinents is significant in the accretionary history of Central Asia.

East-central Kazakhstan. Heinhorst *et al.* (2000) undertook a comprehensive study of mineralization in association with a variety of magmatic rocks in east-central Kazakhstan. Although the types of mineralization (Au, Cu, rare-metal, or REE) may be related to a particular magmatic suite or a lithological variety, most granitic rocks have positive $\varepsilon_{Nd}(T)$ values irrespective of their compositions, here represented by SiO_2 contents (Fig. 8a; Heinhorst *et al.* 2000). The granitoids were intruded in several episodes: 450 and 300 Ma for magmatic suites with gold mineralization, about 300 Ma for granitoids with rare-metal mineralization, and c.250 Ma for A-type granites with REE mineralization. There is a slight tendency for an increase of $\varepsilon_{Nd}(T)$ with younger ages of the rocks (Fig. 8b). Single-stage model ages for all cases are between 400 and 1500 Ma.

The above data for the southern belt of the CAOB–from Kazakhstan, northern Xinjiang, Inner Mongolia to NE China, covering a distance

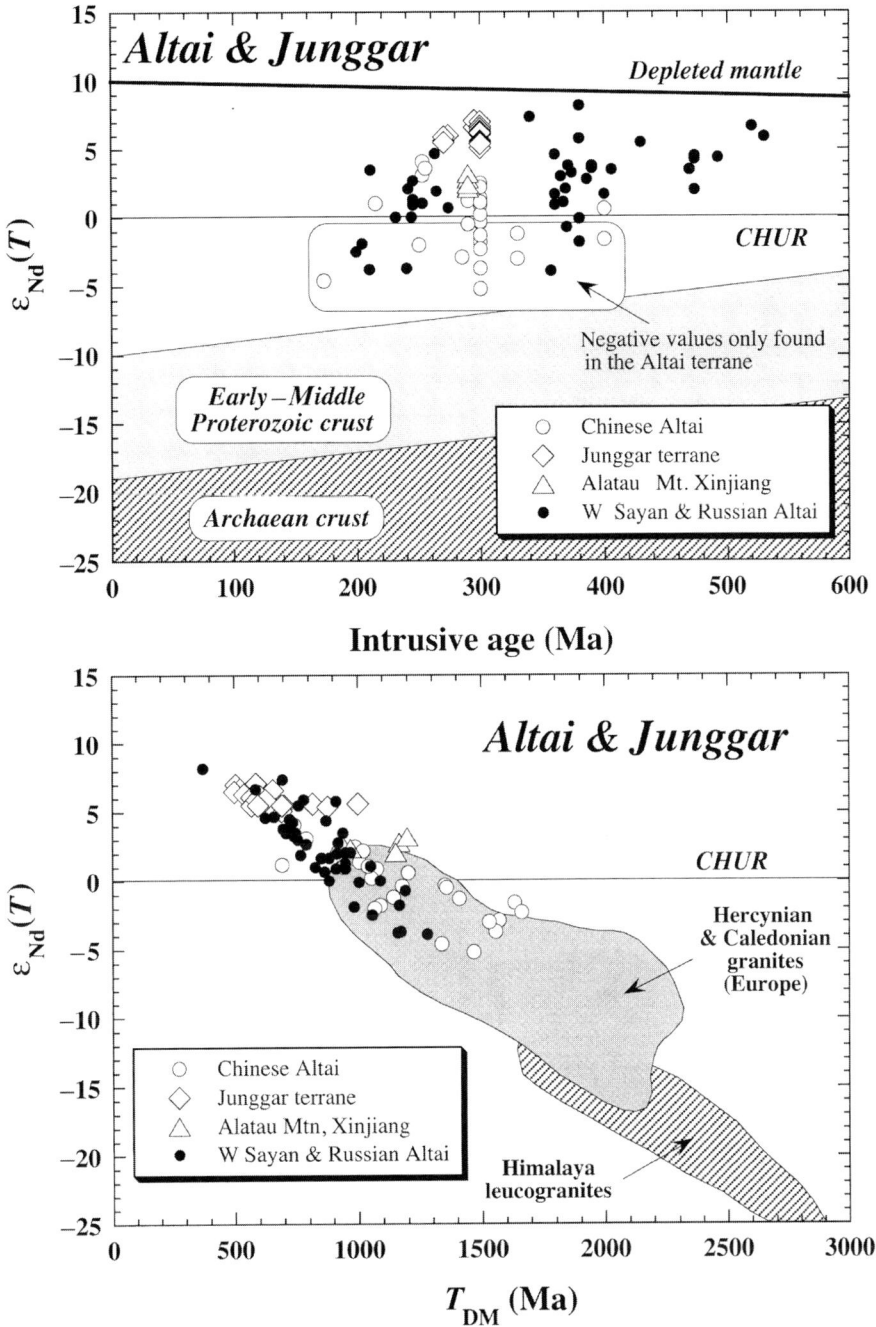

Fig. 7. Isotope diagrams for granitoids from the Altai Mountians (Chinese and Russian) and Junggar terrane.
(**a**) $\varepsilon_{Nd}(T)$ v. intrusive ages, (**b**) $\varepsilon_{Nd}(T)$ v. T_{DM} plot. Data sources: **Chinese Altai**: Zhao (1993), Junggar: Han *et al.*
(1997), Chen and Jahn (2002), **Alatau**: Zhou *et al.* (1995), **Western Sayan and Gorny Altai**: Kruk *et al.* (2001).

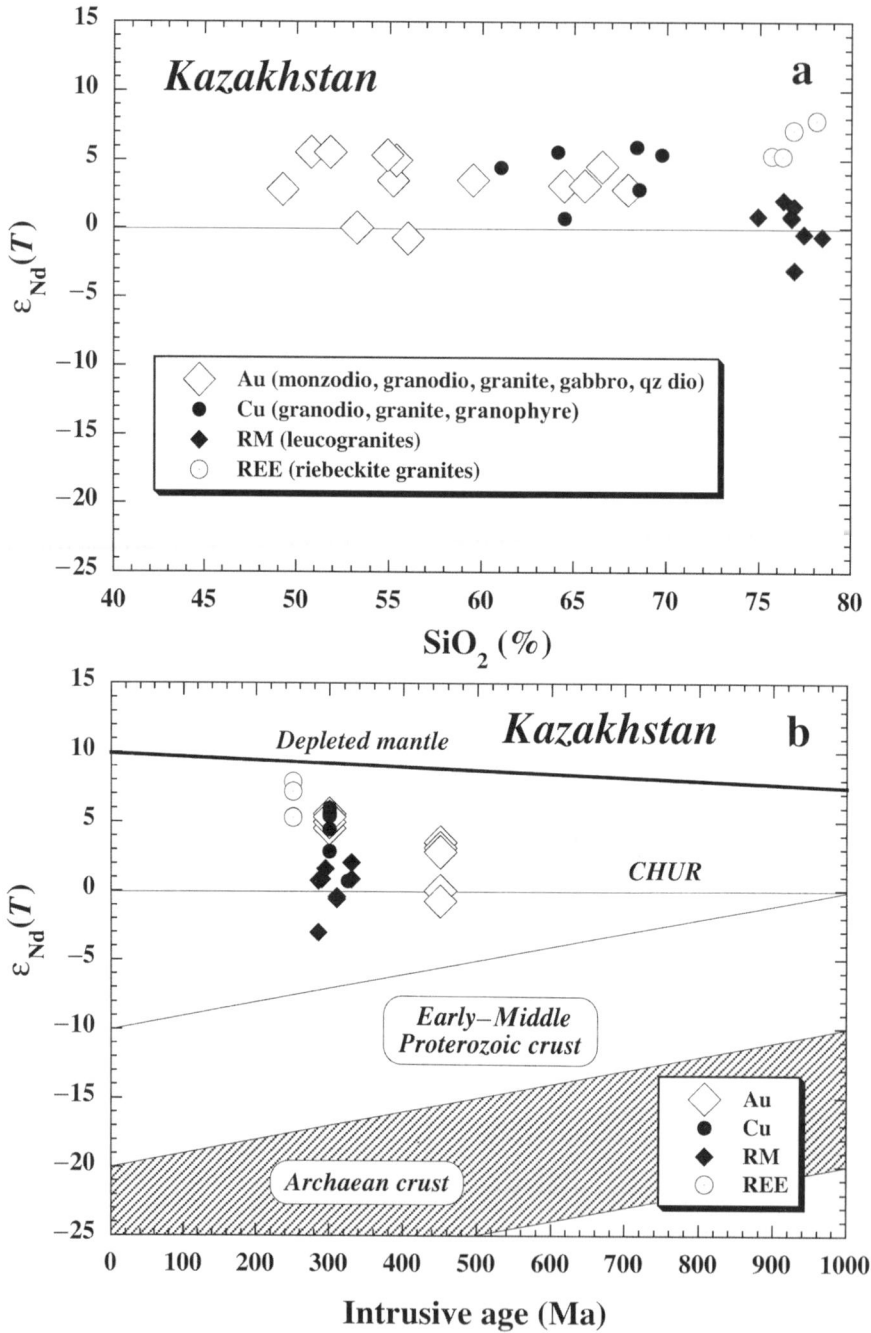

Fig. 8. Isotope diagrams for granitoids from east-central Kazakhstan. (**a**) $\varepsilon_{Nd}(T)$ v. SiO_2, and (**b**) $\varepsilon_{Nd}(T)$ v. intrusive ages. Data source: Heinhorst *et al.* (2000).

of nearly 5000 km, indicate that most of the granitoids, despite their highly differentiated nature and sometimes strong hydrothermal alteration leading to important mineralizations, possess a clear signature of high proportions of the mantle component in their petrogeneses.

Central Mongolia to Transbaikalia. The Nd–Sr isotopic compositions of granitoids from the northern belt of central Mongolia to Transbaikalia have been extensively studied by Kovalenko *et al.* (1992, 1996, 2004). These authors delineated four isotope provinces ('Precambrian', 'Caledonian', 'Hercynian', and 'Indosinian'), which coincide with three tectonic zones of corresponding ages for the northern belt of the CAOB, and with one (Indosinian) in Inner Mongolia and NE China (Kovalenko *et al.* 2004). Presented in Figure 9a, b, c are the published data of Kovalenko *et al.* (1996) and the recently acquired Nd isotopic data for granitoids from three regions:

(1)   the Baydrag and Hangay terranes of west-central Mongolia (Kozakov *et al.* 1997; Jahn *et al.* 2004);

(2)   northern Mongolia (Jahn, unpublished); and

(3)   Transbaikalia (the Bryansky and Tsagan–Khurtei complexes; Litvinovsky *et al.* 2002*b*).

The Transbaikalian and northern Mongolian samples roughly correspond with the 'Barguzin' belt and 'Caledonides' of Kovalenko *et al.* (2004). The Baydrag terrane is the only Precambrian microcontinent containing granulitic and amphibolitic gneisses of Archaean ages. The Barguzin Belt and the Hangay–Hentey Basin are known to be 'composite' terranes comprising Proterozoic and Phanerozoic formations.

The Hangay–Hentey Basin occupies a large area in central to north-central Mongolia. The basin is filled with Cambrian to Carboniferous sediments, mainly of shallow-marine origin, and volcanics. It is intruded by impressive amounts of Early Palaeozoic to Mesozoic granitoids. The granitoid belt extends to Transbaikalia and further to the Sea of Okhotsk. The tectonic significance of the basin is controversial. Zorin (1999) and Parfenov *et al.* (1999) considered the basin as an accretionary wedge, whereas Sengör and Natal'in (1996), following Zonenshain *et al.* (1990), took it as part of the Mongol–Okhotsk oceanic gulf. Alternatively, it has also been suggested to be a back-arc basin formed within an Andean type margin produced by northward subduction of the South Mongolian oceanic plate (Gordienko, 1987), or as a post-orogenic successor basin formed on the Early Palaeozoic basement of northern Mongolia (Ruzhentsev & Mossakovsky 1996). Note that although the Mongol–Okhotsk ocean basin has been frequently hypothesized in tectonic models, its closure history and location of suture zone have never been clearly defined (Badarch *et al.* 2002). Based on Nd isotopic data, Kovalenko *et al.* (1996, 2004) suggested that the southern Hangay Basin is underlain by Precambrian rocks.

As shown in Figure 9a, b, c, Phanerozoic granites emplaced into 'Caledonian' and 'Hercynian' provinces have positive $\varepsilon_{Nd}(T)$ values, suggesting their juvenile characteristics, whereas those intruded into the Baydrag and the composite terranes (Barguzin and Hangay–Hentey) show $\varepsilon_{Nd}(T)$ values from positive to negative, indicating variable but always minor contributions of Precambrian crust in the generation of the granitic rocks. Note that some Late Neoproterozoic to Early Palaeozoic granites (600–500 Ma) have $\varepsilon_{Nd}(T)$ values as high as $+10$ (Fig. 9b), suggesting their derivation from an almost pure depleted mantle component ($=100\%$ basaltic source).

In addition to the Nd isotopic evidence, oxygen isotope analyses of alteration-resistant titanites from anorogenic granites of Transbaikalia (Wickham *et al.* 1995, 1996) show a progressive decrease in $\delta^{18}O$ of titanite (sphene) from $+6.5\%o$ in the earliest suite (*c.*450 Ma) to $+1.5\%o$ in the youngest suite (*c.*220 Ma). This corresponds with a decrease in whole-rock $\delta^{18}O$ from $+11\%o$ to $+6\%o$. It appears that whereas the older granitoids with higher $\delta^{18}O$ values may have a crustal heritage, the younger magmas, particularly the A-type granites, became increasingly mantle-like in terms of their oxygen isotopic composition. This suggests that a series of important crust growth events was taking place in Central Asia in the Late Phanerozoic.

Discussion

Genesis of the Phanerozoic crust in the CAOB

Windley (1993, 1995) distinguishes two types of orogens:

(1)   collisional orogens, formed by collision and amalgamation of two or more large continental blocks (e.g. Himalayas, Alps, Grenville, etc.), and

(2)   accretionary orogens, formed by accretion of island-arc complexes and intervening accretionary prisms, etc. (e.g. CAOB, North American Cordillera, Andes, Birimian, Nubian–Arabian, etc.).

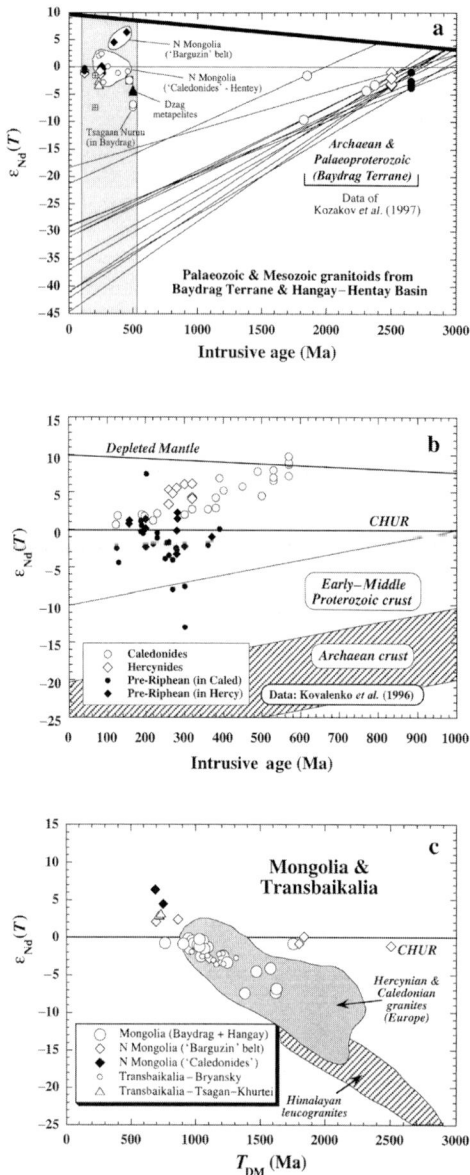

Fig. 9. Isotope diagrams for granitoids from Mongolia and Transbaikalia. (**a**) ε_{Nd} (T) v. intrusive ages for granitoids of west-central Mongolia (data: Jahn *et al.* 2004; Kozakov *et al.* 1997); (**b**) $\varepsilon_{Nd}(T)$ v. intrusive ages for granitoids of Mongolia and Transbaikalia (data: Kovalenko *et al.* 1996). Granites intruded in 'Caledonian and Hercynian' belts (open symbols) are characterized by positive ε_{Nd} (T) values, whereas those intruded in pre-Riphean basement (black symbols) have both positive and negative $\varepsilon_{Nd}(T)$ values. (**c**) ε_{Nd} (T) v. T_{DM} plot for west-central Mongolia and Transbaikalia (data: Jahn *et al.* 2004).

The CAOB is a prime example for accretionary tectonics. The Central Asian orogenic system evolved during a span of about 400–450 Ma since the Vendian. Documented field data clearly indicate that many lithological assemblages represent products of oceanic and subduction-related processes (Sengör *et al.* 1993; Kovalenko *et al.* 1995; Sengör and Natal in 1996; Badarch *et al.* 2002). The lithology includes ophiolite suites, ocean island and seamount fragments, fore-arc/back-arc basin assemblages, and accretionary complexes. These rocks are evidently of mantle derivation, at least for most of them. If the CAOB is entirely accreted from such lithological assemblages, then its juvenile character is easily justified without conformation from isotopic studies. However, the voluminous distribution of granitoids in the CAOB requires further explanation.

Granitoids are the most representative 'continental component'. They probe the lower part of the continental crust. Reputedly, they also have the most diverse and controversial origins. Since the CAOB contains Precambrian microcontinental fragments, probably derived from Gondwana (Zonenshain *et al.* 1990; Buslov *et al.* 2001) as well as from Laurasia (Berzin and Dobretsov 1994), determination of their juvenile v. recycled nature is important for understanding the growth rate of the continental crust and the geodynamic evolution of the orogen. This can only be achieved through using radiogenic isotope tracer techniques. The most striking feature of the granitoids from Central Asia is their dominantly positive $\varepsilon_{Nd}(T)$ values, low I_{Sr} ratios and young T_{DM} model ages (Kovalenko *et al.* 1996, 2004; Jahn *et al.* 2000a, b, 2004; Chen *et al.* 2000; Wu *et al.* 2000, 2002, 2003a, b; Chen & Jahn 2004). That is, much of the crust in the CAOB was made up of 'young' mantle-derived material. Such a conclusion is essentially identical to that reached by Sengör *et al.* (1993) on the basis of their geological and tectonic analyses. The ultimate message is clear: the formation of a large composite continent, like Asia, involves not only amalgamation of broken-up supercontinental fragments (e.g. the Angara and Sino-Korean cratons), but also massive addition of juvenile material from the upper mantle.

Two mechanisms may be envisaged for the growth of the continental crust (e.g. Rudnick 1995):

(1) Lateral growth: melting of mantle-wedge or subducted oceanic crust in convergent plate margins where the most active crust–mantle interaction takes place. In this case, magmas are formed mainly by melting of mantle-wedge peridotites that have been metasomatized by fluids/melts

released from subducted oceanic crust, or, occasionally, by direct melting of subducted basaltic crust when the subducted slab is hot. These magmas, mainly basaltic to andesitic, make up the bulk of the arc complexes, and accretion of arc complexes, including fore-arc/back-arc sediments, contributes to the 'lateral growth' process.

(2) Vertical growth: overplating of volcanic rocks and underplating of mantle-derived basaltic magmas near or at the crust–mantle interface within pre-existing continental blocks or accreted arc complexes. Subsequent differentiation and melting of the lower crustal mixed lithologies triggered by hot underplated basaltic magmas contributes to the formation of new crust. This is the 'vertical growth' process.

Both processes have probably played equally important roles in the making of the CAOB throughout the Phanerozoic. The lateral growth process appears indisputable in the concept of the plate tectonics, and Sengör *et al.* (1993) even used 'magmatic front' as a new tool to delineate the tectonic evolution of the Altaid Collage. The vertical process may be controversial, but it could be the most viable process to explain the impressive belts of post-collisional peralkaline and A-type granitoids in Central Asia. The idea expressed herein is illustrated in Figure 10.

Formation of 'syn-orogenic' granitoids of arcs and accreted arcs

Granitic rocks are commonly abundant in continental arcs, but are generally trivial in oceanic island arcs. Such a contrast is likely to be due to the different nature of the crust overlying the zones of melt generation (Fig. 10a). However, the CAOB is essentially accreted from oceanic arcs and ancient microcontinental mass; the latter is limited in number and size. But, why has there been such a widespread felsic (granitic and rhyolitic) magmatism since the early Palaeozoic?

The granitoids of the CAOB may be considered to fall into two broad categories:

(1) Those formed during the building of an arc (Fig. 10a), following the accretion of arc complexes, and possibly in the arc–microcontinent collision (Fig. 10b); these will be collectively called as 'syn-orogenic' granitoids. This would correspond roughly with the 'early granitoids' of Litvinovsky and Zanvilevich (1998), and the lithology includes the tonalite–granodiorite–granite,

tonalite–plagiogranite, and quartz monzonite–granite series.

(2) Those formed after the accretion of arc complexes and during the post-accretion extensional or rifting tectonic phase (Figs 10c, d). These are understood as 'post-accretionary' and correspond with the often used term of 'post-collisional' granitoids in the literature.

Accretion of arc complexes, especially if subparallel and along-margin transport of the arc terrane is involved (e.g. Patchett and Chase 2002), is likely to result in 'soft collision', which lacks the dynamics to form significant thrust belts or nappes. Thus, the use of terms like 'syncollisional' or 'post-collisional' may be somewhat misleading for the case of Central Asia. In Transbaikalia and Mongolia, the post-accretionary granitoids are represented by quartz syenite–granite, monzonite–syenite–granite, syenite–granite, and alkaline to peralkaline granite series (Litvinovsky and Zanvilevich, 1998). In general, the abundance of synorogenic granitoids in the CAOB is overshadowed by that of post-accretionary granitoids. However, the precise proportion remains to be determined.

For the generation of granitoids in individual arcs, the process is known in the subduction zones to involve melting of the mantle wedge followed by magmatic differentiation, or direct melting of the warm subducted slab, forming tonalite–trondhjemite–granodiorite (TTG) or adakite magmas (Fig. 10a). Arc complexes also comprise rocks of sedimentary origin, including clastic and calcareous sediments. The subsequent accretion of arc complexes might not induce melting without further heat supplies. A good example is that the modern arc–continent collision of Taiwan (at an oblique angle) does not produce significant magmatism. To engender large volumes of granitic magmas, accreted arc complexes might have been underplated by basaltic magma which was buoyantly blocked at the lower crust (Fig. 10c). Basaltic underplating provides the necessary heat source for melting, and also materially participates in felsic magma generation (Fig. 10d). How the basaltic liquids are generated in the first place is quite debatable, but the parallel evolution of chemical composition between the granitoids and contemporaneous basaltic rocks (Litvinovsky and Zanvilevich, 1998) observed in Transbaikalia provides the best evidence for the participation of basaltic magma in the generation of granitoids (Fig. 11).

To achieve a high-temperature melting, as suggested from melt inclusion studies for some granites (Litvinovsky *et al.* 2002*a*, *b*), crust overthickening may be another viable process

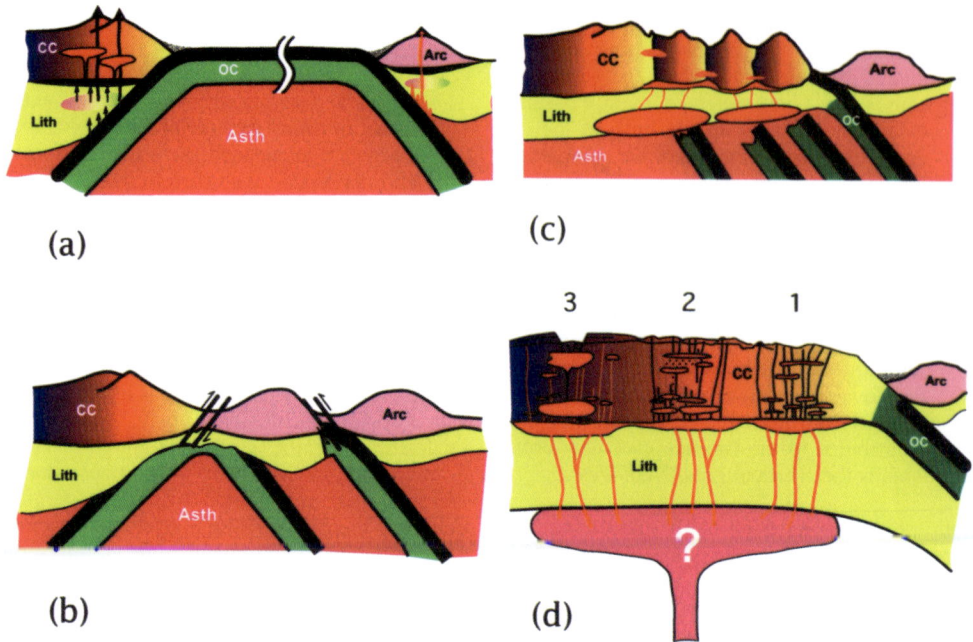

Fig. 10. Schematic diagrams showing the growth of continental crust by lateral and vertical accretion. (**a**) Generation of arc magmas in a continental arc (left side) via partial melting of a mantle wedge, followed by a series of processes (basaltic underplating, assimilation of lower crust, melting of mixed lithologies, magma differentiation), leading to formation of arc granitoids. Subduction of a young and hot oceanic lithosphere may produce TTG (tonalite–trondhjemite–granodiorite) or adakitic magmas in an intra-oceanic arc setting (right side). (**b**) Accretion of island arcs to the continental margin. New arcs are formed by back-stepping of subduction zone. They are then accreted again to the continental margin. Together they enlarge the mass and size of the continental crust by lateral accretion processes. (**c**) Generation of granitic magmas in accreted arc complexes. Basaltic underplating provides heat and material (=mantle component) for the magma generation. Initial basaltic magma could be produced by partial melting of mantle peridotites at the lithosphere/asthenosphere interface after delamination of the oceanic lithosphere. (**d**) Intracrustal melting for the generation of post-orogenic granitoids. This is achieved by a combination of processes, including crustal and lithospheric extension, melting of mantle plume, underplating of basaltic magma at the base of the crust, partial melting of 'young' crust and magmatic differentiation.

in addition to basalt underplating. This process is considered very likely in continent–continent collisional orogens, like the Himalayas, but it is rather ambiguous or may be unfavourable for an accretionary orogen like the CAOB. However, if accreted arc complexes contain sufficient water-saturated sediments, then melting could be promoted by the added water.

In conclusion, the synorogenic granitoids were probably generated in two stages: first during the building of individual arcs via subduction zone magmatism, and then conversion of arc complexes plus underplated basaltic liquids within accreted arc complexes. Melting of accreted arc complexes may be further promoted by water-saturated sediments. Finally, the syn-orogenic granitoids dominate the felsic magmatism of the early Palaeozoic (≥ 300 Ma) when arc complexes were being built.

Origin of post-accretional granitoids – via vertical growth

Although many plutons and batholiths of the CAOB have the calc-alkaline characteristics typical of sub-duction zone magmatism, the emplacement of voluminous granites of the alkaline and peralkaline series deserves attention. Petrogenetically, these rocks are akin to the A-type granites, which are generally known to form in post-collisional extensional (rifting) environments. In the CAOB, two gigantic belts of alkaline granites are recognized:

(1) the northern belt from northern Mongolia to Transbaikalia, and
(2) the southern belt from the Gorny Altai through southern Mongolia, Inner Mongolia to the Great Xing'an and Zhangguanggcai Ranges in NE China.

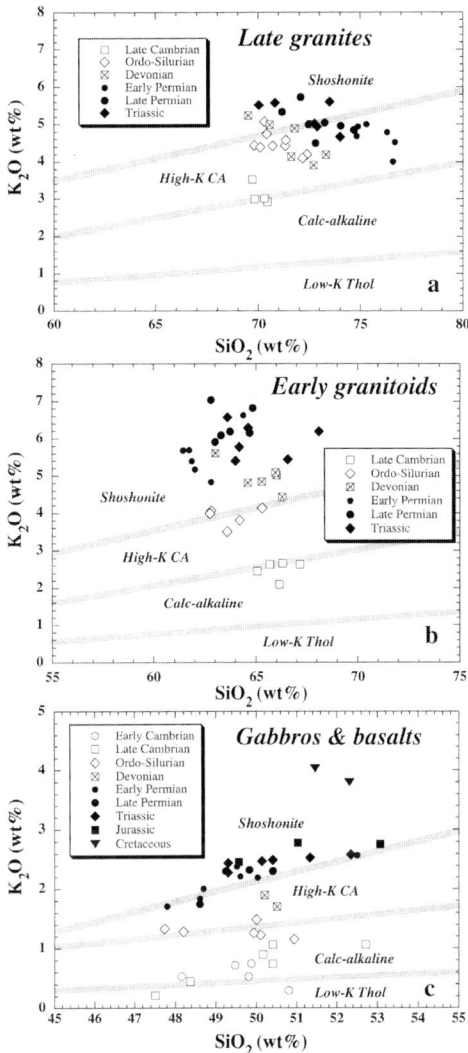

Fig. 11. Chemical evolution of basic and granitic rocks from the Mongolian–Transbaikalian belt, as represented by the K$_2$O v. SiO$_2$ plots (data from Litvinovsky and Zanvilevich, 1998). Increase of K$_2$O as rocks become younger is clearly indicated for late granites ('post-collisional'), early granites (synorogenic) and gabbros/basalts. This suggests that the ultimate mantle source is increasingly enriched in alkali elements through time. The parallel evolution between basic and granitic rocks further implies that basaltic magmas have materially participated in the genesis of granitoid rocks. The boundary lines are taken from Le Maitre et al. (1989) and Rickwood (1989).

The northern belt is probably the largest province of peralkaline granitoids, which extends over 2000 km with a width of 200–300 km, and occupy a total area of ≥500 000 km^2

(Kovalenko et al. 1995; Zanvilevich et al. 1995). It comprises more than 350 massifs of peralkaline granitoids and a number of bimodal volcanic rocks of the basalt–trachyrhyolite–comendite series. The largest Bryansky pluton has a dimension of about 1500 km^2, emplaced at c.280 Ma, and is composed of a syenite-to-granite series (Yarmolyuk et al. 2001; Litvinovsky et al. 2002a, b). The entire belt was developed in several stages from Devonian to Cretaceous, with the main pulses of peralkaline granitic plutons and their volcanic analogues during the Permian and Triassic.

The southern belt is equally impressive in the length of distribution, but the total volume appears less important in comparison with the northern belt. In the Gorny Altai (Russia), A-type granitoids were emplaced at two stages: Permo-Triassic (250 Ma) and Triassic–Jurassic (200 Ma), in probably the largest Permo-Triassic rift system in Asia (Vladimirov et al. 1997). In the Junggar terrane of northern Xinjiang and in Inner Mongolia, A-type granites were formed at 300–280 Ma (Hong et al. 1995; Han et al. 1997; Zhao et al. 2000; Chen & Jahn 2004). In NE China, several hundred granitic bodies have been identified and constitute an area of c.100 000 km^2. A-type granites are mainly distributed in the Great Xing'an and Zhangguangcai Ranges, but they were emplaced in three episodes (Jahn et al. 2001; Wu et al. 2002):

(1) Permian (300–280 Ma);
(2) Late Triassic to Early Jurassic (210–180 Ma), and Cretaceous (c.120 Ma).

It appears that throughout the entire southern belt, post-accretionary A-type granites were emplaced in the Late Palaeozoic, Early Jurassic and Cretaceous.

All these rocks possess positive to slightly negative $\varepsilon_{Nd}(T)$ values (+7 to −4) and young model ages (T_{DM}) of 500–1300 Ma, which are summarized in Figure 12 (Kruk et al. 1998; Jahn et al. 2001; Chen & Jahn 2004; Wu et al. 2002; Litvinovsky et al. 2002a, b; Jahn, unpublished). This advocates that the source of post-accretionary granitoids in the CAOB is dominated by the mantle-derived component, rather than recycled ancient crust, as has been documented for some occurrences in SE China (e.g. Charoy & Raimbault 1994; Darbyshire & Sewell 1997). In most cases, rocks of mantle derivation have also been contaminated by crustal material. Models involving mixing of mantle-derived magmas and crustal components (assimilated crustal rocks or crust-derived

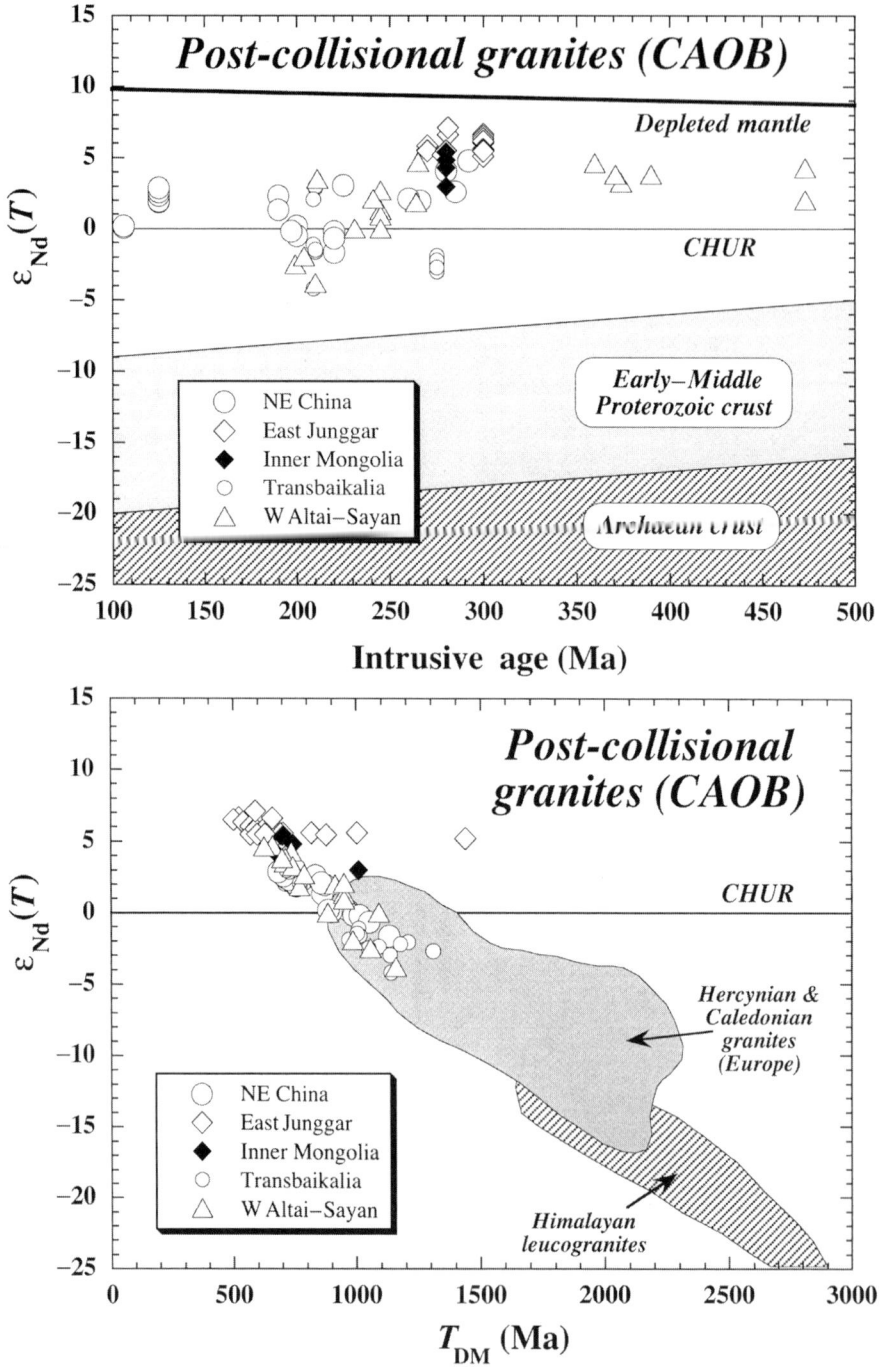

Fig. 12. Isotope diagrams for A-type granitoids from Central Asia. (**a**) ε_{Nd} (T) v. intrusive ages, and (**b**) $\varepsilon_{Nd}(T)$ v. T_{DM} plot. Data source: **NE China**: Wu *et al.* (2002), **eastern Junggar**: Chen and Jahn (2002), **Inner Mongolia**: Jahn (unpubl.), **Transbaikalia**: Jahn (unpubl.), Yarmolyuk *et al.* (2001), western Sayan–Gorny Altai: Kruk *et al.* (2001).

magmas), followed by fractional crystallization, are most acceptable.

The production of a huge amount of alkaline to peralkaline granites was very likely initiated by large-scale crustal extension and accompanied basalt underplating. This has been suggested to be connected with mantle plume activities. A Siberian superplume has been proposed for the Northern Mongolia–Transbaikalia Belt (Yarmolyuk *et al.* 2001; Kovalenko *et al.* 2004). The geological manifestations of a superplume are not expected to have a sharp chronology, particularly when intrusive events are concerned. A sharp chronology can only be observed in basaltic volcanism related to large plume head melting, such as the 65 Ma Deccan Traps (Courtillot *et al.* 1986), the 184 Ma Karoo Traps (Encarnacion *et al.* 1996; Duncan *et al.* 1997), or the 250 Ma Siberian Traps (Renne and Basu 1991; Campbell *et al.* 1992; Renne *et al.* 1995). Magmatic differentiation and cooling of magmatic bodies in plutonic conditions would retard the radiochronometers to different degrees for intrusive rocks. Consequently, intrusive rocks of different ages, up to several tens of millions of years, could be related to the same superplume activity. The hypothesized Siberian superplume could be responsible for the Siberian Trap, as well as all the post-collisional granitic intrusions in the period from 250 to 200 Ma. Note that the magmatism of a superplume need not be spatially contiguous; it would take place where the lithosphere is thin, favouring decompressional melting. Yarmolyuk *et al.* (2001) observed that the intraplate magmatism (peralkaline granites and bimodal basalt–comendite) was centred around the Hangay batholith in Mongolia in the Permian. The centre was then shifted eastward to the Hentay batholith and Transbaikalia in the Triassic. To these authors, this suggests a westward displacement of 800 km for the CAOB relative to a mantle plume.

Stein and Goldstein (1996) argued that the present-day arc magmatism is commonly associated with oceanic plateaux, hence suggesting that generation of large amounts of continental lithosphere (crust and mantle) over short periods is probably associated with plume head magmatism. They further advanced this idea for the formation of the Arabian–Nubian Shield, which was considered to build up initially from an oceanic plateau as a result of melting of a plume head. Upon reaching a convergent margin, the thick oceanic plateau resisted subduction, and plate convergence took place on its own margins, generating calc-alkaline magmas. Later transition from calc-alkaline to alkaline magmatism would mark the end of plate convergence. They considered that alkaline magmatism was associated with melting of enriched lithospheric mantle, and that the transformation from plume head to continental lithosphere has been an important process of crustal growth.

The manifestation of oceanic plateaux in the CAOB remains to be identified. The Permian and Cretaceous A-type granites in NE China seem to be better explained by a delamination model, in which the lithospheric delamination was followed by upwelling and partial melting of the asthenosphere (Wu *et al.* 2002). While the lateral growth mechanism via accretion of arc complexes undeniably played an important part in the making of the CAOB, the vertical growth through intraplate magmatism has probably contributed even more significantly to the formation of the gigantic orogenic belt and the enlargement of the Asian continent. Certainly, a prerequisite of such enlargement is the presence of crustal extension. If not, the underplated mafic rocks could be nullified by likely delamination. In the case of the CAOB, rifting reached its climactic phase during the Permian and Triassic (e.g. Kovalenko *et al.* 2004).

Estimate of the proportions of juvenile crust, and its implications

The proportions of juvenile crust in any given area in the CAOB must be evaluated from detailed knowledge about the distribution of lithological types, of which the granitoids must be further estimated using the isotope tracer technique. This is not an easy task. However, for individual granitoid bodies, this can be done reasonably well using a simplistic two-component mixing calculation. Assuming a fixed depleted mantle ($\varepsilon_{Nd} = +8$) and variable crustal endmembers, the result of calculation is shown in Figure 13. The proportion of the mantle component (or percentage of juvenile crust) for positive $\varepsilon_{Nd}(T)$ granites varies from 60 to 100%, depending on the compositions of the assumed crustal end-members. We take the Nd isotopic composition of the Jiamusi Massif for NE China ($\varepsilon_{Nd} = -12$), the Baidarik Block for Central Mongolia (-30, Fig. 9), the basement gneisses for Altai (-15; Hu *et al.* 2000), Junggar (-4), and the Kazak basement is assumed to be the same as the Altai gneisses. For the Mongolian granitoids with $\varepsilon_{Nd}(T)$ values of -5 or higher, the proportion of juvenile crust is 80% or higher. Kovalenko *et al.* (2004) made an extensive compilation of Nd isotope data, and showed that basaltic rocks of Permian and younger in

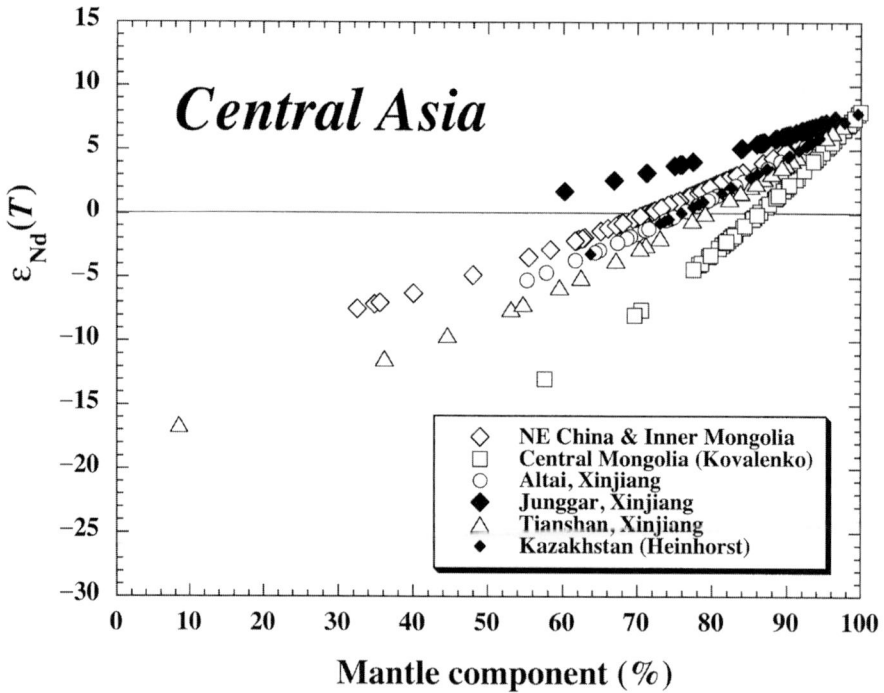

Fig. 13. Estimate of proportions of the mantle or juvenile component in the generation of Central Asian granitoids. The equation used is:

$$X^m = (\varepsilon^c - \varepsilon^r)Nd_c/[\varepsilon^r(Nd_m - Nd_c) - (\varepsilon^m Nd_m - \varepsilon^c Nd_c)]$$

where X^m = % mantle component (represented by basalt). ε^c, ε^r, ε^m = Nd isotope compositions of the crustal component, rock measured, and mantle component, respectively. Nd_c, Nd_m = Nd concentrations in the crustal and mantle components, respectively. Parameters used: $\varepsilon^m = +8$, $\varepsilon^c = -12$ (NE China), -30 (Central Mongolia), -15 (Altai and Kazakhstan), -4 (Junggar), -18 (Tianshan). Nd_m = 15 ppm, Nd_c = 25 ppm.

Mongolia and Transbaikalia are characterized by rather low $\varepsilon_{Nd}(T)$ values ($\leq +3$). Interpretation of the low values could be multiple without added constraints. However, trace-element abundances of mafic rocks and granitoids from Transbaikalia (e.g. Litvinovsky *et al.* 2002*a, b*) strongly suggest that their ultimate mantle sources were enriched. If a lower $\varepsilon_{Nd}(T)$ value is used for the mantle component in the mixing calculation, then the proportion of the mantle component would be further enhanced.

In any case, the general scenario in Central Asia as shown in Figure 13 implies extensive mantle differentiation and rapid crustal growth during the Phanerozoic. However, a significant proportion of recycled crust is also 'isotopically' visible in the granitoids emplaced in the Jiamusi Massif, Altai and Tianshan composite terranes,

and Pre-Riphean zones of Mongolia as exemplified by the present study.

Conclusions

These are as follows:

(1) The Central Asian Orogenic Belt (CAOB) that welded together the Siberian and Sino-Korean cratons is characterized by very abundant granitic rocks of Palaeozoic to Mesozoic ages. The Nd–Sr isotopic data show that most of these granitic rocks are very juvenile. Considering the immense geographical coverage, the CAOB represents undoubtedly the most important site of crustal growth in the Phanerozoic.

(2) The tectonic evolution of the CAOB is mainly related to accretion of arc complexes, as suggested by Sengör *et al.* (1993). This idea is generally supported by the available Sr–Nd–O isotopic data (Wickham *et al.* 1995, 1996 for oxygen isotopes). However, the granites with negative $\varepsilon_{Nd}(T)$ values require the existence of old Precambrian blocks; thus, accretion of old terranes is confirmed.

(3) Conversion of arc assemblages into granitoids (=intracrustal differentiation) was most likely triggered by underplating of basaltic magma, which provided not only the heat source but also the added juvenile material. The process involved mixing of underplated magma and lower crustal rocks, and melting of the mixed sources, and was followed by fractional crystallization.

(4) The origin of A-type granites has long been controversial, but most post-accretionary A-type granites from the CAOB are demonstrably of mantle origin (*sensu lato*). The generation of voluminous Late Permian to Cretaceous alkaline and peralkaline granites could be related to the Siberian superplume and to the lithospheric delamination. In this connection, intraplate magmatism was an important process – in addition to the lateral accretion of arc complexes – of crustal growth in the Phanerozoic.

(5) The rate of apparent crustal growth for the entire period of the Phanerozoic may be estimated, but the growth rate at any given time interval of 50–100 Ma is not possible, due to the paucity of reliable age data for many granites. For the entire Altaid Collage, Sengör *et al.* (1993) estimated that during the 350 Ma of crustal evolution, a total area of about 2.5 million km^2 of juvenile crust was added to Asia. This is translated into a growth rate of about 0.3 km^3/year. Combining this with that of the Canadian Cordillera (about 0.15–0.23 km^3/year, Samson *et al.* 1989; Samson and Patchett, 1991) and western United States (DePaolo *et al.* 1991), the new rate would be at least 50% higher than the global growth rate of *c.*1.1 km^3/year, deduced from arc magmatism by Reymer and Schubert (1984, 1986). Consequently, the recent 'discovery' of juvenile crust in several Phanerozoic orogenic belts, in particular the CAOB, may considerably change our views on the growth rate of the continental crust.

I have benefited from close collaboration with many of my colleagues in Rennes (R. Capdevila), in China (Fuyuan Wu, Bin Chen and Aiqin Hu), and elsewhere (B. Litvinovsky, A. Zanvilevich, S. Wilde) during the last five years since the launch of the IGCP-420 project: *Crustal Growth in the Phanerozoic: Evidence from East-Central Asia.* I have also learned a great deal about the geology of northern Xinjiang, west-central Mongolia, the Gorny Altai in southern Siberia and NE China, during the past field workshops. Unesco's support of this project is highly appreciated. The constructive comments of R. Capdevila, B. Litvinovsky, Shen-su Sun, Fuyuan Wu, Bin Chen, V. I. Kovalenko, W. Collins and an anonymous reviewer have significantly improved the final version. The financial assistance of the Croucher Advanced Studies Institutes made it possible for me to deliver this lecture at the University of Hong Kong. This is a contribution to *IGCP-420: Crustal Growth in the Phanerozoic: Evidence from East-Central Asia.* The National Science Council of Taiwan provided financial support for the final preparation of this article.

References

AJAJI, T., WEIS, D., GIRET, A. & BOUABDELLAH, M. 1998. Coeval potassic and sodic calc-alkaline series in the post-collisional Hercynian Tanncherfi intrusive complex, northeastern Morocco: geochemical, isotopic and geochronological evidence. *Lithos*, **45**, 371–393.

ARAKAWA, Y. & SHIMURA, T. 1995. Nd–Sr isotopic and geochemical characteristics of two contrasting types of calc-alkaline plutons in the Hida belt, Japan. *Chemical Geology*, **124**, 217–232.

ARAKAWA, Y., SAITO, Y. & AMAKAWA, H. 2000. Crustal development of the Hida belt, Japan: evidence from Nd–Sr isotopic and chemical characteristics of igneous and metamorphic rocks. *Tectonophysics*, **328**, 183–204.

ARNDT, N. T. & GOLDSTEIN, S. L. 1987. Use and abuse of crust-formation ages. *Geology*, **15**, 893–895.

AZEVEDO, M. R. & NOLAN, J. 1998. Hercynian late-post-tectonic granitic rocks from the Fornos de Algodres area (Northern Central Portugal). *Lithos*, **44**, 1–20.

BADARCH, G., CUNNINGHAM, W. D. & WINDLEY, B. F. 2002. A new terrane subdivision for Mongolia: implications for the Phanerozoic crustal growth of Central Asia. *Journal of Asian Earth Sciences*, **21**, 87–110.

BAU, M. 1996. Controls on the fractionation of isovalent trace elements in magmatic and aqueous systems: evidence from Y/Ho, Zr/Hf, and lanthanide tetrad effect. *Contributions to Mineralogy and Petrology*, **123**, 323–333.

BEN OTHMAN, D., FOURCADE, S. & ALLEGRÉ, C. J. 1984. Recycling processes in granite–granodiorite complex genesis: the Querigut case studied by Nd–Sr isotope systematics. *Earth and Planetary Science Letters*, **69**, 290–300.

BERNARD-GRIFFITHS, J., PEUCAT, J. J., SHEPPARD, S. & VIDAL, P. 1985. Petrogenesis of Hercynian leucogranites from the southern Armorican Massif: contribution of REE and isotopic (Sr, Nd, Pb, and O) geochemical data to the study of source rock characteristics and ages. *Earth and Planetary Science Letters*, **74**, 235–250.

BERZIN, N. A. & DOBRETSOV, N. L. 1994. Geodynamic evolution of southern Siberia in late Proterozoic–early Paleozoic time. *In*: COLEMAN, R. G. (ed.), Reconstruction of the Paleo-Asian ocean. VSP International Science Publishers, Utrecht, The Netherlands, 53–70.

BUSLOV, M. M., SAPHONOVA, I. Y. *et al.* 2001. Evolution of the Paleo-Asian Ocean (Altaid–Sayan Region, central Asia) and collision of possible Gondwana-derived terranes with the southern marginal part of the Siberian continent. *Geosciences Journal*, **5**, 203–224.

CAMPBELL, I. H., CZAMANSKE, G. K., FEDORENKO, V. A., HILL, R. I. & STEPANOV, V. 1992. Synchronism of the Siberian traps and the Permo-Triassic boundary. *Science*, **258**, 1760–1763.

CARROLL, A. R., LIANG, Y. H., GRAHAM, S. A., XIAO, X., HENDRIX, M. S., CHU, J. & MCKNIGHT, C. L. 1990. Junggar basin, northwest China. trapped Late Paleozoic ocean. *Tectonophysics*, **181**, 1–14.

CHAROY, B. & RAIMBAULT, L. 1994. Zr-, Th- and REE-rich biotite differentiates in the A-type granites pluton of Suzhou (eastern China): the key role of fluorine. *Journal of Petrology*, **35**, 919–962.

CHEN, J. F. & JAHN, B. M. 1998. Crustal evolution of southeastern China: Nd and Sr isotopic evidence. *Tectonophysics*, **284**, 101–133.

CHEN, B. & JAHN, B. M. 2002. Geochemical and isotopic studies of the sedimentary and granitic rocks of the Altai orogen of northwest China and their tectonic implications. *Geological Magazine*, **139**, 1–13.

CHEN, B. & JAHN, B. M. 2004. Genesis of post-collisional granitoids and basement nature of the Junggar Terrane, NW China: Nd–Sr isotope and trace element evidence. *Journal of Asian Earth Sciences*, (in press).

CHEN, B., JAHN, B. M., WILDE, S. & XU, B. 2000. Two contrasting Paleozoic magmatic belts in northern Inner Mongolia, China: petrogenesis and tectonic implications. *Tectonophysics*, **328**, 157–182.

CHEONG, C. S. & CHANG, H. W. 1997. Sr, Nd, and Pb isotope systematics of granitic rocks in the central Ogcheon belt, Korea. *Geochemical Journal*, **31**, 17–36.

COCHERIE, A., ROSSI, PH., FOUILLAC, A. M. & VIDAL, PH. 1994. Crust and mantle contributions to granite genesis: an example from the Variscan batholith of Corsica, France, studied by trace-element and Nd–Sr–O isotope systemetics. *Chemical Geology*, **115**, 173–211.

COLEMAN, R. G. 1989. Continental growth of northwest China. *Tectonics*, **8**, 621–635.

COLLINS, W. J. 1996. Lachlan Fold Belt granitoids: products of three-component mixing. *Transactions of the Royal Society of Edinburgh: Earth Sciences*, **87**, 171–181.

COLLINS, W. J. 1998. Evaluation of petrogenetic models for Lachlan Fold Belt granitoids: implications for crustal architecture and tectonic models.

Australian Journal of Earth Sciences, **45**, 483–500.

COURTILLOT, V., BESSE, J., VANDAMME, D., MONTIGNY, R., JAEGER, J. J. & CAPPETTA, H. 1986. Deccan flood basalts at the Cretaceous/Tertiary boundary? *Earth and Planetary Science Letters*, **80**, 361–374.

DARBYSHIRE, D. P. F. & SEWELL, R. J. 1997. Nd and Sr isotope geochemistry of plutonic rocks from Hong Kong: implications for granite petrogenesis, regional structure and crustal evolution. *Chemical Geology*, **143**, 81–93.

DARBYSHIRE, D. P. F. & SHEPHERD, T. J. 1994. Nd and Sr isotope constraints on the origin of the Cornubian batholith, SW England. *Journal of the Geological Society of London*, **151**, 795–802.

DEMPSEY, C. S., HALLIDAY, A. N. & MEIGHAN, I. G. 1990. Combined Sm–Nd and Rb–Sr isotope systematics in the Donegal granitoids and their petrogenic implications. *Geological Magazine*, **127**, 75–80.

DENIEL, C., VIDAL, P., FERNANDEZ, A., LE FORT, P. & PEUCAT, J. J. 1987. Isotopic study of the Manaslu granite (Himalaya, Nepal): inferences on the age and source of Himalayan leucogranites. *Contributions to Mineralogy and Petrology*, **96**, 78–92.

DEPAOLO, D. J., LINN, A. M. & SCHUBERT, G. (1991). The continental crustal age distribution: methods of determining mantle separation ages from Sm–Nd isotopic data and application to the Southeastern United States. *Journal of Geophysical Research*, **96**, 2071–2088.

DIAS, G. & LETERRIER, J. 1994. The genesis of felsic–mafic plutonic associations: a Sr and Nd isotopic study of the Hercynian Braga granitoid Massif (Northern Portugal). *Lithos*, **32**, 207–223.

DOBRETSOV, N. L., BERZIN, N. A. & BUSLOV, M. M. 1995. Opening and tectonic evolution of the Paleo-Asian ocean. *International Geology Review*, **37**, 335–360.

DOBRETSOV, N. L., BUSLOV, M. M. & UCHIO, YU. 2004. Fragments of oceanic islands in accretion-collision areas of Gorny Altai and Salair, southern Siberia, Russia: early stages of continental crust growth of the Siberian continent in Vendian–Early Cambrian time. *Journal of Asian Earth Sciences* (in press).

DOWNES, H. & DUTHOU, J. L. 1988. Isotopic and trace-element arguments for the lower-crustal origin of Hercynian granitoids and Pre-Hercynian orthogeneisses, Massif Central (France). *Chemical Geology*, **68**, 291–308.

DOWNES, H., SHAW, A., WILLIAMSON, B. J. & THIRLWALL, M. F. 1997. Sr, Nd and Pb isotope geochemistry of the Hercynian granodiorites and monzogranites, Massif Central, France. *Chemical Geology*, **136**, 99–122.

DUNCAN, R. A., HOOPER, P. R., REHACEK, J., MARSH, J. S. & DUNCAN, A. R. 1997. The timing and duration of the Karoo igneous event, southern Gondwana. *Journal of Geophysical Research*, **102**, 18 127–18 139.

EBERZ, G. W., NICHOLLS, I. A., MASS, R., MCCULLOCH, M. T. & WHITEFORD, D. J. 1990. The Nd- and Sr-isotopic composition of I-type microgranitoid enclaves and their host rocks from the Swifts Creek pluton, southeast Australia. *Chemical Geology*, **85**, 119–134.

ENCARNACION, J., FLEMING, T. H., ELLIOT, D. H. & EALES, H. V. 1996. Synchronous emplacement of Ferrar and Karoo dolerites and the early breakup of Gondwana. *Geology*, **24**, 535–538.

FANG, W. C. 1992. *Granitoids and Mineralisation in Jilin Province*. Jilin Science and Technology Publishing House, Jilin, China, 271 pp.

FENG, Y., COLEMAN, R. G. & TILTON, G. 1989. Tectonic evolution of the West Junggar region, Xinjiang, China. *Tectonics*, **8**, 729–752.

FORSTER, H.-J., TISCHENDORF, G., TRUMBULL, R. B. & GOTTESMANN, B. 1999. Late-collisional granites in the Variscan Erzgebirge, Germany. *Journal of Petrology*, **40**, 1613–1645.

FROST, C. D. & O'NIONS, R. K. 1985. Caledonian magma genesis and crustal recycling. *Journal of Petrology*, **26**, 515–544.

GAZIS, C. A., BLUM, J. D., CHAMBERLAIN, C. P. & POAGE, M. 1998. Isotope systematics of granites and gneisses of the Nanga Parbat Massif, Pakistan Himalaya. *American Journal of Science*, **298**, 673–698.

GILDER, S. A., GILL, J. *et al.* 1996. Isotopic and paleomagmatic constraints on the Mesozoic tectonic evolution of south China. *Journal of Geophysical Research*, **101**, 16 137–16 154.

GORDIENKO, I. V. 1987. Paleozoic magmatism and geodynamics of Central-Asian foldbelt. *Nauka*, Moscow, p. 238.

HALLIDAY, A. N. 1984. Coupled Sm–Nd and U–Pb systematics in late Caledonian granites and the basement under northern Britain. *Nature*, **307**, 229–233.

HAMILTON, P. J., O'NIONS, R. K. & PANKHURST, R. J. 1980. Isotopic evidence for the provenance of some Caledonian granites. *Nature*, **287**, 279–284.

HAN, B. F., WANG, S. G., JAHN, B. M., HONG, D. W., KAGAMI, H. & SUN, Y. L. 1997. Depleted-mantle magma source for the Ulungur River A-type granites from north Xinjiang, China: geochemistry and Nd–Sr isotopic evidence, and implication for Phanerozoic crustal growth. *Chemical Geology*, **138**, 135–159.

HARRISON, T. M., GROVE, M., McKEEGAN, K. D., COATH, C. D., LOVERA, O. M. & LE FORT, P. 1999. Origin and episodic emplacement of the Manaslu intrusive complex, Central Himalaya. *Journal of Petrology*, **40**, 3–19.

HEINHORST, J., LEHMANN, B., ERMOLOV, P., SERYKII, V. & ZHURUTIN, S. 2000. Paleozoic crustal growth and metallogeny of Central Asia: evidence from magmatic–hydrothermal ore systems of Central Kazakhstan. *Tectonophysics*, **328**, 69–87.

HENSEL, H.-D., McCULLOCH, M. T. & CHAPPELL, B. W. 1985. The New England Batholith: constraints on its derivation from Nd and Sr isotopic studies of granitoids and country rocks. *Geochimica et Cosmochimica Acta*, **49**, 369–384.

HONG, D. W., HUANG, H. Z., XIAO, Y. J., XU, H. M. & JIN, M. Y. 1995. Permian alkaline granites in central Inner Mongolia and their geodynamic significance. *Acta Geologica Sinica*, **8**, 27–39.

HONG, D. W., WANG, S. G., HAN, B. F. & JIN, M. Y. 1996. Post-orogenic alkaline granites from China and comparisons with anorogenic alkaline granites elsewhere. *Journal of SE Asian Earth Sciences*, **13**, 13–27.

HSÜ, K. J. 1989. Relict back-arc basins: principles of recognition and possible new examples from China. *In*: KLEINSPEHN, K. L. & PAOLA, C. (eds). *New Perspectives in Basin Analysis*. Springer-Verlag, New York, 245–263.

HSÜ, K. J., LI, J. L., CHEN, H. H., WANG, Q. C., SUN, S. & SENGÖR, A. M.C. 1990. Tectonics of South China: key to understanding west Pacific geology. *Tectonophysics*, **183**, 9–39.

HU, A. Q., JAHN, B. M., ZHANG, G. & ZHANG, Q. 2000. Crustal evolution and Phanerozoic crustal growth in Northern Xinjiang: Nd–Sr isotopic evidence. Part I: Isotopic characterization of basement rocks. *Tectonophysics*, **328**, 15–51.

HUANG, X., SUN, S. H., DePAOLO, D. J. & WU, K. L. 1986. Nd–Sr isotope study of Cretaceous magmatic rocks from Fujian province. *Acta Petrologica Sinica*, **2**, 50–63.

IRBER, W. 1999. The lanthanide tetrad effect and its correlation with K/Rb, Eu/Eu*, Sr/Eu, Y/Ho, and Zr/Hf of evolving peraluminous granite suites. *Geochimica et Cosmochimica Acta*, **63**, 489–508.

INGER, S. & HARRIS, N. 1993. Geochemical constraints on leucogranite magmatism in the Langtang Valley, Nepal Himalaya. *Journal of Petrology*, **34**, 345–368.

JAHN, B. M., CAPDEVILA, R., LIU, D. Y., VERNON, A. & BADARCH, G. 2004. Sources of Phanerozoic granitoids in Mongolia: geochemical and Nd isotopic evidence, and implications for Phanerozoic crustal growth. *Journal of Asian Earth Sciences* (in press).

JAHN, B. M., CHEN, P. Y. & YEN, T. P. 1976. Rb–Sr ages of granitic rocks in southeastern China and their tectonic significance. *Bulletin of the Geological Society of America*, **86**, 763–776.

JAHN, B. M., MARTINEAU, F., PEUCAT, J. J. & CORNICHET, J. 1986, Geochronology of the Tananao Schist complex, Taiwan, and its regional tectonic significance. *Tectonophysics*, **125**, 103–124.

JAHN, B. M., WU, F. Y. & CHEN, B. 2000a. Granitoids of the Central Asian Orogenic Belt and continental growth in the Phanerozoic. *Transactions of the Royal Society of Edinburgh: Earth Sciences*, **91**, 181–193.

JAHN, B M., WU, F. Y. & CHEN, B. 2000b. Massive granitoid generation in Central Asia: Nd isotope evidence and implication for continental growth in the Phanerozoic. *Episodes*, **23**, 82–92.

JAHN, B. M., WU, F. Y., CAPDEVILA, R., MARTINEAU, F., ZHAO, Z. H. & WANG, Y. X. 2001. Highly evolved juvenile granites with tetrad REE patterns: the Woduhe and Baerzhe granites from the Great Xing'an Mountain in NE China. *Lithos*, **59**, 171–198.

JAHN, B. M., ZHOU, X. H. & LI, J. L. 1990. Formation and tectonic evolution of southeastern China and

Taiwan: isotopic and geochemical constraints. *Tectonophysics*, **183**, 145–160.

KEAY, S., COLLINS, W. J. & McCULLOCH, M. T. 1997. A three-component Sr–Nd isotopic mixing model for granitoid genesis, Lachlan fold belt, eastern Australia. *Geology*, **25**, 307–310.

KIM, C. K. & KIM, G. S. 1997. Petrogenesis of the early Tertiary A-type Namsan alkali granite in the Kyongsang Basin, Korea. *Geoscience Journal*, **1**, 99–107.

KING, P. L., WHITE, A. J. R., CHAPPELL, B. W. & ALLEN, C. M. 1997. Characterization and origin of aluminous A-type granites from the Lachlan Fold Belt, Southeastern Australia. *Journal of Petrology*, **38**, 371–391.

KOVALENKO, V. I., TSAREVA, G. M., YARMOLYUK, V. V., TROISKY, V. A., FARMER, G. L. & CHERNISHEV, I. V. 1992. Sr–Nd isotopic compositions and the age of rare-metal peralkaline granitoids from western Mongolia. *Dokladi Akademiia Nauka*, **237**, 570–574 (in Russian).

KOVALENKO, V., YARMOLYUK, V. & BOGATIKOV, O. 1995. Magmatism, geodynamics, and metallogeny of Central Asia. Miko Commercial Herald Publishers, Moscow, 272 pp.

KOVALENKO, V. I., YARMOLYUK, V. V., KOVACH, V. P., KOTOV, A. B., KOZAKOV, I. K. & SAL'NIKOVA, E. B. 1996. Sources of Phanerozoic granitoids in Central Asia: Sm–Nd isotope data. *Geochemistry International*, **34**, 628–640.

KOVALENKO, V. I., YARMOLYUK, V. V. *et al.* 2004. Isotope provinces, mechanisms of generation and sources of the continental crust in the Central Asian Mobile Belt: geological and isotopic evidence. *Journal of Asian Earth Sciences* (in press).

KOZAKOV, I. K., KOTOV, A. B., KOVACH, V. P. & SAL'NIKOVA, E. B. 1997. Crustal growth in the geological evolution of the Baidarik block, central Mongolia: evidence from Sm–Nd isotopic systematics. *Petrology*, **5**, 201–207 (translated from *Petrologiya*, **5**, 227–235).

KRUK, N. N., TITOV, A. V., PONOMAREVA, A. P., SHOKALSKII, S. P., VLADIMIROV, A. G. & RUDNEV, S. N. 1998. The internal structure and petrology of the Aya syenite–granosyenite–granite series (Gorny Altai). *Russian Geology & Geophysics*, **39**, 1075–1087.

KRUK, N. N., RUDNEV, S. N., VYSTAVNOI, S. A. & PELEESKIY, S. V. 2001. Sr–Nd isotopic systematics of granitoids and evolution of continental crust of the western part of the Altai–Sayan Fold Region. Abst., *In*: DOBRETSOV, N. L., JAHN, B. M., VLADIMIROV, A. G. (eds), IGCP-420 Workshop III, Abstract volume, 68–72.

KUZMIN, M. I. & ANTIPIN, V. S. 1993. Geochemical types of granitoids of the Mongol–Okhotsk belt and their geodynamic settings. *Chinese Journal of Geochemistry*, **12**, 110–117.

LAN, C, Y., LEE, T., ZHOU, X. H. & KWON, S. T. 1995a. Nd isotopic study of Precambrian basement of South Korea: evidence for early Archean crust? *Geology*, **23**, 249–252.

LAN, C. Y., LEE T., JAHN, B. M. & YUI, T. F. 1995b. Taiwan as a witness of repeated mantle inputs – Sr–Nd–O isotopic geochemistry of Taiwan granitoids and metapelites. *Chemical Geology*, **124**, 287–303.

LEE, J. I., JWA, Y. J., PARK, C. H., LEE, M. J., MOUTTE, J. & KAGAMI, H. 1999. Sr and Nd isotopic compositions of late Paleozoic Youngju and Andong granites in the northeastern Yeongnam massif, Korea. *Geochemical Journal*, **33**, 153–165.

LE MAITRE, R. W., BATEMAN, P. *et al.* 1989. *A Classification of Igneous Rocks and Glossary of Terms.* Blackwell, Oxford.

LIEW, T. C., FINGER, F. & HOCK, V. 1989. The Moldanubian granitoid plutons of Austria: chemical and isotopic studies bearing on their environmental setting. *Chemical Geology*, **76**, 41–55.

LIEW, T. C. & HOFMANN, A. W. 1988. Precambrian crustal components, plutonic association, plate environment of the Hercynian fold belt of central Europe: indications from a Nd and Sr isotopic study. *Contributions to Mineralogy and Petrology*, **98**, 129–138.

LITVINOVSKY, B. A. & ZANVILEVICH, A. N. 1998. Compositional trends of silicic and mafic magmas formed in the course of evolution of the Mongolian–Transbaikalian Mobile Belt. *Russian Geology*, **39**, 155–180.

LITVINOVSKY, B. A., ZANVILEVICH, A. N., ALAKSHIN, A. M., & Podladchikov, YU. 1992. The Angara–Vitim batholith – the largest granitoid pluton. *Novosibirsk, Nauka*, 141 pp.

LITVINOVSKY, B. A., JAHN, B. M. & ZANVILEVICH, A. N. 2002a. Crystal fractionation as the dominant mechanism in the petrogenesis of alkali gabbrosyenite series: the Oshurkovo intrusive complex, Transbaikalia, Russia. *Lithos*, **64**, 97–130.

LITVINOVSKY, B. A., JAHN, B. M., ZANVILEVICH, A. N., SAUNDERS, A., POULAIN, S., KUZMIN, D. V., REICHOW, M. K. & TITOV, A. V. 2002b. Petrogenesis of syenite–granite suites from the Bryansky complex (Transbaikalia, Russia): implications for the origin of A-type granitoid magmas. *Chemical Geology*, **189**, 105–133.

LITVINOVSKY, B. A., ZANVILEVICH, A. N. & WICKHAM, S. M. 1994. Angara–Vitim batholith, Transbaikalia: structure, petrology and petrogenesis. *Russian Geology and Geophysics*, **35**, 190–203.

McCULLOCH, M. T. & CHAPPELL, B. W. 1982. Nd isotopic characteristics of S- and I-type granites. *Earth and Planetary Science Letters*, **58**, 51–64.

MARTIN, H., BONIN, B., CAPDEVILA, R., JAHN, B. M., LAMAYRE, J. & WANG, Y. 1994. The Kuiqi peralkaline granitic complex (SE China): petrology and geochemistry. *Journal of Petrology*, **35**, 983–1015.

MASS, R., NICHOLLS, I. A. & LEGG, C. 1997. Igneous and metamorphic enclaves in the S-type Deddick granodiorite, Lachlan Fold Belt, SE Australia: petrographic, geochemical and Nd–Sr isotopic evidence for crustal melting and magma mixing. *Journal of Petrology*, **38**, 815–841.

MASUDA, A. & AKAGI, T. 1990. Lanthanide tetrad effect observed in leucogranites from China. *Geochemical Journal*, **23**, 245–253.

MASUDA, A., KAWAKAMI, O., DOHMOTO, Y. & TAKE-NAKA, T. 1987. Lanthanide tetrad effects in nature: two mutually opposite types, W and M. *Geochemical Journal*, **21**, 119–124.

MORENO-VENTAS, I., ROGERS, G. & CASTRO, A. 1995. The role of hybridization in the genesis of Hercynian granitoids in the Gredos Massif, Spain: inferences from Sr–Nd isotopes. *Contributions to Mineralogy and Petrology*, **120**, 137–149.

PIN, C. & DUTHOU, J. L. 1990. Sources of Hercynian granitoids from the French Massif Central: Inferences from Nd isotopes and consequences for crustal evolution. *Chemical Geology*, **83**, 281–296.

PARFENOV, L. M., POPEKO, L. I. & TOMURTOGOO, O. 1999. The problems of tectonics of the Mongol–Okhotsk orogenic belt. *Pacific Oceanic Geology*, **18**, 24–43.

PATCHETT, P. J. & CHASE, C. G. 2002. Role of transform continental margins in major crustal growth episodes. *Geology*, **30**, 39–42.

POITRASSON, F., PIN, C., DUTHOU, J. L. & PLATEVOET, B. 1994. Aluminous subsolvus anorogenic granite genesis in the light of Nd isotopic heterogeneity. *Chemical Geology*, **112**, 199–219.

POITRASSON, F., DUTHOU, J. L. & PIN, C. 1995. The relationship between petrology and Nd isotopes as evidence for contrasting anorogenic granite genesis: example of the Corsican Province (SE France). *Journal of Petrology*, **36**, 1251–1274.

RENNE, P. & BASU, A. R. 1991. Rapid eruption of the Siberian traps flood basalts at the Permo-Triassic boundary. *Science*, **253**, 176–179.

RENNE, P., ZICHAO, Z., RICHARDS, M. A., BLACK, M. T. & BASU, A. 1995. Synchrony and causal relations between Permian–Triassic boundary crises and Siberian flood volcanis. *Science*, **269**, 1413–1415.

REYMER, A. & SCHUBERT, G. 1984. Phanerozoic addition rates to the continental crust and crustal growth. *Tectonics*, **3**, 63–77.

REYMER, A. & SCHUBERT, G. 1986. Rapid growth of some major segments of continental crust. *Geology*, **14**, 299–302.

RICKWOOD, P. C. 1989. Boundary lines within petrologic diagrams which use oxides of major and minor elements. *Lithos*, **22**, 247–263.

RUDNICK, R. L. 1995. Making continental crust. *Nature*, **378**, 571–578.

RUZHENTSEV, S. V. & MOSSAKOVSKIY, A. A. 1996. Geodynamics and tectonic evolution of the Central Asian Paleozoic structures as the result of the interaction between the Pacific and Indo-Atlantic segments of the earth. *Geotectonics*, **29**, 211–311.

SAMSON, S. D. & PATCHETT, P. J. 1991. The Canadian Cordillera as a modern analogue of Proterozoic crustal growth. *Australian Journal of Earth Sciences*, **38**, 595–611.

SAMSON, S. D., MCCLELLAND, W. C., PATCHETT, P. J., GEHRELS, G. E. & ANDERSON, R. G. 1989. Evidence from neodymium isotopes for mantle contributions to Phanerozoic crustal genesis in the Canadian Cordillera. *Nature*, **337**, 705–709.

SENGÖR, A. M.C. & NATAL'IN, B. A. 1996. Turkin-type orogeny and its role in the making of the continental crust. *Annual Review of Earth and Planetary Sciences*, **24**, 263–337.

SENGÖR, A. M.C., NATAL'IN, B. A. & BURTMAN, V. S. 1993. Evolution of the Altaid tectonic collage and Paleozoic crustal growth in Eurasia. *Nature*, **364**, 299–307.

SIEBEL, W., HOHNDORF, A. & WENDT, I. 1995. Origin of late Variscan granitoids from NE Bavaria, Germany, exemplified by REE and Nd isotope systematics. *Chemical Geology*, **125**, 249–270.

SKJERLIE, K. P. 1992. Petrogenesis and significance of late Caledonian granitoid magmatism in western Norway. *Contributions to Mineralogy and Petrology*, **110**, 473–487.

STEIN, M. & GOLDSTEIN, S. L. 1996. From plume head to continental lithosphere in the Arabian–Nubian shield. *Nature*, **382**, 773–778.

TANG, K. D. 1990. Tectonic development of Paleozoic foldbelts at the northern margin of the Sino-Korean craton. *Tectonics*, **9**, 249–260.

TOMMASINI, S., POLI, G. & HALLIDAY, A. N. 1995. The role of sediment subduction and crustal growth in Hercynian plutonism: isotopic and trace element evidence from the Sardinia-Corsica batholith. *Journal of Petrology*, **36**, 1305–1332.

TURPIN, L., CUNEY, M., FRIEDRICH, M., BOUCHEZ, J-L. & AUBERTIN, M. 1990. Meta-igneous origin of Hercynian peraluminous granites in N.W. French Massif Central: implications for crustal history reconstructions. *Contributions to Mineralogy and Petrology*, **104**, 163–172.

VIDAL, P., BERNARD-GRIFFITHS, J., COCHERIE, A., LE FORT, P., PEUCAT, J. J. & SHEPPARD, S. 1984. Geochemical comparison between Himalayan and Hercynian leucogranites. *Physics of Earth and Planetary Interiors*, **35**, 179–190.

VLADIMIROV, A. G., PONOMAREVA, A. P. et al. 1997. Late Paleozoic–Early Mesozoic granitoid magmatism in Altai. *Russian Geology and Geophysics*, **38**, 755–770.

VLADIMIROV, A. G., BABIN, G. A., RUDNEV, S. N. & KRUK, N. N. 2001. Geology, magmatism and metamorphism of the western part of Altai-Sayan Fold Region. *IGCP-420 Workshop III Excursion Guidebook*. 139 pp. Institute of Geology, RAS-SB, Novosibirsk, Russia.

WANG, S. G., HAN, B. F., HONG, D. W., XU, B. L. & SUN, Y. L. 1994. Geochemistry and tectonic significance of alkali granites along Ulungur River, Xinjiang (in Chinese with English abstract). *Scientia Geologica Sinica*, **29**, 373–383.

WICKHAM, S. M., LITVINOVSKY, B. A., ZANVILEVICH, A. N. & BINDEMAN, I. N. 1995. Geochemical evolution of Phanerozoic magmatism in Transbaikalia, East Asia: a key constraint on the origin of K-rich silicic magmas and the process of cratonization. *Journal of Geophysical Research*, **100**, 15 641–15 654.

WICKHAM, S. M., ALBERTS, A. D., ZANVILEVICH, A. N., LITVINOVSKY, B. A., BINDEMAN, I. N. & SCHUBLE, E. A. 1996. A stable isotope study of anorogenic magmatism in East Central Asia. *Journal of Petrology*, **37**, 1063–1095.

WILDE, S. A., DORSETT-BAIN, H. L. & LIU, J. L. 1997. The identification of a Late Pan-African granulite facies event in Northeastern China: SHRIMP U–Pb zircon dating of the MashanGroup at Liu Mao, Heilongjiang Province, China. *Proceedings of the 30th IGC: Precambrian Geology and Metamorphic Petrology*, **17**, 59–74.

WILDE, S. A., ZHANG, X. Z. & WU, F. Y. 2000. Extension of a newly-identified 500 Ma metamorphic terrain in NE China: further U–Pb SHRIMP dating of the Mashan Complex, Heilongjiang Province, China. *Tectonophysics*, **328**, 115–130.

WILLIAMSON, B. J., DOWNES, H. & THIRLWALL, M. F. 1992. The relationship between crustal magmatic underplating and granite genesis: an example from the Velay granite complex, Massif Central, France. *Transactions of the Royal Society of Edinburgh: Earth Sciences*, **83**, 235–245.

WINDLEY, B. 1993. Proterozoic anorogenic magmatism and its orogenic connections. *Journal of the Geological Society of London*, **150**, 39–50.

WINDLEY, B. 1995. *The Evolving Continents*, Third edn, John Wiley & Sons, Ltd, Chichester, 526 pp.

WU, F. Y., JAHN, B. M., WILDE, S. & SUN, D. Y. 2000. Phanerozoic crustal growth: Sr–Nd isotopic evidence from the granites in northeastern China. *Tectonophysics*, **328**, 89–113.

WU, F. Y., SUN, D. Y., LI, H. M. & WANG, X. L. 2001. The nature of basement beneath the Songliao Basin in NE China: geochemical and isotopic constraints. *Physics and Chemistry of the Earth (Part A)*, **26**, 793–803.

WU, F. Y., SUN, D. Y., LI, H. M., JAHN, B. M. & WILDE, S. 2002. A-type granites in Northeastern China: age and geochemical constraints on their petrogenesis. *Chemical Geology*, **187**, 143–173.

WU, F. Y., JAHN, B. M. *et al.* 2003a. Highly fractionated I-type granites in NE China (I): geochronology and petrogenesis. *Lithos*, **66**, 241–273.

WU, F. Y., JAHN, B. M. *et al.* 2003b. Highly fractionated I-type granites in NE China (II): isotopic geochemistry and implications for crustal growth in the Phanerozoic. *Lithos*, **67**, 191–204.

WU, Q. F. 1987. The Junggar terrane and its significance in the tectonic evolution of the Kazakhstan Plate. *In*: *Plate Tectonics of Northern China*, No. 2. *Geological Publishing House*, Beijing. pp. 29–38 (in Chinese with English abstract).

YARMOLYUK, V. V., KOVALENKO, V. I., KOTOV, A. B. & SAL'NIKOVA, E. B. 1997. The Angara–Vitim Batholith: on the problem of batholith geodynamics in the Central Asian Foldbelt. *Geotectonics*, **31**, 345–358.

YARMOLYUK, V. V., LITVINOVSKY, B. A. *et al.* 2001. Formation stages and sources of the peralkaline granitoid magmatism of the northern Mongolia–Transbaikalia rift belt during the Permian and Triassic. *Petrology*, **9**, 302–328.

ZANVILEVICH, A. N., LITVINOVSKY, B. A. & WICKHAM, S. M. 1995. Genesis of alkaline and peralkaline syenite–granite series: the Kharitonovo pluton (Transbaikalia, Russia), *Journal of Geology*, **103**, 127–145.

ZHAO, Z. H. 1993. REE and O–Pb–Sr–Nd isotopic compositions and petrogenesis of the Altai granitoids. *In*: *New Development of Solid Earth Science in Northern Xinjiang*. Science Publishing Company, Beijing, 239–266.

ZHAO, Z. H., BAI, Z. H., XIONG, X. L., MEI, H. J. & WANG, Y. X. 2000. Geochemistry of alkali-rich igneous rocks of northern Xinjiang and its implications for geodynamics. *Acta Geologica Sinica*, **74**, 321–328.

ZHOU, T. X., CHEN, J. F., CHEN, D. G. & LI, X. M. 1995. Geochemical characteristics and genesis of granitoids from Alataw Mountain, Xinjiang, China. *Geochimica*, **24**, 32–42 (in Chinese with English abstract).

ZHOU, X. M. & LI, W. X. 2000. Origin of late Mesozoic igneous rocks in southeastern China: implications for lithospheric subduction and underplating of mafic magmas. *Tectonophysics*, **326**, 269–287.

ZONENSHAIN, L. P., KUZMIN, M. I. & NATAPOV, L. M. 1990. Geology of the USSR: a plate tectonic synthesis. *AGU Geodynamics Series*, **21**, 242 pp.

ZORIN, Yu. A. 1999. Geodynamics of the western part of the Mongolia–Okhotsk collisional belt, Transbaikal region (Russia) and Mongolia. *Tectonophysics*, **306**, 33–56.

Tectonic evolution of Palaeozoic terranes in West Junggar, Xinjiang, NW China

SOLOMON BUCKMAN[1] & JONATHAN C. AITCHISON[2]

[1]*School of Geoscience, Minerals and Civil Engineering, University of South Australia, Mawson Lakes Campus, Mawson Lakes, South Australia 5095, Australia (e-mail: solomon.buckman@unisa.edu.au)*

[2]*Department of Earth Sciences, University of Hong Kong, Pokfulam Road, Hong Kong SAR, China (e-mail: jona@hku.hk)*

Abstract: Nine separate Cambrian to Carboniferous terranes are recognized in West Junggar, northwest China. They were amalgamated as part of the Central Asian Orogenic Belt which records accretion of continental, island-arc and oceanic terranes to Archaean–Proterozoic continental nuclei. Tangbale, Kekesayi, Ebinur and Mayila terranes (Cambrian–Silurian) evolved in intra-oceanic settings and docked, along a series of north-dipping subduction zones, on to the Laba terrane to their south. This southern continent was contiguous with lithosphere of the Kulumudi Ocean to the north. Devonian subduction on the northern edge of this ocean resulted in formation of a continental arc (Toli terrane) and accretionary complex (Kulumudi terrane). The Karamay terrane formed as an accretionary complex during the Carboniferous. The ophiolitic Sartuohai terrane was emplaced as mélange between Kulumudi and Karamay terranes during the Late Carboniferous. Subduction migrated southward, continuing beneath these terranes, resulting in the intrusion of I-type granites into the Toli, Kulumudi, Sartuohai and Karamay terranes. These granites are closely associated with epithermal and porphyry-style gold mineralization. Composite terranes either side of the Kulumudi Ocean collided in the Late Carboniferous, marking the final consolidation of Central Asia. Collision was accompanied by anorogenic granite and diabase dyke intrusion, followed by widespread latest Carboniferous to Permian extension, and subsequently the formation of the Junggar Basin. West Junggar has been further disrupted by Cenozoic strike-slip faulting along Junggar and Dalabute faults.

Central Asian Orogenic Belt (CAOB)

The Central Asian Orogenic Belt (CAOB) represents the products of a considerable amount of continental growth throughout the Palaeozoic. It is composed of distinct Palaeozoic blocks or terranes, and the boundaries between these units are commonly identified by the presence of ophiolitic fragments along suture zones (Coleman 1989). The CAOB grew sporadically via the consumption of multiple ocean basins. Sengör *et al.* (1993) and Sengör & Natal'in (1996) proposed a hypothesis invoking a tectonic evolution of the Altaids, involving the development and southward migration through time of a single, huge accretionary complex containing off-scraped ophiolitic fragments in association with a single, long-lived subduction zone that existed along the southern margin of the Angara Craton. This model is not supported by detailed investigations, which reveal considerable complexity such as might be expected of any orogenic collage. Recent detailed structural studies of ophiolite belts within the region, for example, Mongolia (Buchan *et al.* 2001), confirm the complexity of the region.

The West Junggar region comprises a small portion of the central Altaids in Xinjiang Autonomous Region, northwest China (Fig. 1). Relatively little is known, in Western literature, about the Early Palaeozoic geology and tectonic evolution of this area, apart from regional-scale studies (Coleman 1989; Feng *et al.* 1989; Sengör *et al.* 1993; Zhang *et al.* 1993). The area lies between two major intracontinental mountain ranges, the Tian Shan to the south and the Altai Shan to the north, and incorporates a variety of Palaeozoic subduction complex rocks. The Late Palaeozoic–Mesozoic Junggar Basin is a large intracontinental basin unconformably overlying Palaeozoic rocks of West Junggar. Our investigations of these rocks indicate that the region has a very complex history, involving the accretion of multiple intra-oceanic island arcs and

From: MALPAS, J., FLETCHER, C. J. N., ALI, J. R. & AITCHISON, J. C. (eds) 2004. *Aspects of the Tectonic Evolution of China*. Geological Society, London, Special Publications, **226**, 101–129.
0305-8719/04/$15 © The Geological Society of London 2004.

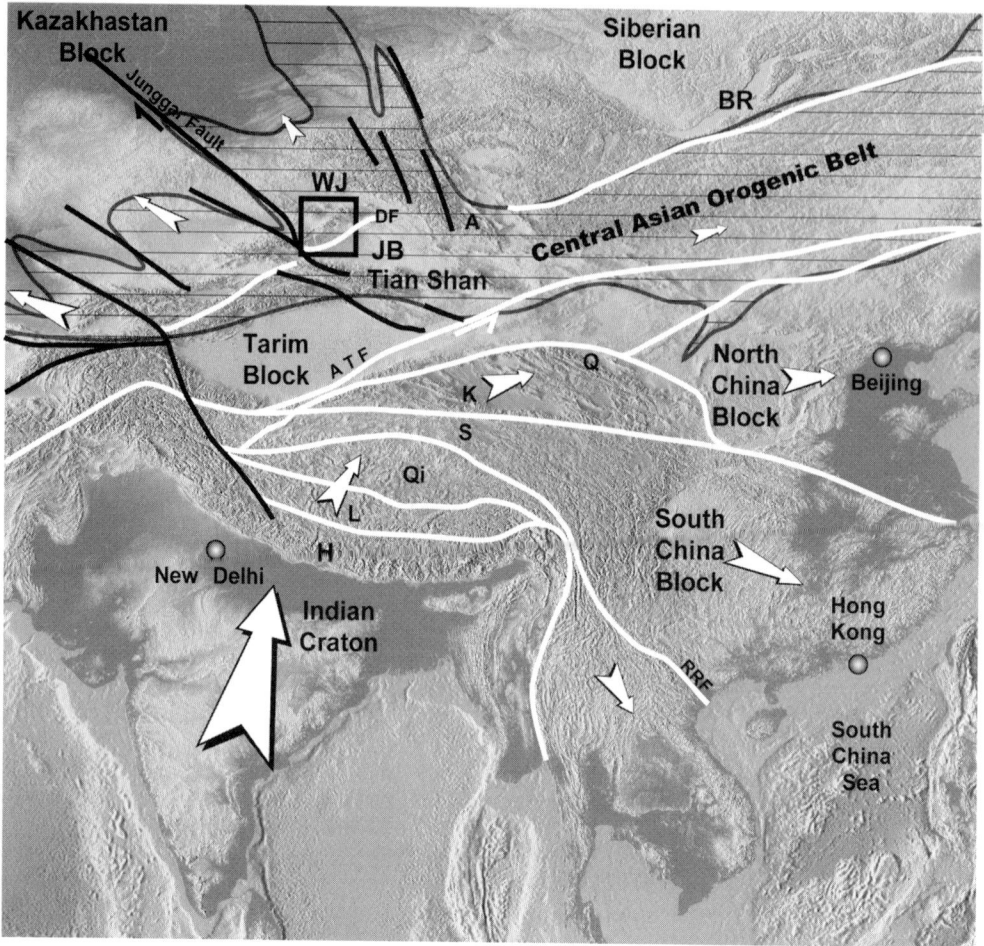

Fig. 1. Location and tectonic setting of West Junggar within the Central Asian Orogenic Belt. Arrows represent relative plate velocities. Information adapted from Coleman (1989), Yin & Nie (1996), and Zhang *et al.* (1984). White lines are major left-lateral faults and red lines are major right-lateral faults associated with the India–Asia collision. Abbreviations include: A, Altay Shan; ATF, Altyn Tagh Fault; BR, Baikal Rift; DF, Dalabute Fault; H, Himalaya frontal thrust system; JB, Junggar Basin; K, Kunlun Block; L, Lhasa Block; Q, Qaidam/Qilian Block; Qi, Qiantang Block; RRF, Red River Fault; S, Songpan Garze Block; TS, Tian Shan; WJ, West Junggar. Topography sourced from NGDC (2001).

continental fragments to one of three main blocks, Siberia, Tarim and Kazakhstan, although it is not apparent as to whether the latter was a single, consolidated continent at this time (Zonenshain *et al.* 1990).

Tectonostratigraphy of West Junggar

The regional framework of West Junggar has been established through 1:100 000-scale geological mapping (1st Regional Geological Survey Team 1985a) and numerous investigations of stratigraphic relations (Feng *et al.* 1983, 1985, 1989; Feng 1986, 1989; Zhou 1986; Fei *et al.* 1987;

Feng & Huo 1987; Gan & Wang 1987; Ma 1987; Zhang & Wei 1989; Wang 1990; Wu & Pan 1991; Gong 1993*a*, *b*; Zhang *et al.* 1993). The regional 1:100 000-scale maps are accurate in their lithological descriptions, but lack critical age, stratigraphic and structural constraints, especially for various ophiolitic rocks and intrusive phases. Formal descriptions of type sections for lithological units within West Junggar are unknown, and some confusion or conflict exists in previous literature as to the exact content of lithological units. In this study we present a compilation and revision of the existing geological data of West Junggar, combined with results of our own mapping and sampling specifically in the areas from Tangbale in the south to Sartuohai in the north (Buckman 2000) (Fig. 2). We adopt a terrane analysis approach (Coney *et al.* 1980; Jones *et al.* 1983; Howell 1989, 1995) and present a summary of the various tectonic units that we have discriminated in the region, in order to develop an integrated model for their evolution through time. We group or divide Palaeozoic units according to their age, composition, cross-cutting or stratigraphic relationships and bounding structures. The ages of various terranes have been determined by utilizing palaeontological data, cross-cutting relations and/or isotopic dates.

Tangbale terrane

Description. A zone of ophiolitic melange, the Tangbale terrane (Guo 1983; Feng *et al.* 1989), crops out in southern West Junggar where it extends east–west for about 100 km (Fig. 2). It includes blocks of peridotite, gabbro and basalt enveloped within a schistose serpentinite-matrix (Huo 1984) (Fig. 3F). The peridotites consist mainly of serpentinized harzburgite and dunite, with minor lherzolite, pyroxenite and podiform chromite. The gabbros include layered differentiates and massive units that outcrop in the

hanging wall of north-dipping thrust faults. Whole-rock geochemistry (Table 1) indicates that they are calc-alkaline in nature and range in composition from gabbroic to highly fractionated plagiogranite. Small volumes of plagiogranite occur as thin (<30 cm) sheets, dykes and pods which intrude the gabbro. An absence of sheeted dykes is notable. Basalt–chert–tuff units incorporated into the serpentinite mélange as blocks have previously been linked to the Tangbale ophiolite (Feng *et al.* 1989; Zhang *et al.* 1993), but there is no evidence that these units were initially stratigraphically continuous, and they appear to be considerably younger than rocks of the Tangbale terrane. For that reason they are considered herein to represent portions of a separate, younger (Kekesayi) terrane.

Age. Titanite extracted from quartz-bearing leucocratic rocks (plagiogranite) at Tangbale has yielded an age of 523.7 ± 7.2 Ma from using Pb/Pb isotope methods (Feng *et al.* 1989). This indicates that the late intrusive phases associated with the formation of the Tangbale ophiolite are Late Cambrian and they represent the oldest dated rocks in West Junggar. The timing of disruption of the Tangbale ophiolite

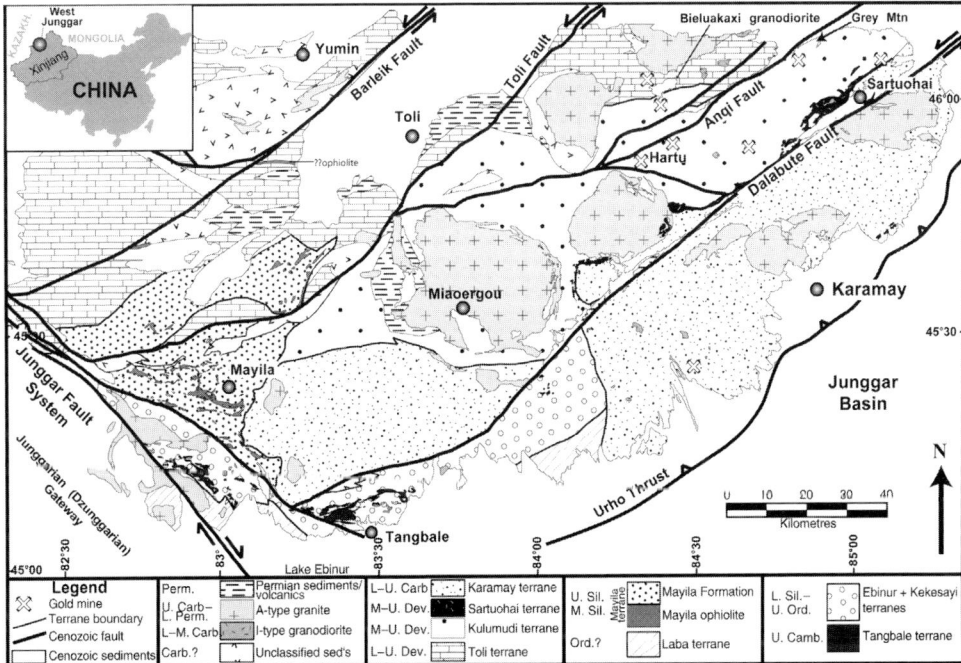

Fig. 2. Regional geology of West Junggar. A terrane map compiled from 1 : 100 000 scale geology maps (1st Regional Geological Survey Team 1985a) and the mapping and field observations of Buckman (2000).

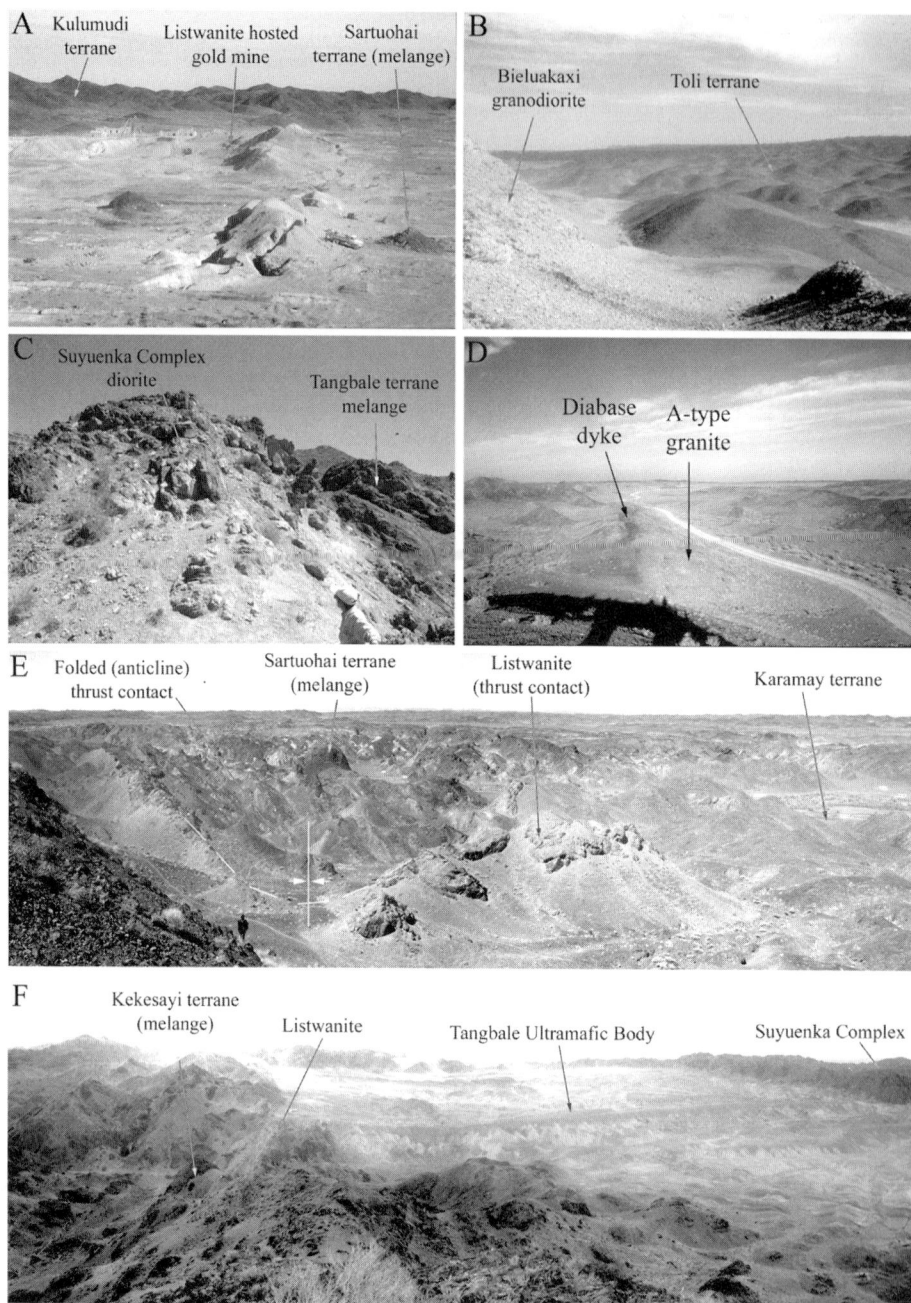

Fig. 3. Images of the various terranes, terrane boundaries and intrusives of West Junggar. (**A**) The Sartuohai gold mine hosted in listwanite alteration is located at the thrust contact between the serpentinite-matrix mélange of the Sartuohai terrane and the Kulumudi terrane; (**B**) the Bieluakaxi granodiorite intruding volcanics and sediments of the Toli terrane; (**C**) a small diorite intrusive of the Suyuenka Complex intruding the serpentinite-matrix mélange of the Tangbale terrane; (**D**) diabase dykes intruding the large A-type granites at Sartuohai; (**E**) the thrust contact between the Sartuohai terrane ophiolitic mélange and the Karamay terrane, marked by the presence of orange, listwanite alteration. The thrust contact has been folded to form an inclined anticline, plunging into the background (south); (**F**) the Tangbale terrane (ophiolite) is faulted against a mélange of chert and basalt of the Kulumudi terrane, which have both been intruded and overlain by igneous rocks of the Suyuenka Complex.

Table 1. *Whole-rock and trace-element analyses for intrusive phases from the Tangbale terrane, Suyuenka Complex and a post-orogenic granitoid from the Sartuohai region (Buckman 2000)*

Sample	TA159	TA161	TA144	TA130	TA70	TA5A	TA39	TA18	TA76	TA57	TA83	TA86	TA82	TA38	SA40
Terrane	T	T	T	T	T	T	T	T	T	T	S	S	S	T	—
Rock	Gab.	Gab.	Gab.	Gab.	Gab.	Dior.	Dior.	Dior.	Dior.	Dior.	Dior.	Dior.	Dior.	Plg gr	Gran.
SiO$_2$	43.23	45.79	50	51.44	55.92	56.24	56.43	56.44	57.69	57.97	59	59.4	68.3	72.5	80.59
TiO$_2$	0.65	0.08	0.9	0.98	0.74	0.7	0.77	0.31	0.37	0.67	0.68	0.56	0.11	0.22	0.07
Al$_2$O$_3$	17.66	23.46	14.64	15.54	17.02	15.63	17.04	14.9	18.23	15.74	16.7		17.9	13	11.57
Fe$_2$O$_3$	1.5	3.13	12.63	9.14	5.43	6.08	6.35	6.65	4.96	6.46	5.47	4.84	1.31	3.02	0.76
MnO	0.03	0.08	0.2	0.13	0.08	0.1	0.09	0.11	0.14	0.12	0.09	0.09	0.03	0.06	0.03
MgO	7.79	5.74	5.01	8.56	3.27	5.61	4.88	7.27	2.57	6.18	4.01	2.97	0.31	0.73	0.14
CaO	23.76	12.96	10.73	5.61	4.13	5.64	5.88	6.74	4.1	3.19	4.07	4.22	1.24	4.12	0.8
Na$_2$O	0.05	0.91	3.48	4.68	6.51	3.9	5.66	5.48	6.75	5.45	5.4	5.03	6.54	4.79	0.19
K$_2$O	0.01	3.35	0.85	1.76	2.25	0.97	0.57	0.37	2.28	1.32	0.62	1.38	2.43	0.09	3.69
P$_2$O$_5$	0.21	0.32	0.29	0.22	0.23	0.17	0.17	0.06	0.55	0.18	0.2	0.21	0.18	0.06	0.02
L.O.I.	5.1	4.49	1.26	1.92	4.37	4.89	2.14	1.64	2.33	2.69	3.67	4.28	1.57	1.31	—
Total	99.81	99.98	99.72	99.58	99.73	99.87	99.69	99.66	99.92	99.33	99.7	99.8	99.7	99.6	99.64
Nb	2	0	2	3	4	2	2	0	3	3	3	3	4	1	3
Zr	42	0	60	65	96	63	64	31	101	82	89	81	103	41	66
Y	18	2	29	15	13	9	9	9	31	12	10	9	4	14	40
Sr	31	386	289	255	267	666	427	167	285	510	1307	768	631	163	13
Rb	1	49	7	26	24	12	7	3	24	13	9	26	52	2	86
Th	1	3	4	0	0	<2	0	0	1	3	3	2	4	1	9
Pb	1	3	0	4	7	5	3	1	3	4	7	5	7	2	6
As	0	0	0	2	55	11	17	0	9	33	23	9	10	3	68
U	1	2	0	0	0	<1	0	0	2	0	0	0	0	0	0
Ga	7	11	16	18	21	18	20	8	14	18	19	19	21	12	18
Zn	0	17	81	74	73	80	69	45	59	73	71	64	38	32	12
Cu	4	17	173	118	63	37	68	97	29	66	52	24	5	28	31
Ni	56	52	30	196	46	132	125	52	9	148	88	43	0	3	5
Cr	74	198	27	646	71	433	296	58	10	410	177	83	3	14	16
Ce	12	10	23	24	28	20	19	11	37	16	26	25	25	8	16
Nd	5	3	19	16	13	10	10	4	20	11	13	13	6	5	8
Ba	7	453	281	405	499	517	164	68	485	386	582	297	747	24	194
V	187	66	488	245	172	171	160	171	160	154	129	106	4	47	2
La	2	0	20	10	10	11	0	8	15	6	12	13	8	2	1
Sc	57	30	48	20	11	24	16	42	21	17	8	13	1	10	3

Abbreviations: T, Tangbale terrane; K, Kulumudi terrane; S, Suyuenka Complex; gabb., gabbro; dior., diorite; plg gr, plagiogranite; gran, granitoid (Sartuohai).

into mélange is uncertain, but must have occurred prior to development of the Suyuenka Complex which both intrudes and overlies the Tangbale terrane.

Interpretation. The depleted nature of the peridotitic (harzburgite) material (Zhu *et al.* 1983; Huo 1984) and the presence of highly fractionated plagiogranites suggest that this residual mantle material has undergone prolonged or multiphase partial melting. The suite of Late Cambrian lithologies present in the ophiolitic mélange of the Tangbale terrane is indicative of formation at a spreading centre in a suprasubduction zone environment, possibly during the earliest stages of subduction initiation (Shervais 2001).

Kekesayi terrane

Description. A disrupted, structurally imbricated, sequence of basalt–chert–tuff rocks crops out in the Tangbale region (Fig. 3F). The entire sequence has a maximum stratigraphic thickness of <100 m but outcrops across-strike for up to 20 km due to structural thickening and multiple thrust repetition of the same sequence. Basalts from the Kekesayi terrane are tholeiitic to alkaline, and have MORB or OIB geochemical signatures (Table 2, Figs. 4, 5 and 6). They are typically overlain by a few metres of red chert that is, in turn, succeeded by fine-grained tuffaceous cherts. Kekesayi terrane lithologies commonly occur as blocks within the serpentinite-matrix mélange associated with the Tangbale terrane, and are locally metamorphosed to amphibolite or blueschist facies. The best-preserved sequences occur on the northern margin of the Tangbale ultramafic body adjacent to the Tangbale River (45° 07.733′ N and 083° 21.974′ E).

Age. Radiolarian assemblages from this terrane are Middle Ordovician (Buckman & Aitchison 2001). Thus a considerable time gap (*c.*50 Ma) exists between the Middle-Late Cambrian formation of the Tangbale ophiolite at 523.7 ± 7.2 Ma (Feng *et al.* 1989) and deposition of cherts in the Kekesayi terrane (Middle Ordovician). If the cherts were deposited conformably on top of the Tangbale ophiolite, little difference might be expected between their ages. As rocks of the Kekesayi terrane can be clearly discriminated from igneous rocks in the Tangbale terrane on the basis of their composition, geological association and age, it is regarded as unlikely that the Kekesayi terrane represents the upper levels of the Tangbale terrane ophiolite. Blueschist metamorphism of elements of the terrane has been

dated at 458.2 ± 2.5 Ma and 470.0 ± 2.3 Ma (Zhang 1997).

Interpretation. The stratigraphic succession preserved within the terrane and a structural style where numerous south-vergent thrusts repeat the thin stratigraphic succession are typical of that seen in an accretionary wedge. The geochemical signatures of tholeiitic to alkaline basalts indicate MORB or OIB affinities (Figs. 4, 5 and 6) suggesting that they represent MORB-type oceanic crust and alkali basalts of intra-oceanic seamounts that originated on the subducting plate. Pillow basalts and hyaloclastic breccia indicate formation at an oceanic spreading centre. Overlying radiolarian cherts and the absence of clastic and tuffaceous material in the lower part of the unit suggest initial deposition in a deep marine environment, isolated from any clastic influx. A gradual influx of tuffaceous material up-section indicates the oceanic crust was slowly advancing towards a volcanic arc. Eventually the influx of tuffaceous and volcaniclastic material overwhelmed the accumulation of biogenic ooze. Most rocks of the Kekesayi terrane were incorporated by off-scraping into a subduction complex (Fig. 7B). As parts of the terrane have been subject to high-pressure/low-temperature metamorphism to blueschist facies, this suggests that underplating was also involved. The blueschists have been dated by (Zhang 1997) and yielded $^{40}Ar/^{39}Ar$ plateau ages of 458.2 ± 2.5 Ma and 470.0 ± 2.3 Ma, indicating that subduction of oceanic material was active throughout the Mid Ordovician – relatively soon after formation of the Kekesayi terrane.

Serpentinite-matrix mélange surrounding the terrane contains large blocks of basalt–chert sequences originating from the Kekesayi terrane. These blocks are highly disrupted and fragmented, but have little internal deformation or alteration, suggesting that the mélange formed as a result of extensional tectonics in a fore-arc region. The serpentinite rose and penetrated pre-existing zones of crustal weakness such as faults, and incorporated fragments of the trapped mantle (Tangbale terrane), accretionary wedge (Kekesayi terrane), and metamorphosed underplated material (blueschist) on its path toward the surface. The mechanisms of mélange development may be analogous to the serpentinite/mud volcanoes recently observed on the seafloor in the actively forming fore-arc region of the Mariana subduction zone (Fryer *et al.* 1985, 1999; Maekawa *et al.* 1993). Mechanisms such as slab-rollback (Stern & Bloomer 1992; Hamilton 1994) and oblique subduction (Howell *et al.* 1980) can create extensional regimes

Table 2. *Whole-rock and trace-element analyses for volcanic rocks from Tangbale*

Sample	TA10	TA84	TA85	TA12	TA34	TA127	TA55	TA50	TA28	TA25	TA75	TA87
Terrane	K	S	S	S	S	S	S	S	S	S	S	S
SiO_2	50.26	51.9	52.1	53.27	55.5	58.04	58.1	60.3	60.9	60.93	61.95	68.5
TiO_2	1.64	0.85	1.08	0.78	0.76	0.71	0.68	0.67	0.62	0.53	0.59	0.13
Al_2O_3	17.24	13.9	15.7	16.95	15.6	17.6	16.1	17.3	18	16.93	15.87	17.9
FeO	7.59	7.18	7.23	7.21	6.46	5.35	5.89	5.06	4.8	4.63	5.51	1.43
MnO	0.13	0.11	0.11	0.11	0.11	0.08	0.1	0.08	0.08	0.09	0.07	0.02
MgO	4.65	7.95	6.72	6.16	6.33	3.18	4.49	3.02	2.67	3.42	1.99	0.41
CaO	5.11	6.03	5.37	3.37	3.97	4	5.23	3.83	3.84	4.52	3.25	0.48
Na_2O	3.96	3.77	4.59	5.67	4.79	4.87	4.4	6.59	5.48	5.21	6.78	6.91
K_2O	4.6	0.36	0.75	0.81	0.56	1.67	1.59	0.14	1.45	1.13	0.47	3.37
P_2O_5	0.24	0.18	0.23	0.18	0.19	0.19	0.21	0.2	0.2	0.16	0.28	0.12
L.O.I.	4.55	7.75	6.08	5.45	5.71	4.28	3.2	2.85	1.95	2.44	3.22	0.72
Total	99.9	99.6	99.6	99.92	99.6	99.6	99.9	99.9	99.8	99.7	99.95	99.7
Nb	14	3	4	2	3	2	3	2	2	4	2	3
Zr	110	87	113	71	86	71	97	84	79	77	84	101
Y	16	13	16	15	14	9	13	8	9	8	27	5
Sr	479	421	665	506	621	949	715	242	843	779	330	402
Rb	39	6	10	8	7	21	24	2	19	11	7	48
Th	<2	1	1	<2	4	<2	3	2	3	3	2	2
Pb	<2	3	6	7	3	4	8	6	8	6	3	6
As	1	10	16	20	20	23	7	16	13	13	5	2
Ga	12	17	18	18	17	20	19	18	20	18	14	20
Zn	64	63	64	78	70	71	73	84	70	68	16	31
Cu	48	10	17	62	66	37	65	49	21	43	51	3
Ni	47	241	156	138	167	52	86	58	36	80	6	4
Cr	150	650	374	424	279	58	228	118	39	88	13	7
Ce	28	27	24	20	18	15	22	19	21	27	39	28
Nd	18	15	15	14	12	9	15	10	8	14	21	13
Ba	649	126	297	661	192	623	673	87	546	344	90	887
V	199	163	164	192	155	130	144	127	104	105	129	5
La	18	6	13	8	11	6	13	7	8	10	17	9
Sc	27	23	18	20	17	13	18	13	8	10	12	1

Abbreviations: T, Tangbale; terrane, K, Kekesayi terrane; and S, Suyuenka Complex.

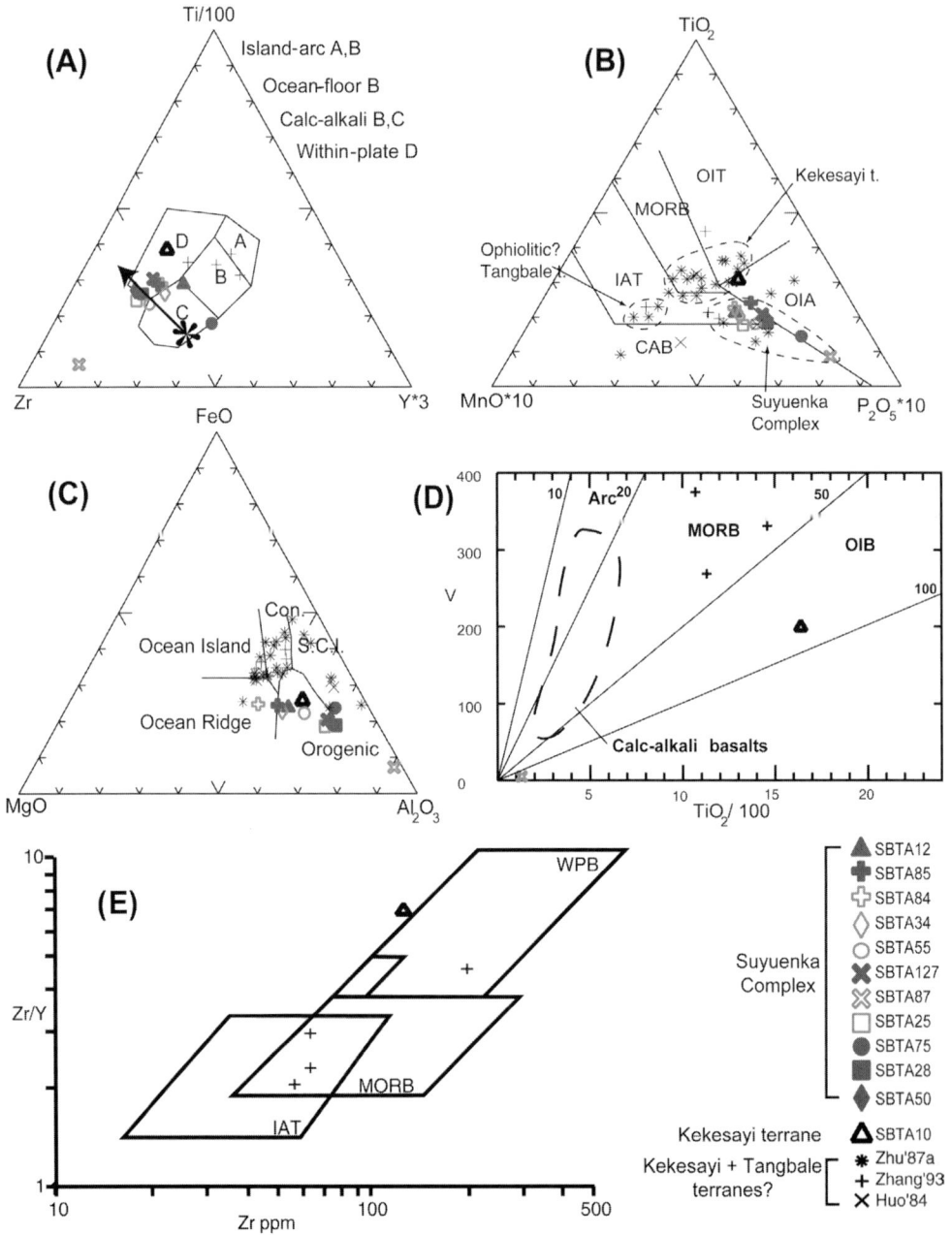

Fig. 4. Tectonic discrimination diagrams for basaltic to andesitic rocks from Tangbale. These include data from Huo (1984), Yang (1993), Zhang *et al.* (1993) and Zhu *et al.* (1987a). Note that in most diagrams the andesitic samples from the Suyuenka Complex have been removed from the plot, as significant shifts from the primary melt appear to have occurred due to fractionation processes (as illustrated in A). (**A**) Zr–Ti–Y diagram (Pearce & Cann 1973). (**B**) MnO–TiO$_2$–P$_2$O$_5$ diagram (Mullen 1983). (**C**) MgO–FeO*–Al$_2$O$_3$ diagram (Pearce *et al.* 1977). (**D**) V/TiO$_2$ diagram (Shervais 1982). (**E**) (Zr/Y)/Zr diagram (Pearce & Norry 1979).

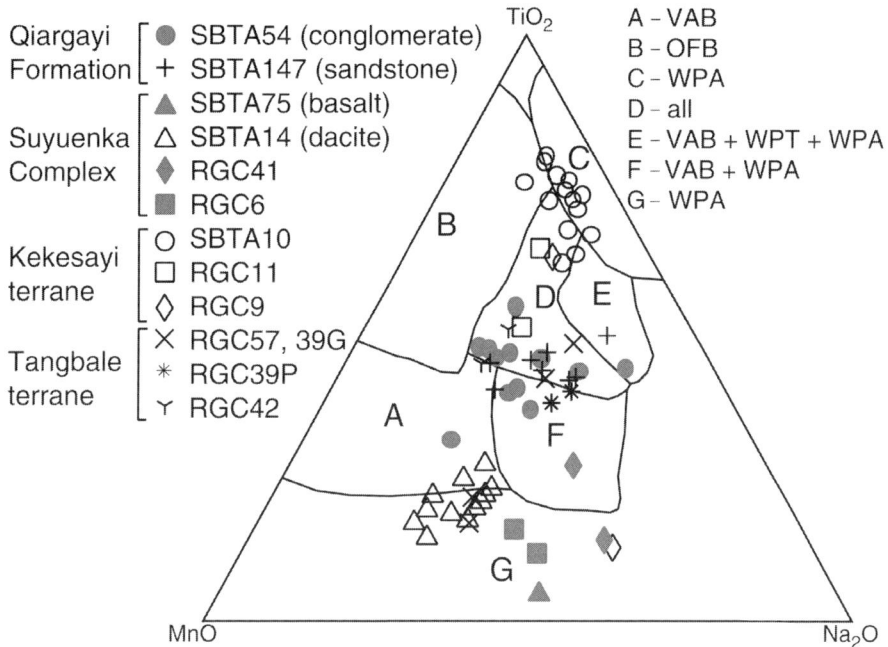

Fig. 5. Tectonic discrimination diagrams for clinopyroxenes from volcanic and sedimentary rocks at Tangbale, including data from Feng *et al.* (1989). Samples include basalts from the Kekesayi terrane (SBTA10, RGC11, RGC9), volcanics from the Suyuenka Complex (SBTA14, SBTA75, RGC6, RGC41, RGC9), and volcaniclastic sediments from the Qiargayi Formation (SBTA54, SBTA147). (**A**) TiO_2–MnO–Na_2O tectonic discrimination triangle (Nisbet & Pearce 1977). (**B**) SiO_2/Al_2O_3 plot which separates subalkaline, alkaline and peralkaline magma types (Nisbet & Pearce 1977). (**C**) SiO_2/Al_2O_3 plot.

directly above the descending oceanic slab, possibly along faults within an accretionary wedge such as the Kekesayi terrane. The ascent of the mélange was sufficiently rapid to prevent re-equilibration of blueschist mineral assemblages at lower metamorphic grades.

Stratigraphic relations indicate that the Tangbale terrane (ophiolite) had amalgamated with the Kekesayi terrane by the Late Ordovician, forming the basement upon which the Suyuenka Complex and Qiargayi Formation formed (Late Ordovician–Early Silurian).

Laba terrane

Description. The Laba terrane is a metamorphosed (greenschist to amphibolite facies) sedimentary/volcanic unit in fault contact against the southern margin of the Ebinur terrane. The lower part of the terrane consists of grey sericite–biotite–quartz schist, tourmaline-bearing chlorite–plagioclase schist and biotite–quartz schist, interbedded with pelitic mudstone, plagioclase–hornblende schist and quartzite. The middle section consist of light-grey sericitized phyllites, which were originally composed of tuffaceous siltstone, tuffaceous sandstone, and acidic fine-grained tuff, interbedded with minor sericite–chlorite phyllite, muddy siltstone, and lithic–dacitic tuff. The upper sequence consists of sericite–chlorite phyllite, interbedded with small amounts of chloritized fine-grained siliceous rock (possibly chert), siltstone, mudstone, sandstone, lithic tuffs, slates and felsic intrusives (1st Regional Geological Survey Team 1985a, b).

Age. Metamorphism must have occurred prior to juxtaposition with the Ebinur terrane (Late Ordovician–Early Silurian), which records no evidence of this event. A Pb isotope age of 471 ± 3 Ma was reported (Wang 1996) for metamorphosed terrigenous clastic rocks from the Tangbale region. Presumably, these rocks are from the Laba terrane; however, the source of

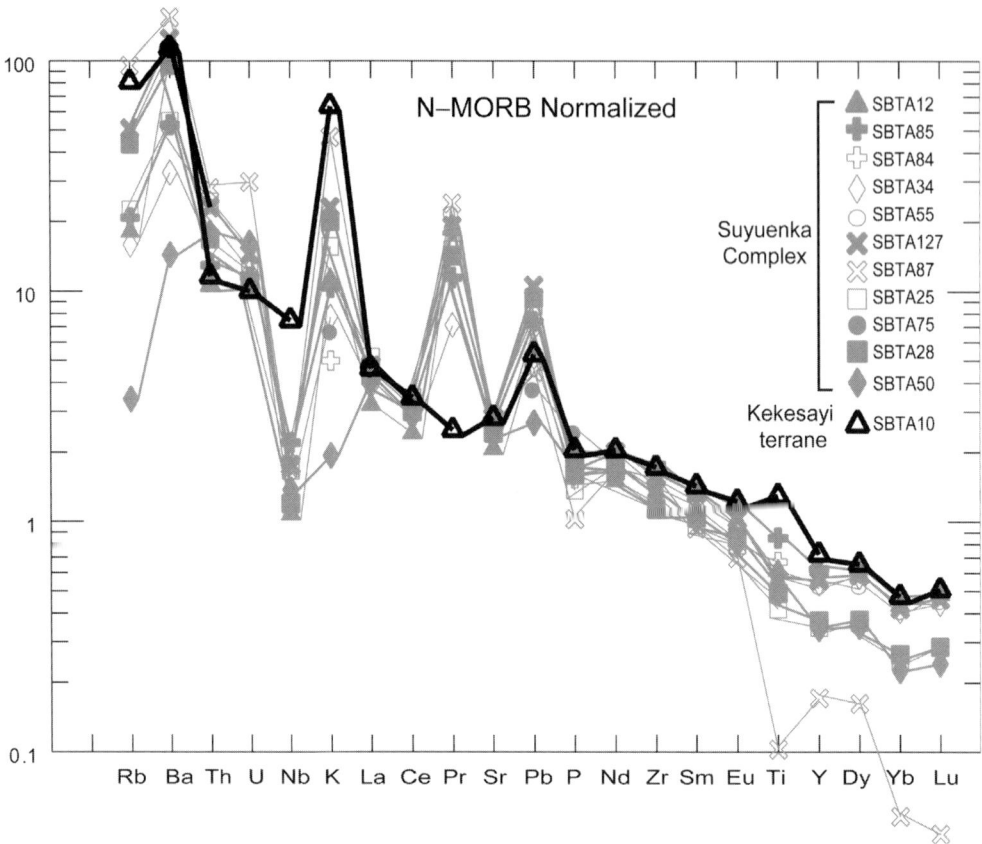

Fig. 6. N-MORB normalized plot (Sun & McDonough 1989) of REE and trace-element abundances for various volcanic rocks at Tangbale. Symbols are the same as for Figure 5.

the original data is unknown and details of the results are undocumented. Zhang *et al.* (1993) point out that an Early Ordovician age is sometimes ascribed to the Laba terrane, but this is merely because they are more deformed than the nearby Middle Ordovician Kekesayi Formation. The contact between the two formations is tectonic, and, based on the presence of more intense deformation and a higher metamorphic grade, we have ascribed terrane status to the original Laba Formation.

Interpretation. The Laba terrane possibly represents a rifted fragment of the trailing edge of the continental margin of either the Tarim or the Kazakhstan block. The pervasive metamorphism observed within the Laba terrane is not observed within any other terrane at Tangbale, and younger intrusive or sedimentary rocks of the Ebinur terrane do not occur within, or overlie, the Laba terrane. This indicates that

these two terranes developed independently, at least until after the deposition of the Lower Silurian Qiargayi Formation. The minimum age of juxtaposition is difficult to constrain, but the fault separating the Laba terrane from the Ebinur terrane is truncated by the Cenozoic Junggar Fault. The Laba terrane was separated from the composite Tangbale + Kekesayi terrane to the north by the Kekesayi Ocean, the lithosphere of which was subducting beneath the composite Tangbale + Kekesayi terrane (Fig. 4C).

Ebinur terrane

Description. The Ebinur terrane is composite and includes several formations or lithological associations. Amalgamation of the Kekesayi and Tangbale terranes created basement for the terrane, over which the overlapping island-arc volcanics of the Suyuenka Complex and volcaniclastics of the Qiargayi Formation accumulated.

Table 3. *Trace- and rare-earth elements for volcanic rocks from the Tangbale district*

Sample	TA10	TA85	TA12	TA34	TA127	TA55	TA50	TA28	TA25	TA87
Terrane	K	S	S	S	S	S	S	S	S	S
Rb	45.3	11.63	10.12	8.87	28.3	23.98	1.9	24.39	12.63	53.28
Y	18.32	17.24	16.51	16.12	16.42	14.26	9.82	10.46	10.22	4.55
Zr	127.1	130.3	84.63	100.9	116.6	104.8	96.79	83.96	88.97	113.2
Nb	17.39	5.12	2.5	4.21	4.1	3.71	3.19	2.78	3.87	3.96
Ba	716.9	331.9	655.95	206	781.2	705.3	90.49	601.6	344.37	969.3
La	11.52	10.96	8.04	9.68	11.83	11.59	10.12	10.71	12.92	11.02
Ce	25.91	24.82	18.16	21.17	24.75	24.74	21.67	22.49	25.83	23.47
Pr	3.73	3.62	2.74	3.12	3.73	3.52	3.1	3.2	3.34	3.28
Nd	14.95	14.35	11	12.46	14.61	13.68	12.18	12.13	12.64	12.11
Sm	3.76	3.55	2.96	3.09	3.51	3.24	2.66	2.75	2.56	2.46
Eu	1.25	1.19	0.85	0.97	1.05	0.98	0.82	0.88	0.81	0.71
Gd	4.08	3.61	3.13	3.08	3.53	3.5	2.2	2.66	2.29	1.68
Tb	0.58	0.51	0.42	0.43	0.43	0.4	0.28	0.31	0.29	0.19
Dy	3.29	2.84	2.53	2.45	2.53	2.34	1.56	1.69	1.58	0.78
Ho	0.66	0.6	0.53	0.52	0.53	0.48	0.3	0.34	0.32	0.12
Er	1.75	1.63	1.5	1.46	1.44	1.33	0.79	0.92	0.86	0.26
Tm	0.26	0.24	0.22	0.22	0.22	0.2	0.12	0.14	0.13	0.03
Yb	1.44	1.4	1.32	1.25	1.28	1.22	0.69	0.81	0.76	0.16
Lu	0.23	0.22	0.22	0.2	0.21	0.2	0.11	0.13	0.13	0.02
Hf	2.79	2.74	2.05	2.39	2.88	2.76	2.36	2.21	2.1	3.38
Ta	1.09	0.45	0.25	0.38	0.46	0.5	0.45	0.46	0.5	0.68
Pb	0.75	3.44	5.57	2.14	5.75	6.33	5.67	4.28	4.54	7.21
Th	1.39	1.55	1.27	1.87	2.77	2.77	2.14	2.02	2.93	3.36
U	0.47	0.54	0.48	0.57	0.68	0.72	0.76	0.52	0.58	1.4

Abbreviations are the same as for Tables 1 and 2.

The most significant feature of the terrane was the development of the Suyuenka Complex. Andesites, diorites, dacites, tuffs (Table 2) and volcaniclastic sediments comprise a suite of island-arc rocks built on top of, and intruded into, the Tangbale and Kekesayi terranes. Small dioritic bodies (Table 1) intrude ophiolitic mélange on the northeast margin of the Tangbale ultramafic body.

A succession of volcaniclastic turbidites referred to as the Qiargayi Formation, fines upward from basal conglomerates into a section dominated by siltstones. This formation is around 3000 m thick (Feng *et al.* 1989) and overlies the Suyuenka Complex. It includes green and purple lithic arkose turbidites, red ferruginous muddy siltstone, coarse sandstone, gritstone, bedded tuffaceous siltstone, tuffs, tuffaceous breccia, cherty siltstone and small limestone blocks or lenses.

The composition of detrital clinopyroxene within Qiargayi Formation (Table 4) plot mostly within the 'all' field of a $MnO-TiO_2-Na_2O$ tectonic discrimination plot (Fig. 5). However, some of the data lies within the volcanic arc and within-plate alkali field. This probably indicates that detrital clinopyroxene in the Qiargayi Formation was sourced mostly from calc-alkaline volcanics of the Suyuenka Complex, but also

from tholeiitic to alkaline rocks of the Kekesayi terrane.

Whereas basalts of the Kekesayi terrane are typically overlain by red chert, and occur as knockers within the mélange, basaltic-andesites and diorites of the Suyuenka Complex intrude Kekesayi terrane lithologies and are interlayered with tuffaceous and volcaniclastic sediments. Suyuenka volcanics are highly porphyritic and generally andesitic in composition compared to the fine-grained tholeiitic–alkaline basalts of the Kekesayi terrane (Table 2). They display geochemical signatures typical of volcanic arc type volcanics as opposed to the MORB or OIBs found in the Kekesayi terrane (Fig. 4).

Age. As diorites of the Suyuenka Complex intrude the serpentinite mélange (Fig. 3C) they must be younger than the blocks of Middle Ordovician Kekesayi terrane cherts incorporated into the mélange (Fig. 3F). Graptolites in shale and in siltstone constrain the unconformably overlying Qiargayi Formation as Lower Silurian (1st Regional Geological Survey Team 1985*a, b*) providing a minimum age constraint. Thus, development of the Suyuenka Complex must have occurred in the Late Ordovician. Fossils reported from interbedded limestone and clastic beds at

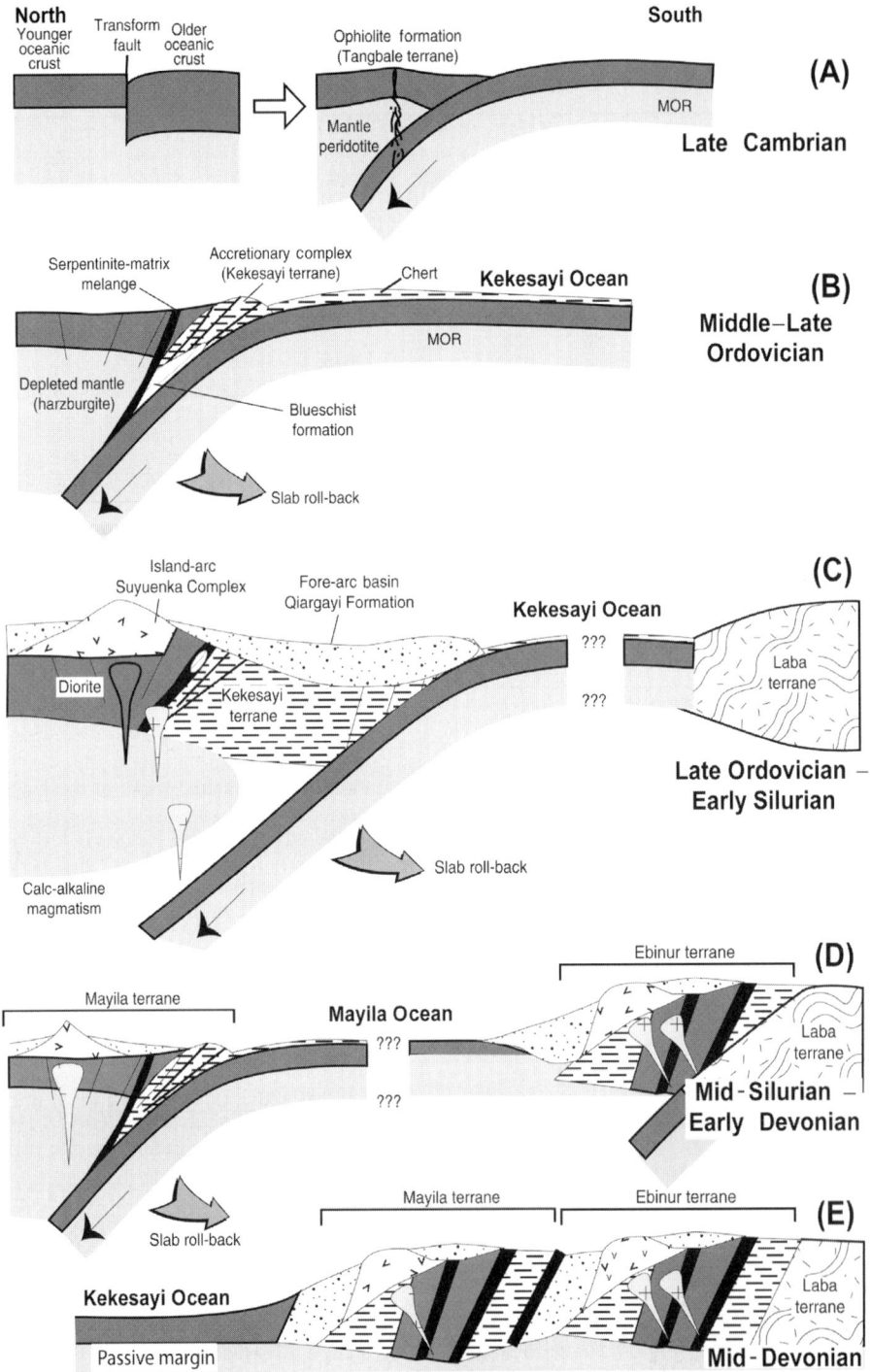

Fig. 7. Interpretation of the tectonic evolution of Early Palaeozoic terranes in West Junggar. Ophiolitic, accretionary complex and island-arc complexes were formed via a series of north-dipping subduction zones and obducted on to a southern continental margin represented by the Laba terrane. By Mid-Devonian the composite Mayila + Ebinur + Laba terrane was a passive continental margin contiguous with the Kulumudi Ocean to the north.

Table 4. *Clinopyroxene compositions (wt%) from volcanic and sedimentary rocks at Tangbale*

Block 1

Samp.	TA147	TA147	TA147	TA54	TA54	TA54	TA54	TA54	TA54	TA54	TA54	TA54	TA54	TA54	TA147	TA147	TA147	TA147	TA147
Terr.	Qi s.s.	Qi s.s.	Qi s.s.	Qi cgl	Qi cgl	Qi cgl	Qi cgl	Qi cgl	Qi cgl	Qi cgl	Qi cgl	Qi cgl	Qi cgl	Qi cgl	Qi s.s.	Qi s.s.	Qi s.s.	Qi s.s.	Qi s.s.
SiO_2	51.29	50.04	51.41	51.32	46.55	51.55	51.47	51.86	52.45	52.02	51.08	51.53	50.98	50.75	51.55	51.65	49.66	49.8	50.13
TiO_2	0.34	0.49	0.43	0.58	1.68	0.42	0.61	0.51	0.43	0.39	0.33	0.38	0.38	0.53	0.25	0.26	0.51	0.48	0.51
Al_2O_3	2.56	2.3	2.71	2.55	7.78	2.45	2.61	2.51	2.78	1.28	4.49	2.52	2.38	3.11	1.84	1.95	2.72	2.72	2.2
Cr_2O_3	0.04	0	0.11	0	0	0	0.02	0.09	0.01	0.01	0.1	0	0	0	0.27	0.14	0.12	0.14	0.01
MgO	14.62	15.6	15.58	14.37	14.15	14.2	15	15.43	15.44	12.96	14.75	14.77	14.71	14.42	16.57	16.3	15.34	15.47	15.4
CaO	19.93	19.63	20.63	18.61	11.11	19.49	19.09	19.3	20.98	19.39	21.31	19.34	19.73	18.93	20.99	21.05	20.18	19.92	20.08
MnO	0.29	0.3	0.24	0.42	0.52	0.3	0.28	0.33	0.3	0.58	0.16	0.33	0.3	0.3	0.19	0.23	0.14	0.25	0.25
FeO^*	9.65	9.81	8.09	11.42	13.12	10.53	10.39	9.93	8.29	12.79	7.16	10.25	10	10.75	6.84	7.12	8.98	9	8.87
Na_2O	0.25	0.36	0.38	0.25	1.72	0.22	0.25	0.28	0.2	0.3	0.29	0.35	0.28	0.36	0.13	0.17	0.4	0.32	0.33
Total	99.08	98.62	99.59	99.53	97.15	99.19	99.78	100.24	100.9	99.72	99.68	99.5	98.76	99.16	98.65	98.89	98.16	98.18	97.86

Block 2

Samp.	TA147	TA147	TA14	TA14	TA14	TA14	TA14	TA14	TA14	TA14	TA14	TA14	TA14	TA14	TA75	TA10
Terr.	Qi s.s.	Qi s.s.	S and.	S and.	S and.	S and.	S and.	S and.	S and.	S and.	S and.	S and.	S and.	S and.	S and.	K bas.
SiO_2	50.15	51.41	50.58	51.04	50.98	50.44	51.12	51.01	50.84	50.73	51.25	51.65	51.1	51.62	52.44	55.69
TiO_2	0.46	0.43	0.16	0.13	0.18	0.18	0.17	0.14	0.2	0.21	0.2	0.12	0.18	0.12	0.04	0.04
Al_2O_3	2.69	2.71	0.86	0.95	1.54	0.98	0.98	0.95	0.84	1.15	1.1	0.81	1.21	0.87	0.15	0.08
Cr_2O_3	0.17	0.11	0	0	0.04	0	0.01	0	0	0.04	0	0	0.01	0	0	0.02
MgO	15.34	15.58	14.04	14.22	13.79	14.05	13.94	14.09	14.47	13.91	14.18	14.3	14.16	14.13	13.66	18.54
CaO	19.97	20.63	21.56	21.7	21.05	21.61	21.58	21.95	21.59	21.71	21.4	21.29	21.67	21.42	23.4	26.11
MnO	0.24	0.24	0.47	0.38	0.45	0.35	0.39	0.4	0.32	0.41	0.5	0.42	0.38	0.49	0.4	0.03
FeO^*	8.91	8.09	9.24	9.1	10.36	9.91	9.53	9.52	9.35	9.72	9.78	9.54	9.89	9.84	8	0
Na_2O	0.41	0.38	0.22	0.25	0.31	0.27	0.28	0.23	0.23	0.25	0.24	0.18	0.28	0.24	0.44	0
Total	98.41	99.59	97.14	97.82	98.72	97.79	98.05	98.13	97.86	98.12	98.67	98.31	98.91	98.75	98.54	100.5

Block 3

Samp.	TA10	TA10	TA10	TA10	TA10	TA10	TA10	TA10	TA10	TA10	TA10	TA10	TA10	TA10	TA10	TA10
Terr.	K bas	K bas.	K bas.	K bas.	K bas.	K bas.	K bas.	K bas.	K bas.	K bas.	K bas.	K bas.	K bas.	K bas.	K bas.	K bas.
SiO_2	55.4	50.96	51.31	51.74	50.89	50.29	49.85	50.67	34.63	50.18	45.32	50.08	49.79	49.66	50.8	50.92
TiO_2	0.03	0.95	0.91	0.91	0.89	1.07	1.45	1.17	0.38	1.12	2.08	1.16	1.24	1.18	1.01	1.03
Al_2O_3	0.07	3.34	2.88	2.68	3.26	3.8	3.91	3.11	14.02	4.04	5.63	4.03	4.37	4.17	3.31	3.33
Cr_2O_3	0.02	0.54	0.21	0.15	0.67	0.69	0.42	0.18	0.45	0.79	0.9	0.75	0.86	0.89	0.29	0.21
MgO	18.74	15.5	15.29	15.82	15.29	15	15.75	16.25	18.24	15.23	13.42	15.08	14.83	14.81	15.09	15.46
CaO	25.93	21.92	21.84	21.88	21.9	21.95	19.9	19.93	5.56	21.85	20.11	22.02	21.75	21.66	21.46	21.84
MnO	0.09	0.07	0.21	0.14	0.16	0.16	0.15	0.2	0.42	0.11	0.19	0.09	0.12	0.12	0.11	0.1
FeO^*	0.04	5.91	6.53	6.37	5.9	6.06	7.59	7.78	16.69	6.01	10.01	6.06	5.8	5.97	6.9	6.36
Na_2O	0.01	0.31	0.38	0.32	0.38	0.27	0.26	0.2	0.08	0.3	0.36	0.3	0.28	0.39	0.42	0.31
Total	100.3	99.52	99.57	100	99.35	99.29	99.31	99.5	90.51	99.64	98.04	99.57	99.05	98.87	99.39	99.56

Abbreviations are the same as for Tables 1 and 2: Qi, Qiargayi Formation (cgl, conglomerate; s.s., sandstone); and., andesite; bas., basalt.

Tangbale (1st Regional Geological Survey Team 1985a, b; Zhu et al. 1987a, b) range from Mid to Upper Ordovician, confirming a Late Ordovician age.

Interpretation. Calc-alkaline igneous rocks of the Suyuenka Complex mark the beginning of island-arc volcanism associated with development of the Ebinur terrane. The basic to intermediate composition of these volcanics (Table 2) possibly indicates arc initiation within an intra-oceanic island-arc setting (Shervais, 2001). Evolution to more dacitic compositions may be the result of crustal thickening resulting in increased fractionation of the arc magmas. The Suyuenka Complex intrudes and overlies subduction complex rocks of the Tangbale and Kekesayi terranes. This requires that the axis of subduction migrated trenchward (south) in order for the arc-related igneous rocks to intrude the former accretionary prism (Fig 7C). Trenchward arc migration is a common phenomenon observed during the development of many modern intra-oceanic island-arc systems such as the Izu–Bonin–Mariana system (Stern & Bloomer 1992).

Basal conglomerates in the Qiargayi Formation contain chert, dolerite, basalt, gabbro, peridotite and blueschist clasts, together with a heavy mineral assemblage that includes garnet, chromite and olivine (Zhang et al. 1993). This indicates that mantle peridotites in the Tangbale terrane were exposed by the Early Silurian. The absence of detrital quartz and abundance of volcanic and lithic fragments in the conglomerates and sandstones (Buckman 2000) suggests that the turbidites were locally sourced from units of basic to intermediate composition, such as the Tangbale terrane, Kekesayi terrane or Suyuenka Complex. The absence of detrital metamorphic fragments may indicate that the Laba terrane had not docked with the Ebinur terrane by the Early Silurian.

The ocean separating the Ebinur terrane from the Laba terrane, the 'Kekesayi Ocean', was eventually consumed by subduction, resulting in the obduction of the Ebinur terrane on to the trailing edge of a continental margin – the Laba terrane. The Ebinur terrane was obducted on to the Laba terrane as subduction was north-directed, beneath the Ebinur terrane (Fig. 7D). The accretion event halted subduction and effectively brought an end to any further arc-related igneous or sedimentary activity in the Ebinur and Laba terranes.

Mayila Terrane

Description. The Mayila terrane is composite and contains ophiolitic peridotite and gabbro enveloped in serpentinite and mixed with basalt–chert associations, both of which are intruded and overlain by diorites and andesites, which are themselves unconformably overlain by volcaniclastic turbidites. The overall lithological composition is similar to that in the Ebinur terrane; however, the Mayila and Ebinur terranes are regarded herein as separate entities, as they are separated by faults and the ages of cherts within the terranes differ considerably. The terrane is a composite of older units, which in previous literature (1st Regional Geological Survey Team 1985a, b) are referred to as the Mayila Ophiolite and the Lower Mayila Formation. These units appear to be part of an accretionary complex containing basalt, chert, tuff and siltstone. They amalgamated before being intruded by calc-alkaline intrusives and overlain by volcanics/volcaniclastics (Upper Mayila Formation). As access to the Mayila terrane is difficult, it was not possible to record the relationships between the various rock types, and no further subdivision is attempted.

Age. Ophiolitic rocks in the terrane are Middle to Late Silurian in age, based on radiolarian biostratigraphy for cherts of the lower Mayila Formation (Li 1991, 1994). Fossils reported from limestones in the Mayila region are predominantly Upper Silurian (1st Regional Geological Survey Team 1985b). The relationship of the cherts and limestones to other rocks in the terrane remains uncertain.

Interpretation. Ophiolitic rocks at Mayila consist of highly depleted harzburgitic material and cumulate gabbros incorporated into a serpentinite-matrix mélange (Zhang et al. 1993) and are host to numerous chromite deposits. Given the highly refractory nature of these rocks, it is likely that they represent fragments of suprasubduction zone oceanic crust, similar to the Tangbale terrane.

The basalt, chert and tuff found within the lower Mayila Formation are either the upper portions of the Mayila ophiolite (Zhang et al. 1993) or fragments of Wenlockian MOR material that were off-scraped into an accretionary complex. If they form part of the original ophiolite stratigraphy, the radiolarian ages constrain the minimum age of ophiolite formation. However, if the basalt, chert and tuff units occur as thrust repetitions, then it may be more likely that they are similar to rocks of the Kekesayi terrane and represent Mid-Silurian MOR material that was off-scraped into an accretionary complex (Fig. 7D). This oceanic lithosphere material is referred to herein as the 'Mayila Ocean' and interpreted to have separated the actively forming Mayila terrane in the north from the passive margin of the composite Ebinur + Laba terrane to the south. Subduction-related igneous and sedimentary processes continued to operate within the Mayila terrane until the Mayila Ocean was completely consumed and the terrane was obducted on to composite Ebinur + Laba terrane in the Early Devonian (Fig. 7E).

There is no record of arc-related igneous rocks younger than Late Silurian within the Mayila, Ebinur or Laba terranes; thus, the accretion of the Mayila terrane is interpreted to represent the last subduction-related event to affect this southern continental margin. By the Mid-Devonian, the Mayila terrane represented the trailing-edge margin of a continental block (composite Laba + Ebinur + Mayila terrane) (Fig. 7E) and subduction had been taken up on the opposite side of the adjoining ocean ('Kulumudi Ocean') where new terranes were developing. The Kulumudi Ocean was the last ocean to exist before the final consolidation of the Siberian, Kazakhstan and Tarim blocks in the Late Carboniferous.

Toli Terrane

Description. The Toli terrane is a stratigraphic terrane comprising Lower Devonian sedimentary rocks overlain by the Middle Devonian Barleik Formation and the Upper Devonian Tielieketi Formation. Unnamed Lower Devonian rocks restricted to areas northeast of Toli constitute the base of the terrane. This unit consists of porphyritic andesites overlain by dark grey-green lithic and vitric tuffs. The upper part of the section consists entirely of vitric tuffs, fine-grained banded siliceous tuffs and lithic tuffs. The overlying Barleik Formation contains grey-purple tuffaceous sandstone, coarse sandstone, sandy conglomerate interbedded with muddy siltstone, calcareous siltstone, lithic tuffs, fine-grained siliceous tuffs and jasper (red chert) in the lower part of the unit. The upper section consists of grey-green tuffaceous sandstone, calcareous sandstone, sandy conglomerate and lenses of limestone and coal. The overlying Tielieketi Formation crops out north of Mayila, where basal sediments include well-rounded purple-grey tuffaceous conglomerate, tuffs, rhyolitic welded tuffs, breccias, and sandy conglomerate interbedded with sandstone and siltstone lenses. The unit fines upward into fine-grained tuffaceous material, siltstone and calcareous sandstones.

We note that a small sliver of rocks of ophiolitic affinity has been recorded within the Toli terrane close to the Barleik Fault. The age, composition and cross-cutting relationships of these to the surrounding lithologies are not known and we have not attempted to incorporate them into our tectonic models. It may be that these are slices of the Mayila terrane that have been displaced along the Barleik Fault.

Age. Corals in limestone lenses from the lowermost unit in the Toli terrane are Lower Devonian (1st Regional Geological Survey Team 1985b). Middle Devonian fossils occur within the Barleik Formation and the Tielieketi Formation contains Upper Devonian brachiopod and plant fossils. The Toli terrane is faulted against the Kulumudi terrane to its SE. This fault is cut by the 369 Ma (Late Devonian) I-type Bieluakaxi biotite-granite (Fig. 3B), (Jin & Xu 1997) indicating juxtaposition soon after formation of the two terranes in the Late Devonian.

Interpretation. Andesites, lithic and vitric tuffs associated with the lowermost, unnamed unit are similar to volcanic lithologies associated with continental convergent margins. Abundant brachiopods and corals indicate a shallow-marine environment. A fining-upward sequence of marine clastic sediments, fossiliferous calcareous units and fine-grained tuffs and chert within the lower Barleik Formation indicates deposition in a deepening marine environment. An overlying, coarsening-upward sequence of tuff, sandstone, conglomerate, limestone and coal, indicates a later shallowing of the depositional environment, which eventually led to the formation of terrestrial coal deposits. Siliciclastic sediments, rhyolitic tuffs and plant fossils in the Tielieketi Formation, indicate formation in terrestrial to shallow-marine conditions. Rhyolitic lavas indicate that by the Late Devonian the arc had evolved a more silicic composition.

We interpret this terrane to represent products of an Andean-type continental margin or at least a mature island arc that developed independently and to the north of the composite Mayila + Ebinur + Laba terrane (Fig. 5A), possibly as part of either the Kazakhstan or Siberian block. Arc volcanism is inferred to be related to the northward subduction of the Kulumudi Ocean. This appears to be an important gold mineralizing event in West Junggar, as most mineralization in the region has a close spatial relationship with I-type granitoids and faults that associated with them.

Kulumudi terrane

Description. The Kulumudi terrane is composed of a series of thrust repetitions of a sequence of basalt, chert, tuff, siltstone and volcaniclastic rocks, which strike approximately NE–SW. It is faulted against the Sartuohai terrane to the east and the Toli terrane to the west. There is some ambiguity concerning the exact content and extent of the unit known as the Kulumudi Formation. Basaltic units in the Anqi, Sartuohai and Karamay regions were originally thought to all represent the upper levels of the Sartuohai/ Dalabute ophiolite (Huo 1984, 1985, 1987; Zhang *et al.* 1993). However, critical age constraints (Buckman 2000) and geochemical data indicate there are three separate suites of basalts belonging to different terranes. The Kulumudi terrane has been revised to include three sub-units:

(1) a basalt–chert–tuff association with distinct MORB characteristics;
(2) thick volcaniclastic sequences; and
(3) black carbonaceous shale and silicic pyroclastic deposits.

A NW-dipping thrust fault characterized by a gradual transition over several hundred metres into a highly disrupted serpentinite-matrix melange marks the SE contact of the Kulumudi terrane with ophiolitic mélange of the Sartuohai terrane (Fig. 3A). In some regions, the serpentinite-matrix mélange is not present, and the Kulumudi terrane is in direct fault contact with Carboniferous sediments of the Karamay terrane. This fault has been folded (Fig. 3E) and intruded by Late Carboniferous to Permian granites, indicating that it is a Palaeozoic structure.

Basalt–chert–tuff associations. Three horizons of the basalt–chert–tuff sequence trending NE–SW can be traced within the terrane. The easternmost unit occurs adjacent to the ophiolitic mélange of the Sartuohai terrane. Another lies *c.*5 km to the NW and the third, a further 5–10 km NW, is referred to as the Dagun Valley or Grey Mountain zone. The Anqi Fault truncates this zone. Zhang *et al.* (1993) presented trace-element and REE data for basalts from the Anqi, Sartuohai and Karamay regions, which indicated a range of tectonic settings, including within-plate basalts (WPB-Karamay terrane), mid-ocean ridge basalts (MORB–Kulumudi terrane), island-arc tholeiites (IAT–Sartuohai terrane) and transitions between these three. Kulumudi terrane is host to predominantly N-MORB type basalts. These basalts probably represent the upper portions of mid-ocean ridge material that was scraped into an accretionary complex. Pyritic black shales occur in close association with the lowermost basalt–chert

sequences, especially in the Hatu region and are economically important, as they are commonly associated with significant gold deposits.

Volcaniclastic turbidites. Turbidite sequences dominate lower parts of the Kulumudi terrane stratigraphy. A unit of volcaniclastic turbidites and high-density mass-flow conglomerates occurs between successive slices of the basalt–chert horizons. Individual beds consist of basal conglomerates grading up into sandstone and siltstone, and exhibit typical fining-upward Bouma sequences. Fine-grained units predominate, indicating a distal source.

Black siltstone, shale and pyroclastic deposits. Black carbonaceous shale horizons and felsic to silicic pyroclastics form a large proportion of the upper portions of Kulumudi terrane (1st Regional Geological Survey Team 1985a). The pyroclastic deposits are predominantly lithic/crystal tuffs composed of feldspar, quartz and glassy tuffaceous lithic fragments.

Age. Poorly preserved conodonts extracted from heavily recrystallized cherts within the Kulumudi terrane (He Wenjun pers. comm.) were identified (Dr Ruth Mawson, MacQuarie University, pers. comm.) as upper Middle Devonian to Upper Devonian/Lower Carboniferous. Middle Devonian brachiopods occur within basal shale and clastic units (1st Regional Geological Survey Team 1985b) but, in general, the volcaniclastic units are sparsely fossiliferous.

Interpretation. The basalt–chert sequences initially formed in deep-marine conditions. The N-MORB affinities of basalts within the Kulumudi terrane support the interpretation that these rocks formed at a mid-ocean ridge spreading centre and were off-scraped into an accretionary complex. There is no evidence to suggest that these basalts represent the upper portion of the Sartuohai terrane (ophiolite). Auriferous pyritic black shales occur in association with the lowermost basalts in the Hatu gold mining district (Kulumudi terrane), and possibly formed at the time of sea-floor volcanism as 'black smoker' type deposits. Increasing felsic tuff up-section indicates a gradual approach towards a volcanic arc, accompanied by an influx of arc-derived ash-fall deposits. The adjacent Toli terrane is of a similar age to the Kulumudi terrane, and is the most likely source of increasingly silicic, pyroclastic material up-section (Fig. 8A). The composition of clastic material, younging direction of chert horizons and structural style of the Kulumudi terrane suggest that it formed as an accretionary prism of a NW-dipping subduction zone in close vicinity to a volcanic arc, which evolved from a mafic to felsic composition. Conodonts from cherts in the NW of the terrane are slightly older (upper Middle Devonian) than those from cherts in the Sartuohai region (Upper Devonian–Lower Carboniferous), possibly indicating a younging progression to the SE. This may indicate that subduction was directed to the NW, and progressively younger basalt–chert–tuff sequences were off-scraped into the accretionary complex as it grew.

We interpret the Kulumudi and Toli terranes to represent a SE-facing accretionary complex and volcanic arc respectively, related to a single NW-subducting slab (Fig. 8A). The ocean subducting beneath the Kulumudi and Toli terranes is referred to as the Kulumudi Ocean. This ocean has previously been referred to as the 'Junggar Ocean' (Carroll *et al.* 1990), but this name has genetic implications to the formation of the Junggar Basin which, in this study, is not thought to be underlain by trapped oceanic lithosphere.

Sartuohai terrane

Description. The Sartuohai terrane is a zone of ophiolitic mélange located 40 km NW of Karamay. Previously this unit has been referred to as the Sartuohai ophiolite, Dalabute ophiolite, Kulumudi Formation (in part), and Tailegula Formation (in part) (Feng *et al.* 1989; Zhang *et al.* 1993). This unit is one of the most studied ophiolitic bodies in West Junggar due to an abundance of chromite mineralization (Hao *et al.* 1989, 1991*a*, *b*, 1992; Bao *et al.* 1990, 1992; Hao 1991; Peng *et al.* 1992, 1995; Zhou & Bai 1994; Peng 1996; Zhou *et al.* 2001). Unfortunately, the nomenclature is confused and loosely defined, with lithological descriptions varying between authors. The mélange contains rocks of ophiolitic character (peridotite, gabbro and basalt?) as well as fragments of the adjacent Kulumudi terrane (basalt, chert, tuff and volcaniclastics) and Karamay terrane (conglomerate, limestone and siliciclastics) within a serpentinite-matrix mélange (Fig. 3E). The extent of this mélange is difficult to determine, as its boundary with the adjacent Kulumudi terrane appears transitional across a NW-dipping thrust fault zone. For practical purposes, the Sartuohai terrane is taken to include all disrupted rocks that are contained within a schistose serpentinite-matrix. The thrust is commonly marked by bright orange silica-carbonate (listwanite) alteration. In some regions the thrust surface is folded by a N–S-directed compressional event (Fig. 3E).

The silica-carbonates are host to gold mineralization in the Sartuohai terrane (Fig. 3A). Rare amphibolite-facies blocks occur in the mélange near the Dalabute Fault at Sartuohai. Feng *et al.* (1989) reported the presence of blueschist-facies minerals within meta-cherts and meta-gabbros in the mélange. This indicates that some lithologies within the mélange were subject to high-pressure/low-temperature metamorphism commonly associated with subduction.

Age. A Sm–Nd isotopic age of 395 ± 12 Ma has been obtained from gabbro within mélange of the Sartuohai terrane (Huang 1990; Zhang *et al.* 1993), indicating ophiolite formation in the Early Devonian. The mélange itself, however, formed later, as blocks within it include fragments of the Kulumudi terrane (Late Devonian–Early Carboniferous) as well as upper Tailegula Formation-derived knockers of quartz–lithic conglomerate. Inclusion of these blocks constrains the maximum age of mélange formation to Mid-Carboniferous. A small anorogenic granite body intrudes the mélange at Sartuohai and fragments of the Sartuohai terrane occur as pendants within the Akbastao granite. These granites are Late Carboniferous (Feng *et al.* 1989; Tilton *et al.* 1986) thus formation and emplacement of the Sartuohai terrane mélange can be constrained as Middle to Late Carboniferous.

Interpretation. Sartuohai terrane peridotites are typical of ophiolitic complexes, and are composed of highly depleted harzburgitic material, which is host to podiform chromite deposits and cumulate gabbro containing small volumes of highly fractionated plagiogranite (Zhou & Bai 1994; Zhou *et al.* 2001). The highly refractory nature of the peridotite (harzburgite) suggests that this mantle material did not give rise to the N-MORB basalts of the Kulumudi terrane or the alkali basalts of the Karamay terrane. Therefore, the ophiolitic peridotites contained within the Sartuohai terrane are considered to have origins distinct from those of the basalt–chert lithologies of the Kulumudi and Karamay terranes, even though they have been tectonically incorporated into the same mélange (Sartuohai terrane).

The high alumina contents of the chromites led Zhou & Robinson (1994) to suggest that the Sartuohai terrane (Dalabute ophiolite) formed in a back-arc environment where relatively aluminous tholeiitic melts could react with lithospheric mantle to form high-Al podiform chromites. This may have occurred as a result of back-arc spreading behind the Kulumudi terrane in the Mid

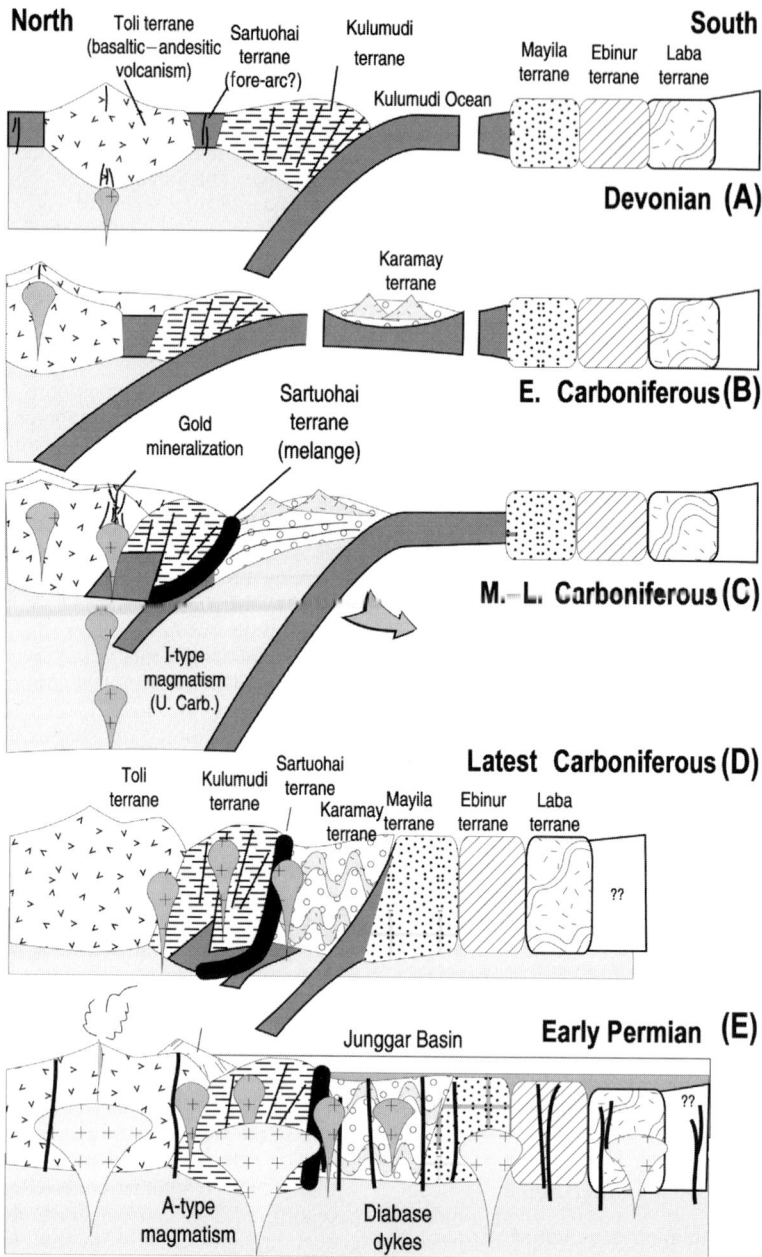

Fig. 8. Interpretation of the tectonic evolution of Late Palaeozoic terranes in West Junggar. The Toli terrane developed as a continental arc on the northern margin of the Kulumudi Ocean, well separated from the composite Mayila + Ebinur + Laba terrane to the south. North-directed subduction beneath either the Kazakhstan or the Siberian block resulted in the formation of a typical Andean-style continental margin and an accretionary complex containing off-scraped fragments of the Kulumudi Ocean (Kulumudi terrane). The Karamay terrane represents a portion of an ocean-island seamount incorporated into the accretionary complex during the Mid–Late Carboniferous. I-type magmatism is related to this subduction event and restricted to the Toli, Kulumudi and possibly Karamay terranes, and closely associated with epithermal and gold mineralization in these terranes. All oceanic lithosphere was consumed by the Late Carboniferous, resulting in complete consolidation of Central Asia. Further igneous activity occurred throughout the latest Carboniferous to Permian as a result of a regional extensional event, and led to the intrusion of large, A-type granites and diabase dykes, followed by the formation of the Junggar Basin.

Devonian. Alternatively, if the Toli terrane represents the arc that evolved behind the Kulumudi terrane accretionary complex, then the relative position of the Sartuohai terrane suggests that it may have formed in the fore-arc region. It was subsequently exhumed as a tectonic mélange along the suture between the Kulumudi and Karamay terranes in the Late Carboniferous.

The Sartuohai terrane separates the Kulumudi terrane from the Karamay terrane. The mélange is interpreted to have formed during a later stage of extension, rather than during the collision of these two terranes (Fig. 8C). Serpentinite-matrix melanges are known to develop in response to extension and diapirism in fore-arc regions (Fryer *et al.* 1985, 1995, 1999). This process involves dewatering of the descending oceanic lithosphere and hydration of overlying mantle material to produce large volumes of warm buoyant serpentinite, which rises diapirically along zones of weakness, such as older faults or fractures, for example the suture between the Kulumudi and Karamay terranes.

Karamay terrane

Description. The Karamay terrane (Buckman 2000) is a stratigraphic terrane dominated by volcaniclastic turbidites intercalated with basaltic volcanic rocks, and incorporates the Xibeikulas, Baogutu and Tailegula formations. It is characterized by well-defined symmetrical folds that have not been extensively disrupted by faulting. Southwest of Karamay the axis of a large anticline plunges gently to the SSE. Ductile deformation is evident along the thrust contact between the Karamay terrane and the Sartuohai or Kulumudi terranes on its NW boundary. Conglomerate pebbles in the upper Tailegula Formation show considerable stretching indicative of ductile deformation with a dextral component of shear. Deformation is confined to rocks close to the thrust fault separating the Karamay terrane from the Sartuohai terrane (Fig. 3E).

Xibeikulas Formation. The lower Xibeikulas Formation is composed of grey-black turbidite facies consisting of polymictic conglomerate, sandstone and siltstone. Zhang *et al.* (1993) reported sharp lateral facies changes, and small lenses of coal and limestone are present (1st Regional Geological Survey Team 1985*b*). The upper section is dominated by grey-black/green tuffs and tuffaceous sandstones, interbedded with basic to intermediate volcanics, siltstones, carbonaceous muddy siltstones, sandy conglomerates and limestones. Greywackes are dominated by basaltic lithics and plagioclase, with minor

clinopyroxene and opaque minerals (Buckman 2000). Detrital quartz is rare ($<5\%$), indicating a provenance that was far removed from any continental source. Angular to subangular basaltic clasts indicate close proximity to a volcanogenic source.

Baogutu Formation. The Baogutu Formation includes grey-black, turbiditic greywackes, siltstones, carbonaceous shale and vitric tuffs, intercalated with silty carbonaceous limestone (marls). These grade into dominantly grey-black carbonaceous siltstones with minor bioclastic sandstones and conglomerates that contain fossils and plant spores (1st Regional Geological Survey Team 1985*b*). Sedimentary rocks are predominantly composed of black carbonaceous shale and vitric tuffs. Rare conglomerate and sandstones of volcanic origin are interbedded with the shales.

Tailegula Formation. The Tailegula Formation consists of grey-green/black tuffs, siltstone and chert intercalated with basaltic volcanics. The upper portions of this formation consist of tuffs, quartzofeldspathic sandstone, conglomerate, and muddy siliceous siltstone intercalated with basic to intermediate volcanics and minor limestone lenses. Volcaniclastic turbidites (lithic greywackes) from the lower Tailegula Formation are dominated by basaltic and carbonaceous shale fragments and plagioclase. The upper parts of the unit are dominated by tuffaceous material, siltstones, quartz and lithic conglomerates and fossiliferous, clastic limestone lenses. The change to a more quartz-rich composition up-section indicates the influx of continental material into the basin.

Basalts from the Tailegula Formation (25 km SW of Karamay) are high-TiO_2, high-Nb alkali basalts (Zhang *et al.* 1993) similar to either intra-plate, ocean-island basalts or E-type MORB. Zhang *et al.* (1993) considered basalts of the Tailegula Formation (Karamay terrane) equivalent to basalts found at Sartuohai and Dagun Valley. However, geochemistry (high-TiO_2, high-Nb alkali basalts – samples 19.1 and 19.5 (Zhang *et al.* 1993) indicates that basalts of the Tailegula Formation (Karamay terrane) are significantly different in composition to the N-MORB type basalts found in the Kulumudi terrane.

Age. Fossils reported from the stratigraphic column accompanying 1:100 000 geological maps (1st Regional Geological Survey Team 1985*a*, *b*) indicate that the Xibeikulas Formation is Lower Carboniferous, the Baogutu Formation is Middle

Carboniferous and the Tailegula Formation is Middle to Upper Carboniferous.

Interpretation. The Karamay terrane accumulated during the Carboniferous in an oceanic setting with periodic alkali basaltic volcanism (Fig. 8B). The alkali basalts are significantly different to the N-MORB basalts of the Kulumudi terrane, and there is a distinct structural contrast between the two terranes, which justifies separation into two distinct lithotectonic entities. We interpret this terrane as a subduction complex, possibly an extension of the Kulumudi terrane accretionary complex, containing off-scraped basalts of ocean-island origin (Fig. 8C).

The Kulumudi Ocean separated the composite Toli + Kulumudi + Karamay + Sartuohaiterrane in the north from the composite Ebinur + Laba + Mayila terrane to the south. This oceanic lithosphere is interpreted to have been subducting beneath the composite Toli + Kulumudi + Karamay + Sartuohai terrane and given rise to I-type magmatism (Fig. 8C). The lack of post-Silurian, arc-related magmatism within the Ebinur, Laba and Mayila terranes, may be why these older rocks are generally better preserved and of slightly lower metamorphic grade than rocks of the Toli and Kulumudi terranes. It may also explain why gold deposits are rare within the older terranes but abundant throughout the Toli, Kulumudi, Sartuohai and Karamay terranes where pervasive I-type magmatism injected auriferous hydrothermal fluids into the surrounding rocks.

The consumption of the Kulumudi Ocean lead to the final consolidation of Central Asia and the collision of two continental blocks in the Late Carboniferous (Fig. 8D). This collision probably resulted in widespread crustal thickening and uplift. Thrust faulting and ductile deformation accompanied the earliest phase of deformation, with later folding of fault surfaces (Fig. 3E). Ductile deformation, as evident by the presence of stretched pebbles close to the Sartuohai/Karamay thrust contact (Buckman 2000), is interpreted to have occurred during the collision in the Late Carboniferous. The collision and uplift were short lived (Late Carboniferous) and followed by a period of widespread extension (latest Carboniferous–Permian) (Fig. 8E).

Stitching plutons and overlap sequences

Terranes in West Junggar were completely consolidated as part of the Eurasian Plate by the latest Carboniferous. The timing of the amalgamation is variously constrained by Mid Carboniferous to Early Permian granites and diabase dykes, which cut terrane boundaries, overlapping units of Early Permian alkaline basalts, andesites, rhyolites and silicic pyroclastics, and provenance links to Upper Permian continent-derived sediments within the Junggar Basin.

Widespread, post-orogenic extension occurred throughout Central Asia during the latest Carboniferous to Early Permian. This is manifest as voluminous A-type granites and diabase dykes in the Junggar region. A-type granites are commonly surrounded by normal ring faults, indicating that they were emplaced in an extensional regime. These intrusions represent a form of continental growth unrelated to subduction, which involves the addition of mantle-derived melts into the upper crust (Han *et al.* 1997; Hu *et al.* 2000; Jahn *et al.* 2000; Wu *et al.* 2000).

Sediments also began to accumulate in the Junggar and Tacheng basins as a result of widespread extension in the latest(?) Carboniferous to Early Permian. These basins record the erosion of surrounding mountain ranges, a process which continues today.

I-type magmatism

Description. Numerous granitic plutons and diabase dykes intrude the Palaeozoic lithologies of West Junggar. The granites are divided into two categories:

(1) small metaluminous (I-type) granodiorite to quartz diorite intrusions (Upper Devonian to Upper Carboniferous); and
(2) large alkali, hypersolvus A-type granites (Upper Carboniferous to Lower Permian).

A suite of relatively small (<10 km diameter) felsic intrusives stitches the Toli and Kulumudi terranes and trends roughly ENE–WSW. The Bieluakaxi granodiorite (Fig. 3B) is the largest of these intrusions, and numerous small gold mines have been established around its periphery. These granites have a wide metamorphic halo (*c.*100 m) and have been cut by diabase dykes. The composition of these intrusives (Jin & Zhang 1993; Jin & Xu 1997) is characteristic of I-type magmatism associated with continental arc-related subduction. The subsolvus nature and the porphyritic texture of the metaluminous (I-type) granitoids in the Hatu region indicates that crystallization of feldspars took place independently and originated at deep crustal levels. They are distinguished from later A-type granites by being slightly more mafic, porphyritic, smaller and closely associated with gold mineralization.

Age. The Bieluakaxi granodiorite intrudes the contact between the Toli and Kulumudi terranes, which constrains the maximum age of intrusion to latest Devonian. The Bieluakaxi granodiorite has been dated at 369 Ma (Jin & Xu 1997; Jin & Zhang 1993).

Interpretation. I-type magmatism took place in West Junggar between mid- to Late Carboniferous. This igneous activity may represent the final stages of subduction of the Kulumudi Ocean before complete continental consolidation of the Kazakhstan, Siberian and Tarim blocks in the Late Carboniferous (Fig. 8A–C).

A-type magmatism

Description. Large granite batholiths (up to 30 km in diameter) (Fig. 3D) intrude all Palaeozoic terranes and terrane boundaries of West Junggar (Fig. 2), and have been referred to as post-collisional, anorogenic or A-type granites (Feng *et al.* 1989). The A-type granites of West Junggar are predominantly composed of K-feldspar (perthite + microcline) and quartz, with lesser amounts of plagioclase, biotite, amphibole and pyroxene. The distinguishing petrographic feature of these granites is their hypersolvus character. Perthite is the dominant feldspar, with only small quantities of plagioclase. Analyses of eight different granites from West Junggar (Kwon *et al.* 1989) show that they are alkaline and quite distinct from any calc-alkaline granites which form above subduction zones.

Age. Feng *et al.* (1983) used K/Ar methods to date anorogenic granites within West Junggar as Permian. More recent Pb-isotope studies of the anorogenic Karamay granite (Kwon *et al.* 1989; Tilton *et al.* 1986; Yang *et al.* 1994) indicate age ranges between 300–328 Ma (mid–Late Carboniferous). However, as the surrounding rocks of the Tailegula Formation are Mid-Carboniferous, the age of the Karamay granite is further constrained as Late Carboniferous.

Interpretation. Highly positive $\sum_{Nd}(T)$ values of +6.1 (Kwon *et al.* 1989) suggest that anorogenic granites of West Junggar were derived from a depleted upper mantle source. Kwon *et al.* (1989) interpreted the positive $\sum_{Nd}(T)$ value to indicate the melt was derived from lower crustal material or oceanic lithosphere underlying West Junggar. However, the partial melting of mafic oceanic crustal material cannot account for the contemporaneous formation of mafic dykes. The melting of lower crustal material, even if it was completely oceanic

lithosphere, should result in low or negative $\sum_{Nd}(T)$ values. If West Junggar were underlain by Precambrian basement rocks (Wang 1986; Wang & Wang 1986; Xu *et al.* 1987; Zhang *et al.* 1993) then the granites should have highly negative $\sum_{Nd}(T)$ values. There are no known outcrops of Precambrian basement in West Junggar, which further suggests that these granites were not derived from the melting of lower crustal rocks, but were derived from the partial melting of upper mantle material.

The close association of A-type granites with later stage diabase dyke swarms is a common feature of many anorogenic granites (Pe-Piper *et al.* 1991; Coleman *et al.* 1992). Positive $\sum_{Nd}(T)$ values of anorogenic granites and the close spatial association with diabase dykes (Buckman 2000) indicates that both have been derived from the partial melting of mantle material. However, the early-stage granites are highly fractionated, and numerous meta-sedimentary xenoliths and pendants around the margins of the granite indicate contamination from surrounding crustal rocks. Both the A-type granites and diabase dykes represent a period of regional crustal extension and thinning from the Late Carboniferous to Early Permian (Fig. 8E) (Allen *et al.* 1995; Han *et al.* 1997; Hu *et al.* 2000; Jahn *et al.* 2000; Wu *et al.* 2000).

The mechanisms responsible for extensional igneous activity in Central Asia are not fully understood. However, the extension may be related to orogenic collapse following the collision of the Siberian, Kazakhstan and Tarim blocks in the Late Carboniferous. Collapse of large mountain belts is a common phenomenon, as observed in the Himalaya and Tibet (Tapponnier *et al.* 1981) with the N–S directed extensional faults. These form as the lithosphere shifts or flows (England & Molnar 1997) laterally in a direction approximately orthogonal to compression in order to obtain thermal and isostatic equilibrium (Dewey 1988; Inger 1994). A similar but more advanced or extensive style of extension may explain the regional extent of igneous and sedimentary activity in West Junggar. Other forms of extension-related igneous activity, such as hot-spots or rifting usually create a single well-defined zone of volcanism rather than the widespread igneous activity observed in the Junggar region.

Diabase dykes

Description. Diabase dyke swarms are common throughout West Junggar (Qi 1993) but have received little attention in previous geological studies. Individual dykes range from 0.3 to

5 m wide and can extend for tens of km. They commonly occur in sets of parallel, linear swarms (Fig. 3D) and also occur as conical (circular) sets within some of the anorogenic granites – notably the Miaoergou and Hong Shan granites.

Age. The diabase dykes intrude the Late Carboniferous to Early Permian A-type granites and the ?Lower Permian Qialebayi Formation but not the Lower–Middle Permian Molaoba Formation or the Upper Permian Kurjirtai Formation. Thus the timing of diabase dyke intrusion is constrained to Early Permian.

Interpretation. The Early Permian diabase dykes of West Junggar represent a period of widespread extension and mantle upwelling (Fig. 8E). They are closely associated with, and intrude into, the large A-type granites. Diabase dyke swarms are not restricted to West Junggar. Similar dyke and granite associations occur in the Altay Mountains (c.270–300 Ma (Han *et al.* 1997)) and in the Kuruktag Mountains (northern Tarim) (c.282 Ma (Zhang *et al.* 1998)). This indicates that West Junggar and possibly the entire Central Asia region underwent a period of widespread extension during the Late Carboniferous to Early Permian. The tectonic significance of the dykes may be found by looking at other well-documented dyke complexes. Ernst & Buchan (1997) suggested that pre-Mesozoic diabase dyke swarms found intruding uplifted orogenic belts may be indicative of ancient large igneous provinces (LIPs). The age of the dykes coincides with the oldest sediments found within the Junggar Basin, and may represent an important mechanism by which large intracontinental basins, such as the Junggar Basin, are formed.

Qialebayi Formation

Description. The Qialebayi Formation crops out on the western side of the Miaoergou Granite. It includes a lower sequence of olivine basalt flows along with olivine andesitic basalt, and basaltic agglomerates. These are overlain by medium to thick-bedded tuffaceous fine-grained sandstone and siltstone, interbedded with terrigenous sandstone and conglomerate (1st Regional Geological Survey Team 1985b).

Age. The age of the Qialebayi Formation is not accurately constrained, due to a lack of fossil or isotopic dating. As the Miaoergou granite intrudes the Qialebayi Formation (1st Regional Geological Survey Team 1985b) it is older than

the 277 Ma isotopic age reported for this body (Yang 1997).

Interpretation. The formation of olivine-rich basalts, as found in the Qialebayi Formation, requires partial melting of a mantle source. Andesitic lavas within the Qialebayi Formation indicate primary basaltic magma was either contaminated by the surrounding Kulumudi terrane rocks, or the melt underwent fractionation on its ascent to the surface. The formation is intruded by the Miaoergou granite and several diabase dykes and their close spatial association may indicate that they were successive stages of the same extensional event, with the Qialebayi Formation representing the earliest phase of within-plate igneous activity. The Qialebayi Formation may have formed where some of the diabase dykes reached the surface and were erupted as flows (Fig. 8E). Elsewhere in the region most of the Qialebayi Formation has been removed by erosion in response to post-Permian uplift, thereby exposing the lower-level diabase dykes.

Molaoba Formation

Description. The Molaoba Formation occurs in close association with anorogenic granitic intrusions in the Toli region. The lower part of the formation consists of porphyritic dacite, rhyolite, porphyritic felsic intrusives, andesitic porphyries, tuffaceous sandy conglomerates, acidic tuffs, ignimbrites and volcanic breccias. The upper part of the formation consists of muddy siltstones and tuffs. Most of the Molaoba Formation has been removed by erosion associated with post-Permian uplift. Preservation of this unit is restricted to localized structural depressions, such as at Toli.

Age. The Molaoba Formation lies unconformably below the Middle Permian Kurjirtai Formation and has been assigned to the Lower Permian (1st Regional Geological Survey Team 1985b).

Interpretation. The silicic volcanic rocks and associated pyroclastics may represent extrusive equivalents of the large A-type granites in West Junggar. The sequence of sediments and tuffs deposited on top of the volcanics indicates that the region was experiencing widespread subsidence during the Early Permian. Similar volcanic/sedimentary associations have been found around the margins of the Junggar Basin (Carroll *et al.* 1990), and are interpreted as the initial deposits of the Junggar Basin. The Molaoba Formation is interpreted to have formed as a result of regional intra-continental extension related to the A-type igneous activity. It represents the initial phase of sedimentation during extension that ultimately led to the formation of the Junggar Basin.

Kurjirtai Formation

Description. The Kurjirtai Formation crops out in small, elongate, fault-bounded packages located along the Dalabute Fault. It also occurs further north, where it unconformably overlies the Molaoba Formation. The lower Kurjirtai Formation consists of Middle Permian, conglomerates, sandy conglomerates and cross-bedded sandstones, interbedded with thin beds of coal, marly limestone and muddy siltstone. Porphyritic andesite flows occur towards the top of the sequence (1st Regional Geological Survey Team 1985b). The upper section is composed almost entirely of conglomerate and thin sandstone lenses. Various splays of the Dalabute Fault bound the Kurjirtai Formation. There are no records of depositional contacts with the underlying Palaeozoic basement.

Age. Fossils constrain the age of the formation as Middle Permian (1st Regional Geological Survey Team 1985*b*).

Interpretation. The Kurjirtai Formation denotes the initial stages of widespread intra-continental sedimentation following consolidation of Palaeozoic terranes within the Eurasian continent. It unconformably overlies the Molaoba Formation and was therefore deposited after the igneous activity associated with the A-type granites. Andesite flows in the sequence may be extrusive equivalents of some of the more fractionated dykes commonly found throughout West Junggar. The upper unit of conglomerate and sandstone lenses was deposited in a fluvial environment.

The Kurjirtai Formation represents either small remnants of the Junggar Basin, or contemporaneous localized depositional sites. The Junggar Basin may have once blanketed the Palaeozoic rocks of West Junggar, with tectonic movements associated with Cenozoic strike-slip faulting exposing some areas of the basin to erosion (West Junggar) and others to rejuvenated sedimentation (Lake Ebinur). In uplifted regions such as West Junggar, the Kurjirtai Formation was almost completely eroded, with the exception of small packages preserved in topographic depressions along the Dalabute Fault.

The mature, well-rounded, well-sorted sediments of the Kurjirtai Formation differ considerably from the texturally and mineralogically immature sediments that would be expected to have developed in strike-slip basins, and is inconsistent with previous interpretations (Allen & Vincent 1997) that it formed in a series of small elongate strike-slip basins along the Dalabute Fault.

Junggar Basin

Description. The Junggar Basin contains ?uppermost Carboniferous to Lower Permian marine sediments which are overlain by thick successions of fluvial/deltaic sediments. This basin is an important petroleum source, and there has been much speculation as to its origins (Watson *et al.* 1987; Carroll *et al.* 1990; Allen *et al.* 1993, 1994, 1995). Few researchers have looked to the underlying or adjacent Palaeozoic rocks for evidence of the extensional mechanisms that gave rise to this large inland basin.

The Junggar Basin overlaps Palaeozoic subduction complex rocks in West Junggar. There is no evidence of strike-slip faults or mass-flow deposits on the margins of the basin to suggest a pull-apart origin. Available evidence indicates that the basin is everywhere underlain by Palaeozoic subduction complex rocks, and formed in response to a regional extensional event linked to the preceding intrusion of anorogenic (A-type) granites and diabase dykes during the latest Carboniferous to Early Permian (Fig. 8E).

Cenozoic deformation associated with the India–Asia collision led to the formation of the Dalabute and Junggar fault systems, resulting in rejuvenated uplift and erosion of the West Junggar mountains. Largely unconsolidated sediments of the Junggar Basin that originally overlay basement rocks of West Junggar would have been rapidly eroded, revealing the underlying Palaeozoic subduction complex rocks. Small remnants of the basin (Kurjirtai Formation) have been preserved in structural depressions within the uplifted regions.

Conclusions

Nine unique, fault-bounded terranes are present in the West Junggar area (Fig. 9). These are:

(1) the Tangbale terrane (disrupted ophiolitic, Upper Cambrian–Lower Ordovician);
(2) the Kekesayi terrane (disrupted accretionary complex, Mid-Ordovician);
(3) the Laba terrane (metamorphic, ?Ordovician)
(4) the Ebinur terrane (composite island-arc, Upper Ordovician–Lower Silurian);

(5) the Mayila terrane (composite ophiolitic, accretionary complex, island-arc, ?Lower Silurian–?Lower Devonian);

(6) the Toli terrane (stratigraphic continental–arc, Lower–Upper Devonian);

(7) Kulumudi terrane (disrupted/stratigraphic accretionary complex, Mid-Devonian– Lower Carboniferous);

(8) Sartuohai terrane (disrupted ophiolitic, Mid-Devonian);

(9) Karamay terrane (stratigraphic, intra-oceanic basin + OIB, Lower–Mid-Carboniferous).

The tectonic evolution of these terranes involved the accretion of the older, Upper Cambrian to Upper Silurian terranes (Tangbale,

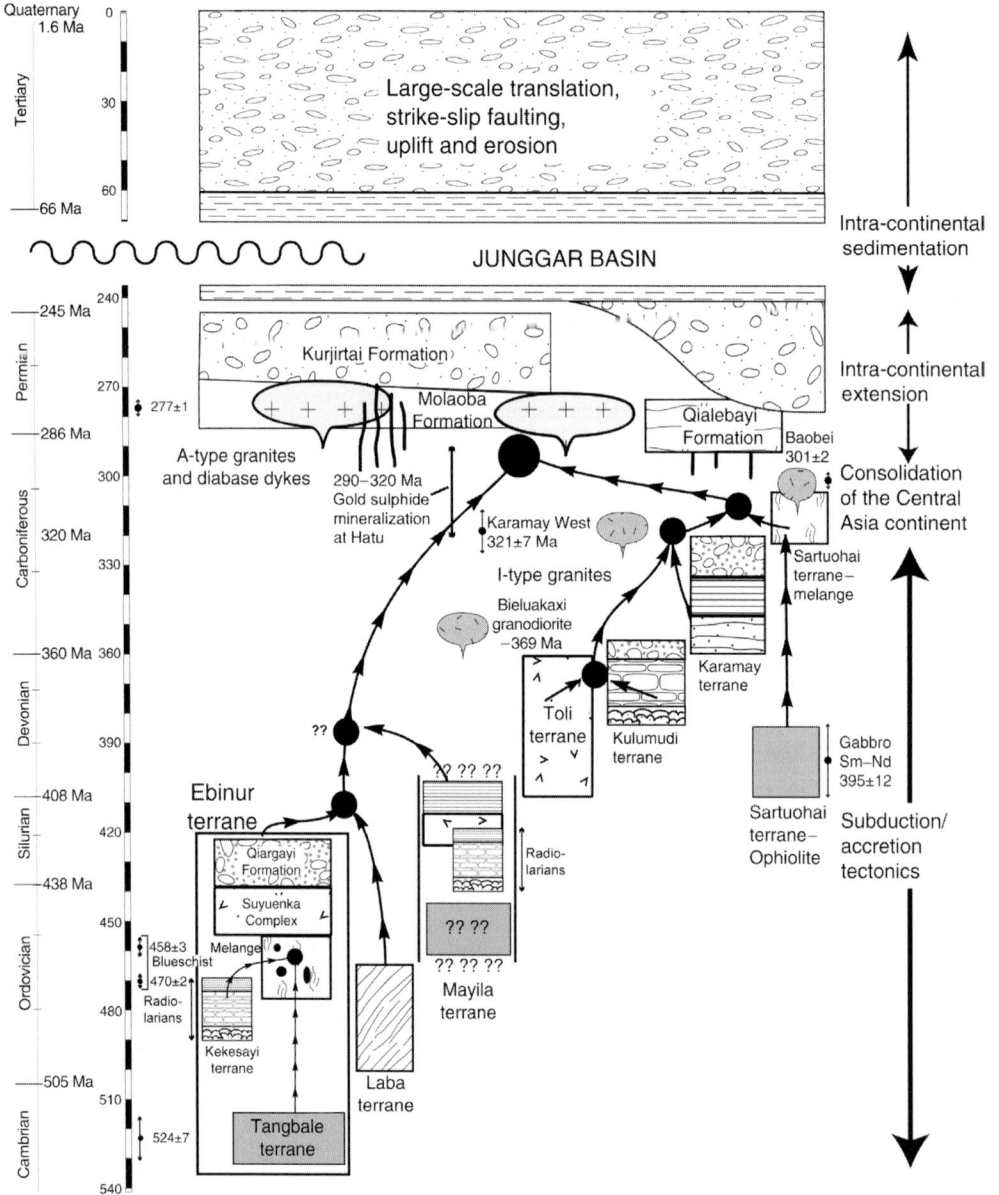

Fig. 9. Time–space diagram illustrating the tectonic evolution of Palaeozoic terranes of West Junggar and the successor basins, overlap sequences and stitching intrusive bodies.

Kekesayi, Laba, Ebinur and Mayila), to a southern continental margin. This involved the closure of at least two oceans (Kekesayi and Mayila), at a series of north-dipping subduction zones (Fig. 7). The closest Archaean continental block at present is the Tarim Block, but large-scale Cenozoic translations may have shuffled terranes from the Kazakhstan Block region, and it is difficult to determine where accretion originally occurred. The other terranes (Toli, Kulumudi, Sartuohai and Karamay) developed on the northern side of the Kulumudi Ocean (Fig. 8). Consumption of oceanic lithosphere at a series of north-dipping subduction zones resulted in the amalgamation of these terranes to a northern continental margin, possibly the Kazakhstan or Siberian block. This left only the Kulumudi Ocean separating two continents, the composite Ebinur + Laba + Mayila terrane to the south, and the composite Toli + Kulumudi + Sartuohai + Karamay terrane to the north. Oceanic lithosphere flooring the Kulumudi Ocean continued to subduct northwards beneath the latter terrane, resulting in widespread continental I-type magmatism. However, the composite Ebinur + Laba + Mayila terrane was contiguous with the Kulumudi Ocean as a passive margin, and did not experience any post-Silurian arc-related igneous activity. This is significant with regard to gold mineralization, as epithermal and porphyry-style gold mineralization in West Junggar is closely associated with I-type granites that formed during the final subduction-related event and rich gold deposits are restricted to the Toli, Kulumudi, Sartuohai and Karamay terranes.

This study shows that the Palaeozoic was an important period of crustal growth in Central Asia. The positive continental crustal budget is largely related to the accretion of island-arc, accretionary wedge and ophiolite complexes to continental nuclei. Field observations and analyses of rocks from West Junggar indicate that Central Asia grew by accretion of distinct packages of island-arc, ophiolitic and accretionary complex rocks on to the margins of continental nuclei. This process involved the interaction of numerous subduction complexes, some of which evolved contemporaneously as outlined in Figure 9, before being amalgamated together to produce the complex tectonic collage seen today. Previous models proposing a single long-lived, north dipping subduction zone (Sengör et al. 1993; Sengör & Natal in 1996) do not adequately account for this, as they exclusively involve an Andean/continental type margin. As the oldest terranes in West Junggar occur to the south of younger terranes, the hypothesis of progressive off-scraping of younger rocks into a single, southward-migrating accretionary complex can be rejected. A modern tectonic analogue for Central Asia may be SE Asia, where numerous micro-continents, island-arcs and oceanic basins are forming, colliding and being driven by several subduction zones, orientated in various directions.

We wish to thank Wenjun He, Meifu Zhou and Jianbing Lui for their assistance in helping organize and carry out the fieldwork for this project. Thank are due to the Chinese 'Project 305' for their assistance and permission to carry out fieldwork in West Junggar. A special thanks goes to the Tangbale chromite miners and nomadic families for the salubrious accommodation and their warm hospitality. The work described in this paper was supported by grants from the Research Grants Council of the Hong Kong Special Administrative Region, China (Project No. HKU 7089/97P).

References

ALLEN, M. B., SENGÖR, A. M. C. & NATAL'IN, B. A. 1995. Junggar, Turfan and Alakol Basins as late Permian to ?early Triassic extensional structures in a sinistral shear zone in the Altaid orogenic collage, Central Asia. *Journal of the Geological Society of London*, **152**, 327–338.

ALLEN, M. B. & VINCENT, S. J. 1997. Fault reactivation in the Junggar region, northwest China: the role of basement structures during Mesozoic–Cenozoic compression. *Journal of the Geological Society of London*, **154**, 151–155.

ALLEN, M. B., WINDLEY, B. F. & ZHANG, C. 1994. Cenozoic tectonics in the Urumqi–Korla region of the Chinese Tien Shan. *Geologische Rundschau*, **83**, 406–416.

ALLEN, M. B., WINDLEY, B. F., ZHANG, C. & GUO, J. 1993. Evolution of the Turfan Basin, Chinese Central Asia. *Tectonics*, **12**, 889–896.

BAO, P., HAO, Z., WANG, X. & PENG, G. 1992. Discussion on formation environments of podiform chromite deposits in western Junggar, Xinjiang. *Bulletin of the Institute of Geology, Chinese Academy of Geological Sciences*, **23**, 48–63.

BAO, P., WANG, X., HAO, Z., PENG, G., ZHANG, R., CHEN, Q. & YANG, T. 1990. A new idea about the genesis of the aluminum-rich podiform chromite deposit; with the Sartuohai chromite deposit of Sinkiang. *Mineral Deposits*, **9**, 97–111.

BUCHAN, C., CUNNINGHAM, D., WINDLEY, B. F. & TOMURHUU, D. 2001. Structural and lithological characteristics of the Bayankhongor ophiolite zone, central Mongolia. *Journal of the Geological Society, London*, **158**, 445–460.

BUCKMAN, S. 2000. *Tectonics and mineralization of West Junggar, northwest China*. PhD, University of Hong Kong.

BUCKMAN, S. & AITCHISON, J. C. 2001. Middle Ordovician (Llandeilan) radiolarians from West Junggar, Xinjiang, China. *Micropaleontology*, **47**, 359–367.

CARROLL, A. R., LIANG, Y., GRAHAM, S. A., XIAO, X., HENDRIX, M. S., CHU, J. & McKNIGHT, C. L. 1990. Junggar basin, northwest China: trapped Late Paleozoic ocean. *Tectonophysics*, **181**, 1–14.

COLEMAN, R. G. 1989. Continental growth of northwest China. *Tectonics*, **8**, 621–635.

COLEMAN, R. G., DeBARI, S. & PETERMAN, Z. 1992. A-type granite and the Red Sea opening. *Tectonophysics*, **204**, 27–40.

CONEY, P. J., JONES, D. L. & MONGER, J. W. H. 1980. Cordilleran suspect terranes. *Nature*, **239**, 329–333.

DEWEY, J. F. 1988. Extensional collapse of orogens. *Tectonics*, **7**, 1123–1139.

ENGLAND, P. & MOLNAR, P. 1997. Active deformation of Asia: from kinematics to dynamics. *Science*, **278**, 647–650.

ERNST, R. E. & BUCHAN, K. L. 1997. Giant radiating dyke swarms: their use in identifying pre-Mesozoic large igneous provinces and mantle plumes. *In:* MAHONEY, J. J. & COFFIN, M. F. (eds) Large igneous provinces; continental, oceanic, and planetary flood volcanism. *American Geophysical Union, Geophysical Monograph*, **100**, 297–333.

FEI, D., ZHANG, X. S., FENG, Y., COLEMAN, R. G., TILTON, G. & XIAO, X, 1987 Tectonic evolution of the West Junggar region, Xinjiang, China. *Acta Geophysica Sinica*, **30**, 459–468.

FENG, Y. 1986. Genetic environments and original types of ophiolites in West Junggar. [*Bulletin of the Xi'an Institute of Geological Sciences*], **13**, 37–45.

FENG, Y. 1989. Tectonic framework and evolution of the northern Xinjiang and its surroundings. [*Bulletin of the Xi'an Institute of Geological Sciences*], **25**, 115–126.

FENG, Y. & HUO, Y. 1987. Some advances in research of ancient plate tectonics of West Junggar. *In:* ZHOU, W. (ed.) *International Symposium on Tectonic Evolution and Dynamics of Continental Lithosphere; the Third All-China Conference on Tectonics; abstracts.* **1**, 16, Chinese Academy of Geological Sciences, Beijing, China.

FENG, Y., COLEMAN, R. G., TILTON, G. & XIAO, X. 1989. Tectonic evolution of the West Junggar region, Xinjiang, China. *Tectonics*, **8**, 729–752.

FENG, Y., XIAO, X., WANG, G., JU, B., YEN, B., ZHANG, Z. M. & COLEMAN, R. G. 1985. Tectonic evolution of the Tangbale Ophiolite melange, in West Junggar of Xinjiang region, China. *In:* Anon. (eds) *AGU 1985 Fall Meeting.* **66** (46), 1129, American Geophysical Union, Washington, DC.

FENG, Y., ZHANG, Y. & HUO, Y. 1983. Tectonic evolution of the western part of northern Xinjiang. *Contribution Project Plate Tectonics, Northern China*, **1**, 17–33.

1ST REGIONAL GEOLOGICAL SURVEY TEAM 1985a. *Geology and Mineral Resources of Southern West Junggar, Xinjiang Province (Map Series: L-44-84, 93, 94, 96; L-45-61, 62, 63, 73, 74, 75, 85, 86, 87, 97; & Stratigraphic Column of Southern West Junggar)*, 1:100 000. Geological Bureau of Xinjiang, Urumqi.

1ST REGIONAL GEOLOGICAL SURVEY TEAM 1985b. *Stratigraphic Column of Southern West Junggar*, Geological Bureau of Xinjiang, Urumqi.

FRYER, P., AMBOS, E. L. & HUSSONG, D. M. 1985. Origin and emplacement of Mariana forearc seamounts (Pacific). *Geology*, **13**, 774–777.

FRYER, P., MOTTL, M. J., JOHNSON, L. E., HAGGERTY, J. A., PHIPPS, S. & MAEKAWA, H. 1995. Serpentinite bodies in the forearcs of Western Pacific convergent margins: origin and associated fluids. *In:* TAYLOR, B. & NATLAND, J. (eds) Active margins and marginal basins of the Western Pacific. *American Geophysical Union, Geophysical Monograph*, **88**, 259–279.

FRYER, P., WHEAT, C. G. & MOTTL, M. J. 1999. Mariana blueschist mud volcanism; implications for conditions within the subduction zone. *Geology*, **27**, 103–106.

GAN, Q. & WANG, X. 1987. On Paleo-Junggar oceanic plate and ophiolite zones in West Junggar, Xinjiang, China. In: ZHOU, W. (eds) *International Symposium on Tectonic Evolution and Dynamics of Continental Lithosphere; the Third All-China Conference on Tectonics; Abstracts.* **1**, 16–17, Chinese Academy of Geological Sciences, Beijing, China.

GONG, Y. 1993a. Process-facies types and sequences of Devonian volcanic–sedimentary successions in northern Xinjiang and their relation to plate tectonics. *Acta Geologica Sinica*, **67**, 37–51.

GONG, Y. 1993b. Types, characteristics and geologic significance of the Devonian pyroclastic tempestites in east and west Junggar, Xinjiang. [*Sedimentary Facies and Palaeogeography*], **13**, 18–25.

GUO, Y. H. 1983. The glaucophane schist belt in Tangbale of Xinjiang. In: Anon. (eds) *Papers on the Tectonic Plate in North China*. Geological Publishing House, Beijing, 89–104.

HAMILTON, W. B. 1994. Subduction systems and magmatism. *In:* SMELLIE, J. (ed) *Volcanism Associated with Extension at Consuming Plate Margins*. Geological Society, London Special Publications, **81**, 3–28.

HAN, B., WANG, S., JAHN, B.-M., HONG, D., KAGAMI, H. & SUN, Y. 1997. Depleted-mantle source for the Ulungur River A-type granites from North Xinjiang, China: geochemistry and Nd-Sr isotopic evidence, and implications for Phanerozoic crustal growth. *Chemical Geology*, **138**, 135–159.

HAO, Z. 1991. Study on the genesis of ophiolites and podiform chromite deposits of the western Junggar area, Xinjiang. *Bulletin of the Chinese Academy of Geological Sciences*, **23**, 73–83.

HAO, Z., WANG, X. & BAO, P. 1991a. Genesis of enriched upper mantle in an orogenic zone: a case study of the Saertuohai ophiolitic block. *Bulletin of the Chinese Academy of Geological Sciences*, **22**, 159–166.

HAO, Z., WANG, X., BAO, P., PENG, G. & JIN, Y. 1989. Geological characteristics and genetic study on ophiolites of the two types in western Zhungeer, Sinkiang Uighur. [*Acta Petrologica et Mineralogica*], **8**, 299–310.

HAO, Z., WANG, X., BAO, P., PENG, G. & JIN, Y. 1991b. Study of mineralogy of the metamorphic peridotite in the two types of ophiolite in the western Zhungeer, Xinjiang. *Journal of Changchun College of Geology*, **21**, 33–38.

HAO, Z., WANG, X., BAO, P., PENG, G. & JIN, Y. 1992. Geochemical characteristics and petrologic significance of REE in ophiolites in the western Junggar area, Xinjiang. *Bulletin of the Institute of Geology, Chinese Academy of Geological Sciences*, **23**, 36–47.

HOWELL, D. G. 1989. *Tectonics of Suspect Terranes: Mountain Building and Continental Growth.* Topics in the Earth Sciences, **3**, Chapman and Hall, London.

HOWELL, D. G. 1995. *Principles of Terrane Analysis; New Applications for Global Tectonics.* Topics in the Earth Sciences, **8**, Chapman and Hall, London.

HOWELL, D. G., CROUCH, J. K., GREENE, H. G., McCULLOCH, D. S. & VEDDER, J. G. 1980. Basin development along the late Mesozoic and Cainozoic California margin: a plate tectonic margin of subduction, oblique subduction and transform tectonics. *In*: BALANCE, P. F. & READING, H. G. (eds) Sedimentation in oblique-slip mobile zones. *Special Publication of the International Association of Sedimentologists*, **4**, 43–62.

HU, A., JAHN, B.-M., ZHANG, G., CHEN, Y. & ZHANG, Q. 2000. Crustal evolution and Phanerozoic crustal growth in northern Xinjiang: Nd isotopic evidence. Part I. Isotopic characterization of basement rocks. *Tectonophysics*, **328**, 15–51.

HUANG, X. 1990. Isotope geochronology and geochemistry. *In*: ZHANG, C. (ed.) *Geological Character and Mineralization of the Western Junggar Ophiolites in Xinjiang, China*, 104–183.

HUO, Y. 1984. Petrochemical and REE characteristics of the ophiolites at Tangbale, West Junggar, Xinjiang and their geological significance. [*Bulletin of the Xi'an Institute of Geological Sciences*], **1984**, 82–94.

HUO, Y. 1985. Characteristics of REE and petrochemistry of the ophiolites and their geological significance in Darbut, West Junggar, Xinjiang. [*Bulletin of the Xi'an Institute of Geological Sciences*], **10**, 55–68.

HUO, Y. 1987. Petrochemistry and geological evolution of island arc and back-arc basin, Hoboksair of Junggar, Xinjiang. [*Bulletin of the Xi'an Institute of Geological Sciences*], **15**, 57–68.

INGER, S. 1994. Magmagenesis associated with extension in orogenic belts: examples from the Himalaya and Tibet *Tectonophysics*, **238(1–4)**, 183–197.

JAHN, B.-M., GRIFFIN, W. L. & WINDLEY, B. 2000. Continental growth in the Phanerozoic: evidence from Central Asia. *Tectonophysics*, **328**, vii–x.

JIN, C. & XU, Y. 1997. Petrology and genesis of the Bieluagaxi granitoids in Tuoli, Xinjiang, China. *Acta Petrologica Sinica*, **13**, 529–537.

JIN, C. & ZHANG, X. 1993. Geochronology and genesis of the western Junggar granitoids, Xinjiang, China. *Scientia Geologica Sinica*, **28**, 28–36.

JONES, D. L., HOWELL, D. G., CONEY, P. J. & MONGER, J. W. H. 1983. Recognition, character and analysis of tectonostratigraphic terranes in western North America. *Journal of Geological Education*, **31**, 295–303.

KWON, S. T., TILTON, G. R., COLEMAN, R. G. & FENG, Y. 1989. Isotopic studies bearing on the tectonics of the West Junggar region, Xinjiang, China. *Tectonics*, **8**, 719–727.

LI, H. 1991. First discovery of Middle Silurian Radiolaria fossils in Xinjiang. [*Scientia Geologica Sinica*], **1991**, 75.

LI, H. 1994. Middle Silurian radiolarians from Keerhada, Xinjiang, China. *Acta Micropalaeontologica Sinica*, **11**, 259–272.

MA, X. 1987. *Lithospheric Dynamics Map of China and Adjacent Areas*, 1:4 000 000. Geological Publishing House, Beijing.

MAEKAWA, H., SHOUZUI, M., ISHII, T., FRYER, P. & PEARCE, J. A. 1993. Blueschist metamorphism in an active subduction zone. *Nature*, **364**, 520–523.

MULLEN, E. D. 1983. $MnO/TiO_2/P_2O_3$: a minor element discrimination for basaltic rocks of oceanic environments and its implications for petrogenesis. *Earth and Planetary Science Letters*, **62**, 53–62.

NGDC 2001. *Surface of the Earth, 2 Minute Color Relief Images.* World Wide Web Address: http://www.ngdc.noaa.gov/mgg/image/2minrelief.html.

NISBET, E. G. & PEARCE, J. A. 1977. Clinopyroxene composition in mafic lavas from different tectonic settings. *Contributions to Mineralogy and Petrology*, **63**, 149–160.

PEARCE, J. A. & CANN, J. R. 1973. Tectonic setting of basic volcanic rocks determined using trace element analysis. *Earth and Planetary Science Letters*, **19**, 290–300.

PEARCE, J. A. & NORRY, M. J. 1979. Petrogenic implications of Ti, Zr, Y and Nb variations in volcanic rocks. *Contributions to Mineralogy and Petrology*, **69**, 33–47.

PEARCE, T. H., GORMAN, B. E. & BIRKETT, T. C. 1977. The relationship between major element chemistry and tectonic environment of basic and intermediate volcanic rocks. *Earth and Planetary Science Letters*, **36**, 121–132.

PENG, G. 1996. Podiform chromite and associated ophiolitic rocks in West Junggar, Xinjiang, NW China. *American Association of Petroleum Geologists Bulletin*, **80**, 984–985.

PENG, G., BAO, P., WANG, X., JIN, Y. & HAO, Z. 1992. Genesis of chromite deposit in the Hongguleleng Ophiolite, Xinjiang. *Bulletin of the Institute of Geology, Chinese Academy of Geological Sciences*, **23**, 64–72.

PENG, G., LEWIS, J. F., LIPIN, B. R., McGEE, J. J., BAO, P. & WANG, X. 1995. Inclusions of phlogopites and their hydrates in chromite from the Hongguleleng Ophiolite in Xinjiang, NW China. *American Mineralogist*, **80**, 1307–1316.

PE-PIPER, G., PIPER, D. J. W. & CLERK, S. B. 1991. Persistent mafic igneous activity in an A-type granite pluton, Cobequid Highlands, Nova Scotia. *Canadian Journal of Earth Sciences*, **28**, 1058–1072.

Qı, J. 1993. Geology and genesis of dike swarms in western Junggar, Xinjiang, China. [*Acta Petrologica Sinica*], **9**, 288–299.

SENGÖR, A. M. C. & NATAL'IN, B. A. 1996. Turkic-type orogeny and its role in the making of the continental crust. *Annual Review of Earth and Planetary Sciences*, **24**, 263–337.

SENGÖR, A. M. C., NATAL'IN, B. A. & BURTMAN, V. S. 1993. Evolution of the Altaid tectonic collage and Palaeozoic crustal growth in Eurasia. *Nature*, **364**, 299–307.

SHERVAIS, J. W. 1982. Ti–V plots and the petrogenesis of modern and ophiolitic lavas. *Earth and Planetary Science Letters*, **59**, 101–118.

SHERVAIS, J. W. 2001. Birth, death, and resurrection; the life cycle of suprasubduction zone ophiolites. *Geochem. Geophys. Geosyst.*, **2**, paper number 2000GC000080, 1–45.

STERN, R. J. & BLOOMER, S. H. 1992. Subduction zone infancy: examples from the Eocene Izu–Bonin–Mariana and Jurassic California arcs. *Geological Society of America Bulletin*, **104**, 1621–1636.

SUN, S. S. & McDONOUGH, W. F. 1989. Chemical and isotopic sytematics of oceanic basalts: implications for mantle composition and processes. *In*: SAUNDERS, A. D. & NORRY, M. J. (eds) *Magmatism in the Ocean Basins*. Geological Society, London, Special Publications, **42**, 313–345.

TAPPONNIER, P., MERCIER, J. L., ARMIJO, R., HAN, T. & ZHOU, J. 1981. Field evidence for active normal faulting in Tibet. *Nature*, **294**, 410–414.

TILTON, G. R., KWON, S. T., COLEMAN, R. G. & XIAO, X. 1986. Isotopic studies from the West Junggar Mountains, NW China. *Geological Society of America Abstracts and Programs*, **18**, 773.

WANG, B. 1990. The late Ordovician Tabulata and heliolitidea in Xinjiang and their biostratigraphical significance. [*Xinjiang Geology*], **8**, 61–79.

WANG, G. 1996. Classification of tectonic units and geologic evolution in northern Xinjiang and neighboring areas. [*Xinjiang Geology*], **14**, 12–27.

WANG, H. 1986. Tectonic evolution and oil pools of Junggar Basin, Xinjiang. *China Oil*, **3**, 26–30.

WANG, S. & WANG, Y. 1986. Formation of the intermontane basins of the Tianshan System in China. *Mountain Research*, **4**, 287–294.

WATSON, M. P., HAYWARD, A. B., PARKINSON, D. N. & ZHANG, Z. H. 1987. Plate tectonic history, basin development and petroleum source rock deposition onshore China. *Marine and Petroleum Geology*, **4**, 205–225.

WU, F. Y., JAHN, B. M., WILDE, S. & SUN, D. Y. 2000. Phanerozoic crustal growth: U–Pb and Sr–Nd isotopic evidence from the granites in northeastern China. *Tectonophysics*, **328**, 89–113.

WU, H. & PAN, Z. 1991. 'Structural complex' and its geological significance; in the case of western Junggarian complexes. [*Scientia Geologica Sinica*], **1991**, 1–8.

XU, W., ZHANG, K., GAO, M. & YAO, H. 1987. Evolution of plate tectonics of northeastern Junggar Basin and its control over oil and gas. [*Oil & Gas Geology*], **8**, 163–170.

YANG, F. 1997. *Geochemistry and magmatic dynamics of alkaline granites in North and West Junggar regions*. PhD, Chinese Academy of Science.

YANG, J. S., ROBINSON, P. T., ZHOU, M. F., CHEN, Y., BAI, W. & HU, X. 1994. Ophiolite-related gold deposits in West Junggar, NW China. *Geological Association of Canada; Mineralogical Association of Canada; Annual Meeting; Program with Abstracts*, **19**, 122.

YANG, R. 1993. Geochemical characteristics of the Middle Ordovician volcanic rocks, western Junggar, Xinjiang, China. [*Geochimica*], **1993**, 399–408.

YIN, A. & NIE, S. 1996. A Phanerozoic palinspastic reconstruction of China and its neighbouring regions. *In*: YIN, A. & HARRISON, T. M. (eds) *Tectonic Evolution of Asia*. Cambridge University Press, New York, 442–485.

ZHANG, C., ZHAI, M., ALLEN, M. B., SAUNDERS, A. D., WANG GUANG, R. & HUANG, X. 1993. Implications of Palaeozoic ophiolites from western Junggar, NW China, for the tectonics of central Asia. *Journal of the Geological Society of London*, **150**, 551–561.

ZHANG, L. 1997. The $^{40}Ar/^{39}Ar$ metamorphic ages of Tangbale blueschists and their geological significance in West Junggar of Xinjiang. *Chinese Science Bulletin*, **42**, 1902–1904.

ZHANG, Q. & WEI, Z. 1989. Formation age of the Darbut Fault belt, West Junggar. *Xinjiang* [*Xinjiang Petroleum Geology*], **10**, 35–38.

ZHANG, Z., ZUO, Z. & LIU, S. 1998. Age and tectonic significance of the mafic dyke swarm in the Kuruktag region, Xinjiang. *Acta Geologica Sinica (English Edition)*, **72**, 29–36.

ZHANG, Z. M., LIOU, J. G. & COLEMAN, R. G. 1984. An outline of the plate tectonics of China. *Geological Society of America Bulletin*, **95**, 295–312.

ZHOU, L. 1986. The polycyclic development and migration of the West Junggar Geosyncline. [*Bulletin of the Xi'an Institute of Geological Sciences*], **13**, 25–36.

ZHOU, M. & BAI, W. 1994. The origin of the podiform chromite deposits. *Mineral Deposits*, **13**, 242–249.

ZHOU, M. F. & ROBINSON, P. T. 1994. High-Cr and high-Al podiform chromitites, western China: relationship to partial melting and melt/rock reaction in the upper mantle. *International Geology Review*, **36**, 678–686.

ZHOU, M. F., ROBINSON, P. T., MALPAS, J., AITCHISON, J., SUN, M., BAI, W. J., HU, X. F. & YANG, J. S. 2001. Melt/mantle interaction and melt evolution in the Sartohay high-Al chromite deposits of the Dalabute Ophiolite (NW China). *Journal of Asian Earth Sciences*, **19**, 517–534.

ZHU, B., FENG, Y. & YE, L. 1987a. Paleozoic ophiolites in western Junggar and their geological significance. *Contributions to the Project of Plate Tectonics in Northern China*, **2**, 19–28.

ZHU, B., WANG, L. & WANG, L. 1987b. Paleozoic era ophiolite of the southwestern part of western Junggar, Xinjiang, Sinkiang Uighur, China. *Zhongguo Dizhi Kexueyuan Xian* [*Bulletin of the Xi'an Institute of Geological Sciences*], **17**, 1–64.

ZHU, B., ZHU, S., REN, Y. & WANG, L. 1983. The characteristics and origin of Middle Ordovician ophiolite in West Junggar, Xingjiang. *Contributions to the Project on Plate Tectonics, Northern China*, **1**, 64–88.

ZONENSHAIN, L. P., KUZMIN, M. I., NATAPOV, L. M. & PAGE, B. M. 1990. *Geology of the USSR; a Plate-Tectonic Synthesis*. American Geophysical Union Geodynamics Series, **21**. American Geophysical Union, Washington, DC.

Nb-depleted, continental rift-related Akaz metavolcanic rocks (West Kunlun): implication for the rifting of the Tarim Craton from Gondwana

CHAO YUAN[1,2], MIN SUN[2], JINGSUI YANG[3], HUI ZHOU[4] & MEI-FU ZHOU[2]

[1]*Guangzhou Institute of Geochemistry, Chinese Academy of Sciences, Guangzhou 510640, China*

[2]*Department of Earth Sciences, The University of Hong Kong, Pokfulam Road, Hong Kong, China (e-mail: minsun@hkucc.hku.hk)*

[3]*Institute of Geology, Chinese Academy of Geological Sciences, 26 Baiwanzhuang Road, Beijing 100073*

[4]*Department of Geology, Peking University, Beijing 100871, China*

Abstract: The Akaz metavolcanic rocks of the West Kunlun Mountains possess low to intermediate SiO_2 (42.3–64.7 wt%) and MgO (2.69–7.54 wt%) and high TiO_2 (0.94–3.05 wt%) and $Fe_2O_3^T$ (7.64–18.47 wt%), indicating a basaltic to andesitic protolith. These rocks have high contents of Zr (89.6–470 ppm), Nb (10.0–40.3 ppm), Y (19.7–52.7 ppm), Th (0.86–15.96 ppm) and total REE (67.7–407 ppm), and are characterized by relatively high Ti/Y (183–649), Th/Yb (0.5–3.9), and low Hf/Ta (3.0–8.6) ratios. They are LREE-enriched (La/Yb = 5.4–20) and most have small negative Nb anomalies (Nb/Nb* = 0.20–1.16). These characteristics are transitional between within-plate and subduction-related basalts. The relatively high Gd/Yb ratios (1.4–2.9) distinguish these rocks from island-arc tholeiites and the high Zr/Y (3–12), Ta/Yb (0.3–0.7) and low Zr/Nb (<12) ratios strongly support a continental affinity. The protoliths for the Akaz metavolcanic rocks are interpreted to be continental rift basalts formed during rifting of the Tarim Craton from Gondwana. Stratigraphic and palaeontological data indicate that the rifting occurred in Sinian to Cambrian times, roughly contemporaneously with rifting in the East Kunlun and North Qilian orogenic belts farther to the east.

Understanding the tectonic evolution of the Western Kunlun mountain range is important for unravelling the early history of the Tibetan Plateau. A previously proposed collisional model envisaged the West Kunlun as a tectonic collage composed of a continental block (North Kunlun Block) and an accreted arc terrane (South Kunlun Block) (Yao & Hsü 1994; Hsü *et al.* 1995; Sengör & Natal'in 1996; Li *et al.* 1999; Xiao *et al.* 2002). However, an alternative model considers both the North and South Kunlun blocks to be parts of the Tarim Craton (Pan *et al.* 1994; Ding *et al.* 1996; Mattern *et al.* 1996; Mattern & Schneider 2000). The Tarim Craton itself, as a major block of the Chinese continent (Fig. 1), was originally part of Gondwana. The Tarim Craton was rifted from Gondwana by the opening of the Tethyan ocean (*sensu lato*) (Li *et al.* 1991; Li *et al.* 1996; Metcalfe 1996; Zhao *et al.* 1996; Li 1998; Stampfli & Borel 2002). The Akaz meta-volcanic sequence in the West Kunlun crops out along the south margin of the North Kunlun Block, and may provide critical constraints on the above tectonic models. The sequence is considered to be the result of Neoproterozoic continental rifting (Pan *et al.* 1994), or to be part of a seamount accreted during subduction of the Tethyan ocean basin (Xiao *et al.* 2002). These different interpretations stem from the lack of a comprehensive geochemical study of the volcanic rocks. In this paper, we present new systematic geochemical data for the Akaz metavolcanic rocks, and use these data to constrain their petrogenesis and to shed light on the tectonic evolution of the West Kunlun.

Geological setting

The Tibetan Plateau was formed by the successive accretion of several microcontinental and arc blocks from Gondwana to the Laurasian continent (Chang *et al.* 1986; Molnar *et al.* 1987;

From: MALPAS, J., FLETCHER, C. J. N., ALI, J. R. & AITCHISON, J. C. (eds) 2004. *Aspects of the Tectonic Evolution of China*. Geological Society, London, Special Publications, **226**, 131–143.

Dewey *et al.* 1988; Yin & Harrison 2000) (Fig. 1). These accreted blocks are separated by several suture zones that young progressively to the south. The Kunlun Mountains in the northern-most part of the Plateau are divided by the Altyn Tagh Fault into eastern and western segments (Fig. 1). The West Kunlun consists of the North Kunlun and South Kunlun Blocks, separated by the early Paleozoic Kudi–Subashi suture (Pan *et al.* 1994; Matte *et al.* 1996; Yang *et al.* 1996) (Fig. 1). The North Kunlun Block is in fault contact with the Tarim Craton in the north, whereas the South Kunlun Block is bounded by the strike-slip Karakash Fault to the south (Matte *et al.* 1996), which is coincident with one of the sutures of the Palaeo-Tethys (Pan *et al.* 1994; Mattern *et al.* 1996) (Fig. 1).

The basement of the North Kunlun Block consists of the Precambrian Ailiankate Complex, made up of metaclastic and carbonate rocks with subordinate metavolcanic rocks (Wen *et al.* 2000) (Figs 2 and 3). These rocks were intruded by 2.2 Ga granitic plutons (Xu *et al.* 1994) suggesting that the Precambrian basement of the North Kunlun Block and the Tarim Craton are the same (Pan *et al.* 1994; Hsü *et al.* 1995; Ding *et al.* 1996; Matte *et al.* 1996; Mattern & Schneider 2000; Yuan *et al.* 2002, 2003; Xiao *et al.* 2002).

The Sailajaz Tagh Group, a metavolcanic sequence with some interlayered metasedimentary rocks, unconformably overlies the Ailiankate Complex (Fig. 2). The metavolcanic sequence occurs in the lower part of the Sailajaz Tagh Group, and is referred to as the Akaz greenschist in the literature (Deng 1989; Pan *et al.* 1994; Wen *et al.* 2000) (Fig. 3). These metavolcanic rocks include basalt, spilite, quartz-keratophyre, felsite-porphyry, potash keratophyre and rhyolite, intercalated with silicic and mafic tuff. Thin

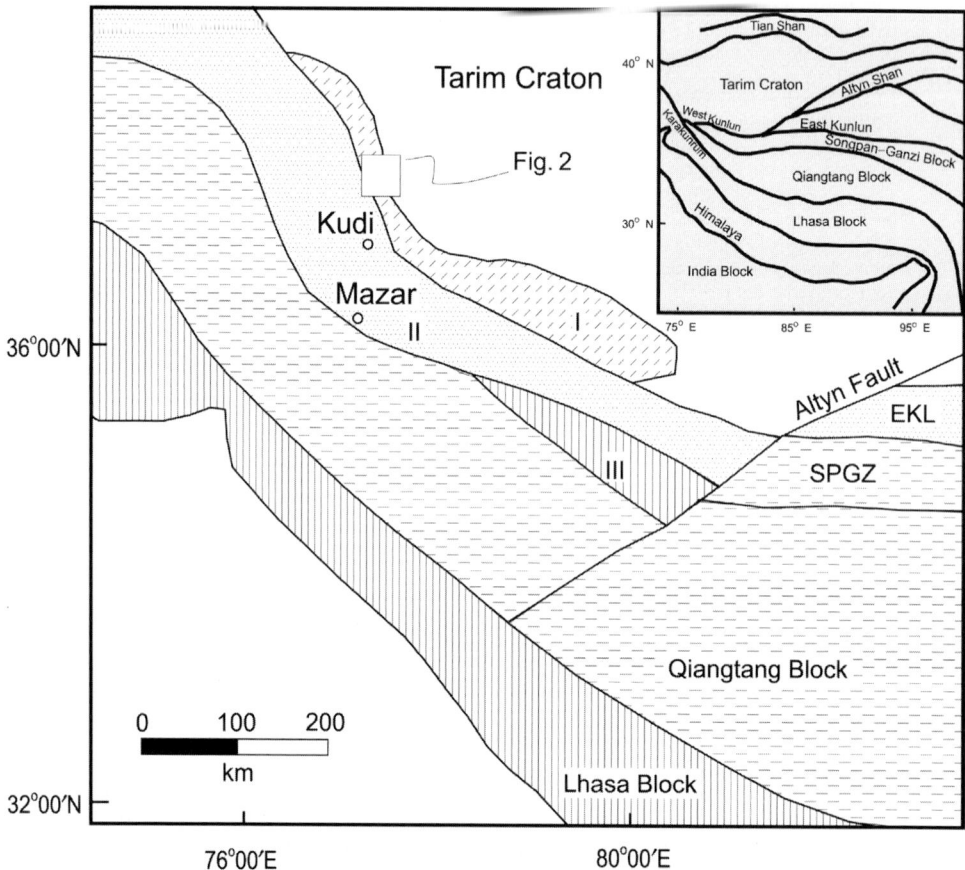

Fig. 1. Main tectonic blocks of the Tibet Plateau and adjacent regions. I, North Kunlun Block; II, South Kunlun Block; III, Tianshuihai Block; EKL, East Kunlun; SPGZ, Songpan Ganzi.

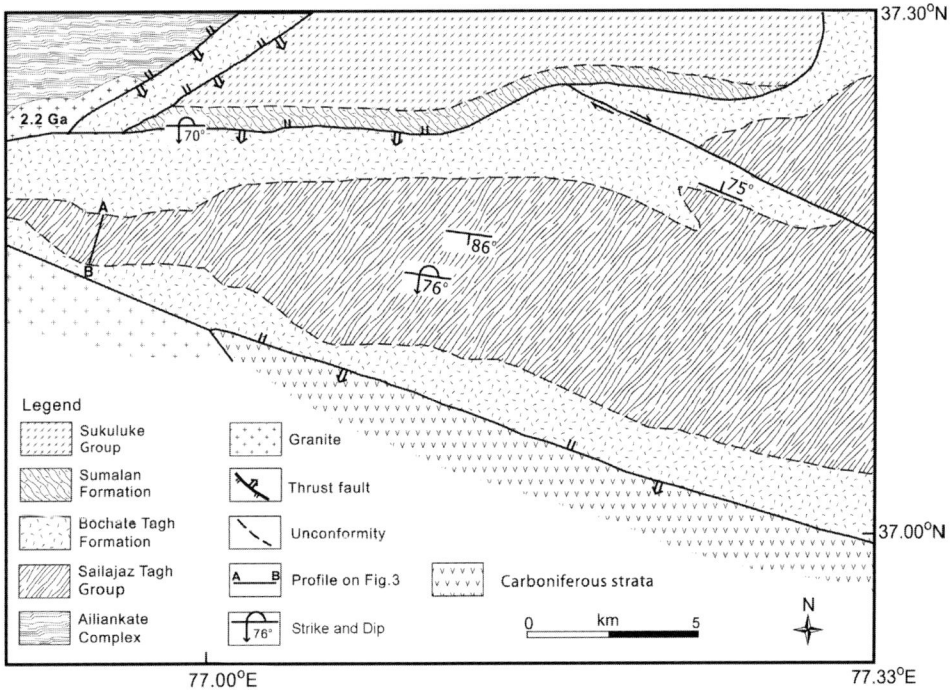

Fig. 2. Geological map of the Akaz pass area, western Kunlun Mountains (modified from XGIT-II 1985).

layers or lenses of metamorphosed sandstone, siltstone and limestone are intercalated with the metavolcanic rocks (XGIT-II 1985) and the entire sequence is conformably overlain by dolomite, marble, phyllite, sandstone and schist. The relative abundance of metavolcanic rocks in the Sailajaz Group increases from east to west, where they become dominant.

Samples for this study were collected in the Akaz Pass, where the Xinjiang–Tibet road transects the western part of the Sailajaz Tagh Group and exposes the entire metavolcanic sequence (Figs 2 and 3). The metavolcanic rocks in this area are dominantly basalt and basaltic andesite flows with sparse tuff (Deng 1989). Six layers of metavolcanic rocks, from 20 to more than 100 m thick, crop out south of the Akaz Pass. The bottom layer is massive and relatively fresh, whereas the other layers are variably fractured, altered and decomposed due to neotectonic activity and weathering. Twenty samples were collected at regular intervals across the bottom layer at milestone 124.6 (Fig. 2).

Isotope geochronological data are not currently available for the Sailajaz Tagh Group. The unit has been tentatively placed in the Sinian (Late Neoproterozoic) or Cambrian, because it contains stromatolites and crinoid fossils and is unconformably overlain by Devonian to Triassic strata (Pan *et al.* 1994; Xiao *et al.* 2002). To the east, Cambrian rift-related volcanic rocks have been recognized in East Kunlun (Fig. 1) (Pan *et al.* 1996). Even farther east, in the North Qilian Mountains, single-grain zircon dating and whole-rock Sm–Nd dating of rift-related volcanic rocks yielded ages of 738–604 Ma and 522 to 593 Ma, respectively (Xia *et al.* 1996; Mao *et al.* 1998; Xia *et al.* 1999). These ages are thought to be roughly consistent with that of the Akaz metavolcanic rocks in the West Kunlun.

Petrography

The Akaz metavolcanic samples consist of chlorite, epidote, albite, quartz, calcite, and magnetite with, or without, biotite. Some samples (GS-13, 14, 18, 19 and 21) contain zoisite instead of epidote. Three samples (GS-13, 15 and 19) have relatively few mafic minerals, but more quartz and feldspar, and contain biotite that does not occur in other samples. The chlorite, epidote and albite are anhedral to subhedral, and most albite crystals do not show polysynthetic twinning. Muscovite was identified in one sample (GS-13).

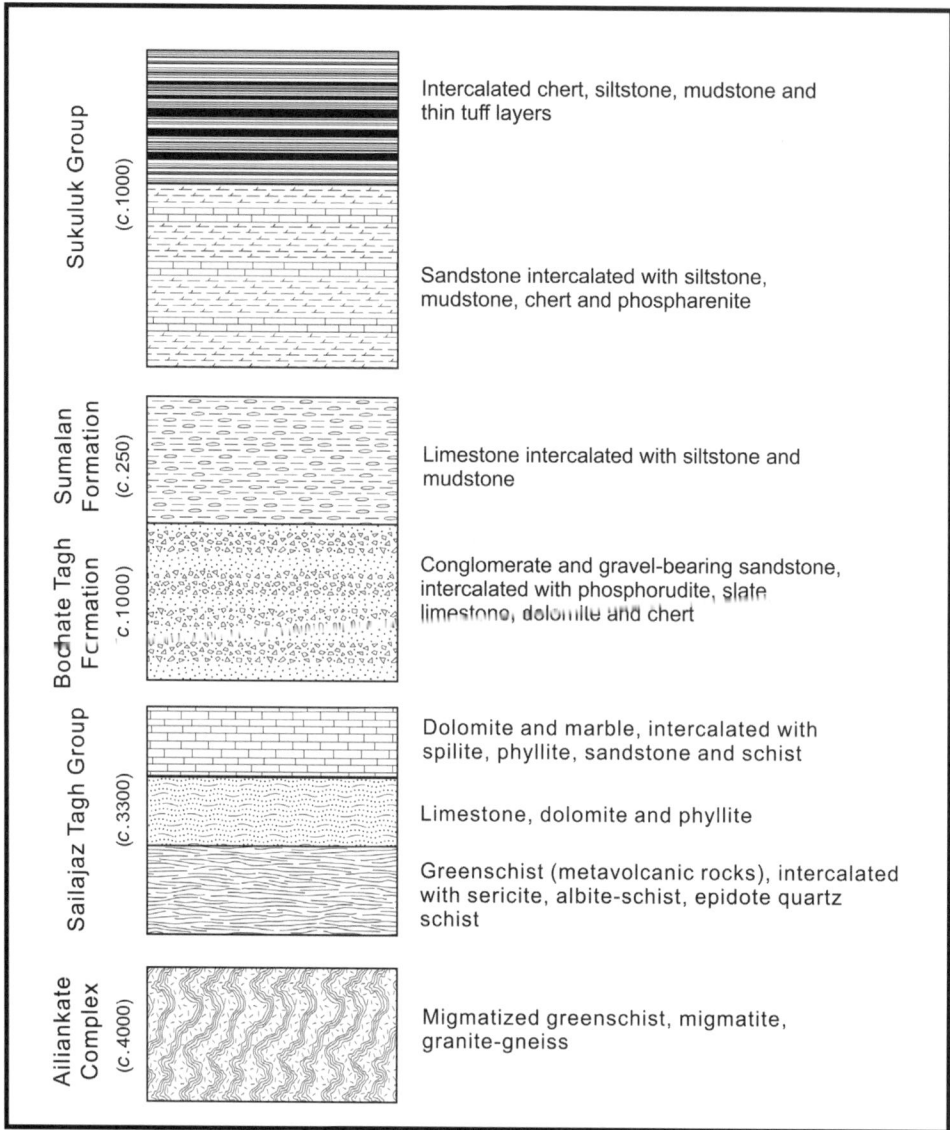

Fig. 3. Stratigraphic column of the North Kunlun Block, West Kunlun Mountains (XGIT-II, 1985 and Wen *et al.* 2000). Estimated thickness of units in metres given in brackets.

The Akaz metavolcanic samples are strongly deformed, and the original textures and structures are mostly obliterated. These metavolcanic rocks exhibit schistosity, comprised of green bands of chlorite and epidote and light-coloured bands of albite and quartz. The metamorphic mineral compositions and textures of the rocks indicate low-temperature and moderate-pressure greens-chist-facies metamorphism (Williams *et al.* 1989).

Analytical methods

Whole-rock samples were crushed using a jaw crusher, and the resulting chips were cleaned three times with de-ionized water in an ultrasonic vessel (15 minutes each time), dried and then ground into powder using an agate mill. Samples were dissolved with mixed acid ($HF + HNO_3$) in Teflon screw-capped vials to ensure complete digestion. Major oxides were measured on a Varian

VISTA-PRO ICP–AES in the Guangzhou Institute of Geochemistry, Chinese Academy of Sciences, whereas trace elements were analysed on a Perkin Elmer Elan 6000 ICP–MS, installed in the same institute. Chinese (GSR1-5) and international rock standards (W-2, MRG-1, G-1, SY-4 and GSP-1) were used as either reference materials or external standards to monitor the analytical accuracy. Samples were separately dissolved using the NaOH sinter method and measured for SiO_2 content with the ICP–AES. The precision for major oxides is within 0.5–1% RSD (relative standard deviation), whereas those for trace elements are better than 5% (for Rb, Sr, Cs, Ba, Y, Zr, Nb, Ta, U, Th, and Hf or 3% RSD rare-earth elements (REEs). The results are presented in Table 1.

Analytical results

The Akaz metavolcanic rocks vary widely in SiO_2 contents (42–64 wt%), but the majority fall between 45 and 55 wt%. The Al_2O_3 (12.18–16.91 wt%) and $Fe_2O_3^T$ (c.18.47 wt%) contents fall within the range of MORB compositions, but they have relatively low MgO (2.69–7.54 wt%) and P_2O_5 (0.08–0.36 wt%), and relatively high TiO_2 (0.94–3.05 wt%) concentrations.

All the samples in this study exhibit LREE-enriched patterns (La/Yb = 5.4–20), with various Eu anomalies (Eu/Eu* = 0.74–1.4) and slightly negative Ce (Ce/Ce* = 0.87–0.97) anomalies (Table 1; Fig. 4). Most samples have LILE and HFSE concentrations higher than those of E-MORB (Sun & McDonough 1989), but very similar to those of the Deccan continental flood basalts (e.g. Wilson 1989) (Table 1; Fig. 5). Their high La–Nb (0.92–2.49) and Th–Nb (0.09–1.29) ratios indicate considerable HFSE depletion relative to LILE (Table 1). Most samples show various degree of Nb depletion (Nb/Nb* = 0.20–0.65) in spider diagrams (Fig. 5) and the most Nb-depleted sample GS-19 (Nb/Nb* = 0.20) has the highest SiO_2 (64.7 wt%) and the lowest P_2O_5 (0.08 wt%) and TiO_2 (0.94 wt%) contents. Two samples (GS-7 and GS-18), with the lowest contents of LILE and HFSE have slightly positive Nb anomalies (Nb/Nb* c.1.16) (Fig. 5, Table 1).

Discussion

Nature of the Akaz metavolcanic rocks

Most Akaz metavolcanic samples are mafic in composition, although three samples are intermediate (SiO_2 = 54.2–55.1 wt%) and one

sample (GS-19) is relatively silicic, with a SiO_2 content of 64.7 wt%. The relatively high TiO_2, $Fe_2O_3^T$ and MgO contents, although variable, are consistent with their mafic compositions. The rocks have variable CaO (2.05–9.96 wt%), K_2O (0.09–5.29 wt%), Na_2O (0.15–5.3 wt%) and LOI contents (1.94–8.53 wt%), reflecting the effects of hydrothermal alteration. Accordingly, only relatively immobile elements are reliable indicators of the protolith composition (Lightfoot 1993). Because Nb, Zr, Ti and Y are relatively insensitive to alteration, the Nb/Y v. Zr/TiO_2 diagram was used to classify the rocks. Most of the Akaz metavolcanic rocks plot in the field of subalkaline basalt, whereas the three samples with relatively high SiO_2 contents plot in the alkaline basalt (GS-16), trachyandesite (GS-13) and dacite (GS-19) fields (Fig. 6a), suggesting that the protoliths of the metavolcanic rocks are dominantly tholeiitic basalts. In the Mg# v. TiO_2 diagram, almost all the samples plot in the high-Ti tholeiite field (Fig. 6b). Their low Cr (77.3–238 ppm) and Ni (28.6–72.3) contents, suggest that all these samples were derived from an evolved magma.

Origin and tectonic setting of the Akaz metavolcanic rocks

Tholeiitic basalts can occur in a variety of tectonic settings, e.g. oceanic floor, oceanic plateau, oceanic island, seamount, island arc, back-arc basin and continental interior (Wilson 1989; Flower 1991; Floyd 1991; Saunders & Tarney 1991). The intermediate Nb/Y ratios and relatively high LILE (Th, U) and HFSE (Nb, Ta, Zr, Hf, Ti, Y) contents of the Akaz metavolcanic rocks suggest that the basaltic protolith might have been derived from an enriched mantle source.

The Akaz metavolcanic rocks have transitional geochemical signatures of between within-plate and subduction-related basalts. On one hand, the high contents of HFSE (e.g. Ti, Nb and Ta), relatively high Ti/Y (mostly >350) and low Hf/Ta (mostly <5) ratios make the rocks akin to within-plate basalts (Condie 1989). On the other hand, the negative Nb anomalies, relatively high Th/Nb (0.1–1.3), La/Nb (0.9–2.5) and Th/Yb (0.5–3.9) ratios show an arc-related signature for the metavolcanic rocks. Therefore, different tectonic discrimination diagrams give different results. For example, in the Zr-Zr/Y diagram, most of the Akaz samples plot in the within-plate basalt field (Pearce & Cann 1973) (Fig. 7a); whereas in the Ta/Yb v. Th/Yb

Table 1. *Representative chemical analyses of Akaz metavolcanics, West Kunlun*

Sample	GS-1	GS-2	GS-3	GS-4	GS-5	GS-6	GS-7	GS-8	GS-9	GS-10	GS-11	GS-12	GS-13	GS-14	GS-15	GS-16	GS-17	GS-18	GS-19	GS-21
SiO_2	48.59	45.69	54.23	50.43	44.96	43.51	46.37	46.00	46.23	49.92	47.30	49.84	55.13	50.16	47.10	54.33	42.33	46.00	64.70	45.15
TiO_2	1.95	3.05	2.28	2.06	2.67	2.37	2.13	2.18	2.18	2.13	2.19	1.98	1.83	2.25	2.12	1.98	2.98	1.96	0.94	2.37
Al_2O_3	12.63	16.91	14.38	13.06	16.85	14.92	13.04	14.00	14.03	13.21	13.91	12.63	14.77	12.18	13.75	13.72	13.6	14.05	12.53	14.85
$Fe_2O_3^T$	13.16	16.29	14.93	13.17	18.47	15.65	13.72	15.11	15.56	16.38	15.44	13.78	8.88	12.78	15.10	12.30	15.79	13.73	7.64	15.88
MnO	0.25	0.17	0.12	0.21	0.12	0.23	0.19	0.25	0.23	0.18	0.20	0.27	0.07	0.14	0.24	0.11	0.18	0.22	0.09	0.18
MgO	4.88	5.38	3.69	4.59	3.97	5.51	7.40	7.06	4.88	3.56	6.33	4.28	2.69	4.81	6.85	5.24	6.14	7.54	2.75	6.68
CaO	7.41	2.58	2.21	5.99	3.23	5.92	7.46	5.75	5.80	4.84	7.38	5.48	2.05	6.79	5.42	3.67	7.61	9.96	3.87	4.84
Na_2O	4.35	3.59	3.63	4.64	2.84	5.14	0.98	3.59	5.30	4.99	3.79	4.46	0.15	2.31	3.52	2.42	1.97	1.55	3.17	2.90
K_2O	0.09	2.42	2.00	0.14	3.44	0.34	2.21	0.11	0.22	0.60	0.29	0.33	5.29	3.47	0.12	3.12	3.59	1.53	2.27	1.66
P_2O_5	0.25	0.19	0.22	0.24	0.34	0.26	0.17	0.31	0.34	0.28	0.30	0.29	0.36	0.30	0.30	0.28	0.35	0.20	0.08	0.26
LOI	6.47	3.62	2.33	5.41	3.11	5.69	6.26	5.61	5.08	3.92	2.82	5.62	8.53	4.58	5.51	2.84	5.34	3.23	1.94	5.22
Total	100.03	99.89	100.02	99.94	100.00	99.54	99.93	99.97	99.85	100.01	99.95	99.96	99.75	99.77	100.03	100.01	99.88	99.97	99.98	99.77
Mg#*	0.48	0.45	0.38	0.47	0.35	0.47	0.58	0.54	0.44	0.35	0.51	0.44	0.43	0.49	0.53	0.52	0.49	0.58	0.48	0.51
Sc	32.8	37.9	26.5	27.6	31.0	49.7	38.5	29.6	31.1	28.0	26.8	24.9	19.7	27.4	29.6	22.1	31.9	12.6	11.2	35.0
V	214	205	121	237	218	411	329	304	246	286	283	239	209	282	309	285	346	163	157	263
Cr	112.0	113.0	105.0	108.0	131.0	130.0	156.0	117.0	107.0	77.3	87.8	12..0	237.0	181.0	123.0	129.0	152.0	115.0	97.0	140.0
Ni	54.5	62.0	45.9	45.8	64.1	72.2	66.8	50.8	50.5	45.9	53.5	5..7	30.1	45.1	50.7	44.7	53.4	28.9	28.6	72.2
Cu	26.7	39.3	16.1	38.2	8.99	35.5	96.1	67.1	20.5	9.64	58.9	6..8	6.08	5.08	66.3	10.4	5.63	59.8	56.6	62.5
Zn	168.0	125.0	89.8	162.0	112.0	198.0	113.0	207.0	113.0	90.6	133.0	125.0	71.5	122.0	202.0	124.0	154.0	58.4	57.6	125.0
Rb	2.52	87.90	72.70	4.00	118.00	11.30	111.00	3.22	6.30	20.80	7.15	11.00	118.00	109.00	3.18	96.60	113.00	70.80	112.40	73.70
Sr	124.0	58.0	67.6	113.0	77.6	106.0	316.0	153.0	112.0	91.4	557.0	94.2	75.5	162.0	160.0	171.0	171.0	365.0	413.0	155.0
Ba	44.6	708.0	672.0	61.9	921.0	85.7	132.0	39.2	99.5	175.0	78.7	97.8	966.0	526.0	40.0	703.0	554.0	87.0	241.0	303.0
U	0.656	0.880	0.676	0.557	1.120	0.969	0.498	0.475	0.662	0.593	0.583	0.5.2	2.140	0.974	0.466	1.790	1.220	0.448	2.290	0.587
Th	2.46	4.03	3.06	2.51	3.30	2.92	0.94	2.74	2.67	2.72	2.90	2.5.	16.0	9.19	2.70	15.5	6.90	0.86	13.40	2.25
Y	26.2	28.3	29.2	26.5	31.4	30.5	19.7	27.3	26.6	25.1	26.4	26.0	39.9	44.3	27.2	36.7	52.7	29.2	30.7	24.6
Zr	126	184	145	130	175	142	94	140	136	142	140	11.	470	189	138	363	250	89	307	133
Nb	12.3	18.6	14.5	12.9	17.5	14.9	10.2	13.8	13.6	14.0	14.5	12.5	40.3	22.6	13.9	37.3	28.7	10.0	10.3	14.1
Ga	18.3	20.5	20.1	18.0	24.5	23.1	18.5	22.3	18.5	18.6	18.6	18.5	24.9	16.3	21.6	21.4	19.8	19.3	19.1	18.1

Hf	3.16	4.35	3.86	3.28	4.42	3.76	2.48	3.51	3.55	3.56	3.93	3.33	13.20	5.28	3.62	9.73	6.72	2.41	8.95	3.62
Ta	0.88	1.36	1.06	0.90	1.22	1.02	0.76	0.96	1.19	0.98	1.21	0.92	3.03	1.74	0.98	3.24	1.85	0.76	1.04	1.03
La	18.6	39.3	27.1	20.3	30.6	25.0	9.4	22.7	23.4	23.4	23.9	21.6	57.0	28.7	23.2	93.0	37.0	9.8	23.1	23.6
Ce	38.2	73.7	57.6	42.0	61.8	49.9	21.9	45.8	45.3	47.5	48.8	42.6	111.0	56.0	47.3	168.0	72.0	22.2	49.4	48.9
Pr	5.17	8.56	7.16	5.38	8.00	6.59	3.21	5.94	5.93	6.11	6.47	5.65	13.6	7.92	6.22	21.3	9.97	3.33	6.26	6.28
Nd	22.3	34.2	29.6	22.3	33.1	27.5	14.7	24.6	24.7	25.4	26.9	23.8	51.4	32.8	25.8	78.5	40.7	15.2	24.2	27.7
Sm	4.39	6.54	5.97	4.56	6.69	5.59	3.56	4.97	4.90	4.95	5.45	4.75	9.20	6.98	5.21	12.20	8.53	3.63	4.96	5.63
Eu	1.44	2.44	1.93	1.57	2.07	1.87	1.30	1.71	1.59	1.63	1.80	1.61	2.18	3.45	1.80	2.63	4.11	1.39	1.47	1.94
Gd	4.89	7.29	6.60	5.13	7.59	6.05	4.35	5.63	5.33	5.40	6.04	5.19	8.50	8.47	5.67	9.71	10.00	4.32	5.14	6.31
Tb	0.731	1.360	1.020	0.756	1.103	0.916	0.650	0.855	0.816	0.826	0.914	0.781	1.360	1.400	0.872	1.440	1.560	0.667	0.918	0.928
Dy	4.23	5.91	5.84	4.43	6.36	5.23	3.75	4.87	4.75	4.72	5.24	4.50	7.79	8.28	5.07	8.01	9.25	3.78	5.90	5.26
Ho	0.851	1.14	1.21	0.876	1.28	1.04	0.743	0.959	0.954	0.936	1.03	0.896	1.59	1.64	1.02	1.61	1.85	0.752	1.29	0.980
Er	2.20	2.80	3.22	2.26	3.32	2.66	1.88	2.52	2.47	2.44	2.71	2.33	4.16	4.01	2.62	4.40	4.66	1.90	3.63	2.44
Tm	0.342	0.411	0.504	0.347	0.508	0.408	0.290	0.382	0.382	0.370	0.402	0.355	0.652	0.572	0.408	0.705	0.691	0.288	0.607	0.354
Yb	2.17	2.49	3.10	2.21	3.22	2.55	1.74	2.39	2.39	2.30	2.52	2.28	4.14	3.34	2.47	4.57	4.03	1.77	3.79	2.19
Lu	0.352	0.384	0.521	0.360	0.513	0.414	0.272	0.374	0.382	0.361	0.402	0.361	0.672	0.462	0.395	0.791	0.583	0.273	0.609	0.331
ΣREE	106	186	151	112	166	136	67.7	124	123	126	133	117	273	164	128	407	205	69.4	131	133
Gd/Yb	2.3	2.9	2.1	2.3	2.4	2.4	2.4	2.4	2.2	2.3	2.4	2.3	2.1	2.5	2.3	2.1	2.5	2.4	1.4	2.9
La/Nb	1.5	2.1	1.9	1.6	1.7	1.7	0.9	1.6	1.7	1.7	1.7	1.7	1.4	1.3	1.7	2.5	1.3	1.0	2.2	1.7
Th/Nb	0.20	0.22	0.21	0.20	0.19	0.20	0.09	0.20	0.20	0.19	0.20	0.20	0.40	0.41	0.19	0.42	0.24	0.09	1.29	0.16
Zr/Nb	10	10	10	10	10	9.5	9.2	10	10	10	9.6	9.5	12	8.4	10	9.7	8.7	9.0	30	9.4
Ti/Zr	93	99	94	95	91	100	136	93	96	90	94	100	23	71	92	33	71	131	18	107
Ti/Y	446	649	468	466	510	465	649	478	491	510	498	456	275	304	467	324	339	403	183	577
La/Yb	8.6	15.8	8.7	9.2	9.5	9.8	5.4	9.5	9.8	10.1	9.5	9.4	13.8	8.6	9.4	20.4	9.2	5.6	6.1	10.8
Zr/Y	4.8	6.5	5.0	4.9	5.6	4.6	4.8	5.1	5.1	5.7	5.3	4.6	11.8	4.3	5.1	9.9	4.7	3.1	10.0	3.5
Hf/Ta	3.6	3.6	3.7	3.6	3.6	3.7	3.2	3.6	3.0	3.6	3.3	3.6	4.4	3.0	3.7	3.0	3.6	3.2	8.6	3.5
Ta/Yb	0.40	0.55	0.34	0.41	0.38	0.40	0.44	0.40	0.50	0.43	0.48	0.40	0.73	0.52	0.40	0.71	0.46	0.43	0.27	0.47
Th/Yb	1.1	1.6	1.0	1.1	1.0	1.1	0.5	1.1	1.1	1.2	1.2	1.1	3.9	2.8	1.1	3.4	1.7	0.5	3.5	1.0
Eu/Eu*[†]	0.95	1.08	0.94	1.00	0.89	0.98	1.01	0.99	0.95	0.96	0.96	0.99	0.75	1.37	1.01	0.74	1.36	1.07	0.89	1.00
Ce/Ce*[‡]	0.91	0.94	0.97	0.94	0.93	0.91	0.94	0.93	0.90	0.93	0.92	0.90	0.93	0.87	0.92	0.88	0.88	0.91	0.96	0.94
Nb/Nb*[§]	0.62	0.50	0.54	0.61	0.59	0.59	1.16	0.59	0.58	0.60	0.59	0.58	0.45	0.47	0.59	0.33	0.61	1.16	0.20	0.65

*Mg# (Mg number) = (MgO/40.4)/[(MgO/40.4 + 0.85*$Fe_2O_3^T$*0.70/(0.78*77.8)].

[†]Eu/Eu* = (Eu/0.087)/(Sm*Gd/0.306/0.231)$^{1/2}$.

[‡]Ce/Ce* = (Ce/0.957)/(La*Pr/0.367/0.137)$^{1/2}$.

[§]Nb/Nb* = 0.3384*Nb/(Th*La)$^{1/2}$.

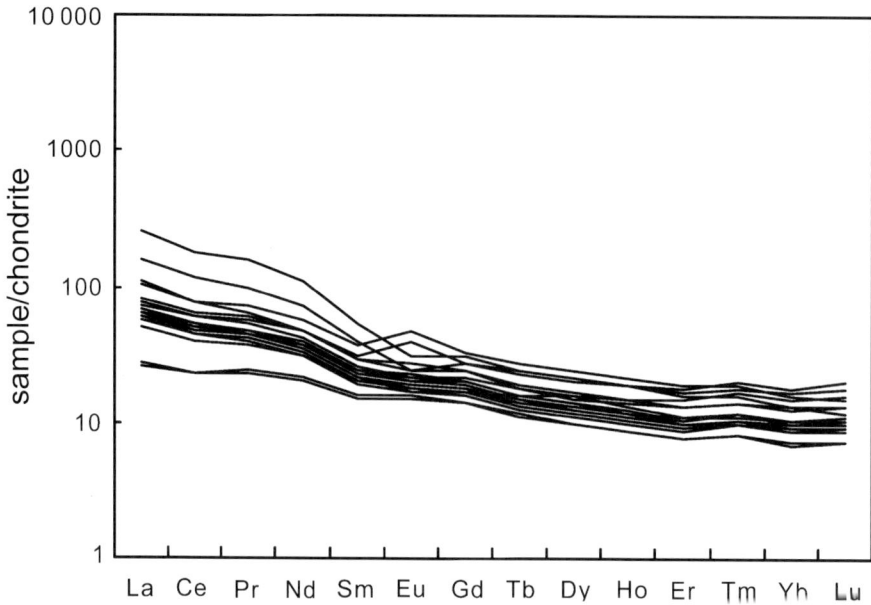

Fig. 4 Chondrite normalized rare-earth element patterns for the Akaz metavolcanic rocks (chondrite values from Taylor & McLennan 1985).

diagram (Pearce 1983) (most rocks plot in the active continental margin field; Fig. 7b).

Although the Akaz metavolcanics exhibit some characteristics of within-plate basalts, they cannot have been derived from ocean floor, oceanic plateau or mature back-arc basin basalts,

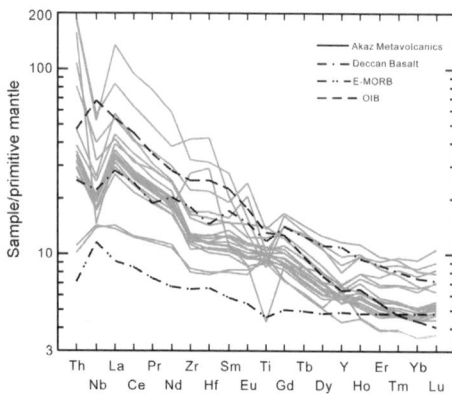

Fig. 5. Primitive-mantle-normalized spider diagram for the Akaz metavolcanic rocks (primitive mantle, E-MORB and OIB values from Sun & McDonough 1989; Deccan basalt data from Wilson 1989).

because they are more enriched than these lavas and have significant negative Nb anomalies (Floyd 1989; Wilson 1989; Saunders & Tarney 1991) (Figs 4 and 5). Ocean island basalts are generally considered to originate from plume-related sources (Wilson 1989; Floyd 1991), whereas seamount basalts have complicated geochemical characteristics that strongly depend on the tectonic settings. Seamounts in within-oceanic plate settings have been explained as the products of hot-spots (e.g. Clague & Dalrymple 1987); however, some seamounts occur in suprasubduction zone environments, and their lavas can be imprinted with a strong subduction-related signature (e.g. Kamenetsky *et al.* 1997). The within-plate characteristics of the Akaz metavolcanic rocks and their association with limestone have led some researchers to suggest that these rocks formed as part of a seamount (Xiao *et al.* 2002). Although these metavolcanic rocks have trace-element concentrations and some element ratios close to those of E-MORB or OIB, their Nb/La ratios (0.4–1.1) are significantly lower than modern OIB or E-MORB (*c.* 1.3) (Sun & McDonough 1989).

The relatively LREE- and LILE-enriched compositions and negative Nb anomalies are characteristics of subduction-related basalts or

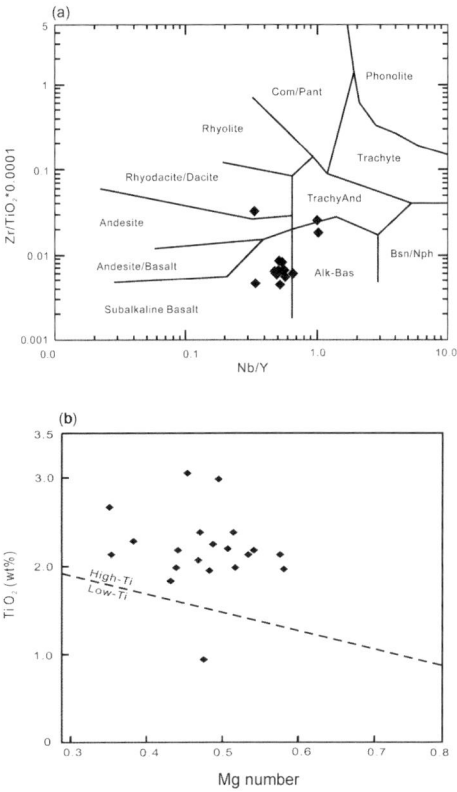

Fig. 6. Classification diagrams for the Akaz metavolcanic rocks. (a) Nomenclature of the Akaz metavolcanic rocks (after Winchester & Floyd 1977); (b) Correlation diagram of Mg number and TiO$_2$ for the Akaz metavolcanic rocks (after Lightfoot 1993).

geochemical characteristics between within-plate basalts and arc-related basalts indicate that the most likely protolith is either subduction-related basalts in an active continental margin, or continental flood/rift basalts contaminated with crustal materials. Continental flood basalts usually have steeply sloping heavy rare-earth element patterns that are rarely seen in arc tholeiites (Arndt, pers. comm.). Data for the Akaz metavolcanic rocks, continental flood basalts and island arc tholeiites are compared in a Gd/Yb v. Gd diagram (Fig. 8). Except for one sample

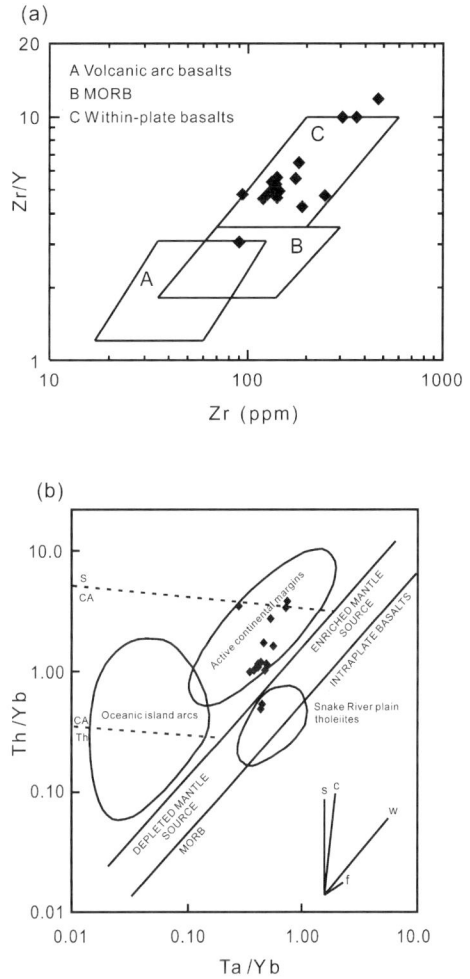

contaminated continental flood/rift basalts. Partial melting induced by dehydration of a subducting slab in the mantle (Pearce & Parkinson 1993; Thirlwall et al. 1994; Tatsumi & Eggins 1995), and contamination of continental crust (Dupuy & Dostal 1984; Cox & Hawkesworth 1985; Arndt et al. 1993; Cadman et al. 1995) may produce basalts with Nb–Ta depletions. The subduction-related basalts generated in intra-oceanic island arcs and active continental margins can be effectively distinguished by their incompatible element ratios. When compared with typical intra-oceanic arc basalts (Zr/Y <3; Ta/Yb <0.1 and 25 <Zr/Nb <70) (Condie 1989; McCulloch & Gamble 1991), the Akaz rocks have high Zr/Y (3–12), Ta/Yb (0.3–0.7) and low Zr/Nb (mostly <12) ratios, strongly supporting a continental affinity (Table 1, Fig. 7b). This precludes intra-oceanic arc basalts as the protolith of these rocks, and the transitional

Fig. 7. Immobile-element-based tectonic discrimination diagrams for the Akaz metavolcanic rocks: (a) Zr v. Zr/Y diagram (after Pearce 1983); (b) Th/Yb versus Ta/Yb diagram (after Pearce 1983). Vectors shown indicate the influence of a subduction component (s), within-plate enrichment (w), crustal contamination (c) and fractional crystallization (f).

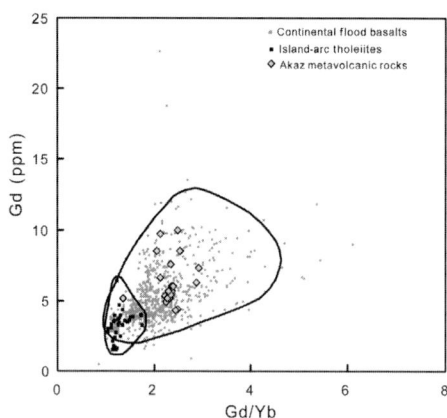

Fig. 8. Discrimination diagram of Gd/Yb v. Gd for the Akaz metavolcanic rocks. CFB, continental flood basalts; IAT, island-arc tholeiites (fields are drawn based on data from the GEOROC database, Max-Planck-Institut für Chemie).

(GS-19, $SiO_2 = 64.7$ wt%), most of the Akaz metavolcanic rocks plot in the CFB field (Fig. 8). The existence of E-MORB-type samples and the transitional nature of the other samples suggest that the Akaz metavolcanic rocks were derived from contaminated continental rift basalts. Some tholeiitic samples in Fig. 6a contain relatively high K_2O contents (Table 1), which also suggests assimilation of crustal materials. The most Nb-depleted sample GS-19 ($Nb/Nb^{\#} = 0.2$) has the highest SiO_2 (64.7 wt%) and lowest TiO_2 (0.94 wt%) contents and displays a Zr–Hf peak ($Zr/Sm = 62$) in its trace-element pattern (Table 1; Fig. 5), reflecting the most intensive crustal contamination. Samples GS-7 and GS-18 exhibit slightly positive Nb-anomalies, and are characterized by the lowest LILE and LREE contents (Fig. 5), and may represent the uncontaminated primary magma.

Further constraints on the tectonic setting of the Akaz metavolcanic rocks come from field evidence. The Sailajaz Tagh Group has a total thickness of about 3500 m, only about one-tenth of which is occupied by the metavolcanic rocks (XGIT-II 1985). Typical continental flood basalts, e.g. Deccan and Siberian traps (Mahoney 1988; Zolotukhin & Al'mukhamedov 1988) are much thicker than this, suggesting a continental rift environment for the Akaz lavas.

Implications for the separation of Tarim from Gondwana

No consensus has been reached on the tectonic evolution of the Kunlun Mountains. A number

of authors have accepted the arc–continental collisional model to explain the accretion of the Kunlun Mountains along the southern margin of the Tarim craton (Yao & Hsü 1994; Hsü *et al.* 1995; Sengör & Natal'in 1996; Li *et al.* 1999; Xiao *et al.* 2002), whereas others consider the South Kunlun Block as a microcontinental block, rifted from the Tarim craton (Pan *et al.* 1994; Ding *et al.* 1996; Jiang *et al.* 2000). Granitoids of the South Kunlun Block, regardless of their ages and tectonic setting, all have young T_{DM} ages (1.0 to 1.5 Ga) (Yuan *et al.* 2003). These young T_{DM} ages are consistent with those of the metamorphic complex in the South Kunlun Block (Zhou 1998), but significantly different from those of the North Kunlun Block (2.8 Ga) (Arnaud & Vidal 1990). These features suggest that the South Kunlun Block does not have an Archean basement and is probably an ancient accretionary prism (Yuan *et al.* 2003).

The available palaeomagnetic and stratigraphic data show that the Tarim Craton was originally part of Gondwana (e.g. Li *et al.* 1991; Li *et al.* 1996; Metcalfe 1996; Zhao *et al.* 1996; Li, 1998; Stampfli & Borel 2002). However, there is little agreement as to when the Tarim Craton rifted from Gondwana. Some workers suggest that rifting took place in the Neoproterozoic as the Rodinia supercontinent dispersed (Li *et al.* 1996; Li 1998), whereas others propose a Devonian or later rifting event (Metcalfe 1996; Zhao *et al.* 1996; Li 1998; Stampfli & Borel 2002). This study indicates that the Akaz metavolcanic rocks formed in a continental rift environment along the south margin of the Tarim Craton. Although precise radiometric data are not available, the Akaz metavolcanic rocks are certainly Sinian to Early Cambrian based on palaeontological and stratigraphic studies of the associated sedimentary rocks (Pan *et al.* 1994; Xiao *et al.* 2002). Coeval continental rift volcanism has also been recently recognized in the East Kunlun (Pan *et al.* 1996) and North Qilian (Xia *et al.* 1996; Mao *et al.* 1998; Zuo *et al.* 1999) orogenic belts. These findings support a pre-Devonian rifting age.

Conclusions

The Akaz metavolcanic rocks were metamorphosed from a high-Ti tholeiitic protolith, with characteristics transitional between within-plate and subduction-related basalts. LREE-enriched patterns and negative Nb anomalies in most samples preclude oceanic environments (ocean floor, within-plate seamount, oceanic island or plateau basalts) for the protolith. The high Zr/Y, Ta/Yb, Gd/Yb and low Zr/Nb ratios strongly support a continental affinity and make the Akaz metavolcanic rocks distinct from island–arc tholeiites. The existence of a few E-MORB-like samples, undepleted in Nb, indicate that the Akaz metavolcanic rocks were originally continental rift basalts contaminated by crustal materials. The Akaz metavolcanic rocks, together with other continental rift volcanic rocks in the East Kunlun and North Qilian orogenic belts, provide evidence for rifting of the Tarim Craton from Gondwana in the Sinian or Early Cambrian.

We are grateful to Li Jiliang, Zhang Yuquan, Pan Yusheng, Xu Ronghua, Xiao Wenjiao and Hou Quanlin for their fruitful discussions about the tectonic evolution of the Kunlun Mountains. Special thanks are given to Brian Windley for his enlightening suggestions. N. T. Arndt and P. Black are greatly thanked for their constructive and encouraging reviews. We gratefully thank P. Robinson for his kind help in the revision of the manuscript. This research was supported by Chinese Project 973 (G1998040800), NSF of China (project 40003005), and an Outstanding Researcher Award from the University of Hong Kong.

References

ARNAUD, N. & VIDAL, PH. 1990. Geochronology and geochemistry of the magmatic rocks from the Kunlun–Karakorum geotraverse. *Colloque Kunlun–Karakorum, IGP' Paris*, 52.

ARNDT, N. T., CZAMANSKE, G. K., WOODEN, J. L. & FEDERENKO, V. A. 1993. Mantle and crustal contributions to continental flood volcanism. *Tectonophysics*, **223**, 39–52.

CADMAN, A. C., TARNEY, J. & BARAGAR, W. R. A. 1995. Nature of mantle source contributions and the role of contamination and *in situ* crystallization in the petrogenesis of Proterozoic mafic dykes and flood basalts, Labrador. *Contributions to Mineralogy and Petrology*, **122**, 213–229.

CHANG, C. F., CHEN, N. S. *et al.* 1986. Preliminary conclusions of the Royal Society and Academia Sinica 1985 geotraverse of Tibet. *Nature*, **323**, 501–507.

CLAGUE, D. A. & DALRYMPLE G. B., 1987. The Hawaiian-Emperor volcanic chain, Part I, Geological evolution. *In:* DECKER, R. W., WRIGHT, T. & STAUFFER, P. H. (eds). Volcanism in Hawaii: *US Geological Survey Professional Paper*, **1350**, 5–54.

CONDIE, K. C. 1989. Geochemical changes in basalts and andesites across the Archean–Proterozoic boundary: identification and significance. *Lithos*, **23**, 1–18.

COX, K. G. & HAWKESWORTH, C. J. 1985. Geochemical stratigraphy of the Deccan Traps, at Mahabaleshwar, Western Ghats, India, with implications for open system magmatic processes. *Journal of Petrology*, **26**, 355–377.

DENG, W. 1989. A preliminary study on the basic–ultrabasic rocks of the Karakorum–western Kunlun

Mts. *Journal of Natural Resources*, **4**, 204–211 (in Chinese with English abstract).

DEWEY, J. F., SHACKLETON, R. M., CHANG, C. F. & SUN, Y. Y. 1988. The tectonic evolution of the Tibetan Plateau. *Philosophical Transactions of the Royal Society of London*, **327**, 379–413.

DING, D., SHAN, X. & ZHANG, Y. 1996. The basin protype and sedimentary–tectonic subdivision of South Tarim and West Kunlun Orogens. *In:* DING, D., WANG, D., LIU, W. & SUN, S. (eds) *The Western Kunlun Orogenic Belt and Basin*. Geological Publishing House, Beijing, China, 9–35 (in Chinese, with English summary).

DUPUY, C. & DOSTAL, J. 1984. Trace element geochemistry of some continental tholeiites. *Earth and Planetary Sciences Letters*, **67**, 61–69.

FLOWER, M. 1991. Magmatic processes in oceanic ridge and intraplate settings. *In:* FLOYD, P. A. (ed.) *Oceanic Basalts*. Blackie, Glasgow & London and Van Nostrand Reinhold, New York, 116–147.

FLOYD, P. 1989. Geochemical features of intraplate oceanic plateau basalts. *In:* SAUNDERS, A. D. & NORRY, M. J. (eds) *Magmatism in the Ocean Basin*. Geological Society, London, Special Publications, **42**, 215–230.

FLOYD, P. 1991. Oceanic islands and seamounts. *In:* FLOYD, P. A. (ed.) *Oceanic Basalts*. Blackie, Glasgow and London, and Van Norstrand Reinhold, New York, 174–218.

HSÜ, K. J., PAN, G. *et al.* 1995. Tectonic evolution of the Tibetan Plateau: a working hypothesis based on the archipelago model of orogenesis. *International Geology Review*, **37**, 473–508.

JIANG, C. F., WANG, Z. Q. & LI, J. Y. 2000. *Opening–closure tectonics of Chinese central orogenic belt*. Geological Publishing House, Beijing, 1–153 (in Chinese with English abstract).

KAMENETSKY, V. S., CRAWFORD, A. J., EGGINS, S. & MÜHE, R. 1997. Phenocryst and melt inclusion chemistry of near axis seamounts, Valu Fa Ridge, Lao Basin: insight into mantle wedge melting and the addition of subduction components. *Earth and Planetary Science Letters*, **151**, 205–223.

LI, J., SUN, S., HAO, J., CHEN, H., HOU, Q. & XIAO, W. 1999. On the classification of collision orogenic belts. *Scientia Geologica Sinica*, **34**, 129–138 (in Chinese with English abstract).

LI, Y. P., LI, Y. A., SHARPS, R., MCWILLIAMS, M., & GAO, Z. J. 1991. Sinian palaeomagnetic results from the Tarim block, western China. *Precambrian Research*, **49**, 61–67.

LI, Z. X. 1998. Tectonic history of the major East Asian lithospheric blocks since the Mid-Proterozoic–a synthesis. *In:* FLOWER, M. F. J., CHUNG, S. L., LO, C. H. & LEE, T. Y. (eds) *Mantle Dynamics and Plate Interactions in East Asia*. American Geophysical Union, Geodynamics Series, **27**, 221–243.

LI, Z. X., ZHANG, L. & POWELL, C. MCA. 1996. Positions of the East Asian cratons in the Neoproterozoic supercontinent Rodinia. *Australian Journal of Earth Sciences* **43**, 593–604.

LIGHTFOOT, P. C. 1993. The interpretation of geoanalytical data. *In:* RIDDLE, C. (ed.) *Analysis of*

Geological Materials. Marcel Dekker, New York, 377–455.

McCULLOCH, E. M. & GAMBLE, J. A. 1991. Geochemical and geodynamic constraints on subduction magmatism. *Earth and Planetary Science Letters*, **102**, 358–374.

MAHONEY, J. J. 1988. Deccan traps. *In:* MACDOUGALL, J. D. (ed.) *Continental Flood Basalts*. Kluwer Academic Publishers, Dordrecht, Netherlands, 151–194.

MAO, J., ZHANG, Z., YANG, J., SONG, B., WU, M. & ZUO, G. 1998. Single-zircon dating of Precambrian strata in the west sector of the northern Qilian Mountains and its geological significance. *Chinese Science Bulletin*, **43**, 1289–1294.

MATTE, PH., TAPPONNIER, P., ARNAUD, N., BOURJOT, L., AVOUAC, VIDAL, PH., LIU, Q., PAN, Y. S. & Wang, Y. 1996. Tectonics of Western Tibet, between the Tarim and the Indus. *Earth and Planetary Science Letters*, **142**, 311–330.

MATTERN, F. & SCHNEIDER, W. 2000. Suturing of the Proto- and Paleo-Tethys oceans in the western Kunlun (Xinjiang, China). *Journal of Asian Earth Sciences*, **18**, 637–650.

MATTERN, F., SCHNEIDER, W., LI, Y. & LI, X. 1996. A traverse through the western Kunlun (Xinjiang, China): tentative geodynamic implications for the Paleozoic and Mesozoic. *Geologische Rundschau*, **85**, 705–722.

METCALFE, I. 1996. Gondwanaland dispersion, Asian accretion and evolution of eastern Tethys. *Australia Journal of Earth Sciences*, **43**, 605–623.

MOLNAR, P., BURCHFIEL, B. C., ZHAO, Z., LIANG, K., WANG, S. & HUANG, M. 1987. Geological evolution of northern Tibet: results of an expedition to Ulugh Muztagh. *Science*, **235**, 299–304.

PAN, Y. S., WANG, Y., MATTE, PH. & TAPPONNIER, P. 1994. Tectonic evolution along the geotraverse from Yecheng to Shiquanhe. *Acta Geologica Sinica*, **68**, 295–307 (in Chinese with English abstract).

PAN, Y., ZHOU, W. *et al.* 1996. Early Paleozoic geological characteristics and tectonic evolution. *Science in China (Series D)*, **26**, 302–307 (in Chinese).

PEARCE, J. A. 1983. Role of the sub-continental lithosphere in magma genesis at active continental margins. *In:* HAWKESWORTH, C. J. & NORRY, M. J. (eds) *Continental Basalts and Mantle Xenoliths*. Shiva, Nantwich, 230–249.

PEARCE, J. A. & CANN, J. R. 1973. Tectonic setting of basic volcanic rocks determined using trace element analyses. *Earth and Planetary Science Letters*, **19**, 290–300.

PEARCE, J. A. & PARKINSON, I. J. 1993. Trace element models for mantle melting: application to volcanic arc petrogenesis. *In:* PRICHARD, H. M., ALABASTER, T., HARRIS, N. B. W. & NEARY, C. R. (eds) *Magmatism Processes and Plate Tectonics*. Geological Society, London, Special Publications, **76**, 373–403.

SAUNDERS, A. & TARNEY, J. 1991. Back-arc basins. *In:* FLOYD, P. A. (ed.) *Oceanic Basalts*. Blackie, Glasgow and London; Van Nostrand Reinhold, New York, 219–263.

SENGÖR, A. M. C. & NATAL'IN, B. A. 1996. Paleotectonics of Asia: fragments of a synthesis. *In:* YIN, A. & HARRISON, T. M. (eds) *The Tectonic Evolution of Asia*. Cambridge University Press, 486–640.

STAMPFLI, G. M. & BOREL, G. D. 2002. A plate tectonic model for the Paleozoic and Mesozoic constrained by dynamic plate boundaries and restored synthetic oceanic isochrones. *Earth and Planetary Science Letters*, **196**, 17–33.

SUN, S-S. & McDONOUGH, W. F. 1989. Chemical and isotopic systematics of oceanic basalts: implications for mantle composition and processes. *In:* SAUNDERS, A. D. & NORRY, M. J. (eds) *Magmatism in the Ocean Basins*. Geological Society, London, Special Publications, **42**, 313–346.

TATSUMI, Y. & EGGINS, S. 1995. *Subduction Zone Magmatism*. Blackwell Science, Cambridge, MA.

TAYLOR, S. R. & McLENNAN, S. M. 1985. *The Continental Crust: its Composition and Evolution: an Examination of the Geochemical Record Preserved in Sedimentary Rocks*. Blackwell Scientific Publications, Oxford.

THIRLWALL, M. F., SMITH, T. E., GRAHAM, A. M., THEODOROU, N., HOLLINGS, P., DAVIDSON, J. P. & ARCULUS, R. J. 1994. High field strength element anomalies in arc lavas: source or process? *Journal of Petrology*, **35**, 819–838.

WEN, S., SUN, D., YIN, J., CHEN, T. & LUO, H. 2000. Stratigraphy and paleontology. *In:* PAN, Y. (ed.) *Geological Evolution of the Karakorum and Kunlun Mountains*. Science Press, Beijing, 6–92 (in Chinese).

WILLIAMS, H., TURNER, F. J. & GILBERT, C. M. 1989. *Petrography – An Introduction to the Study of Rocks in Thin Sections, 2nd edn*, W. H. Freeman, San Francisco.

WILSON, M. 1989. *Igneous Petrogenesis*. Unwin Hyman, London.

WINCHESTER, J. A. & FLOYD, P. A. 1977. Geochemical discrimination of different magma series and their differentiation products using immobile elements. *Chemical Geology*, **20**, 325–343.

XGIT-II 1985. *Stratigraphy, 1/500,000 geological map and explanation of Southwestern Xinjiang, China*. Geological Publishing House, Beijing.

XIA, L., XIA, Z., ZHAO, J., XU, X., YANG, H. & ZHAO, D. 1999. Determination of properties of Proterozoic continental flood basalts of western part from North Qilian Mountains. *Science in China (Series D)*, **42**, 506–514.

XIA, Z., XIA, L. & XU, X. 1996. The Late Proterozoic–Cambrian active continental rift volcanism in northern Qilian mountains. *Acta Geoscientia Sinica*, **17**, 282–291 (in Chinese with English abstract).

XIAO, W. J., WINDLEY, B. F., HAO, J. & LI, J. L. 2002. Arc-ophiolite obduction in the Western Kunlun Range (China): implications for the Paleozoic evolution of central Asia. *Journal of the Geological Society of London*, **159**, 517–528.

XU, R., ZHANG, *et al.* 1994. A discovery of an early Palaeozoic tectono-magmatic belt in the Northern part of west Kunlun Shan. *Scientia Geologica*

Sinica, **29**, 313–328 (in Chinese with English abstract).

YANG, J. S., ROBINSON, R. T., JIANG, C. F. & XU, Z. Q. 1996. Ophiolites of the Kunlun Mountains, China and their tectonic implications. *Tectonophysics*, **258**, 215–231.

YAO, Y. & HSÜ, K. J. 1994. Origin of the Kunlun Mountains by arc–arc and arc–continent collisions. *The Island Arc*, **3**, 75–89.

YIN, A. & HARRISON, T. M. 2000. Geologic evolution of the Himalayan–Tibetan Orogen, *Annual Review of Earth and Planetary Sciences*, **28**, 211–280.

YUAN, C., SUN, M., ZHOU, M. F., ZHOU, H., XIAO, W. J. & LI, J. I. 2002. Tectonic evolution of the West Kunlun: geochronologic and geochemical constraints from Kudi Granitoids. *International Geology Review*, **44**, 653–669.

YUAN, C., SUN, M., ZHOU, M. F., ZHOU, H., XIAO, W. J. & LI, J. L. 2003. Absence of Archean basement in the South Kunlun Block: Nd–Sr–O isotopic evidence from granitoids. *The Island Arc*, **12**, 13–21.

ZHAO, X., COE, R. S., GILDER, S. A. & FROST, G. M. 1996. Paleomagnetic constraints on the palaeogeography of China: implications for Gondwanaland. *Australian Journal of Earth Sciences*, **43**, 643–672.

ZHOU, H. 1998. *The main ductile shear zone and the lithosphere effective elastic thickness of west Kunlun orogenic belt.* PhD thesis, Institute of Geology, Chinese Academy of Sciences, Beijing, China (in Chinese with English abstract).

ZOLOTUKHIN, V. V. & AL'MUKHAMEDOV, A. I. 1988. Traps of the Siberian Platform. *In:* MACDOUGALL, J. D. (ed.) *Continental Flood Basalts*, 273–310, Kluwer Academic Publishers, Dordrecht, Netherlands.

ZUO, G., WU, M., MAO, J. & ZHANG, Z. 1999. Structural evolution of Early Paleozoic tectonic belt in the west section of northern Qiliang area. *Acta Geologica Gansu*, **8**, 6–13 (in Chinese with English abstract).

Basement heterogeneity in the Cathaysia crustal block, southeast China

CHRIS J. N. FLETCHER[1], LUNG. S. CHAN[1], RODERICK J. SEWELL[2],
S. DIARMAD G. CAMPBELL[2], DONALD W. DAVIS[3] & JIESHOU ZHU[4]

[1]*Department of Earth Sciences, The University of Hong Kong, Pokfulam Road, Hong Kong*
[2]*Civil Engineering Department, The Government of the Hong Kong Special Administrative Region, 101, Princess Margaret Road, Homantin, Hong Kong*
[3]*Earth Science Department, Royal Ontario Museum, 100 Queen's Park, Toronto, Canada*
[4]*Chengdu University of Technology, Chengdu, China*

Abstract: Isotope signatures and T_{DM} model ages in Hong Kong and neighbouring Guangdong Province have indicated that the basement of the Cathaysia Block is probably an amalgamation of narrow crustal slices, ranging in age from latest Archaean to Mesoproterozoic. Inheritance ages from zircons contained within Mesozoic volcanic and plutonic rocks also show Proterozoic and Archaean components. Regional gravity survey studies display NNE- to NE-trending Bouguer anomalies that are indicative of sharp changes in rock densities at middle and lower crustal levels. The anomalies displayed on the gravity profile from Guangdong to Hong Kong have been modelled as narrow slices of Archaean and Proterozoic crust. A substantial E–W-trending Bouguer anomaly, which largely parallels the trend of the foliation in the Proterozoic schists of the region, is present to the east of Guangzhou. It is proposed that the basement of the Cathaysia Block consists of an amalgamation of NE- to NNE-trending Palaeo- to Mesoproterozoic and Archaean crustal terranes, which in places have retained the pre-amalgamation E–W-trending tectonic fabric. The discontinuities between the basement terranes, and the E–W structures have strongly influenced the geological evolution of the Phanerozoic sequences and igneous complexes in southeast China. These are most obviously manifest in the regional NE-trending fault and shear zones that displace the cover sequences.

Southeast China is composed of two major crustal blocks (Yang *et al.* 1986): the Yangtze Block in the north, which forms a stable cratonic area, and the Cathaysia Block in the south, which is made up of several Phanerozoic mobile belts (Fig. 1). The nature of the crystalline basement to the Yangtze and Cathaysia crustal blocks is largely conjectural (Hsü *et al.* 1988, 1990; Li 1997), since until recently few outcrops of proven Palaeoproterozoic or older rocks have been found. However, ion microprobe analyses of zircons from trondhjemites within the Yangtze Block, close the suture zone with the North China Block, have yielded U–Pb ages between 2.90 and 2.95 Ga (Qiu *et al.* 2000). The U–Pb ages of detrital zircons from metapelites in the same area range from 2.87 to 3.28 Ga. Thus, at least the northern part of the Yangtze Block is of Archaean age, but the age of the basement to the Cathaysia Block is still an enigma. It has been commonly assumed that the Cathaysia Block is underlain by Palaeo- to Mesoproterozoic continental crust (Jahn *et al.* 1990; Li *et al.* 1992; Li 1994; Li & McCulloch 1996; Li 1998; Chen & Jahn 1998). It has been postulated that the eastern part of the Cathaysia Block is composed of several microcontinental fragments (Zhang *et al.* 1984; Guo *et al.* 1989; Chen *et al.* 1993; Gilder *et al.* 1995, 1996) some of which may be narrow Archaean terranes (Fletcher *et al.* 1997; Sewell *et al.* 2000*a*). This paper will examine the evidence for the heterogeneity and age of the crystalline basement to the southeastern part of the Cathaysia Block that can be derived from the Nd and Sr isotope data and zircon inheritance ages from the Mesozoic magmatic rocks, and interpretations of the regional gravity data-sets.

Regional geological setting

Palaeozoic platform sedimentary rocks are widespread across the inland regions of the Cathaysia Block, and in a few places the underlying Neoproterozoic metasedimentary sequences are exposed. The predominant trend of the foliation in the Proterozoic schists in Guangdong Province

From: MALPAS, J., FLETCHER, C. J. N., ALI, J. R. & AITCHISON, J. C. (eds) 2004. *Aspects of the Tectonic Evolution of China*. Geological Society, London, Special Publications, **226**, 145–155.
0305-8719/04/$15 © The Geological Society of London 2004.

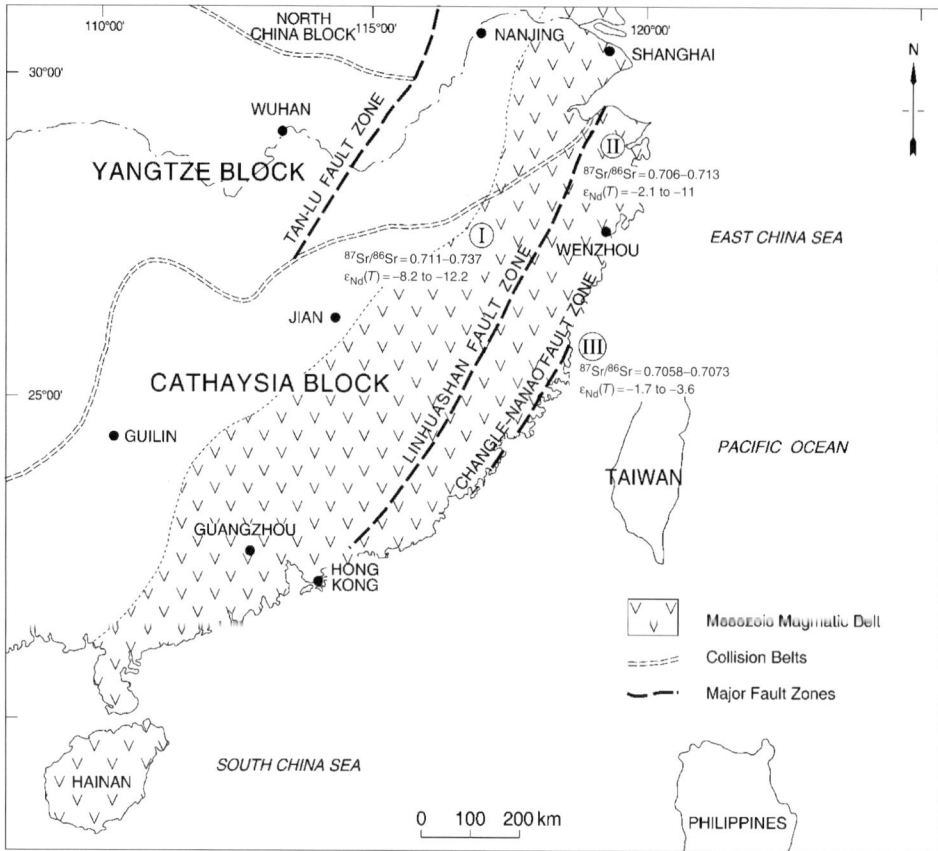

Fig. 1. Regional geological setting of southeast China (isotope zones after Huang *et al.* 1986 and Sewell *et al.* 2000a).

(Fig. 2) is east–west, except close to the NE-trending faults where it is oriented parallel to these faults (Bureau of Geology and Mineral Resources of Guangdong 1998). No Mesoproterozoic or older rocks have been positively identified at surface within the Cathaysia Block. The coastal region of southeast China is dominated by an extensive NE-trending belt of Mesozoic magmatic rocks that extends from Hainan in the south to Korea and eastern Russia in the north. In China, the belt, which is on average 200 km wide but in places extends up to 400 km, consists predominantly of thick accumulations of pyroclastic deposits and numerous sub-volcanic granite intrusions (Fig. 1). The nature of the belt is exemplified by the geology of SE Guangdong (Fig. 2) where over 80% of the exposures are Mesozoic magmatic rocks. Some of the granite plutons display east–west elongation, indicative of basement control to their orientation. Two major NE-trending fault zones transect the

region: the Changle–Nanao Fault Zone, located close to the coast, defines the western margin of the Mesozoic magmatic belt and juxtaposes metamorphosed, Early Palaeozoic, volcanic arc assemblages from the Mesozoic magmatic rocks; and the Lianhuashan Fault Zone (Chen 1987) that parallels the continental margin from Hong Kong to just south of Shanghai (Fig. 1). These faults have been interpreted to be the surface expressions of deep-seated basement structures that have at times controlled the Mesozoic magmatic activity (Campbell & Sewell 1997; Darbyshire & Sewell 1997; Sewell *et al.* 2000b). Small post-volcanic, Cretaceous sedimentary basins are commonly also controlled by reactivation of the NE-trending faults (Liu & Fu 1988).

Isotope signatures

The Nd and Sr isotope data for the Mesozoic granites of the coastal region of southeast China

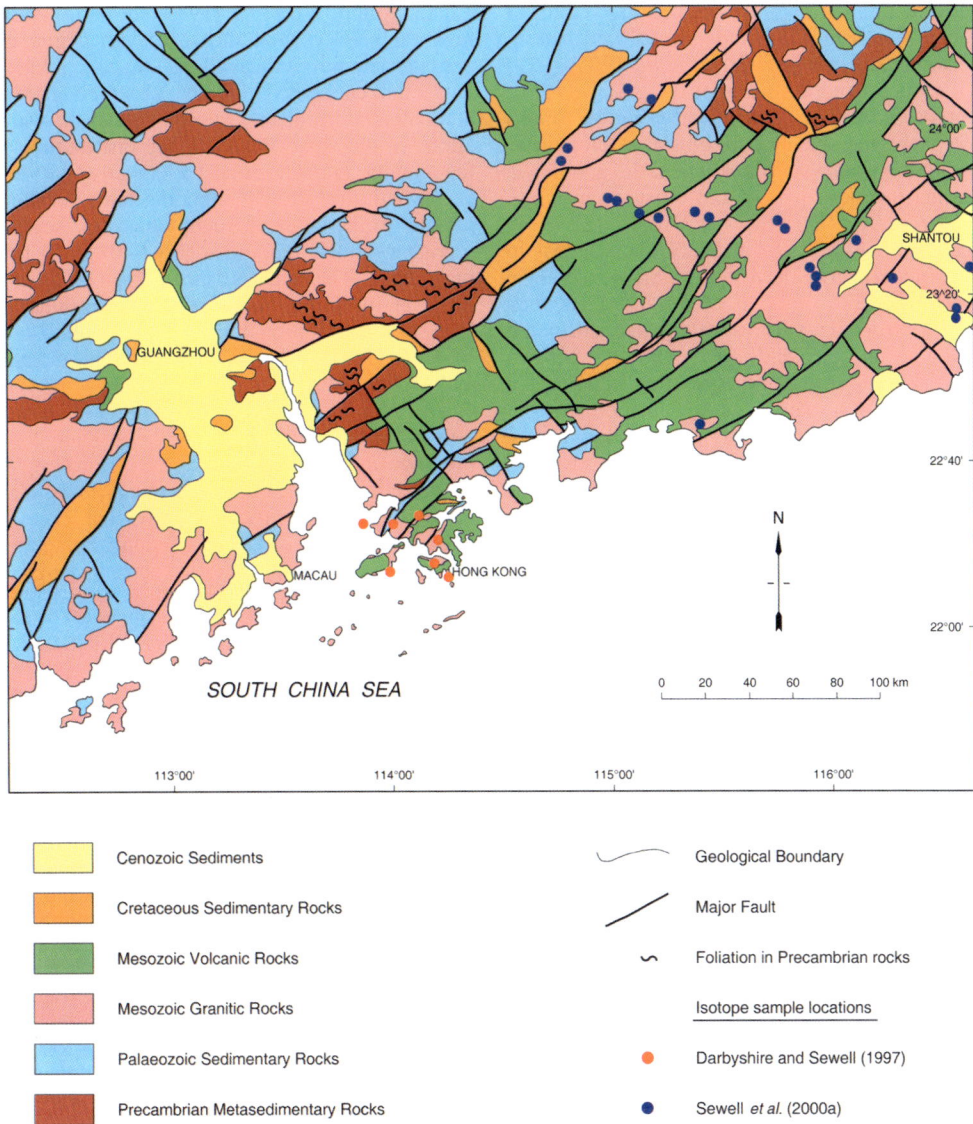

Fig. 2. Simplified geological map of Guangdong Province (based on Bureau of Geology and Mineral Resources of Guangdong Province 1988) and location of isotope sample points (Darbyshire & Sewell 1997; Sewell *et al.* 2000a).

have suggested that the composition of the crustal basement is extremely varied (Huang *et al.* 1986; Huang & DePaolo 1989; Pei & Hong 1995; Zhou *et al.* 1996). Regional Nd–Sr studies in Fujian Province distinguished three isotopic zones within the Mid-Jurassic to Early Cretaceous granites (Fig. 1): a western zone (I) with $^{87}Sr/^{86}Sr_i = 0.711$ to 0.737 and $\varepsilon_{Nd}(T) = -8.2$ to -12.2, a central zone (II) with $^{87}Sr/^{86}Sr_i = 0.706$ to 0.713 and $\varepsilon_{Nd}(T) = -2.1$ to -11, and a coastal

zone (III) with $^{87}Sr/^{86}Sr_i = 0.7058$ to 0.7073 and $\varepsilon_{Nd}(T) = -1.7$ to -3.6 (Huang *et al.* 1986). This trend of increasing ε_{Nd} and decreasing initial Sr ratios from west to east was interpreted to reflect a higher mantle component within the granitoid magma, and it was suggested that the boundary between the coastal and central zones marked the eastern limit of crystalline basement. In contrast, it has been argued that the mean depleted-mantle (T_{DM}) model ages from this region agree within error and therefore the basement could be considered homogeneous, and heterogeneity was unproven (Gilder *et al.* 1995).

Detailed isotope studies of the granites to the southwest in Hong Kong (Darbyshire & Sewell 1997) provide evidence for crustal heterogeneity within the Cathaysia Block. Here, the granites in the northwestern part of the territory are characterized by $^{87}Sr/^{86}Sr_i > 0.710$, $\varepsilon_{Nd}(T) < -9$ and T_{DM} between 1.67 and 2.02 Ga, whereas in the southeast they have $^{87}Sr/^{86}Sr_i = 0.7071$ to 0.7109, $\varepsilon_{Nd}(T) = -5.5$ to -6.5 and T_{DM} between 1.39 and 1.47 Ga. Such variations were attributed to the mixing of mantle-derived melts with two distinct compositions of crystalline basement: a dominantly Late Archaean to Palaeoproterozoic source in the northwest and a dominantly Proterozoic crust in the southeast. The strongest mantle-derived imprint is found in the granites lying between these two zones, and these were considered to delineate a NE-trending boundary between distinct Precambrian terranes.

Additional evidence for basement heterogeneity comes from the granites sampled along a traverse across eastern Guangdong Province undertaken by Sewell *et al.* (2000a). In this region, four isotopically distinct zones were recognized (Fig. 3): Zone I (northwest) $^{87}Sr/^{86}Sr_i = 0.70656$ to 0.71594, $\varepsilon_{Nd}(T) = -7.9$ to -10.8 and T_{DM} between 1.58 and 1.78 Ga; Zone II (west central): $^{87}Sr/^{86}Sr_i = 0.70905$ to 0.70623, $\varepsilon_{Nd}(T) = -7.5$ to -0.8 and T_{DM} between 1.54 and 1.02 Ga; Zone III (east central): $^{87}Sr/^{86}Sr_i = 0.70619$ to 0.70646, $\varepsilon_{Nd}(T) = -7.5$ to -2.4 and T_{DM} between 1.55 and 1.16 Ga; Zone IV (coastal): $^{87}Sr/^{86}Sr_i = 0.70857$ to 0.70324, $\varepsilon_{Nd}(T) = -9.1$ to -2.3 and T_{DM} between 1.67 and 1.14 Ga. Evidence of mantle influence occurs in narrow belts, some of which lie close to the zonal boundaries (Fig. 3). The boundaries between these zones have been inferred to correspond to mapped NE-trending faults and shear zones at surface. The widths of these zones may be as narrow as 20 km.

Geophysical evidence

The relation between the tectonic framework and geophysical data of South China has been discussed in a qualitative manner by Cheng (1987), Yuan (1987), Wu & Gao (1985) and Wang (1985). However, no detailed gravity anomaly maps were presented in these reports. Relatively more comprehensive Bouguer gravity data were given in a 1 : 5 000 000-scale map by Liu (1992) for the coastal areas of China. Zeng *et al.* (1997) provided a tectonic interpretation of China based on a 1:1 000 000 Bouguer gravity

anomaly map. Wu & Lu (1989) showed that the regional gravity anomalies of South China do not correlate well with available crustal thickness data, and concluded that the observed gravity anomalies are probably caused by density variations in the crust. Based on the distribution of the gravity gradient zones, Zeng *et al.* (1997) delineated a major ENE-trending fault that runs parallel to the coast of South China and several secondary faults trending at about 40° to the primary fault. The gravity data of Liu (1992) and Zeng *et al.* (1997) have been re-presented in Fig. 4 with additional data from unpublished sources (Gravity Centre Map, Ministry of Land and Mineral Resources, China, 1997).

The main features of the regional Bouguer gravity anomaly map of the southern Guangdong Province and offshore areas of SE China (Fig. 4) are a series of NE-trending anomalies and an E–W trending positive anomaly that extends through Guangzhou. The NE-trending anomalies are less conspicuous over the onshore parts of the area, due to the overprinting of this latter anomaly. Additional gravity data have been acquired by the authors along several traverses between Hong Kong and Guangzhou. Gravity measurements were taken at about 1 km intervals along these traverses. The new data were combined with the high-resolution results previously published and interpreted by Fletcher *et al.* (1997) and are shown together in Fig. 5. This map illustrates the precise nature of the NE-trending gravity gradient through Hong Kong, and shows that the E–W anomaly to the south of Guangzhou has several NE-trending components to it. The gravity data from this area have been projected on to a NNW-trending section (A–B on Fig. 5) that passes through Hong Kong (Fig. 6).

The variations of gravity values along this traverse reveal the presence of two major gravity anomalies to the north of Hong Kong: a negative anomaly at approximately 45 km and a positive anomaly at approximately 120 km. The gravity variations along the southern part of the section have been studied in detail by Busby *et al.* (1992) and Fletcher *et al.* (1997), where the gravity contours cut across the mapped distribution of the volcanic and granitic rocks. They concluded that the short-wavelength gravity anomalies in this area reflect variations in the upper crust, particularly the presence of basic and granodiorite intrusions, whereas longer wavelength anomalies are due to a narrow felsic segment flanked by more mafic segments in the middle to lower crust. They suggested that the model densities of the segments were consistent with Archaean and Proterozoic terranes respectively. In addition, a detailed analysis of the

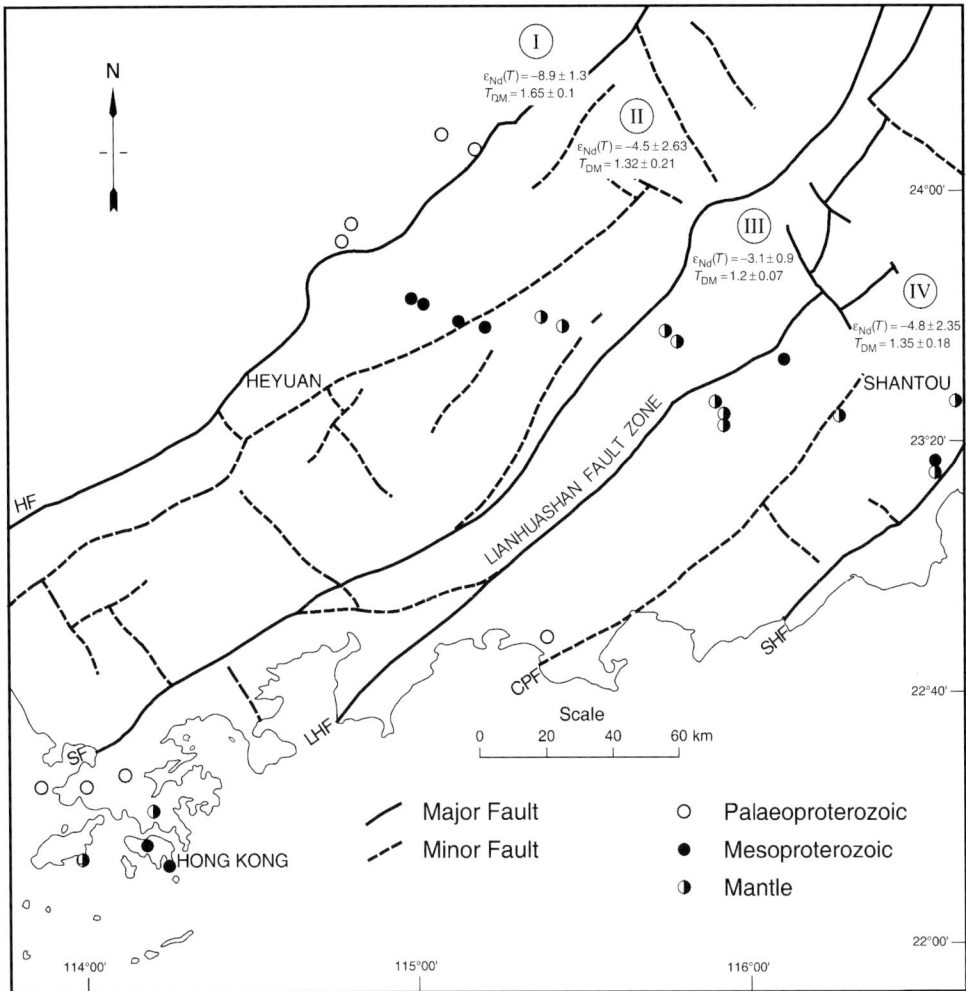

Fig. 3. Isotope signatures of Mesozoic granites from Guangdong Province and Hong Kong (based on Sewell *et al.* 2000*a*). HF – Heynan Fault; SF – Shenzhen Fault; LHF – Lianhuashan Fault; CPF - Chao'an-Puing Fault; SHF – Shanton-Heilei Fault.

gravity data-set from Hong Kong (Fletcher *et al.* l997) using the Euler deconvolution technique has revealed four sets of linear anomalies that relate to known fault trends in Hong Kong. The most prominent set has a NE trend, corresponding to the dominant structural trend in the region. However, the deepest Euler solutions define short, discontinuous E–W-trending linear anomalies, which are truncated by the other linear anomaly sets. Solution depths for these anomalies range up to 8 km, indicating that they are generated from inferred faults in the upper part of the middle crust, and their presence elsewhere has been masked by shallower anomalies.

The northern part of the section cross-cuts the east-trending positive anomaly through Guangzhou. It is proposed that this anomaly is mainly generated by E-trending Mesozoic granites in the upper crust that overprint and modify a NE-trending high-density segment in the middle to lower crust. East-trending geological features are also found in other areas of the region, for example the elongate granite to the NE of Guangzhou and the dominant foliation in the exposed Neoproterozoic rocks to the east of Guangzhou (Fig. 2). This would suggest that there are fundamental east-trending structures in the basement rocks that have been reactivated in Mesozoic times. More detailed gravity surveys

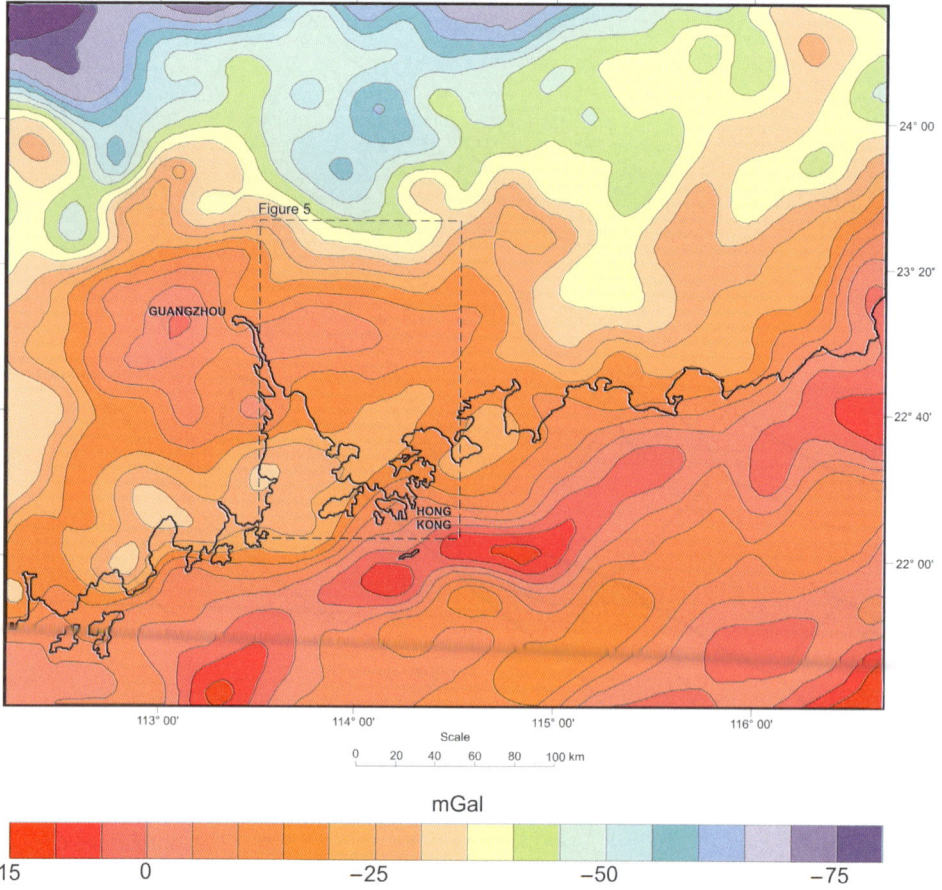

Fig. 4. Gravity map of the region around Guangzhou and Hong Kong (based on Liu *et al.* 1992; Zeng *et al.* 1997; Gravity Centre Map, Ministry of Land and Mineral Resources, China, 1997 (unpublished)).

will have to be undertaken in this area to substantiate this interpretation. However, taken in conjunction with the regional gravity anomaly patterns and the detailed gravity modelling in the Hong Kong region, it is concluded that the best-fitting crustal model for the whole section is the juxtaposition of several distinct NE-trending middle to lower crust blocks, about 20–50 km wide and with density contrasts of up to about 0.1 Mg m^{-3}. This would be consistent with the accretion of a series of narrow low-density felsic Archaean and higher-density mafic Proterozoic terranes along the southeastern margin of the Cathaysia Block.

Zircon inheritance

The most important evidence for the age of the crystalline Precambrian basement, in the absence of xenoliths or surface exposures of these rocks,

comes from inherited zircon contained within the Mesozoic granites and volcanic rocks. Zircon can survive crustal anatexis, and its blocking temperatures for diffusion of Pb and U are sufficiently high to preserve a record of its primary age (Lee *et al.* 1997).

The systematic U–Pb dating of the plutonic and volcanic rocks of Hong Kong, using zircons and monazites, was undertaken to determine accurately the Mesozoic magmatic evolution of a transect across part of the magmatic belt of southeastern China. The analyses, undertaken by the Royal Ontario Museum, Canada, on fresh, colourless, generally euhedral zircon prisms, enabled the Mesozoic magmatic events to be dated precisely (Davis *et al.* 1997; Sewell *et al.* 2000a). Most of the analyses were carried out on fractions containing several grains, because of the young age and generally low U concentration of the zircons. Single-grain analysis was only possible on

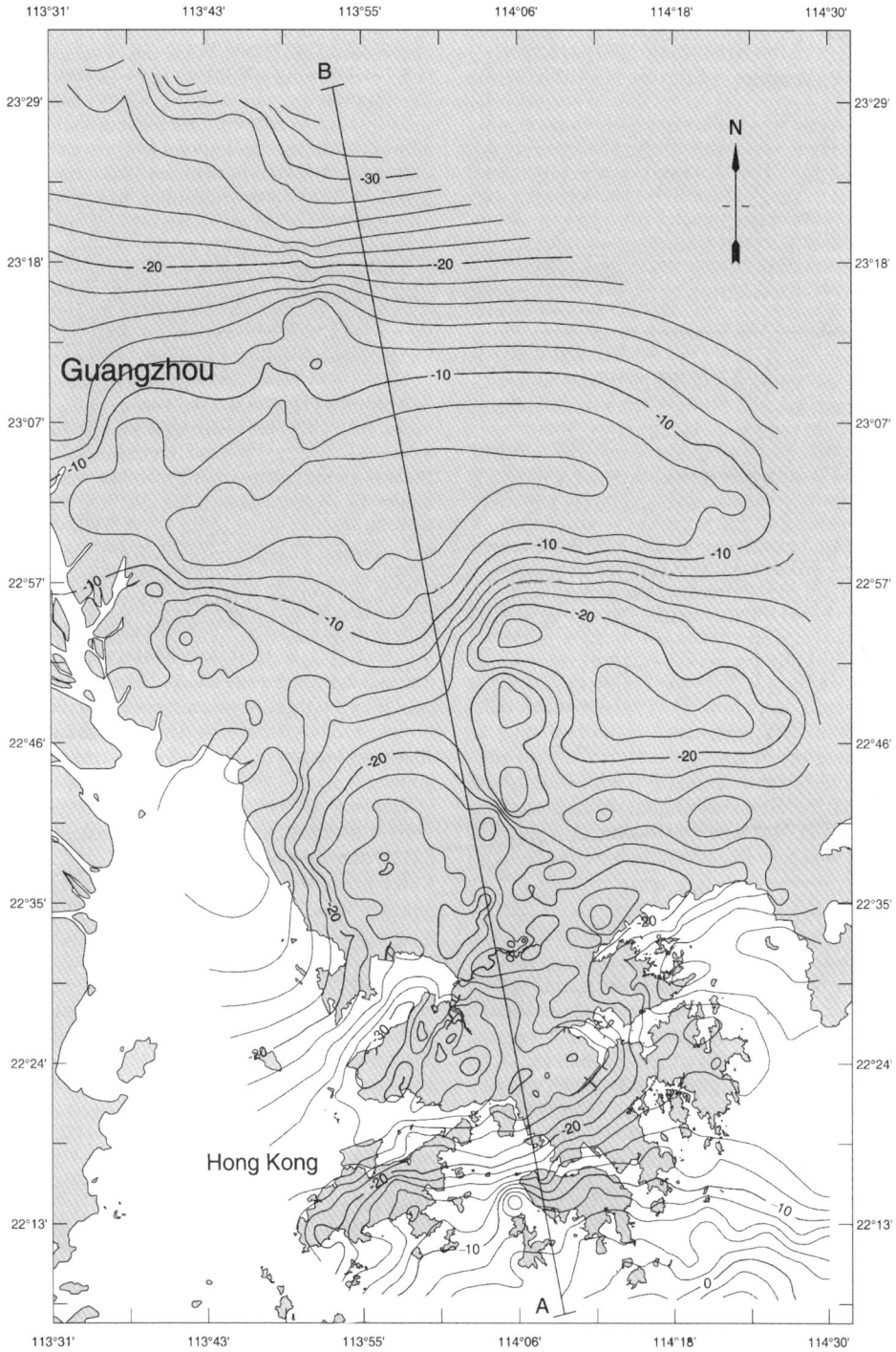

Fig. 5. Extended Bouguer gravity anomaly map of the transect from Guangzhou to Hong Kong (based on Fletcher *et al.* 1997 and new data). The contours are at 2 mGal intervals. The position of the N–S gravity profile (A–B, in Fig. 6) is indicated. Land area shaded.

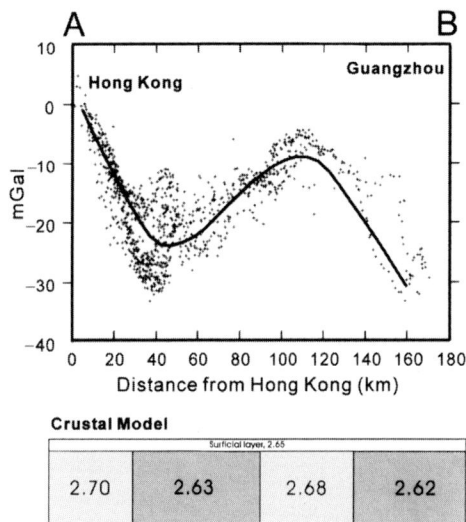

Fig. 6. Gravity model along the profile A–B in Figure 5. Gravity values within 5 km of the line of profile have been projected on to the section. The numbers in the crustal model refer to densities (Mg m^{-3}) assigned to each block.

exceptionally large single crystals or much older xenocrysts. The U–Pb dates confirmed the four main phases of magma generation that had been previously inferred through field relationships, geochemistry and petrographic characterization (Campbell & Swell, 1997; Sewell & Campbell 1997). However, slight to moderate amounts of inheritance were found in samples from two of the Mid-Jurassic to Early Cretaceous magmatic periods. This inheritance provides important evidence for the probable age of the crystalline basement beneath this part of the Cathaysia Block.

Inheritance ages were apparent in granites and volcanic rocks from three of the five Mesozoic magmatic events. Data are tabulated in Davis *et al.* (1997). Precambrian inherited zircons were detected in HK11640, a 236 Ma granite (projections to 1269 Ma), as well as 507 Ma. For at least five of the seven samples from the 159–164 Ma magmatic event, projection of lines from the best fit ages of the concordant data through older discordant data yield Precambrian upper concordia intercept ages. Inherited data for HK11837 (1000 Ma to 3000 Ma), and HK10277 (1050 Ma to 2200 Ma), as well as for HK11640 above are presented in Davis *et al.* (Figs 3B, 3G, 3A, 1997). Inherited data for samples HK11822, HK11025 and HK11821 with upper concordia intercepts of 713 ± 61 Ma, 1136 ± 64 Ma and 2719 ± 4, respectively, are presented in Figure 7A, B and C. Most of the fractions that

showed inheritance contained several grains, therefore some of the upper intercept ages may represent averages of different older components. Small amounts of inheritance are probably due to the presence of small cores, since a Precambrian xenocryst would be likely have a much higher radiogenic Pb content than the Mesozoic zircons, resulting in a highly discordant mixed age. Rare pink or violet rounded grains are present in some samples and probably represent xenocrysts. Analysis of one such zircon from HK11837 gave a concordant datum with an age of 1872 ± 3 Ma (Fig. 3B of Davis *et al.* 1997). The U–Pb age of this crystal was apparently undisturbed during the emplacement of the Mesozoic tuff.

Multi-grain zircon fractions from fine ash vitric tuffs of the 142.7 ± 0.2 Ma Repulse Bay Volcanic Group, which represents the fourth Mesozoic magmatic event, provide further evidence of Palaeoproterozoic to Late Archaean inheritance. One fraction from HK11840, a fine-grained rhyolite, produced a highly discordant datum that, together with the concordant data, defines a line with an upper concordia intercept age of 2426 ± 5 Ma (Fig. 7D). It is very probable that at least one of the zircons in the multi-grain fraction is an Archaean xenocryst or contains a large Archaean core. Zircons from HK11835, a coarse ash crystal tuff, define a short mixing line, giving an imprecise Archaean upper intercept age of 3000 ± 700 Ma (Fig. 6D of Davis *et al.* 1997).

Even if upper intercept ages of partially reset data from multi-grain fractions are averages from different-aged sources, the youngest and oldest precise upper intercept ages should represent minimum estimates for the age range of the inherited components. Upper intercept ages are probably due to a combination of xenocrysts derived from wall-rock contamination, and invisible cores derived from older rocks at the site of melt generation.

Discussion

Recent U–Pb zircon ages from the northern boundary of the Yangtze Block show that it is, in part, composed of Archaean crystalline basement (Qiu *et al.* 2000). These ages support the previous inherited zircon date of 2.52 Ga from the Tanghu Granite exposed in the southeastern part of the Yangtze Block (Li *et al.* 1989). Thus, the common assumption that the Yangtze Block is Proterozoic and the North China Block is characterized by Archaean basement cannot be strictly sustained. This paper has further shown that the Cathaysia Block, which probably amalgamated with the Yangtze Block during the

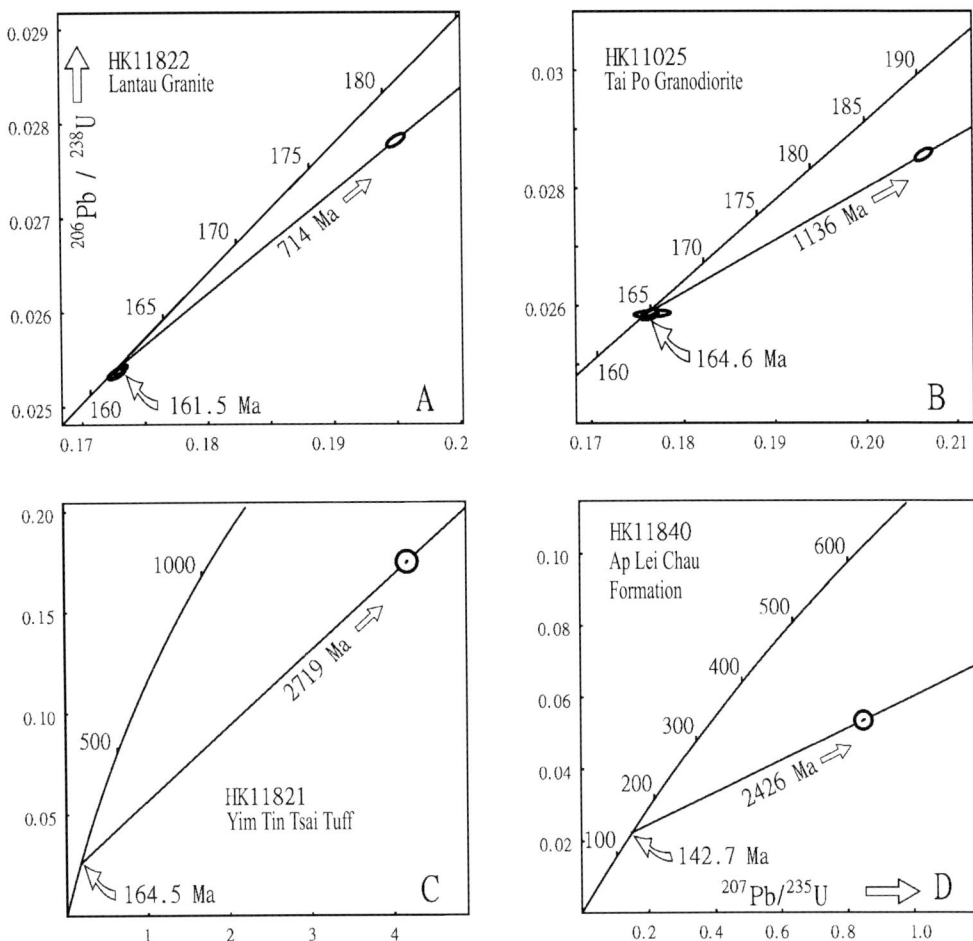

Fig. 7. Zircon inheritance ages (based on data from Davis *et al.* 1997).

Late Mesoproterozoic or Early Neoproterozoic (Chen *et al.* 1991; Xing *et al.* 1992), also probably contains Archaean elements. The zircons that yielded Archaean ages could have been derived from younger sedimentary sequences through which the granite conduits passed, rather than from an Archaean crystalline basement. However, this would still necessitate the proximity of a significant Archaean source area.

The Nd–Sr signatures and T_{DM} model ages of the Mesozoic granites from the southeastern part of the Cathaysia Block also support the hypothesis that the crystalline basement is probably an amalgamation of narrow crustal slices, ranging in age from Late Archaean to Mesoproterozoic. Isotope studies on zircons contained within these granites and the associated volcanic rocks suggest that some of the zircons within both rock types have an inherited Archaean component. The nature,

depth and orientation of these crustal slices or terranes have been ascertained using the results of regional gravity and magnetic surveys. In Hong Kong and neighbouring Guangdong Province, NNE- to NE-trending Bouguer anomalies are indicative of sharp changes in rock densities at middle and lower crustal levels. These have been modelled as narrow slices of Archaean and Proterozoic rocks. Subordinate E–W trending Bouguer anomalies, which largely parallel the trend of the foliation in the Neoproterozoic schists of the region and some Mesozoic granite intrusions, have overprinted and modified the regional gravity anomaly patterns near Guangzhou.

In summary, the crystalline basement of the southeastern part of the Cathaysia Block consists of an amalgamation of NE- to NNE-trending Palaeo- to Mesoproterozoic and possibly Late Archaean crustal terranes. The discontinuities between the basement terranes have strongly influenced the geological evolution of the Phanerozioc sequences and igneous complexes in southeast China. These are most obviously manifest in the regional NE-trending fault and shear zones that have, in particular, controlled the Mesozoic magmatism and framework for the Late Mesozoic sedimentary basins.

This paper is published with the permission of the Director of Civil Engineering and Head of the Geotechnical Engineering Office, Hong Kong Special Administrative Region of China. The authors would like to thank the Gravity Centre, Ministry of Land and Mineral Resources, China, who gave permission to use some of their unpublished gravity data.

References

BUREAU OF GEOLOGY AND MINERAL RESOURCES OF GUANGDONG PROVINCE 1988. *Regional Geology of Guangdong Province.* Geological Memoirs Series, **1**, Geological Publishing House, Beijing, China.

BUSBY, J. P., EVANS, R. B., LAM, M. S., RIDLEY-THOMAS, W. N. & LANGFORD, R. L. 1992. The gravity base station network and regional gravity survey of Hong Kong. *Geological Society of Hong Kong Newsletter*, **10**, 549–556.

CAMPBELL, S. D. G. & SEWELL, R. J. 1997. Structural control and tectonic setting of Mesozoic volcanism in Hong Kong. *Journal of the Geological Society of London*, **154**, 1039–1052.

CHEN, H. H., SUN, S., LI, J. L., HELLER, F., DOBSON, J., HAAG, M. & HSÜ, K. J. 1993. Early Triassic palaeomagnetism and tectonics, South China. *Journal of Southeast Asian Earth Sciences*, **8**, 269–276.

CHEN, T. G. 1987. Basic features of the Lianhuashan Fault Zone in Hong Kong and Shenzhen area. *Journal of Guangdong Geology*, **2**, 57–68.

CHEN, J. F. & JAHN, B. M. 1998. Crustal evolution of southeastern China: Nd and Sr isotopic evidence. *Tectonophysics*, **284**, 101–133.

CHEN, J., FOLAND, K. A., XING, F., XU, X. & ZHOU, T. 1991. Magmatism along the southeastern margin of the Yangtze block: Precambrian collision of the Yangtze and Cathaysian blocks of China. *Geology*, **19**, 815–818.

CHENG, J. 1987. The relationship of regional gravity and magnetic fields to granitoids and tungsten–tin metallizations in South China. *In*: WANG, M. & CHENG, J. (eds) Contributions to the Exploration Geophysics and Geochemistry, **6**, Geological Publishing House, Beijing, 161–168 (in Chinese).

DARBYSHIRE, D. P. F. & SEWELL, R. J. 1997. Nd and Sr isotope geochemistry of plutonic rocks from Hong Kong: implications for granite petrogenesis, regional structure, and crustal evolution. *Chemical Geology*, **143**, 81–93.

DAVIS, D. W., SEWELL, R. J. & CAMPBELL, S. D. G. 1997. U–Pb dating of Mesozoic igneous rocks

from Hong Kong. *Journal of the Geological Society, London*, **154**, 1067–1076.

FLETCHER, C. J. N., CAMPBELL, S. D. G, CARRUTHERS, R. M., BUSBY, J. P. & LAI, K. W. 1997. Regional tectonic setting of Hong Kong: implications of new gravity models. *Journal of the Geological Society, London*, **154**, 1021–1030.

GILDER, S. A., COE, R. S., WU, H., KUANG, G., ZHAO, X. & WU, Q. 1995. Triassic paleomagnetic data from south China and their bearing on the tectonic evolution of the western circum-Pacific region. *Earth and Planetary Science Letters*, **131**, 269–287.

GILDER, S. A., GILL, J. A., COE, R. S. *et al.* 1996. Isotopic and paleomagnetic constraints on the Mesozoic tectonic evolution of South China. *Journal of Geophysical Research*, **101**, 16 137–16 154.

GUO, L. Z., SHI, Y. S., LU, H. F., MA, R. S. & DONG, H. G. 1989. The pre-Devonian tectonic patterns and evolution of South China. *Journal of Southeast Asian Earth Sciences*, **3**, 87–93.

HUANG, X., SUN, S. H., DEPAOLO, D. J. & WU, K. L. 1986. Nd–Sr isotope study of Cretaceous magmatic rocks from Fujian Province. *Acta Petrologica Sinica*, **2**, 50–63.

HUANG, X. & DEPAOLO, D. J. 1989. Study of sources of Paleozoic granitoids and the basement of south China by means of Nd–Sr isotopes. *Acta Petrologica Sinica*, **1**, 28–36.

HSÜ, K. J., JILIANG, L., QINGCHEN, W., SHU, S. & SENGOR, A. M. C. 1990. Tectonics of South China: key to understanding West Pacific geology. *Tectonophysics*, **183**, 9–39.

HSÜ, K. J., SHU, S., JILIANG, L., HAIHONG, C., HAIPO, P. & SENGOR, A. M. C. 1988. Mesozoic overthrust tectonics in south China. *Geology*, **16**, 418–421.

JAHN, B. M., ZHOU, X. H. & LI, J. L. 1990. Formation and tectonic evolution of southeastern China and Taiwan: isotopic and geochemical constraints. *Tectonophysics*, **183**, 145–160.

LEE, J. K. W., WILLIAMS, I. S. & ELLIS, D. J. 1997. Pb, U and Th diffusion in natural zircon. *Nature*, **390**, 159–162.

LI, X. H., 1994. A comprehensive U–Pb, Sm–Nd, Rb–Sr and ^{40}Ar–^{39}Ar geochronological study on Guidong Granodiorite, southeast China: records of multiple tectonothermal events in a single pluton. *Chemical Geology*, **115**, 283–295.

LI, X. H., 1997. Timing of the Cathaysia Block formation: constraints from SHRIMP U–Pb zircon geochronology. *Episodes*, **20**, 188–192.

LI, X. H. & McCULLOCH, M. T. 1996. Secular variations in the Nd isotopic composition of the Neoproterozoic sediments from the southern margin of the Yangtze Block: evidence for a Proterozoic continental collision in southeast China. *Precambrian Research*, **76**, 67–76.

LI, X. H., TATSUMOTO, M., PREMO, W. R. & GUI X. T. 1989. Age and origin of the Tanghu granite, southeast China: results from U–Pb single zircon and Nd isotopes. *Geology*, **17**, 395–399.

LI, X. H., ZHAO, Z., GUI, X. & YU, J. L. 1992. Sm–Nd and zircon U–Pb isotopic constraints on the age of formation of the Precambrian crust in Southeast China. *Chinese Journal of Geochemistry*, **11**, 111–120.

LI, Z. X. 1998. Tectonic history of the major East Asia lithospheric blocks since the mid-Proterozoic – a synthesis. *In*: FLOWER, M. J., CHUNG, S.-L., LEE, T. Y. & LO, C.-H. (eds) *Mantle Dynamics and Plate Interactions in East Asia.* American Geophysical Union Geodynamic Series, **27**, 221–244.

LIU, G., 1992. *Map Series of Geology and Geophysics of China Seas and Adjacent Regions,* 1 : 5 000 000, Geological Publishing House, Beijing.

LIU, X. & FU, D. R. 1988. *The Sedimentary Association and Tectonic Evolution of Meso–Cenozoic Basins of eastern China.* Environmental Science Press of China, Beijing, 189 pp.

PEI, R. & HONG, D. 1995. The granites of South China and their metallogeny. *Episodes*, **18**, 77–82.

QIU, Y. M., GAO, S., MCNAUGHTON, N. J., GROVES, D. I. & LING, W. 2000. First evidence of >3.2 Ga continental crust in the Yangtze Craton of south China and its implications for Archaean crustal evolution and Phanerozoic tectonics. *Geology*, **28**, 11–14.

SEWELL, R. J. & CAMPBELL, S. D. G. 1997. Geochemistry of coeval Mesozoic plutonic and volcanic suites in Hong Kong. *Journal of the Geological Society, London*, **154**, 1053–1066.

SEWELL, R. J., CAMPBELL, S. D. G., FLETCHER, C. J. N., LAI, K. W. & KIRK, P. A. 2000a. *The Pre-Quaternary Geology of Hong Kong.* Geotechnical Engineering Office, Civil Engineering Department, Hong Kong SAR.

SEWELL, R. J., CHAN, L. S., FLETCHER, C. J. N., BREWER, T. S. & ZHU, J. C. 2000b. Isotope zona-tion in basement crustal blocks of southeastern China: evidence for multiple terrane amalgamation. *Episodes*, **23**, 257–261.

WANG, M. 1985. An investigation into the crustal structure of South China in terms of gravity field. *Geophysical and Geochemical Exploration*, **9**, 161–169 (in Chinese).

WU, R. & LU, J. 1989. The correlation between Bouguer gravity anomaly and crustal thickness of North China plate and South China plate. *Geophysical and Geochemical Exploration*, **13**, 7–14.

XING, F., XU, X., CHEN, J., ZHOU, T. & FOLAND, K. A. 1992. The late Proterozoic continental accretionary history of the southeastern margin of the Yangtze Platform. *Acta Geologica Sinica*, **2**, 27–32.

YANG, Z., CHENG, Y. & WANG, H. 1986. *The Geology of China.* Oxford Monographs on Geology and Geophysics, **3**, Clarendon Press, Oxford.

YUAN, Z. 1987. Relationships between geotectonic characteristics of Guangdong China and the geophysical fields. *Geotectonica et Metallogenia*, **11(1)**, 55–64.

ZENG, H., ZHANG, Q., LI, Y. & LIU, J. 1997. Crustal structure inferred from gravity anomalies in south China. *Tectonophysics*, **283**, 189–203.

ZHANG Z. H., LIOU, J. G. & COLEMAN, R. G. 1984. An outline of the plate tectonics of China. *Geological Society of America Bulletin*, **95**, 295–312.

ZHOU, Z., LAO, Q., CHEN, H., DING, S. & LIAO, Z. 1996. Early Mesozoic orogeny in Fujian, southeast Asia. *In*: HALL, R. & BLUNDELL, D. (eds) *Tectonic Evolution of Southeast Asia.* Geological Society, London, Special Publications, **106**, 549–556.

Subduction, collision and exhumation in the ultrahigh-pressure Qinling–Dabie orogen

BRADLEY R. HACKER[1], LOTHAR RATSCHBACHER[2] & J. G. LIOU[3]

[1]Geological Sciences, University of California, Santa Barbara, CA 93106, USA

[2]Institut für Geologie, Technische Universität Bergakademie Freiberg,
Freiberg, D-09599, Germany

[3]Geological and Environmental Sciences, Stanford University,
Stanford, CA 94305, USA

Abstract: High-pressure metamorphism and ophiolite emplacement (Songshugou ophiolite) attended suturing of the Yangtze craton to Rodinia during the $c.1.0$ Ga Grenvillian orogeny. The Qinling microcontinent then rifted from the Yangtze craton at $c.750$ Ma. The Erlangping intraoceanic arc formed in the Early Ordovician, was emplaced onto the Qinling microcontinent in the Ordovician–Silurian, and then both units were accreted to the Sino-Korea craton before being stitched together by the $c.400$ Ma Andean-style Qinling arc. Subsequent subduction beneath the Qinling–Sino-Korean plate created a Devonian–Triassic accretionary wedge that includes eclogites, and formed a coeval volcano-plutonic arc that stretches from the Longmen Shan to Korea. In the Late Permian–Early Triassic, the northern edge of the South China Block was subducted to >150 km depth, creating the diamond- and coesite-bearing eclogites of the Dabie and Sulu areas. Exhumation from the mantle by lithosphere-scale extension occurred between 245 and 195 Ma during clockwise rotation of the craton. The Yangtze–Sino-Korea suture locally lies tens of km north of the exhumed UHP–HP part of the South China Block, implying perhaps that the very tip of the South China Block was not subducted, or that the UHP–HP rocks rose as a wedge that peeled the upper crust of the unsubducted South China Block from the lower crust. The Tan–Lu fault is an Early Cretaceous to Cenozoic feature. The apparent offset of the Dabie and Sulu UHP terranes by the Tan–Lu fault is a result of this Cretaceous to Cenozoic faulting combined with post-collisional extension north of Dabie.

The Sino-Korean craton and the Qinling microcontinent collided in the Ordovician–Silurian. The North China Block (NCB) and the South China Block (SCB) collided in the Permo-Triassic. The orogenic belt associated with these collisions and intervening tectonism extends $c.2000$ km west–east through the Qinling, Tongbai, Dabie Shan and Sulu areas and into Korea (Fig. 1). It is of special interest because it contains high-pressure (HP) rocks of four different ages: Grenvillian, Devonian, Carboniferous and Triassic. This paper aims to provide a review of the collision zone as both a summary of what conclusions have been drawn where and with which data, and as a guide to future research.

Sino-Korean craton

In our usage, the Sino-Korean and Yangtze cratons comprise Precambrian basement and cover as described below; in principle, all Precambrian basement outcrops can be assigned either a Sino-Korean or Yangtze craton affinity, whereas the affinity of areally extensive domains of sedimentary strata or of volcano-plutonic complexes can be ambiguous. 'North China Block' and 'South China Block' are used to refer to the continental blocks north and south of the Qinling–Dabie–Sulu suture.

The Sino-Korean craton is subdivided into three units: the western, eastern and central blocks. The western and eastern blocks are Archean cratons linked by a central suture belt that formed during collision at $c.1.8$ Ga (Zhao et al. 2000). The U–Pb zircon ages from the Sino-Korean craton cluster at $c.3.8, 3.3, 3.0, 2.5$ and 1.7–1.8 Ga (Song et al. 1996; Yu et al. 1996; Zhao et al. 2000). The craton consists of a basement of Archaean to Proterozoic metamorphic rock overlain by a relatively uninterrupted 4–8 km thick superjacent section of Sinian (Late Proterozoic to Early Cambrian) to

From: Malpas, J., Fletcher, C. J. N., Ali, J. R. & Aitchison, J. C. (eds) 2004. *Aspects of the Tectonic Evolution of China*. Geological Society, London, Special Publications, **226**, 157–175.
0305-8719/04/$15 © The Geological Society of London 2004.

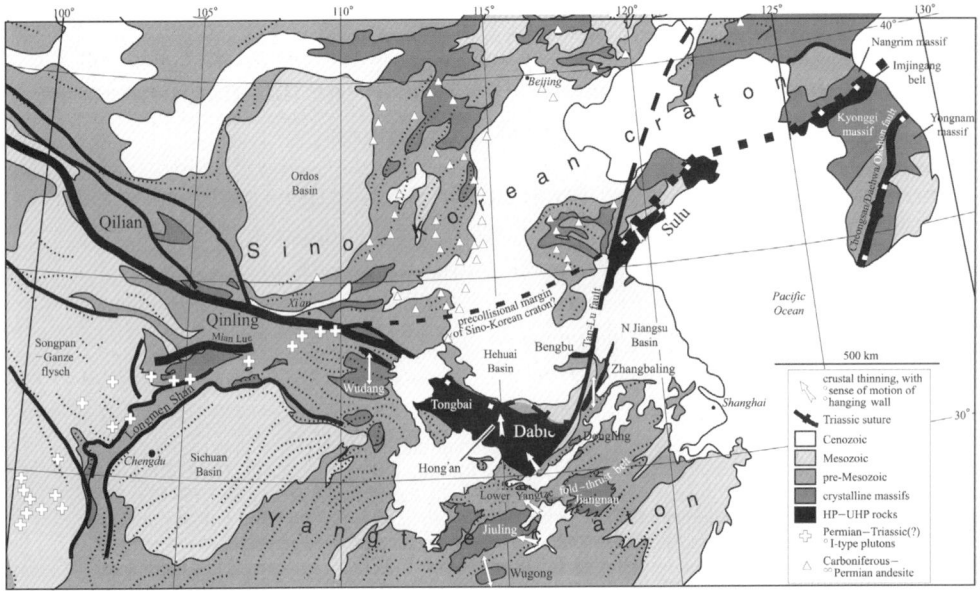

Fig. 1. The Qilian–Qinling–Tongbai–Hong'an–Dabie–Sulu–Imjingang collisional orogen in eastern Asia (modified after Cluzel *et al.* 1991; Ree *et al.* 1996; Wallis *et al.* 1997; Hacker *et al.* 2000). Structural interpretations from previous studies (Faure *et al.* 1996; Wallis *et al.* 1997; Schmid *et al.* 1999; Hacker *et al.* 2000; Lin *et al.* 2000; Ratschbacher *et al.* 2000; Webb *et al.* 2001). Andesitic volcanic rocks from Zhang (1997) and Permian–Triassic plutons from Regional Geological Survey of Sichuan (1991) and Regional Geological Survey of Shaanxi (1989) suggest a pre Late Permian arc related to subduction.

Triassic age (Hsü *et al.* 1987; Ma 1989; Regional Geological Survey of Henan 1989). Sinian through Late Ordovician rocks are platform-facies sandstone, stromatolitic dolomite, limestone, mudstone and rare evaporites, basaltic flows and pyroclastic rocks. Late Ordovician through Early Carboniferous rocks are locally absent in the Qinling orogen, probably indicating tectonism (as discussed below). Devonian to Permian (and locally Triassic in Henan Province) rocks are shallow-marine limestone and dolomite or lacustrine sandstone, mudstone, limestone, gypsum and coal. The Middle Carboniferous to Lower Permian coal-bearing series contains andesitic volcanic rocks (Zhang 1997). Sediment deposition changed markedly on the Sino-Korean craton in the Early Triassic, when lacustrine-to-alluvial conglomerate, arkosic sandstone, and siltstone were laid down; this continental environment persisted throughout the Mesozoic (Regional Geological Survey of Anhui 1987; Regional Geological Survey of Henan 1989; Regional Geological Survey of Hubei 1990).

Some crystalline materials within the core of the Qinling orogenic belt can also be assigned to the Sino-Korean craton (Figs 2 and 3). The Kuanping unit, chiefly amphibolite- to greenschist-facies marbles and two-mica quartz schists, contains detrital zircon with a Pb/Pb age of 638 Ma, is intruded and metamorphosed by *c.*434 Ma diorite and is unconformably overlain by middle Carboniferous to Permian sedimentary rocks (see summary in Ratschbacher *et al.* in press). It could represent the metamorphosed south-facing passive margin of the Sino-Korean craton (Wang 1989). The upper part of the Qinling unit consists of marble with minor amphibolite and garnet–sillimanite gneiss (You *et al.* 1993) that might be correlative with the Kuanping unit (Huang & Wu 1992).

Yangtze craton

The oldest rocks in the Yangtze craton, exposed in the classic Yangtze Gorge section in the Shennong and Huangling areas (Fig. 2), comprise gneissic trondhjemites with *c.*2.9 Ga zircons (U–Pb SHRIMP ages) and paragneisses with 2.9–3.3 Ga detrital zircons (Ames *et al.* 1996; Qiu *et al.* 2000). Zircons from elsewhere in the Yangtze craton, including the Zhangbaling and Dongling areas (Fig. 1), range from *c.*2.5 Ga to *c.*700 Ma (see review in Grimmer *et al.* in press). The youngest zircon signature that is

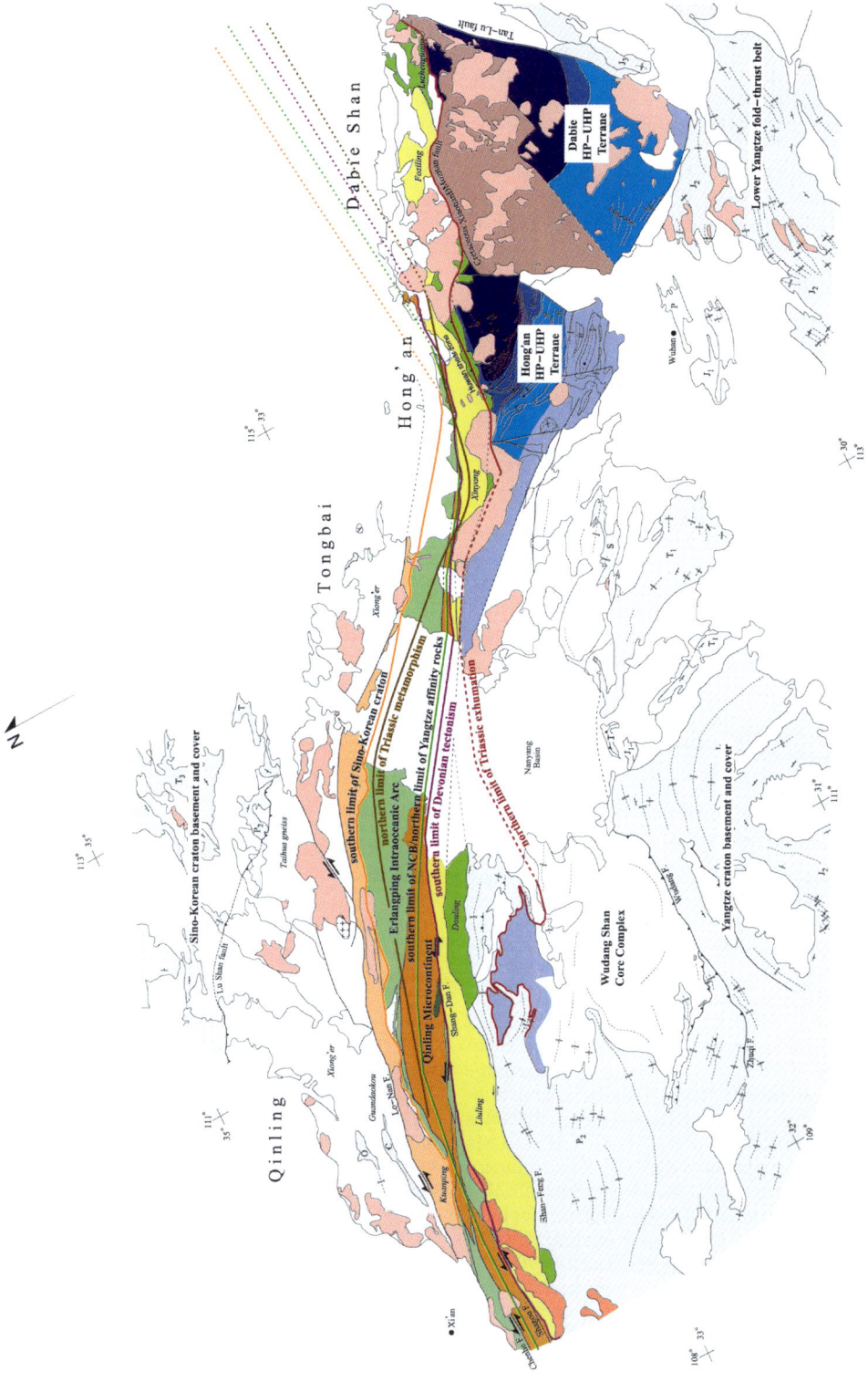

Fig. 4. Major units and unit boundaries of the Qinling–Dabie orogen (see Fig. 2 and its caption).

North China Block–South China Block suture in the Hong'an–Dabie areas.

High-pressure metagabbro, amphibolitized eclogite and felsic granulite associated with *c*.1.0–1.2 Ga ophiolitic rocks were emplaced on to the lower part of the Qinling unit in the Songshugou area, following high-pressure metamorphism in an oceanic setting (Liu *et al.* 1996; Song *et al.* 1998; Zhang 1999). The amphibolitized eclogite was derived from MORB, is in fault contact with mantle peridotite and contains decompression textures common in many eclogites, including symplectites of plagioclase and diopside derived from omphacite estimated to have formed at pressures ≥1.5 GPa (Zhang 1999). A Sm–Nd mineral isochron of 983 Ma is taken to date later cooling. The high-*P* felsic granulite includes garnet + kyanite + microperthite + quartz + rutile assemblages formed at 800–900 °C and 1.3–1.6 GPa (Liu *et al.* 1996).

Palaeoclimatic, palaeobiogeographical and palaeomagnetic data imply that the North China and South China blocks were close to or part of near-equatorial East Gondwana through the Late Devonian, and that rifting of the NCB from Gondwana took place after the Devonian (Zhao *et al.* 1996; Huang *et al.* 2000).

Ophiolites and accretionary complexes

Early Ordovician Erlangping Ophiolite, the Qinling Microcontinent and their amalgamation with the Sino-Korean Craton

The distribution of ophiolites within the Qinling–Dabie–Sulu orogen provides important information about the (partial) closure of ocean basins. The oldest Palaeozoic orogenic event in the Qinling orogen was the formation of the Erlangping intra-oceanic arc, which includes the Heihe, Danfeng and Erlangping units (Fig. 3). Sediments associated with the arc contain Cambrian–Ordovician through Ludlovian–Wenlockian (419–428 Ma) radiolaria, and trondjhemites, tonalites, gabbros, and rare pyroxenites with single-zircon Pb/Pb ages of 470–488 Ma that intrude a volcanic sequence (see summary in Ratschbacher *et al.* in press). Ratschbacher *et al.* (in press) proposed south-directed emplacement of this arc on to the lower Qinling unit (the Qinling microcontinent) between *c*.470 and 435 Ma, based on the present geographical positions of the arc, the upper Qinling unit and the lower Qinling unit. The Erlangping–Qinling collision may have been related to the collision of the Pamir–South Tarim–Qaidam continent with the

NCB (Yin & Nie 1993). This intra-oceanic arc can be traced eastward as far as Hong'an, where it disappears beneath Cenozoic sediments (Figs 2 and 4). No such ophiolite is known from Sulu or Korea, but there are several candidates for Palaeozoic ophiolites in the Kunlun–Altyn Tagh–Qaidam–Qilian area (Matte *et al.* 1996; Sobel & Arnaud 1999).

Silurian–Early Devonian Qinling arc

Following emplacement of the Early Ordovician Erlangping intra-oceanic arc, a continental margin arc was built on the Kuanping, Erlangping and Qinling units from *c*.438–395 Ma; this implies that the Erlangping and Qinling units were amalgamated with the Kuanping unit (and thus the Sino-Korean craton) prior to *c*.440 Ma (Fig. 3) (Ratschbacher *et al.* in press). Significantly, none of these Silurian–Early Devonian plutons crop out within the Liuling unit. However, the northern part of the Liuling unit, which Ratschbacher *et al.* (in press) correlated with the Qinling unit, experienced the regional contact metamorphism produced by this batholith, and Silurian to Lower Devonian strata in the northern Liuling unit received metamorphic detritus (Mattauer *et al.* 1985) from the Qinling arc and *c*.780 Ma and *c*.1.0 Ga detrital zircons (see above). We thus interpret the southern NCB, and units as far south as the northern Liuling unit, as having been stitched together by the 400 Ma magmatic–metamorphic event, and suggest that an Andean-type continental margin arc was built along the southern margin of the NCB at this time; we place the subduction zone producing the Silurian–Devonian arc south of the northern Liuling unit (Ratschbacher *et al.* in press). The local absence of Upper Ordovician (*c*.445 Ma) through Lower Carboniferous (*c*.350 Ma) rocks from the NCB in the Qinling orogen probably reflects uplift related to the formation of this Andean-style Qinling arc.

High-pressure rocks that may be associated with the Qinling arc have been found by Hu *et al.* (1995, 1996) in the northern Qinling area (Fig. 2, offset). Hu *et al.* reported lenses and blocks of eclogite, coesite-bearing eclogite and retrogressed amphibolite within garnet-bearing quartz and phengitic mica schist. While the individual outcrops are no more than a few metres wide, the belt of eclogites extends more than 10 km (Fig. 2). The eclogites consist of garnet + omphacite + rutile + quartz + zoisite ± phengite (3.5 Si atoms per formula unit) ± amphibole; many have been extensively retrogressed. Inclusions of coesite and its pseudomorphs in garnet and omphacite were identified in a few samples, based on optical properties and the presence of radial fractures around the inclusions; this identification, however, should be confirmed by Raman spectroscopy. Pressure–temperature estimates based on mineral composition range from 1.3 to 1.5 GPa and 590 to 758 °C; coesite of course would indicate $P > 2.6$ GPa. A Sm–Nd isochron for garnet, omphacite, rutile, amphibole and whole rock is reportedly 400 ± 16 Ma (although we are unable to reproduce this age from the reported isotopic ratios).

The Qinling continental arc can be traced eastward as far as Hong'an, where it disappears beneath Cenozoic sedimentary rocks (Figs 2 and 4). No such arc is known from Sulu or Korea, but possible correlatives exist in the Kunlun Mountains of northern Tibet (Matte *et al.* 1996; Yang *et al.* 1996).

Devonian–Carboniferous accretionary wedge

A probable fossil accretionary complex, also with high-pressure rocks, lies south of the Silurian–Early Devonian arc in the Liuling and Sujiahe units (Fig. 3). The Liuling unit is a mixture of siliciclastic and volcaniclastic rocks, amphibolite and minor carbonate (You *et al.* 1993). Correlative rocks include the Xinyang Group in Tongbai–Hong'an and the Foziling unit in Dabie. Devonian fossils in the Tongbai area (Du 1986; Niu *et al.* 1993) have been used to infer a Devonian age for the entire Liuling; however, Yu and Meng (1995) showed that several conglomerate-bearing, volcaniclastic deposits, locally containing metabasalts and metacarbonates, have proven Upper Devonian and Lower Carboniferous and suspected Carboniferous to Permian ages. The Sujiahe unit mainly contains volcaniclastic rocks and 'mélange' (Ratschbacher *et al.* in press).

Blocks of eclogitic rocks and retrograde amphibolites occur in a shear zone 1–3 km wide near the southern edge of the Sujiahe unit 5–20 km northwest of Tongbai. The eclogites contain garnet, omphacite, quartz, rutile, phengite and barroisitic amphibole, and have yielded P–T estimates of 480–550 °C and 1.3–1.8 GPa (Wei *et al.* 1999). Ye *et al.* (1994) reported Sm–Nd mineral/rock isochrons of 533 ± 13 Ma and 544 ± 14 Ma and a $^{40}Ar/^{39}Ar$ barroisite age of 399 ± 4 Ma. This belt of HP rocks in the Sujiahe unit extends as far east as the Xiongdian area of Hong'an, where eclogite lenses or layers crop out in quartzite and felsic gneiss. These eclogites contain minor glaucophane and phengite in addition to garnet, omphacite and rutile, and formed at $P \geq 1.3$–1.5 GPa, $T \geq 590$–680 °C, similar to tectonic blocks in the Franciscan Complex

of California (Ye *et al.* 1994; Liu *et al.* 1996). They have yielded SHRIMP U–Pb zircon ages of *c.*310 Ma (Sun *et al.* 2002), 400 Ma (Jian *et al.* 1997), a Sm/Nd isochron of 422 Ma (Li *et al.* 1995), an amphibole $^{40}Ar/^{39}Ar$ age of 400 Ma (Jian *et al.* 1997), and a muscovite $^{40}Ar/^{39}Ar$ age of 400 Ma (Xu *et al.* 2000a). Sun *et al.* (2002) directly tied their *c.*310 Ma age to high-pressure recrystallization with laser ICP-MS trace-element analyses that demonstrated zircon growth while garnet was stable and plagioclase was not stable.

During and after the formation of the Qinling Andean-style volcano-plutonic arc, the southern margin of the NCB was dissected by sinistral wrenching along the Lo-Nan and Shang-Dan shear systems at 420–380 Ma (Fig. 2) (Ratschbacher *et al.* in press). We assume that oblique subduction imposed these spectacular transpressive wrench zones. Muscovite and biotite $^{40}Ar/^{39}Ar$ ages ranging from 348 to 314 Ma have been interpreted to reflect a continuation of this strike-slip motion into the Permian (Mattauer *et al.* 1985). On a larger scale, the NCB and SCB rifted from Gondwana in the Late Devonian–Early Carboniferous (Li & Powell 2001).

Devonian–Triassic volcano-plutonic arc

It is unclear whether the inferred Devonian–Permian subduction beneath the NCB produced a continental margin arc. Younger, Permian and Triassic metaluminous, probably I-type plutons with low Sr_i ratios (Xue *et al.* 1996a) intrude the Kuanping, Erlangping, Qinling and Liuling units as far east as Xi'an (Figs 2 and 4). Coeval intrusions pierce the western edge of the Yangtze Craton in the Xue Shan, Longmen Shan, the Songpan–Ganze flysch and the SCB cover south of the Qinling mountains (Fig. 1, Regional Geological Survey of Gansu 1989, Regional Geological Survey of Shaanxi 1989). The plutons intruding the SCB cover raise an important problem, as they appear to be south of the inferred north-dipping subduction zone. East of Xi'an this arc may end, disappear under Cenozoic sedimentary rocks, or be covered by younger thrust slices. Middle Carboniferous to Early Permian andesites form a NE-trending belt across the eastern half of the NCB (Fig. 1) (Zhang 1997), but the belt is surprisingly far north of and oblique to the inferred NCB–SCB suture.

We suggest that the pre-collisional, Permo-Carboniferous NE trend of the NCB–SCB suture in Tongbai–Hong'an–Dabie and the suture–arc distance were modified by the Triassic, syncollisional clockwise rotation of the SCB described below (e.g. Zhao & Coe 1987), and the Late Jurassic–Cenozoic extension within the Hehuai Basin (e.g. Han *et al.* 1989; Ren *et al.* 2002). The apparent intrusion of subduction-related plutons into the SCB cover can most easily be reconciled by attributing them to Late Triassic subduction along the cryptic Mianlue 'suture' of the southwestern Qinling orogen (Fig. 1) (Meng & Zhang 2000). The Mianlue rock assemblage, containing volcaniclastic and ophiolitic remnants with proven or suspected Early–Middle Triassic metamorphic ages (Li *et al.* 1999a), apparently terminates in the southwestern Qinling orogen, but might connect with the Liuling unit through the central and eastern Qinling orogen.

Triassic collision

Palaeomagnetic data (Zhao & Coe 1987; Lin & Fuller 1990; Enkin *et al.* 1992) suggest that the NCB and SCB moved farther apart during the Middle Permian to Middle–Late Triassic, but then approached each other and underwent 60° of relative rotation between Middle–Late Triassic and Middle–Early Jurassic times (Fig. 5) (Zhao & Coe 1987; Gilder *et al.* 1999). A regional unconformity and a cusp on the NCB apparent polar wander path imply that collision ended at the Middle to Late Jurassic boundary (Gilder & Courtillot 1997).

Collisional metamorphism

The UHP–HP metamorphism of the Qinling–Dabie Orogen constitutes the most readily identifiable feature of the North China Block–South China Block collision. Triassic HP rocks occur in four distinct areas: (1) the northern Wudang core complex, which tapers eastward into the Suixian and Yaolinghe areas; (2) the Hong'an–Dabie area, which forms an eastward-thickening wedge of UHP rocks (including eclogite relits recently discovered within northeastern Dabie

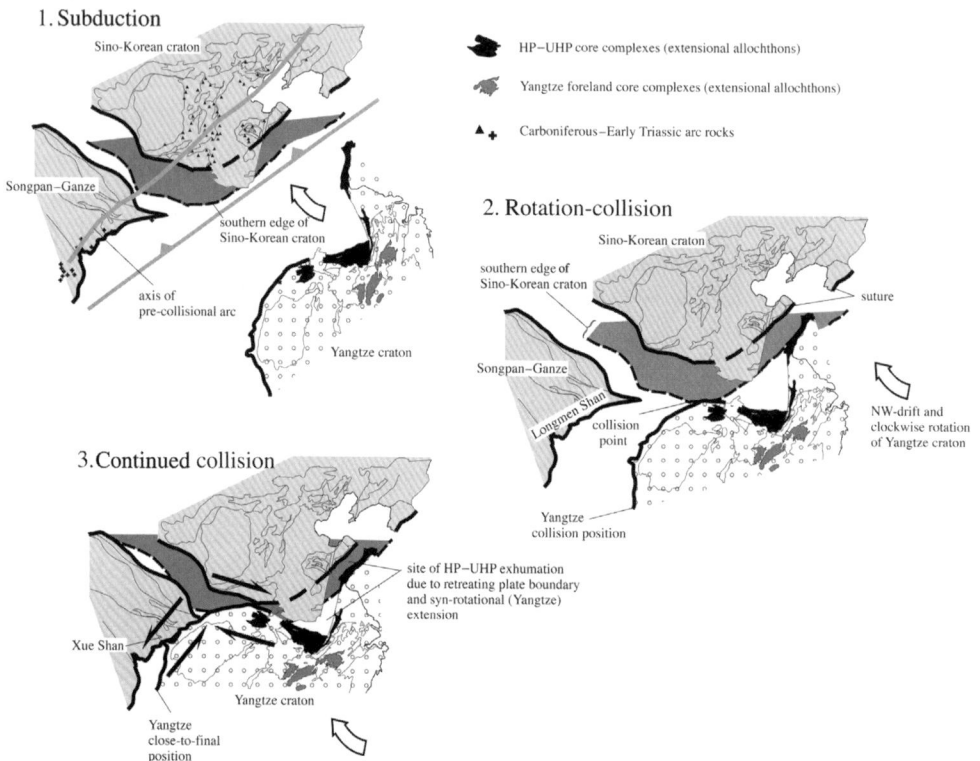

Fig. 5. Carboniferous–Triassic tectonic evolution of eastern Asia. (1) Formation of a Carboniferous–Permian magmatic arc, oblique to the presently east–west-trending Triassic suture along the Qinling–Dabie belt. (2–3) Palaeomagnetically supported rotation–collision scenario for the Permian–Triassic Sino-Korean–Yangtze approach, illustrating transpressive wrenching along the Qinling–Dabie belt and extensional exhumation of the Dabie–Sulu HP–UHP rocks by retreat of the Sino-Korean plate boundary due to rotation during collision.

(Wei *et al.* 1998; Tsai & Liou 2000; Xu *et al.* 2000b; Zhou *et al.* 2000; Liu *et al.* 2001); (3) the Zhangbaling and Bengbu areas, a mostly blueschist-facies extensional complex (our observations); and (4) the Sulu area (Wallis *et al.* 1999) (Fig. 1).

The evidence of Triassic HP and UHP metamorphism comes chiefly from a few per cent of eclogite and garnet peridotite boudins that are hosted by paragneiss plus less granodioritic–tonalitic orthogneiss (Fig. 6) (Cong 1996; Liou *et al.* 2000). The highest temperatures and pressures attained in the coesite- and, locally, diamond-bearing Dabie eclogites were 825–850 °C and 3.3–4.0 GPa (Carswell *et al.* 1997). The coesite-free eclogites reached somewhat lower peak conditions of 625–700 °C and 2.2–2.4 GPa in the Dabie Shan (Okay 1993; Liou *et al.* 1996; Carswell *et al.* 1997), and have been divided into kyanite-bearing and kyanite-absent eclogites in Hong'an, with estimated physical conditions of 550–650 °C, 1.6–2.5 GPa and 450–550 °C, 0.8–1.2 GPa, respectively (Eide & Liou 2000). The amphibolite unit is a hornblende-rich orthogneiss; peak pressure and temperature estimated from one locality are >1.0 GPa and ≥650 °C (Liu & Liou 1995). Blueschists reached conditions of

only 400–800 MPa at 350–450 °C (Eide & Liou 2000). While most of the evidence of UHP derives from eclogites, the paragneiss also contains local unambiguous indications of metamorphism at similar pressure and temperature, such that the eclogites clearly were metamorphosed *in situ* (Liou *et al.* 1996), the same may not hold true for ultramafic blocks, some of which record recrystallization pressures >4 GPa (Okay 1994; Hacker *et al.* 1997; Liou & Zhang 1998). The spatial distribution and metamorphic conditions of these HP through UHP rocks indicate subduction-zone metamorphism on a regional scale.

The age of UHP metamorphism in the Dabie Shan is now well constrained to be Middle–Late Triassic (e.g. Hacker *et al.* 2000). The most recent data indicate that UHP recrystallization may have begun as early as 245 Ma and extended through c.225–230 Ma (Okay 1993; Hacker *et al.* 1998; Hacker *et al.* 2000; Li *et al.* 2000; Chavagnac *et al.* 2001; Ayers *et al.* 2002). High-pressure rocks in Hong'an (the area least affected by Cretaceous reheating) indicate that the regional amphibolite-facies overprint at 500–650 °C and 0.8 GPa occurred during the 195–225 Ma time span ([40]Ar/[39]Ar phengite

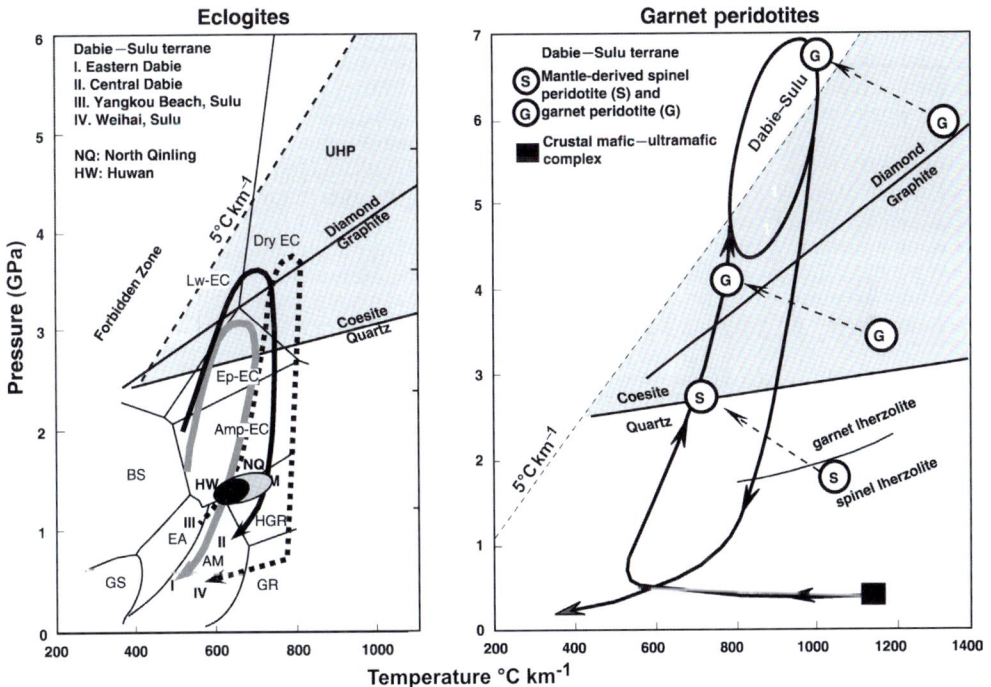

Fig. 6. Metamorphic pressures and temperatures for eclogites and peridotites of the Qinling–Dabie–Sulu orogen (modified after Liou *et al.* 1999; Zhang *et al.* 2001).

ages). Most of the K-feldspar $^{40}Ar/^{39}Ar$ spectra have inflections indicating cooling below $c.200\,°C$ near 170 Ma, although some rocks were this cool as early as 200 Ma. Such $P–T$ paths, with concomitant decompression and cooling are only possible if exhumation was rapid or the slab was refrigerated by deeper level subduction (Hacker & Peacock 1995; Ernst & Peacock 1996). The exhumation rate is only crudely constrained to $c.3–15$ mm/year.

Geochronology shows that the Middle–Late Triassic thermal event associated with the UHP metamorphism in the eastern Dabie Shan and Hong'an areas extended as far north as the Erlangping unit in the Qinling area, the Xinyang unit in the Tongbai and Hong'an areas, and the Foziling and Luzhenguang units in the Dabie Shan (Hacker *et al.* 1998; Hacker *et al.* 2000). The absence of HP metamorphism means that these northernmost units were not subducted during the collision, but the Late Triassic metamorphism is certainly related to the collision.

Collisional deformation

The Triassic NCB–SCB collision produced a distinct set of structures throughout the Qinling-Dabie orogen: NW–SE to N–S contraction by folding and thrusting throughout the belt, subhorizontal NW–SE to N–S extension within core complexes along the northern margin of the South China Block, and dextral transpressive reactivation of existing shear zones in the Qinling area (Fig. 5).

The UHP rocks in both Dabie–Hong'an (Faure *et al.* 1999; Hacker *et al.* 2000) and Sulu (Wallis *et al.* 1999; Faure *et al.* 2001) form the cores of structural domes, exhibit a top-NW sense of shear and are overlain by extensional faults that exhumed the UHP rocks. At least in Hong'an–Dabie, the entire crystalline core of the orogen constitutes a normal-sense shear zone $c.15$ km thick; the Huwan shear zone, a normal-sense detachment that reactivated the plate suture, tops the extensional allochthon in Hong'an (Hacker *et al.* 2000; Webb *et al.* 2001). Associated Triassic metamorphic core complexes in the northern part of the SCB include the Wudang Shan (Ratschbacher *et al.* in press) and Zhangbaling–Bengbu (our unpublished data); whether the Lu Shan (Lin *et al.* 2000), Wugong Shan (Faure *et al.* 1996), Dongling (Grimmer *et al.* in press), and Jiuling Shan (Lin *et al.* 2001) basement uplifts within the foreland SCB fold-thrust belt south and east of Hong'an–Dabie, which show a Triassic to Cretaceous extensional overprint, are related to

extensional exhumation of the HP–UHP rocks remains unclear.

The HP to UHP rocks of Dabie–Hong'an are mostly a structural homocline with SE-dipping foliation, SE-plunging lineation, and overall top-to-NW flow that includes significant coaxial stretching (Fig. 7). In northern Hong'an, however, the S-oriented structures roll over through horizontal into a 5-km-thick zone of N(W)-dipping foliations, N(W)-plunging lineations, and a N(W)-directed sense of shear. This, the Huwan shear zone (Webb *et al.* 2001), effectively straddles the Triassic suture. It developed out of the high-strain gneisses of typical Yangtze affinity in northern Hong'an, contains strongly retrogressed eclogite boudins of pre-Triassic age in the Sujiahe complex (see above), and dies out northward in the phyllitic quartzites typical of the southern Liuling unit. The Huwan shear zone can be followed – with interruptions by Cretaceous plutons – into northern Dabie, where it is truncated by a large Cretaceous pluton. The pseudostratigraphy of the Hong'an area also defines a series of km-scale NW-trending synforms and antiforms that are overturned to the north in central Hong'an, and are upright to south facing in southern Hong'an. The coesite-bearing eclogite unit forms the core of the northernmost antiform; the lowest pressure rocks, blueschist, are present only on the south limb of this orogen-scale fold. The westward decrease in peak metamorphic pressures reveals that the antiform plunges west. This antiform extends eastward into Dabie where it is partially overprinted by the dominantly Cretaceous intrusions and structures of the Northern Orthogneiss (Ratschbacher *et al.* 2000), and then terminates against the Tan–Lu fault.

The large-scale Wudang Shan dome (Fig. 4) that lies west of the UHP Hong'an–Dabie area is an extensional core complex overprinted by foreland folding and thrusting. Deformation in the Wudang Shan began with an early unresolved, but possibly contractional deformation, was followed by sub-vertical contraction and sub-horizontal N–S extension during blueschist–greenschist-facies metamorphism of the basement and basement–cover contact zone; $^{40}Ar/^{39}Ar$ dating of syn- to post-kinematic hornblende and muscovite puts the extension at, or prior to, 230–235 Ma. Subsequent folding and thrusting of the basement and Palaeozoic cover sequence occurred during N–S contraction.

The northern limit of the Triassic extensional deformation (Fig. 4) is located at different positions along the Qinling orogen. In the Qinling mountains, the northern limit of extension is the Wudang basement; the Douling unit

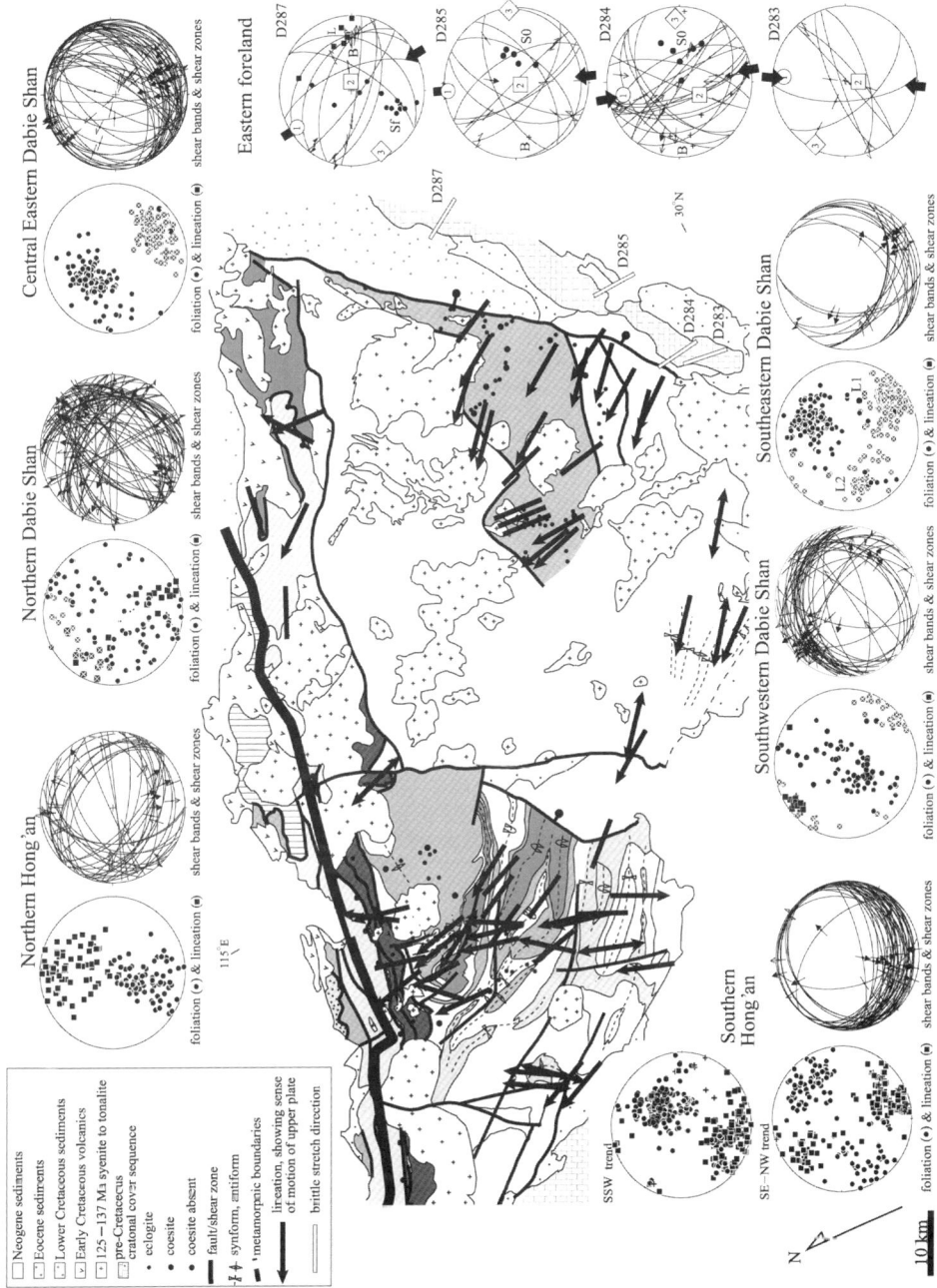

Fig. 7. Structural overview of the Dabie–Hong'an area (after Schmid *et al.* 1999; Hacker *et al.* 2000; Ratschbacher *et al.* 2000; Webb *et al.* 2001). Areas disturbed by Cretaceous plutons are not included in the synoptic stereonets. Stereonets from the eastern foreland show dominantly brittle faults: (**1**), (**2**), and (**3**) indicate principal stress directions; B, fold axis; S0, bedding; and Sf, foliation.

is non-mylonitic and lacks significant Triassic metamorphism. In northern Hong'an, the Huwan shear zone constitutes a lithosphere-scale reactivation of the Devonian–Triassic subduction complex. In Dabie, the Huwan detachment cannot be mapped, as Cretaceous tectonism has obliterated earlier fabrics, but the northern limit of extension must be south of the Foziling and Luzhenguang units, which have clear Yangtze craton affinity and were unaffected by HP metamorphism, very likely coinciding with the Early Cretaceous Xiaotian–Mozitang crustal-scale shear zone (Hacker *et al.* 2000; Ratschbacher *et al.* 2000).

Much of the area shown in Figure 2 has WNW-trending folds and craton-directed thrusts of millimeter to kilometre scale that formed during the collision (Mattauer *et al.* 1985; Regional Geological Survey of Henan 1989; Regional Geological Survey of Shaanxi 1989; Regional Geological Survey of Hubei 1990). On the NCB, the north-directed Lu Shan thrust, placing crystalline basement over Palaeozoic cover, was active in the Middle Triassic and Late Jurassic (Huang & Wu 1992). Huang and Wu (1992) also identified a series of Mesozoic S-directed thrusts imbricating the Sino-Korean basement, Kuanping, Qinling, Erlangping and Douling units. The fold–thrust belt along the Lower Yangtze river began to develop through NW–SE contraction in the Middle Triassic and continued through the Early Jurassic, by which time the shortening direction had rotated to NNE–SSW (Schmid *et al.* 1999).

The Early Palaeozoic core of the Qinling orogen was reactivated during the relatively early stages of the NCB–SCB collision by overall top-south imbrication during a change from NW–SE (dextral transpression) to NE–SW shortening. Ratschbacher *et al.* (in press) documented *c.*200–250 Ma low-grade metamorphism and dextral transpression along the Lo-Nan, Shang-Xian and Shang-Dan faults, and N–S shortening within the Liuling and Douling units.

Syn- to post-collisional overlap assemblage

Changes in sedimentation patterns are often among the best guides to collisional timing. Depositional facies up through the Permian trend E–W in the southern NCB, and NNE in the northern SCB (Han *et al.* 1989; Sun *et al.* 1989; Wang *et al.* 1989; Zhang *et al.* 1989). Although collision-related (?) shortening began on the SCB in the Early to Middle Triassic, the major sedimentological change occurred in the Middle Triassic, which was marked locally by either a depositional hiatus, continental

sedimentation (Huang & Opdyke 2000) or unconformable deposition of coarse carbonate breccias (Breitkreuz *et al.* 1994). Yin & Nie (1993) proposed that the transition from marine to continental sedimentation was diachronous: Early Permian in Shandong and Korea, Lower/Upper Permian south of Sulu, Lower/Middle Triassic SE of Dabie.

Nie *et al.* (1994) and Zhou & Graham (1996) proposed that detritus eroded from the active Qinling–Dabie mountain range was channeled westward and deposited to form the Songpan–Ganze flysch from the Anisian through the Norian (*c.*240–210 Ma). Bruguier *et al.* (1997) found detrital zircons with U/Pb ages of 233 and 231 Ma within the Songpan–Ganze flysch that match the ages of zircons found in Dabie (Hacker *et al.* 2000).

Post-collisional, Jurassic, continental clastic sedimentation on both cratons was accompanied by deposition of up to 5 km of calc-alkaline, crustal-derived, intermediate-composition volcanic rocks: tuff, volcanogenic sandstone, and some lava in the Late Jurassic (Regional Geological Survey of Anhui 1987; Regional Geological Survey of Henan 1989; Regional Geological Survey of Hubei 1990). Gneiss cobbles, presumably derived from the Dabie Mountains, first appeared in the Zhuji Formation on the northern slope of the Dabie Mountains in Middle Jurassic time (Ma 1991). Middle Jurassic sedimentary rocks in the Lower Yangtze fold–thrust belt show clear evidence of erosion of the Dabie UHP core, including Triassic-Jurassic detrital micas with high-Si contents and zircon grains as young as *c.*218 Ma (Grimmer *et al.* in press) their oldest mica ages indicate that exhumation in Hong'an-Dabie might have started at 240 ± 5 Ma.

Plate-scale collision model

While there is still much that we do not know, some key structural observations constrain the exhumation mechanism of the UHP rocks (Hacker *et al.* 2000):

(1) Unfolding the 'Hong'an antiform' – the orogen-scale fold trending NW–SE across Hong'an–Dabie – yields a N-dipping lithospheric slab with coesite eclogite in the north and lower pressure rocks progressively farther south and closer to the interior of the SCB. The upper boundary of the slab is the Huwan shear zone, which encompasses the suture between the NCB and the SCB, implying that it is essentially a plate boundary reactivated as a lithosphere-scale, normal-sense shear zone.

(2) The size of the UHP outcrop and the peak pressures diminish westward, indicating that the depth of exhumation increased eastward.

(3) Stretching lineations within Hong'an and Dabie show a clockwise rotation with depth of exhumation (Fig. 5).

These features are most easily reconciled with a model involving subduction of an triangular promontory of the SCB that reached its greatest depth in the east (Fig. 8, Hacker *et al.* 2000). At some point, the buoyant, subducted continental crust tore away from the oceanic part of the plate and began to rise within the channel between the two cratons. The tear began at the deepest point of subduction of the buoyant slab and ripped upward along the promontory. The shape and orientation of the subducted crust, combined with the dependence of buoyancy on

depth below the Moho, imply that the promontory might have pivoted about its shallow end, producing the curved lineations and an extruded wedge of formerly subducted SCB plate. The E–W trend of the southern margin of the NCB in the Qinling–Dabie area, coupled with the observed motion within the extrusion channel, implies that the UHP slab was extruded eastward along the plate margin during exhumation. We suggest that this occurred toward an eastern re-entrant in the plate margin that imposed a weak constraint on the extruding lithosphere (Ratschbacher *et al.* 1991).

Such a model makes no significant predictions about the apparent sinistral offset of the Qinling–Dabie–Sulu suture along the Tan–Lu fault. Yin and Nie (1993) explained the left-lateral offset along the Tan–Lu fault and the orientations of sedimentary facies patterns on the

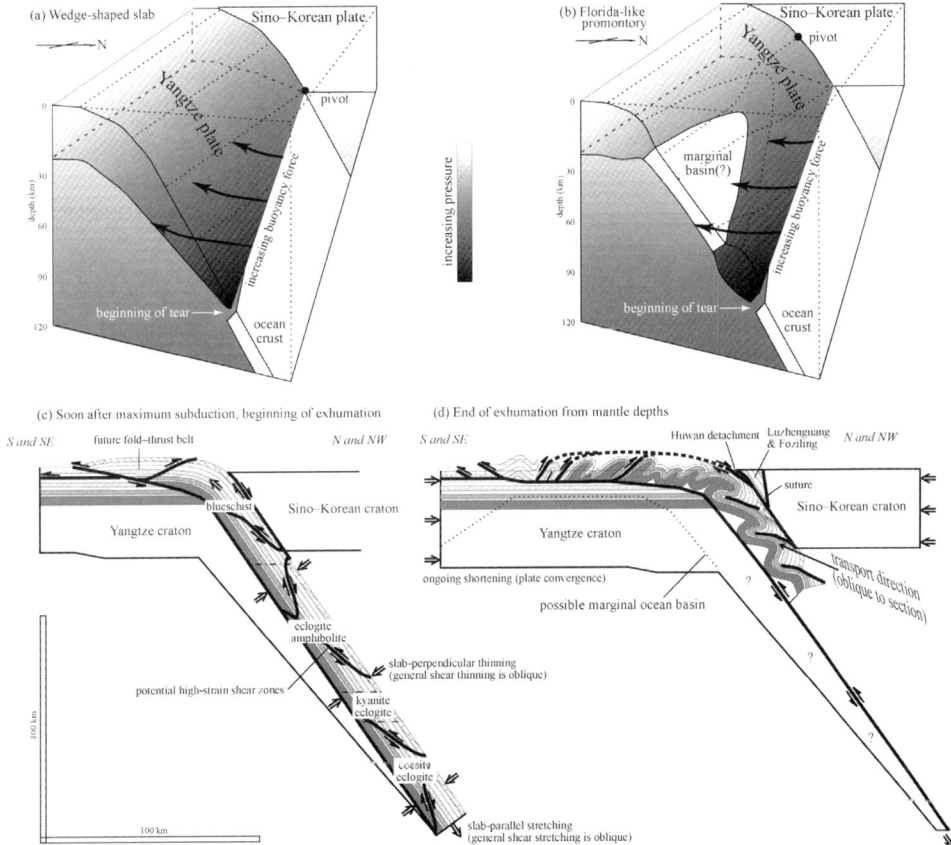

Fig. 8. Exhumation model. (**a**) Subduction of wedge-shaped continental promontory; arrow shows pivoting of slab during exhumation. (**b**) Subduction of Florida-like continental promontory (dark) attached to oceanic crust (white). (**c**) 2D representation of subducted continental crust, with detached sliver beginning to exhume. (**d**) 2D representation of continental crust exhumed by buoyancy plus subhorizontal extrusion.

two plates as the result of Late Permian–Early Triassic indentation of the NCB by the SCB; they rationalized that the active margin of the NCB should have been relatively straight and that the passive margin of the SCB could have been relatively complex. Gilder *et al.* (1999) criticized this idea because of a 'lack of significant folding of Phanerozoic rocks . . . north of the Sulu belt', and proposed instead that the older NCB acted as a rigid indentor to deform the SCB in the Early to Middle Jurassic – leading to the palaeomagnetically recorded bending of the Lower Yangtze fold–thrust belt (Gilder *et al.* 1999), formation of the Tan–Lu fault, and 60° of relative rotation between the NCB and SCB (Zhao & Coe 1987). While there is structural and thermochronological evidence of Late Cretaceous and Cenozoic strike-slip and normal faulting (Ratschbacher *et al.* 2000; e.g. Grimmer *et al.* 2002), no well-documented structural data demonstrate sinistral Triassic or Jurassic motion along the Tan–Lu fault. However, Mid-Triassic NW–SE contraction and Late Triassic–Early Jurassic N–S contraction in the eastern Dabie foreland (Schmid *et al.* 1999) implies that if the Tan–Lu fault existed at that time then it would have been sinistral. Note that the age of Triassic metamorphism in the Qinling area is similar to that in Dabie; the proposed younging and westward migration of collision (e.g. Zhao & Coe 1987; Yin & Nie 1993; Zhang 1997) is not supported by extant geochronology (Hacker & Wang 1995).

Problems and possible solutions

The preceding sections have outlined a number of important problems within the Qinling–Dabie orogen.

(1) The boundary between rocks of Sino-Korean and Yangtze affinity appears to be a suture of Ordovician–Silurian age. This interpretation could benefit from further analysis of the tectonic histories of key units in the Qinling orogen.

(2) Devonian–Triassic rocks of the Liuling, Xinyang, and Foziling units separate the NCB from the SCB. These relatively monotonous units thus potentially experienced quite a varied history. Can this history be read from these rocks?

(3) The northern boundary of Triassic exhumation is generally south of the northern edge of the SCB. In the Qinling mountains, it lies south of the Douling unit and south of the Yangtze craton cover. In Tongbai and Hong'an it coincides with the northern edge of the SCB. In Dabie, it lies south of the Luzhenguang unit. This implies that the Douling, Luzhenguang and other units (i) represent the leading edge of the SCB but were not subducted; (ii) were originally south of the HP–UHP rocks and ended up in their present position as a result of exhumation of the HP–UHP rocks; or (iii) are part of the NCB (although of Yangtze affinity, like the Qinling unit), but are now mysteriously south of the Liuling, Xinyang and Foziling units and escaped intrusion by the Qinling arc.

(4) A weakness of the exhumation model outlined in the previous section is that we have not identified a sole thrust along the southern and eastern edges of Dabie–Hong'an. SE-directed extrusion of the wedge of UHP–HP rocks should have induced equivalent shortening at the tip of the extruding wedge, but one of the characteristics of the Triassic/Jurassic foreland deformation east of Dabie is upright folding and a

combination of hinterland- and foreland-directed thrusting with relatively moderate shortening (Schmid *et al.* 1999; Grimmer *et al.* in press). To explain the apparent lack of the mega-thrust and the distinctive structural style, we can envision at least three scenarios. (i) The leading edge of the UHP–HP rocks forms a wedge that is buried beneath the cover strata of the foreland. This implies that the crust of the foreland was detached at middle to upper crustal levels and thrust northwestward on to the crystalline core. (ii) The UHP–HP rocks were exhumed by buoyancy into the middle/lower crust, and the overlying rocks were removed by extension and erosion. A plate rotation model proposes that after initial collision in the eastern part of the orogen, continued closure of the 'Songpan–Ganze Sea' by subduction of oceanic lithosphere caused the SCB to rotate 60° clockwise (required by palaeomagnetic data), pulling parts of the subducted SCB back toward the surface.

Summary

Grenvillian orogeny, involving oceanic subduction with HP metamorphism and ophiolite emplacement (Songshugou ophiolite), assembled the Yangtze craton, including the Qinling microcontinent, into Rodinia. Rifting at *c.*750 Ma separated the Qinling microcontinent from the Yangtze craton. Intra-oceanic arc formation (Erlangping–Danfeng–Heihe) between $c.490$–470 Ma was followed by the accretion of the lower Qinling unit to the intra-oceanic arc and the North China Block. Oceanward (northward in present coordinates) subduction beneath the northern Liuling unit imprinted the $c.400$ Ma Andean-type magmatic arc on to the North China Block. Oblique subduction imposed the spectacular Early Devonian left-lateral transpressive wrench zones. A subduction signature is again evident during the mid-Carboniferous to Late Permian on the North China Block, when the Palaeo-Tethys was subducted northward, producing the andesitic magmatism on the North China Block. In the Late Permian–Early Triassic the leading edge of the South China Block was subducted to >150 km and subsequently exhumed by crustal extension during clockwise rotation of the craton.

Thanks to Yongjun Yue, Stanford, for assistance with Chinese publications; Shuwen Dong, Yueqiao Zhang and many other Chinese colleagues for field guidance and discussions; our colleagues J. C. Grimmer, L. Franz, R. Schmid, E. Enkelmann and R. Oberhänsli for discussions; and C. Blake, P. Robinson and C. Fletcher for useful criticism. Funded by projects Ra442/14, 19 of the Deutsche Forschungsgemeinschaft and NSF grants EAR-9796119, EAR-9725667 and EAR-0003568.

References

AMES, L. 1995. *Geochronology and isotopic character of ultrahigh-pressure metamorphism with implications for collision of the Sino-Korean and Yangtze cratons, central China.* PhD thesis, University of California.

AMES, L., ZHOU, G. & XIONG, B. 1996. Geochronology and geochemistry of ultrahigh-pressure metamorphism with implications for collision of the Sino-Korean and Yangtze cratons, central China. *Tectonics*, **15**, 472–489.

AYERS, J. C., DUNKLE, S., GAO, S. & MILLER, C. 2002. Constraints on timing of peak and retrograde metamorphism in the Dabie Shan ultrahigh-pressure metamorphic belt, east-central China, from U–Th–Pb dating of zircon and monazite. *Chemical Geology*, **186**, 315–331.

BREITKREUZ, H., MATTERN, F. & SCHNEIDER, W. 1994. Mid-Triassic shallow marine evaporites and a genetic model for evaporation during regional compression (lower Yangtze basin, Anhui, China). *Carbonates and Evaporites*, **9**, 211–222.

BRUGUIER, O., LANCELOT, J. R. & MALAVIELLE, J. 1997. U–Pb dating on single detrital zircon grains from the Triassic Songpan–Ganze flysch (central China): provenance and tectonic correlations. *Earth and Planetary Science Letters*, **152**, 217–231.

CAO, R. & ZHU, S. 1995. The Dabieshan coesite-bearing eclogite terrain – a neo-Archean ultra-high pressure metamorphic belt. *Acta Geologica Sinica*, **69**, 232–242.

CARSWELL, D. A., O'BRIEN, P. J., WILSON, R. N. & ZHAI, M. 1997. Thermobarometry of phengite-bearing eclogites in the Dabie Mountains of central China. *Journal of Metamorphic Geology*, **15**, 239–252.

CHAVAGNAC, V. & JAHN, B.-M. 1996. Coesite-bearing eclogites from the Bixiling Complex, Dabie Mountains, China; Sm–Nd ages, geochemical characteristics and tectonic implications. *Chemical Geology*, **133**, 29–51.

CHAVAGNAC, V., JAHN, B.-M. VILLA, I. M., WHITEHOUSE, M. & LIU, D. 2001. Multichronometric evidence for an in situ origin of the ultrahigh-pressure metamorphic terrane of Dabieshan, China. *Journal of Geology*, **109**, 633–646.

CHEN, D., WU, Y., WANG, Y.-J., ZHI, X., XIA, Q. & YANG, J. 1998. Ages, Nd and Sr isotopic compositions of the Jiaoziyan gabbroic intrusion from the northern Dabie terrain. *Scientia Geologica Sinica*, **7**, 29–35.

CHEN, J., XIE, Z., LIU, S., LI, X. & FOLAND, K. A. 1995. Cooling age of Dabie orogen, China, determined by ^{40}Ar–^{39}Ar and fission track techniques. *Science in China, Series B*, **38**, 749–757.

CHEN, N., HAN, Y., YOU, Z. & SUN, M. 1992. Whole-rock Sm–Nd, Rb–Sr and zircon Pb–Pb dating of the metamorphic complex in the interior of Qinling orogenic belt, western Henan, and their implications with regard to crustal evolution. *Chinese Journal of Geochemistry*, **11**, 168–177.

CHEN, N.-S., YOU, Z.-D., SUO, S.-T., YANG, Y. & LI, H.-M. 1996. U–Pb zircon ages of intermediate granulites and deformed granites in Dabie Mountains, central China. *Chinese Science Bulletin*, **41**, 1886–1890.

CLUZEL, D., JOLIVET, L. & CADET, J.-P. 1991. Early middle Paleozoic intraplate orogeny in the Ogcheon belt (South Korea): a new insight on the Paleozoic buildup of east Asia. *Tectonics*, **10**, 1130–1151.

CONG, B. 1996, *Ultrahigh-pressure metamorphic rocks in the Dabieshan–Sulu region of China.* Science Press, Beijing.

DU, D. 1986. *Research on the Devonian System of the Qin-Ba Region Within the Territory of Shaanxi.* Jiaotang University Publishing House, Xi'an.

EIDE, E. A. & LIOU, J. G. 2000. High-pressure blueschists and eclogites in Hong'an: a framework for addressing the evolution of ultrahigh-pressure rocks in central China. *Lithos*, **52**, 1–22.

EIDE, E. A, McWILLIAMS, M. O. & LIOU, J. G. 1994. ^{40}Ar/^{39}Ar geochronologic constraints on the exhumation of HP–UHP metamorphic rocks in east-central China. *Geology*, **22**, 601–604.

ENKIN, R. J., YANG, Z., CHEN, Y. & COURTILLOT, V. 1992. Paleomagnetic constraints on the geodynamic history of the major blocks of China from the Permian to the present. *Journal of Geophysical Research*, **97**, 13953–13989.

ERNST, W. G. & PEACOCK, S. M. 1996. A thermotectonic model for preservation of ultrahigh-pressure phases in metamorphosed continental crust. *In*: BEBOUT, G. E., SCHOLL, D. W., KIRBY, S. H. & PLATT, J. P. (eds) *Subduction Top to Bottom*, American Geophysical Union, Washington, D.C., 171–178.

FAURE, M., LIN, W. & LE BRETON, N. 2001. Where is the North China–South China block boundary in eastern China? *Geology*, **29**, 119–122.

FAURE, M., LIN, W., SHU, L., SUN, Y. & SCHÄRER, U. 1999. Tectonics of the Dabieshan (eastern China) and possible exhumation mechanism of ultra high-pressure rocks. *Terra Nova*, **11**, 251–258.

FAURE, M., SUN, Y., SHU, L., MONIÉ, P. & CHARVET, J. 1996. Extensional tectonics within a subduction-type orogen: the case study of the Wugongshan dome (Jiangxi Province, SE China). *Tectonophysics*, **263**, 77–108.

GILDER, S. A. & COURTILLOT, V. 1997. Timing of the North–South China collision from new middle to late Mesozoic paleomagnetic data from the North China Block. *Journal of Geophysical Research*, **102**, 17 713–17 727.

GILDER, S. A., LELOUP, P. H., COURTILLOT, V., CHEN, Y., COE, R., ZHAO, X., XIAO, W., HALIM, N., COGNE, J.-P. & ZHU, R. 1999. Tectonic evolution of the Tancheng–Lujiang (Tan-Lu) fault via Middle Triassic to Early Cenozoic paleomagnetic data. *Journal of Geophysical Research*, **104**, 15 365–15 390.

GRIMMER, J. C., JONCKHEERE, R., ENKELMANN, E., RATSCHBACHER, L., HACKER, B. R., BLYTHE, A. E., WAGNER, G. A., WU, Q., LIU, S. & DONG, S. 2002. Late Cretaceous–Cenozoic history of the southern Tan–Lu fault zone: Apatite fission-track and structural constraints from the Dabie Shan (eastern China). *Tectonophysics*, **359**, 225–253.

GRIMMER, J. C., RATSCHBACHER, L., FRANZ, L., GAITZSCH, I., TICHOMIROWA, M., MCWILLIAMS, M. & HACKER, B. R. in press. When did the ultrahigh-pressure rocks reach the surface? A $^{207}Pb/^{206}Pb$ zircon, $^{40}Ar/^{39}Ar$ white mica, Si-in-phengite single grain study of Dabie Shan synorogenic foreland sediments. *Chemical Geology*, **197**, 87–110.

HACKER, B. R. & PEACOCK, S. M. 1995. Creation, preservation, and exhumation of coesite-bearing, ultrahigh-pressure metamorphic rocks. *In*: COLEMAN, R. G. & WANG, X. (eds) *Ultrahigh Pressure Metamorphism*, Cambridge University Press, 159–181.

HACKER, B. R., RATSCHBACHER, L., WEBB, L., IRELAND, T., WALKER, D. & DONG, S. 1998. U/Pb zircon ages constrain the architecture of the ultrahigh-pressure Qinling–Dabie Orogen, China. *Earth and Planetary Science Letters*, **161**, 215–230.

HACKER, B. R., RATSCHBACHER, L., WEBB, L. E., IRELAND, T. R., CALVERT, A., DONG, S., WENK, H.-R. & CHATEIGNER, D. 2000. Exhumation of ultrahigh-pressure continental crust in east-central China: Late Triassic–Early Jurassic tectonic unroofing. *Journal of Geophysical Research*, **105**, 13 339–13 364.

HACKER, B. R., SHARP, T., ZHANG, R. Y., LIOU, J. G. & HERVIG, R. L. 1997. Determining the origin of ultrahigh-pressure lherzolites. *Science*, **258**, 702–704.

HACKER, B. R. & WANG, Q. C. 1995. Ar/Ar geochronology of ultrahigh-pressure metamorphism in central China. *Tectonics*, **14**, 994–1006.

HAN, J., ZHU, S. & XU, S. 1989. The generation and evolution of the Hehuai basin. *In*: ZHU, X. (ed.) *Sedimentary Basins of the World*, Elsevier, Amsterdam, 125–135.

HSÜ, K. J., WANG, Q., LI, J., ZHOU, D. & SUN, S. 1987. Tectonic evolution of the Qinling Mountains, China. *Eclogae Geologicae Helvetiae*, **80**, 735–752.

HU, N., YANG, J. & ZHAO, D. 1996. Sm–Nd isochron age of eclogite from northern Qinling Mountains. *Acta Mineralogica Sinica*, **16**, 349–352.

HU, N., ZHAO, D., XU, B. & WANG, T. 1995. Petrography and metamorphism study on high–ultrahigh pressure eclogite from Guanpo area, northern Qinling Mountain. *Journal of Mineralogy and Petrology*, **15**, 1–9.

HUANG, B., OTOFUJI, Y.-I., YANG, Z. & ZHU, R. 2000. New Silurian and Devonian palaeomagnetic results from the Hexi Corridor terrane, northwest China, and their tectonic implications. *Geophysical Journal International*, **140**, 132–146.

HUANG, K. & OPDYKE, N. D. 2000. Magnetostratigraphic investigations of the Middle Triassic Badong Formation in South China. *Geophysical Journal International*, **142**, 74–82.

HUANG, W. & WU, Z. W. 1992. Evolution of the Qinling orogenic belt. *Tectonics*, **11**, 371–380.

JIAN, P., YANG, W., LI, Z. C. & ZHOU, H. 1997. Isotopic geochronological evidence for the Caledonian Xiongdian eclogite in the western Dabie mountains, China. *Acta Geologica Sinica*, **10**, 455–465.

KRÖNER, A., COMPSTON, W., ZHANG, G. W., GUO, A. L. & TODT, W. 1988. Age and tectonic setting of late Archean greenstone–gneiss terrane in Henan province, China, as revealed by single-grain zircon dating. *Geology*, **16**, 211–215.

KRÖNER, A., ZHANG, G. W. & SUN, Y. 1993. Granulites in the Tongbai area, Qinling belt, China: geochemistry, petrology, single zircon geochronology, and implications for the tectonic evolution of eastern Asia. *Tectonics*, **12**, 245–255.

LERCH, M. F., XUE, F., KRÖNER, A., ZHANG, G. W. & TODT, W. 1995. A Middle Silurian–Early Devonian magmatic arc in the Qinling Mountains of central China. *Journal of Geology*, **103**, 437–449.

LI, S. G., CHEN, Y., ZHANG, F. & ZHANG, Z. 1991. A 1 Ga B.P. alpine peridotite body emplaced into the Qinling Group: evidence for the existence of the Late Proterozoic plate tectonics in the north Qinling area. *Geological Review*, **37**, 235–242.

LI, S. G., HAN, W.-L., HUANG, F. & ZHENG, Y.-F. 1998. Sm–Nd and Rb–Sr ages and geochemistry of volcanics from the Dingyuan Formation in Dabie Mountains, central China: evidence to the Paleozoic magmatic arc. *Scientia Geologica Sinica*, **7**, 461–470.

LI, S. G., HART, S. R., ZHENG, S. G., LIU, D. L., ZHANG, G. W. & GUO, A. L. 1989. Timing of collision between the north and south China blocks – the Sm–Nd isotopic age evidence. *Science in China, Series B*, **32**, 1393–1400.

LI, S. G., HUANG, F. *et al.* 2001. Geochemical and geochronological constraints on the suture location between the North and South China blocks in the Dabie Orogen, Central China. *Physics and Chemistry of the Earth, Part A: Solid Earth and Geodesy*, **26**, 655–672.

LI, S. G., JAGOUTZ, E., ZHANG, Z. Q., CHEN, W. & LO, Q. H. 1995. Structure of high-pressure metamorphic belt in the Dabie mountains and its tectonic implications. *Chinese Science Bulletin*, **40**, 138–140.

LI, J., JAGOUTZ, E., CHEN, Y. & LI, Q. 2000. Sm–Nd and Rb–Sr isotopic chronology and cooling history of ultrahigh pressure metamorphic rocks and their country rocks at Shuanghe in the Dabie Mountains, Central China. *Geochimica et Cosmochimica Acta*, **64**, 1077–1093.

LI, J., WANG, Z. & ZHAO, M. 1999a. $^{40}Ar/^{39}Ar$ thermochronological constraints on the timing of collisional orogeny in the Mian–Lüe collision belt, southern Qinling Mountains. *Acta Geologica Sinica*, **73**, 208–215.

LI, X. H. 1999. U–Pb zircon ages of granites from the southern margin of the Yangtze Block: timing of Neoproterozoic Jinning Orogeny in SE China and implications for Rodinia assembly. *Precambrian Research*, **97**, 43–57.

LI, X. X., YAN, Z. & LU, X. X. 1992. *Granitoids of Qinling-Dabie mountains.* Geological Press, Beijing.

LI, X. X., LI, X. H., KINNY, P. D. & WANG, J. 1999b. The breakup of Rodinia: did it start with a mantle plume beneath South China? *Earth and Planetary Science Letters*, **173**, 171–181.

LI, X. X. & POWELL, C. M. 2001. An outline of the palaeogeographic evolution of the Australasian region since the beginning of the Neoproterozoic. *Earth-Science Reviews*, **53**, 237–277.

LIN, J. L. & FULLER, M. 1990. Palaeomagnetism, North China and South China collision, and the Tan–Lu fault. *Philosophical Transactions of the Royal Society of London*, A331, 589–598.

LIN, W., FAURE, M., MONIÉ, P., SCHÄRER, U., ZHANG, L. & SUN, Y. 2000. Tectonics of SE China: new insights from the Lushan massif (Jiangxi Province). *Tectonics*, **19**, 852–871.

LIN, W., FAURE, M., SUN, Y., SHU, L. & WANG, Q. 2001. Compression to extension switch during the Middle Triassic orogeny of Eastern China: the case study of the Jiulingshan massif in the southern foreland of the Dabieshan. *Journal of Asian Earth Sciences*, **20**, 31–43.

LIU, J. & LIOU, J. G. 1995. Kyanite anthophyllite schist and the southwest extension of the Dabie Mountains ultrahigh to high pressure belt. *The Island Arc*, **4**, 334–346.

LIOU, J. G., & ZHANG, R. Y. 1998. Petrogenesis of an ultrahigh-pressure garnet-bearing ultramafic body from Maowu, Dabie Mountains, east-central China. *The Island Arc*, **7**, 115–134.

LIOU, J. G., ZHANG, R. Y., EIDE, E. A., MARUYAMA, S., WANG, X. & ERNST, W. G. 1996. Metamorphism and tectonics of high-P and ultrahigh-P belts in Dabie–Sulu Regions, eastern central China. *In:* YIN, A. & HARRISON, T. M., (eds) *The Tectonic Evolution of Asia.* Cambridge University Press, 300–343.

LIOU, J. G., ZHANG, R. Y., & JAHN, B. M. 2000. Petrological and geochemical characteristics of ultrahigh-pressure metamorphic rocks from the Dabie–Sulu terrane, east-central China. *International Geology Review*, **12**, 328–352.

LIU, L., ZHOU, D., WANG, Y., CHEN, D. & LIU, Y. 1996. Study and implication of the high-pressure felsic granulite in the Qinling complex of East Qinling. *Science in China*, **39**, 60–68.

LIU, Y.-C., LI, S.-G. *et al.* 2001. Sm–Nd dating of eclogites from North Dabie and its constraints on the timing of granulite-facies retrogression. *Geochimica*, **30**, 79–87.

MA, W. 1989. Tectonics of the Tongbai–Dabie fold belt. *Journal of Southeast Asian Earth Sciences*, **3**, 77–85.

MA, W. 1991. The Carboniferous at the northern foot of the Dabie Mountains and its tectonic implications. *Acta Geologica Sinica*, **4**, 237–249.

MARUYAMA, S., TABATA, H., NUTMAN, A. P., MORIKAWA, T. & LIOU, J. G. 1998. SHRIMP U–Pb geochronology of ultrahigh-pressure metamorphic rocks of the Dabie Mountains, central China. *Continental Dynamics*, **3**, 72–85.

MATTAUER, M., MATTE, P. *et al.* 1985. Tectonics of the Qinling belt: build-up and evolution of eastern Asia. *Nature*, **317**, 496–500.

MATTE, P., TAPPONNIER, P. *et al.* 1996. Tectonics of Western Tibet, between the Tarim and the Indus. *Earth and Planetary Science Letters*, **142**, 311–330.

MENG, Q.-R. & ZHANG, G.-W. 2000. Geologic framework and tectonic evolution of the Qinling orogen, central China. *Tectonophysics*, **323**, 183–195.

NIE, S., YIN, A., ROWLEY, D. B. & JIN, Y. 1994. Exhumation of the Dabie Shan ultrahigh-pressure rocks and accumulation of the Songpan–Ganzi flysch sequence, central China. *Geology*, **22**, 999–1002.

NIU, B., FU, Y., LIU, Z., REN, J. & CHEN, W. 1994. Main tectonothermal events and $^{40}Ar/^{39}Ar$ dating of the Tongbai–Dabie Mts. *Acta Geoscientia Sinica*, **1994**, 20–34.

NIU, B., LIU, Z. & REN, J. 1993. The tectonic relationship between the Qinling Mountains and Tongbai–Dabie Mountains with notes on the tectonic evolution of the Hehuai Basin. *Bulletin of the Chinese Academy of Geological Sciences*, **26**, 1–12.

OKAY, A. I. 1993. Petrology of a diamond and coesite-bearing metamorphic terrain: Dabie Shan, China. *European Journal of Mineralogy*, **5**, 659–675.

OKAY, A. I. 1994. Sapphirine and Ti-clinohumite in ultra-high-pressure garnet-pyroxenite and eclogite from Dabie Shan, China. *Contributions to Mineralogy and Petrology*, **116**, 145–155.

QIU, Y. M., GAO, S., MCNAUGHTON, N. J., GROVES, D. I. & LING, W. 2000. First evidence of >3.2 Ga continental crust in the Yangtze craton of south China and its implications for Archean crustal evolution and Phanerozoic tectonics. *Geology*, **28**, 11–14.

RATSCHBACHER, L., FRISCH, W., LINZER, H.-G. & MERLE, O. 1991. Lateral extrusion in the eastern Alps, Part 2: structural analysis. *Tectonics*, **10**, 257–271.

RATSCHBACHER, L., HACKER, B. R. *et al.* 2004. Tectonics of the Qinling (central China): tectonostratigraphy, geochronology, and deformation kinematics. *Tectonophysics*, **366**, 1–53.

RATSCHBACHER, L., HACKER, B. R. *et al.* 2000. Exhumation of ultrahigh-pressure continental crust in east central China: Cretaceous and Cenozoic unroofing and the Tan–Lu fault. *Journal of Geophysical Research*, **105**, 13303–13338.

REE, J.-H., CHO, M., KWON, S.-T. & NAKAMURA, E. 1996. Possible eastward extension of Chinese collision belt in South Korea; the Imjingang Belt. *Geology*, **24**, 1071–1074.

REGIONAL GEOLOGICAL SURVEY OF ANHUI, 1987. *Regional Geology of Anhui Province.* Geological Publishing House, Beijing (in Chinese).

REGIONAL GEOLOGICAL SURVEY OF GANSU, 1989, *Regional Geology of Gansu Province.* Geological Publishing House, Beijing.

REGIONAL GEOLOGICAL SURVEY OF HENAN, 1989. *Regional Geology of Henan Province.* Geological Publishing House, Beijing (in Chinese).

REGIONAL GEOLOGICAL SURVEY OF HUBEI, 1990. *Regional Geology of Hubei Province.* Geological Publishing House, Beijing (in Chinese).

REGIONAL GEOLOGICAL SURVEY OF SHAANXI, 1989. *Regional Geology of Shaanxi Province*, **13**, Geological Publishing House, Beijing.

REGIONAL GEOLOGICAL SURVEY OF SICHUAN, 1991. *Regional Geology of Sichuan Province.* Geological Publishing House, Beijing.

REN, J., TAMAKI, K., LI, S. & JUNXI, Z. 2002. Late Mesozoic and Cenozoic rifting and its dynamic setting in Eastern China and adjacent areas. *Tectonophysics*, **344**, 175–205.

ROWLEY, D. B., XUE, F., TUCKER, R. D., PENG, Z. X., BAKER, J. & DAVIS, A. 1997. Ages of ultrahigh pressure metamorphism and protolith orthogneisses from the eastern Dabie Shan: U/Pb zircon geochronology. *Earth and Planetary Science Letters*, **151**, 191–203.

SCHMID, J. C., RATSCHBACHER, L., GAITZSCH, I., HACKER, B. R. & DONG, S. 1999. How did the foreland react? Yangtze foreland fold-and-thrust belt deformation related to exhumation of the Dabie Shan ultrahigh-pressure continental crust (eastern China). *Terra Nova*, **11**, 266–272.

SHEN, J., ZHANG, Z. & LIU, D. 1997. Sm–Nd, Rb–Sr, ^{40}Ar/^{39}Ar, ^{207}Pb/^{206}Pb age of the Douling metamorphic complex from eastern Qinling orogenic belt. *Acta Geoscientia Sinica*, **18**, 248–254.

SOBEL, E. R. & ARNAUD, N. 1999. A possible middle Paleozoic suture in the Altyn Tagh, NW China. *Tectonics*, **18**, 64–73.

SONG, B., NUTMAN, A. P., LIU, D. & WU, J. 1996. 3800 to 2500 Ma crustal evolution in the Anshan are of Liaoning Province, northeastern China. *Precambrian Research*, **78**, 79–94.

SONG, S., SU, L., YANG, H. & WANG, Y. 1998. Petrogenesis and emplacement of the Songshugou Peridotite in Shagnan, Shaanxi. *Acta Petrologica Sinica*, **14**, 212–221.

SUN, W., LI, S., SUN, Y., ZHANG, G. & ZHANG, Z. 1996. Chronology and geochemistry of a lava pillow in the Erlangping Group at Xixia in the northern Qinling Mountains. *Geological Review*, **42**, 144–153.

SUN, W., WILLIAMS, I. S. & LI, S. 2002. Carboniferous and Triassic eclogites in the western Dabie Mountains, east-central China: evidence for protracted convergence of the North and South China Blocks. *Journal of Metamorphic Geology*, **20**, 873–886.

SUN, Z., XIE, Q. & YANG, J. 1989. Ordos Basin – a typical example of an unstable cratonic interior superimposed basin. *In:* ZHU, X. (ed.) *Sedimentary Basins of the World.* Elsevier, Amsterdam, 63–75.

TSAI, C.-H. & LIOU, J. G. 2000. Eclogite facies relics and inferred ultrahigh-pressure metamorphism in the North Dabie Complex, central-eastern China. *American Mineralogist*, **85**, 1–8.

WALLIS, S. R., ENAMI, M. & BANNO, S. 1999. The Sulu UHP terrane: a review of the petrology and structural geology. *International Geology Review*, **41**, 906–920.

WALLIS, S. R., ISHIWATARI, A. *et al.* 1997. Occurrence and field relationships of ultrahigh-pressure metagranitoid and coesite eclogite in the Su–Lu terrane, eastern China. *Journal of the Geological Society of London*, **154**, 45–54.

WANG, N. 1989. Micropaleontological study of lower Paleozoic siliceous sequences of the Yangtze platform and eastern Qinling Range. *Journal of Southeast Asian Earth Sciences*, **3**, 141–161.

WANG, Q., BAO, C., LOU, Z. & GUO, Z. 1989. Formation and development of the Sichuan basin. *In.* ZHU, X. (ed.) *Sedimentary Basins of the World*, Elsevier, Amsterdam, 147–163.

WEBB, L. E., HACKER, B. R., RATSCHBACHER, L. & DONG, S. 1999. Thermochronologic constraints on deformation and cooling history of high and ultrahigh-pressure rocks in the Qinling–Dabie orogen, eastern China. *Tectonics*, **18**, 621–637.

WEBB, L. E., RATSCHBACHER, L., HACKER, B. R. & DONG, S. 2001. Kinematics of exhumation of high- and ultrahigh-pressure rocks in the Hong'an and Tongbai Shan of the Qinling–Dabie collisional orogen, eastern China. *In:* HENDRIX, M. S. & DAVIS, G. A. (eds) *Paleozoic and Mesozoic Tectonic Evolution of Central Asia – From Continental Assembly to Intracontinental Deformation.* Geological Society of America Special Memoirs. **194**, 231–245.

WEI, C.-J., SHAN, Z.-G., ZHANG, L.-F., WANG, S.-G. & CHANG, Z.-G. 1998. Determination and geological significance of the eclogites from the northern Dabie Mountains, central China. *Chinese Science Bulletin*, **43**, 253–256.

WEI, C.-J., WU, Y. H., MI, Y. Y., CHEN, B. & WANG, S. G. 1999. Characteristics of Tongbai eclogites in Hennain and their geologic significance. *Science Report*, **44**, 1882–1885.

XIE, Z., CHEN, J.-F., ZHANG, X., GAO, T.-S., DAI, Z.-Q., ZHOU, T.-X. & LI, H.-M. 2001. Zircon U–Pb dating of gneiss from Shizhuhe in North Dabie and its geological implications. *Acta Petrologica Sinica*, **17**, 139–144.

XIE, Z., CHEN, J.-F. & ZHOU, T.-X. 1998. U–Pb zircon ages of the rocks in the North Dabie terrain, China. *Scientia Geologica Sinica*, **7**, 501–511.

XU, B., GROVE, M., WANG, C., ZHANG, L. & LIU, S. 2000a. ^{40}Ar/^{39}Ar thermochronology from the northwestern Dabie Shan: constraints on the evolution of Qinling–Dabie orogenic belt, east-central China. *Tectonophysics*, **322**, 279–301.

XU, S.-T., LIU, Y.-C., SU, W., WANG, R.-C., JIANG, L.-L. & WU, W.-P. 2000b. Discovery of the eclogite and its petrography in Northern Dabie Mountains. *Chinese Science Bulletin*, **45**, 273–278.

XUE, F., KRÖNER, A., REISCHMANN, T. & LERCH, F. 1996a. Palaeozoic pre- and post-collision calc-alkaline magmatism in the Qinling orogenic belt, central China, as documented by zircon ages on granitoid rocks. *Journal of the Geological Society of London*, **153**, 409–417.

XUE, F., LERCH, F., KRÖNER, A. & REISCHMANN, T. 1996b. Tectonic evolution of the East Qinling Mountains, China, in the Paleozoic: a review and a new tectonic model. *Tectonophysics*, **253**, 271–284.

XUE, F., ROWLEY, D. B., TUCKER, R. D. & PENG, Z. X. 1997. U–Pb zircon ages of granitoid rocks in the north Dabie complex, eastern Dabie Shan, China. *Journal of Geology*, **105**, 744–753.

YANG, J. S., ROBINSON, P. T., JIANG, C.-F. & XU, Z. Q. 1996. Ophiolites of the Kunlun Mountains, China and their tectonic implications. *Tectonophysics*, **258**, 215–231.

YE, B. D., JIAN, P., XU, J. W., CUI, F., LI, Z. C. & ZHANG, Z. H. 1994. Geochronological study on Huwan Group in Tongbai–Dabie orogenic belt. *In*: CHEN, H. S. (ed.) *Isotopic Geochemistry Research*, Zhejiang University Press, Hangzhou, 187–204.

YIN, A. 1996. A Phanerozoic palinspastic reconstruction of China and its neighboring regions. *In:* YIN, A. & HARRISON, T. M. (eds) *The Tectonic Evolution of Asia*, Cambridge University Press, 442–485.

YIN, A. & NIE, S. 1993. An indentation model for the North and South China collision and the development of the Tanlu and Honam fault systems, eastern Asia. *Tectonics*, **12**, 801–813.

YOU, Z., HAN, Y., SUO, S., CHEN, N. & ZHONG, Z. 1993. Metamorphic history and tectonic evolution of the Qinling Complex, eastern Qinling Mountains, China. *Journal of Metamorphic Geology*, **11**, 549–560.

YU, J., FU, H., HAAPALA, I., RAMO, T. O., VAASJOKI, M. & MORTENSEN, J. K. 1996. Paleoproterozoic anorogenic Rapakivi granite–anorthosite suite in the north part of the North China Craton. *30th International Geological Congress, Beijing*, 104–108.

YU, Z. & MENG, Q. 1995. Late Paleozoic sedimentary and tectonic evolution of the Shangdan suture zone, eastern Qinling, China. *Journal of Southeast Asian Earth Sciences*, **11**, 237–242.

ZHAI, X., DAY, H. W., HACKER, B. R. & YOU, Z. 1998. Paleozoic metamorphism in the Qinling orogen, Tongbai Mountains, central China. *Geology*, **26**, 371–374.

ZHANG, H.-F., GAO, S., ZHANG, B.-R., LUO, T.-C. & LIN, W.-L. 1997. Pb isotopes of granites suggest Devonian accretion of Yangtze (South China) craton to North China craton. *Geology*, **25**, 1015–1018.

ZHANG, K.-J. 1997. North and South China collision along the eastern and southern North China margins. *Tectonophysics*, **270**, 145–156.

ZHANG, R. Y., LIOU, J. G., YANG, J. S., YUI, T. F. 2001. Petrochemical constraints for dual origin of garnet peridotites from the Dobie-Sulu VHP terranes, eastern-central China. *Journal of Metamorphic Geology*, **18**, 149–166.

ZHANG, Y., WEI, Z., XU, W., TAO, R. & CHEN, R. 1989. The North Jiangsu–South Yellow Sea basin. *In*: ZHU, X. (eds.) *Sedimentary Basins of the World*, Elsevier, Amsterdam, 107–123.

ZHANG, Z.-J. 1999. Metamorphic evolution of garnet–clinopyroxene–amphibole rocks from the Proterozoic Songshugou mafic–ultramafic complex, Qinling mountains, central China. *Island Arc*, **8**, 259–280.

ZHANG, Z. Q., LIU, D. Y. & FU, G. M. 1991. Ages of the Qinling, Kuanping, and Taowan Groups in the North Qinling orogenic belt, middle Chiina and their iimplications. *In*: YE, L. J., QIAN, X. L. & ZHANG, G. W. (ed.) *A Selection of Papers Presented at the Conference on the Qinling Orogenic Belt*. Northwest University Press, Xi'an. 214–228.

ZHAO, G., CAWOOD, P. A., WILDE, S. A., SUN, M. & LU, L. 2000. Metamorphism of basement rocks in the Central Zone of the North China Craton: implications for Paleoproterozoic tectonic evolution. *Precambrian Research*, **103**, 55–88.

ZHAO, X. X. & COE, R. S. 1987. Palaeomagnetic constraints on the collision and rotation of North and South China. *Nature*, **327**, 141–144.

ZHAO, X. X., COE, R. S., GILDER, S. A. & FROST, G. M. 1996. Palaeo-magnetic constraints on the palaeogeography of China: implications for Gondwanaland. *Australian Journal of Earth Sciences*, **43**, 643–672.

ZHOU, C. T., GAO, T. S., TANG, J. F., SHEN, H. S. & HU, Y. Q. 2000. Distribution and main characteristics of eclogite in the northern Dabie Mountains, Anhui. *Regional Geology of China*, **19**, 253–257.

ZHOU, D. & GRAHAM, S. A. 1996. Songpan–Ganzi complex of the west Qinling Shan as a Triassic remnant ocean basin. *In:* YIN, A. & HARRISON, T. M. (eds) *The Tectonic Evolution of Asia*. Cambridge University Press, 281–299.

UHP rocks and the Dabieshan Orogenic Belt

QINGCHEN WANG & BOLIN CONG

Laboratory of Lithosphere Tectonic Evolution, Institute of Geology and Geophysics, Chinese Academy of Sciences, Beijing 100029, China

Abstract: The Dabieshan Orogenic Belt, which contains ultra-high-pressure (UHP) metamorphic rocks, is the Mesozoic collision zone between the Sino-Korean and Yangtze cratons. With respect to the exhumation of the UHP rocks, the Dabieshan Orogenic Belt can be divided into four units, i.e. allochthonous, parautochthonous, autochthonous and reworked units. The allochthonous unit is composed of UHP rocks. The parautochthonous unit is represented by the non-UHP rocks of the Yangtze sedimentary cover and crystalline basement, as well as an accretionary wedge. The autochthonous unit includes Jurassic and Cretaceous sedimentary and igneous rocks. The reworked unit is characterized by migmatization. The deep structure of the Dabieshan Orogenic Belt is characterized by a Moho offset and dome structure in the middle and upper crust, as recording a compressional state. A northward subduction of the Yangtze craton is evidenced by geological, geophysical and geochemical data. In the formation of the Dabieshan orogenic belt, a precondition is the low density of the subducted continental materials. If the UHP unit is the *piston* to build up the orogenic belt, then the continuous compression between the Sino-Korean and Yangtze cratons is the *motor* to trigger the mountain-building processes.

The discovery of ultra-high-pressure (UHP) metamorphic rocks could be seen as the most remarkable contribution of petrologists in the twentieth century. With the increasing discoveries of UHP rocks world-wide, attention was shifted gradually from the problems of formation and preservation of UHP rocks to the questions regarding the nature of the mechanisms of subduction and exhumation of the UHP terranes.

Although the first report of coesite from China (Xu, Z. 1987) was made three years later than finds from Europe (Chopin 1984; Smith 1984), the Chinese coesite has attracted world-wide attention since the very beginning. With international efforts, many new outcrops of coesite-bearing eclogite were discovered from the Dabieshan (Dabie Mountains) Orogenic Belt and the Sulu region (*Su* is short for the Jiangsu Province, and *Lu* for Shandong Province), Central China, in 1989 (Okay *et al.* 1989; Wang, X. *et al.* 1989; Yang & Smith 1989; Zhang *et al.* 1989). The third International Field Symposium was held in the Dabieshan in 1995, and the Chinese Continental Scientific Drilling (CCSD) project was approved by the International Continental Scientific Drilling Program (ICDP) in 1999. A planned 5-km-deep borehole in Donghai County, east China, is now penetrating the Dabieshan–Sulu UHP terrane. Like the Himalaya, the Dabieshan Orogenic Belt has become an international natural laboratory for the study of continental dynamics. Inside China, the study of UHP rocks has gained

increasing financial support. With grants from the National Natural Science Foundation of China (NSFC), Ministry of Land and Resources of China, Ministry of Science and Technology (MOST) of China, and Chinese Academy of Sciences, studies of UHP rocks have developed over the past 15 years from small and isolated projects to high-rank and multidisciplinary ones.

The Dabieshan Orogenic Belt is probably the best place to resolve many questions, e.g. what is the greatest depth to which continental crustal materials could be subducted; what is the role of fluids in UHP metamorphism; and what is the mechanism by which UHP rocks could be exhumed from mantle depth to the Earth's surface? Among these questions, some have been partly answered, whereas others have not. In the present paper, we will summarize the recent developments in studies of UHP rocks from the Dabieshan–Sulu region, and then discuss the architecture of the Dabieshan Orogenic Belt in the light of the exhumation of UHP rocks.

UHP rocks from the Dabieshan–Sulu region

The largest UHP-rock-bearing belt in the world

In the past 15 years, many UHP rocks have been discovered in the Dabieshan–Sulu region (Fig. 1).

From: MALPAS, J., FLETCHER, C. J. N., ALI, J. R. & AITCHISON, J. C. (eds) 2004. *Aspects of the Tectonic Evolution of China*. Geological Society, London, Special Publications, **226**, 177–192.
0305-8719/04/$15 © The Geological Society of London 2004.

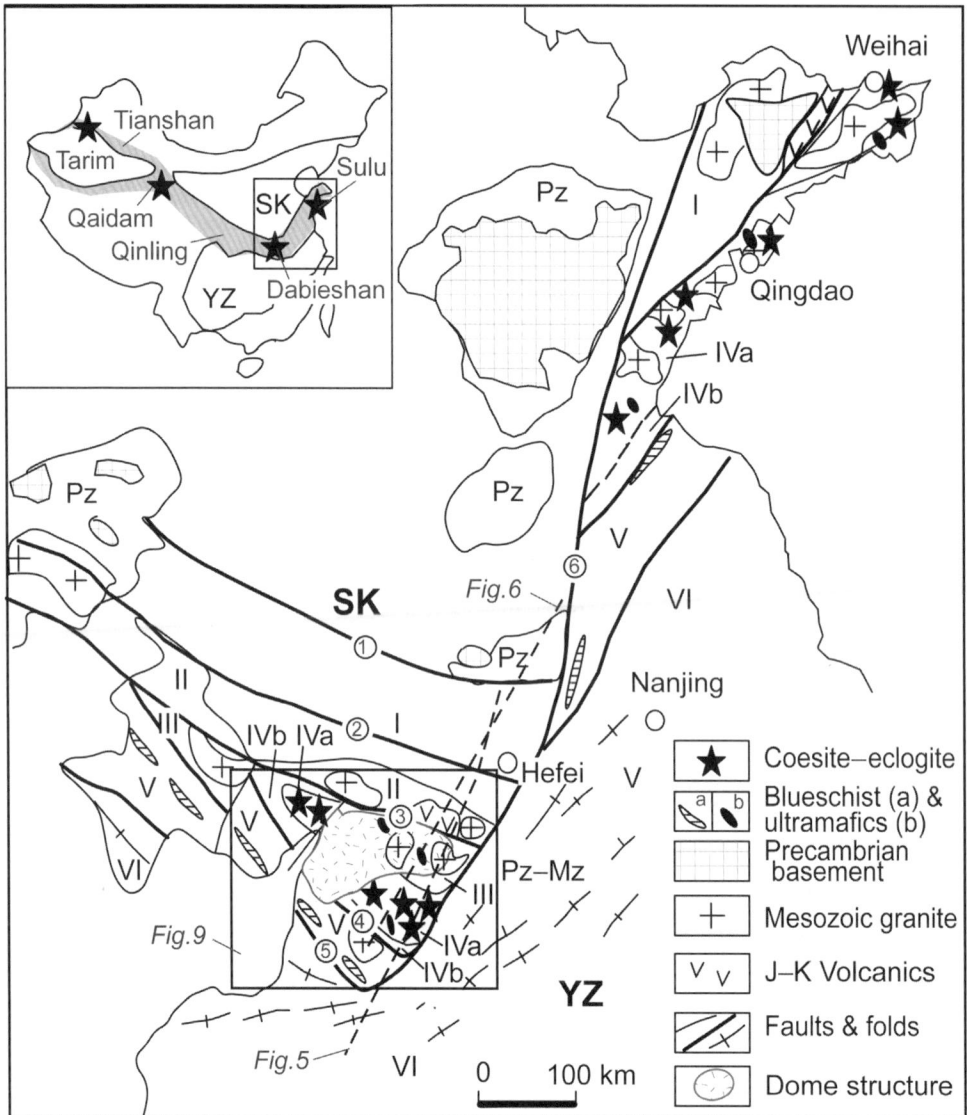

Fig. 1. The Dabieshan–Sulu Orogenic Belt in China. Shaded area in the index map stands for orogenic belts. Faults: (**1**) Shouxian–Dingyuan Fault (SDF); (**2**) Li'an Fault (LAF); (**3**) Xiaotian–Mozitan Fault (XMF); (**4**) Taihu–Mamiao Fault (TMF); (**5**) Xianfan–Guangji Fault (XGF); (**6**) Tancheng–Lujiang Fault (TLF). Abbreviations: SK, Sino-Korean Craton; YZ, Yangtze Craton; Mz, Mesozoic; Pz, Palaeozoic. Tectonic subdivision in the Dabieshan: **I**, the Jurassic–Cretaceous basin; **II**, the North Huaiyang flysch belt; **III**, the NDT gneisses; **IV**, the SDT UHP (**IVa**) and HP (**IVb**) rocks; **V**, the Susong blueschist belt; **VI**, the fold–thrust foreland belt. The subdivisions in the Sulu region can be correlated more or less to those in the Dabieshan.

Lithologically, these UHP rocks include not only eclogite, but also jadeitic quartzite, marble, schist, granitic gneiss and ultramafic rocks (Hirajima *et al.* 1990; Xu, S. *et al.* 1992; Wang, Q. *et al.* 1993; Wang, X. & Liou 1993; Yang *et al.* 1993; Zhang *et al.* 1994, 1995; Cong *et al.* 1995; Cong 1996; Zhang & Liou 1996). Geographically, these UHP rocks are scattered in a belt, about 1000 km in length, from Xinxian County in the west to Rongcheng County in the east. Recently, UHP rocks were discovered in the North Qaidam Mountain (Yang *et al.* 2002, oral report in the *Croucher Advanced Studies Institute*), and even in the Western Tianshan (Zhang, L. *et al.* 2000).

The North Qaidam Mountain, Western Tianshan, and the Dabieshan–Sulu all belong to the Central Orogenic Belt of China (COBC). Therefore, the Dabieshan–Sulu UHP belt and its western extension, which totals about 4000 km in length, could be the largest UHP-rock-bearing belt so far recognized. However, not all UHP rocks in the belt are the same age. For example, UHP rocks from the Dabieshan–Sulu region are Triassic (Li *et al.* 1993, 2000), whereas those from the Qaidam Mountain are Early Palaozoic (Zhang, J. *et al.* 2000). A common feature of these UHP rocks is that they all formed along the margin of an Archaean craton.

The *in situ* v. *foreign* debate has lasted for many years in the study of the Dabieshan eclogite terrain (Cong 1996). The key point to solve the problem is to find evidence for UHP metamorphism in the eclogite-hosting gneisses. The pioneer works, which identified coesite inclusions in zircons separated from UHP-eclogite-hosting gneisses (Tabata *et al.* 1998; Ye *et al.* 2000*b*) have shown an approach to the answer. Because of its mechan-

ical strength, zircon was considered as an excellent container to preserve coesite (Sobolev *et al.* 1994; Chopin & Sobolev 1995). Recently, from the high-pressure (HP) 'cold' eclogite zone (Okay 1993), coesite, jadeite and phengite inclusions were found in zircons (Fig. 2). These zircons were separated from the magnetite–phengite gneisses (Liu, J. *et al.* 2001), which were once considered as non-UHP rocks. Therefore, the volume of UHP rocks in the Dabieshan Orogenic Belt is dramatically larger than previously realized.

These detailed studies in petrology and mineralogy have demonstrated that continental crust with low density could be subducted down to mantle depths. For example, the highest P and T of the coesite- and diamond-bearing eclogites were estimated as 3.0–4.0 GPa and 800–900 °C (Xu *et al.* 1992; Okay 1993; Carswell *et al.* 1997; Xu & Su 1997), implying a subduction depth of 120 km. The exsolution of clinopyroxene, rutile and apatite in garnet point to the possible previous existence of a majoritic component in garnet. The calculated Na_2O content (0.34 wt%) and octahedral silicon values (0.06) of the original garnet in the Yangkou eclogites implied further the subduction of continental materials to depths greater than 200 km (Ye *et al.* 2000*a*).

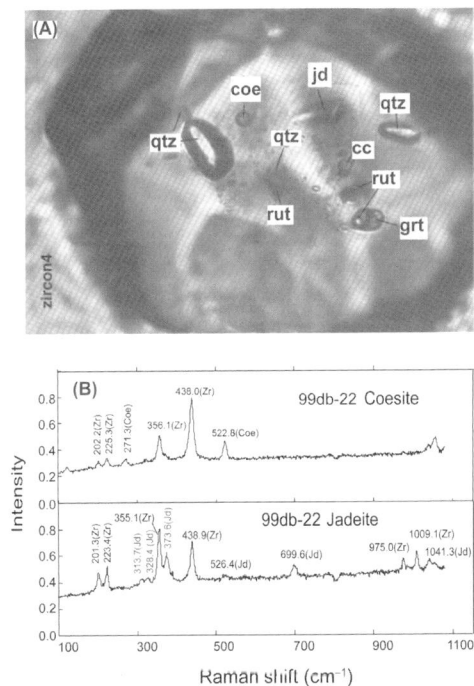

Fig. 2. (**A**) Coesite and jadeite inclusions in a zircon grain with a long axis of about 220 μm (sample 99db-22). (**B**) Raman spectrum of coesite and jadeite. The mineral inclusions were identified by Dr Liu Jingbo with the Jasco NRS Laser Raman spectrophotometer at the Department of Earth and Planetary Sciences, Tokyo Institute of Technology.

Fluid behaviour during formation and exhumation of UHP rocks

The presence of water during the formation of UHP rocks was evidenced by the presence of coesite as an inclusion in epidote (Zhang *et al.* 1994), as well as by other hydrous UHP mineral assemblages, such as clinohumite in UHP ultramafic rocks (Zhang *et al.* 1995; Liou & Zhang 1996), magnesian staurolite in eclogite (Enami & Zang 1988), phengite + talc + kyanite in white schist (Rolfo *et al.* 2000), and phengite in many eclogites. Recently, even hydroxyl and free water (H_2O) were identified in garnet under transmission electron microscopy (Su *et al.* 2002). All of this evidence indicates that continental crustal materials were not totally dehydrated before and during their deep subduction. On the other hand, however, O-, C- and H- isotope studies indicated that no fluid reacted with the subducted supracrustal rocks when they arrived at mantle depth. Their isotope patterns still preserve the character of meteoric water (Yui *et al.* 1995; Zheng *et al.* 1996; Baker *et al.* 1997), and the isotopic composition appeared highly inhomogeneous even at the metre scale (Wang & Rumble 1999; Rumble *et al.* 2000). Therefore, it seems that fluids in the supracrustal rocks, if any, had been limited to certain isolated portions during deep subduction. The positive $\delta^{13}C$ values of the UHP marble, up to 4–6‰ and similar to those of the Precambrian dolomitic limestone deposited on the north margin of the Yangtze craton (Wang & Rumble 1999; Rumble *et al.* 2000), further indicated that the UHP rocks were exhumed rapidly from the mantle environment.

Discovery of tiny vein of Ky + Ab + Qtz (mineral abbreviations following Kretz 1983) in eclogite and inclusions of Kfs + Qtz + Mgt in omphacite (Fig. 3) indicate dehydration melting (Ye *et al.* in press). The vein could be a product of dehydration of phengite at pressures of 1.5–2.2 GPa:

$$Jd \text{ (in omph)} + Ms + Qtz$$
$$\rightarrow Ky + Grt + Ab + Kfs + H_2O.$$

At high pressure, breakdown of omphacite produced Na-enriched feldspar. With a decrease in pressure, more and more phengite grains

A

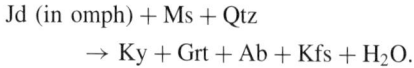

B

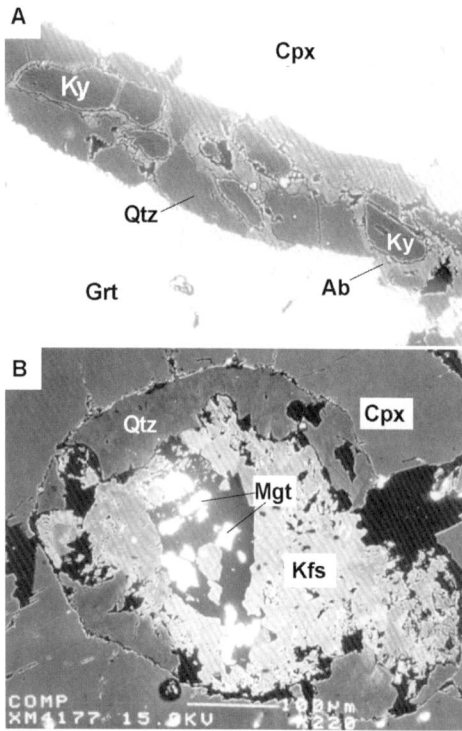

Fig. 3. Back-scatter images showing (**A**) Tiny vein of Ky + Ab + Qtz in eclogite (the width of the vein is about 0.05 mm) and (**B**) inclusion of Kfs + Qtz + Mgt in omphacite.

were broken down to produce K-feldspar. As a result, granitic melts of Ms + Kfs + Qtz + H_2O could form. The inclusion of Kfs + Qtz + Mgt in omphacite can be interpreted as the recrystallization of the hydrous fluid. Furthermore, the accumulation of the fluid-saturated melts could eventually produce migmatite in the gneiss terranes.

Metamorphic events in the Dabieshan Orogenic Belt

Various methods have been employed to date UHP–HP metamorphic events in the Dabieshan Orogenic Belt. Almost all methods (i.e. U–Pb, Sm–Nd, and Ar/Ar) indicated that the UHP metamorphism proceeded at about 240 Ma, when the Yangtze Block collided with the Sino-Korean Block (Li *et al.* 1989, 1992, 1993, 2000; Ames *et al.* 1993; Eide *et al.* 1994; Maruyama *et al.* 1994; Chavagnac *et al.* 1996; Rowley *et al.* 1997, Hacker *et al.* 1998, 2000). A detailed geochronological study showed that UHP rocks from the Dabieshan experienced two rapid cooling stages, with the first stage being 226–219 Ma and the second being 180–167 Ma (Li *et al.* 2000).

Recently, zircons separated from the Sujiahe eclogites at Xiongdian and Hujiawan in the north of the Dabieshan Orogenic Belt yielded SHRIMP ages of 370–430 Ma in the core and 310–320 Ma in the rim (Li, 2001, pers. comm.). The rim was interpreted as the product of HP metamorphism. The $\Sigma_{Nd}(T)$ values (>+5.1) of the Sujiahe eclogites points to a protolith of oceanic crust (Li *et al.* 1996). Therefore, it is implied that the HP eclogites in the north portion formed during oceanic subduction that predates the collision between the Sino-Korean and Yangtze cratons.

At least two post-UHP metamorphic events happened during the exhumation of UHP rocks. One event is the Barrovian amphibolite-facies overprint, starting at about 200 Ma and lasting to 180 Ma, as dated by Sm–Nd and Rb–Sr methods (Li *et al.* 2000), as well as the Ar/Ar method (Hacker *et al.* 2000). It was implied that the UHP rocks were already exhumed to crustal depths by 200 Ma. The other event is the reheating at amphibolite-facies temperatures and crustal pressures at about 180 Ma (Hacker *et al.* 2000), with no tectonic interpretation. Migmatite occurs in the north Dabieshan terrane (NDT) and its formation should define a metamorphic event. Although the migmatization was considered as a Cretaceous event, the field evidence that migmatite occurs as xenoliths in Cretaceous granite (Fig. 4) implies an event

Fig. 4. Photograph showing migmatite as an inclusion in the Cretaceous Baimajian granite (taken by Lin Wei). The compass at centre right is about 10 cm in diameter.

earlier than Cretaceous. Perhaps the migmatization and the Jurassic reheating are two features of the same reheating event.

The protolith ages of the UHP–HP rocks have also been dated by using the U–Pb method, represented either by the upper intercept age in the concordia diagram, or SHRIMP ages obtained from the zircon core. Most of the data range from 700 Ma to 800 Ma (Rowley *et al.* 1997; Hacker *et al.* 1998). Gneiss in the north Dabieshan terrane (NDT) also yielded Late Precambrian ages (Xie *et al.* 2001). It is implied that these rocks are derived from the Yangtze craton, which has a distinctive Late Precambrian age signature.

Architecture of the Dabieshan Orogenic Belt

The tectonic framework and structure of the Dabieshan Orogenic Belt have been described in several papers (Faure *et al.* 1998, 1999; Wang, Q. *et al.* 1998; Hacker *et al.* 2000; Ratschbacher *et al.* 2000). Although tectonic interpretations varied among the researchers, several petrologically unique terranes could be recognized. They are, from north to south, the Jurassic–Cretaceous Hefei Basin, the North Huaiyang Belt composed of low-grade metamorphic flysch, the north Dabieshan terrane (NDT) containing orthogneiss with metamorphosed mafic–ultramafic blocks, the south

Dabieshan terrane (SDT) that consists of UHP–HP rocks, the Susong terrane which includes blueschist, and the foreland of the Yangtze fold–thrust belt (Fig. 1). Among these units, the NDT was once at the centre of a debate as to whether it contained eclogites or not. The debate has been settled by the discovery of eclogite with a U–Pb age of 230 Ma at Huazhuang in the NDT (Xu, S. *et al.* 2000; Liu, Y. *et al.* 2001).

The present Dabieshan Orogenic Belt appears as an asymmetrical dome structure, with foliation dipping north in the narrow north wing, and dipping south in the wide south wing. The axis of the dome extends ENE–WSW across the NDT in the eastern Dabieshan, and in the northern portion of the Xinxian coesite–eclogite terrane in the western Dabieshan (also called the Hong'an block by some researchers). Detailed measurements of foliation and lineation revealed a NW–SE lineation and top-to-NW shear sense (Wang *et al.* 1995; Faure *et al.* 1998, 1999; Hacker *et al.* 2000; Lin 2000), that was overprinted later by top to SE shear sense in the southern NDT, and the UHP–HP units of the SDT (Lin 2000; Ratschbacher *et al.* 2000).

Deep crustal and upper mantle structures of the Dabieshan Orogenic Belt were studied by using deep seismic refraction and reflection, as well as seismic tomography (Wang, C. *et al.* 2000; Xu, P. *et al.* 2000, 2001). The deep seismic refraction data (Wang, C. *et al.* 2000) reveal that the Dabieshan Orogenic Belt is composed of a 35-km-thick crust, with average seismic velocities of $6.0 \, km \, s^{-1}$ in the upper crust, $6.5 \, km \, s^{-1}$ in the middle crust and $6.8 \, km \, s^{-1}$ in the lower crust (Fig. 5). The isolines of P-wave velocity outlined an arch-shaped structure implying a dome in the upper and middle crust. The crust reaches a maximum thickness of 41.5 km beneath the Xiaotian–Mozitan Fault, where the Moho was offset. The Moho offset was also revealed by recent deep seismic reflection data (Yuan 2002, pers. comm.). A north-dipping high-velocity ($6.3 \, km \, s^{-1}$) zone, as thick as about 10 km, in the uppermost crust beneath Yuexi County was identified and interpreted as a UHP lithology.

The seismic tomography yielded high-resolution images of the P-wave velocity structures to a depth of 150 km (Xu, P. *et al.* 2000, 2001). The images revealed that the slab-like high-velocity anomaly occurs directly beneath the UHP terrane of the Dabieshan Orogenic Belt, and extends northwards down to a depth of 100 km, where it is interrupted by a low-velocity anomaly (Fig. 6). A horizontal high-velocity anomaly was imaged under the Sino-Korean (SK) block above a depth of about 40 km, and was interpreted as the SK crust. The high-velocity anomaly beneath the UHP terrane was interpreted as the Yangtze crust, whereas that beneath the Hefei Basin was considered as the remnant of the subducted Yangtze slab (Xu, P. *et al.* 2000, 2001).

Exhumation history of UHP rocks

The most amazing puzzle might be how the UHP rocks were exhumed from mantle depth. To explain the exhumation of UHP rocks from the Dabieshan Orogenic Belt, at least three mechanisms have been invoked, i.e. buoyancy, erosion, and wedge extrusion (Ernst *et al.* 1991; Yin & Nie 1993; Maruyama *et al.* 1994). Since individually none of the above mechanisms could properly explain the exhumation process

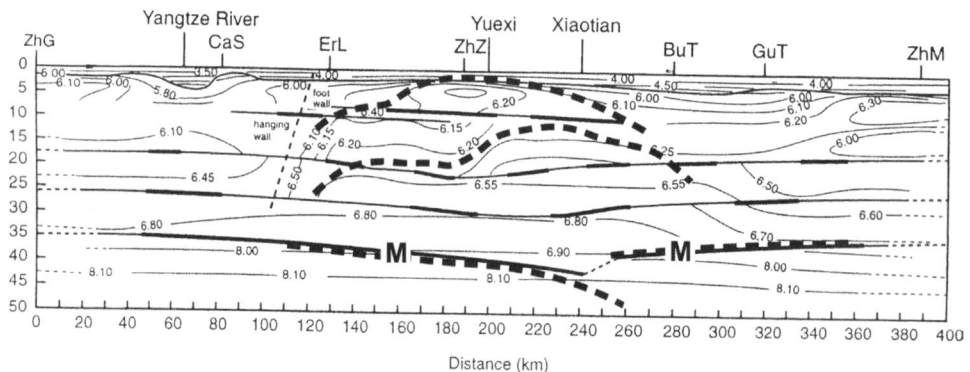

Fig. 5. P-wave velocities in a deep seismic profile (after Wang, C. *et al.* 2000) outlining an arch structure in the middle and upper crust of the Dabieshan Orogenic Belt. The position of the profile is shown in Figure 1. The Moho (M) offset is manifested by the contour of the P-wave velocity $= 8.0 \, km \, s^{-1}$. Abbreviations of place names: ZhG, Zhanggongdu; CaS, Caishan; ErL, Erlanghe; ZhZ, Zhuangzhong; BuT, Butasi; GuT, Guanting; ZhM, Zhuangmu.

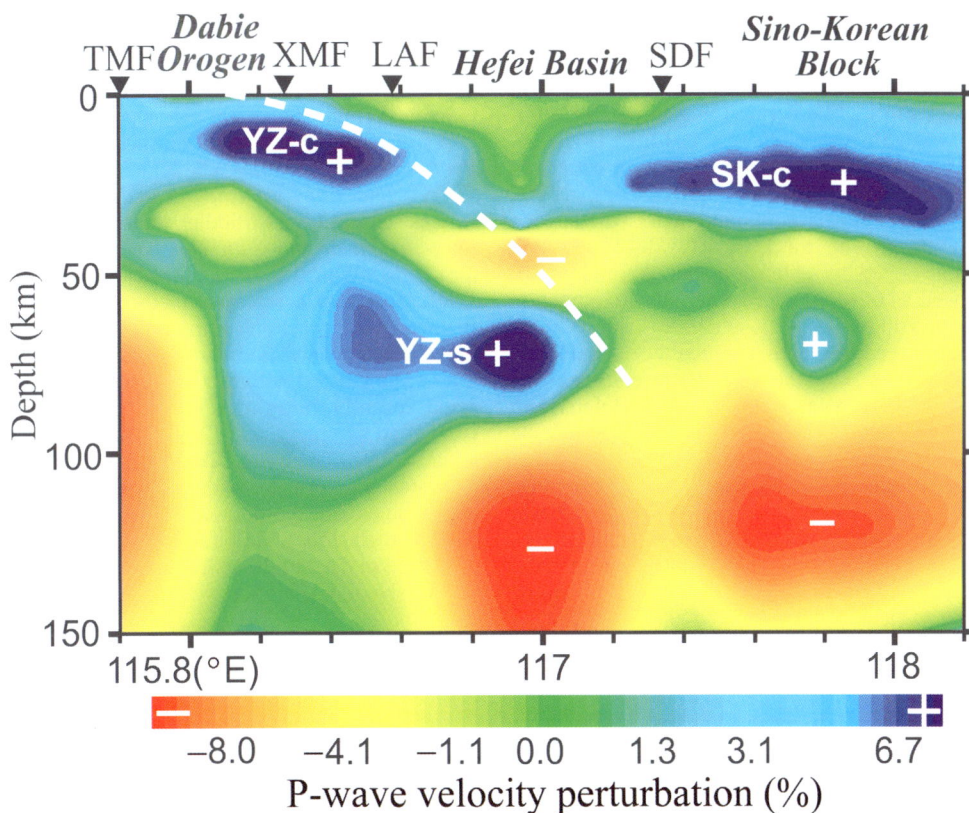

Fig. 6. Seismic tomographic image across the Dabieshan Orogenic Belt, showing that the Yangtze craton has subducted beneath the Sino-Korean craton (after Xu, P. *et al.* 2000, 2001). The position of the profile is shown in Figure 1. Abbreviation of fault names: TMF, Taihu–Mamiao Fault; XMF, Xiaotian–Mozitan Fault; LAF, Li'an Fault; SDF, Shouxian–Dingyuan Fault. SK-c, crust of the Sino-Korean craton; YZ-c, crust of the Yangtze Craton; YZ-s, the remnant of the subducted Yangtze slab.

of UHP rocks from the Dabieshan–Sulu belt, a three-stage model (Wang & Cong 1999) was suggested, based on the $P–T–t–D$ path of UHP rocks and the architecture of the UHP terrane displayed by the geophysical survey. The first stage (230–210 Ma) is characterized by a low geothermal gradient ($10\,°C\,km^{-1}$). The $P–T$ estimation of the peak UHP metamorphic stage is 2.7–4.0 GPa and 700–900 °C and that of the eclogite recrystallization stage is 1.2–2.4 GPa and 600–700 °C. A high exhumation rate (3.3–3.6 mm/year) or rapid cooling (about 40 °C/Ma) of the UHP rocks in the period implies a syn subduction exhumation (Wang & Cong 1996; Li *et al.* 2000). A syn-exhumation ductile deformation at eclogite facies was implied by the shape fabric of garnet and omphacite in eclogite (Fig. 7) and dislocation features in garnet (Wang & Cong 1996). In the second stage (210–170 Ma), the geothermal gradient was

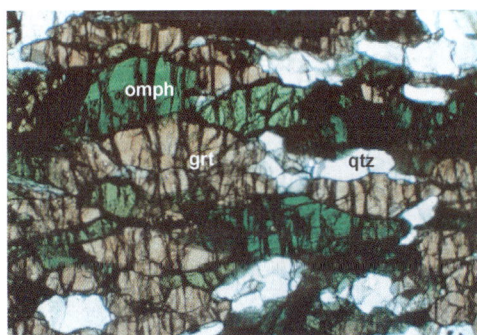

Fig. 7. Microphotograph of shape-fabric showing deformation at eclogite facies (width of view = 1.2 mm).

enhanced to about $20\,^{\circ}\text{C}\,\text{km}^{-1}$, as indicated by overprint of the Barrovian amphibolite-facies at pressure of 0.4 GPa and temperature of $400 \pm 50\,^{\circ}\text{C}$. The cooling rate in the period of 180–170 Ma was about $15\,^{\circ}\text{C}/\text{Ma}^{-1}$ (Li et al. 2000), and the exhumation rate was estimated as 0.7–1.1 mm/year (Wang & Cong 1996), implying a slower exhumation than that in the first stage. The third stage (170–120 Ma) is characterized by extension and thermal uplift, as well as erosion, with a very slow exhumation rate at 0.15 mm/year (Chen 1995).

However, the above suggested exhumation histories are 'shortened' by the recent discovery of eclogite pebbles in the Jurassic Hefei Basin (Wang, D. et al. 2001). The presence of coesite pseudomorphs in the eclogite pebbles implies an UHP metamorphism for some of these eclogites. The sedimentary sequences in the Hefei Basin are, from bottom to top, the Fanghushan, Sanjianpu, Fenghuangtai, Maotanchang and Xiaotian formations. Conglomerates containing eclogite pebbles are found in the Fenghuangtai and Maotanchang formations. Sandstones con-

taining phengite with a high content of Si (>3.5 p.f.u.) occur in the horizon stratigraphically lower than the conglomerates in the Fenghuangtai Formation (Fig. 8).

A flora in the Fanghushan Formation includes *Podozamites lanceolatu* (T–K), *Neocalamites* (T–J₂), and *Cycadocarpidium erdmanni* (T₃), whereas the Lower–Middle Jurassic bivalves *Ferganoconcha* and *Sibireconcha* occur in the formation (Wang, S. et al. 1985; Yu & Yao 1989). *Eosestheria,* representing the Upper Jurassic (Wang, S. et al. 1985; Hao et al. 1986), occurs in the Maotanchang Formation. Several volcanic layers intercalated with the conglomerates were dated recently using the K–Ar method, and yielded an age range of 149–138 Ma (Wang, Y. et al. 2002). Therefore, the Fenghuangtai and Maotanchang formations that contain the eclogite pebbles and the detrital phengite with a Si content higher than 3.5 p.f.u. should have been deposited in the Middle to Late Jurassic (Fig. 8). Obviously, the UHP rocks had been exhumed and exposed on the Earth's surface at least by the later part of the Middle Jurassic (about

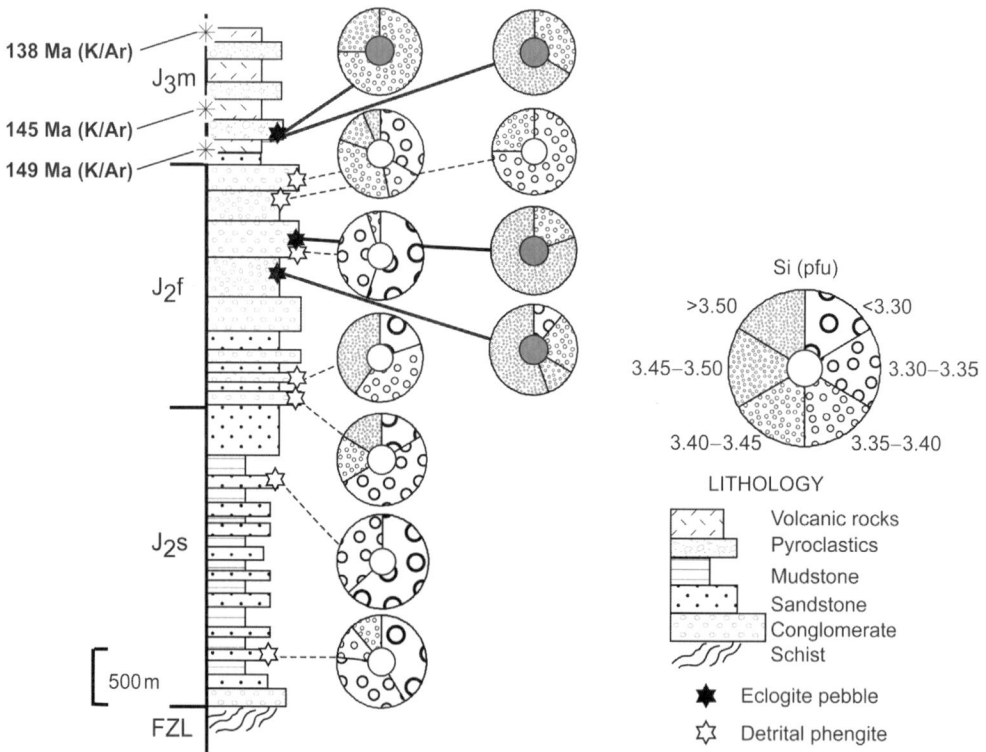

Fig. 8. Sedimentary sequence from the Jurassic Hefei Basin (Wang, D. et al. 2001) showing the horizons containing eclogite pebbles and detrital phengite.

160 Ma). Therefore, Cretaceous tectonic events played a lesser role in the exhumation of the UHP rocks.

Dynamic pattern of the Dabieshan Orogenic Belt

Many researchers are used to interpreting the Dabieshan Orogenic Belt in terms of subduction, i.e. trying to look for a narrow suture zone between the Sino-Korean and Yangtze blocks. This effort has resulted in controversies (Ernst *et al.* 1991; Xu *et al.* 1992; Okay 1993; Maruyama *et al.* 1994; Hacker *et al.* 1996). In fact, a suture zone should be represented by a three-dimensional geological body, the subduction complex. At least two boundary faults should be recognized when the subduction complex is truncated by the erosion surface. Furthermore, the fact that the UHP rocks occupy a significant volume of the Dabieshan Orogenic Belt indicates that exhumation has played a more important role than subduction in forming the Dabieshan Orogenic Belt. Therefore, our suggestion is to differentiate the exhumation complex from the subduction complex. The exhumation complex could have partly destroyed the subduction complex. As a result, the original suture zone could have been reworked to a great extent. Here we suggest a dynamic subdivision of the Dabieshan Orogenic Belt from the viewpoint of exhumation (Fig. 9).

The key point in our new subdivision is to consider the upward travel distance, with the end of the Jurassic as the reference time and the Jurassic ground-level as the reference surface. The UHP rocks are all put into the allochthonous unit, because they were exhumed from mantle depths and travelled more than 100 km vertically. The HP rocks are also included in the allochthonous unit, although they were exhumed from depths of only several tens of kilometres. The non-UHP rocks of the Yangtze sedimentary cover and crystalline basement, as well as the accretionary wedge, are considered as parautochthonous units. They were metamorphosed at only amphibolite facies or even greenschist facies, implying a shallow subduction and short distance of exhumation. The autochthonous units include Jurassic and Cretaceous sedimentary and igneous rocks. A reworked unit characterized by migmatite is differentiated to denote the strong tectonic–thermal reworking before the Cretaceous. The reworked unit occurs as a dome structure, as was revealed by both surface geological survey (Faure *et al.* 1998, 1999; Hacker *et al.* 2000; Ratschbacher *et al.* 2000) and geophysical survey

(Wang, C. *et al.* 2000). We should emphasize here that the terms *allochthonous*, *autochthonous* and *parautochthonous* are borrowed from Alpine geology. In the Alps, the term *allochthonous* usually applies to nappes that have horizontally travelled a long distance, whereas *autochthonous* describes *in situ* units. Here we use the terms *allochthonous* and *parautochthonous* to qualitatively describe the upward travel. However, relative to the present-day boundaries between the units, these terms are the same, because the contacts between units are shallow-dipping.

The new pattern of the architecture has focused on exhumation processes and will help us to understand the geodynamics involved in the formation of the Dabieshan Orogenic Belt. The allochthonous unit can be correlated to UHP and HP rocks in the units III, IV-a, IV-b and V in Fig. 1 and comprises a 10-km-thick slab, estimated by Hacker *et al.* (2000). Its side view looks like a piston in a chamber (Fig. 10). Its upward extrusion built up the backbone of the orogenic belt. The parautochthonous unit can be correlated to units II and VI, as well as non-UHP gneiss in unit V in Figure 1 and serves as both the hanging wall and footwall of the extruded unit. The hanging wall is composed of metamorphic flysch and Carboniferous rocks in the North Huaiyang Belt. They represent a tectonic mélange zone that formed during oceanic subduction. The suture line drawn there by some researchers is just a northern boundary of the subduction complex. The Xiaotian–Mozitan Fault, on the other hand, represents the boundary between the subduction complex and the exhumation complex, which is characterized by UHP rocks. The footwall is composed of non-UHP gneiss, representing the mobilized Yangtze basement. Low-angle ductile faults represent the bottom boundary of the exhumation complex, and separate the UHP rocks and the underlying non-UHP gneiss (Figs 9 & 10). The non-UHP rocks of the footwall, in turn, overlie the deeply buried Yangtze basement that is not exposed.

A very important feature in the sequence of evolutionary maps (Fig. 10b) is the reversal of shear sense on faults. For example, the Xiaotain–Mozitan Fault (XMF in Fig. 10) once had a top-to-the-south thrust shear sense during the northward subduction of the NDT, as is evidenced by fabrics quartz *c*-axis data (Wang, Q. *et al.* 1996*b*). However, during the exhumation of the NDT, the XMF changed to a normal fault with a top-to-the-north shear sense (Wang, Q. *et al.* 1996). Another example is the Shuihou–Wuhe Fault (SWF in Fig. 10) between the allochthonous unit and its lower unit. It should also have a top-to-the-south thrust shear sense

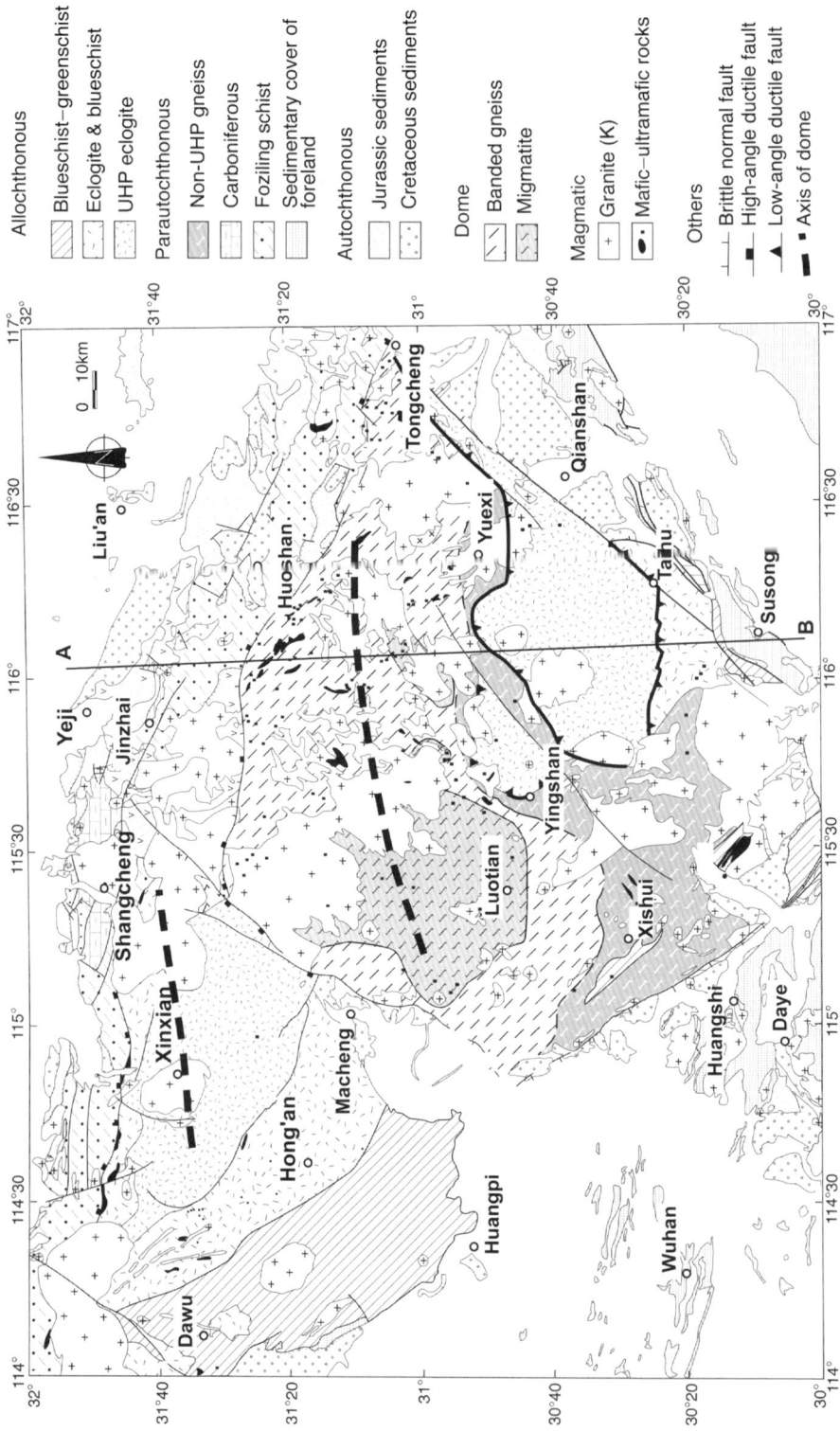

Fig. 9. Dynamic architecture of the Dabieshan Orogenic Belt.

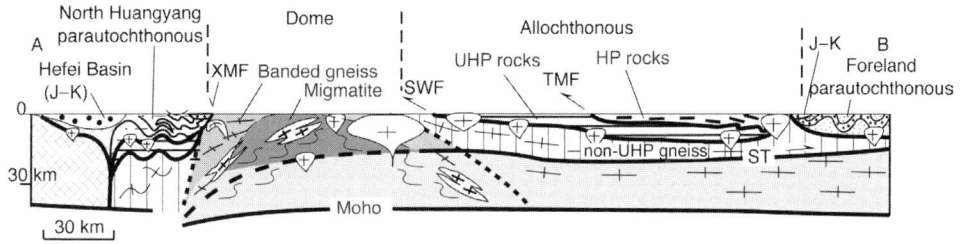

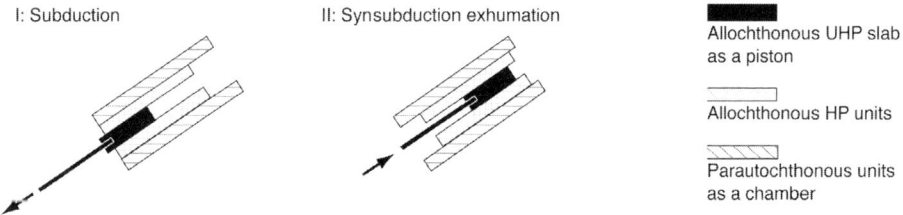

Fig. 10. (**a**) Profile of the Dabieshan Orogenic Belt (the position is shown in Fig. 9); (**b**) the interpreted evolutionary stages; and (**c**) a *piston* model. Faults: XMF, Xiaotian–Mozitan Fault; SWF, Shuihou–Wuhe Fault; TMF, Taihu–Mamiao Fault; TLF, Tancheng–Lujiang Fault; ST, Sole thrust. The filled dots, diamonds, triangles and crosses in (**b**) stand for HP eclogite, UHP rocks, non-UHP gneiss and the Yangtze basement. See text for explanation.

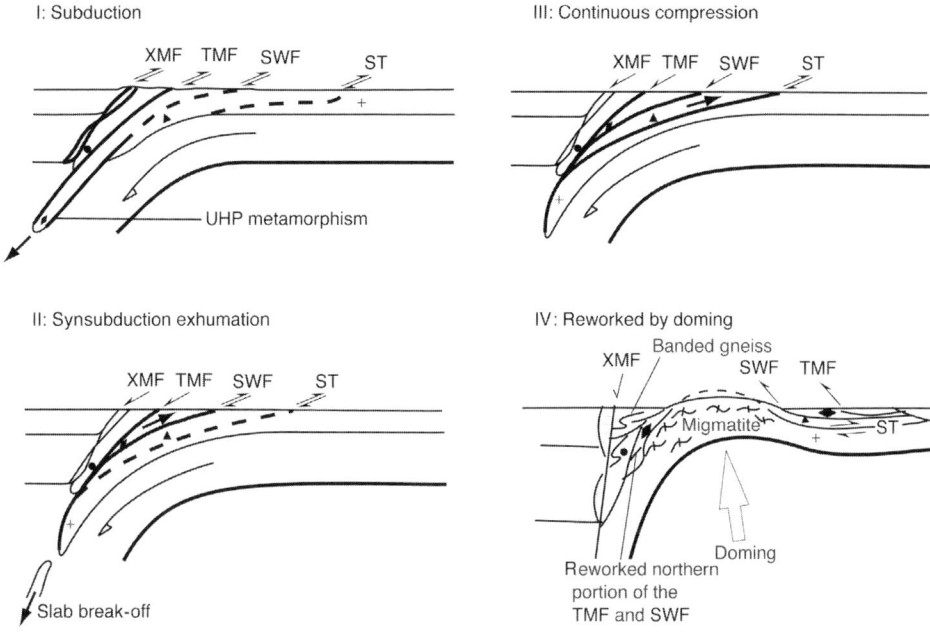

during the northward subduction of its lower unit. During later exhumation it reversed to a normal fault with a top-to-the-north shear sense. During the doming event, however, it was reworked to a thrust-like fault with a top-to-the-north shear sense (Wang et al. 1995).

As previously indicated, the peak metamorphic conditions of the rocks from various units are not the same (Wang, Q. & Cong, B. 1996). The $P-T-t$ estimations and exhumation rates of the UHP rocks in the allochthonous unit have been given in the foregoing section (p. 183). However, the HP rocks differ from UHP rocks in both peak metamorphism and exhumation rate. For example, the HP eclogite in the southern part of the Dabie-shan has an exhumation rate of 1.4 mm/year between the peak metamorphism (1.8–2.5 GPa at 640 °C) during 220–230 Ma and retrograde metamorphism (0.5 GPa at 450 °C) around 170 Ma (Wang, Q. et al. 1996a). The eclogite in the northern Dabieshan has an exhumation rate of 1.5 mm/year from eclogite metamorphic stage (2.0 GPa at 850 °C) at 230 Ma to granulite metamorphic stage (1.1 GPa at 870 °C) at 210 Ma, and an exhumation rate of 0.7 mm/year to amphibolite metamorphic stage (0.5 GPa at 550 °C), according to the $P-T-t$ path given by Xu, S. et al. (2000) and Liu, Y. et al. (2001). Obviously, the UHP slab has been subducted deeper than, and also exhumed faster than, its neighbouring HP rocks.

Discussion and conclusions

The Dabieshan Orogenic Belt has attracted world-wide attention. One of the important reasons is that it can serve as a window into deep dynamic processes involved in continent–continent collision and orogenesis. In contrast to previous works, we provide a dynamic pattern for evolution of the Dabieshan Orogenic Belt from the viewpoint of exhumation of UHP rocks based on published data and our own work. With this new model, we try to convey our understanding of several important phenomena in the orogenic belt.

Moho offset and dome structure in the mountain root

The Moho offset and the dome structure beneath the Dabieshan Orogenic Belt were revealed by deep crustal geophysical data. Although the geophysical data usually outline the present-day structural pattern, the shear sense and ductile deformation history indicate that the dome, or antiform at least, is a Jurassic structure. The

Moho offset beneath the dome indicates a compressional structure, and a similar feature can be seen also beneath the Alps (Fig. 2b of Michard et al. 1993), where the offset marks the gap between the Moho of the subducting Adriatic Plate and that of the overriding European plate. On the other hand, a vertically extruding slab can bend itself when it meets a hard obstacle, as shown in models of Chemenda et al. (1996). In contrast to the core complexes in the western United States, in the Dabieshan this type of dome formed in a compression environment.

Subduction complex and exhumation complex

We have tried to differentiate the subduction complex from the exhumation complex. Theoretically, they are different in their formational environments and mechanisms. The subduction complex formed during subduction of oceanic lithosphere, while the exhumation complex formed during exhumation of the deeply subducted materials. However, it is not easy to differentiate them in an orogenic belt. This is because a subduction complex itself can also experience exhumation, such as the Franciscan Complex occurring on the California coastline. Our practice in the Dabieshan Orogenic Belt is to put all UHP–HP rocks into the exhumation complex. The subduction complex includes the Foziling Flysch and Carboniferous strata in the North Huaiyang Belt. The Foziling Flysch represents the accretionary wedge and has been partly metamorphosed at greenschist-facies, whereas the Carboniferous strata have been strongly deformed during the continental collision.

Subduction polarity

One important question pertains to subduction polarity in the Dabieshan Orogenic Belt. To answer this question one needs to consider the geological, geophysical and geochemical data. Our dynamic model (Figs 9 & 10) hints at a geological answer. A northward subduction of the Yangtze craton is implied by the pattern of the subduction complex occuring in the north and the non-UHP rocks representing the mobilized Yangtze basement lying to the south. Our interpretation of northward subduction is supported by the north-dipping high-velocity bodies in the crust and the upper mantle (Figs 5 & 6). The north-dipping high-velocity bodies in the crust have been interpreted as an UHP slab within the crust (Wang, C. et al. 2000), and those in the upper mantle have been interpreted as the subducted Yangtze craton (Xu, P. et al. 2001). Geochemical evidence for northward subduction comes from the similarity in the C-isotopes between the UHP marble and the unmetamorphosed carbonate rocks on the north margin of the Yangtze craton (Wang & Rumble 1999; Rumble *et al.* 2000).

The motor to build up the Dabieshan Orogenic Belt

In our dynamic architecture of the Dabieshan Orogenic Belt, the allochthonous UHP unit is considered as an extrudant, which can be viewed as a *piston* side-on. Then, what has pushed it out? Buoyancy has undoubtedly played an important role. However, whenever subduction is proceeding, buoyancy is only a potential force, otherwise the continental crust could not be pulled down to mantle depths. To trigger buoyancy effects, the force pulling down the crust should be removed. The break-off of the subducted slab can provide a chance for this to occur (Davies & von Blanckenburg 1995). The slab-like high-velocity anomaly is truncated by a low-velocity anomaly at a depth of >100 km beneath the Dabieshan Orogenic Belt (Fig. 6) and can be interpreted to be a result of slab break-off. Therefore, in forming the Dabieshan Orogenic Belt, the low density of the subducted continental materials is but a precondition. However, the *motor* that pushed out the extruded UHP unit as a *piston* was not the buoyancy itself. The slab break-off only triggered buoyancy to work by decoupling of the UHP continental materials of low density from the lithosphere moving down. Based on the P–T–t path and exhumation rates of various units, as well as the kinematic analysis above, we prefer a multistage exhumation process in which the syn-subduction compression between the Sino-Korean craton and the Yangtze craton has played a major role. The UHP allochthonous unit was pushed out as a *piston* by the synsubduction compression. The main architecture of the Dabieshan Orogenic Belt was built up by the compression, and reworked by a late doming event.

This study is funded by the Chinese Ministry of Science and Technology (G19990755) and the National Natural Science Foundation of China (No. 49794042). We are grateful to Li Suguang, Lin Wei, Liu Jingbo, Ye Kai and Xu Peifen, who took part in our projects and provided their new data. We thank also B. Windley, L. S. Teng and Bor-Ming Jahn for constructive discussion in the Croucher Advanced Studies Institute (Hong Kong), that was chaired by Sun Shu and J. Malpas during 23–28 April 2002. Our special thanks go to two reviewers, R. Coleman and P. O'Brien, for their critical and helpful comments that have greatly improved the present paper.

References

AMES, L., TILTON, G. R. & ZHOU, G. 1993. Timing of collision of the Sino-Korean and Yangtze cratons: U–Pb zircon dating of coesite-bearing eclogites. *Geology*, **21**, 339–342.

BAKER, J., MATTHEWS, A., MATTEY, D., ROWLEY, D. & XUE, F. 1997. Fluid–rock interactions during ultra-high pressure metamorphism, Dabie Shan, China. *Geochimica et Cosmochimica Acta*, **61**, 1685–1696.

CARSWELL, D. A., O'BRIEN, P. J., WILSON, R. N. & ZHAI, M. 1997. Thermobarometry of phengite-bearing eclogites in the Dabie Mountains of central China. *Journal of Metamorphic Geology*, **15**, 239–252.

CHAVAGNAC, V. & JAHN, B. M. 1996. Coesite-bearing eclogites from the Bixiling Complex, Dabie Mountains, China: Sm–Nd ages, geochemical characteristics and tectonic implications. *Chemical Geology*, **133**, 29–51.

CHEMENDA, A. I., MATTAUER, M. & BOKUN, A. N. 1996. Continental subduction and a mechanism for exhumation of high-pressure metamorphic rocks: new modeling and field data from Oman. *Earth and Planetary Science Letters*, **143**, 173–182.

CHEN, J. 1995. Cooling age of Dabie orogen, China, determined by $^{40}Ar–^{39}Ar$ and fission track techniques. *Science in China (Series B)*, **38**, 749–757.

CHOPIN, C. 1984. Coesite and pure pyrope in high grade pelitic blueschists of the Western Alps: a first record and some consequences. *Contributions to Mineralogy and Petrology*, **86**, 107–118.

CHOPIN, C. & SOBOLEV, N. V. 1995. Principal mineralogic indicators of UHP in crustal rocks. *In*: COLEMAN, R. G. & WANG, X. (eds) 1995, *Ultrahigh-Pressure Metamorphism*. Cambridge University Press, Cambridge, 96–131.

CONG, B. 1996. (ed.) *Ultrahigh-Pressure Metamorphic Rocks in the Dabieshan–Sulu Region of China*. Science Press, Beijing, China, Kluwer Academic Publishers, Dordrecht.

CONG, B., ZHAI, M., CARSWELL, D. A., WILSON, R. N., WANG, Q., ZHAO, Z. & WINDLEY, B. F. 1995. Petrogenesis of the ultrahigh-pressure rocks and their country rocks at Shuanghe in Dabieshan, central China. *European Journal of Mineralogy*, **7**, 119–138.

DAVIES, J. H. & VON BLANCKENBURG, F. 1995. Slab breakoff: a model of lithosphere detachment and its test in the magmatism and deformation of collisional orogens. *Earth and Planetary Science Letters*, **129**, 85–102.

EIDE, E., MCWILLIAMS, M. O. & LIOU, J. G. 1994. $^{40}Ar/^{39}Ar$ geochronology and exhumation of high-pressure to ultrahigh-pressure metamorphic rocks in east-central China. *Geology*, **22**, 601–604.

ENAMI, M. & ZANG, Q. 1988. Magnesian staurolite in garnet–corundum rocks and eclogite from the Donghai District, Jiangsu Province, East China. *American Mineralogist*, **73**, 48–56.

ERNST, W. G., ZHOU, G., LIOU, J.G., EIDE, E. & WANG, X. 1991. High-pressure and superhighpressure metamorphic terranes in the Qinling-Dabie mountain belt, central China: early- to mid-Phanerozoic accretion of the western paleo-Pacific Rim. *Pacific Scientific Association of Information Bulletin*, **43**, 6–15.

FAURE, M., LIN, W. & SUN, Y. 1998. Doming in the southern foreland of the Dabieshan (Yangtze block, China). *Terra Nova*, **10**, 307–311.

FAURE, M., LIN, W., SHU, L. & SUN, Y. 1999. Tectonics of the Dabieshan and its bearing on the exhumation of the ultra high-pressure rocks. *Terra Nova*, **11**, 251–258.

HACKER, B. R., RATSCHBACHER, L., WEBB, L., IRELAND, T., WALKER, D. & DONG, S. 1998. U/Pb zircon ages constrain the architecture of the ultrahigh-pressure Qinling–Dabie Orogen, China. *Earth and Planetary Science Letters*, **161**, 215–230.

HACKER, B. R., RATSCHBACHER, L. *et al.* 2000. Exhumation of ultrahigh-pressure continental crust in east central China: Late Triassic–Early Jurassic tectonic unroofing. *Journal of Geophysical Research*, **105**, 13 339–13 364.

HACKER, B. R., WANG, X., EIDE, E. A. & RATSCHBACHER, L. 1996. The Qinling–Dabie ultrahigh pressure collisional orogen. *In*: HARRISON, M.T. & YIN, A. 1996. (eds) *Tectonic Development of Asia*. Cambridge University Press, Cambridge, 345–370.

HAO, Y., SU, D. *et al.* 1986. *The Cretaceous System of China*. Geological Publishing House, Beijing (in Chinese with English abstract).

HIRAJIMA, T., ISHIWATARI, A., CONG, B., ZHANG, R., BANNO, S. & NOZAKA, T. 1990. Coesite from Mengzhong eclogite at Donghai county, northeastern Jiangsu province, China. *Mineralogical Magazine*, **54**, 579–583.

KRETZ, R. 1983. Symbols for rock-forming minerals. *American Mineralogist*, **68**, 277–279.

LI, S., CHEN, Y. *et al.* 1993. Collision of the North China and Yangtze Blocks and formation of coesite-bearing eclogites: timing and processes. *Chemical Geology*, **109**, 70–89.

LI, S., HART, S. R., ZHENG, S., LIOU, D., ZHANG, G. & GUO, A. 1989. Timing of collision between the North and South China Blocks – Sm–Nd isotopic age evidence. *Science in China (Series B)*, **32**, 1391–1400.

LI, S., JAGOUTZ, E., XIAO, Y., GE, N. & CHEN, Y. 1996. Chronology of ultrahigh-pressure metamorphism in the Dabie Mountains and Su–Lu terrane: I. Sm–Nd isotope system. *Science in China (Series D)*, **39**, 597–609.

LI, S., JAGOUTZ, E., CHEN, Y. & LI, Q. 2000. Sm/Nd and Rb/Sr isotopic chronology and cooling history of ultrahigh pressure metamorphic rocks and their country rocks at Shuanghe in the Dabie Mountains, Central China. *Geochimica et Cosmochimica Acta*, **64**, 1077–1093.

LI, S., LIOU, D., CHEN, Y. & GE, N. 1992. The Sm–Nd isotopic age of coesite-bearing eclogite from the southern Dabie Mountains. *Chinese Science Bulletin*, **37**, 1638–1641.

LIN, W. 2000. *Etude tectonique de l'avant-pays méridional de la chaîne Debie–Qinling (Nord du bloc de Chine du Sud)*. Docteur Thèse, de l'Université d'Orléans.

LIOU, J. G. & ZHANG, R. Y. 1996. Occurrence of intergranular coesite in Sulu ultrahigh-P rocks from China: implications for fluid activity during exhumation. *American Mineralogist*, **81**, 1217–1221.

LIU, J., YE, K., MARUYAMA, S., CONG, B. & FAN, H. 2001. Mineral inclusions in zircon from gneisses in the ultrahigh-pressure zone of the Dabie Mountains, China. *The Journal of Geology*, **109**, 523–535.

LIU, Y., LI, S. & XU, S. 2001. Zircon U–Pb ages of eclogite and tonolitic gneiss from the North Dabie Terrane and multi-stage metamorphism. *Geological Journal of Chinese Universities*, **6**, 417–423. (in Chinese with English abstract).

MARUYAMA, S., LIOU, J. G. & ZHANG, R. 1994. Tectonic evolution of the ultrahigh-pressure and high-pressure metamorphic belts from central China. *The Island Arc*, **3**, 112–121.

MICHARD, A., CHOPIN, C. & HENRY, C. 1993. Compression versus extension in the exhumation of the Dora Maira coesite-bearing unit, Western Alps, Italy. *Tectonophysics*, **221**, 173–193.

OKAY, A. I. 1993. Petrology of a diamond and coesite-bearing metamorphic terrain: Sabie Shan, China. *European Journal of Mineralogy*, **5**, 659–675.

OKAY, A. I., XU, S. & SENGÖR, A. M. C. 1989. Coesite from the Dabie Shan eclogites, central China. *European Journal of Mineralogy*, **1**, 595–598.

RATSCHBACHER, L., HACKER, B. *et al.* 2000. Exhumation of the ultrahigh-pressure continental crust in east central China: Cretaceous and Cenozoic unroofing and the Tan–Lu fault. *Journal of Geophysical Research*, **105**, 13 303–13 338.

ROLFO, F., COMPAGNONI, R., XU, S. & JIANG, L. 2000. First report of felsic whiteschist in the ultrahigh-pressure metamorphic belt of Dabie Shan, China. *European Journal of Mineralogy*, **12**, 883–898.

ROWLEY, D. B., XUE, F., TUCKER, B. D., PENG, Z. X. & DAVIES, A. M. 1997. Ages of ultrahigh-pressure metamorphism and protolith orthogneisses from the eastern Dabie Shan: U/Pb zircon geochronology. *Earth and Planetary Science Letters*, **151**, 191–203.

RUMBLE, D., WANG, Q. & ZHANG, R. 2000. Stable isotope geochemistry of marbles from the coesite UHP terranes of Dabieshan and Sulu, China. *Lithos*, **52**, 79–98.

SMITH, D. C. 1984. Coesite in clinopyroxene in the Caledonides and its implications for geodynamics. *Nature*, **310**, 641–644.

SOBOLEV, N. V., SHATSKY, V. S., VAVILOV, M. A. & GORYAINOV, S. V. 1994. Zircon from ultrahigh-pressure metamorphic rocks of folded regions as an unique container of inclusions of diamond, coesite and coexisting minerals. *Dokladi Akademii Nauk*, **334**, 488–492.

SU, W., YOU, Z., CONG, B., YE, K. & ZHONG, Z. 2002. Cluster of water molecules in garnet from ultrahigh-pressure eclogite. *Geology*, **30**, 611–614.

TABATA, H., YAMAUCHI, K., MARUYAMA, S. & LIOU, J. G. 1998. Tracing the extent of a UHP metamorphic terrane: mineral-inclusion study of zircons in gneiss from the Dabieshan. *In*: HACKER, B. & LIOU, J. G. (eds) *When Continents Collide: Geodynamics and Geochemistry of Ultrahigh-Pressure Rocks*, Kluwer Academic Publishers, Dordrecht, 261–273.

WANG, C., ZENG, R., MOONEY, W. D. & HACKER, B. R. 2000. A crustal model of the ultrahigh-pressure Dabie Shan orogenic belt, China, derived from deep seismic refraction profiling. *Journal of Geophysical Research*, **105 (B5)**, 10 857–10 869.

WANG, D., LIU, Y., LI, S. & JIN, F. 2001. Lower time limit on the UHPM rock exhumation: discovery of eclogite pebbles in the Late Jurassic conglomerates from the northern foot of the Dabie Mountains, eastern China. *Chinese Science Bulletin*, **47**, 231–235.

WANG, Q. & CONG, B. 1996. Tectonic implication of UHP rocks from the Dabie Mountains. *Science in China (Series D)*, **39**, 311–318.

WANG, Q. & CONG, B. 1999. Exhumation of UHP terranes: a case study from the Dabie Mountains, Eastern China. *International Geology Review*, **41**, 994–1004.

WANG, Q. & RUMBLE, D. 1999. Oxygen and carbon isotope composition from the UHP Shuanghe marbles, Dabie Mountains, China. *Science in China (Series D)*, **42**, 88–96.

WANG, Q., CONG, B. & ZHAI, M., 1996a. Tectonic evolution of UHPM rocks. *In*: CONG, B. (ed.) *Ultrahigh-Pressure Metamorphic Rocks in the Dabieshan–Sulu Region of China*. Science Press, Beijing; Kluwer Academic Publishers; Dordrecht, 161–170.

WANG, Q., CONG, B. & ZHU, R. 1998. Geodynamics of UHP-rock-bearing continental collision zone in Central China. *In*: FLOWER, M. F. J., CHUNG, S. L., LO, C. H. & LEE, T.-Y. (eds) *Mantle Dynamics and Plate Interactions in East Asia*. American Geophysical Union, Geodynamics Series, **27**, 259–268.

WANG, Q., ISHIWATARI, A. *et al.* 1993. Coesite-bearing granulite retrograded from eclogite in Weihai, eastern China, *European Journal of Mineralogy*, **5**, 141–152.

WANG, Q., LIU, X., MARUYAMA, S. & CONG, B. 1995. Top Boundary of the Dabie UHPM rocks, Central China. *Journal of Southeast Asian Earth Sciences*, **11**, 295–300.

WANG, Q., LIU, X., ZHAO, Z. & CHEN, J. 1996b. Structural geology. *In*: CONG, B. (ed.) *Ultrahigh-Pressure Metamorphic Rocks in the Dabieshan–Sulu Region of China*. Science Press, Beijing; Kluwer Academic Publishers, Dordrecht, 27 48.

WANG, S., CHENG, Z. & WANG, N. 1985. *The Jurassic System of China*. Geological Publishing House, Beijing (in Chinese with English abstract).

WANG, X. & LIOU, J. G. 1993. Ultrahigh-pressure metamorphism of carbonate rocks in the Dabie Mountains, central China. *Journal of Metamorphic Geology*, **11**, 575–588.

WANG, X., LIOU, J. & MAO, H. K. 1989. Coesite-bearing eclogite from the Dabie mountains in central China, *Geology*, **17**, 1085–1088.

WANG, Y., FAN, W. & GUO, F. 2002. K/Ar ages of the Mesozoic volcanic rocks and geochemistry of pyroclastic rocks in the North Huaiyang. *Chinese Science Bulletin*, **47**, 1688–1695.

XIE, Z., CHEN, J., ZHANG, X., GAO, T., DAI, S., ZHOU, T. & LI, H. 2001. Zircon U–Pb dating of gneiss from Shizhuhe in North Dabie and its geologic implications. *Acta Petrologica Sinica*, **17**, 139–144.

XU, S. & SU, W. 1997. Raman determination on microdiamond in eclogite from the Dabie Mountains, eastern China. *Chinese Science Bulletin*, **42**, p. 87.

XU, S., LIU, Y., SU, W., WANG, R., JIANG, L. & WU, W. 2000. Discovery of the eclogite and its petrography in the Northern Dabie Mountain. *Chinese Science Bulletin*, **45**, 273–278.

XU, P., LIU, F., WANG, Q., CONG, B. & CHEN, H. 2001. Slab-like high velocity anomaly in the uppermost mantle beneath the Dabie–Sulu orogen. *Geophysical Research Letters*, **28**, 1847–1850.

XU, S., OKAY, A. I., JI, S., SENGÖR, A. M. C., SU, W., LIU, Y. & JIANG, L. 1992. Diamond from the Dabie Shan metamorphic rocks and its implication for tectonic setting. *Science*, **256**, 80–82.

XU, P., SUN, R., LIU, F., WANG, Q. & CONG, B. 2000. Seismic tomography showing subduction and slab breakoff of the Yangtze block beneath the Dabie–Sulu orogen. *Chinese Science Bulletin*, **45**, 70–73.

XU, Z. 1987. *Étude tectonique et microtectonique de la chaîne paleozoique et triasique des Quilings (Chine)*. Thèse de doctorat, Université des Sciences et Techniques du Languedoc, Montpellier.

YANG, J. & SMITH, D. C. 1989. Evidence for a former sanidine–coesite–eclogite at Lanshantou, east China, and the recognition of the Chinese 'Su-Lu coesite–eclogite province'. *Third International Eclogite Conference Abstracts*, Blackwell Scientific Publications, Oxford, p. 26.

YANG, J., GODARD, G., KIENSAT, J. R., LU, Y. & SUN, J. 1993. Ultrahigh-pressure (60 kbar) magnesite-bearing garnet peridotites from northeastern Jiangsu, China. *Journal of Geology*, **101**, 541–554.

YE, K., CONG, B. & YE, D. 2000a. The possible subduction of continental material to depths greater than 200 km. *Nature*, **407**, 734–736.

YE, K., LIU, J., YAO, Y. & CONG, B. in press. Dehydration melting of phengite and fluid generation during exhumation of ultrahigh-pressure (UHP) metamorphic rocks (Northern Sulu, eastern China): possible implications for rapid exhumation and exhumation related magmatism. *Journal of Petrology*.

YE, K., YAO, Y., KATAYAMA, I., CONG, B., WANG, Q. & MARUYAMA, S. 2000b. Large areal extent of ultrahighpressure metamorphism in the Sulu ultra-high-pressure terrane of East China: new implica-tions from coesite and omphacite invlusions in zircon granitic gneiss. *Lithos*, **52**, 157–164.

YIN, A. & NIE, S. 1993. An indentation model for the North and South China collision and the development of the Tan–Lu and Honam fault system, eastern Asia. *Tectonics*, **12**, 801–803.

YU, J. & YAO, P. 1989. Assemblage succession of Mesozoic non-marine bivalves in Yanliao region. *In*: SU, D. & WANG, S. (eds) *The Palaeontology and Stratigraphy of the Jurassic and Cretaceous in Eastern China*. Geological Publishing House, Beijing, 52–72 (in Chinese with English abstract).

YUI, T. F., RUMBLE, D. & LO, C. H. 1995. Unusually low $\delta^{18}O$ ultra-high-pressure metamorphic rocks from the Sulu Terrain, eastern China. *Geochimica et Cosmochimica Acta*, **59**, 2859–2864.

ZHANG, L., GAO, J., AI, B. & WANG, Z. 2000. Metamorphism of low temperature eclogite facies in the west Tianshan. *Science in China (Series D)*, **30**, 345–354.

ZHANG, R. Y. & LIOU, J. G. 1996. Significance of coesite inclusions in dolomite from eclogite in the southern Dabie mountains, China. *American Mineralogist*, **80**, 181–186.

ZHANG, R. Y., HIRAJIMA, T., BANNO, S., ISHIWATARI, A., LI, J., CONG, B. & NOZAKA, T. 1989. Coesite-eclogite from Donghai area, Jiangsu Province in China. *The 15th General Meeting of the International Mineralogical Association Abstracts*, **2**, 923–924.

ZHANG, R. Y., LIOU, J. G. & CONG, B. 1994. Petrogenesis of garnet-bearing ultramafic rocks and associated eclogites in the Su–Lu ultrahigh-P metamorphic terrane, eastern China. *Journal of Metamorphic Geology*, **12**, 169–186.

ZHANG, R. Y., LIOU, J. G. & CONG, L. 1995. Ultrahigh-pressure metamorphosed talc-, magnesite- and Ti-clinohumite-bearing mafic–ultramafic complex, Dabie Mountains, east central China. *Journal of Petrology*, **36**, 1011–1038.

ZHENG, Y., FU, B., GONG, B. & LI, S. 1996. Extreme ^{18}O depletion in eclogite from the Su–Lu terrane in East China. *European Journal of Mineralogy*, **8**, 317–323.

Jurassic intraplate magmatism in southern Hunan–eastern Guangxi: ^{40}Ar/^{39}Ar dating, geochemistry, Sr–Nd isotopes and implications for the tectonic evolution of SE China

XIAN-HUA LI[1], SUN-LIN CHUNG[2], HANWEN ZHOU[1,3], CHING-HUA LO[2],
YING LIU[1], & CHANG-HWA CHEN[4]

[1]*Guangzhou Institute of Geochemistry, Chinese Academy of Sciences, Guangzhou 510640,
Guangdong, China (e-mail: lixh@gig.ac.cn)*
[2]*Department of Geosciences, National Taiwan University, Taipei, Taiwan*
[3]*Faculty of Earth Sciences, China University of Geosciences, Wuhan 430074, Hubei*
[4]*Institute of Earth Sciences, Academia Sinica, Taipei, Taiwan*

Abstract: The Mesozoic geology of SE China is characterized by intensive and widespread magmatism. However, the tectonic regime that accounted for the Mesozoic magmatism has been an issue with little consensus. A comprehensive study of ^{40}Ar–^{39}Ar dating, geochemistry and Sr–Nd isotopes has been conducted on basalts from southern Hunan and syenite intrusions from eastern Guangxi. Three episodes of Jurassic magmatism, i.e. alkaline basalts of $c.175$ Ma in age, syenitic intrusions of $c.160$ Ma and high-Mg basalts of $c.150$ Ma, are identified. The older, $c.175$ Ma alkaline basalts are characterized by low Sr ($I_{Sr} = 0.7035–0.7040$) and high Nd ($\varepsilon_{Nd}(T) = 5$ to 6) isotopic compositions and OIB-like trace-element patterns (e.g. Nb/La > 1). In contrast, the younger, $c.150$ Ma high-Mg basalts have high Sr ($I_{Sr} c.0.7054$) and low Nd ($\varepsilon_{Nd}(T) c.-2$) isotopic compositions and incompatible trace-element patterns of arc affinity. The $c.160$ Ma syenitic intrusions display a relatively large range of Sr and Nd isotopic compositions ($I_{Sr} = 0.7032–0.7082$, $\varepsilon_{Nd}(T) = 5.5$ to -4.1), with the Qinghu syenites having the lowest I_{Sr}, highest $\varepsilon_{Nd}(T)$ and OIB-type incompatible trace-element patterns analogous to the $c.175$ Ma alkaline basalts. Such a secular variation in rock types and geochemical and isotopic characteristics reveals changes in melt segregation depth and mantle sources, which are inferred to have resulted from the post-Indosinian orogenic lithosphere extension and thinning. The $c.175$ Ma alkaline basalts are suggested to have formed by small degrees of decompression melting of the asthenosphere or an enriched lithospheric mantle source accreted by asthenosphere-derived melts during the initial extension. The $c.160$ Ma syenitic and $c.150$ Ma high-Mg basaltic rocks mainly originated from the enriched lithospheric mantle that melted owing to a raised geotherm caused by lithosphere thinning. This interpretation is at odds with the active continental margin related to the subduction of palaeo-Pacific plate, but consistent with continental rifting and extension for the Mesozoic of SE China.

The Mesozoic geology of SE China is characterized by intensive and widespread magmatism (Fig. 1). Among the igneous rocks formed, granites and rhyolites are volumetrically predominant, with subordinate basalts and rare intermediate lithologies. There appears to be a southeastward-younging trend for the Mesozoic magmatism, i.e. with the oldest Triassic intrusions occurring mainly in the southwest, the Jurassic intrusions in the middle, and then the Cretaceous granitoids and volcanic rocks in the southeast along the coastal areas. The tectonic regime that accounted for the Mesozoic magmatism in SE China has been an issue of long-term debate. The different models that have been proposed in

the last two decades can be generally grouped into four categories:

(1) An active continental margin related to the subduction of the palaeo-Pacific plate (e.g. Jahn 1974; Jahn *et al.* 1976; Holloway 1982). This model is favoured by many authors for development of calc-alkaline volcanic and intrusive rocks during the Jurassic to Cretaceous in the coastal region of SE China (e.g. Jahn 1974; Jahn *et al.* 1976, 1990; Huang *et al.* 1986; Charvet *et al.* 1994; Martin *et al.* 1994; Lan *et al.* 1996; Fletcher *et al.* 1997; Lapierre *et al.* 1997; Sewell & Campbell 1997). The subduction-related regime was considered to be dominant until the Late Cretaceous, when alkaline

From: MALPAS, J., FLETCHER, C. J. N., ALI, J. R. & AITCHISON, J. C. (eds) 2004. *Aspects of the Tectonic Evolution of China*. Geological Society, London, Special Publications, **226**, 193–215.
0305-8719/04/$15 © The Geological Society of London 2004.

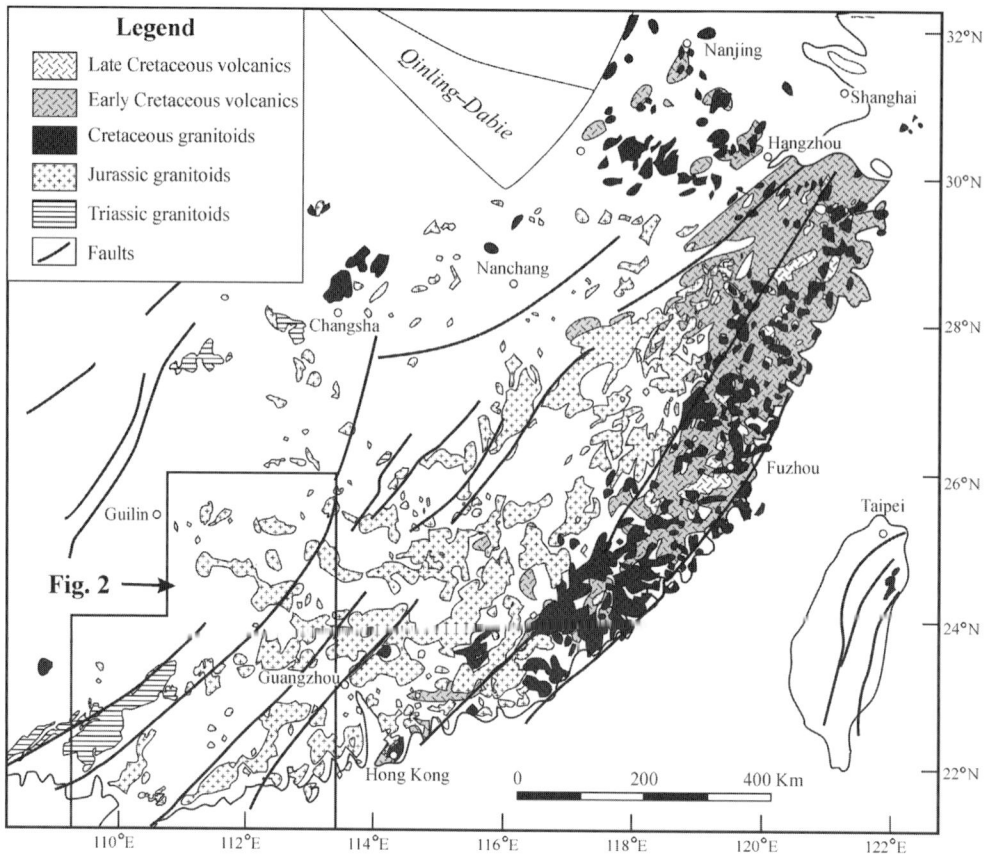

Fig. 1. Distribution of Mesozoic igneous rocks in southeastern China (modified after Wang *et al.* 1985)

granites were intruded along the Fujian coastal areas (e.g. Martin *et al.* 1994). Recently, a modified subduction model was proposed by Zhou & Li (2000) to account for the exceptional wide magmatic arc and the migration of magmatic activity oceanward to the southeast. They suggested an increase in the slab dip angle of the palaeo-Pacific plate subduction underneath SE China, from a very low angle in the Early Jurassic to a median angle in the Late Cretaceous.

(2) *An Alpine-type continental collision model* (Hsü *et al.* 1988, 1990). In this model, Hsü *et al.* proposed that the Huanan collided with the Yangtze during the Triassic, and the Huanan collided with the postulated Dongnanya block during the Triassic/Jurassic to Cretaceous. In this collision model, the Neoproterozoic Banxi Group on the Yangtze side was re-interpreted as a Mesozoic mélange and a long-displaced thrust sheet, and the Cretaceous granites in SE China as the products of a collision between the Huanan block and the Dongnanya microcontinents.

However, this collision model has been refuted by many studies (e.g. Rodgers 1989; Rowley *et al.* 1989; Chen *et al.* 1991; Li & McCulloch 1996) since it was proposed. For instance, a number of studies demonstrated that the Banxi Group and the overlying Sinian System are normal sedimentary sequences (e.g. Li & McCulloch 1996; Wang & Li 2001, 2003), which most likely formed within the Neoproterozoic Nanhua Rift basin, related to the initial rifting of the supercontinent Rodinia (Li *et al.* 1999; Wang & Li 2001, 2003). The U–Pb zircon and geochemical data (Li *et al.* 1995; Zou, 1995) indicate that the Quanzhou Gabbro in the coastal Fujian province crystallized within the continent at 107 Ma, rather than an ophiolitic member of the 'Quanzhou Klippe' proposed by Hsü *et al.* (1990).

(3) *Wrench fault system* (Xu *et al.* 1987, 1993). Xu *et al.* proposed in this model that the Changle–Nanao shear zone along the coastal area of SE China belongs to the huge Tan–Lu Fault system. They inferred that there was no oceanic

plate subduction beneath the East Asia continent before the Late Cretaceous, and the Mesozoic Yanshanian granites in SE China were considered as the products of faulting-induced re-melting. This faulting-induced re-melting model could be a plausible way of explaining the formation of deformed granites in the Dongshan and Chinmen regions along the Changle–Nanao shear zone, as syntectonic granites were emplaced coeval with major phases of shearing (Tong & Tobisch 1996; Yui *et al.* 1996). However, such a faulting-induced re-melting model is difficult to apply to the widespread Mesozoic Yanshanian granites in SE China, because most granites are neither deformed nor associated with fractures.

(4) *Continental rifting and extension.* Gilder *et al.* (1991) proposed that a system of roughly parallel, NE-trending grabens formed in SE China during the Mesozoic, and that they are similar to the present-day Basin and Range Province in the western United States. They referred to this area as the 'SE China Basin and Range Province', which underwent extension primarily in the Late Jurassic to Cretaceous. This model, that is at odds with the traditional Andean-type active continental margin model, was favoured by Li (2000) in view of the association of A-type granitic and intraplate basaltic magmatism (140–90 Ma) with coeval high-K calc-alkaline magmatism. In the interior of SE China, some granites have anomalously high REE, and low-Nd model ages (T_{DM}), possibly due to the input of mantle materials (Gilder *et al.* 1996; Chen & Jahn 1998). These granites form a NE-trending zone, referred to as the 'Shi-Hang Zone' (Gilder *et al.* 1996), which coincides with two prominent Mesozoic basins (the Gan-Hang and Shiwandashan Basins). It is noted that a number of extensional domes, possible metamorphic core complexes, were documented at Wugongshan (Faure *et al.* 1996), Lushan (Lin *et al.* 2000) and Jiulingshan (Lin *et al.* 2001) in SE China. Although these domes experienced multiple tectonometamorphic and magmatic events, their final doming, and possibly formation of core complexes, dated at *c.*130 Ma, suggests a major crustal extension in the Early Cretaceous in SE China. This is not inconsistent with the model of continental rifting and extension, or 'SE China Basin and Range Province' mentioned above, although different mechanisms were advocated for the formation and evolution of these domes (Faure *et al.* 1996; Lin *et al.* 2000, 2001; Ratschbacher *et al.* 2000).

Relative to the numerous studies of the Cretaceous magmatism in the coastal areas, less attention has been paid to the Jurassic (early Yanshanian) magmatism in the interior of SE China. In this paper, a comprehensive study of $^{40}Ar/^{39}Ar$ dating, geochemistry and Sr–Nd isotopes for basalts and syenite intrusions from southern Hunan–eastern Guangxi (the interior of SE China) is presented. These new data are used to explore the petrogenesis and provide constraints on the evolution of tectonic regime of SE China in the Mesozoic.

Geological setting

The southern Hunan–eastern Guangxi region is located at the southwestern flank of the Mesozoic igneous province of SE China (Figs 1 and 2). It demarcates the majority of Mesozoic igneous rocks and extensional basins to the southeast from relatively stable land, which are covered by the Lower Palaeozoic shallow-marine carbonate and clastic deposits and the Upper Palaeozoic to Middle Triassic flysch sedimentation, to the northwest (Guangxi 1985). The Indosinian Shiwandashan granites, with a total exposed area of more than 8000 km^2, occur in southeastern Guangxi. Late Triassic to Cretaceous basins, represented by the NE-trending Shiwandashan Basin, developed on the folded Paleozoic basement and the Indosinian granites.

A number of syenite intrusions occur in eastern Guangxi, and are distributed roughly in a NE direction, with a total exposed area of *c.*634 km^2 (Fig. 2). All plutons are undeformed. Most syenites are medium- to coarse-grained and grey to dark pink in colour, consisting mainly of clinopyroxene, hornblende, K-feldspar, plagioclase and a minor amount of quartz and biotite. They have intruded the Palaeozoic strata, with the Mashan pluton intruding the Triassic Darongshan (eastern Shiwandashan) Granite. Thus, they are generally considered to belong to the Yanshanian magmatism (Guangxi 1985). Xu & Yuan (1992) reported a U–Pb zircon age of 158 ± 2 Ma for the Qinghu pluton occurring on the border of Guangdong and Guangxi provinces.

Basaltic rocks occur sporadically in southern Hunan. They are generally considered as being Cretaceous in age, as some basalts have extruded in the Cretaceous basins (Hunan 1988). However, some basaltic rocks occur as cones, pipes, sills and dykes within the Palaeozoic to Lower Jurassic strata in the Ningyuan and Daoxian regions (Fig. 2), and their age of formation is not well constrained. These basaltic rocks in Ningyuan and Daoxian commonly contain a variety of lower crustal and mantle xenoliths. The major xenolith types are spinel lherzolite, dunite,

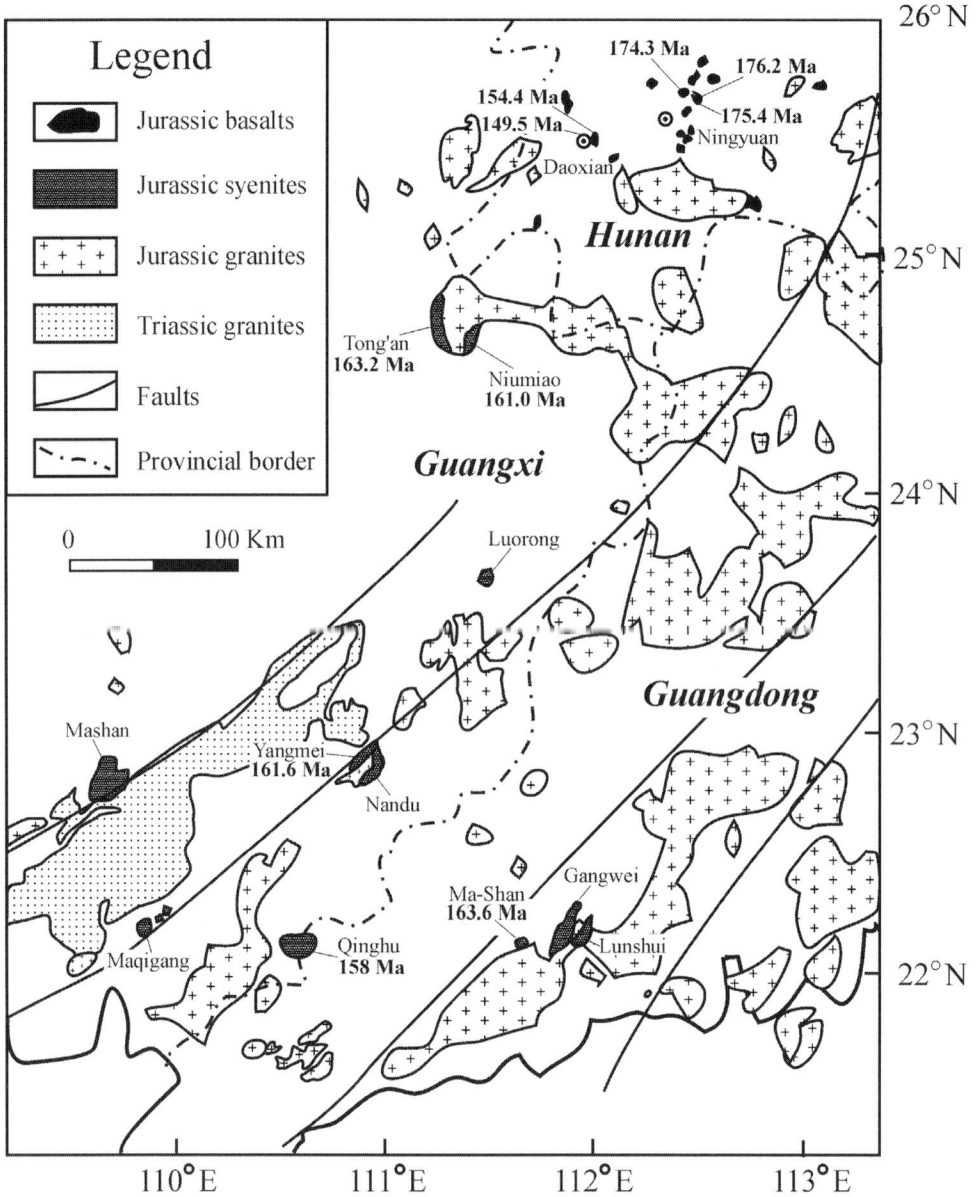

Fig. 2. Distribution of the studied Mesozoic basaltic rocks in southern Hunan and syenite intrusions in eastern Guangxi, as well as syenite intrusions in western Guangdong (modified after Guangxi 1985; Hunan 1988; Li *et al.* 2000). See text for sources of ages.

pyroxenite, gabbro and garnet-bearing biotite gneisses (Zhu *et al.* 1997).

Analytical procedures

Five whole-rock samples and three hornblendes were selected and dated by using the ^{40}Ar/^{39}Ar conventional step-heating method. Whole-rock chips in the size range of 140–250 μm were ultrasonically cleaned in distilled water and dried, and then handpicked to remove visible contamination. Hornblende separates were extracted from syenite rock samples using hand picking under the microscope. Samples (*c.*0.5–1 g) were wrapped in aluminium foil packets, stacked in an aluminium canister along with the LP-6 Biotite standard (Odin *et al.* 1982) and then irradiated in the VT-C position at the THOR Reactor in Taiwan for 30 hours. After irradiation, standards and samples were degassed incrementally from 400 to 1200°C, following a 30 minute/step schedule, using a resistance furnace. The purified gas was analysed with a Varian-MAT GD150 mass spectrometer at the Department of Geosciences, National Taiwan University. The concentrations of ^{36}Ar, ^{37}Ar, ^{38}Ar, ^{39}Ar and ^{40}Ar were corrected for system blanks, mass discrimination, radioactive decay of the nucleogenic isotopes, and minor interference reactions involving Ca, K and Cl, following procedures outlined in detail by Lo & Lee (1994). The mean of *J*-values obtained from the monitor standards was adopted in an age calculation, because the gradient of neutron flux across the canister appeared to be 0.52%; which is rather small. Ages were calculated from the Ar isotopic ratios measured. The results of ^{40}Ar/^{39}Ar analyses are plotted as Cl/K, Ca/K and age spectrum diagrams in Figure 3, and isotope correlation diagrams were also plotted. A summary of the dating results is given in Table 1.

Major-element oxides were determined using a Rigaku RIX 2000 X-ray fluorescence spectrometer (XRF) at the National Taiwan University. Analytical precision was around 1–5%. The details of the analytical procedures are given by Lee *et al.* (1997). Trace elements were analysed using a Perkin-Elmer Sciex ELAN 6000 inductively coupled plasma mass spectrometer (ICP–MS) at the Guangzhou Institute of Geochemistry, Chinese Academy of Sciences. About 50 mg of powdered sample was dissolved in high-pressure Teflon bombs using a $HF + HNO_3$ mixture. An internal standard solution containing the single element Rh was used to monitor signal drift during counting. The USGS standards BCR-1, W-2 and G-2 and the Chinese National Standard GSR-3 were chosen for calibrating element concentrations of measured samples. In-run analytical precision for most elements was generally better than 2%. The detailed procedures for trace-element analysis by ICP–MS are described by Liu *et al.* (1996) and Li (1997).

The Sr and Nd isotopic compositions were determined using a Finnigan MAT-261 mass spectrometer operated in static multi-collector mode at the China University of Geosciences (Wuhan), and a VG-354 mass spectrometer operated in dynamic multi-collector mode at the Institute of Earth Sciences, Academia Sinica (Taipei). Measured ^{87}Sr/^{86}Sr and ^{143}Nd/^{144}Nd ratios were normalized to ^{86}Sr/^{88}Sr = 0.1194 and ^{146}Nd/^{144}Nd = 0.7219, respectively. The reported ^{87}Sr/^{86}Sr and ^{143}Nd/^{144}Nd ratios were further adjusted to the NBS SRM 987 standard ^{87}Sr/^{86}Sr = 0.71025 and the La Jolla standard ^{143}Nd/^{144}Nd = 0.511860, respectively.

Results

^{40}Ar/^{39}Ar ages

The ^{40}Ar/^{39}Ar age spectra for five basalts in southern Hunan and three hornblendes separated from syenites in eastern Guangxi are plotted in Fig. 3. Three basalt samples XPA-1, PA-03 and XTB-1 from Ningyuan yielded consistent plateau ages of *c.*175 Ma (174.3 ± 0.8 Ma, 176.2 ± 0.9 Ma and 175.4 ± 0.9 Ma, respectively), whereas two basalt samples DXB-1 and HTY-1 from Daoxian gave younger plateau ages of 149.5 ± 0.8 Ma and 154.4 ± 1.0 Ma, respectively. Except for sample HTY-1, all of the obtained plateau ages are consistent with their respective intercept ages obtained from regressions of data in an isotope correlation diagram (Table 1). Therefore, there seem to have been two main episodes of Jurassic basalt extrusion at *c.*175 Ma and *c.*150 Ma in southern Hunan.

Hornblende concentrations from the Yangmei (HK-28), Niumiao (HK-61) and Tong'an (HK-65) syenite plutons in eastern Guangxi yielded plateau ages of 161.6 ± 0.9 Ma, 161.0 ± 0.9 Ma and 163.2 ± 0.9 Ma, respectively. These consistent ^{40}Ar/^{39}Ar ages of *c.*161–163 Ma are indistinguishable within analytical error from the U–Pb zircon age of 158 ± 2 Ma for the Qinghu pluton (Xu & Yuan 1992) and the hornblende ^{40}Ar/^{39}Ar age of 163.6 ± 2.0 Ma for the Ma-Shan pluton in western Guangdong (Li *et al.* 2000). Thus, the syenites were likely to have been intruded contemporaneously at *c.*160 Ma in eastern Guangxi and western Guangdong. Taken together, the ^{40}Ar/^{39}Ar dating results document three main episodes of Jurassic magmatism of mantle origin at *c.*175 Ma, *c.*160 Ma and *c.*150 Ma.

Geochemistry

Nine basalt samples from southern Hunan and 18 syenite samples from eastern Guangxi were analysed for major and trace elements, and the data are listed in Table 2. Although all basalt samples have a relatively restricted range of SiO_2 (44–50%), the older, *c.*175 Ma Ningyuan basalts and the younger, *c.*150 Ma Daoxian basalts show contrasting chemical compositions (Fig. 4). The Ningyuan basalts have relative primitive to evolved compositions (Mg# (Mg/

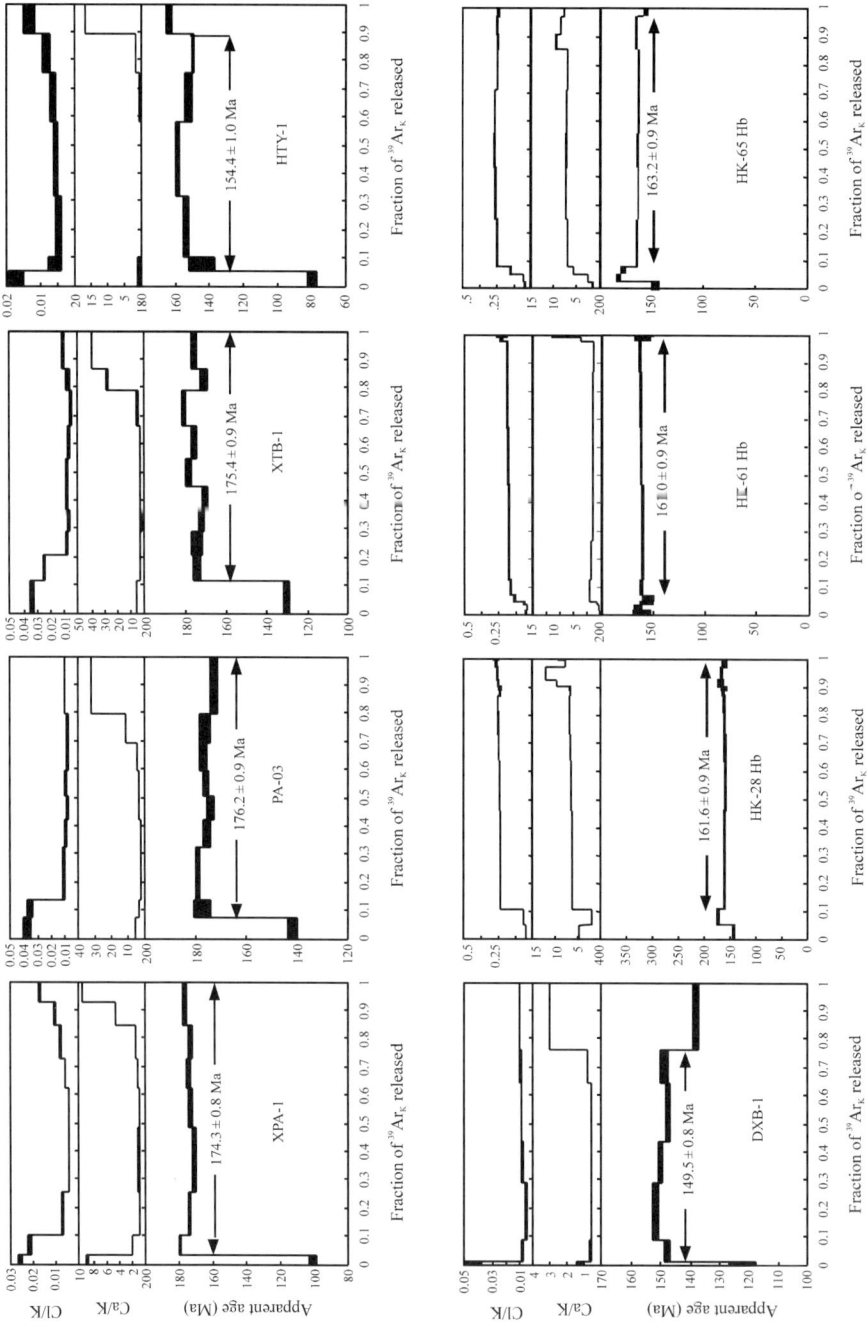

Fig. 3. The ^{40}Ar/^{39}Ar age spectra diagrams for basalts in southern Hunan and hornblendes separated from syen tes in eastern Guangxi. The plateau dates are indicated by the arrows in the spectrum plots. The length of the steps represents the relative volume of ^{39}Ar$_K$ released, and the height of the solid rectangles indicates the 2σ relative uncertainties.

Table 1. *Summary of $^{40}Ar/^{39}Ar$ results*

Sample	Integrated date (Ma)	Plateau plot[a]			$^{40}Ar/^{39}Ar$–$^{39}Ar/^{36}Ar$ isotope correlation plot[b]		
		Temp. step (°C)	%^{39}Ar	Date (Ma)	Date (Ma)	MSWD	$(^{40}Ar/^{36}Ar)_i$
XPA-1	171.8 ± 0.8	600–1200	89.1	174.3 ± 0.8	172.0 ± 1.0	2.8	371 ± 16
PA-03	173.8 ± 0.9	600–1200	72.9	176.2 ± 0.9	176.4 ± 1.5	2.1	334 ± 24
XTB-3	170.3 ± 0.9	600–1200	88.7	175.4 ± 0.9	178.2 ± 2.0	4.2	285 ± 31
HK28Hb	161.6 ± 0.9	900–1160	83.9	161.6 ± 0.9	162.2 ± 1.0	45	269 ± 08
HK61Hb	160.6 ± 1.0	930–1120	82.6	160.6 ± 0.9	161.9 ± 0.9	4.1	247 ± 16
HK65Hb	163.5 ± 0.8	920–1140	84.1	163.2 ± 0.9	162.7 ± 1.3	14	347 ± 53
DXB-1	147.3 ± 0.3	650–1100	63.0	150.9 ± 0.8	150.3 ± 3.7	2.1	314 ± 56
HTY-1	151.6 ± 1.0	620–1110	84.1	154.5 ± 1.0			

(a) Plateau date is calculated from the sum total gas from the continuous steps with radiometric dates that fall within 2σ errors with each other.
(b) Result of $^{40}Ar/^{39}Ar$-$^{39}Ar/^{36}Ar$ isotope correlation plot for the plateau steps indicates the trapped $(^{40}Ar/^{36}Ar)_i$ ratio and the radiogenic age for the samples. MSWD represents the mean sum of weighted deviations for the regression.

$(Mg + Fe^{2+}) = 0.68–0.52)$. They are high in TiO_2 (1.8–3.1%), Al_2O_3 (14–16%), total Fe_2O_3 (11–13%) and Na_2O (2–4%), plotting into the alkaline basalt field on a total alkalis v. silica plot (Fig. 5a). In contrast, the Daoxian basalts have uniquely high and restricted MgO (15–16%) and high Mg# = 0.82–0.86, variably high CaO (8–15%) and K_2O (1–3.4%), but are low in other major elements, with TiO_2 (0.55–0.67%), Al_2O_3 (9–11%), total Fe_2O_3 (7–8%) and Na_2O (0.5–1.7%). They straddle the boundary between the sub-alkaline and alkaline fields (Fig. 5a). On the Zr/P_2O_5 v. Nb/Y diagram of Winchester & Floyd (1976), the Ningyuan samples, having a high Nb/Y ratio of 2.2–2.6, plot exclusively in the alkaline field, whereas the Daoxian samples, with a low Nb/Y ratio of 0.38–0.42, fall in the sub-alkaline field (Fig. 5b).

The eastern Guangxi syenites show variable SiO_2 contents of 51–65% and high alkalis ($K_2O + Na_2O > 6.7\%$). They have low MgO (0.5–3.9%) and Mg# (0.25–0.53), indicating a significant crystal fractionation. On the TAS diagram, the analysed rocks mostly fall into the syeno-diorite and syenite fields (Fig. 5a). They are characterized by high Nb/Y ratios (1.1–4.2) of typical alkaline igneous rocks (Fig. 5b). All samples are high in K_2O at given SiO_2, with K_2O/Na_2O ratios of 0.8–1.7 – typical of shoshonitic affinity.

Figure 6 shows the chondrite-normalized REE patterns for the studied samples. The alkaline basalts have LREE-enriched patterns with $La_N = 167–204$, $(La/Yb)_N = 12.5–14.2$, and no Eu anomalies. The high-Mg basalts display similar LREE-enriched REE patterns with $La_N = 173–214$, but have relatively low HREE, high $(La/Yb)_N = 17.0–23.1$ and clear negative Eu anomalies ($Eu/Eu^* = 0.77–0.82$). The Guangxi syenites display REE patterns similar to the high-Mg basalts, having $La_N = 225–330$, $(La/Yb)_N = 14–26$ and varying degrees of negative Eu anomalies ($Eu/Eu^* = 0.48–0.93$).

On the primitive mantle-normalized incompatible trace-element spidergrams (Fig. 7), the alkaline basalts have 'humped' patterns characterized by variable enrichment in all the trace elements with respect to the primitive mantle. They display significant enrichments in Nb–Ta relative to La (i.e. Nb/La = 1.6–1.7) similar to many alkali basalts formed in continental rifts or oceanic islands without appreciable crustal contamination. In contrast, the high-Mg basalts display 'spiky' patterns with significant depletion in Nb–Ta (Nb–La = 0.18–0.25) and Zr–Hf and Ti relative to the neighbouring elements, characteristic of arc-related magma, although they possess an abundance of REE, Th, Sr and P similar to the alkaline basalts. The Guangxi syenites are characterized by variable enrichment in all trace elements apart from Sr, P and Ti, showing different degrees of negative anomaly due to crystal fractionation. It is noted that, unlike most potassic alkaline rocks occurring in arc-related settings (e.g. Müller & Groves 1995), the Guangxi syenites have no evident Nb–Ta depletion relative to La (Nb/La = 0.56–1.45). A few samples have high Nb/La ratios (1.2–1.45), resembling many alkali basalts formed in oceanic islands and continental rift areas (Sun & McDonough 1989).

Sr–Nd isotopes

Five basalt and eleven syenite samples were analysed for Sr and Nd isotopes, and the results along with calculated initial $^{87}Sr/^{86}Sr$ ratios (I_{Sr}) and $\varepsilon_{Nd}(T)$ values are listed in Table 3. In general, the basaltic rocks in southern Hunan have a restricted range of Sr and Nd isotopic compositions, with the alkaline basalts having low $I_{Sr} = 0.7035–0.7040$ and high $\varepsilon_{Nd}(T) = 4.6–5.1$, whilst the high-Mg basalts display slightly higher $I_{Sr} = 0.7050$ and lower $\varepsilon_{Nd}(T) = -1.6$ to -1.9. In contrast, the Guangxi

syenites have a relatively large range of $I_{Sr} = 0.7032–0.7082$ and a $\varepsilon_{Nd}(T)$ value of 5.5 to -4.1. Among the syenites, two Qinghu samples (HK-16 and HK-21) have the lowest Sr and highest Nd isotopic compositions ($I_{Sr} = 0.7032–0.7042$ and $\varepsilon_{Nd}(T) = 4.2–5.5$), very similar to those of the Ningyuan alkaline basalts, whereas the Maqigang sample HK-12 possesses the most evolved compositions ($I_{Sr} = 0.7082$ and $\varepsilon_{Nd}(T) = -4.1$), possibly caused by significant involvement of continental crustal components. Taken together, all samples are negatively correlated on an I_{Sr} v. $\varepsilon_{Nd}(T)$ plot (Fig. 8).

Table 2. *Major- and trace-element data for the syenites and basalts*

Sample locality	HK-3 Mashan	HK-10 Mashan	HK-12 Maqigang	HK-16 Qinghu	HK-21 Qinghu	HK-22 Qinghu
Major (%)						
SiO_2	65.16	51.27	63.60	63.27	58.68	60.08
TiO_2	0.44	1.34	0.97	0.68	1.02	0.92
Al_2O_3	17.07	17.72	15.10	16.98	17.57	17.97
Fe_2O_3	3.47	5.83	6.09	3.33	5.76	5.25
MnO	0.09	0.18	0.10	0.06	0.12	0.11
MgO	0.50	1.98	1.26	0.60	1.35	1.21
CaO	1.75	4.91	3.23	2.43	4.20	3.72
Na_2O	5.17	5.85	2.90	4.82	5.17	5.05
K_2O	5.34	2.93	4.60	4.80	4.33	4.74
P_2O_5	0.14	0.39	0.30	0.18	0.30	0.26
Trace (ppm)						
V	15.9	92.1	58.4	43.4	68.2	53.0
Cr	3.22	67.8	15.2	13.6	10.0	8.85
Co	0.54	8.81	11.1	40.5	6.45	5.43
Ni	1.97	19.0	11.4	7.22	7.30	6.28
Ga	21.8	24.2	18.8	23.8	24.7	24.8
Rb	171	52.9	167	232	183	201
Sr	414	693	330	458	652	645
Y	27.2	23.6	37.3	19.8	29.2	29.5
Zr	258	423	447	385	422	367
Nb	61.5	99.2	35.4	76.0	77.4	67.8
Ba	906	712	1118	670	645	784
La	56.4	77.9	63.9	60.5	53.3	60.6
Ce	97.0	123	121	123	100	107
Pr	11.7	12.7	13.6	11.2	11.3	11.6
Nd	43.1	44.4	49.1	36.9	39.9	39.8
Sm	7.36	6.75	9.11	5.87	7.07	6.85
Eu	2.12	2.07	2.10	1.33	2.01	1.96
Gd	6.86	6.68	8.24	5.41	6.41	6.24
Tb	0.90	0.78	1.17	0.71	0.89	0.84
Dy	4.95	4.19	6.83	4.05	4.89	4.68
Ho	0.91	0.80	1.28	0.77	0.93	0.89
Er	2.55	2.20	3.70	2.11	2.53	2.47
Tm	0.39	0.37	0.53	0.36	0.40	0.40
Yb	2.27	2.16	3.39	2.16	2.40	2.47
Lu	0.35	0.33	0.52	0.32	0.34	0.38
Hf	6.27	11.3	10.6	9.63	10.9	9.47
Ta	4.13	5.58	2.11	5.68	4.92	5.40
Th	13.3	9.34	18.2	33.4	24.7	37.1
U	5.91	3.64	3.77	25.1	9.01	11.4

(continued)

Discussion

Petrogenesis

The $c.175$ Ma Ningyuan alkaline basalts exhibit OIB-type incompatible trace-element characteristics, and are relatively uniform in Sr and Nd isotopes over variable (but low) SiO_2 (Fig. 9), indicating that their geochemical and isotopic characters are reflective of the mantle source compositions. It is noted that the Ningyuan alkaline basalts have Sr and Nd isotopic compositions similar to the extension/rift-related Cenozoic intraplate basalts in SE China (Tu *et al.* 1991; Chung *et al.* 1994, 1995, 1997) and the post-spreading seamount basalts in the South China Sea (SCS) (Tu *et al.* 1992). These Cenozoic basalts have been widely advocated as having originated from an upper mantle source without appreciable crustal contamination during magma ascent. By analogy, we interpret that the alkaline basalts in southern Hunan, although >150 Ma older than the Cenozoic basalts, are derived from similar petrogenetic processes, in view of their trace-element and Sr–Nd isotope similarities. Two kinds of upper mantle sources,

Table 2. *Continued*

Sample locality	HK-26 Yangmei	HK-28 Yangmei	HK-34 Nandu	HK-37 Nandu	HK-40 Nandu	HK-43 Nandu
Major (%)						
SiO_2	53.79	63.83	58.02	52.10	55.33	56.04
TiO_2	2.11	0.81	1.65	1.36	1.72	1.69
Al_2O_3	16.47	15.89	15.83	14.45	16.27	16.16
Fe_2O_3	9.30	5.03	7.64	6.64	8.08	7.92
MnO	0.13	0.11	0.11	0.10	0.13	0.17
MgO	2.87	1.06	2.29	2.28	2.55	2.25
CaO	5.47	2.28	4.48	8.15	5.65	4.08
Na_2O	4.14	3.58	3.21	4.26	3.45	4.06
K_2O	3.42	5.54	4.78	4.36	4.31	4.84
P_2O_5	0.59	0.32	0.49	0.44	0.53	0.53
Trace (ppm)						
V	163	39.9	119	110	136	131
Cr	1.64	9.02	12.6	8.74	7.75	14.2
Co	14.7	2.26	16.4	7.65	11.9	10.6
Ni	3.31	0.91	12.0	11.3	7.92	6.47
Ga	22.3	21.6	21.5	19.2	21.3	22.4
Rb	133	225	175	139	142	171
Sr	769	450	609	359	638	692
Y	26.9	32.8	32.2	25.9	30.9	34.5
Zr	310	334	309	251	284	252
Nb	62.3	57.6	81.7	59.4	71.1	78.6
Ba	1007	708	888	828	795	897
La	59.8	56.5	77.0	57.2	65.4	65.7
Ce	111	89.0	141	106	122	124
Pr	12.8	10.5	16.2	12.2	14.5	14.8
Nd	47.1	41.8	58.0	45.3	53.3	54.8
Sm	8.28	8.13	10.3	7.73	9.32	9.79
Eu	2.43	1.56	2.24	1.93	2.24	2.18
Gd	7.49	6.57	8.79	7.06	8.76	9.00
Tb	0.92	1.02	1.19	0.89	1.10	1.16
Dy	4.80	5.61	6.37	4.62	5.58	6.08
Ho	0.87	1.05	1.12	0.86	1.05	1.11
Er	2.28	2.86	3.14	2.27	2.85	2.97
Tm	0.35	0.46	0.44	0.34	0.41	0.45
Yb	1.97	2.67	2.66	1.92	2.40	2.55
Lu	0.28	0.39	0.40	0.28	0.35	0.37
Hf	7.28	8.22	7.45	5.73	6.69	5.90
Ta	3.42	4.74	4.72	3.38	3.94	4.10
Th	16.8	8.12	17.7	8.87	9.00	11.0
U	2.91	4.07	3.69	2.25	2.23	3.70

(*continued*)

i.e. the asthenosphere and the continental litho-sphere mantle (CLM), could have been involved in the magma generation. It is noted that the SCS seamount basalts exhibit Sr and Nd isotopic com-positions that are more depleted than, despite somewhat overlapping with, the Cenozoic intra-plate basalts from South China (Fig. 8). The for-mer were erupted mostly within the oceanic lithosphere, and may have originated exclusively from the convecting asthenosphere (Tu *et al.* 1992). The latter are thought to be the products of lithosphere–asthenosphere interaction (Chung *et al.* 1994), or generated from recently accreted

CLM overprinted by more radiogenic melts involving subducted sediment (Tu *et al.* 1991). Considering the radiogenic isotopic growth for the mantle source from the Jurassic to Cenozoic, the Ningyuan alkaline basalts should have been derived from a mantle source with Nd–Sr isotope compositions less enriched than the source of the Cenozoic intraplate basalts in SE China, but very similar to those of the SCS seamount basalts. It may therefore be reasonable to conclude that the *c.*175 Ma alkaline basalts in southern Hunan were generated by small degrees of decompression melting of the asthenosphere

Table 2. *Continued*

Sample locality	HK-59 Niumiao	HK-61 Niumiao	HK-62 Niumiao	HK-65 Tong'an	HK-67 Tong'an
Major (%)					
SiO_2	58.78	61.22	51.79	62.43	64.40
TiO_2	1.55	1.28	2.10	1.12	1.14
Al_2O_3	14.50	15.28	16.16	15.78	14.84
Fe_2O_3	8.05	6.89	10.04	6.45	6.35
MnO	0.11	0.10	0.16	0.09	0.09
MgO	3.86	2.50	3.63	1.40	1.45
CaO	5.67	4.67	6.38	3.38	3.06
Na_2O	2.76	3.01	3.15	3.47	3.09
K_2O	3.95	3.94	4.64	4.76	4.69
P_2O_5	0.34	0.31	0.58	0.27	0.26
Trace (ppm)					
V	143	117	181	81.6	76.9
Cr	96.3	36.0	15.72	7.25	8.06
Co	17.0	12.2	32.2	7.52	7.55
Ni	42.1	17.3	20.5	3.23	4.04
Ga	20.2	20.8	22.0	22.3	21.6
Rb	213	188	243	265	268
Sr	468	465	452	384	327
Y	31.4	27.5	40.7	35.6	40.7
Zr	275	249	252	338	356
Nb	42.8	42.1	70.9	51.9	50.9
Ba	721	653	477	918	571
La	59.7	59.9	63.2	61.4	71.0
Ce	112	122	115	120	133
Pr	12.6	12.7	13.9	14.0	15.6
Nd	45.9	45.3	52.0	50.1	55.5
Sm	8.21	7.86	10.3	8.85	10.0
Eu	1.87	1.63	1.72	1.85	1.56
Gd	7.45	7.21	9.07	8.37	9.41
Tb	0.99	0.94	1.33	1.09	1.24
Dy	5.41	5.06	7.18	5.93	6.79
Ho	1.02	0.98	1.34	1.13	1.26
Er	2.71	2.58	3.68	3.07	3.56
Tm	0.42	0.41	0.55	0.47	0.51
Yb	2.38	2.37	3.19	2.80	2.97
Lu	0.36	0.35	0.45	0.40	0.43
Hf	7.01	6.24	6.01	8.47	8.95
Ta	3.07	3.10	4.76	3.37	3.21
Th	28.4	22.6	12.2	30.5	33.7
U	9.28	9.16	5.74	10.2	7.80

(continued)

and/or the young, enriched CLM accreted by the asthenosphere-derived melts.

The $c.160$ Ma syenite intrusions exhibit variable trace-element and Sr–Nd isotope compositions. Most samples reveal a general increase in I_{Sr} and decrease in $\varepsilon_{Nd}(T)$ with increasing SiO_2 (Fig. 9), indicating an AFC process for the genesis of syenites, i.e. fractional crystallization of pyroxene, plagioclase, apatite and Ti–Fe oxides of a parent alkaline basaltic magma, accompanied by various degrees of crustal contamination. Among these syenite samples, the Maqigang sample (HK-12) shows the most

evolved, crust-like isotopic signature, suggesting significant involvement of crustal components. Other syenite samples have geochemical and isotopic compositions between those of the Ningyuan alkaline basalt, Maqigang syenite and the $c.150$ Ma high-Mg basalt, as illustrated on I_{Sr} v. $\varepsilon_{Nd}(T)$, SiO_2 and Nb/La plots (Figs 8–10). The Qinghu syenites are constantly low in I_{Sr} and high in $\varepsilon_{Nd}(T)$, coupled with an OIB-type trace-element distribution pattern and high Nb/La ratios of 1.26–1.45 (Figs 7 and 10). Their overall geochemical and isotopic features are very similar to those of the $c.175$ Ma alkaline

Table 2. *Continued*

Sample locality	HK-68 Tong'an	XPA-1 Ningyua	PA-01 Ningyua	PA-02* Ningyua	PA-03* Ningyua
Major (%)					
SiO_2	63.37	49.64	44.81	45.07	44.97
TiO_2	1.00	1.79	2.68	2.69	2.68
Al_2O_3	15.67	16.15	14.77	14.53	14.61
Fe_2O_3	5.95	10.82	12.65	12.63	12.65
MnO	0.09	0.19	0.19	0.20	0.20
MgO	1.27	6.62	8.23	8.17	8.15
CaO	2.80	6.63	10.18	10.21	10.19
Na_2O	3.24	3.81	3.19	3.19	3.25
K_2O	5.58	2.16	0.66	0.76	0.80
P_2O_5	0.24	0.76	0.66	0.67	0.66
Trace (ppm)					
V	65.2	140	274		
Cr	9.57	204	216		
Co	6.55	32.9	50.2	48.2	48.8
Ni	3.46	149	149		
Ga	21.4	24.9	22.8		
Rb	295	47.1	11		
Sr	380	743	971		
Y	34.8	30.5	30.2		
Zr	328	354	267		
Nb	46.5	78.9	67.3		
Ba	1208	537	483		
La	61.4	48.4	39.6	40.1	39.5
Ce	116	95.1	78.2	76.2	78.8
Pr	13.3	11.3	9.5		
Nd	47.7	43.2	37.6	35.1	33.8
Sm	8.37	8.23	7.32	7.76	7.61
Eu	1.75	2.62	2.49	2.42	2.43
Gd	7.65	7.27	7.19		
Tb	1.02	1.04	1.06	1.01	1.10
Dy	5.57	5.61	5.63		
Ho	1.05	1.02	1.03		
Er	2.84	2.79	2.73		
Tm	0.45	0.40	0.38		
Yb	2.56	2.45	2.27	2.09	2.02
Lu	0.38	0.36	0.33	0.32	0.30
Hf	7.87	7.83	5.90	5.73	6.10
Ta	2.96	4.78	4.64	3.95	4.05
Th	38.5	7.99	6.20	5.55	5.48
U	8.79	2.45	2.01	1.53	1.69

(*continued*)

basalts, implying derivation of their parent magma from a similar mantle source with insignificant crustal contamination. It has been generally considered that potassic alkaline magmas form either by small-degree melting of a subduction-related metasomatized mantle source enriched in LIL and LREE (e.g. Wyllie & Sekine 1982), or by interactions of an asthenosphere-derived melt with the overlying lithospheric mantle (e.g. Menzies 1987). Müller & Groves (1995) have classified potassic alkaline igneous rocks into two main categories:

(1) the majority of potassic alkaline rocks formed in various subduction- or arc-related settings ('arc-type'), such as continental arc, oceanic arc and post-collisional arc environments, and

(2) others occurring within continental and oceanic settings ('intraplate-type'), including those affiliated with rifts and hotspots.

The former and latter types generally correspond with the high-K/Ti–low-Ti-type and low-K/Ti–high-Ti-type potassic magmas (Rogers 1992), respectively. The salient feature observed in

Table 2. *Continued*

Sample locality	XTB-2 Ningyua	XTB-3* Ningyua	DXB-1 Daoxian	HTY-1 Daoxian	HTY-2* Daoxian
Major (%)					
SiO_2	44.60	44.80	48.99	46.39	46.49
TiO_2	2.67	2.68	0.67	0.56	0.55
Al_2O_3	14.76	14.86	11.27	9.03	8.89
Fe_2O_3	12.53	12.52	8.11	6.93	6.96
MnO	0.19	0.20	0.16	0.16	0.16
MgO	8.18	8.04	16.24	16.16	16.52
CaO	10.28	10.11	8.23	13.77	14.00
Na_2O	3.25	3.17	1.52	0.80	1.73
K_2O	0.69	0.74	3.35	2.19	1.23
P_2O_5	0.66	0.66	0.47	0.61	0.60
Trace (ppm)					
V	273		208	182	
Cr	218		1200	970	
Co	49.7	45.7	57.7	53.0	52.2
Ni	149		503	619	
Ga	22.5		13.2	9.2	
Rb	7.1		89	95	
Sr	985		772	1282	
Y	29.8		24.2	24.5	
Zr	258		104	104	
Nb	67.0		10.1	9.3	
Ba	517		1326	2629	
La	39.5	39.7	41.0	50.8	50.0
Ce	77.7	76.0	79.4	104	94.3
Pr	9.5		10.0	12.6	
Nd	37.7	31.9	34.5	48.2	39.7
Sm	7.70	7.23	7.96	9.90	9.70
Eu	2.53	2.37	2.07	2.43	2.40
Gd	7.26		7.19	9.04	
Tb	1.05	1.05	0.95	1.12	1.40
Dy	5.58		4.60	5.09	
Ho	1.02		0.79	0.83	
Er	2.66		2.07	2.05	
Tm	0.37		0.29	0.27	
Yb	2.21	2.04	1.73	1.58	1.48
Lu	0.32	0.31	0.25	0.23	0.23
Hf	5.63	5.88	2.52	2.42	2.33
Ta	3.95	3.72	0.55	0.53	0.34
Th	6.15	5.47	10.1	10.2	9.40
U	1.99	1.46	3.76	3.97	4.04

*Trace elements were analysed using INAA at the National Taiwan University. Total iron as Fe_2O_3.

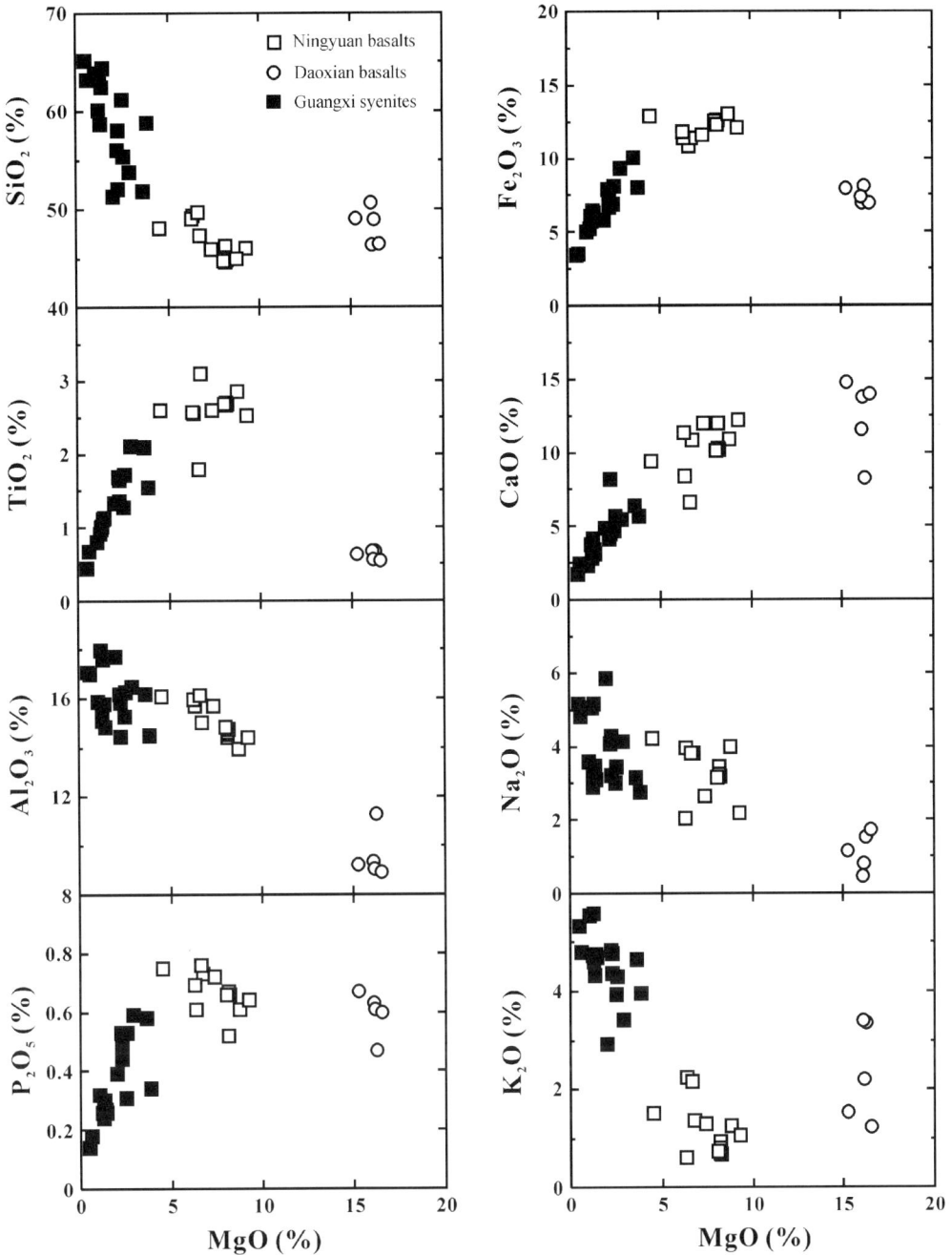

Fig. 4. Chemical variation diagrams for the basalts in southern Hunan and syenites in eastern Guangxi, including chemical data of basalts by Zhu *et al.* 1997).

most Guangxi syenites is their OIB-type trace-element compositions with a high Nb/La ratio (>1). All syenite samples are highly fractionated, with TiO_2 decreasing and K_2O increasing with decreasing MgO (Fig. 4). The less-fractionated samples (HK-26 and HK-62) are high in TiO_2 (c.2%) and low in K/Ti (c.0.7), which are comparable with the geochemical features of

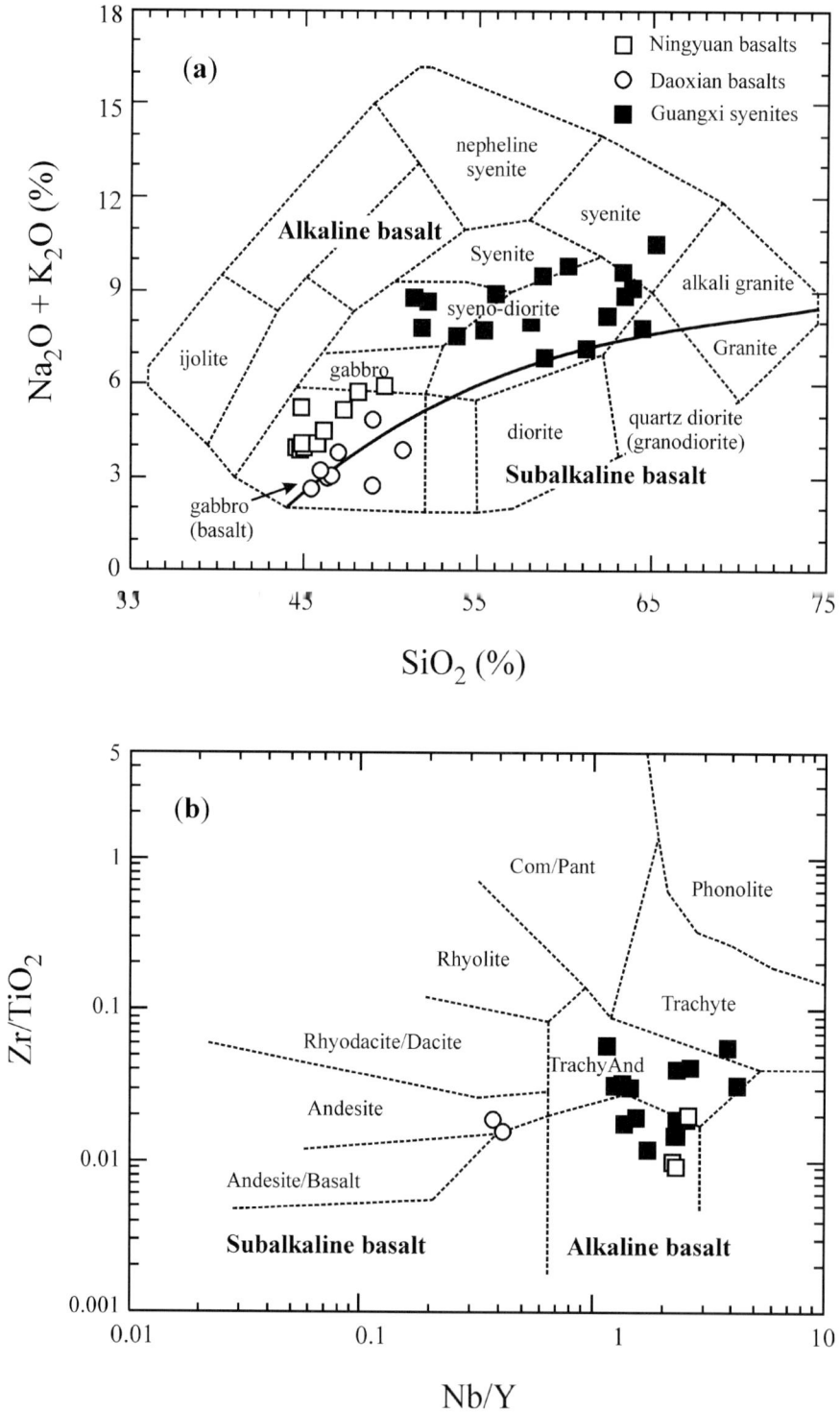

Fig. 5. (a) SiO$_2$ v. Na$_2$O + K$_2$O diagram of Cox *et al.* (1979); (b) Nb/Y v. Zr/P$_2$O$_5$ diagram of Winchester & Floyd (1976). The thick trend line subdivides the alkaline from subalkaline rocks. Data sources as in Figure 4.

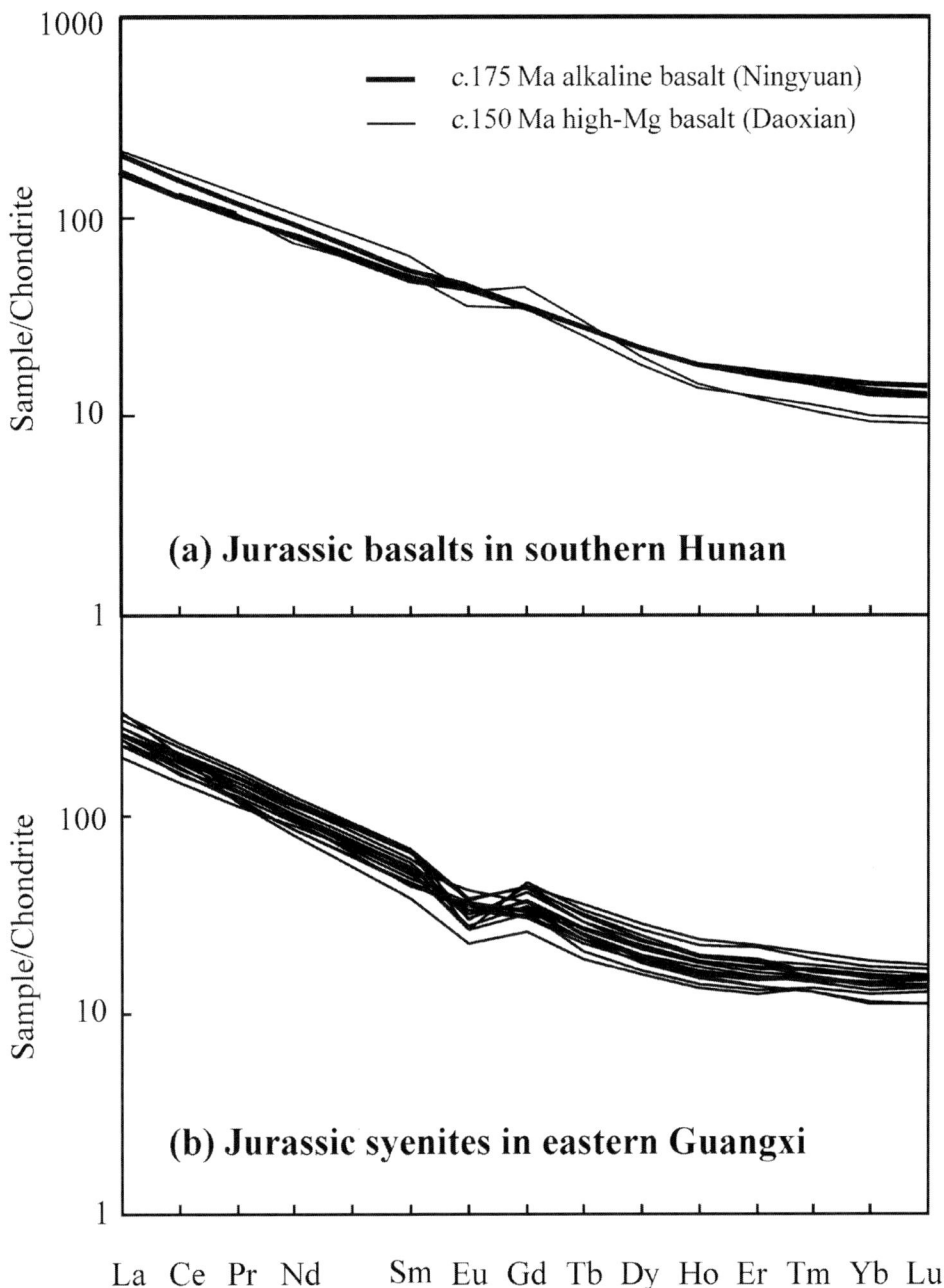

Fig. 6. Chondrite-normalized REE diagrams for (**a**) basalts in southern Hunan, and (**b**) syenites in eastern Guangxi. The normalization values are from Sun & McDonough (1989).

the intraplate, low-K/Ti–high-Ti-type potassic rocks (Nb/La ≥ 1, TiO_2 > 2% and K/Ti < 2), such as the potassic rocks from Navajo (Roden 1981), and the western branch of East African Rift (Thompson *et al.* 1984; Rogers *et al.* 1998), but significantly different from the arc-related, high-K/Ti–low-Ti-type potassic rocks (Nb/La < 1, TiO_2 < 1.5% and K/Ti > 3). Thus, the parent magma of Guangxi syenites is most likely to have been derived from an

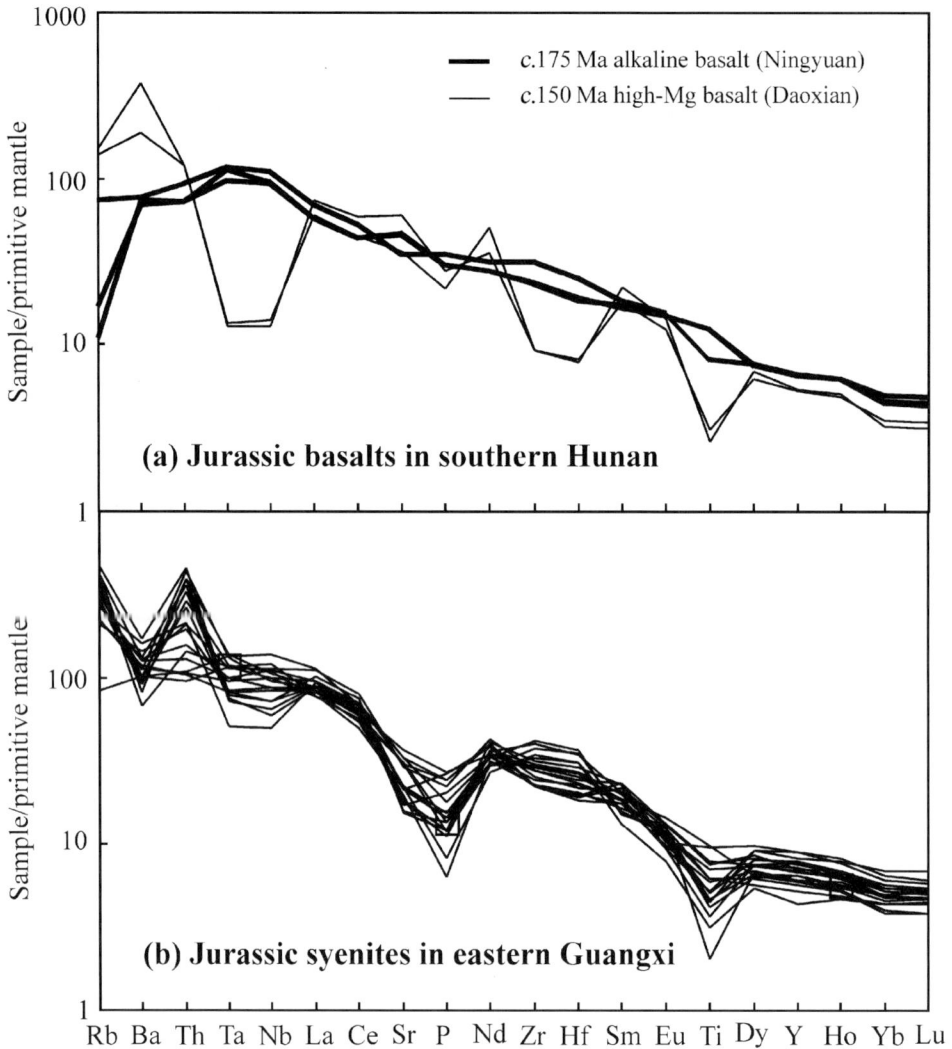

Fig. 7. Primitive-mantle-normalized incompatible element spidergrams for (**a**) basalts in southern Hunan, and (**b**) syenites in eastern Guangxi. The normalization values are from Sun & McDonough (1989).

enriched lithospheric mantle that had been meta-somatized, very shortly before the generation of potassic magmas, by an asthenosphere-derived, OIB-type melt. The Qinghu and Mashan sye-nites, showing the 'least-evolved' isotope ratios, might have been derived dominantly from a newly asthenosphere-derived, OIB-type enriched component, whilst the others were generated by mixing of variable proportions of newly astheno-sphere-derived and previously subduction-related metasomatized lithosphere components, plus variable degrees of crustal contamination (Figs 9 and 10).

The high Mg# values of 0.80–0.86 and high Ni (500–620 ppm) and Cr (970–1200 ppm) for the *c.*150 Ma Daoxian basalts indicate that they may represent primary melts of the mantle peridotite. Their TiO_2 (0.55–0.67%) and total Fe_2O_3 (7–8%) contents fall within the field defined by experimental melts of depleted peridotite (Falloon *et al.* 1988). Thus, the Daoxian high-Mg basalts are most likely derived from a refractory mantle source that could be related to the important extraction of basaltic melts during the Neoproter-ozoic mantle plume or superplume activities (Li *et al.* 1999, 2003). These high-Mg basalts have

Table 3. Rb–Sr and Sm–Nd isotopic data for the syenitic intrusions in eastern Guangxi and the basalts in southern Hunan

	Rb (ppm)	Sr (ppm)	^{87}Rb/^{86}Sr	^{87}Sr/^{86}Sr $\pm 2\sigma_{m}$	Sm (ppm)	Nd (ppm)	^{147}Sm/^{144}Nd	^{143}Nd/^{144}Nd $\pm 2\sigma_{m}$	I_{Sr}	εNd(T)
Syenitic intrusions (T c.160 Ma)										
HK-3	171	436	1.132	0.706456 ± 20	7.36	43.1	0.1033	0.512636 ± 06	0.70388	1.84
HK-12	167	299	1.618	0.711876 ± 21	9.11	49.1	0.1121	0.512342 ± 07	0.70820	−4.07
HK-16	251	467	1.550	0.707680 ± 12	5.87	36.9	0.0962	0.512747 ± 05	0.70415	4.15
HK-21	185	678	0.788	0.704979 ± 17	7.07	39.9	0.1070	0.512825 ± 06	0.70319	5.45
HK-26	148	791	0.539	0.707234 ± 16	8.28	47.1	0.1062	0.512533 ± 07	0.70601	−0.23
HK-28	235	455	1.491	0.709338 ± 16	8.13	41.8	0.1175	0.512483 ± 05	0.70595	−1.43
HK-34	175	609	0.826	0.706669 ± 18	10.3	58.0	0.1069	0.512547 ± 06	0.70479	0.03
HK-40	150	511	0.846	0.706736 ± 14	9.32	53.3	0.1057	0.512487 ± 10	0.70481	−1.11
HK-59	229	432	1.530	0.708218 ± 21	8.21	45.9	0.1081	0.512477 ± 06	0.70474	−1.36
HK-62	260	427	1.760	0.709109 ± 19	10.3	52.1	0.1199	0.512529 ± 11	0.70511	−0.58
HK-65	274	317	2.488	0.711488 ± 19	8.85	50.1	0.1068	0.512523 ± 09	0.70583	−0.43
Alkaline basalts (T c.175 Ma)										
XPA-1*	47.1	743	0.183	0.703977 ± 14	8.23	43.2	0.1703	0.512846 ± 13	0.70352	4.64
PA-01*	11.0	971	0.0327	0.703986 ± 16	7.32	37.6	0.1703	0.512859 ± 21	0.70353	4.90
XTB-2*	7.13	985	0.0210	0.704076 ± 17	7.70	37.7	0.1703	0.512867 ± 14	0.70402	5.05
High-Mg basalts (T c.150 Ma)										
DXB-1*	88.6	772	0.332	0.706115 ± 16	7.96	34.5	0.1703	0.512514 ± 16	0.70542	−1.92
HTY-1*	95.0	1282	0.215	0.705865 ± 14	9.90	48.2	0.1703	0.512530 ± 17	0.70541	−1.60

Samples with asterisks were analysed at the Institute of Earth Sciences, Academia Sinica (Taipei), and the others at the China University of Geosciences (Wuhan).

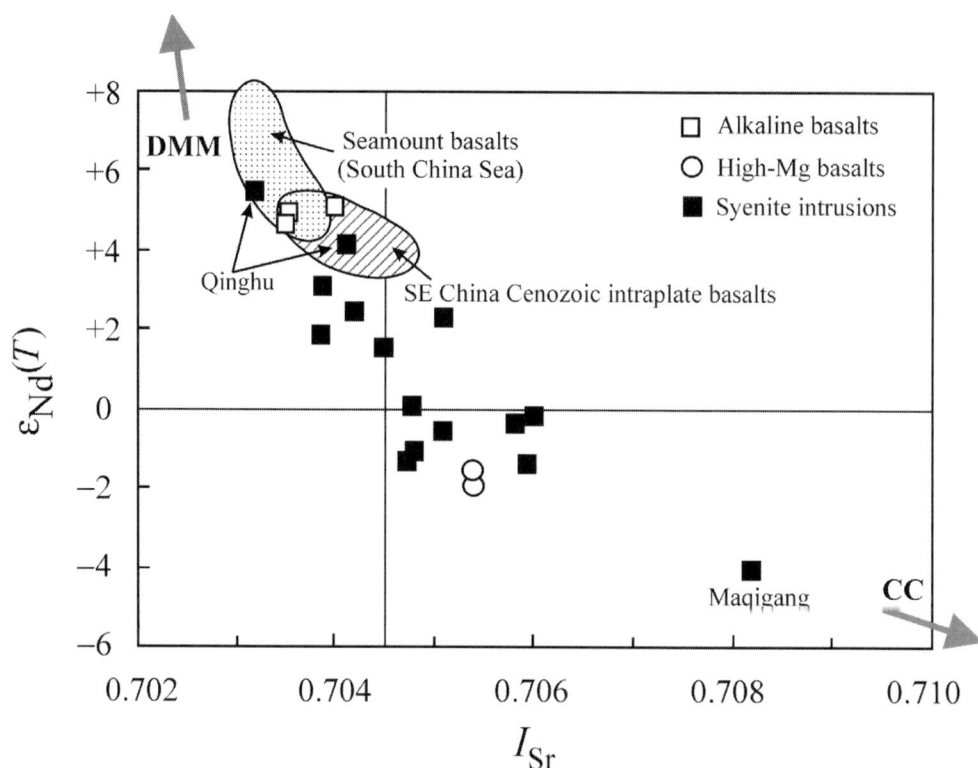

Fig. 8. Initial $\varepsilon_{Nd}(T)$ v. I_{Sr} diagram for basalts in southern Hunan and syenites in eastern Guangxi, including isotopic data for the Manshan and Luorong plutons from Guo *et al.* (2001). Also illustrated for comparison are the fields of the post-spreading seamount basalts in the South China Sea (Tu *et al.* 1992) and the Cenozoic intraplate basalts in SE China (Tu *et al.* 1992; Chung *et al.* 1994, 1995, 1997). DMM, depleted MORB-type mantle; CC, continental crust. Symbols as in Figure 4.

restricted Sr–Nd isotope ratios, and a pronounced, Nb–Ta negative anomaly of typical arc signature. Hence, their refractory mantle source domain, most likely residing in the continental lithospheric mantle, must have been metasomatized by subduction-related enrichment before being melted to produce the high-Mg basalt. This subduction-related metasomatism event has not been well documented, but possibly occurred during the Early Palaeozoic (Caledonian-age) orogeny (Li 1998). It is noted that the high-Mg basalts have a wide range of CaO and K_2O. The K_2O appears to correlate positively with Al_2O_3, TiO_2, Fe_2O_3 and LREE, but negatively with CaO, Sr and Ba, possibly implying a peridotite-plus-vein mantle source (Foley 1992). Alternative interpretations involving significant crustal contamination to the asthenosphere-derived magmas are not favoured because of their very high MgO contents (15–16%) and low Nb/La ratios (0.18–0.25), which are much lower than the con-

tinental crust values and average between 0.5 and 0.8 (Rudnick & Fountain 1995). A detailed geochemical and isotopic study is needed to provide further constraints on the mantle source of these high-Mg basalts.

Tectonic implications

As mentioned above, the tectonic regime that accounted for the Mesozoic (particularly the Early Mesozoic) magmatism in SE China has been an issue with little consensus. In the model of an active continental margin, a dominantly compressive regime related to the palaeo-Pacific subduction has been thought to have prevailed until the Late Cretaceous, when bimodal volcanic rocks and associated alkaline granites were emplaced in the coastal areas (e.g. Charvet *et al.* 1994; Martin *et al.* 1994). The early Yanshanian (Jurassic) magmatism in the interior of SE China was also considered to

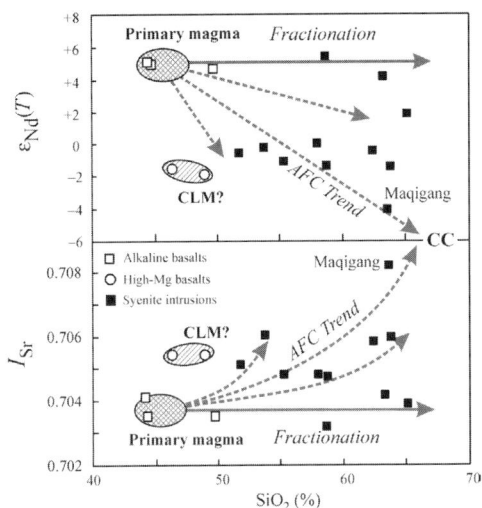

Fig. 9. Plots of (**a**) initial $\varepsilon_{Nd}(T)$ and (**b**) I_{Sr} v. SiO_2 for basalts in southern Hunan and syenites in eastern Guangxi. The Ningyuan alkaline basalts are relatively uniform in Sr and Nd isotopes over variable and low SiO_2, whilst the syenite samples reveal a wildly varying range of I_{Sr} and $\varepsilon_{Nd}(T)$. See text for discussion.

relate to a very low-angle subduction (Zhou & Li 2000). It is noted that such a subduction-related interpretation, which rests mainly on the prevailing arc-signatures observed in the Cretaceous calc-alkaline igneous rocks, is inconsistent with recent observations. The association of A-type granitic and intraplate basaltic magmatism of 140–90 Ma suggests that extension has been dominant in SE China during that time period (Li 2000). Note that the early two episodes during 140–120 Ma of the Cretaceous intraplate magmatism correspond with the timing of the final doming of the core complexes in Wugongshan, Lushan and Jiulingshan (Faure *et al.* 1996; Lin *et al.* 2000, 2001). Furthermore, Chen *et al.* (1999) reported a *c.*175–160 Ma bimodal magmatic suite in southern Jiangxi, *c.*300 km east to our study area. This implies that the lithosphere extension may have been active since the Middle Jurassic. Pulses of rapid extension during the Late Jurassic are also reported from the Hong Kong region (Fletcher *et al.* 1997). The Upper Triassic–Cretaceous extensional basins, such as the Shiwandashan basin in SE Guangxi (Gilder *et al.* 1996), mostly developed on the Palaeozoic to Middle Triassic basement that was ubiquitously folded by the Indosinian Orogeny (Li 1998). Thus, the rifting and extension are likely to have commenced in the Late Triassic–Early Jurassic. However, prior to our work here, this continental rifting

and extension in the Late Triassic to Jurassic had gained little support from magmatic evidence.

Our new data of $^{40}Ar/^{39}Ar$ ages, geochemistry and Sr–Nd isotopes for the intraplate basaltic and syenitic rocks provide important petrogenetic and tectonic evolution constraints on the Jurassic of SE China. First, the data make it possible to propose an intraplate rifting and extension environment for the genesis of these alkaline basalts and 'intraplate-type' syenites. Second, a secular variation in rock types and geochemical and isotopic characteristics reveals changes in melt segregation depth and mantle sources in response to the lithosphere extension and thinning. The *c.*175 Ma alkaline basalts are suggested to have formed by small degrees of decompression melting of the asthenosphere, or enriched CLM by asthenosphere-derived OIB-type melts during the early stage of extension. The *c.*160 Ma syenites and *c.*150 Ma high-Mg basalts were mainly derived from the asthenosphere-derived OIB-type and subduction-related metasomatised CLM, respectively, that melted owing to a raised geotherm caused by subsequent lithosphere extension and thinning.

Continental rifts may be classified into two types, namely, the low-volcanicity (LV) and high-volcanicity (HV) types (Barberi *et al.* 1982), on the basis of the relative volume of volcanic products. The Jurassic alkaline magmatism in southern Hunan suggests that at this time the tectonic regime of SE China interior was analogous to the LV-type rifts, which are characterized by relatively small volumes of eruptive products, low rates of crustal extension, discontinuous volcanic activity along the rift, highly alkaline silica-undersaturated mafic and ultramafic magmas, and small volumes of felsic differentiates, as represented by the western branch of the East African Rift, the Rhine Graben and the Baikal Rift (Wilson 1989). In such a LV-type rift, deep lithosphere fractures could permit upward migration of small-degree melts from the asthenosphere to form minor alkaline basaltic eruptions and volatile/melt fluxing from the asthenosphere, that would cause metasomatism of the overlying lithosphere mantle. The metasomatized lithosphere mantle may, in turn, undergo partial melting to generate more substantial alkaline magmatism. The intrusion (underplating) could cause subsequent interaction between crustal melts and LIL- and LREE-enriched alkaline magmas, which is borne out by those granites with anomalously high REE and low T_{DM} model ages reflecting juvenile mantle inputs (Gilder *et al.* 1996; Chen & Jahn 1998).

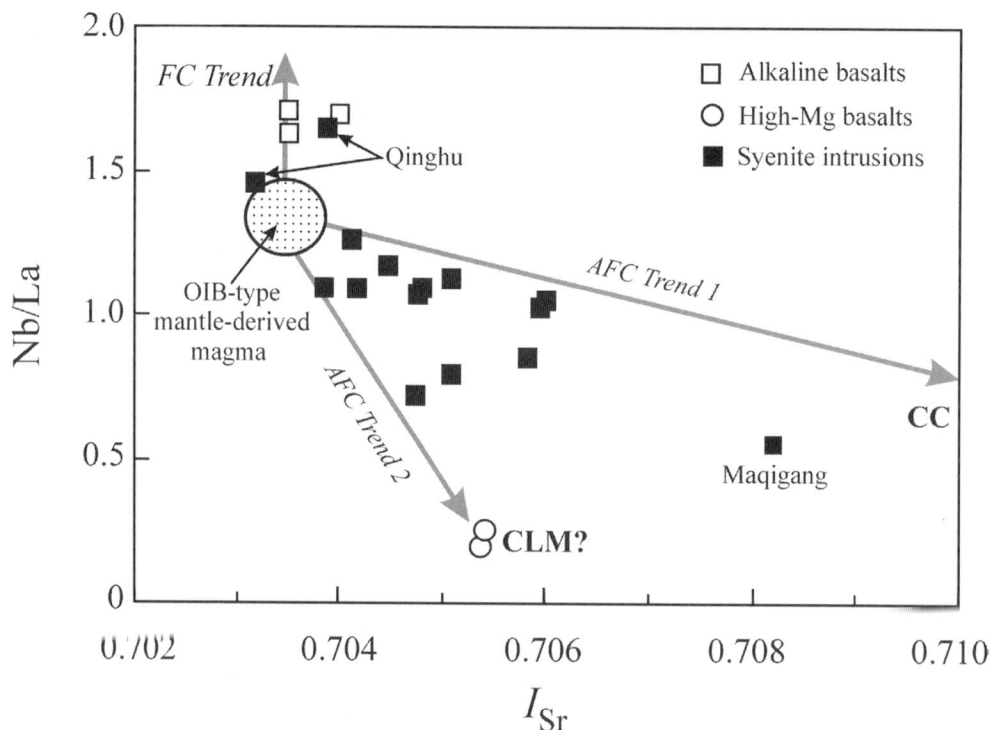

Fig. 10. A Nb–La v. $\varepsilon_{Nd}(T)$ v. I_{Sr} diagram. Note that the alkaline basalts and the Qinghu syenites have low I_{Sr} and high Nb/La ratios showing a fractional crystallization (FC) trend. The high-Mg basalts have moderately enriched I_{Sr} but a very low Nb–La ratio, indicative of the enriched continental lithosphere mantle (CLM) metasomatized by previous subduction processes. The continental crust (CC) is characterized by high I_{Sr} (>0.71) and a moderate Nb–La ratio of 0.5–0.8 (Rudnick & Fountain 1995). Most syenites plot between the two trends of AFC 1 and AFC 2 (assimilation with fractional crystallization), indicative of variable involvement of CLM and CC into the OIB-type mantle-derived magma. Four samples from the Mashan and Luorong plutons are cited from Guo *et al.* (2001). Symbols as in Figure 4.

In summary, we propose here a tectonomagmatic model (Fig. 11) to describe the geodynamic evolution of the interior of SE China in the Early Mesozoic. The Permo-Triassic Indosinian Orogeny, whose activity probably ceased in the latest Triassic to Early Jurassic, resulted in substantial crustal thickening, as witnessed by the folded and thrust belt in SE China (Li 1998). Then, an LV-type, passive rifting started developing during the early Middle Jurassic, in response to the post-Indosinian orogenic collapse and lithosphere extension. In the meantime, minor alkaline basalts that originated from small degrees of decompression melting of the upwelling asthenosphere mantle (or OIB-type enriched CLM located at the base of CLM) erupted rapidly through deep lithosphere fractures. The overlying lithospheric mantle was hence metasomatized by these alkaline basaltic melts and/or volatiles with OIB-type geo-chemical and isotopic signatures. The region experienced further lithospheric extension and thinning during the Late Jurassic, associated with the geotherm raised due to upwelling of the asthenosphere and partial melting of newly and pre-existing metasomatized lithosphere mantle domains caused the formation of syenite intrusions and high-Mg basalts, respectively.

Conclusion

The $^{40}Ar/^{39}Ar$ age determinations and geochemical and Sr–Nd isotopic analyses document three episodes of Jurassic intraplate magmatism, i.e. alkaline basalts of *c.*175 Ma, syenitic intrusions of *c.*160 Ma, and high-Mg basalts of *c.*150 Ma, in southern Hunan–eastern Guangxi in the interior of SE China. A secular variation in geochemical and isotopic characteristics of these

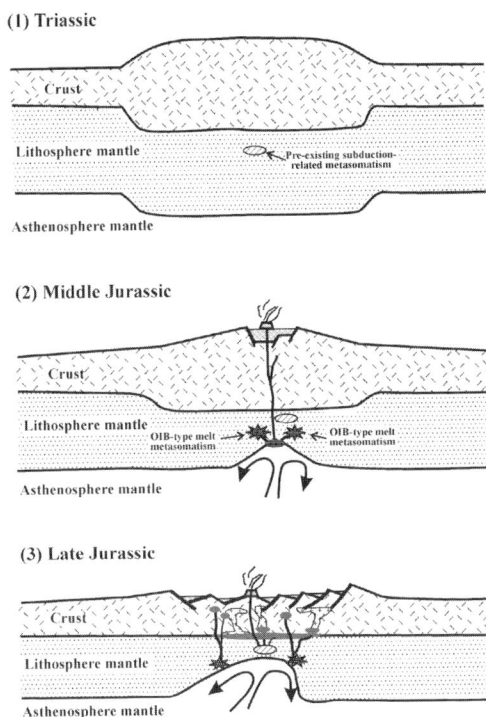

Fig. 11. Cartoon illustrating the Early Mesozoic tectonic and magmatic evolution of the interior of SE China. (**1**) The Permo-Triassic Indosinian orogeny resulted in a thickened crust and lithosphere mantle; (**2**) an LV-type, passive rift started developing in response to the post-Indosinian orogenic collapse and lithosphere extension. Small degrees of decompression melting of the upwelling asthenosphere produced minor alkaline basaltic melts which either erupted through deep lithosphere fractures, or metasomatized the overlying lithospheric mantle. (**3**) Syenites and high-Mg basalts formed during further lithospheric extension and thinning, associated with the formation of Late Jurassic granitoids.

intraplate igneous rocks is interpreted as being related to changes in melt segregation depth and mantle sources. The Jurassic tectonic regime in SE China interior is regarded as continental rifting and extension in response to the post-Indosinian orogenic collapse and lithosphere extension.

We appreciate B.-Y Zhang for assistance in fieldwork, C.-Y. Lee for major-element analysis, and W. L. Ling for Sr–Nd isotope analyses. Discussions with Shen-su Sun and Yigang Xu were very helpful. Reviews and constructive comments by M. Menzies, R. Sewell and C. Fletcher have substantially improved the manuscript. This work was supported by the Chinese Academy of Sciences (grants KZCX2-102) and NFSC (grants 49725309 and 49702020).

References

BARBERI, F., SANTACROE, R. & VARET, J. 1982. Chemical aspects of rift magmatism. *In*: PALMSON, G. (ed.) *Continental and Oceanic Rifts*, American Geophysical Union, Washington D.C., 223–258.

CHARVET, J., LAPIERRE, H. & YU, Y. 1994. Geodynamic significance of the Mesozoic volcanism of southeastern China. *Journal of Southeast Asian Earth Sciences*, **68**, 387–396.

CHEN, J. F. & JAHN B. M. 1998. Crustal evolution of southeastern China: evidence from Sr, Nd and Pb isotopic compositions of granitoids and sedimentary rocks. *Tectonophysics*, **284**, 101–133.

CHEN, J. F., FOLAND, K. A., XING, F., XU, X. & ZHOU, T.X. 1991. Magmatism along the southeast margin of the Yangtze block: Precambrian collision of the Yangtze and Cathaysia blocks of China. *Geology*, **19**, 815–818.

CHEN, P., KONG, X., WHAN, Y., NI, Q., ZHANG, B. & LING, H. 1999. Rb–Sr isotopic dating and significance of Early Yanshanian bimodel volcanic–intrusive complex from south Jiangxi Province. *Geological Journal of China Universities*, **5**, 378–383 (in Chinese with English abstract).

CHUNG, S.-L., CHENG, H., JAHN, B. M., O'REILLY, S. Y. & ZHU B. Q. 1997. Major and trace element, and Sr–Nd isotope constraints on the origin of Paleogene volcanism in South China prior to the South China Sea opening. *Lithos*, **40**, 203–220.

CHUNG, S.-L., JAHN, B. M., CHEN, S. J., LEE, T. & CHEN, C.-H. 1995. Miocene basalts in northwestern Taiwan: evidence for EM-type mantle source in continental lithosphere. *Geochimica et Cosmochimica Acta*, **59**, 549–555.

CHUNG, S.-L., SUN, S.-S., TU, K., CHEN, C.-H. & LEE C.-Y. 1994. Late Cenozoic basaltic volcanism around the Taiwan Strait, SE China: product of lithosphere–asthenosphere interaction during continental extension. *Chemical Geology*, **112**, 1–20.

COX, K. G., BELL, J. D. & PANKHURST, R. J. 1979. *The Interpretation of Igneous rocks*. Allen and Unwin, London, 450 pp.

FALLOON, T. J., GREEN, D. H., HATTON, C. J. & HARRIS, K. L. 1988. Anhydrous partial melting of a fertile and depleted peridotite from 2 to 30 kb and application to basalt petrogenesis. *Journal of Petrology*, **29**, 1257–1282.

FAURE, M., SUN, Y., SHU, L., MONIÉ, P. & CHARVET, J. 1996. Extensional tectonics within a subduction-type orogen: the case study of the Wugongshan dome (Jiangxi Province, SE China). *Tectonophysics*, **263**, 77–108.

FLETCHER, C. J. N., CAMPBELL, S. D. G., BUSBY, J. P., CARRUTHERS, R. M. & LAI, K. W. 1997. Regional tectonic setting of Hong Kong: implications of new gravity models. *Journal of the Geological Society of London*, **154**, 1021–1030.

FOLEY, S. F. 1992. Vein-plus-wall-rock melting mechanisms in the lithosphere and the origin of potassic alkaline magmas. *Lithos*, **28**, 435–453.

GILDER, S. A., GILL, J. *et al.* 1996. Isotopic and paleomagnetic constraints on the Mesozoic tectonic evolution of south China. *Journal of Geophysical Research*, **101(B7)**, 16 137–16 154.

GILDER, S. A., KELLER, G. R., LUO, M. & GOODELL, P. C. 1991. Timing and spatial distribution of rifting in China. *Tectonophysics*, **197**, 225–243.

GUANGXI (BUREAU OF GEOLOGY AND MINERAL RESOURCES OF GUANGXI PROVINCE) 1985. *Regional Geology of Hunan Province*. Geological Publishing House, Beijing, 853 pp.

GUO, X. H., CHEN, J. F., ZHANG, X., TANG, J. F., XIE, Z., ZHOU, T. X. & LIU, Y. L. 2001. Nd isotopic ratios of K-enriched magmatic complexes from southeastern Guangxi Province. *Acta Petrologica Sinica*, **17**, 19–27 (in Chinese with English abstract).

HOLLOWAY, N. H. 1982. North Palawan Block, Philippines – its relation to Asian Mainland and role in evolution of South China Sea. *American Association of Petroleum Geologists Bulletin*, **66**, 1355–1383.

HSÜ, K. J., LI, J. L., CHEN, H. H., PEN, H. P. & SENGOR, A. M. C. 1990. Tectonics of south China: key to understanding west Pacific geology. *Tectonophysics*, **183**, 9–39.

HSÜ, K. J., SUN, S., LI, J. L., CHEN, H. H., PEN, H. P. & SENGOR, A. M. C. 1988. Mesozoic overthrust tectonics in south China. *Geology*, **16**, 418–421.

HUANG, X., SUN, S. H., DEPAOLO, D. J. & WU, K. L. 1986. Nd–Sr isotope study of Cretaceous magmatic rocks from Fujian province. *Acta Petrologica Sinica*, **2**, 28–36 (in Chinese with English abstract).

HUNAN (BUREAU OF GEOLOGY AND MINERAL RESOURCES OF HUNAN PROVINCE) 1988. *Regional Geology of Hunan Province*. Geological Publishing House, Beijing, 719 pp.

JAHN, B. M. 1974. Mesozoic thermal events in southeast China. *Nature*, **248**, 480–483.

JAHN, B. M., CHEN, P. Y. & YEN, T. P. 1976. Rb–Sr ages of granitic rocks in southeastern China and their tectonic significance. *Geological Society of America Bulletin*, **86**, 763–776.

JAHN, B. M., ZHOU, X. H. & LI, J. L. 1990. Formation and tectonic evolution of southeast China: isotopic and geochemical constraints. *Tectonophysics*, **183**, 145–160.

LAN, C. Y., JAHN, B. M., MERTZMAN, S. A. & WU, T. W. 1996. Subduction-related granitic rocks of Taiwan. *Journal of Southeast Asian Earth Sciences*, **14**, 11–28.

LAPIERRE, H., JAHN, B. M., CHARVET, J. & YU, Y. W. 1997. Mesozoic magmatism in Zhejiang Province and its relation with the tectonic activities in SE China. *Tectonophysics*, **274**, 321–338.

LEE, C.-Y., TSAI, J.-H., HO, H.-H., YANG, T. F., CHUNG, S.-L. & CHEN, C.-H. 1997. Quantitative analysis in rock samples by an X-ray fluorescence spectrometer, (I) major elements. *Annual Meeting of Geological Society of China*, Abstract Volume, 418–420 (in Chinese).

LI, H. M., DONG, C. W., XU, X. S. & ZHOU, X. M. 1995. U–Pb single-grain zircon dating of the Quanzhou Gabbro: implications to the origin of the basic magmatism in SE Fujian. *Chinese Science Bulletin*, **40**, 158–160 (in Chinese).

LI, X. H. 1997. Geochemistry of the Longsheng Ophiolite from the southern margin of Yangtze Craton, SE China. *Geochemical Journal*, **31**, 323–337.

LI, X. H. 2000. Cretaceous magmatism and lithospheric extension in Southeast China. *Journal of Asian Earth Sciences*, **18**, 293–305.

LI, X. H. & MCCULLOCH, M. T. 1996. Secular variations in the Nd isotopic composition of Late Proterozoic sediments from the southern margin of the Yangtze Block: evidence for a Proterozoic continental collision in SE China. *Precambrian Research*, **76**, 67–76.

LI, X. H., ZHOU, H., LIU, Y., LEE, C.-Y., CHEN, C.-H., YU, J. & GUI, X. 2000. Mesozoic shoshonitic intrusives in the Yangchun Basin, Western Guangdong, and their tectonic significance: 1. petrology and isotope geochronology. *Geochimica*, **29**, 513–520 (in Chinese with English abstract).

LI, Z. X. 1998. Tectonic evolution of the major East Asian lithospheric blocks since mid-Proterozoic – a synthesis. *In*: MARTIN, F. J., CHUNG, S.-L., LO, C.-H. & LEE, T.-Y. (Eds) *Mantle Dynamics and Plate Interactions in East Asia*, American Geophysical Union, Washington, D.C., 221–243.

LI, Z. X., LI, X. H., KINNY, P. D. & WANG, J. 1999. The breakup of Rodinia: did it start with a mantle plume beneath South China? *Earth and Planetary and Science Letters*, **173**, 171–181.

LI, Z. X., LI, X. H., KINNY, P. D., WANG, J., ZHANG, S. & ZHOU, H. 2003. Geochronology of Neoproterozoic syn-rift magmatism in the Yangtze Craton, South China and correlations with other continents: evidence for a mantle superplume that broke up Rodinia. *Precambrian Research*, **122**, 85–109.

LIN, W., FAURE, M., MONIÉ, P., SCHÄRER, U., ZHANG, L. & SUN, Y. 2000. Tectonics of SE China: new insights from the Lushan massif (Jiangxi Province). *Tectonics*, **19**, 852–871.

LIN, W., FAURE, M., SUN, Y., SHU, L. & WANG, Q. 2001. Compression to extension switch during the middle Triassic orogeny in eastern China: the case study of the Jiulingshan massif in the southern foreland of the Dabieshan. *Journal of Asian Earth Sciences*, **20**, 31–43.

LIU, Y., LIU, H. C. & LI, X. H. 1996. Simultaneous and precise determination of 40 trace elements in rock samples by ICP–MS. *Geochimica*, **25**, 552–558 (in Chinese with English abstract).

LO, C.-H. & LEE, C.-Y. 1994. $^{40}Ar/^{39}Ar$ method of K–Ar age determination of geological samples using the Tsing-Hua Open-Pool Reactor (THOR). *Journal of the Geological Society of China*, **37**, 143–164.

MARTIN, H., BONIN, B., CAPDEVILA, R., JAHN, B. M., LAMEYRE, J. & WANG, Y. 1994. The Kuiqi peralkaline granitic complex (SE China): petrology and geochemistry. *Journal of Petrology*, **35**, 983–1015.

MENZIES, M. 1987. Alkaline rocks and their inclusion: a window on the Earth's interior. *In*: FITTON, J. G. & UPTON, B. G. J (eds) *Alkaline Igneous Rocks*, Geological Society, London, Special Publications, **30**, 15–27.

MÜLLER, D. & GROVES, D. I. 1995. *Potassic Igneous Rocks and Associated Gold-Copper Mineralization*. Springer-Verlag, Berlin, 144 pp.

ODIN, G. S. *et al.* 1982. Interlaboratory standards for dating purposes. *In*: ODIN, G. S. (ed.), *Numerical Dating in Stratigraphy*. John Wiley & Sons, Ltd, Chichester, 123–149.

RATSCHBACHER, L., HACKER, B. R. *et al.* 2000. Exhumation of the ultrahigh-pressure continental crust in east central China: Cretaceous and Cenozoic unroofing and the Tan–Lu fault. *Journal of Geophysical Research*, **105(B6)**, 13 303–13 338.

RODEN, M. F. 1981. Origin of coexisting minette and ultramafic breccia, Navajo volcanic field. *Contributions to Mineralogy and Petrology*, **77**, 195–206.

RODGERS, J. 1989. Comment on 'Mesozoic overthrust tectonics in south China'. *Geology*, **17**, 671–672.

ROGERS, N. W. 1992. Potassic magmatism as a key to trace-element enrichment processes in the upper mantle. *Journal of Volcanology and Geothermal Research*, **50**, 85–99.

ROGERS, N. W., JAMES, D., KELLEY, S. P. & DE MULDER, M. 1998. The generation of potassic lavas from the eastern Virunga province, Rwanda. *Journal of Petrology*, **39**, 1223–1247.

ROWLEY, D. B., ZIEGLER, A. M. & GYOU, N. 1989. Comment on 'Mesozoic overthrust tectonics in south China'. *Geology*, **17**, 384–386.

RUDNICK, R. & FOUNTAIN, D. M. 1995. Nature and composition of the continental crust: a lower crustal perspective. *Reviews of Geophysics*, **33**, 267–309.

SEWELL, R. J. & CAMPBELL, S. D. G. 1997. Geochemistry of coeval Mesozoic plutonic and volcanic suites in Hong Kong. *Journal of the Geological Society, London*, **154**, 1053–1066.

SUN, S.-S. & MCDONOUGH, W. F. 1989. Chemical and isotopic systematics of oceanic basalt: implications for mantle composition and processes. *In*: SAUNDERS, A.D. & NORRY, M. J. (eds), *Magmatism in the Ocean Basins*, Geological Society, London, Special Publications, **42**, 528–548.

THOMPSON, R. N., MORRISON, M. A., HENDRY, G. L. & PARRY, S. J. 1984. An assessment of the relative roles of a crust and mantle in magma genesis: an element approach. *Philosophical Transactions of the Royal Society of London*, **A310**, 549–590.

TONG, W. X. & TOBISCH, O. T. 1996. Deformation of granitoid plutons in the Dongshan area, southeast China: constraints on the physical conditions and timing of movement along the Changle–Nanao shear zone. *Tectonophysics*, **267**, 303–316.

TU, K., FLOWER, M. F. J., CARLSON, R. W., ZHANG, M. & XIE, G. H. 1991. Sr, Nd, and Pb isotopic compositions of Hainan basalts (South China): implications for a subcontinental lithosphere Dupal source. *Geology*, **19**, 567–569.

TU, K., FLOWER, M. F. J., CARLSON, R. W., XIE, G. H., CHEN, C.-Y. & ZHANG, M. 1992. Magmatism in the South China basin, 1. Isotopic and trace element

evidence for an endogenous Dupal mantle component. *Chemical Geology*, **97**, 47–63.

WANG, L. K., YANG, W. J., ZHANG, S. L. & XU, W. X. 1985. *Distribution Map of Granitoids of Two Petrogenetic Series in SE China*. Institute of Geochemistry, Chinese Academy of Sciences, Guiyang.

WANG, J. & LI, Z. X. 2001. Sequence stratigraphy and evolution of the Neoproterozoic marginal basins along southeastern Yangtze Craton, South China. *Gondwana Research*, **4**, 17–26.

WANG, J. & LI, Z. X. 2003. History of Neoproterozoic rift basins in South China: implications for Rodinia breakup. *Precambrian Research*, in press.

WYLLIE, P. J & SEKINE, T. 1982. The formation of mantle phlogopite in subduction zone hybridization. *Contributions to Mineralogy and Petrology*, **79**, 375–380.

WILSON, M. 1989. *Igneous Petrogenesis*. Unwin Hyman, London, 466 pp.

WINCHESTER, J. A. & FLOYD, P. A. 1976. Geochemical magma type discrimination: application to altered and metamorphosed igneous rocks. *Earth and Planetary Science Letters*, **28**, 459–469.

XU, J., MA, G., TONG, W. X., ZHU, G., & LIIN, S. 1993. Displacement of the Tancheng–Lujiang wrench fault system and its geodynamic setting in the northwestern Circum-Pacific. *In*: XU, J. (ed.) *The Tancheng–Lujiang Wrench Fault System*. John Wiley & Sons, Ltd., 51–74.

XU, J., ZHU, G., TONG, W. X., GUI, K. R. & LIU, Q. 1987. Formation and evolution of the Tancheng–Lujiang wrench fault system: a major shear system to the northwest of the Pacific Ocean. *Tectonophysics*, **134**, 273–310.

XU, M. L. & YUAN, Z. X. 1992. The U-Pb zircon isotopic age and its geological significance of the Qinghu Monzonite in Guangxi. *Geology of Guangxi*, **5**, 33–36 (in Chinese with English abstract).

YUI, T. F., HEAMAN, L. & LAN, C. Y. 1996. U–Pb and Sr isotopic studies on granitoids from Taiwan and Chinmen–Lieyu and their tectonic implications. *Tectonophysics*, **263**, 61–76.

ZHOU, X. M. & LI, W. X. 2000. Origin of Late Mesozoic igneous rocks in Southeastern China: implications for lithosphere subduction and underplating of mafic magmas. *Tectonophysics*, **326**, 269–287.

ZHU, Q., WANG, F., LU, F. & ZHONG, Z. 1997. Petrogenesis research and tectonic environment analysis about Mesozoic–Cenozoic basalts from southern Hunan. *Earth Science: Journal of China University of Geosciences*, **22**, 584–588 (in Chinese with English abstract).

ZOU, H. B. 1995. A mafic–ultramafic rock belt in the Fujian coastal area, southeastern China: a geochemical study. *Journal of Southeast Asia Earth Sciences*, **12**, 121–127.

Evidence for the multiphase nature of the India–Asia collision from the Yarlung Tsangpo suture zone, Tibet

JONATHAN C. AITCHISON & AILEEN M. DAVIS

Tibet Research Group, Department of Earth Sciences, University of Hong Kong, Pokfulam Road, Hong Kong SAR, China (e-mail: jona@hku.hk)

Abstract: Recent investigations in southern Tibet enable the testing and refinement of existing models for India–Asia collision. Presently available data indicate that marine deposition continued in the southern central portion of Tibet until at least the end of the Eocene. Subduction-related magmatism continued until the Mid-Oligocene, after which rapid uplift of the plateau was initiated. Mass-wasting of sediments into molasse basins did not commence until the latest Oligocene. The implications are that existing models, based on less-precise age constraints, invoking India–Asia collision at 55 Ma, are either flawed, or collision began at a different time. Recent work has produced sufficient data to allow the recognition of two different collisional events along the suture between India and Asia. Features related to each event require separate interpretation, and no collisional continuum should be assumed. In southern Tibet, a collision between the northern margin of India and a south-facing intra-oceanic island arc occurred at around 55 Ma, whereas continent–continent collision between India and Asia did not occur until at least 20 million years later.

The Tibet–Qinghai Plateau is the type area for a continent–continent collision zone and contains some of the most exciting geology on Earth. It consists of a series of fault-bounded continental blocks separated by oceanic sutures (Fig. 1). These blocks mostly originated elsewhere, migrated across oceans and were accreted to Asia at various times. The present-day elevated topography of the plateau is a response of the region to the ongoing collision between Asia and Indian continental landmasses. Continental collision is an efficient process in the formation of supercontinents, and the mountain belts created by such collisions are the most dominant geological features of the Earth's surface. The Himalayan–Tibetan Orogen is the youngest and most spectacular collisional belt existing today. It is part of the greater Himalayan–Alpine system stretching from Europe through East Asia. Because of its immense size and high elevation, the geological evolution of the Himalayan–Tibetan Orogen is also thought to have played a critical role in controlling global climate change.

General aspects of the overall geological framework of Tibet are widely known. Much of the plateau has been mapped at reconnaissance scale, and several key traverses have allowed description of the key geological elements. Several important summaries of the geology of the Tibet–Qinghai Plateau are already published (Chang *et al.* 1988; Dewey *et al.* 1989; Chang

1996). Key elements of the geology of the Tibet–Qinghai Plateau include several terranes, which from north to south are the: Qilian, Kunlun–Qaidam, Songpan–Ganzi, Qiangtang, Lhasa and Himalayan (Indian) terranes. Faults which bound these terranes include sutures such as the Kokoxili or Qilian suture (Early Pz), Kunlun suture–AKMS–Xidatan Fault (Pz), Jinsha suture (Early J), Bangong–Nujiang suture (J/K) and Yarlung Tsangpo suture (Cz). Significant displacement has also occurred along numerous more recent and still active faults, such as the Late Cenozoic Himalayan Fault systems and Neotectonic Altyn Tagh and Karakoram faults. A detailed description of the geology of the Tibet–Qinghai Plateau is given by Chang (1996). Because of intense interest in the evolution of this orogenic system, the crustal structure of the region has also been explored in some detail (Nelson *et al.* 1996). As further detailed work is yet to be undertaken in northern parts of the plateau, it is difficult to rigorously test and improve on the simple models for tectonic evolution of the region that are presented in the works cited above.

Access to field areas is somewhat easier in southern Tibet, and considerably more work has been undertaken in this region. Nevertheless, much work is still reconnaissance in nature when compared with investigations carried out south of the Himalayan divide. Major Sino-Western

From: MALPAS, J., FLETCHER, C. J. N., ALI, J. R. & AITCHISON, J. C. (eds) 2004. *Aspects of the Tectonic Evolution of China*. Geological Society, London, Special Publications, **226**, 217–233.
0305-8719/04/$15 © The Geological Society of London 2004.

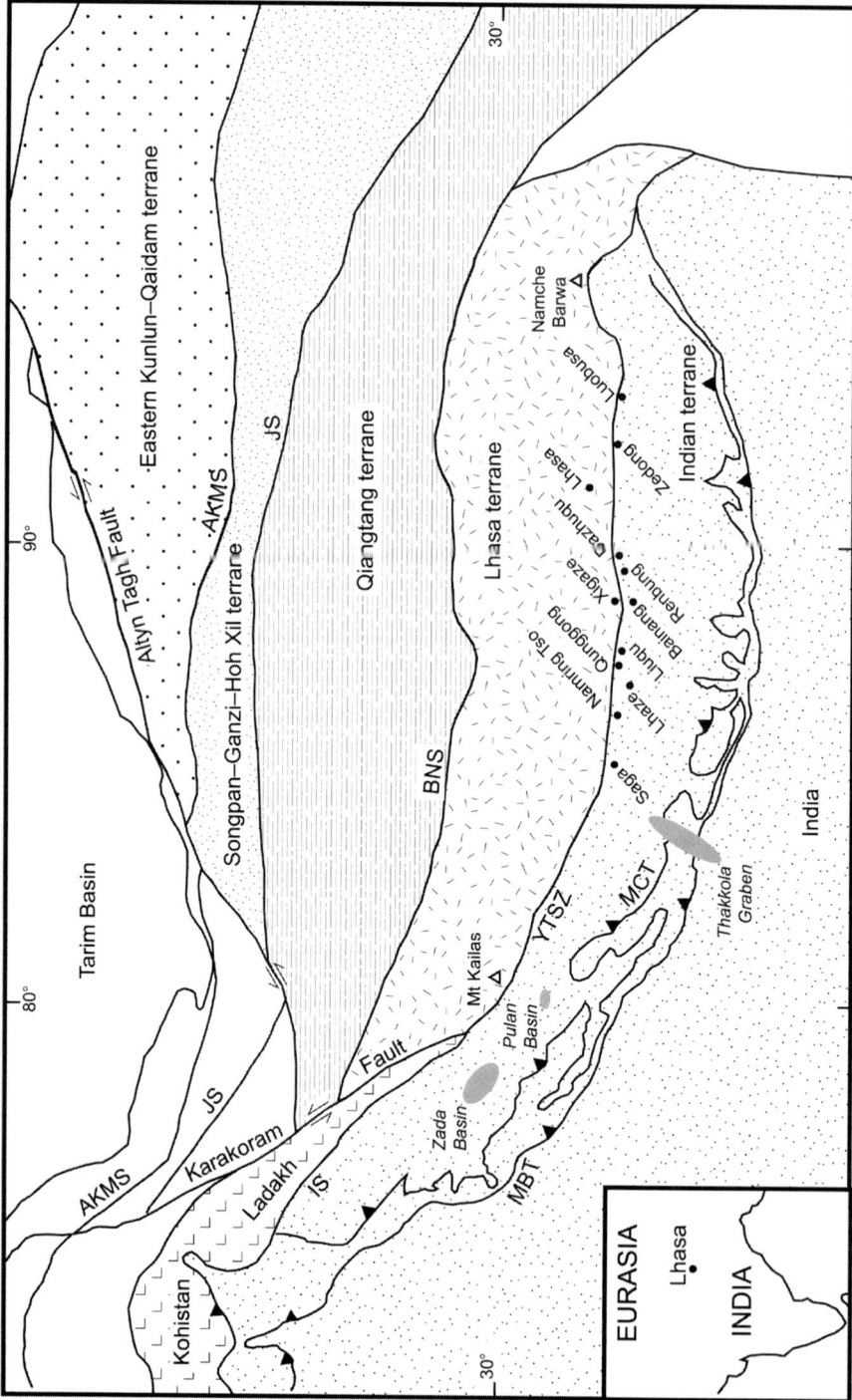

Fig. 1. Tectonic units and boundaries within the Tibet–Qinghai Plateau. AKMS, Anyimaqen–Kunlun–Muztagh suture; JS, Jinsha suture; BNS, Bangong–Nujiang suture; IS, Indus suture; YTSZ, Yarlung Tsangpo suture zone; MCT, Main Central Thrust; MBT, Main Boundary Thrust.

collaborations in the early 1980s along the Yarlung Tsangpo suture zone (YTSZ) led to the recognition that this zone was where Tethyan oceanic lithosphere was subducted (Nicolas *et al.* 1981). Several simple models for the tectonic evolution of this suture were proposed (Tapponnier *et al.* 1981*b*; Burg & Chen 1984; Mercier & Li 1984; Searle *et al.* 1987). The India–Asia collision and the role of the YTSZ in this event have resulted in considerable interest being shown in this region. Southern Tibet has also witnessed numerous investigations into the on-going events associated with India–Asia collision (see Hodges 2000 and references therein). Over the past few years the Tibet Research Group based at Hong Kong University has been carrying out research in Tibet, which has mostly been concentrated in southern Tibet, in particular along the YTSZ. Data emanating from this work are summarized herein. They provide a substantial body of new information, which can be used to test earlier models for tectonic evolution of the India–Asia collision in this region and as a basis for developing new models.

Geological features of the (Neotethyan) Yarlung–Tsangpo suture zone

The Lhasa Terrane, which bounds the north of the suture, is reported to have a Proterozoic to Early Cambrian basement in the Amdo region (Harris *et al.* 1988; Yin *et al.* 1988). Elsewhere Ordovician and Carboniferous to Triassic shallow-marine clastic sediments are overlain by Jurassic turbidites, which are themselves interlayered with volcanic flows and tuffs. Cretaceous limestone and marine deposits are widespread. In the Late Mesozoic, the northern margin of the terrane experienced collision with the Qiangtang terrane along the Bangong–Nujiang suture.

The southern Lhasa terrane has a variably deformed and metamorphosed basement represented by the Sangri Group (Badengzhu 1979; Yin *et al.* 1988). These andesitic meta-volcanics and associated sediments crop out from Xigaze to Zedong and provide evidence of Late Jurassic to Early Cretaceous northward subduction of Tethyan ocean crust beneath the southern Lhasa terrane. Mesozoic sequences along the southern margin of the terrane are intruded by the well-known, if not extensively studied, Cretaceous to Palaeogene Gangdese batholith, correlatives of which extend west beyond Mount Kailas.

The southern margin of the Lhasa terrane lies at the YTSZ. The history of convergence and the nature of what once lay between India and Asia are not simple, and the suture includes tectonically disrupted fragments of at least one Mesozoic intra-oceanic island arc system, together with remnants of the Tethyan Ocean. Early models recognized that subduction had occurred, and suggested the development at a simple south-facing continental margin arc–fore-arc–trench system along the southern margin of the Lhasa terrane. Recognition of a south-facing intra-oceanic island arc system within the Neo-Tethys (Aitchison *et al.* 2000), leads to a new model invoking Palaeogene collision of this arc system with the Indian margin prior to the main India–Asia collision.

Within the suture, several key terranes (Xigaze, Zedong, Dazhuqu and Bainang) can be recognized. Other key units along this zone include various conglomerate units; Liuqu Conglomerate and the 'Gangrinboche facies' conglomerates of the Luobusa, Dazhuqu, Qiuwu and Kailas formations, which record important tectonic events (Aitchison *et al.* 2002*b*; Davis *et al.* 2002, 2003, 2004). South of the suture, the Yamdrok mélange (Liu 2001; Liu & Aitchison 2002) represents another key unit for understanding the tectonic evolution of the region.

The Xigaze terrane contains turbidites and other deep-sea fan deposits sourced from coeval Lhasa terrane volcanic rocks associated with north-directed subduction of Neotethyan oceanic lithosphere (Einsele *et al.* 1994; Dürr 1996). The terrane is possibly the best-exposed fore-arc basin in the world. The base of the succession is faulted and the top has been removed by erosion. Macrofossils, are rare, but those present are mid to Upper Cretaceous.

The Zedong terrane in the east of the suture is dominated by tholeiitic arc volcanic rocks and shoshonites. Bajocian radiolarian fossils and geochronological data (U–Pb and Ar/Ar) both indicate that these are remnants of a late Middle to early Late Jurassic intra-oceanic island arc (McDermid 2002; McDermid *et al.* 2002). Lower Cretaceous fossils are also reported from the terrane (Badengzhu 1979).

The Dazhuqu terrane contains all elements of an ophiolite including harzburgite, dunite, gabbro, diabase, basalt and chert in a disrupted sequence. It is exposed along a strike length of >250 km near Xigaze (Nicolas *et al.* 1981; Girardeau *et al.* 1984*a*, 1984*b*, 1985*a*, 1985*b*, 1985*c*; Girardeau & Mercier 1988). Correlatives of these rocks are known from along the entire length of the suture within Tibet. The sequence near Xigaze is up to 25 km wide but has been tectonically attenuated. Petrology and geochemistry indicate generation in a suprasubduction zone setting (Hébert *et al.* 2000; Aitchison *et al.* 2002*a*). Detailed studies of radiolarians from several localities, where unequivocal depositional contacts of chert upon ophiolitic pillow basalts can be observed, indicate ophiolite formation in the Barremian to Albian (mid-Cretaceous) (Ziabrev 2001; Ziabrev *et al.* 2003). Dazhuqu terrane sediments comprise a distinctly different assemblage to the fore-arc sediments of the Xigaze terrane, and the contact between these two units is everywhere faulted. Amphibolites in serpentinite-matrix mélange at the southern margin of the terrane indicate that it experienced a Late Cretaceous 70–90 Ma (based on Ar/Ar dating) metamorphic event (Wang *et al.* 1987).

In the Zedong–Luobusa area, ophiolitic rocks constitute part of a northerly-directed thrust stack dominated by ultramafic rocks in which Cr-mineralization is abundant (Zhou *et al.* 1996). Gabbros and diabase dykes are locally preserved. The only available age data (Zhou *et al.* 2002) suggest that these ophiolitic rocks may be of early Middle Jurassic (177 ± 31 Ma) rather than Mid-Cretaceous age, as they are near Xigaze. The Luobusa ophiolite may thus be part of a different terrane and possibly has greater similarity to the Spontang ophiolite in Ladakh (Reuber *et al.* 1992; Corfield *et al.* 2001; Pedersen *et al.* 2001) than to the Dazhuqu terrane. Alternatively, these results may indicate that there is only partial preservation of any record of a long-lived intra-oceanic subduction system that once existed within the Tethys.

The Bainang terrane is an imbricate south-vergent thrust stack of Tethyan oceanic rocks (Ziabrev 2001; Ziabrev *et al.* 2004). Numerous north-facing, south-verging imbricated slices are present, and the terrane is interpreted as once being part of a subduction complex. Individual thrust slices incorporate various lithologies such as basalt, chert, siliceous and tuffaceous mudstones, limestone and calcareous shales. The terrane developed in response to north-directed subduction, and five mappable units, each of which has a distinctive composition and structural style, are recognized at Bainang (Ziabrev 2001). The oldest fossils occur in accreted cherts and are Upper Triassic. The youngest radiolarian faunas are Mid-Cretaceous and they occur in units which possibly constrain the timing of the subduction event.

Several terrane-sealing conglomerate units occur along the suture and appear to indicate two different collision events (Davis *et al.* 2004). The Palaeogene Liuqu Conglomerate was derived from Indian, Tethyan and intra-oceanic terranes, but does not contain any detritus from the Xigaze and Lhasa terranes. This unit is interpreted to record Palaeogene collision

of an intra-oceanic island arc with the northern margin of India (Davis *et al.* 2002).

Upper Oligocene to Lower Miocene conglomerates of the Gangrinboche facies include the correlative Luobusa, Dazhuqu, Qiuwu and Kailas formations. This facies is a molasse that records aspects of the India/Asia collision history (Aitchison *et al.* 2002*b*). Derivation of clasts was initially from the north, then, as the collision progressed, an increasing volume of clasts was shed from sources to the south.

South of the YTSZ, the Indian (northern Himalayan) terrane is composed of continent-derived siliciclastic sedimentary rocks and carbonates. The northern Himalaya comprise a succession of rocks recording passive margin sedimentation from Ordovician to at least the end of the Eocene. Sediments are increasingly distal to the north, and pass into Tethyan oceanic sediments. Uppermost Cretaceous clastic sedimentary rocks indicate tectonic disturbance of the Indian continental margin. Marine nannofossils from Himalayan sediments north of Qomolangma indicate that marine conditions prevailed until at least the Eocene/Oligocene boundary (NP20) (Wang *et al.* 2002) and thereby place a maximum age constraint on the timing of India–Asia collision in the region.

Within the Indian terrane south of, and subparallel to, the YTSZ, mud-matrix mélange crops out on a regional scale. Various degrees of stratal disruption are recorded (Liu 2001) in the Yamdrok mélange, which contains blocks of Indian passive margin and more distal marine rocks of Permian to Cretaceous age. Radiolarians recovered from the mélange matrix include Palaeocene/Eocene boundary faunas (Liu & Aitchison 2002).

Further south, within the Indian terrane, the Greater (or High) Himalaya are bounded on their northern side by the South Tibetan Detachment Fault and to their south by the Main Central Thrust. A protolith of Upper Proterozoic to Lower Cambrian sedimentary rocks has experienced high-grade Late Cenozoic metamorphism related to India–Asia collision (see Hodges 2000 and references therein). The South Tibetan Detachment System is a north-side down, low-angle normal fault system, traceable along Himalaya, that places low-grade Tethyan metasediments against Greater Himalayan gneisses. Radiometric data indicate that it was active from *c.*17 Ma until 8–9 Ma.

The Himalayan leucogranites are syn-collisional anatectic igneous rocks that crop out in two roughly parallel granite belts across southern Tibet. The High Himalayan leucogranites formed between 24.0 and 17.2 Ma, and the North Himalayan granite belt, which is parallel to, and c.80 km to the north of, the High Himalaya, formed at 17.6 to 9.5 Ma.

India–Asia collision and establishment of the Tibet–Qinghai Plateau

Development of the Tibet–Qinghai Plateau is intimately related to the India–Asia collision, which has powered this phenomenon. Thus, it is important to understand the sequence of events involved in this collision. Most models, for the tectonic evolution of Tibet, consider that the first contact between India and Asia occurred in the earliest Eocene (Molnar 1984; Patriat & Achache 1984; Searle *et al.* 1987; Dewey *et al.* 1989; Klootwijk *et al.* 1991; Harrison *et al.* 1992; Klootwijk *et al.* 1992; Molnar *et al.* 1993; Beck *et al.* 1995; Butler 1995; Le Fort 1996; Rowley 1996*a*; Hodges 2000; Yin & Harrison 2000). Estimates of the timing of initial India–Asia contact have been refined on the basis of several lines of evidence (Searle *et al.* 1987). These include: the slowdown in the convergence rate between India and Asia; the initiation of compressional tectonics along and south of the Indus–Yarlung–Tsangpo suture; a lack of data indicating continuing subduction-related magmatic activity north of the suture between India and Asia; and the accumulation of molasse deposits along the suture. The existing consensus model that invokes earliest Eocene (c.55 Ma) collision between India and Asia is widely believed and usually largely unquestioned. However, since early estimates of the timing of this collision were made two decades ago, a considerable body of new data, emanating in large part from research in previously unstudied parts of Tibet, has accumulated. It is therefore timely to now reconsider the 55 Ma collision hypothesis. We suggest that the first two collision-related features mentioned above may be ascribed to a different collision event and the last two require reassessment because new data provide improved age constraints for critical units in question.

The elegance of the above-mentioned simple models notwithstanding, interpretations of rocks in the greatest collisional system known on Earth suggest, somewhat curiously, that there is a time lag of >20 million years between the proposed (55 Ma) timing of initial contact of buoyant continental lithospheric masses and the development of the attendant effects of that collision. However, in modern observable systems involving tectonic units of lesser magnitude, collision appears to have immediate effect (e.g. Taiwan, northern New Guinea). We suggest that although an end-Cretaceous/Palaeogene tectonic event is clearly recorded throughout the Tibet–Himalayan region, it was collision between an intra-oceanic island arc and the Indian continent, and not that between India and Asia. The continental lithosphere of northern Indian was conveyed, as part of the northward-subducting slab to which it was attached, into a south-facing intra-oceanic subduction system. The unsubductable nature of continental crust caused the failure of this convergent plate margin. The larger collision between India and Asia proper occurred later than previously suggested and was, as might be expected, synchronous with many orogenic events.

Cretaceous–Palaeogene events and products: an India/intra-Tethyan island-arc collision

The ages of numerous significant rock units and events recorded in Tibet reflect a significant Late Cretaceous–Palaeogene event. Recent field investigations along the Indus–Yarlung Tsangpo suture zone, combined with a reassessment of the interpretation of some well-known rock units in this zone, have led to the recognition of an Early Cretaceous intra-oceanic island arc (Aitchison *et al.* 2000). A model consistent with observed field data and interpretations of seismic tomography (Van der Voo *et al.* 1999) suggests the existence of a south-facing intra-oceanic island-arc system that developed over a northward-subducting slab and probably collided with the northern margin of India in the latest Cretaceous/Palaeogene (Aitchison *et al.* 2000). As this event did not involve terranes north of the present-day YTSZ, it cannot be implicated in development of the Tibet–Qinghai Plateau uplift, which did not occur until India collided with Asia.

Since early (1970s to 1980s) models for India–Asia collision were developed, the end-Cretaceous obduction of ophiolitic rocks on to the northern margin of India has become an increasingly widely recognized event (Beck *et al.* 1995; Searle 1996; Corfield *et al.* 1999; Aitchison *et al.* 2000; Maheo *et al.* 2000). Advances in the understanding of the likely genesis of ophiolites emanating from the past decade of Ocean Drilling Program research clearly show that many ophiolites are generated in intra-oceanic supra-subduction zone settings (Bloomer *et al.* 1995; Shervais 2001). Not only are the chemistries of well-studied ophiolites in accord with their generation above a subducting slab (Miyashiro 1973), a tectonic position in the upper plate of a convergent margin favours their preservation and obduction during arc–continent collision. On the other hand, as normal mid-ocean ridge-generated oceanic crust involved in a convergent plate system is typically located on the down-going slab, it is likely to be subducted except in exceptional circumstances. Thus, the presence of an ophiolite might be regarded as a good indicator that collision between an intra-oceanic island arc and a continental margin has occurred.

Ophiolitic successions along the YTSZ formed in the mid Jurassic (Luobusa–Zhou et al. 2002; Yungbwa in western Tibet – Miller et al. 2003) to mid-Cretaceous (Xigaze – Ziabrev 2001; Ziabrev et al. 2003). Together with other elements of a south-facing intra-oceanic island arc, they were obducted on to the northern Indian margin sometime around the end of the Cretaceous–early Palaeogene (Aitchison et al. 2000). Similar obduction events have been recognized from further west in India and Pakistan (Beck et al. 1995; Corfield et al. 1999; Maheo et al. 2000), and ophiolite obduction in Oman (Searle 1996) may even have been part of this same regional event. Metamorphic ages of amphibolites and blueschists (Honegger et al. 1989; Searle 1996; Anczkiewicz et al. 2000) within serpentinite-matrix mélanges along the southern margins of the YTSZ ophiolites all indicate an end-Cretaceous event. However, whether this event was obduction on to the Indian margin, or reflects the subduction of a spreading centre in the manner described by Shervais (2001), remains unclear. Similarities in the nature and timing of ophiolite obduction events over such an extensive strike-length must have major significance for any interpretation of regional tectonics.

Other evidence for a Late Cretaceous–Palaeogene collision event is seen in southern Tibet, with the initiation of tectonic shortening accompanied by the widespread development of south-verging structures in the Tethyan Himalayan zone (Burg & Chen 1984; Burg et al. 1984b; Ratschbacher et al. 1994). Structures that developed during this event are truncated by later Oligocene-Miocene structures.

Mud-matrix mélanges as young as latest Palaeocene (Burg et al. 1987; Liu & Aitchison 2002) are extensive across Tibet south of the YTSZ, extending from at least Mount Kailas in the west to near Lhasa Airport. These mélanges developed tectonically in response to a widespread over-pressuring event. They are analogous to Lichi and Kenting mélanges that have developed in association with contemporary arc-continent collision in Taiwan (Chang et al. 2000).

Uplift and mass-wastage of coarse clastic sediments into molasse basins axial to this arc–continent collision zone resulted in deposition of the Liuqu Formation (Davis et al. 2002). Fossils indicate accumulation of this unit in the Palaeogene (Tao 1988a, b), rather than during the Miocene as has previously been inferred (Searle et al. 1987). Although these conglomerates occur along the YTSZ, they contain no clasts of Asian affinity, such as Lhasa terrane igneous material that could have been derived from terranes north of the suture. This presumably reflects the terranes that were present and thus available as sediment sources at the time of deposition. Given the upper plate position of Asia during the India–Asia collision, it is reasonable to expect large quantities of such material to have been eroded and make an appearance in the sedimentary record of associated foreland basin deposits.

Sedimentary successions south of the suture within the Indian–Himalayan zone record a Late Cretaceous hiatus in sedimentation (Rowley 1996a; Shi et al. 1996; Willems et al. 1996), possibly representing the transit of a fore-bulge through the region. Distinctly coarser-grained sediments in the form of intraformational conglomerates accumulated in the Early Palaeocene (Wang et al. 2002). There is no good reason, however, why the fore-bulge could not have been associated with an earlier collision that occurred prior to that between India and Asia. Accumulation of orogenic molasse shed from such a collision indeed appears to be recorded elsewhere in a fore-deep basin located to the south of the collision zone. The uppermost Palaeocene–Middle Eocene Subathu Formation in northern India contains coarse-grained detritus derived from an uplifted ophiolitic and volcanic arc source terrane to its north (Najman & Garzanti 2000). Subsequent to deposition of this unit, suture zone input was drastically reduced and no sediments derived from any collision zone are seen again until the end of the Oligocene. The Upper Palaeocene to Lower Eocene Ghazij Formation in central Pakistan (Warwick et al. 1998) also contains volcanic rock fragments and detrital chromite grains indicative of arc (including ophiolite)–continent collision. These clastic sediments are overlain by Middle Eocene limestones, and no further influx of collision-derived coarse clastic material is seen until the Miocene.

A high $P–T$ metamorphic event at 55 Ma is indicated by eclogites that crop out at Tso Morari in the western Himalaya (Tonarini et al. 1993; De Sigoyer et al. 2000; O'Brien et al. 2001). We speculate that these eclogites were not necessarily generated in the India–Asia collision, as suggested by some workers (O'Brien et al. 2001;

Kohn & Parkinson 2002; Treloar *et al.* 2003). They may have been part of the continental crust on the northern edge of the Indian Plate that was dragged, by an attached down-going slab of oceanic lithosphere preceding it, into an intra-oceanic subduction zone during arc–continent collision. Continental crust may have entered sufficiently far into the subduction zone to achieve eclogite-facies metamorphism before isostatically rebounding to higher crustal levels. These Eohimalayan eclogites remained at mid- to upper crustal levels until the later India–Asia collision provided final uplift to near surface levels.

Palaeomagnetic studies indicate an equatorial latitude for India at the proposed (55 Ma) time of arc-continent collision (Klootwijk *et al.* 1992). Data from the Dazhuqu terrane ophiolite in the YTSZ indicate that it formed at an equatorial latitude (Abrajevitch *et al.* 2001), and should therefore have accreted to India in the Palaeocene. Paleomagnetic results (Lin & Watts 1988) from coeval rocks in Tibet indicate a more northerly position (6–20°N) and are, notably, not in accord with the suggestion of earliest Eocene India–Asia collision. However, India should have arrived at exactly the same palaeolatitude as that suggested for Tibet in the Eocene by the end of the Oligocene.

Studies of seismic tomography (Van der Voo *et al.* 1999) reveal that remnants of subducted Tethyan oceanic lithosphere can be clearly imaged beneath India and other parts of Asia. What is most intriguing is the existence of multiple slabs suggesting the former presence of more than one subduction system within Tethys. The greater depth to which a slab that lies south of the one attributed to subduction along the southern margin of Asia has descended implies that it was part of a system, which terminated somewhat earlier. Interpretation of seismic tomography (Van der Voo *et al.* 1999) is entirely consistent with that of surface geology (Aitchison *et al.* 2000).

The relative plate motion between India and Asia prior to the suggested 55 Ma collision had a convergence rate significantly greater than the presently observed global range of plate convergence rates (minimum 2 cm/year to maximum 11 cm/year, with an average of 7 cm/year; Gordon & Stein 1992). A rate of 20 cm/year is approximately twice that which might be expected from a single plate margin, even where convergence is very rapid. We therefore suggest that the entire 55 Ma slowdown in India/Asia convergence can be reconciled in a three-plate scenario in which an additional oceanic plate lay between India and Asia. Because of the removal of an entire destructive plate boundary at an intra-Tethyan intra-oceanic zone of plate convergence, and the resultant conversion to a two-plate situation, the net convergence rate between India and Asia slowed dramatically to around 5 cm/year. This rate is more consistent with the presence of a single plate boundary. Indeed, given the presence of intra-oceanic island-arc fragments obducted on to India, such a slowdown in the overall rate of convergence is entirely predictable. Fragments of the now extinct intra-oceanic island arc and associated ophiolites, which collided with India in the Palaeogene, were swept along with the continental bow-wave at the north of India, and experienced a second collision when Indian and Asian continental lithosphere met towards the end of the Oligocene. The response to this second collision was immediate, with cessation of arc magmatism along the southern margin of Asia, uplift of the Tibetan Plateau, collisional orogenesis and readjustment of plate boundaries throughout eastern Asia.

Oligocene–Miocene events and products: the India/Asia collision

Dramatic uplift, which is the typical, immediate and most obvious manifestation of any active collision occurring on Earth today, apparently did not affect the Tibetan Plateau until the Late Oligocene–Early Miocene (Harrison *et al.* 1992). At that time, in both Tibet and surrounding regions, numerous important tectonic phenomena that can clearly be related to collision were initiated. Despite little evidence of any collision-related events in older (Mid-Eocene to Lower Oligocene) rocks, collision is, however, still inferred to have begun at least 20 million years earlier.

Thermochronological investigations (Copeland *et al.* 1987; Richter *et al.* 1991; Harrison *et al.* 1992, 1993) of variations in mineral ages with elevation indicate that late Middle Eocene subduction-related granites in the Lhasa terrane experienced relatively low denudation rates for around 20 million years prior to their rapid unroofing beginning at around 21 Ma. Detrital K-feldspars collected from sediments within the modern Yarlung Tsangpo yield ages consistent with widespread Miocene uplift of Tibet (Copeland & Harrison 1990; Harrison *et al.* 1992). Widespread uplift throughout the Tibet–Himalayan region in the Late Oligocene to Early Miocene must have occurred in response to continental collision.

Modern systems where collision results in immediate concomitant orogenesis are readily observable (Richardson & Blundell 1996;

Hill & Raza 1999; Niitsuma 1999; Huang *et al.* 2000). Although the early stages of collision between two major fragments of continental lithosphere cannot be observed anywhere at present, we are able to observe smaller-scale collisions between accreting island arcs and continents in locations such as Taiwan, Timor, Japan and Papua New Guinea. These areas have all experienced orogenesis with juvenile topographic relief in the order of 4000 metres. Uplift accompanies collision and, in each of these locations, there has been a geologically immediate response to the entry of continental lithosphere into the subduction system. Collision of a relatively minor intra-oceanic island arc, the Luzon arc, with the continental margin of Asia has given rise to more than 4 km of topographic relief in Taiwan over the past five million years. The deformation front is propagating across the leading edge of the continental margin of eastern China just as the last of the remaining oceanic lithosphere is being subducted and arc magmatism is ceasing (Lundberg *et al.* 1997; Tang & Chemenda 2000). In northern parts of Taiwan where collision has finished, orogenic collapse with attendant extrusion parallel to the axis of the collision belt is now well advanced (Crespi *et al.* 1996), and topographic relief is actively being reduced and post-collisional shoshonitic volcanism is occurring (Chung *et al.* 2001). It therefore seems somewhat improbable that uplift should significantly (25 Ma) lag behind a collision event of considerably greater magnitude.

The timing of the cessation of subduction along the continental margin at the southern edge of the Lhasa terrane where oceanic lithosphere of the Tethys was being consumed is also crucial. Until recently, there was a lack of data to indicate subduction in Tibet after the Early Eocene (38 Ma). Recent work, however, changes this, as the youngest evidence now known, from the Lhasa terrane, for magmatism related to the subduction of Tethyan oceanic lithosphere comes from the 30.4 ± 0.4 Ma Yaja granodiorite exposed SE of Lhasa (Harrison *et al.* 2000). As dating of rocks within the Lhasa terrane is still far from exhaustive, other young subduction-related rocks may await discovery.

As lithospheric material is subducted, it should only take around one million years to reach depths at which the dehydration of water-saturated and hydrated oceanic lithosphere generates a sufficient fluid flux into the overlying mantle wedge such that magma generation is possible (Ernst 1999). The nature of continental crust is such that it is unlikely to produce a melt in response to subduction. As the buoyancy of continental crust is such that once it reaches depths of 100–150 km it is no longer dense enough to sink into the asthenosphere, the dehydrated oceanic lithosphere breaks off and continues to sink into the asthenosphere. Arc magmas are no longer generated and the collision zone rises isostatically. In modern collision settings such as Taiwan, Timor and Papua New Guinea, a cessation of magmatism is attendant on the subduction of continental crust. Notably, even the youngest previously reported subduction-related rocks of 38 Ma substantially post-date previously suggested timings for collision initiation. The new age data (Harrison *et al.* 2000) seem to require that subduction must have continued in some form until the Late Oligocene.

The sedimentary record is highly sensitive to change, and is widely used as a proxy for phenomena such as climatic variation on a millennial scale. It is therefore reasonable to suggest that the sedimentary record within, and peripheral to, Tibet should closely reflect the tectonic evolution of area. It may also be expected that the sedimentary response to any change in tectonic environment, such as continent–continent collision, should be geologically immediate.

Cenozoic sedimentary units within foreland basin sediments of western and central Nepal directly record tectonic events associated with the India–Asia collision (Najman *et al.* 1997; DeCelles *et al.* 1998a, 1998b). Notably, Upper Eocene to Oligocene sediments are mostly carbonates and reflect tectonic quiescence. In the south of Nepal, fluvial sedimentation in the Siwalik Basin began during the latest Oligocene to Miocene, and associated sediments represent the first influx of coarse sediment into a foredeep migrating southward from the developing collision zone (DeCelles *et al.* 1998a). Up-section evolution of the Siwalik Group through the Miocene and Pliocene corresponds closely to the development of major thrust systems within the Himalaya and the exposure of progressively deeper crustal levels along these faults.

In the Indian Himalayan zone, molasse was also deposited in a fore-deep south of a collision zone. The Subathu Formation in northern India contains coarse-grained Upper Paleocene–Middle Eocene sediments derived from an uplifted ophiolitic and volcanic arc source to the north (Najman & Garzanti 2000; Najman *et al.* 2000). Suture zone input was drastically reduced subsequent to deposition of this unit, and no sediments were derived from any collision zone again until the end of the Oligocene. Sedimentary units of the Balakot Formation, which occur in a foreland basin in Pakistan, contain orogen-derived detritus. These rocks were previously thought to be uppermost Palaeocene–Mid-Eocene (Bossart & Ottiger 1989). Recent studies show that they contain 36–40 Ma detrital micas (Najman *et al.* 2001). Therefore, they must be younger than previously believed. The Upper Palaeocene to Lower Eocene Ghazij Formation of central Pakistan contains volcanic rock fragments and detrital chromite grains indicative of collision between an arc (including ophiolite) and a continent. This unit is, however, overlain by Middle Eocene limestones which indicate tectonic quiescence, and no further influx of collision-derived coarse clastic material is seen until the Miocene to Pliocene.

Sedimentary rocks in the northern Himalaya have long been regarded (Rowley 1996*a*, *b*) as a key indicator of possible timing of collision, and the youngest previously reported marine strata were of Eocene age. New data from north of Qomolangma (Mount Everest) (Wang *et al.* 2002) clearly indicate that marine conditions prevailed until at least the latest Eocene/earliest Oligocene (*c.*34 Ma NP20). The top of the section has been removed by erosion. It seems improbable that marine deposition could continue along the axis of collision if it had initiated 20 Ma earlier.

Extensive molasse deposits of the Gangrinboche facies (Aitchison *et al.* 2002*b*), which crop out along the Indus–Yarlung Tsangpo suture zone, are clearly likely to have developed in association with a collisional event. Previous interpretations of the timing of collision (Searle *et al.* 1987) have, at least in part, been based on ages assigned to some of these units. As mentioned above, however, at least two distinctive molasse units of different ages are present, and each requires separate interpretation. The Kailas, Qiuwu, Dazhuqu and Luobusa formations are correlative units that all record a similar history of sedimentation (Aitchison *et al.* 2002*b*). The best available age constraints are considerably younger than the previously inferred Eocene estimate and unequivocally indicate that these units are post mid-Oligocene. Clasts within lower portions of the Kailas Formation were not exposed at the surface prior to 19–18 Ma and were thus unavailable as a source of sediment (Harrison *et al.* 1993). In each unit, initial sediments appear to have been derived from the north, and locally are deposited upon a basement of Lhasa terrane rocks. Up-section, clasts of Tethyan and Indian affinity, derived from south of the Indus–Yarlung Tsangpo suture zone appear then to gradually become the dominant components. In the Kailas region this has been interpreted to reflect initial south-directed and later north-directed thrusting (Yin *et al.* 1999). This pattern of evolution of clast sources is consistent along at least 2000 km of strike length. Similar molasse sediments are

also known from the Indus suture (Garzanti & Van Haver 1988; Searle *et al.* 1990). However, biostratigraphic control is generally poor to absent in these continental molasse deposits (Garzanti & Van Haver 1988), which can only be accorded an age younger than Early Eocene.

A significant sediment flux is widely observed in basins within the broader region surrounding the Tibet–Qinghai Plateau during the Late Oligocene to Early Miocene (France *et al.* 1993; Métivier *et al.* 1999). This too can presumably be directly correlated to the collision of India and Asia and uplift associated with this event.

The oldest Eohimalayan metamorphism reported from the Everest region is a peak Barrovian event dated at 32.2 ± 0.4 Ma, and post-dates the presently inferred Palaeocene/Eocene boundary age timing for initiation of collision by 20–25 Ma (Simpson *et al.* 2000). Similar Oligocene ages are reported from elsewhere in the Greater Himalaya (Hodges *et al.* 1996; Coleman 1998; Godin *et al.* 1999), while data that indicate Eocene metamorphism are lacking. Widespread collision-related crustal melting gave rise to generation of the High Himalayan Crystalline Series leucogranites and Neohimalayan metamorphism in the Early and Mid-Miocene 21–17 Ma ago (Le Fort *et al.* 1987; Harrison *et al.* 1995*c*; Hodges *et al.* 1996; Searle 1996, 1999*a*, *b*; Edwards & Harrison 1997; Coleman 1998; Wu *et al.* 1998). Leucogranites were emplaced along the top of the High Himalayan slab (Searle 1999*b*) during extension at 21–17 Ma. The North Himalayan granite-gneiss dome belt is a slightly younger series of high-grade metamorphic and igneous rocks that crop out to the north of the High Himalaya. Protolith ages for these rocks are consistent with their once having been part of the Indian continental crust. Melting, formation of these granite bodies and their emplacement occurred at 17.6–9.5 Ma (Harrison *et al.* 1997*a*, 1998).

Significant structural shortening, such as might be expected to accompany continent–continent collision, occurred with the development of major Late Cenozoic thrust systems along the Indus–Yarlung Tsangpo suture zone and further to the south within the Tethyan Himalaya, Greater Himalaya, Lesser Himalaya and in the front of the Himalaya. North-directed backthrusting along the Great Counter thrust or Renbu–Zedong thrust system occurred between 19 and 10 Ma (Yin *et al.* 1999), coeval with development of the North Himalayan granite–gneiss dome belt and the South Tibetan Detachment System. Additional south-directed thrusting occurred within the Himalayan zone along major structures such as the Main Central Thrust.

Amphibolite-facies shear zones are the oldest dated structures associated with this thrust system, and developed between 23 and 20 Ma (Hodges *et al.* 1996; Harrison *et al.* 1997*b*). Ion microprobe dating of synkinematic monazites indicates significant Late Miocene-Pliocene displacement along this structure (Harrison *et al.* 1997*b*). South-directed thrust systems become progressively younger further south across the Himalaya. Sedimentation patterns have been used to infer initial movement on the Main Boundary Thrust system between 11 and 9 Ma (Meigs *et al.* 1995). This fault may still be active as it cuts Pliocene molasse deposits (DeCelles *et al.* 1998*b*). The toe of the Himalayan orogenic wedge is marked by the active Main Frontal Thrust system (Molnar 1984). These structures and their age progression are exactly what

might be expected of a collisional system. The Early Miocene initiation of such thrusting might be expected to have been concomitant with the beginning of collision.

The South Tibetan Detachment System is a Miocene structure for which radiometric ages constrain displacement to between 17 and 9 Ma (Yin & Harrison 2000). This structure is a north-directed low-angle normal fault system traceable along the length of the Himalaya (Burg *et al.* 1984*a*; Burchfiel *et al.* 1992), along which low-grade Tethyan metasediments in the hanging wall are placed against a footwall of Himalayan granites and gneisses.

A series of regularly spaced north–south-trending rifts occurs across the Tibetan Plateau (Tapponnier & Molnar 1977; Molnar & Tapponnier 1978; Ni & York 1978; Tapponnier

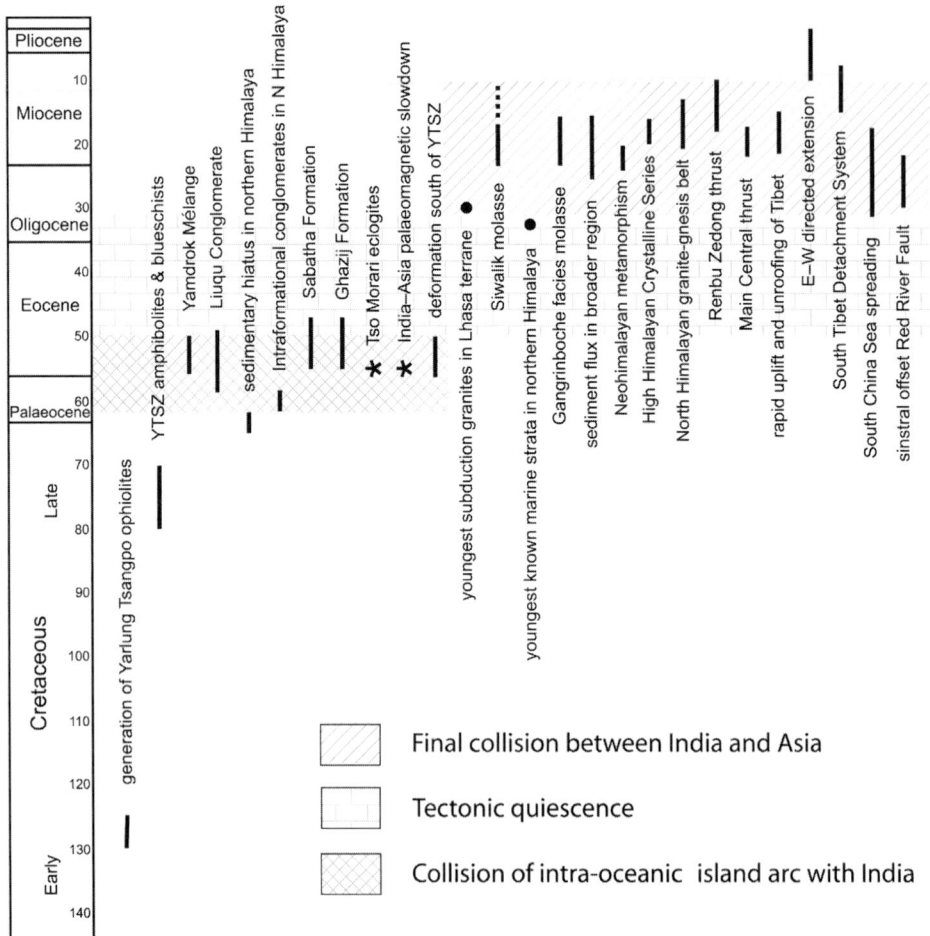

Fig. 2. Time–space plot of geological phenomena potentially related to collision(s) along the Yarlung Tsangpo suture zone.

et al. 1981a; Armijo *et al.* 1986, 1989; Mercier *et al.* 1987; Burchfiel *et al.* 1991; Yin 2000). High-grade metamorphic rocks are locally exposed along these rifts and, in some areas, rifts, such as the Thakkola Graben, were a locus for sedimentation (Garzione *et al.* 2000). In the Nyainquentanghla region the onset of east–west extension is interpreted to be associated with these rifts and is constrained at 8 ± 1 Ma (Pan & Kidd 1992; Harrison *et al.* 1995a). The prevailing interpretation of these rifts is that they developed in response to east–west collapse of orogenically over-thickened Tibetan crust (Molnar & Tapponnier 1978; England & Houseman 1989; Harrison *et al.* 1992), although recent studies suggest that a more complex explanation may be required (Yin 2000). Notably, orogen-parallel extension is not uncommon in modern collisions such as Taiwan where extensional collapse has followed shortly after collision (Crespi *et al.* 1996).

One model suggesting a likely tectonic response to the India–Asia collision is that of extrusion, or escape, tectonics (Tapponnier *et al.* 1982; Peltzer & Tapponnier 1988). Phenomena that may have developed in this manner are all Late Oligocene to Miocene and younger. Considerable left-lateral slip on the Red River Fault occurred during the Early Miocene (Harrison *et al.* 1995b; Leloup *et al.* 1995) synchronous with South China Sea spreading at 32–17 Ma (Briais *et al.* 1993).

Summary

Prevailing models for the India–Asia collision, the greatest on-going collision event in existence, suggest an anomalous situation where collision between two major continental fragments was apparently not followed by significant mountain-building orogenesis for at least 20 million years. Data now available appear to indicate that, in the southern central portion of Tibet at least, marine deposition continued until at least the end of the Eocene. Subduction-related magmatism, which continued until the Mid-Oligocene, was followed by rapid uplift of the Tibet–Qinghai Plateau. Mass-wasting of sediments into molasse basins began in the latest Oligocene. Either existing models that invoke early Eocene India–Asia collision at 55 Ma are flawed, or collision occurred at a different time. Recent work has produced sufficient new data to allow the recognition of two different tectonic events (Fig. 2). Features related to each event require separate interpretation and no collisional continuum should be assumed. One event occurred around the Palaeocene/Eocene boundary, another at some time in the Oligocene (Fig. 3).

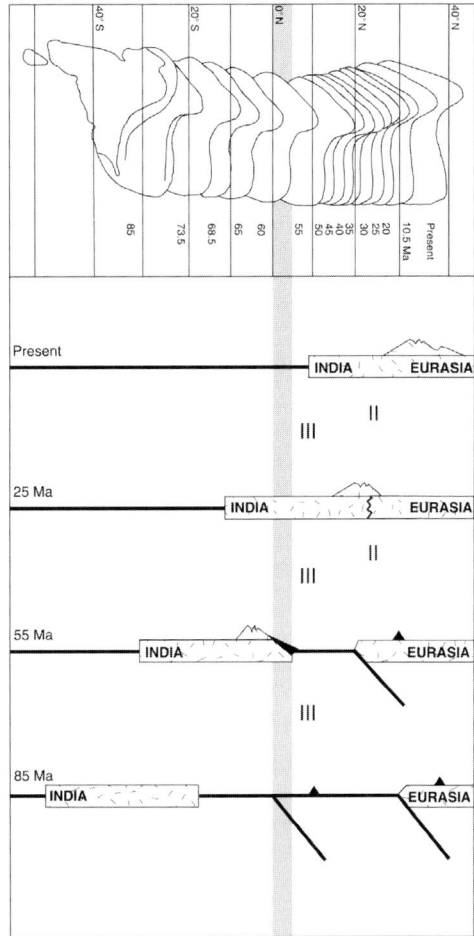

Fig. 3. Northward migration of India towards Eurasia. The outline of the Indian continental mass and its migration, shown at the top of the diagram, are based on Klootwijk *et al.* (1992). Cartoon indicates tectonic evolution of the area including the development of an intra-oceanic island arc within Tethys. This arc collided with India in the zone of shaded latitudes during the Palaeocene. Culmination of the collision of this arc with India and removal of the associated convergent plate boundary at around 55 Ma effectively reduced the net convergence rate between India and Eurasia to a rate more realistically associated with a single convergent plate margin. Approximate locations of different subduction zone remnants (II and III) recognized on seismic tomographs by Van der Voo *et al.* (1999) are indicated. Final collision between India and Asia probably occurred in the Oligocene.

Many models for the response of continental lithosphere to the India-Asia collision have been proposed. They include: lithospheric thickening (England & Thompson 1986); lateral extrusion

(Tapponnier *et al.* 1982; Peltzer & Tapponnier 1988); continental subduction (Argand 1924; Powell & Conaghan 1973; Zhao & Morgan 1987; Willett & Beaumont 1994; Jin *et al.* 1996; Owens & Zandt 1997); lower-crustal channel flow (Bird 1978; Zhao & Morgan 1987; Avouac & Burov 1996; Royden *et al.* 1997) and stepwise growth of the plateau (Tapponnier *et al.* 2001). All of these models are predicated on the assumption of a 55 Ma collision. Estimates, built into these models, of the total shortening observed in Tibet and the Himalaya, always seem to fall short of what might be required of an earliest Eocene India–Asia collision. Reassessment of available data and the recognition of an early arc–continent collision along the northern edge of India significantly reduce this apparent paradox, and provide an interpretation consistent with observed phenomena. The requirement for phenomena such as crustal delamination, in order to explain why uplift occurred 20–25 Ma after collision is removed if uplift accompanied collision, as it appears to do everywhere else where collision has occurred. In the present situation, the 55 Ma hypothesis appears to have become more important than the data (or lack thereof) upon which it is based. A better understanding of the timing and multi-phase nature of the India–Asia collision should provide constraints for the development of more actualistic models for development of the Tibet–Qinghai Plateau and surrounding areas.

On-going Tibet research at The University of Hong Kong is supported by grants from the Research Grants Council of the Hong Kong Special Administrative Region, China (Project Nos HKU7102/98P, 7299/99P and 7069/01P. Although they do not necessarily agree with us, development of ideas presented in this paper has benefited from discussions with Badengzhu, T. M. Harrison, P. Tapponnier, A. Yin, and our colleagues and students here at HKU. Tony Barber is thanked for his helpful and constructive review, which greatly improved the manuscript.

References

ABRAJEVITCH, A., AITCHISON, J. C. & ALI, J. R. 2001. Paleomagnetism of the Dazhuqu Terrane, Yarlung Zangbo Suture Zone, Southern Tibet, *EOS, Transactions of the American Geophysical Union*, **82(47) Suppl.**, Fall Meeting, Abstract GP11A-0188, p. F314.

AITCHISON, J. C., BADENGZHU *et al.* 2000. Remnants of a Cretaceous intra-oceanic subduction system within the Yarlung–Zangbo suture (southern Tibet). *Earth and Planetary Science Letters*, **183**, 231–244.

AITCHISON, J. C., ABRAJEVITCH, A. *et al.* 2002*a*. New insights into the evolution of the Yarlung Tsangpo suture zone, Xizang (Tibet), China. *Episodes*, **25(3)**, 90–94.

AITCHISON, J. C., DAVIS, A. M., BADENGZHU & LUO, H. 2002*b*. New constraints on the India-Asia collision: the Lower Miocene Gangrinboche conglomerates, Yarlung Tsangpo suture zone, SE Tibet. *Journal of Asian Earth Sciences*, **21**, 253–265.

ANCZKIEWICZ, R., BURG, J. P., VILLA, I. M. & MEIER, M. 2000. Late Cretaceous blueschist metamorphism in the Indus Suture Zone, Shangla region, Pakistan Himalaya. *Tectonophysics*, **324**, 111–134.

ARGAND, E. 1924. La tectonique de l'Asie. *Proceedings of the 13th International Geological Congress*, **7**, 171–372.

ARMIJO, R., TAPPONNIER, P., MERCIER, J. L. & HAN, T. L. 1986. Quaternary extension in southern Tibet; field observations and tectonic implications. *Journal of Geophysical Research*, **91B**, 13 803–13 872.

ARMIJO, R., TAPPONNIER, P. & HAN, T. 1989. Late Cenozoic right-lateral strike-slip faulting in southern Tibet. *Journal of Geophysical Research*, **94B**, 2787–2838.

AVOUAC, J. P. & BUROV, E. B. 1996. Erosion as a driving mechanism of intracontinental mountain growth. *Journal of Geophysical Research*, **101B**, 17 747–17 769.

BADENGZHU. 1979. (Compiler) *Xizang Autonomous Region Zhanang–Sangri Regional Geology Reconnaissance Map 1 : 50 000*. Xizang Geological Survey Geological Team #2, Group 2, Lhasa.

BECK, R. A., BURBANK, D. W. 1995. Stratigraphic evidence for an early collision between northwest India and Asia. *Nature*, **373**, 55–58.

BIRD, P. 1978. Initiation of intracontinental subduction in the Himalaya. *Journal of Geophysical Research*, **83B**, 4975–4987.

BLOOMER, S. H., TAYLOR, B., MACLEOD, C. J., STERN, R. J., FRYER, P., HAWKINS, J. W. & JOHNSON, L. 1995. Early arc volcanism and the ophiolite problem; a perspective from drilling in the western Pacific. *American Geophysical Union Geophysical Monograph*, **88**, 1–30.

BOSSART, P. & OTTIGER, R. 1989. Rocks of the Murree Formation in Northern Pakistan: indicators of a descending foreland basin of Late Palaeocene to middle Eocene age. *Eclogae Geologicae Helvetiae*, **82**, 133–165.

BRIAIS, A., PATRIAT, P. & TAPPONNIER, P. 1993. Updated interpretation of magnetic anomalies and seafloor spreading stages in the South China Sea; implications for the Tertiary tectonics of Southeast Asia. *Journal of Geophysical Research*, **98B**, 6299–6328.

BURCHFIEL, B. C., CHEN, Z., ROYDEN, L. H., LIU, Y. & DENG, C. 1991. Extensional development of Gabo Valley, southern Tibet. *Tectonophysics*, **194**, 187–193.

BURCHFIEL, B. C., CHEN, Z., HODGES, K. V., LIU, Y., ROYDEN, L. H., DENG, C. & XU, J. 1992. The South

Tibetan detachment system, Himalayan Orogen; extension contemporaneous with and parallel to shortening in a collisional mountain belt. *Geological Society of America Special Paper*, **269**, 1–41.

BURG, J. P., BRUNEL, M., GAPAIS, D., CHEN, G. M. & LIU, G. H. 1984a. Deformation of leucogranites of the crystalline Main Central Sheet in southern Tibet (China). *Journal of Structural Geology*, **6**, 535–542.

BURG, J. P. & CHEN, G. M. 1984. Tectonics and structural zonation of southern Tibet, China. *Nature*, **311**, 219–223.

BURG, J. P., GUIRAUD, M., CHEN, G. M. & LI, G. C. 1984b. Himalayan metamorphism and deformations in the North Himalayan Belt (southern Tibet, China). *Earth and Planetary Science Letters*, **69**, 391–400.

BURG, J. P., LEYRELOUP, A., GIRARDEAU, J. & CHEN, G. M. 1987. Structure and metamorphism of a tectonically thickened continental crust; the Yalu Tsangpo suture zone (Tibet). *Philosophical Transactions of the Royal Society of London, Series A: Mathematical and Physical Sciences*, **321**, 67–86.

BUTLER, R. 1995. When did India hit Asia? *Nature*, **373**, 20–21.

CHANG, C., SHACKLETON, R. M., DEWEY, J. F. & YIN, J. (eds) 1988. *The Geological Evolution of Tibet; Report of the 1985 Royal Society–Academia Sinica Geotraverse of the Qinghai–Xizang Plateau*. Royal Society, London.

CHANG, C. F. 1996. *Geology and Tectonics of Qinghai–Xizang Plateau*. Science Press, Beijing.

CHANG, C. P., ANGELIER, J. & HUANG, C. Y. 2000. Origin and evolution of a mélange: the active plate boundary and suture zone of the Longitudinal Valley, Taiwan. *Tectonophysics*, **325**, 43–62.

CHUNG, S.-L., WANG, K.-L., CRAWFORD, A. J., KAMENETSKY, V. S., CHEN, C.-H., LAN, C.-Y. & CHEN, C.-H. 2001. High-Mg potassic rocks from Taiwan: implications for the genesis of orogenic potassic lavas. *Lithos*, **59**(4), 153–170.

COLEMAN, M. E. 1998. U–Pb constraints on Oligocene–Miocene deformation and anatexis within the central Himalaya, Marsyandi Valley, Nepal. *American Journal of Science*, **298**, 553–571.

COPELAND, P., HARRISON, T. M., KIDD, W. S. F., XU, R. & ZHANG, Y. 1987. Rapid early Miocene acceleration of uplift in the Gangdese Belt, Xizang (southern Tibet), and its bearing on accommodation mechanisms of the India–Asia collision. *Earth and Planetary Science Letters*, **86**, 240–252.

COPELAND, P. & HARRISON, T. M. 1990. Episodic rapid uplift in the Himalaya revealed by ^{40}Ar/^{39}Ar analysis of detrital K-feldspar and muscovite, Bengal Fan. *Geology*, **18**, 354–357.

CORFIELD, R. I., SEARLE, M. P. & GREEN, O. R. 1999. Photang thrust sheet; an accretionary complex structurally below the Spontang Ophiolite constraining timing and tectonic environment of ophiolite obduction, Ladakh Himalaya, NW India. *Journal of the Geological Society of London*, **156** 1031–1044.

CORFIELD, R. I., SEARLE, M. P. & PEDERSEN, R. B. 2001. Tectonic setting, origin, and obduction history of the Spontang Ophiolite, Ladakh Himalaya, NW India. *Journal of Geology*, **109**, 715–736.

CRESPI, J. M., CHAN, Y. C. & SWAIM, M. S. 1996. Synorogenic extension and exhumation of the Taiwan hinterland. *Geology*, **24**, 247–250.

DAVIS, A. M., AITCHISON, J. C., BADENGZHU, LUO, H. & ZYABREV, S. 2002. Paleogene island arc collision-related conglomerates, Yarlung–Tsangpo suture zone, Tibet. *Sedimentary Geology*, **150**, 247–273.

DAVIS, A. M., AITCHISON, J. C., BADENGZHU & HUI, L. 2004. Conglomerates of the Yarlung Tsangpo suture zone, southern Tibet. *In:* MALPAS, J. G., FLETCHER, C. J. N., ALI, J. R. & AITCHISON, J. C. *Aspects of the Tectonic Evolution of China*. Geological Society, London, Special Publications, **226**, 235–246.

DE SIGOYER, J., CHAVAGNAC, V., BLICHERT-TOFT, J., VILLA, I. M., LUAIS, B., GUILLOT, S., COSCA, M. & MASCLE, G. 2000. Dating the Indian continental subduction and collisional thickening in the northwest Himalaya: multichronology of the Tso Marari eclogites. *Geology*, **28**, 487–490.

DECELLES, P. G., GEHRELS, G. E., QUADE, J. & OHJA, T. P. 1998a. Eocene–early Miocene foreland basin development and the history of Himalayan thrusting, western and central Nepal. *Tectonics*, **17**, 741–765.

DECELLES, P. G., GEHRELS, G. E., QUADE, J., OJHA, T. P., KAPP, P. A. & UPRETI, B. N. 1998b. Neogene foreland basin deposits, erosional unroofing, and the kinematic history of the Himalayan fold–thrust belt, western Nepal. *Geological Society of America Bulletin*, **110**, 2–21.

DEWEY, J. F., CANDE, S. C. & PITMAN, W. C., III. 1989. Tectonic evolution of the India/Eurasia collision zone. *Eclogae Geologicae Helvetiae*, **82**, 717–734.

DÜRR, S.B. 1996. Provenance of Xigaze fore-arc basin clastic rocks (Cretaceous, South Tibet). *Geological Society of America Bulletin*, **108**, 669–684.

EDWARDS, M. A. & HARRISON, T. M. 1997. When did the roof collapse? Late Miocene north–south extension in the high Himalaya revealed by Th–Pb monazite dating of the Khula Kangri Granite. *Geology*, **25**, 543–546.

ENGLAND, P. C. & THOMPSON, A. 1986. Some thermal and tectonic models for crustal melting in continental collision zones. *In:* COWARD, M. P. & RIES ALISON, C. (eds) *Collision Tectonics*. Geological Society, London, Special Publications, **19**, 83–94.

ENGLAND, P. C. & HOUSEMAN, G. 1989. Extension during continental convergence, with application to the Tibetan Plateau. *Journal of Geophysical Research*, **94B**, 17 561–17 579.

EINSELE, G., LIU, B. *et al.* 1994. The Xigaze forearc basin; evolution and facies architecture (Cretaceous, Tibet). *Sedimentary Geology*, **90**, 1–32.

ERNST, W. G. 1999. Hornblende, the continent maker–evolution of H_2O during circum-Pacific subduction versus continental collision. *Geology*, **27**, 675–678.

FRANCE, L. C., DERRY, L. & MICHARD, A. 1993. Evolution of the Himalaya since Miocene time; isotopic and sedimentological evidence from the Bengal Fan. *In:* TRELOAR, P. J. & SEARLE, M. P. (eds) *Himalayan Tectonics.* Geological Society, London, Special Publications, **74**, 603–621.

GARZANTI, E. & VAN HAVER, T. 1988. The Indus clastics: forearc basin sedimentation in the Ladakh Himalaya (India). *Sedimentary Geology*, **59**, 237–249.

GARZIONE, C. N., DETTMAN, D. L., QUADE, J., DECELLES, P. G. & BUTLER, R. F. 2000. High times on the Tibetan Plateau; paleoelevation of the Thakkhola Graben, Nepal. *Geology*, **28**, 339–342.

GIRARDEAU, J. & MERCIER, J. C. C. 1988. Petrology and texture of the ultramafic rocks of the Xigaze Ophiolite (Tibet); constraints for mantle structure beneath slow-spreading ridges. *Tectonophysics*, **147**, 33–58.

GIRARDEAU, J., MARCOUX, J. & ZAO, Y. 1984a. Lithologic and tectonic environment of the Xigaze ophiolite (Yarlung Zangbo suture zone, southern Tibet, China), and kinematics of its emplacement. *Eclogae Geologicae Helvetiae*, **77**, 153–170.

GIRARDEAU, J., NICOLAS, A. *et al.* 1984b. Les ophiolites de Xigaze et la suture du Yarlung Zangbo (Tibet). *In:* MERCIER, J. L. & GUANGCEN, L. (eds) *Mission Franco-Chinoise au Tibet 1980. Etude Géologique et Géophysique de la Croûte Terrestre et du Manteau Supérieur du Tibet et de l'Himalaya.* CNRS, Paris, 189–197.

GIRARDEAU, J., MERCIER, J. C. & XIBIN, W. 1985a. Petrology of the mafic rocks of the Xigaze ophiolite, Tibet; implications for the genesis of the oceanic lithosphere. *Contributions to Mineralogy and Petrology*, **90**, 309–321.

GIRARDEAU, J., MERCIER, J. C. C. & ZAO, Y. G. 1985b. Origin of the Xigaze Ophiolite, Yarlung Zangbo suture zone, southern Tibet. *Tectonophysics*, **119**, 407–433.

GIRARDEAU, J., MERCIER, J. C. C. & ZAO, Y. G. 1985c. Structure of the Xigaze Ophiolite, Yarlung Zangbo suture zone, southern Tibet, China; genetic implications. *Tectonics*, **4**, 267–288.

GODIN, L., BROWN, R. L., HANMER, S. & PARRISH, R. 1999b. Back folds in the core of the Himalaya Orogen: An alternative interpretation. *Geology*, **27**, 151–154.

GORDON, R. G. & STEIN, S. 1992. Global tectonics and space geodesy. *Science*, **259**, 333–342.

HARRIS, N. B. W., HOLLAND, T. J. B. & TINDLE, A. G. 1988. Metamorphic rocks of the 1985 Tibet Geotraverse, Lhasa to Golmud. *Philosophical Transactions of the Royal Society of London, Series A: Mathematical and Physical Sciences*, **327**, 203–213.

HARRISON, T. M., COPELAND, P., KIDD, W. S. F. & YIN, A. 1992. Raising Tibet. *Science*, **255**, 1663–1670.

HARRISON, T. M., COPELAND, P., HALL, S. A., QUADE, J., BURNER, S., OHJA, T. P. & KIDD, W. S. F. 1993. Isotopic preservation of Himalayan/Tibetan uplift, denudation, and climate histories of two molasse deposits. *Journal of Geology*, **100**, 157–175.

HARRISON, T. M., COPELAND, P., KIDD, W. S. F. & LOVERA, O. M. 1995a. Activation of the Nyain-

quentanghla shear zone: Implications for uplift of the southern Tibetan Plateau. *Tectonics*, **14**, 658–676.

HARRISON, T. M., GROVE, M., LOVERA, O. M. & CATLOS, E. J. 1998. A model for the origin of Himalayan anatexis and inverted metamorphism. *Journal of Geophysical Research*, **103B**, 27 017–27 032.

HARRISON, T. M., LELOUP, P. H., RYERSON, F. J., TAPPONNIER, P., LACASSIN, R. & CHEN, W. 1995b. Diachronous initiation of transtension along the Ailao Shan–Red River Shear zone Yunnan and Vietnam. *In:* YIN, A. & HARRISON, T. M. (eds) *The Tectonic Evolution of Asia.* Cambridge University Press, New York, 208–226.

HARRISON, T. M., LOVERA, O. M. & GROVE, M. 1997a. New insights into the origin of two contrasting Himalayan granite belts. *Geology*, **25**, 899–902.

HARRISON, T. M., MAHON, K. I., GUILLOT, S., HODGES, K., LE FORT, P. & PECHER, A. 1995c. New constraints on the age of the Manaslu leucogranite: Evidence for episodic tectonic denudation in the central Himalaya; discussion and reply. *Geology*, **23**, 478–480.

HARRISON, T. M., RYERSON, F. J., LE FORT, P., YIN, A., LOVERA, O. M. & CATLOS, E. J. 1997b. A late Miocene–Pliocene origin for the central Himalayan inverted metamorphism. *Earth and Planetary Science Letters*, **146**, E1–E7.

HARRISON, T. M., YIN, A., GROVE, M., LOVERA, O. M., RYERSON, F. J. & ZHOU, X. 2000. The Zedong Window: a record of superposed Tertiary convergence in southeastern Tibet. *Journal of Geophysical Research*, **105(8)**, 19 211–19 230.

HÉBERT, R., BEAUDOIN, G., VARFALVY, V., HUOT, F., WANG, C. S. & LIU, Z. F. 2000. Yarlung Zangbo ophiolites, southern Tibet revisited, 15th Himalaya–Karakorum–Tibet workshop, Chengdu. *Earth Science Frontiers*, **7**, 124–126.

HILL, K. C. & RAZA, A. 1999. Arc–continent collision in Papua New Guinea: constraints from fission track thermochronology. *Tectonics*, **18**, 950–966.

HODGES, K. V. 2000. Tectonics of the Himalaya and southern Tibet from two perspectives. *Geological Society of America Bulletin*, **112**, 324–350.

HODGES, K. V., PARRISH, R. R. & SEARLE, M. P. 1996. Tectonic evolution of the central Annapurna Range, Nepalese Himalayas. *Tectonics*, **15**, 1264–1291.

HONEGGER, K., LE FORT, P., MASCLE, G. & ZIMMERMANN, J. L. 1989. The blueschists along the Indus suture zone in Ladakh, NW Himalaya. *Journal of Metamorphic Petrology*, **7**, 57–73.

HUANG, C. Y., YUAN, P. B., LIN, C. W., WANG, T. K. & CHANG, C. P. 2000. Geodynamic processes of Taiwan arc–continent collision and comparison with analogs in Timor, Papua New Guinea, Urals and Corsica. *Tectonophysics*, **325**, 1–21.

JIN, Y., MCNUTT, M. K. & ZHU, Y. 1996. Mapping the descent of Indian and Eurasian plates beneath the Tibetan Plateau from gravity anomalies. *Journal of Geophysical Research*, **101B**, 11 275–11 290.

KLOOTWIJK, C. T., GEE, J. S., PEIRCE, J. W. & SMITH, G. M. 1991. Constraints on the India–Asia convergence: Paleomagnetic results from Ninetyeast Ridge. *In:* PEIRCE, J. W., WEISSEL, J. K. *et al.* (eds) *Proceedings of the Ocean Drilling Program, Scientific Results,* **Leg 121**, 777–882.

KLOOTWIJK, C. T., GEE, J. S., PEIRCE, J. W., SMITH, G. M. & MCFADDEN, P. L. 1992. An early India–Asia contact; paleomagnetic constraints from Ninetyeast Ridge, ODP Leg 121; with Suppl. Data 92–15. *Geology,* **20**, 395–398.

KOHN, M. J. & PARKINSON, C. D. 2002. Petrologic case for Eocene slab breakoff during the Indo-Asian collision. *Geology,* **30**, 591–594.

LE FORT, P. 1996. Evolution of the Himalaya. *In:* YIN, A. & HARRISON, T. M. (eds) *The Tectonics of Asia.* Cambridge University Press, New York, 95–106.

LE FORT, P., CUNEY, M., DENIEL, C., FRANCE-LANORD, C., SHEPPARD, S. M. F., UPRETI, B. N. & VIDAL, P. 1987. Crustal generation of Himalayan leucogranites. *Tectonophysics,* **134**, 39–57.

LELOUP, P. H., LACASSIN, R. *et al.* 1995. The Ailao Shan–Red River shear zone (Yunnan, China), Tertiary transform boundary of Indochina. *Tectonophysics,* **251**, 3–84.

LIN, J. & WATTS, D. R. 1988. Palaeomagnetic results from the Tibetan Plateau. *In:* Chang, C., SHACKLETON, R. M., DEWEY J. F. & YIN, J. (eds) The geological evolution of Tibet: Report of the 1985 Royal Society–Academia Sinica geotraverse of the Qinghai–Xizang Plateau. *Philosophical Transactions of the Royal Society of London, Series A: Mathematical and Physical Sciences,* **327**, 239–262.

LIU, J. B. 2001. *Yamdrok Mélange, Gyantze district, Xizang (Tibet), China.* M. Phil. thesis, University of Hong Kong.

LIU, J. B. & AITCHISON, J. C. 2002. Upper Paleocene radiolarians from the Yamdrok mélange, south Xizang (Tibet), China. *Micropaleontology,* **48 (Supplement 1)**, 145–154.

LUNDBERG, N., REED, D. L., LIU, C. S. & LIESKE JR, J. 1997. Forearc-basin closure and arc accretion in the submarine suture zone south of Taiwan. *Tectonophysics,* **274**, 5–23.

MCDERMID, I. R. C. 2002. *Zedong terrane, south Tibet.* PhD thesis, University of Hong Kong.

MCDERMID, I. R. C., AITCHISON, J. C., DAVIS, A. M., HARRISON, T. M. & GROVE, M. 2002. The Zedong terrane: a Late Jurassic intra-oceanic magmatic arc within the Yarlung–Zangbo suture zone, southeastern Tibet. *Chemical Geology,* **187**, 267–277.

MAHEO, G., BERTRAND, H., GUILLOT, S., MASCLE, G., PECHER, A., PICARD, C. & DE, S. J. 2000. Evidence of a Tethyan immature arc within the South Ladakh ophiolites (NW Himalaya, India). *Comptes Rendus de l'Académie Sciences Paris, Sciences de la Terre et des Planètes,* **330**, 289–295.

MEIGS, A. J., BURBANK, D. W. & BECK, R. A. 1995. Middle–late Miocene (>10 Ma) formation of the Main Boundary Thrust in the western Himalaya. *Geology,* **23**, 423–426.

MERCIER, J. L. & LI, G. C. (eds) 1984. *La collision Inde-Asie côté Tibet Mission Franco-Chinoise au Tibet 1980. Etude géologique et géophysique de la croûte terrestre et du manteau supérieur du Tibet et de l'Himalaya.* CNRS, Paris.

MERCIER, J. L., ARMIJO, R., TAPPONNIER, P., CAREY, G. E. & HAN, T. L. 1987. Change from late Tertiary compression to Quaternary extension in southern Tibet during the India–Asia collision. *Tectonics,* **6**, 275–304.

MÉTIVIER, F., GAUDEMER, Y., TAPPONNIER, P. & KLEIN, M. 1999. Mass accumulation rates in Asia during the Cenozoic. *Geophysical Journal International,* **137**, 280–318.

MILLER, C., THÖNI, M., FRANK, W., SCHUSTER, R., MELCHER, F., MEISEL, T. & ZANETTI, A. 2003. Geochemistry and tectonomagmatic affinity of the Yungbwa ophiolite, SW Tibet. *Lithos,* **66**, 155–172.

MIYASHIRO, A. 1973. The Troodos ophiolitic complex was probably formed in an island arc. *Earth and Planetary Science Letters,* **19**, 218–224.

MOLNAR, P. 1984. Structure and tectonics of the Himalaya: Constraints and implications of geophysical data. *Annual Review of Earth and Planetary Sciences,* **12**, 489–518.

MOLNAR, P. & TAPPONNIER, P. 1978. Active tectonics of Tibet. *Journal of Geophysical Research,* **83B**, 5361–5375.

MOLNAR, P., ENGLAND, P. & MARTINOD, J. 1993. Mantle dynamics, uplift of the Tibetan Plateau, and the Indian monsoon. *Reviews of Geophysics,* **31**, 357–396.

NAJMAN, Y. & GARZANTI, E. 2000. Reconstructing early Himalayan tectonic evolution and paleogeography from Tertiary foreland basin sedimentary rocks, northern India. *Geological Society of America Bulletin,* **112**, 435–449.

NAJMAN, Y., BICKLE, M. & CHAPMAN, H. 2000. Early Himalayan exhumation: isotopic constraints from the Indian foreland basin. *Terra Nova,* **12**, 28–34.

NAJMAN, Y., PRINGLE, M., GODIN, L. & OLIVER, G. 2001. Dating of the oldest continental sediments from the Himalayan foreland basin. *Nature,* **410**, 194–197.

NAJMAN, Y. M. R., PRINGLE, M. S., JOHNSON, M. R. W., ROBERTSON, A. H. F. & WIJBRANS, J. R. 1997. Laser ^{40}Ar/^{39}Ar dating of single detrital muscovite grains from early foreland-basin sedimentary deposits in India: implications for early Himalayan evolution. *Geology,* **25**, 535–538.

NELSON, K. D., ZHAO, W. 1996. Partially molten middle crust beneath southern Tibet; synthesis of Project INDEPTH results. *Science,* **274**, 1684–1688.

NI, J. & YORK, J. E. 1978. Late Cenozoic tectonics of the Tibetan Plateau. *Journal of Geophysical Research,* **83B**, 5377–5384.

NICOLAS, A., GIRARDEAU, J. *et al.* 1981. The Xigaze ophiolite (Tibet); a peculiar oceanic lithosphere. *Nature,* **294**, 414–417.

NIITSUMA, N. 1999. Rupture and delamination of arc crust due to the arc–arc collision in the South Fossa magna, central Japan. *The Island Arc,* **8**, 441–458.

O'BRIEN, P. J., ZOTOV, N., LAW, R., KHAN, M. A. & JAN, M. Q. 2001. Coesite in Himalayan eclogite and implications for models of India–Asia collision. *Geology,* **29**, 435–438.

OWENS, T. J. & ZANDT, G. 1997. Implications of crustal property variations for models of Tibetan Plateau evolution. *Nature*, **387**, 37–43.

PAN, Y. & KIDD, W. S. F. 1992. Nyainqentanglha shear zone: a late Miocene extensional detachment in the southern Tibetan Plateau. *Geology*, **20**, 775–778.

PATRIAT, P. & ACHACHE, J. 1984. India–Eurasia collision chronology has implications for crustal shortening and driving mechanism of plates. *Nature*, **311**, 615–621.

PEDERSEN, R. B., SEARLE, M. P. & CORFIELD, R. I. 2001. U–Pb zircon ages from the Spontang Ophiolite, Ladakh Himalaya. *Journal of the Geological Society of London*, **158**, 513–520.

PELTZER, G. & TAPPONNIER, P. 1988. Formation and evolution of strike-slip faults, rifts, and basins during India–Asia collision: an experimental approach. *Journal of Geophysical Research*, **93B**, 15 085–15 117.

POWELL, C. M. & CONAGHAN, P. J. 1973. Plate tectonics and the Himalayas. *Earth and Planetary Science Letters*, **84**, 87–99.

RATSCHBACHER, L., FRISCH, W. & LIU, G. 1994. Distributed deformation in southern and western Tibet during and after the India–Asia collision. *Journal of Geophysical Research*, **99B**, 19 917–19 945.

REUBER, I., COLCHEN, M. & MEVEL, C. 1992. The Spontang ophiolite and ophiolitic mélanges of the Zanskar, NW Himalaya, tracing the evolution of the closing Tethys in the Upper Cretaceous to the Early Tertiary. *In:* SINHA, A. K. (eds) *Himalayan Orogen and Global Tectonics.* A. A. Balkema, Rotterdam, 235–266.

RICHARDSON, A. N. & BLUNDELL, D. J. 1996. Continental collision in the Banda Arc. *In:* HALL, R. & BLUNDELL, D. (eds) *Tectonic Evolution of Southeast Asia.* Geological Society, London, Special Publications, **106**, 47–60.

RICHTER, F. M., LOVERA, O. M., HARRISON, T. M. & COPELAND, P. 1991. Tibetan tectonics from $^{40}Ar/^{39}Ar$ analysis of a single K-feldspar sample. *Earth and Planetary Science Letters*, **105**, 266–278.

ROWLEY, D. B. 1996a. Age of initiation of collision between India and Asia; a review of stratigraphic data. *Earth and Planetary Science Letters*, **145**, 1–13.

ROWLEY, D. B. 1996b. Minimum age of initiation of collision between India and Asia north of Everest based on the subsidence history of the Zhepure Mountain Section. *Journal of Geology*, **106**, 1–13.

ROYDEN, L. H., BURCHFIEL, B. C., KING, R. W., WANG, E., CHEN, Z., SHEN, F. & LIU, Y. 1997. Surface deformation and lower crustal flow in eastern Tibet. *Science*, **276**, 788–790.

SEARLE, M. P. 1996. Cooling history, erosion, exhumation, and kinematics of the Himalaya–Karakorum–Tibet orogenic belt. *In:* YIN, A. & HARRISON, M. (eds) *The Tectonic Evolution of Asia.* Cambridge University Press, New York, 110–137.

SEARLE, M. P. 1999a. Extensional and compressional faults in the Everest–Lhotse Massif, Khumbu Himalaya, Nepal. *Journal of the Geological Society of London*, **156**, 227–240.

SEARLE, M. P. 1999b. Emplacement of Himalayan leucogranite by magma injection along giant sill complexes: examples from Cho Oyu, Gyachung Kang and Everest leucogranites (Nepal Himalaya). *Journal of Asian Earth Sciences*, **17**, 773–783.

SEARLE, M. P., PICKERING, K. T. & COOPER, D. J. W. 1990. Restoration and evolution of the intermontane Indus molasse basin, Ladakh Himalaya, India. *Tectonophysics*, **174**, 301–314.

SEARLE, M. P., WINDLEY, B. F. *et al.* 1987. The closing of Tethys and the tectonics of the Himalaya. *Geological Society of America Bulletin*, **98**, 678–701.

SHERVAIS, J. W. 2001. Birth, death, and resurrection: the life cycle of suprasubduction zone ophiolites. *Geochemistry Geophysics Geosystems*, **2**, 2000GC000080.

SHI, X. Y., YIN, J. R. & JIA, C. P. 1996. Mesozoic to Cenozoic sequence stratigraphy and sea-level changes in the northern Himalaya, southern Tibet, China. *Newsletters on Stratigraphy*, **33**, 15–61.

SIMPSON, R. L., PARRISH, R. R., SEARLE, M. P. & WATERS, D. J. 2000. Two episodes of monazite crystallization during metamorphism and crustal melting in the Everest region of the Nepalese Himalaya. *Geology*, **28**, 403–406.

TANG, J. C. & CHEMENDA, A. I. 2000. Numerical modelling of arc–continent collision: application to Taiwan. *Tectonophysics*, **325**, 23–42.

TAO, J. 1988a. The Paleogene flora and palaeoclimate of Liuqu Formation in Xizang. *In:* WHYTE, P., AIGNER, J. S., JABLONSKI, N. G., TAYLOR, G., WALKER, D. & WANG, P. (eds) *The Paleoenvironment of East Asia from the Mid-Tertiary.* Occasional Papers and Monographs–Centre of Asian Studies, **77**, Centre of Asian Studies, University of Hong Kong, Hong Kong, 520–522.

TAO, J. 1988b. Plant fossils from Lepequ formation in Lhaze County, Xizang and their palaeoclimatological singificances. *Academia Sinica Geological Institute Memoir*, **3**, 223–238.

TAPPONNIER, P. & MOLNAR, P. 1977. Active faulting and tectonics in China. *Journal of Geophysical Research*, **82**, 2905–2930.

TAPPONNIER, P., MERCIER, J. L., ARMIJO, R., HAN, T. & ZHOU, J. 1981a. Field evidence for active normal faulting in Tibet. *Nature*, **294**, 410–414.

TAPPONNIER, P., MERCIER, J. L. *et al.* 1981b. The Tibetan side of the India–Eurasia collision. *Nature*, **294**, 405–410.

TAPPONNIER, P., PELTZER, G., Le DAIN, A. Y., ARMIJO, R. & COBBOLD, P. 1982. Propagating extrusion tectonics in Asia: new insights from simple experiments with plasticine. *Geology*, **10**, 611–616.

TAPPONNIER, P., ZHIQIN, X., ROGER, F., MEYER, B., ARNAUD, N., WITTLINGER, G. & JINGSUI, Y. 2001. Oblique stepwise rise and growth of the Tibet Plateau. *Science*, **294**, 1671–1677.

TONARINI, S., VILLA, I. M., OBERLI, F., MEIER, M., SPENCER, D. A., POGNANTE, U. & RAMSAY, J. G. 1993. Eocene age of eclogite metamorphism in Pakistan Himalaya; implications for India–Eurasia collision. *Terra Nova*, **5**, 13–20.

TRELOAR, P. J., O'BRIEN, P. J., PARRISH, R. R. & KHAN, M. A. 2003. Exhumation of early Tertiary, coesite-bearing eclogites from the Pakistan Himalaya. *Journal of the Geological Society*, **160**, 367–376.

VAN DER VOO, R., SPAKMAN, W. & BIJWAARD, H. 1999. Tethyan subducted slabs under India. *Earth and Planetary Science Letters*, **171**, 7–20.

WANG, C. S., LI, X. H., HU, X. M. & JANSA, L. F. 2002. Latest marine horizon north of Qomolangma (Mt Everest): implications for closure of Tethys seaway and collision tectonics. *Terra Nova*, **14**, 114–120.

WANG, X. B., BAO, P. S. & XIAO, X. C. 1987. *Ophiolites of the Yarlung Zangbo (Tsangbo) River, Xizang (Tibet)*. Publishing House of Surveying and Mapping, Beijing.

WARWICK, P. D., JOHNSON, E. A. & KHAN, I. H. 1998. Collision-induced tectonism along the northwestern margin of the Indian subcontinent as recorded in the Upper Paleocene to Middle Eocene strata of central Pakistan (Kirthar and Sulaiman Ranges). *Palaeogeography, Palaeoclimatology, Palaeoecology*, **142**, 201–216.

WILLEMS, H., ZHOU, Z., ZHANG, B. & GRAEFE, K. U. 1996. Stratigraphy of the Upper Cretaceous and lower Tertiary strata in the Tethyan Himalayas of Tibet (Tingri area, China). *Geologische Rundschau*, **85(4)**, 723–754.

WILLETT, S. D. & BEAUMONT, C. 1994. Subduction of Asian lithospheric mantle beneath Tibet inferred from models of continental collision. *Nature*, **369**, 642–645.

WU, C., NELSON, K. D. *et al.* 1998. Yadong cross structure and South Tibetan detachment in the east central Himalaya (89°–90° E). *Tectonics*, **17**, 28–45.

YIN, A., HARRISON, T. M. 1999. Tertiary deformation history of southeastern and southwestern Tibet during the Indo-Asian collision. *Geological Society of America Bulletin*, **111**, 1644–1664.

YIN, A. 2000. Mode of Cenozoic east-west extension in Tibet suggesting a common origin of rifts in Asia during the Indo-Asian collision. *Journal of Geophysical Research*, **105B**, 21 745–21 759.

YIN, A. & HARRISON, T. M. 2000. Geologic evolution of the Himalayan–Tibetan orogen. *Annual Reviews of Earth and Planetary Science*, **28**, 211–280.

YIN, J., XU, J., LIU, C. & LI, H. 1988. The Tibetan Plateau: regional stratigraphic context and previous work. *Philosophical Transactions of the Royal Society of London, Series A: Mathematical and Physical Sciences*, **327**, 5–52.

ZHAO, W. L. & MORGAN, W. J. P. 1987. Injection of Indian crust into Tibetan lower crust: A two-dimensional finite element model study. *Tectonics*, **6**, 489–504.

ZHOU, M. F., ROBINSON, P. T., MALPAS, J. & LI, Z. 1996. Podiform chromitites in the Luobusa Ophiolite (southern Tibet): implications for melt–rock interaction and chromite segregation in the upper mantle. *Journal of Petrology*, **37**, 3–21.

ZHOU, S., MO, X. X., MAHONEY, J. J., ZHANG, S. Q., GUO, T. J. & ZHAO, Z. D. 2002. Geochronology and Nd and Pb isotope characteristics of gabbro dikes in the Luobusha ophiolite, Tibet. *Chinese Science Bulletin*, **47**, 143–146.

ZIABREV, S. 2001. *Tectonic evolution of Dazhuqu and Bainang terranes, Yarlung Zangbo suture, Tibet as constrained by radiolarian biostratigraphy*. PhD thesis, University of Hong Kong.

ZIABREV, S. V., AITCHISON, J. C., ABRAJEVITCH, A., BADENGZHU, DAVIS, A. M. & LUO, H. 2003. Precise radiolarian age constraints on the timing of ophiolite generation and sedimentation in the Dazhuqu terrane, Yarlung–Tsangpo suture zone, Tibet. *Journal of the Geological Society*, **160**, 591–600.

ZIABREV, S. V., AITCHISON, J. C., ABRAJEVITCH, A., BADENGZHU, DAVIS, A. M. & LUO, H. 2004 in press. Bainang Terrane, Yarlung-Tsangpo suture, southern Tibet: a record of intra-Tethyan subduction on the Roof of the World. *Journal of the Geological Society*, **161**(3), 523–538.

Conglomerates record the tectonic evolution of the Yarlung–Tsangpo suture zone in southern Tibet

AILEEN M. DAVIS[1], JONATHAN C. AITCHISON[1], BADENGZHU[2] & LUO HUI[3]

[1]*Tibet Research Group, Department of Earth Sciences, University of Hong Kong, Hong Kong SAR, China (e-mail: davisaileen@hotmail.com)*
[2]*Tibetan Geological Survey, Lhasa, Tibet, China*
[3]*Nanjing Institute of Geology and Palaeontology, Academia Sinica, Nanjing, China*

Abstract: The histories of individual conglomeratic units along the Yarlung–Tsangpo (River) suture zone in southern Tibet reflect significant phases in the Mesozoic to Cenozoic tectonic evolution of this area. Several temporally distinct conglomerate units are recognized along the suture, and their detailed examination permits analysis of the collision between India and Asia.

Upper Jurassic to Lower Cretaceous conglomerates crop out within the Sangri Group along the southern Lhasa terrane. They are dominated by limestone and andesitic volcanic cobbles derived entirely from the Lhasa terrane. These rocks have experienced amphibolite facies metamorphism, and exhibit a strong penetrative regional foliation.

Thick successions of the Palaeocene Liuqu Conglomerate crop out within the suture from Xigaze to Lhaze. They contain detritus sourced from intra-oceanic terranes associated with the suture zone, as well as clasts of Indian affinity, while Lhasa and Xigaze terrane-derived material is notably absent. These conglomerates record an early suture zone event prior to India–Asia collision.

Uppermost Oligocene to Lower Miocene 'Gangrinboche facies' conglomerates crop out on the southern edge of the Lhasa terrane along the length of the suture. Several correlative units within this facies exhibit broadly similar stratigraphic histories. A basal depositional contact upon an eroded Lhasa terrane surface is ubiquitous with initial clast derivation from the north. Up-section, the first arrival of coarse-grained, suture-zone and India-derived clasts, is abrupt. These southerly derived clasts predominate by the top of most sections.

An areally restricted succession of gently dipping Late Neogene ultramafic breccias unconformably overlies folded Liuqu Conglomerate near Quanggong. Other Neogene sediments are extensive west of Mount Kailas. Deposition of coarse clastic sediments is presently continuing along the length of the Yarlung Tsangpo.

Discrimination and detailed investigation of each of these units will improve our understanding of the evolution of the India–Asia collision.

The Cenozoic Yarlung–Tsangpo Suture Zone (YTSZ) is the youngest and most southerly of the sutures that divide the Tibetan Plateau into a number of east–west-trending blocks. It separates Indian rocks from those of the Lhasa terrane to the north, and incorporates Neotethyan oceanic material that once lay between these two areas (Fig. 1). Rocks that crop out along the YTSZ have been studied extensively in the past in an effort to reconstruct the geological history of the India–Asia collision and the demise of Tethys (Tapponnier *et al.* 1981; Burg & Chen 1984; Mercier & Li 1984; Searle *et al.* 1987). Aitchison *et al.* (2000) recognize a number of terranes of oceanic affinity within and bounding the suture zone. These terranes, together with the conglomerates described in this paper, record

the history of the subduction of Tethys and the India–Asia collision.

Several distinct conglomerate units can be recognized along the YTSZ (Aitchison *et al.* 2002a, b; Davis *et al.* 2002). The development of this type of coarse clastic sedimentary unit typically reflects periods of orogenesis, uplift and basin development. Thus, detailed examination of such units along the YTSZ permits an analysis of different phases in the closure of the Neotethyan Ocean and eventual collision between India and Asia. Some units are extensive and record regional tectonic events, whereas others, such as Neogene breccia, only have a local distribution. Nonetheless, each of these units is significant, as they all provide constraints on the timing and sequence of tectonic events in the region.

From: MALPAS, J., FLETCHER, C. J. N., ALI, J. R. & AITCHISON, J. C. (eds) 2004. *Aspects of the Tectonic Evolution of China*. Geological Society, London, Special Publications, **226**, 235–246.
0305-8719/04/$15 © The Geological Society of London 2004.

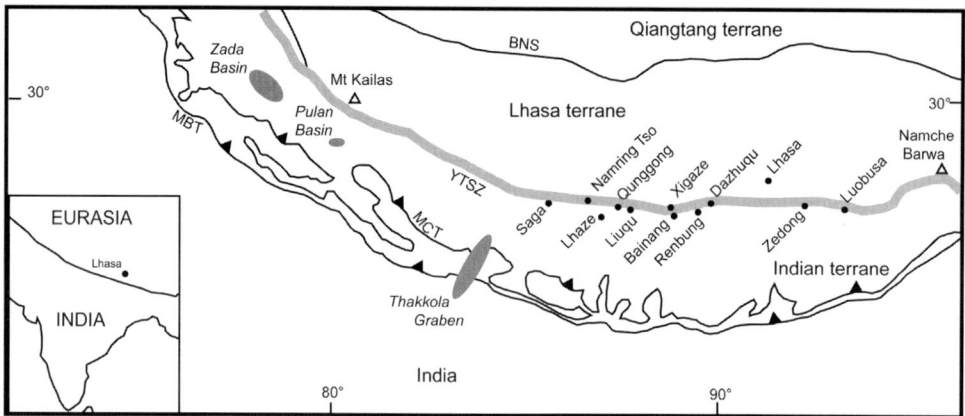

Fig. 1. Location map indicating major tectonic units and boundaries within the Tibet–Qinghai Plateau. Locations mentioned in the text are indicated. The conglomerates discussed in this paper crop out along, and near, the Yarlung Tsangpo suture zone (YTSZ). Individual terranes and conglomerate units along the suture cannot be discriminated at this scale, and readers are referred to more detailed maps published in Davis *et al.* (2002) and Aitchison *et al.* (2002a). BNS, Bangong–Nujiang suture; MCT, Main Central thrust and MBT, Main Boundary thrust of the Himalaya.

Until recently, the existence and significance of these conglomerates have largely been overlooked, as many researchers have concentrated on Tethyan oceanic rocks within the suture zone. Misinterpretation of the age, or inappropriate correlation between some of these conglomerate units, has led to the development of errors regarding the timing and significance of various events. For example, an incorrect age assignment for the Qiuwu Conglomerate (Gangrinboche facies) may be a contributing reason as to why an Eocene age has been inferred for the initial collision between India and Asia. By carefully examining each of these conglomerates over the past few field seasons, we can now suggest a more refined tectonic model for the development of the YTSZ.

As conglomerates within the suture record aspects of its history, it is pertinent to mention here some key units within, and bounding, the suture, and to provide a short description of the tectonic evolution of these rocks. Consumption of oceanic lithosphere on the northern side of the Neotethyan Ocean occurred at a continental convergent plate margin along the southern margin of Asia. The basement of the southern Lhasa terrane is represented by the Sangri Group (discussed below) and north-directed subduction of Tethyan crust beneath these rocks resulted in voluminous magma production, represented by the Gangdese batholith and its eruptive equivalents along the southern Lhasa terrane (Allegre *et al.* 1984; Searle *et al.* 1987). Radiometric age data from the granites indicate subduction

from Late Jurassic until at least the Oligocene (Harrison *et al.* 2000; Yin & Harrison 2000). The Sangri Group comprises regionally deformed and metamorphosed volcanic and sediment lithologies and is locally subdivided into three formations, the Bima, Mamusha and Tiensutin formations (Badengzhu 1979, 1981). With the exception of the Gangdese batholith, relatively little detailed investigation has been undertaken on any other rocks in the southern part of the Lhasa terrane (Burg *et al.* 1983). During the Late Cretaceous, predominantly deep-water marine sedimentation occurred to the south of the Lhasa terrane in an associated continental margin fore-arc basin (Xigaze terrane; Aitchison *et al.* 2000). Sedimentation in the Xigaze terrane was dominated by turbidite accumulation and occurred from Albian until at least Coniacian (Einsele *et al.* 1994; Dürr 1996; Wang *et al.* 1999). The base of the terrane is everywhere faulted, and erosion has removed younger rocks. Clastic sediment within the Xigaze terrane was largely sourced from coeval volcanic rocks developing upon the Lhasa terrane.

In the Late Jurassic to Mid-Cretaceous, additional subduction of oceanic lithosphere occurred along an intra-oceanic island-arc system that developed within Tethys. The history of subduction related to this arc is recorded in rocks preserved along the YTSZ within the Dazhuqu, Bainang and Zedong terranes. These terranes are presently juxtaposed along a series of north-vergent back-thrusts associated with the Miocene Great Counter thrust (Gansser 1964, syn. Renbu–Zedong thrust system of Yin *et al.* 1999). A discontinuous belt of ophiolitic rocks, which crop out along the suture zone (Nicolas *et al.* 1981; Girardeau *et al.* 1984*a*, *b*, 1985*a*, *b*, *c*; Girardeau & Mercier 1988; Mercier & Li 1984) is assigned to the Dazhuqu terrane (Aitchison *et al.* 2003 in press). Detailed radiolarian biostratigraphy (Ziabrev 2001, 2003 in press) indicates a Barremian age for ophiolite formation. The terrane has been tectonically attenuated, and many of the ophiolitic massifs present have somewhat different individual histories (Hébert *et al.* 2000, 2001). The Bainang terrane (Aitchison *et al.* 2000), which crops out south of the ophiolite, preserves a south-verging imbricate thrust stack of ocean-floor rocks. Cherts, basalts and thinly bedded deep-water marine sediments dominate the sequence. The terrane is interpreted as the subduction complex that developed in association with the intra-oceanic island arc. Triassic to Early Cretaceous radiolarians occur in cherts which were accreted into the subduction complex during the Mid-Cretaceous (Ziabrev 2001; Ziabrev *et al.* 2004). Late Jurassic and Cretaceous igneous rocks preserved in the Zedong terrane represent remnants of the intra-oceanic island arc (McDermid *et al.* 2002) and crop out SE of Lhasa near Zedong.

This intra-oceanic arc system was obducted on to India prior to Tethys closure and collision between India and Asia. Siliciclastic passive margin sediments that make up the northern margin of the India were locally disrupted into extensive zones of mud-matrix mélange (Liu 1992; Liu & Einsele 1996; Liu 2001), known as the Yamdrok mélange, as the intra-oceanic arc was being obducted on to India. Radiolarians, extracted from the matrix of this unit (Liu & Aitchison 2002), constrain the timing of its formation to around the Palaeocene/Eocene boundary. Deposition of the Liuqu Conglomerate was coeval with formation of the Yamdrok mélange, and its origins are also linked to this collision (Davis *et al.* 2002).

Sangri Group conglomerates

The Upper Jurassic to Lower Cretaceous Sangri Group crops out along the southern margin of the Lhasa terrane, where it extends from at least Luobusa to near Dazhuqu in the west (Fig. 1). It consists of sedimentary and volcanic lithologies which have experienced amphibolite facies metamorphism (Badengzhu 1979, 1981; Burg *et al.* 1983; Yin *et al.* 1988). Coarse-grained intraformational conglomerate units occur within the Bima and Mamusha formations near Zedong (Fig. 2). Stretched pebble to cobble size clasts are compositionally dominated by locally sourced

Fig. 2. Coarse-grained, intraformational, metaconglomerate units within the Upper Jurassic–Lower Cretaceous Bima Formation, Sangri Group, at the type locality of the phantom Gangdese thrust to the east of Zedong. Note the penetrative regional foliation dipping to the lower right-hand side (NE) of the photograph.

limestone (marble) and andesitic lithologies. They are deformed, together with the matrix, within a strongly developed penetrative regional foliation, which predates intrusion of the Gangdese batholith. These conglomerates are intraformational and do not represent part of any overlap assemblage (Burg *et al.* 1983). However, their mistaken correlation with the uppermost Oligocene–Lower Miocene Luobusa Formation in the Zedong area has contributed to suggestions of the existence of a south-directed Gangdese thrust in the southern Lhasa terrane (Yin *et al.* 1994, 1999; Aitchison *et al.* 2002a, b, 2003) and the development of associated hypotheses for regional structural evolution. The conglomerates are distinctive because, like the rest of the Sangri Group, they have been affected by regional deformation, resulting in steep isoclinal folding and amphibolite-facies metamorphism. Nearby unmetamorphosed Tertiary conglomerates belonging to the Gangrinboche facies unconformably overlie the Sangri Group, and are locally mapped as the Luobusa Formation. Rare fossils from limestones within the Sangri Group indicate Upper Jurassic to Lower Cretaceous ages (Yin *et al.* 1988). Further study of these conglomerates is presently being undertaken.

Liuqu Conglomerate

Thick successions of rapidly deposited coarse clastic rocks known as the Liuqu Conglomerate crop out within the YTSZ (Fig. 3) near Xigaze, and represent an overlap assemblage that links several terranes. The Liuqu Conglomerate varies between outcrops but is essentially a coherent mappable unit over a strike length of 150 km. Coarse-grained mineralogically and texturally immature sediments indicate source proximal deposition. Numerous sections were measured across exposures during the summers of 1998–2001, in order to establish the depositional setting and tectonic significance of this unit (Davis *et al.* 1999, 2001, 2002).

Liuqu Conglomerate accumulated in a variety of proximal fluvial settings with some localized subaqueous deposition. Lithofacies indicate that sedimentation was voluminous, and suggest high erosion and sedimentation rates. Individual basin geometries are presently elongate (east–west) and narrow, with basin margins delineated by faults along which there are likely to have been several phases of strike-slip displacement. As the nature of these coarse-grained sediments varies little across the strike of these remnant basins, we consider that their present geometry

Fig. 3. Palaeogene Liuqu Formation exposed on the true left-hand side of the Yarlung Tsangpo where it passes through the YTSZ to the northeast of Lhaze. The conglomerates are dominated by red radiolarian chert clasts, and the bedding dips to the southwest.

mimicks the original basin shape. Many clast types in the Liuqu Conglomerate are terrane specific and can therefore be used to give a clear indication of the likely source. Red radiolarian chert dominates the Bainang terrane, which also contributed volcaniclastic psammites and phyllitic clasts. Serpentinite, ultramafics, gabbro, basalt and other ophiolitic detritus were clearly derived from the Dazhuqu terrane – the only potential source for such clasts within the region. Quartzites, sublitharenites, rare limestones, and siliceous mudstones are all of Indian affinity and derived from sources that lie to the south of the Liuqu Conglomerate. Notably, despite the present-day proximity of these conglomerates to the Lhasa and Xigaze terranes, no detritus that could be attributed to any source north of the Dazhuqu terrane was observed. Strikingly obvious mismatches of clast petrography with lithologies in adjacent source terranes are common. The rapidly deposited nature of the conglomerates in narrow elongate basins, where there is a mismatch in the petrography of proximal clasts with that of the rocks exposed immediately adjacent to the basin margins, is comparable with features observed in oblique-slip basins (Ballance & Reading 1980; Biddle & Christie-Blick 1985) elsewhere.

Several genera and numerous plant fossil species are described from fluvio-lacustrine sediments located 15 km NE of Lhaze (Tao 1988a, b). They indicate deposition between latest Cretaceous to earliest Eocene. Formation of the Liuqu Conglomerate thus post-dates mid-Cretaceous intra-oceanic arc activity (Aitchison et al. 2000, 2002a), but must predate the India–Asia collision. Our preferred model for accumulation of the Liuqu Conglomerate is within a valley, or series of valleys, that most probably developed along the collisional axis between the northern margin of the Indian continent and an intra-oceanic arc as it was being accreted southwards on to this margin (Davis et al. 2002).

Gangrinboche facies conglomerates

A distinctive and regionally extensive overlap assemblage of conglomerates crops out on the northern edge of the YTSZ for over 2000 km from west of Mount Kailas (Gangrinboche in Tibetan) to near the eastern Himalayan syntaxis at Namche Barwa. These conglomerates include various units assigned local geographical names (the Luobusa, Dazhuqu, Qiuwu and Kailas formations) during the course of geological mapping in different areas. These units are correlated together under a single name as the 'Gangrinboche facies conglomerates' by Aitchison

et al. (2002b). Each unit records the Early Miocene deposition of coarse clastic sediments, and all record broadly similar stratigraphic evolution. Sedimentary rocks in all sections exhibit characteristics compatible with interpretation of their having accumulated in alluvial fan and braid-plain environments (Wang et al. 1999, 2000; Yin et al. 1999). Most of the conglomerates present are extremely proximal, and clast petrography is influenced strongly by the nature of rocks in immediately adjacent source terranes. The conglomerates formed after closure of the Tethyan Ocean in this region and indicate the establishment of considerable relief. Each formation is subdivisible into three members, and the broad pattern of up-section evolution can be correlated regionally. A basal depositional contact where alluvial fan deposits rest unconformably upon an eroded basement of Lhasa terrane rocks, which initially was the sole source of sediment, is ubiquitous. Up-section braided river deposits dominate, with the first appearance of clastic detritus derived from the south. This material gradually becomes dominant. Major changes in overall source petrography may indicate the history of activity on north-directed thrust faults during collision. Local variations in clast content are likely to be related to the immediate proximity of distinctive rock types in localized source areas. The top of the sequence is dominated by material derived from south of areas where the conglomerates crop out. This includes various suture zone terranes and the northern margin of India. In many areas, the conglomerates are folded and dip steeply, indicating that they have been further affected by ongoing Tibetan orogenesis.

The *Luobusa Formation* crops out along the YTSZ southeast of Lhasa where there is a thick (1300+ m at Luobusa) succession of coarse clastic strata. Distinct changes in sediment sources are observed up-section. A depositional contact upon a Lhasa terrane surface, represented by Late Cretaceous to Oligocene intrusions of the Gangdese batholith or Late Jurassic to Early Cretaceous meta-sedimentary and meta-volcanic units of the Sangri Group, can be observed at many localities (Badengzhu 1979, 1981; Aitchison et al. 2003). All known fossil occurrences are near Luobusa and lie within the middle member (R2). They are regarded as indicating an Oligocene–Miocene depositional age (Badengzhu 1981). Stratigraphic relations can be further used to constrain the age. The formation has an unconformable depositional contact upon the Yaja granodiorite. Emplacement of the Yaja granodiorite at an estimated crustal depth of 13 km is constrained by radiometric dates of

30.4 ± 0.4 Ma (Harrison *et al.* 2000). Deposition of the basal horizons of the Luobusa Formation directly upon this Mid-Oligocene granodiorite clearly indicates surface exposure of the grano-diorite prior to sediment accumulation. Additional radiometric constraints on north-directed thrusting along the Renbu Zedong thrust, a structure that truncates the Luobusa Formation, indicate that displacement occurred between 18 and 10 Ma in this region (Yin *et al.* 1999). Fossil and radiometric ages are therefore in accord with one another, and the Luobusa Formation is most likely to be Lower Miocene.

The *Dazhuqu Formation* crops out to the southwest of Lhasa, and has its type area in the Yarlung Tsangpo gorge near Dazhuqu, where it is over 1500 m thick. The unit extends eastwards past Renbung to the summit of Peak 6126 m near Chara at the downstream end of the Yarlung Tsangpo gorge. It is dominated by coarse clastic strata, and was deposited directly on an eroded surface of Lhasa terrane rocks, which includes intrusions of the Gangdese batholith into meta-morphic rocks of the Sangri Group. Initial depo-sition of lowermost sediments was very proximal, occurring on alluvial fans. Up-section, braided river deposits incorporating material derived from the south are a notable feature of this unit. Near the top of the preserved section, the Dazhuqu Formation grades upwards into red and brown pebbly mudstones. No fossils have been reported from the Dazhuqu Formation, but the nature of its relationship to the Lhasa terrane and the sedimentary succession within this unit strongly support correlation with the Luobusa, Qiuwu and Kailas units.

The *Qiuwu Formation* crops out from about 15 km east of Xigaze and extends westwards for 100s of km beyond Saga. The base of the for-mation lies at a depositional contact upon rocks of the Lhasa terrane. Uppermost levels of the for-mation lie in the footwall of a major north-directed thrust fault, with Cretaceous turbidites of the Xigaze terrane in the hanging wall thrust over the Qiuwu Formation. The formation can be subdivided into three distinctive members, with the succession very similar to that present elsewhere in the Gangrinboche facies. Sedi-ments, immediately above the basal unconfor-mity, are locally derived and were sourced entirely from the underlying Lhasa terrane. Braided river deposits characterize the middle unit, which records the first arrival and then the increasing up-section abundance of material derived from the suture zone. Uppermost units were sourced from both north and south of the suture, and in the Xigaze to Ngamring areas sedi-ments are relatively finer-grained than those in

the middle unit, reflecting maturation of the sediment distribution system.

Fossil data are somewhat questionable and confusing, as many of the fossils reported are derived – coming from clasts within the con-glomerate (Aitchison *et al.* 2002*b*). Thus, these fossils date the age of the clasts and not the tim-ing of deposition of the conglomerate. Felsic dykes which intrude both the Qiuwu Formation and the Xigaze terrane have been radiometrically (^{40}Ar/^{39}Ar) dated at 18.3 ± 0.5 Ma (Yin *et al.* 1994). The dykes cross-cut folds in the Qiuwu Formation, as well as the major north-directed thrust that truncates the top of the formation, indicating deformation of the sediments prior to intrusion of igneous rocks in the late Early Mio-cene. Thus, based on correlation with Gangrin-boche conglomerates elsewhere, the Qiuwu Formation is also likely to be Lower Miocene.

The *Kailas Formation* includes conglomerates similar to those of the Qiuwu Formation, which can be traced near-continuously westwards from Xigaze, at least as far (1500 km) as the 6714 m peak of Gangrinboche (Mount Kailas), where they outcrop spectacularly as subhorizontal beds on the upper slopes of the mountain (Gansser 1964) (Fig. 4). Evidence for distinct changes in sediment sources is observable up-section. Although lowermost sediments were derived from the north, clasts from the suture zone and further south (Indian margin) become increas-ingly dominant up-section. The age of the Kailas Formation is both palaeontologically and radio-metrically constrained. Non-marine bivalves (Unionidae) reported from the middle part of the formation (Miller *et al.* 2000) indicate Mio-cene deposition. Eocene ages previously reported for this unit are from fossils within limestone clasts, and therefore date the clasts not the con-glomerate. This has led to misinterpretation of the age of the unit in this area. Kailas conglomer-ates were deposited directly upon rocks of the Kailas Igneous Complex which is dated radio-metrically at 38 ± 1.3 Ma Rb–Sr; (Honegger *et al.* 1982). Radiometric age constraints (Harri-son *et al.* 1993) indicate that these igneous rocks were not uplifted, and thus not exposed and avail-able for erosion, until the Late Oligocene. Dis-placement on the South Kailas Thrust, which truncates the upper levels of the conglomerates, is bracketed between 20 and 4 Ma (Yin *et al.* 1999), thereby providing a minimum age con-straint. Thus, the Kailas Formation must also be Lower Miocene.

Available age constraints indicate that rocks of the Gangrinboche facies were deposited between the Late Oligocene and Early Miocene over a strike length of more than 1500 km. They clearly

Fig. 4. Thick beds of coarse conglomeratic units that are part of the Lower Miocene Kailas Formation (Gangrinboche facies conglomerates) exposed in a valley approximately 20 km east of Mount Kailas in western Tibet.

indicate an event of regional significance. Deposition probably occurred along a valley system axial to two sub-parallel, east–west-trending, active mountain ranges. Relief was initially greatest on the northern side of the suture, and sediments accumulated along the southern flank of a mountain range in the Lhasa terrane. Later India–Asia-collision-related back-thrusting uplifted areas to the south. As mountains south of the suture shed detritus, it was transported and then deposited in a braided river system that developed axial to the mountain chains.

Neogene sediments

Development of coarse clastic units in Tibet has continued until the present day. A late Neogene breccia unit crops out near the village of Quanggong on the 'Friendship Highway' between Liuqu and Lhaze. Thick westward-dipping beds of breccia unconformably overlie folded units of the Liuqu Conglomerate (Fig. 5). This breccia consists entirely of angular clasts of serpentinized harzburgite set in a coarse-grained matrix of calcite and fine-grained ultramafic detritus. Clast size ranges from cobbles up to $1 \, m^3$ boulders. Sedimentation was obviously proximal to the source, with channels up to 2 m deep scoured in the base of some units. Deposition of this unit was restricted to a rather small area,

and may have occurred in response to local uplift along the margin of a north–south-trending extensional zone. Development of extensional rift zones was widespread throughout Tibet from the Late Miocene onwards (Molnar & Tapponnier 1977; Tapponnier & Molnar 1977; Tapponnier et al. 1981; Burchfiel et al. 1991; Yin 2000). Other significant rift-related zones of sedimentation are seen in the Thakkola Graben of central southern Tibet, which extends southwards into Nepal where it has been studied in more detail (Fort et al. 1982; Fort 1989; Colchen 1999; Garzione et al. 2000a, b), and the Pulan Basin south of Mount Kailas (Murphy et al. 2002). An extensive zone of fluvio-lacustrine sediments crops out in the Zada Basin (Bureau of Geology and Mineral Resources of Xizang Autonomous Region 1993), where the YTSZ is cut by the Karakoram Fault, to the west of Mount Kailas.

Discussion

The conglomerates described herein are lithologically, temporally and structurally distinct (Fig. 6). By recognizing each succession as a discrete entity, several important assertions can be made. Sangri Group conglomerates are the oldest coarse clastic sediments studied within the YTSZ, and were wholly derived from within the Lhasa

Fig. 5. Neogene breccias crop out near the village of Quanggong on the 'Friendship Highway' between Liuqu and Lhaze. These westward-dipping breccia beds unconformably overlie folded Liuqu Conglomerate and Triassic quartzose sandstones of Indian affinity. This breccia consists entirely of angular clasts of serpentinized harzburgite set in a coarse-grained matrix of calcite and fine-grained ultramafic detritus. Clast size ranges from cobbles up to 1 m^3 boulders. Note the channels up to 2 m deep that are scoured in the base of some units.

terrane. They are part of an extensive regionally foliated metamorphosed unit that makes up the basement of the Lhasa terrane (Burg 1983). Previous misinterpretation of these conglomerates as part of the Lower Miocene Luobusa Formation has led to suggestions of the existence of a south-directed Gangdese Thrust (Yin *et al.* 1994, 1999; Harrison *et al.* 2000; Aitchison *et al.* 2002*a, b*, 2003) in the southern Lhasa terrane.

Investigation of the Liuqu and Gangrinboche conglomerates in particular indicates two distinct phases of molasse sedimentation along the suture. The first occurred during the Late Cretaceous to Palaeocene, and a second in the Late Oligocene to Early Miocene. The Palaeocene event records the collision of an intra-oceanic arc with India, while the second phase indicates the close proximity of sources of northern Indian and Lhasa terrane rocks either during, or immediately after, India–Asia collision.

The Liuqu Conglomerate is the oldest coarse clastic sedimentary unit, containing detritus from various Tethyan terranes that lie within the suture zone. This formation records the Late Cretaceous to Palaeocene collision of an intra-oceanic arc with the northern margin of India,

and predates later collision of India with Asia. Rapid deposition occurred in a series of oblique-slip basins, with numerous mismatches between clast petrology and adjacent terranes (Davis *et al.* 2002). Notably, no material from the Lhasa or Xigaze terranes occurs in these conglomerates. This likely indicates that deposition predated consumption of the remainder of the Tethyan oceanic crust that lay north of the Mid-Cretaceous intra-oceanic arc and south of the Lhasa terrane (Davis *et al.* 2002).

The Lower Miocene Gangrinboche-facies conglomerates (Aitchison *et al.* 2002*a*) include rocks of the Kailas, Luobusa, Dazhuqu and Qiuwu formations. Deposition was dominated by alluvial fan and braided river sedimentation with high erosion rates. The timing of deposition for Gangrinboche-facies conglomerate units is of great significance to regional tectonic models. Erosion accompanied Late Oligocene uplift of the Lhasa terrane. Yin *et al.* (1999) suggest that this uplift was a result of south-directed thrusting along the 'Gangdese Thrust', which they inferred to place rocks of the Lhasa terrane over Tertiary conglomerates of the Luobusa Formation. However, observations along 1500 km strike length of the Gangrinboche facies conglomerates

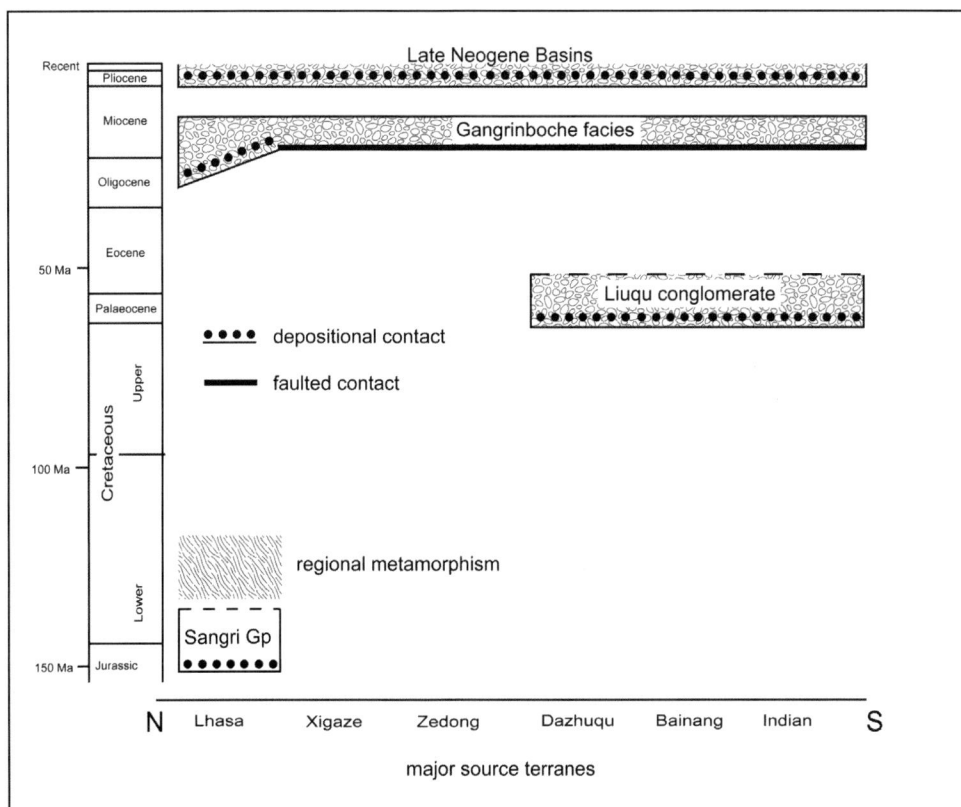

Fig. 6. Plot showing the temporal distribution of conglomeratic units that crop out along the region of the YTSZ in southern Tibet. The major sources (terranes) for clasts within each unit are shown at the base of the diagram.

indicate that they rest unconformably upon a basement of Lhasa terrane rocks at all localities. Detailed investigations in the Zedong–Luobusa area indicate that this depositional contact can be mapped along the entire strike length of the proposed Gangdese thrust, and at no locality are any Lhasa terrane rocks thrust over Gangrinboche conglomerates (Aitchison *et al.* 2003). This observation is in accord with numerous other regional investigations (Gansser 1964; Badengzhu 1979, 1981; Zhang & Fu 1982; Burg 1983; Wei & Peng 1984; Bureau of Geology and Mineral Resources of Xizang Autonomous Region 1993) along the entire length of the suture zone. The 'Tertiary' conglomerates in question are in fact misidentified Sangri Group conglomerates described elsewhere in this paper. Thus, although considerable relief must have been present within the southern Lhasa terrane when Gangrinboche conglomerates accumulated, the mechanism for uplift remains enigmatic.

Gangrinboche-facies rocks have previously been accorded an Eocene age, with this cited (Searle *et al.* 1987) as one of the main lines of evidence in support of models invoking Early Eocene collision between India and Asia. However, all localities with fossil, radiometric or structural constraints indicate latest Oligocene to Early Miocene deposition. There is no evidence in the conglomerates studied to suggest deposition of any molasse-type sediments during the Eocene, despite this classically being the preferred time for collision between India and Asia (Searle *et al.* 1987). Recently, Wang *et al.* (2002) documented latest Eocene/Oligocene boundary (NP20) marine nannofossils from Himalayan sediments north of Qomolangma. Their work thus places a maximum age constraint on the timing of India–Asia collision in the region, which is significantly younger than Early Eocene. While the Gangrinboche conglomerates do not necessarily constrain the precise timing of the initiation of collision, they certainly provide unequivocal evidence that it had happened. Their maximum age is considerably younger than previously thought, as are the ages of the youngest known subduction-related magmatic rocks in the Lhasa terrane, and marine sediments deposited along the northern continental margin of India. The possibility that India–Asia collision was itself later than presently considered therefore warrants consideration. If the conglomerates are key evidence for the collision, then a latest Oligocene to Early Miocene timing for their accumulation is significant. Improved understanding of the timing of molasse accumulation has attendant implications for studies of collision timing, and provides a very simple potential answer as to the cause of large-scale regional uplift at this time.

The development of coarse clastic sedimentary units in north–south-oriented basins associated with region extension began in the Late Neogene and continues today. Huge volumes of coarse sediment are presently being transported from western Tibet all the way east along the Yarlung Tsangpo (River) system. These sediments will eventually make their way to Bangladesh and the Bengal fan. As the India–Asia collision and Tibet–Qinghai Plateau uplift has progressed northeastwards in a stepwise manner, numerous fluvial sedimentary basins have developed (Tapponnier *et al.* 2001). The youngest of these basins are presently developing on the NE margin of the plateau. Numerous other basins of coarse clastic sedimentary rocks exist within the Tibet–Qinghai Plateau region. Each basin records discrete aspects of the history of development of this region, and all these rocks are worthy of further detailed investigation.

We thank members of the Tibetan Geological Survey (Team 2) and Tibetan Geological Society, whose efforts have helped to make this research possible. Many of these friends have assisted with arranging logistics and permission. The work described in this paper was supported by grants from the Research Grants Council of the Hong Kong Special Administrative Region, China (Project Nos HKU7102/98P, HKU 7299/99P and HKU 7069/01P). The authors wish to thank C. Blake and H. Williams for their constructive review comments, which helped to improve the manuscript.

References

AITCHISON, J. C., ABRAJEVITCH, A. *et al.* 2002*a*. New insights into the evolution of the Yarlung Tsangpo suture zone, Xizang (Tibet), China. *Episodes*, **25**(3), 90–94.

AITCHISON, J. C., BADENGZHU *et al.* 2000. Remnants of a Cretaceous intra-oceanic subduction system within the Yarlung–Zangbo suture (southern Tibet). *Earth and Planetary Science Letters*, **183**, 231–244.

AITCHISON, J. C., DAVIS, A. M., BADENGZHU & LUO, H. 2002*b*. New constraints on the India–Asia collision: the Lower Miocene Gangrinboche conglomerates, Yarlung Tsangpo suture zone, SE Tibet. *Journal of Asian Earth Sciences*, **21**(3), 253–265.

AITCHISON, J. C., DAVIS, A. M., BADENGZHU & LUO, H. 2003. The Gangdese Thrust: a phantom structure that did not raise Tibet. *Terra Nova*, **15**, 155–162.

AITCHISON, J. C., DAVIS, A. M. *et al.* 2004. Stratigraphic and sedimentological constraints on the age and tectonic evolution of the Neotethyan ophiolites along the Yarlung Tsangpo suture zone, Tibet. *In*: DILEK, Y. & ROBINSON, P. T. (eds) *Ophiolites in Earth History*, Geological Society, London, Special Publications, **218**, 147–164.

ALLEGRÉ, C. J., COURTILLOT, V. *ET AL.* 1984. Structure and evolution of the Himalaya–Tibet orogenic belt. *Nature (London)*, **307**(5946), 17–22.

BADENGZHU 1979. (Compiler) *Xizang Autonomous Region Zhanang–Sangri Regional Geology Reconnaissance Map*, 1 : 50 000. Xizang Geological Survey Geological Team #2, Group 2. Lhasa.

BADENGZHU 1981. (Compiler) *Xizang Autonomous Region Sangri–Jiacha Regional Geology Reconnaissance Map*, 1:50 000. Xizang Geological Survey Geological Team #2, Lhasa.

BALLANCE, P. F. & READING, H. G. (eds) 1980. *Sedimentation in Oblique-slip Mobile Zones*. Special Publications of the International Association of Sedimentologists, **4**, Blackwell, Oxford.

BIDDLE, K. T. & CHRISTIE-BLICK, N. (eds) 1985. *Strike-slip Deformation, Basin Formation, and Sedimentation*. Special Publications of the Society of Economic Paleontologists and Mineralogists, **37**, Tulsa, Oklahoma.

BURCHFIEL, B. C., CHEN, Z., ROYDEN, L. H., LIU, Y. & DENG, C. 1991. Extensional development of Gabo Valley, southern Tibet. *Tectonophysics*, **194**(1–2), 187–193.

BUREAU OF GEOLOGY AND MINERAL RESOURCES OF XIZANG AUTONOMOUS REGION 1993. *Regional Geology of Xizang (Tibet) Autonomous Region*. Geological Publishing House, Beijing.

BURG, J. P. 1983. *Tectogénèse comparée de deux segments de chaîn de collision: – Le Sud du Tibet (Suture du Tsangpo) – La Chaîn Hercynienne en Europe (sutures du Massif Central)*. Docteur d'État thesis, l'Université des Sciences et Techniques du Languedoc.

BURG, J. P. & CHEN, G. M. 1984. Tectonics and structural zonation of southern Tibet, China. *Nature (London)*, **311**(5983), 219–223.

BURG, J. P., PROUST, F., TAPPONNIER, P. & CHEN, G. M. 1983. Deformation phases and tectonic evolution of the Lhasa Block (southern Tibet, China). *Eclogae Geologicae Helvetiae*, **76**(3), 643–665.

COLCHEN, M. 1999. The Thakkhola–Mustang graben in Nepal and the late Cenozoic extension in the Higher Himalayas. *Journal of Asian Earth Sciences*, **17**, 683–702.

DAVIS, A. M., AITCHISON, J. C., BADENGZHU, LUO, H., MALPAS, J. & ZYABREV, S. 1999. Eocene oblique-slip basin development, Tibet: terrane tracks on the roof of the world. *In*: EVENCHICK, C. A., WOODSWORTH, G. J. & JONGENS, R. (eds) *Circum Pacific Terrane Conference, Okanagan Valley, B.C., Canada,* Terrane Paths 99, Circum Pacific Terrane Conference Abstracts and Program, p. 28.

DAVIS, A. M., AITCHISON, J. C., BADENGZHU & LUI, H. 2001. Late Cretaceous–Paleocene island arc collision related conglomerates, Yarlung–Tsangpo suture zone, Tibet. *In*: GRASEMANN, B. & STÜWE, K. (eds), *16th Himalaya–Karakoram–Tibet Workshop, Graz, Austria,* Journal of Asian Earth Sciences 19/3A, 16th HKT abstracts, p. 13.

DAVIS, A. M., AITCHISON, J. C., BADENGZHU, LUO, H. & ZYABREV, S. 2002. Paleogene island arc collision-related conglomerates, Yarlung–Tsangpo suture zone, Tibet. *Sedimentary Geology,* **150** (3–4), 247–273.

DÜRR, S. B. 1996. Provenance of Xigaze fore-arc basin clastic rocks (Cretaceous, South Tibet). *Geological Society of America Bulletin,* **108**(6), 669–684.

EINSELE, G., LIU, B. *et al.* 1994. The Xigaze forearc basin: evolution and facies architecture (Cretaceous, Tibet). *Sedimentary Geology,* **90**, 1–32.

FORT, M. 1989. The Gongba conglomerates: glacial or tectonic? *Zeitschrift für Geomorphologie, Supplementband,* **76**, 181–194.

FORT, M., FREYTET, P. & COLCHEN, M. 1982. Structural and sedimentological evolution of the Thakkhola Mustang graben (Nepal Himalayas). *Zeitschrift für Geomorphologie, Supplementband,* **42**, 75–98.

GANSSER, A. 1964. *The Geology of the Himalayas.* Wiley-Interscience, New York.

GARZIONE, C. N., DETTMAN, D. L., QUADE, J., DECELLES, P. G. & BUTLER, R. F. 2000a. High times on the Tibetan Plateau; paleoelevation of the Thakkhola Graben, Nepal. *Geology (Boulder),* **28**(4), 339–342.

GARZIONE, C. N., QUADE, J., DECELLES, P. G. & ENGLISH, N. B. 2000b. Predicting paleoelevation of Tibet and the Himalaya from $\delta^{18}O$ vs. altitude gradients on meteoric water across the Nepal Himalaya. *Earth and Planetary Science Letters,* **183**, 215–229.

GIRARDEAU, J. & MERCIER, J. C. C. 1988. Petrology and texture of the ultramafic rocks of the Xigaze Ophiolite (Tibet); constraints for mantle structure beneath slow-spreading ridges. *Tectonophysics,* **147**(1–2), 33–58.

GIRARDEAU, J., MARCOUX, J. & ZAO, Y. 1984a. Lithologic and tectonic environment of the Xigaze ophiolite (Yarlung Zangbo suture zone, southern Tibet, China), and kinematics of its emplacement. *Eclogae Geologicae Helvetiae,* 77(**1**), 153–170.

GIRARDEAU, J., NICOLAS, A. *et al.* 1984b. Les ophiolites de Xigaze et al suture du Yarlung Zangbo (Tibet). *In:* MERCIER, J. L. & GUANGCEN, L. (eds). *Mission Franco-Chinoise au Tibet 1980. Étude Géologique et Géophysique de la Croûte Terrestre et du Manteau Supérieur du Tibet et de l' Himalaya.* CNRS, Paris, 189–197.

GIRARDEAU, J., MERCIER, J. C. & XIBIN, W. 1985a. Petrology of the mafic rocks of the Xigaze ophiolite, Tibet: implications for the genesis of the oceanic lithosphere. *Contributions to Mineralogy and Petrology,* **90**(4), 309–321.

GIRARDEAU, J., MERCIER, J. C. C. & ZAO, Y. G. 1985b. Origin of the Xigaze Ophiolite, Yarlung Zangbo suture zone, southern Tibet. *Tectonophysics,* **119**, 407–433.

GIRARDEAU, J., MERCIER, J. C. C. & ZAO, Y. G. 1985c. Structure of the Xigaze Ophiolite, Yarlung Zangbo suture zone, southern Tibet, China; genetic implications. *Tectonics,* **4**(3), 267–788.

HARRISON, T. M., COPELAND, P., HALL, S. A., QUADE, J., BURNER, S., OHJA, T. P. & KIDD, W. S. F. 1993. Isotopic preservation of Himalayan/Tibetan uplift, denudation, and climate histories of two molasse deposits. *Journal of Geology,* **100**, 157–175.

HARRISON, T. M., YIN, A., GROVE, M., LOVERA, O. M., RYERSON, F. J. & ZHOU, X. 2000. The Zedong Window: a record of superposed Tertiary convergence in southeastern Tibet. *Journal of Geophysical Research,* **105**(B8), 19 211–19 230.

HÉBERT, R., BEAUDOIN, G., VARFALVY, V., HUOT, F., WANG, C. S. & LIU, Z. F. 2000. Yarlung Zangbo ophiolites, southern Tibet revisited. WAN, X. Q. (ed.) *15th Himalaya–Karakoram–Tibet Workshop, Chengdu, China,* Abstract Volume 15th Himalaya–Karakoram–Tibet Workshop Earth Science Frontiers, vol. 7, suppl., 124–126.

HÉBERT, R., WANG, C., VARFALVY, V., HUOT, F., LIU, Z., BEAUDOIN, G. & DOSTAL, J. 2001. Yarlung Zangbo Suture Zone ophiolites and their suprasubduction zone setting (eds) *16th Himalaya–Karakoram–Tibet Workshop, Graz, Austria,* 19/3A, 27–28.

HONEGGER, K., DIETRICH, V., FRANK, W., GANSSER, A., THÖNI, M. & TROMMSDORFF, V. 1982. Magmatism and metamorphism in the Ladakh Himalayas (the Indus–Tsangpo suture zone). *Earth and Planetary Science Letters,* **60**, 253–292.

LIU, G. 1992. Permian to Eocene sediments and Indian passive margin evolution in the Tibetan Himalayas. *Tübinger Geowissenschaftliche Arbeiten,* **13**, 1–268.

LIU, G. H. & EINSELE, G. 1996. Various types of olistostromes in a closing ocean basin, Tethyan Himalaya (Cretaceous, Tibet). *Sedimentary Geology,* **104**, 203–226.

LIU, J. B. 2001. *Yamdrok Melange, Gyantze district, Xizang (Tibet), China.* M.Phil. thesis, University of Hong Kong.

LIU, J. B. & AITCHISON, J. C. 2002. Upper Paleocene radiolarians from the Yamdrok mélange, south Xizang (Tibet), China. *Micropaleontology,* **48** (Supplement 1), 145–154.

MCDERMID, I. R. C., AITCHISON, J. C., DAVIS, A. M., HARRISON, T. M. & GROVE, M. 2002. The Zedong terrane: a Late Jurassic intra-oceanic magmatic arc within the Yarlung–Zangbo suture zone, southeastern Tibet. *Chemical Geology,* **187**(3–4), 267–277.

MERCIER, J. L. & LI, G. C. (eds) 1984. La collision Inde–Asie côté Tibet. *Mission Franco-Chinoise au Tibet 1980. Étude Géologique et Géophysique de la Croûte Terrestre et du Manteau Supérieur du Tibet et de l'Himalaya.* CNRS, Paris, p. 433.

MILLER, C., SCHUSTER, R., KLÖTZLI, U., FRANK, W. & GRASEMANN, B. 2000. Late Cretaceous–Tertiary magmatic and tectonic events in the Transhimalaya batholith (Kailas area, SW Tibet). *Schweizerische Mineralogische und Petrographische Mitteilungen*, **80**, 1–20.

MOLNAR, P. & TAPPONNIER, P. 1977. The collision between India and Eurasia. *Scientific American*, **236(4)**, 30–41.

MURPHY, M. A., YIN, A. *et al.* 2002. Structural evolution of the Gurla Mandhata detachment system, southwest Tibet: implications for the eastward extent of the Karakoram fault system. *Geological Society of America Bulletin*, **114(4)**, 428–447.

NICOLAS, A., GIRARDEAU, J. *et al.* 1981. The Xigaze ophiolite (Tibet): a peculiar oceanic lithosphere. *Nature (London)*, **294(5840)**, 414–417.

SEARLE, M. P., WINDLEY, B. F. *et al.* 1987. The closing of Tethys and the tectonics of the Himalaya. *Geological Society of America Bulletin*, **98(6)**, 678–701.

TAO, J. 1988*a*. The Paleogene flora and palaeoclimate of Liuqu Formation in Xizang. *In:* WHYTE, P., AIGNER, J. S., JABLONSKI, N. G., TAYLOR, G., WALKER, D. & WANG, P. (eds) *The Paleoenvironment of East Asia from the Mid-Tertiary.* Occasional Papers and Monographs – Centre of Asian Studies, **77**, Centre of Asian Studies, University of Hong Kong, Hong Kong, 520–522.

TAO, J. 1988*b*. Plant fossils from Lepequ formation in Lhaze County, Xizang and their palaeoclimatological significances. *Academia Sinica, Geological Institute Memoir*, **3**, 223–238.

TAPPONNIER, P. & MOLNAR, P. 1977. Active faulting and tectonics in China. *Journal of Geophysical Research*, **82(20)**, 2905–2930.

TAPPONNIER, P., MERCIER, J. L. *et al.* 1981. The Tibetan side of the India–Eurasia collision. *Nature (London)*, **294(5840)**, 405–410.

TAPPONNIER, P., ZHIQIN, X., ROGER, F., MEYER, B., ARNAUD, N., WITTLINGER, G. & JINGSUI, Y. 2001. Oblique stepwise rise and growth of the Tibet Plateau. *Science*, **294(5547)**, 1671–1677.

WANG, C. S., LIU, Z. F. *et al.* 1999. *Xigaze Forearc Basin and Yarlung Zangbo Suture Zone, Tibet.* Geological Publishing House, Beijing.

WANG, C. S., LIU, Z. F. & RÉJEAN, H. 2000. The Yarlung–Zangbo paleo-ophiolite, southern Tibet: implications for dynamic evolution of the Yar-

lung–Zangbo Suture Zone. *Journal of Asian Earth Sciences*, **18**, 651–661.

WANG, C. S., LI, X. H., HU, X. M. & JANSA, L. F. 2002. Latest marine horizon north of Qomolangma (Mt Everest): implications for closure of Tethys seaway and collision tectonics. *Terra Nova*, **14(2)**, 114–120.

WEI, B. J. & PENG, Y. H. 1984. (Compilers) *Xizang Autonomous Region Xigaze – Quxu Geological Traverse Map*, 1:200 000. Xizang Geological Survey Geological Team #2, Lhasa.

YIN, A. 2000. Mode of Cenozoic east–west extension in Tibet suggesting a common origin of rifts in Asia during the Indo-Asian collision. *Journal of Geophysical Research*, **105(B9)**, 21 745–21 759.

YIN, A. & HARRISON, T. M. 2000. Geologic evolution of the Himalayan–Tibetan Orogen. *Annual Review of Earth and Planetary Sciences*, **28**, 211–280.

YIN, A., HARRISON, T. M. *et al.* 1999. Tertiary deformation history of southeastern and southwestern Tibet during the Indo-Asian collision. *Geological Society of America Bulletin*, **111(11)**, 1644–1664.

YIN, A., HARRISON, T. M., RYERSON, F. J., CHEN, W. J., KIDD, W. S. F. & COPELAND, P. 1994. Tertiary structural evolution of the Gangdese thrust system in southeastern Tibet. *Journal of Geophysical Research*, **99**, 18 175–18 201.

YIN, J., XU, J., LIU, C. & LI, H. 1988. The Tibetan Plateau: regional stratigraphic context and previous work. *Philosophical Transactions of the Royal Society of London, Series A: Mathematical and Physical Sciences*, **327(1594)**, 5–52.

ZHANG, S. M. & FU, X. L. 1982. (Compilers) *Xizang Autonomous Region Xigaze – Saga Geological Traverse Map 1 : 200 000.* Xizang Geological Survey Geological Team #2, Lhasa.

ZIABREV, S. 2001. *Tectonic evolution of Dazhuqu and Bainang terranes, Yarlung Zangbo suture, Tibet as constrained by radiolarian biostratigraphy.* PhD thesis, University of Hong Kong.

ZIABREV, S. V., AITCHISON, J. C., ABRAJEVITCH, A., BADENGZHU, DAVIS, A. M. & LUO, H. 2003. Precise radiolarian age constraints on the timing of Dazhuqu terrane ophiolite generation and sedimentation, Yarlung–Tsangpo suture zone, Tibet. *Journal of the Geological Society*, **160(4)**, 591–600.

ZIABREV, S. V., AITCHISON, J. C., ABRAJEVITCH, A., BADENGZHU, DAVIS, A. M. & LUO, H. 2004. Bainang Terrane, Yarlung-Tsangpo suture, southern Tibet: a record of intra-Tethyan subduction on the Roof of the World. *Journal of the Geological Society*, **161(3)**, 523–538.

Ultra-high pressure minerals in the Luobusa Ophiolite, Tibet, and their tectonic implications

PAUL T. ROBINSON[1,2], WEN-JI BAI[2], JOHN MALPAS[1], JING-SUI YANG[2], MEI-FU ZHOU[1], QING-SONG FANG[2], XU-FENG HU[3], STANLEY CAMERON[4] & HUBERT STAUDIGEL[5]

[1]*Department of Earth Sciences. The University of Hong Kong, Pokfulam Road, Hong Kong, China (e-mail: probins@hkucc.hku.hk)*

[2]*Laboratory of Continental Dynamics, Institute of Geology, Chinese Academy of Geological Sciences, Beijing, 100037, China*

[3]*Department of Chemistry, Dalhousie University, Halifax, Nova Scotia, Canada*

[4]*Department of Earth Sciences, Dalhousie University, Halifax, Nova Scotia, Canada B3H 3J5*

[5]*Institute of Geophysics and Planetary Physics, Scripps Institution of Oceanography, University of California, San Diego, La Jolla, California 92093-0225, USA*

Abstract: Numerous ultra-high-pressure minerals have been recovered from podiform chromitites in the Luobusa ophiolite, Tibet. Recovered minerals include diamond, moissanite, Fe-silicides, wüstite, $Ni-Fe-Cr-C$ alloys, PGE alloys and octahedral $Mg-Fe$ silicates. These are accompanied by a variety of native elements, including Si, Fe, Ni, Cr and graphite. All of the minerals were hand-picked from heavy-mineral separates of the chromitites and care was taken to prevent natural or anthropogenic contamination of the samples. Many of the minerals and alloys are either enclosed in, or attached to, chromite grains, leaving no doubt as to their provenance. The ophiolite formed originally at a mid-ocean ridge (MOR) spreading centre at 177 ± 33 Ma, and was later modified by suprasubduction zone magmatism at about 126 Ma. The chromitites were formed in the suprasubduction zone environment from boninitic melts reacting with the host peridotites. The UHP minerals are believed to have been transported from the lower mantle by a plume and incorporated in the ophiolite during seafloor spreading at 176 Ma. Blocks of the mantle containing the UHP minerals were presumably picked up by the later boninitic melts, transported to shallow depth and incorporated in the chromitites during crystallization.

Until recently, studies of ultra-high pressure (UHP) minerals were focused mainly on kimberlites and meteorites. However, many new discoveries in recent years have shown that UHP minerals can occur in a variety of rocks. For example, diamonds have been reported from meteor craters (Smith 1984), from a variety of metamorphic rocks (Okay *et al.* 1989; Hirajima *et al.* 1990; Sobolev & Shatsky 1990; Schmaedieke 1991; Xu *et al.* 1992; Jacob *et al.* 1994; Dobrzhinetskaya *et al.* 1995; Hough *et al.* 1995; Shatsky *et al.* 1995; Massonne 1999, 2001; Mukherjee & Sachan 2001; Mposkos & Kostopoulos 2001; Yang, J.-S. *et al.* 2001; Ghiribelli *et al.* 2002), from Alpine ultramafic rocks (Kaminsky & Vaganov 1977; Pearson *et al.* 1989; Davies *et al.* 1992) and from picrites of the Kamenusha massif in the Urals (Lukyanova *et al.* 1980). Fresh diamonds and several other UHP minerals were previously reported from chromitites of the Luobusa and Donqiao ophiolites, Tibet (Fang & Bai 1981, 1986; Bai *et al.* 1993; Hu 1999).

In order to investigate the occurrence of UHP minerals in ophiolites, we undertook a detailed re-examination of podiform chromitites in the Luobusa ophiolite, Tibet. Heavy minerals were separated from several 500 kg samples of chromitite collected directly from orebody 31 and hand-picked under a binocular microscope. A wide variety of HP and UHP minerals, native elements and alloys have been confirmed from the chromitites. Minerals considered to be UHP in origin include diamond, moissanite, Fe-silicide, wüstite,

From: MALPAS, J., FLETCHER, C. J. N., ALI, J. R. & AITCHISON, J. C. (eds) 2004. *Aspects of the Tectonic Evolution of China*. Geological Society, London, Special Publications, **226**, 247–271.
0305-8719/04/$15 © The Geological Society of London 2004.

octahedral Mg–Fe silicates, Os–Ir–Ru alloys and Ni–Fe–Cr–C alloys. Several of these are associated with native elements such as Si, Fe and Cr, which may have had a similar origin. Base-metal alloys composed primarily of Fe, Ni and Co are considered to be secondary, probably formed during serpentization. In this paper, we describe the textures, compositions and modes of occurrence of the UHP minerals in Luobusa, discuss their likely modes of formation and consider the tectonic significance of such minerals in ophiolites.

Geological setting

The Luobusa ophiolite is located about 200 km east–southeast of Lhasa in the Indus–Yarlung Zangbo suture zone – a major tectonic boundary that separates the Lhasa Block to the north from the Indian continent to the south. It is a fault-bounded slab 1–2 km thick, which has been thrust northward on to Tertiary molasse deposits

of the Luobusa Formation and the Gangdese batholith. To the south it is separated from Triassic flysch deposits by a steep reverse fault (Fig. 1). Structurally, the ophiolite consists of several inverted thrust slices, such that the 'pseudostratigraphy' is upside-down (Fig. 1). In the reconstructed section, the deepest rocks are clinopyroxene-bearing harzburgites that contain a number of dunite dykes, some of which have stringers or bands of chromite. The overlying depleted harzburgites contain abundant pods of chromitite, most of which have well-developed dunite envelopes (Zhou, M.-F., *et al.* 1996). At the top of the mantle section is a massive, transition-zone dunite, a few metres to 300 m thick.

The age of the ophiolite is equivocal, but it is believed to have formed in two stages. Stage 1 is represented by MORB-like mantle and gabbroic dykes with a Sm–Nd isochron age of 177 ± 31 Ma (Zhou, S., *et al.* 2002). During stage 2, boninitic melts formed in a suprasubduction zone environment, intruded the mantle

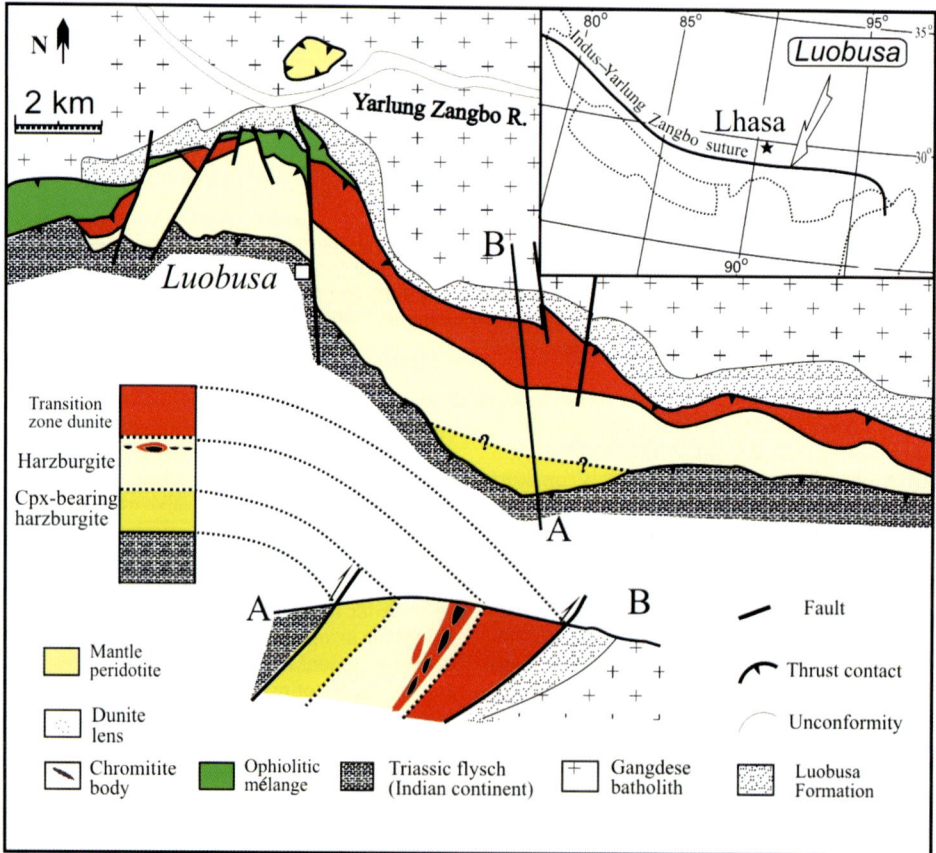

Fig. 1. Location and geological map of the Luobusa ophiolite, Tibet (after Malpas *et al.*, 2003).

section and formed the podiform chromitites. A zircon SHRIMP date of 126 ± 2 Ma has been obtained from another ophiolite in the belt with suprasubduction zone affinity, suggesting a Cretaceous age for the second stage of formation (Malpas *et al.* 2003). The Ar–Ar dates on amphibolites from the mélange zone range from 90–80 Ma and are interpreted as the age of intraoceanic thrusting (Aitchison & Badengzhu *et al.* 2000; Aitchison 2002; Abrajevitch *et al.* 2002; Malpas *et al.* 2003). Final exhumation and emplacement probably took place in the Early Neogene (Malpas *et al.* 2003).

The chromitites occur as tabular, lenticular and podiform bodies in the depleted harzburgites (Fig. 2A) not far from the boundary with the massive dunite. Their textures and mineralogies indicate that they are cumulate magmatic bodies (Fig. 2B) crystallized from a highly depleted, boninitic melt (Zhou *et al.* 1996). Most of the chromitites are surrounded by envelopes of dunite formed by melt–rock reaction in the upper mantle. The chromitites lie at a relatively uniform depth within the mantle (Fig. 1), suggesting the formation of numerous pockets of melt accumulation at the point where mantle flow changed from vertical to horizontal. Small bands of chromitite are also present in some of the dunite dykes.

Microstructures in the harzburgites, Cpx-bearing harzburgites and dunites suggest that the peridotites have undergone at least three stages of deformation:

(1) a high-temperature recrystallization;
(2) a high-pressure deformation; and
(3) a mid-crustal level deformation (Zhou 1995).

High-temperature recrystallization textures, characterized by coarse- to very coarse-grained olivine in the dunite, exaggerated grain growth of olivine 'porphyroblasts' over chromite foliations and layering, and smoothly curved boundaries between coarse-grained olivine–olivine and olivine–orthopyroxene crystal pairs, are common in the dunites and some of the harzburgites. These features are typical of very high-temperature, very low differential stress environments, as seen in spinel lherzolite mantle xenoliths and orogenic lherzolite complexes, such as the Ronda, Lherzo and Beni Bousera massifs (Nicolas & Poirier 1976). Deformation of pyroxene in stage 2 is characterized by the orthopyroxene slip system involving exsolution along the slip planes, kink-band development, and dynamic recrystallization. This deformation was later overprinted by a low-temperature/high-differential stress event (stage 3), characterized by deformation lamellae along slip planes in olivine, mini-kinking of olivine crystals, and low-temperature, recovery-controlled extinction band configurations. Discrete low-temperature plastic shear zones are locally present.

The stage 2 deformation may be a deep crustal overprint on the annealed (stage 1) mantle microstructures, whereas stage 3 deformation appears to be a mid-crustal level orogenic overprint that took place under relatively anhydrous conditions. The succession of deformation events is very similar to what has been described from orogenic garnet peridotites regionally associated with eclogites (Calon 1979).

Sampling, separation procedures and analytical methods

Chromitite has been mined at Luobusa for over a decade, and a number of seams and pods are exposed in open pits. Several samples of 500 kg each were removed directly from orebody 31, and care was taken to avoid any contamination from surrounding materials. The samples were then hand-washed, air-dried and crushed to pass a 1-cm sieve, before being transported to the Institute for Multipurpose Mineral Separation, Zheng Zhou, China for mineral separation. In Zheng Zhou, the samples were crushed to pass a 0.5-mm sieve, and mineral separation was carried out using a variety of magnetic, electric and heavy-liquid techniques (Hu 1999). The UHP minerals and alloys were hand picked from separates of various grain sizes. Selected grains were mounted in epoxy, polished, and then analysed using a JXA8800R microprobe and a S-3500N Scanning Electron Microscope equipped with an energy-dispersive spectrometer. The operating conditions were: voltage 20 kV and beam current 15 nA. SW9100 NIST multiple element standards were used for calibration. The SEM analyses were normalized to 100%. The classification of the alloys is based on CNMMN-INA (Harris & Cabri 1991).

Because the UHP minerals were hand-picked from heavy-mineral separates, natural or anthropogenic contamination is always a possibility. Natural contamination can be ruled out, because the samples were taken directly from the outcrop and carefully cleaned before processing. Anthropogenic contamination also appears unlikely because the same collection of minerals was recovered twice using completely different laboratories for mineral separation (cf. Fang and Bai 1981). All equipment was dismantled and carefully cleaned before the samples were crushed and processed. A 200-kg sample of granite from the Gangdese batholith was processed

A

B

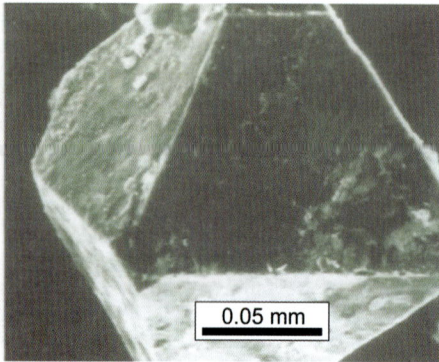

C

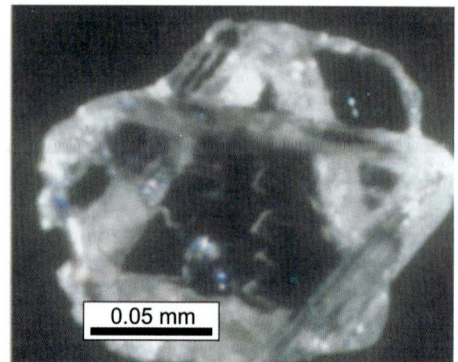

D

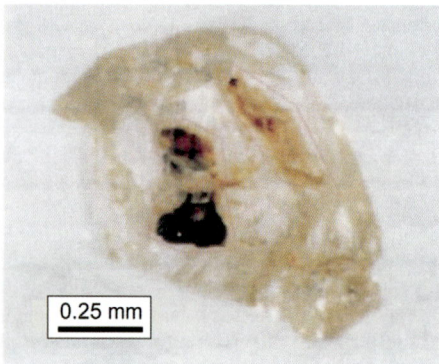

E

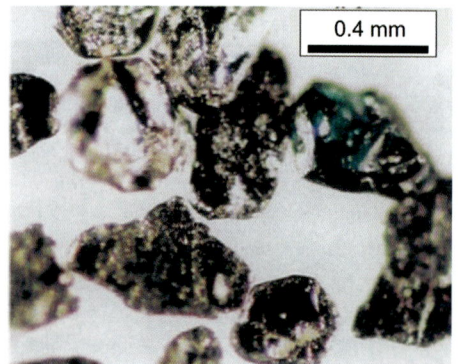

F

Fig. 2. Photographs of podiform chromitites of Luobusa and their UHP minerals. (**A**) Orebody 31 showing seams of massive chromitite sampled during this study. (**B**) Orebody 11 showing magmatic textures of podiform chromitite and lack of high-pressure plastic deformation. (**C**) Diamond grain from orebody 31. (**D**) Diamond grain from orebody 31. (**E**) Relatively large diamond grain from orebody 31, with mafic silicate inclusions. (**F**) Grains of graphite from orebody 31. Green grain is moissanite.

first to test for any contamination during the separation process. Only quartz, K-feldspar, plagioclase, biotite and zircon and apatite were recovered from the granite. Most of the UHP minerals recovered from the chromitites are black, dark grey or brightly coloured and would have been easily recognized in the granite sample had any contamination taken place.

The diamonds recovered in this study are clearly natural in origin (Taylor *et al.* 1995), and the morphology and colour of the moissanite suggest that it is also natural. Many of the grains of Os–Ir and octahedral silicates are either enclosed in (Bai *et al.* 2000), or attached to, chromite grains, leaving no question as to their provenance. Thus, we are confident that the UHP minerals reported here are naturally occurring grains in the Luobusa chromitites.

UHP minerals

A wide variety of minerals have been hand-picked from the chromitite separates, including diamond, other native elements, carbides, PGE and base-metal alloys, sulphides, silicates and oxides (Hu 1999; Bai *et al.* 2000). Here we focus on those minerals and alloys considered to have formed under UHP conditions, including diamond, moissanite (SiC), Fe-silicides, wüstite, CrC and Ni–Fe–Cr–C alloys, Os–Ir alloys and octahedral Mg–Fe silicates. Native Si, Fe, Ti, Ni, Cr, W, Au, Sn, Zn and Pb have also been recovered from the heavy-mineral separates, but only Si, Fe, Ti and Cr are clearly associated with the UHP grains, and thus are believed to have had a similar origin.

Native carbon: diamond and graphite

Diamonds were first reported from the Luobusa ophiolite during an investigation of heavy minerals in the chromitites (Fang & Bai 1981). Most of these are euhedral grains, 0.1 to 0.5 mm across, with octahedral and cubo-octahedral morphologies, but broken fragments are also present (Yan *et al.* 1986). The grains are clear and colourless, without evidence of corrosion or etching as is the case for many kimberlite diamonds (Fig. 2C, D).

In the current work, we have recovered an additional 25 grains of diamond from the chromite samples collected from orebody 31. Unbroken diamond crystals are typically colourless, euhedral octahedra ranging in size from 0.2 to 0.7 mm across, but some irregular grains (Fig. 2E) and broken fragments have also been recovered. X-ray diffraction analysis and Raman spectroscopy have been used to confirm the visual identifications. The *d*-spacings of the diamond crystals match those of standard samples very closely (Table 1).

A few grains contain small inclusions, and one grain contains three, relatively large, dark-green inclusions (Fig. 2E). Microprobe analyses of the inclusions indicate that they are Mg–Fe silicates with a composition close to that of clinoenstatite, but with somewhat higher SiO_2 (Table 2, Analysis 1 and 2).

IR spectra indicate that the Luobusa diamonds are type IaAB (Taylor *et al.* 1995). They have total nitrogen contents of 20–700 atomic ppm and nitrogen aggregation states up to 75% (Taylor *et al.* 1995). These aggregation states are relatively high, and are similar to many diamonds from kimberlites. Temperatures for these nitrogen aggregation states range from about 1250 °C to 1170 °C for mantle residence times between 50 and 1000 Ma, respectively (Taylor *et al.* 1995). The age of the ophiolite ranges from 177 to 126 Ma; however, the diamonds may have resided in the mantle for a much longer period.

Based on their colour, IR spectra, high nitrogen aggregation states and silicate inclusions, the Luobusa diamonds are definitely natural in origin (Taylor *et al.* 1995). However, Taylor *et al.* (1995) argued that their presence in the chromitites resulted from either anthropogenic

Table 1. *X-ray diffraction data for Luobusa diamonds*

1		2		3		4		JCPDS 6-0675		
d(A)	*I/Io*	*d(A)*	*I/Io*	*d(A)*	*I/Io*	*d(A)*	*I/Io*	*d(A)*	*I/Io*	*hkl*
2.042	100	2.05	100	2.042	100	2.048	100	2.06	100	111
1.256	57	1.258	44	1.256	52	1.256	58	1.261	25	220
1.072	15	1.072	13	1.071	17	1.073	30	1.075	16	311
0.8897	17	0.8913	17	0.889	19	0.8914	34	0.8916	8	400
0.8165	9	0.817	7	0.8176	13	0.8168	25	0.8182	16	311
				0.603	6					

MoKα radiation; 47.5 kV; 20 mA; Scan 4–75° 2θ; *I/Io*–relative peak intensity.

Table 2. *Compositions of silicate inclusions in UHP minerals from the Luobusa chromitites*

Number	1	2	3	4	5	6	7	8	9	10	11	12
No. of analyses	1	1	1	1	1	5	2	2	3	2	2	1
SiO_2	63.95	64.15	22.70	23.70	22.90	99.45	0.00	17.70	32.67	34.99	53.43	50.31
TiO_2	0.00	0.00	0.00	0.00	0.00	0.00	0.99	1.24	3.85	8.06	0.00	0.00
Al_2O_3	0.36	0.46	29.70	29.20	32.10	0.00	0.00	0.00	0.50	1.52	11.18	11.09
FeO	4.69	4.52	0.00	0.00	3.00	0.55	99.01	74.30	44.51	37.01	8.56	14.34
MnO	0.00	0.00	0.00	0.00	0.00	0.00	0.00	6.77	18.47	18.42	0.00	0.00
MgO	30.57	30.78	0.00	0.90	0.85	0.00	0.00	0.00	0.00	0.00	16.53	13.33
CaO	0.00	0.24	30.00	33.10	41.30	0.00	0.00	0.00	0.00	0.00	10.32	9.03
Na_2O	0.42	0.00	0.00	0.00	0.00	0.00	0.00	0.00	0.00	0.00	0.00	1.90
K_2O	0.00	0.00	0.00	0.00	0.00	0.00	0.00	0.00	0.00	0.00	0.00	0.00
Total	99.99	100.05	82.40	86.90	100.15	100.00	100.00	100.00	100.00	100.00	100.00	100.00

1, 2, inclusions in diamond shown in Fig. 2E; 3, 4, white inclusions in SiC; 5, gehlenite composition from Anthony *et al.* (1990); 6, SiO_2 grain in Fig. 5B; 7, light-coloured FeO replacing SiO_2 in Fig. 5B; 8, Fe-silicate inclusion in native Fe (Fig. 5D); 9,10, dark patch in silicate inclusion (Fig. 5D); 11,12, silicate inclusions in O.–Ir alloy (Fig. 7C).

or natural contamination. The recovery of exactly the same type of diamonds twice from the same sample location, when the samples were processed in two different laboratories, makes anthropogenic contamination highly unlikely. No diamonds or other UHP minerals were found in the granite control sample, which was processed first. Natural contamination can be ruled out, because our samples were taken as blocks of chromitite directly from the outcrop and cleaned carefully before processing. In addition, there are no kimberlites or other diamond-bearing rocks in the region from which these grains could have been derived.

Graphite is common in the Luobusa chromitites and occurs as grey, tabular prisms and irregular grains, 0.1 to 0.7 mm long (Fig. 2F). Most of the grains have rounded corners, probably due to abrasion during the separation process, but their hexagonal morphology is commonly still apparent. None of the graphite grains have a diamond morphology like those in Ronda and Beni Bousera (Pearson *et al.* 1989; Davies *et al.* 1992).

Moissanite, Fe-silicide, native silicon and silicon rutile

Natural SiC occurs as hexagonal and trigonal polyforms (α-SiC) known as moissanite (Kunz 1905; Leung 1990) and as a cubic polyform (β-SiC). Moissanite (SiC) is a naturally occurring mineral closely associated with diamonds. It occurs both as inclusions within diamonds (Moore *et al.* 1986; Jaques & Hall *et al.* 1989; Moore & Gurney 1989; Otter & Gurney 1989; Leung 1990; Leung *et al.* 1990, 1996) and as an accessory mineral in diamond-bearing rocks (He 1987; Mathez *et al.* 1995; Leung *et al.* 1996).

Many grains and fragments of moissanite have been recovered from the Luobusa chromitites. They are easily recognized, because most are strongly coloured and have an adamantine lustre. The grains range in colour from dark-blue to greyish-blue to nearly colourless, or from pale green to yellow to yellowish-blue to bluish-green (Fig. 3A, B). Many are colour-zoned, with gradual transitions between colours. These colour variations are believed to be caused by small amounts of impurities such as nitrogen (Mathez *et al.* 1995) and aluminium (Leung *et al.* 1996).

The moissanite grains from Luobusa range from 0.1 to 1.1 mm, and occur as either single rounded or pinacoidal idiomorphic crystals (Fig. 3C) or fragments of idiomorphic crystals (Fig. 3A, B). Hexagonal and trigonal forms are

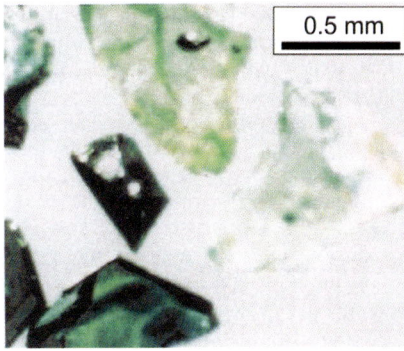

A

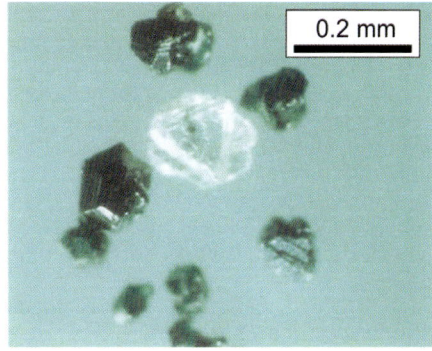

B

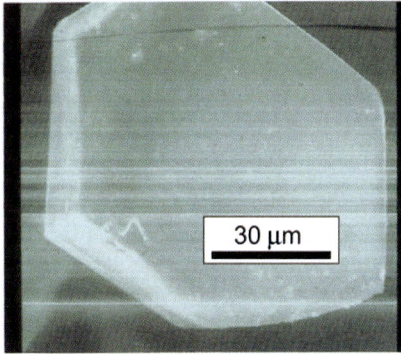

C

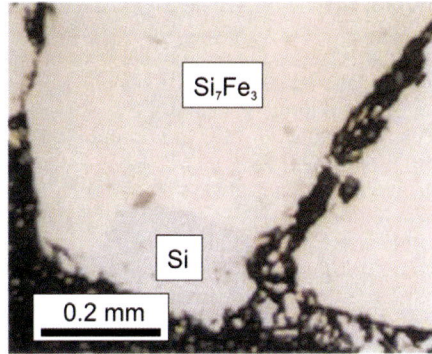

D

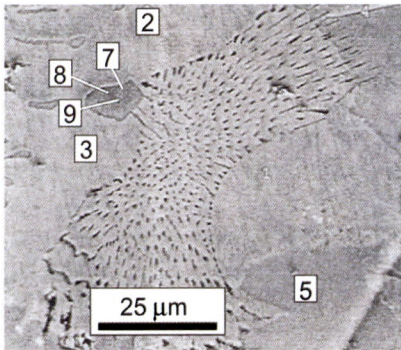

E

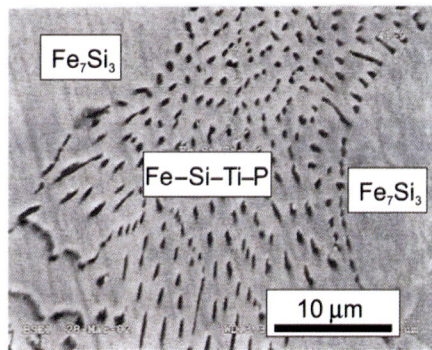

F

Fig. 3. Photographs of UHP minerals from the Luobusa chromitites. (**A**) Fragments of euhedral moissanite grains. Note the small dark inclusions of native Si. (**B**) Twinned grains of dark-blue moissanite. Colourless grain in centre is a diamond. (**C**) SEM photograph of a euhedral moissanite grain from the Luobusa chromitites. (**D**) Grain of Fe-silicide containing a darker inclusion of native Si (after Bai *et al.*, 2000). (**E**) Grain of Fe-silicide (41-9) containing numerous inclusions of different composition. The host grain has a composition of $Fe_{58}Ti_4Si_{38}$, whereas the large dark patch in the upper left corner (points 7, 8, and 9) has an average composition of $Fe_{70.5}Ti_{2.5}Si_{27}$. The cluster of small, dark blebs are also Fe-silicides with abundant Ti and P. (**F**) Enlargement of a section of Fig. 3E, showing the tubular nature of the small, dark inclusions in the Fe-silicide. An average composition of the inclusions is $Fe_{48}Ti_{31}P_{7.5}Si_{13.5}$ (Table 3, small dark blebs).

common, but the many polytypes characteristic of synthetic SiC are absent. Syntactic intergrowths of different crystal forms are common on or across the pinacoidal face. Rarely the crystals have x-shaped intergrowths or twins (Fig. 3B)

Optical identification of the moissanite was confirmed by both energy-dispersive SEM and single-crystal X-ray diffraction analyses. X-ray diffraction analysis on a single, idiomorphic crystal shows that the grain is of the 30H form of the mineral.

A few euhedral moissanite crystals contain small inclusions of a white to pale yellow, powdery mineral, and one broken euhedral grain of moissanite has a spherical cavity half-filled with solid white material. Microprobe analyses of these inclusions reveal a composition close to that of gehlenite ($Ca_2Al_2SiO_7$) (Table 2, Analyses 3 and 4) (cf. Table 2, Analysis 5). Some idiomorphic crystals contain minute oval or needle-like opaque inclusions (Fig. 3A) identified as native Si by microprobe analysis.

Because synthetic SiC (carborundum) is widely used in polishing and drilling, anthropogenic contamination is always a possibility (Milton & Vitaliano 1985). Possible sources of contamination by industrial SiC are carborundum grains used in polishing rock samples, drill bits containing euhedral carborundum crystals, and heavy-mineral concentrates polluted with industrial waste (Milton & Vitaliano 1985). However, it is extremely unlikely that our samples were contaminated by industrial SiC, because samples were taken directly from the outcrop and carefully washed before crushing. None of the samples were polished, and any contamination during the heavy-mineral separation should have been observed in the granite control sample, where the brightly coloured grains of SiC would be easily recognized. In addition, the moissanite grains from Luobusa are relatively large, have a simple morphology and exhibit a wide range of colour, whereas most synthetic grains of SiC are small, have many polytypes and are blue in colour.

Moissanite and diamond are both carbon-bearing and share a common atomic structure. They have similar chemical/physical properties (i.e. similar tetrahedrally oriented sp3 hybrid bonds and identical glide planes) (Leung *et al.* 1996). Clearly, moissanite has a close relationship with diamond and is a more common accessory mineral in diamond-bearing rocks than was previously known (Mathez *et al.* 1995; Leung 1990; Leung *et al.* 1990, 1996).

A recently discovered pebble from the Mediterranean coast of Turkey contains abundant moissanite very similar to that reported here (Di Pierro *et al.* 2003). The newly discovered pebble with moissanite also contains brucite, calcite, magnesite, phlogopite, magnesiochromite, Al-rich orthopyroxene, Mg–Fe silicates and quartz. The moissanite grains are dark-blue to colourless, euhedral, hexagonal crystals that contain inclusions of native Si and Fe-silicide. The moissanite from the Turkish sample is nearly identical in form, colour, structure, composition and inclusions to the grains reported here. Based on the association of SiC and Fe-bearing silicates and the presence of metallic inclusions, Di Pierro *et al.* (2003) suggest a deep-mantle origin for their sample.

The Fe-silicides are also common in the Luobusa chromitites, and occur as grey, irregular grains up to about 1.5 mm across. Most are uniform in composition and character, but some contain small inclusions of native Si (Fig. 3D). The Fe-silicides in Luobusa fall into five main compositional groups, reflecting different proportions of Fe and Si and the addition of significant amounts of other elements, particularly Ti and P (Table 3). Most abundant is a Si-rich variety with a Fe–Si ratio of approximately three to seven. Twenty analyses on six grains of this silicide yielded an average composition of $Fe_{29.4}Si_{70.6}$, with a standard deviation of 0.37. The energy spectrum for a typical grain (Sample 13-2) is shown in Fig. 4A. This silicide typically hosts inclusions of native Si (Fig. 3D).

The other silicides are all Fe-rich and contain variable percentages of other elements. The second group is characterized by variable but significant proportions of Ti, and an absence of either Mn or Al. The most common variety contains approximately 72.2–75.2 wt% Fe, 21.8–23.3 wt% Si and 2.9–4.8 wt% Ti, and has an average composition of $Fe_{59.4}Ti_{3.9}Si_{36.7}$, giving an Fe–Si ratio of approximately six to four if the Ti and Si are combined (Table 3). An energy spectrum of this material from Sample 41-11 is given in Fig. 4B. Under the SEM these grains have a relatively light-grey appearance and some contain irregular patches of darker-coloured material about 5–25 μm across (Fig. 3E). These darker patches are subrounded to subangular in shape and have sharp, distinct boundaries. The darker material is similar in composition to the host grain, but has higher Fe and lower Si and Ti contents (Group 3) (see Table 3, dark patches in grain 41-9). It has an average composition of $Fe_{70.3}Si_{26.9}Ti_{2.8}$, giving an approximate Fe–Si ratio of 7 : 3 when the Ti and Si are combined. Also included in the host grain are irregular, feather- or fan-shaped clusters, up to about 75 μm across, of many minute droplets or blebs of even darker material. These

Table 3. *Chemical compositions of Fe-silicides and native Si of the Luobusa ophiolite*

Sample no.	No. of analyses	Wt%							At%							Formula
		Si	Fe	Mn	Al	Ti	P	Total	Si	Fe	Mn	Al	Ti	P	Total	
Fe-silicides—Fe₃Si₇																
8-1	1	53.78	46.22	0.00	0.00	0.00	0.00	100.00	69.8	30.2	0.00	0.00	0.00	0.00	100.00	Fe_3Si_7
12-22	1	55.26	44.74	0.00	0.00	0.00	0.00	100.00	71.1	28.9	0.00	0.00	0.00	0.00	100.00	Fe_3Si_7
13-2	3	54.33	45.67	0.00	0.00	0.00	0.00	100.00	70.3	29.7	0.00	0.00	0.00	0.00	100.00	Fe_3Si_7
38-2-1	1	54.19	45.81	0.00	0.00	0.00	0.00	100.00	70.17	29.83	0.00	0.00	0.00	0.00	100.00	Fe_3Si_7
23-3	4	55.18	44.82	0.00	0.00	0.00	0.00	100.00	71.0	29.0	0.00	0.00	0.00	0.00	100.00	Fe_3Si_7
23-6	1	53.50	46.50	0.00	0.00	0.00	0.00	100.00	69.6	30.4	0.00	0.00	0.00	0.00	100.00	Fe_3Si_7
23-8	3	54.09	45.91	0.00	0.00	0.00	0.00	100.00	70.1	29.9	0.00	0.00	0.00	0.00	100.00	Fe_3Si_7
23-11	3	54.94	45.06	0.00	0.00	0.00	0.00	100.00	70.8	29.2	0.00	0.00	0.00	0.00	100.00	Fe_3Si_7
32-7	2	54.63	45.38	0.00	0.00	0.00	0.00	100.00	70.5	29.5	0.00	0.00	0.00	0.00	100.00	Fe_3Si_7
32-9	1	54.39	45.61	0.00	0.00	0.00	0.00	100.00	70.3	29.7	0.00	0.00	0.00	0.00	100.00	Fe_3Si_7
38-3	1	54.68	45.32	0.00	0.00	0.00	0.00	100.00	70.6	29.4	0.00	0.00	0.00	0.00	100.00	Fe_3Si_7
Fe-silicides—Fe₆Si₄																
41-9-1	1	23.27	72.16	0.00	0.00	4.57	0.00	100.00	37.39	58.31	0.00	0.00	4.31	0.00	100.00	$Fe_{0.59}Si_{0.37}Ti_{0.04}$
41-9-2	1	23.09	72.52	0.00	0.00	4.39	0.00	100.00	37.16	58.70	0.00	0.00	4.14	0.00	100.00	$Fe_{0.59}Si_{0.37}Ti_{0.04}$
41-9-4	1	23.30	72.38	0.00	0.00	4.32	0.00	100.00	37.44	58.49	0.00	0.00	4.07	0.00	100.00	$Fe_{0.59}Si_{0.37}Ti_{0.04}$
41-9	Average	23.22	72.35	0.00	0.00	4.43	0.00	100.00	37.33	58.50	0.00	0.00	4.17	0.00	100.00	$Fe_{0.59}Si_{0.37}Ti_{0.04}$
41-11-9	1	22.57	72.79	0.00	0.00	4.64	0.00	100.00	36.46	59.14	0.00	0.00	4.40	0.00	100.00	$Fe_{0.59}Si_{0.37}Ti_{0.04}$
41-11-10	1	22.83	73.13	0.00	0.00	4.04	0.00	100.00	36.83	59.34	0.00	0.00	3.82	0.00	100.00	$Fe_{0.59}Si_{0.37}Ti_{0.04}$
41-11-11	1	22.71	74.44	0.00	0.00	2.85	0.00	100.00	36.73	60.56	0.00	0.00	2.70	0.00	100.00	$Fe_{0.60}Si_{0.37}Ti_{0.03}$
41-11	Average	22.70	73.45	0.00	0.00	3.84	0.00	100.00	36.68	59.68	0.00	0.00	3.64	0.00	100.00	$Fe_{0.59}Si_{0.37}Ti_{0.04}$
41-10-3	1	22.72	72.53	0.00	0.00	4.75	0.00	100.00	36.65	58.85	0.00	0.00	4.49	0.00	100.00	$Fe_{0.59}Si_{0.37}Ti_{0.04}$
41-10-4	1	22.87	72.84	0.00	0.00	4.29	0.00	100.00	36.87	59.07	0.00	0.00	4.06	0.00	100.00	$Fe_{0.59}Si_{0.37}Ti_{0.04}$
41-10-5	1	21.77	74.33	0.00	0.00	3.90	0.00	100.00	35.43	60.85	0.00	0.00	3.72	0.00	100.00	$Fe_{0.61}Si_{0.35}Ti_{0.04}$
41-10-6	1	22.00	73.90	0.00	0.00	4.10	0.00	100.00	35.73	60.36	0.00	0.00	3.90	0.00	100.00	$Fe_{0.60}Si_{0.36}Ti_{0.04}$
41-10	Average	22.34	73.40	0.00	0.00	4.26	0.00	100.00	36.17	59.78	0.00	0.00	4.04	0.00	100.00	$Fe_{0.60}Si_{0.36}Ti_{0.04}$

(continued)

Table 3. Continued

Sample no.	No. of analyses	Wt%							At%							Formula
		Si	Fe	Mn	Al	Ti	P	Total	Si	Fe	Mn	Al	Ti	P	Total	
Large dark patches in grain 41-9 (Fig. 3E)																
41-9-5	1	15.38	80.27	0.00	0.00	4.35	0.00	100.00	26.38	69.25	0.00	0.00	4.38	0.00	100.00	$Fe_{0.69}Si_{0.27}Ti_{0.04}$
41-9-7	1	15.85	82.43	0.00	0.00	1.72	0.00	100.00	27.18	71.09	0.00	0.00	1.73	0.00	100.00	$Fe_{0.71}Si_{0.27}Ti_{0.02}$
41-9-8	1	15.97	81.20	0.00	0.00	2.83	0.00	100.00	27.31	69.85	0.00	0.00	2.84	0.00	100.00	$Fe_{0.71}Si_{0.27}Ti_{0.03}$
41-9-8	1	15.53	82.20	0.00	0.00	2.27	0.00	100.00	26.68	71.03	0.00	0.00	2.29	0.00	100.00	$Fe_{0.71}Si_{0.27}Ti_{0.02}$
41-9-9	Average	15.68	81.53	0.00	0.00	2.79	0.00	100.00	26.89	70.30	0.00	0.00	2.81	0.00	100.00	$Fe_{0.70}Si_{0.27}Ti_{0.03}$
Small dark blebs in grain 41-11 (similar to those in Fig. 3F)																
41-11-4	1	13.70	48.05	0.00	0.00	30.99	7.26	100.00	21.87	38.58	0.00	0.00	29.03	10.51	100.00	$Fe_{0.39}Ti_{0.29}Si_{0.22}P_{0.10}$
41-11-5	1	13.79	51.99	0.00	0.00	27.49	6.73	100.00	22.18	42.06	0.00	0.00	25.95	9.82	100.00	$Fe_{0.42}Ti_{0.26}Si_{0.22}P_{0.10}$
41-11-6	1	13.47	48.72	0.00	0.00	30.22	7.59	100.00	21.52	39.15	0.00	0.00	28.33	11.00	100.00	$Fe_{0.39}Ti_{0.28}Si_{0.22}P_{0.11}$
41-11-7	1	13.72	48.18	0.00	0.00	30.61	7.49	100.00	21.88	38.54	0.00	0.00	28.64	10.83	100.00	$Fe_{0.39}Ti_{0.28}Si_{0.22}P_{0.11}$
41-11-8	1	13.42	47.93	0.00	0.00	31.10	7.55	100.00	21.43	38.49	0.00	0.00	29.14	10.93	100.00	$Fe_{0.39}Ti_{0.29}Si_{0.21}P_{0.11}$
41-11	Average	13.62	48.97	0.00	0.00	30.08	7.32	100.00	21.78	39.29	0.00	0.00	28.22	10.62	100.00	$Fe_{0.39}Ti_{0.28}Si_{0.22}P_{0.11}$
Fe-silicides–Fe₃Si																
38-17-3	3	13.38	83.56	0.56	2.50	0.00	0.00	100.00	22.95	72.09	0.49	4.46	0.00	0.00	100.00	$Fe_{0.72}Si_{0.23}Al_{0.045}Mn_{0.005}$
41-1-1	3	15.12	84.88	0.00	0.00	0.00	0.00	100.00	26.15	73.85	0.00	0.00	0.00	0.00	100.00	$Fe_{0.74}Si_{0.26}$
Native silica																
4-6	1	100.00	0.00	0.00	0.00	0.00	0.00	100.00	100.00	0.00	0.00	0.00	0.00	0.00	100.00	Si_{100}
30-2-8	3	99.89	0.11	0.00	0.00	0.00	0.00	100.00	99.94	0.05	0.00	0.00	0.00	0.00	100.00	$Si_{0.99}Fe_{0.01}$
31-2-10	1	100.00	0.00	0.00	0.00	0.00	0.00	100.00	100.00	0.00	0.00	0.00	0.00	0.00	100.00	Si_{100}
32-3-9	2	100.00	0.00	0.00	0.00	0.00	0.00	100.00	100.00	0.00	0.00	0.00	0.00	0.00	100.00	Si_{100}
34-3-1	1	100.00	0.00	0.00	0.00	0.00	0.00	100.00	100.0	0.00	0.00	0.00	0.00	0.00	100.00	Si_{100}
35-31	1	100.00	0.00	0.00	0.00	0.00	0.00	100.00	100.0	0.00	0.00	0.00	0.00	0.00	100.00	Si_{100}
38-2	3	100.00	0.00	0.00	0.00	0.00	0.00	100.00	100.00	0.00	0.00	0.00	0.00	0.00	100.00	Si_{100}
38-34	1	100.00	0.00	0.00	0.00	0.00	0.00	100.00	100.00	0.00	0.00	0.00	0.00	0.00	100.00	Si_{100}

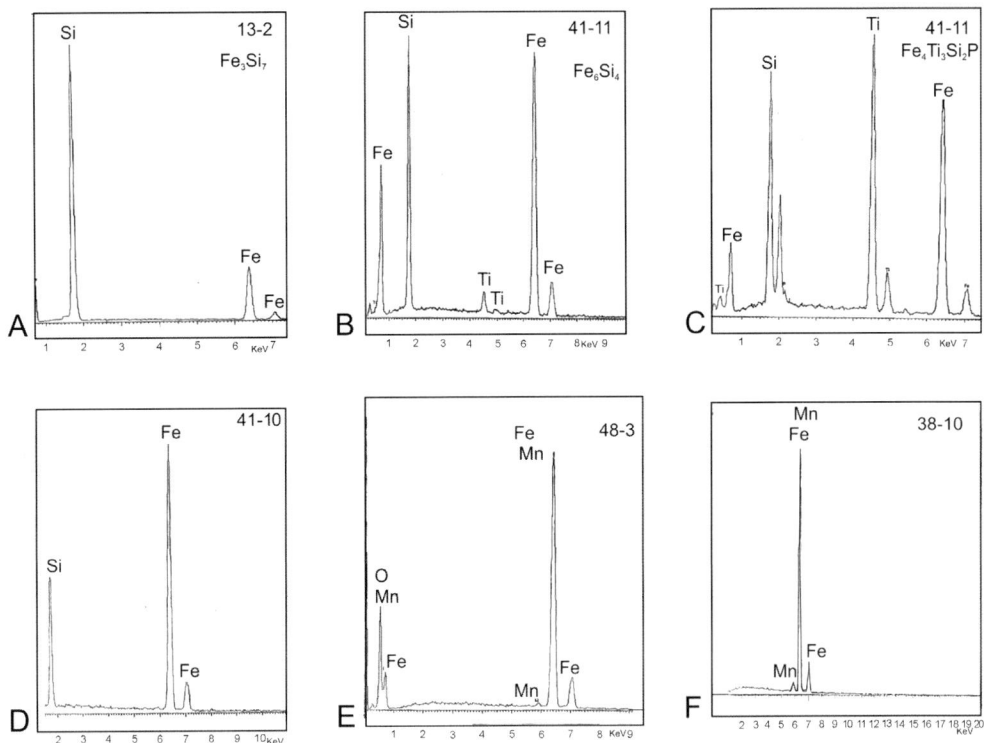

Fig. 4. Energy-dispersive spectra of various minerals from the Luobusa chromitites. (**A**) Fe-silicide (Sample 13-2). (**B**) Fe-silicide (Sample 41-11). (**C**) Fe-silicide with Ti and P. Dark patches in sample 41-11. (**D**) Fe-silicide (Sample 41-1-0). (**E**) Wüstite (Sample 48-3-0). (**F**) Native Fe in wüstite (Sample 38-10).

blebs appear to be elongate tubes, up to 15 μm long, with subcircular cross-sections $1-2$ μm across (Fig. 3E, F). A number of them have teardrop or tubular shapes oriented away from point sources. The blebs have a distinct and unusual composition characterized by approximately 48.5 wt% Fe, 30.5 wt% Ti, 13.5 wt% Si, and 7.5 wt% P (see Table 3, small dark blebs in grain 41-11), which yields an average formula of $Fe_{40}Ti_{28}P_{10}Si_{22}$ (Group 4). An energy spectrum for this material (Sample 41-11) is shown in Fig. 4C. The texture and composition of these blebs strongly suggest that they were immiscible liquids moving through the host material before it solidified.

A few iron-rich grains containing approximately 85 wt% Fe and 15 wt% Si make up Group 5. Some of these grains contain $2-3$ wt% Al, which appears to substitute for Si, and up to 0.5 wt% Mn (Table 3). This silicide has an average composition of $Fe_{73.0}Mn_{0.2}Al_{2.3}Si_{24.5}$, which gives a Fe–Si ratio of approximately 3 : 1 if Mn is combined with Fe and Al with Si. An energy spectrum for Sample 41-10, a grain without Al or Mn, is given in Fig. 4D.

Similar Fe–Ti–Si alloys from the Yakutia kimberlite pipe, Siberia, with compositions close to Fe_3Si_7, have been termed ferrosilicite (Mathez *et al.* 1995). Mathez *et al.* (1995) also reported a ferrosilicite (Fe_3Si_7) very similar in composition to those reported here occurring as an inclusion in SiC.

Native Si occurs typically as irregular inclusions in the Luobusa moissanite (Fig. 3A) and Fe-silicides (Fig. 3D). The inclusions in the Fe-silicides are subcircular in shape, up to about 0.25 mm across, and have sharp boundaries with the host material. In SEM photographs the native Si is darker grey and easily distinguished from the host Fe-silicide (Fig. 3D). The Si has been identified by Raman analysis, which yields a typical spectrum with a peak at 520.5 cm^{-1}, and by energy-dispersive and microprobe analysis. The energy spectrum shows nearly pure Si, sometimes with small amounts of Fe (Table 3). Native Si has been previously reported from the Josephine ophiolite of Oregon (Bird & Weathers 1975) and as inclusions in diamonds from the Yakutia kimberlites of Siberia (Sobolev *et al.* 1997).

One grain of silicon rutile has been recovered from the Luobusa chromitites (Yang *et al.* 2003). The grain is prismatic in shape, about $50 \times 300 \, \mu m$ in size, and has weak cleavage. X-ray powder diffraction data obtained from the sample matches very closely with the JCPD standard 21-1276, showing that it has a tetragonal structure and is isostructural with stishovite. An average composition based on three SEM energy-dispersive analyses is: $SiO_2 = 13.8 \, wt\%$, $TiO_2 = 85.9 \, wt\%$ and $Cr_2O_3 = 0.3 \, wt\%$. The Si substitutes for Ti in the six-fold coordination position, indicating formation under very high pressures at depths equivalent to the mantle transition zone or lower mantle (Yang *et al.* 2003).

Wüstite (FeO) and native iron (Fe)

In Luobusa, wüstite occurs as light-grey, subrounded grains typically 0.1–1.5 mm in diameter, commonly hosting spherical inclusions of native Fe (Fig. 5A). Most of the grains are nearly pure FeO with small amounts of Mn or Ti, but one grain has up to 20.68 wt% MnO (Table 4). This grain also contains a few per cent each of SiO_2 and Al_2O_3 and traces of Cr_2O_3 and MgO. An energy spectrum for sample 48-3 is given in Fig. 4E. X-ray diffraction data for these grains confirm their identification as wüstite (Table 5).

One very interesting grain appears to be an intergrowth of SiO_2 and FeO. This grain has a prismatic form, is about 0.3 mm long, and consists primarily of SiO_2 (Fig. 5B) (Bai *et al.* 2001). Around the margins of the grain and extending toward the centre, the SiO_2 appears to have been replaced by FeO. Minute blebs or lamellae of FeO are also scattered through the grain. The compositions of these two phases are given in Table 2 (Analyses 6 and 7).

Native Fe is relatively common in the Luobusa chromitites, and typically occurs as small round globules, 50–100 μm in diameter, enclosed in wüstite (Fig. 5A) or rarely as irregular clusters of acicular grains (Bai *et al.* 2000). These grains are typically pure Fe, although a few contain small amounts of Mn, Si and Al, up to a total of about 4 at% (Table 6). The energy spectrum for native Fe in sample 38-10 is shown in Fig. 4F. The perfectly round shape of the Fe grains in wüstite and their sharp boundaries indicate that they were originally molten. Because the grains are completely enclosed in wüstite, the two probably represent immiscible liquids

One grain of native Fe (50-2) contains small, round inclusions of an Fe–Mn silicate (Fig. 5C). These inclusions range from about 1 to 7 μm in diameter, are perfectly circular in cross-section and have sharp boundaries with the host Fe.

SEM photographs of the inclusions reveal small, irregular dark patches within a homogeneous, lighter grey background (Fig. 5D). The background phase in the inclusions is an iron-rich silicate (Fig. 5E) (approximately 69–74 wt% FeO and 17.5–22 wt% SiO_2) (Table 2, Analysis 8), whereas the small dark patches are lower in FeO and SiO_2 and higher in MnO (up to about 20 wt%) and TiO_2 (Table 2, Analyses 9 and 10). A few of the dark patches also contain small quantities of Al_2O_3. The distribution of SiO_2 in these inclusions is relatively uniform (Fig. 5F). The Raman spectrum of these inclusions does not match any known mineral, but the grains are similar in composition to Mn-rich fayalite (e.g. Table 2, Analysis 9). Likewise, the composition of the lighter grey, high-FeO, low-SiO_2 background material does not match a known mineral, but it could be a type of Si-rich wüstite. The shape and composition of the inclusions in the native Fe leave little doubt that they represent immiscible silicate liquids that underwent partial crystallization upon cooling.

Native Fe similar to that described here has been reported from alpine ultramafic rocks of the Koryak Peninsula (Rudashevsky *et al.* 1983), where it has a composition of 98.2–99.6 wt% Fe, 0.01–1.25 wt% Cu, 0.02–0.71 wt% Mn and 0–0.09 wt% Si. Wüstite inclusions in diamonds from Tanzania kimberlites contain small round beads of native Fe with a texture very similar to that of the samples from Luobusa (Stachel *et al.* 1998). The samples from Tanzania are estimated to have formed at mantle depths of more than 670 km.

Ni–Fe–Cr–C alloys

These carbide alloys are compositionally variable but fall into three main groups: Cr–C, Fe–Ni–C and Ni–Fe–Cr–C based on their atomic proportions (Fig. 6A). The CrC alloy occurs as steel-grey, acicular crystals with well-developed crystal faces (Fig. 6B). Microprobe analyses for several grains are given in Table 7 (Analyses 1–4). Carbon was not analysed quantitatively but it has a significantly higher peak in these grains than in silicates that have only a carbon coating. In addition to Cr, these alloys have small amounts of Fe and Ni and traces of Ti. Native Cr has also been reported from the Luobusa chromitites (Fang & Bai 1981; Zhang *et al.* 1996).

The Fe–Ni–Cr–C alloys occur as subrounded grains about 200 nm across, with a silver white, metallic lustre. Several have distinct compositional bands or zones. Samples 1-17, 1-23, and 1-18, and are typical Fe–Ni–Cr–C alloys,

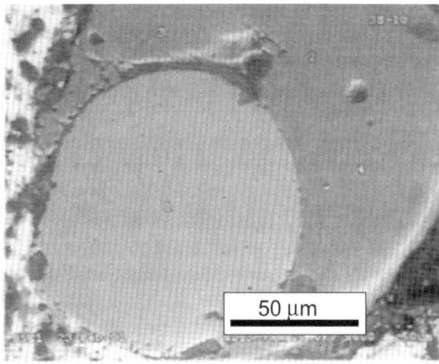

A

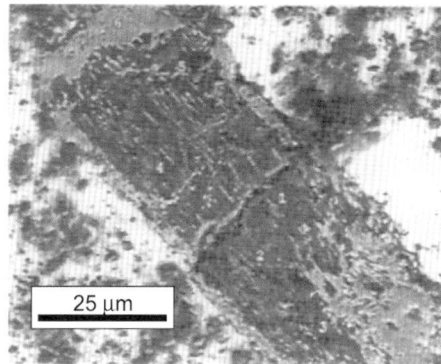

B

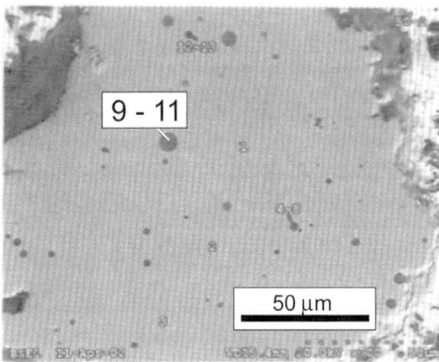

C

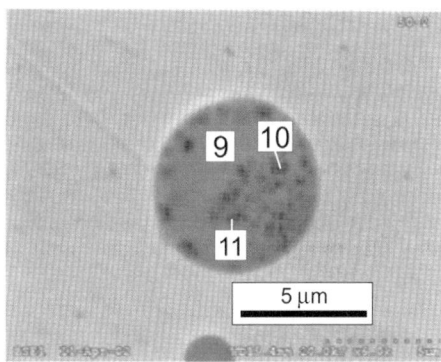

D

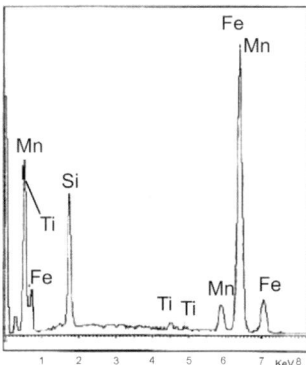

E

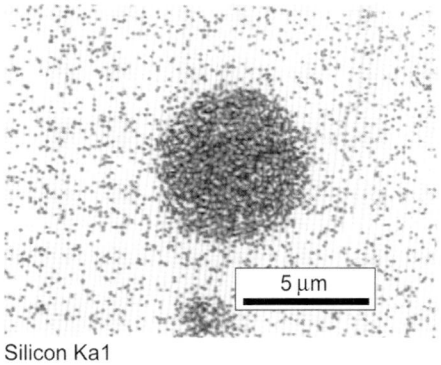

F

Fig. 5. Photos of wüstite, native iron and silicate inclusions. (**A**) Wüstite grain (38-10) containing perfectly round inclusion of native Fe. (**B**) Elongate grain of SiO_2 (dark material) showing replacement by FeO (light material). (**C**) Grain of native Fe (50-2) containing many small, round inclusions of Fe–Mn silicate. (**D**) Enlarged view of silicate in inclusion (9-11) in Fig. 5C showing dark patches of Mn-rich silicate mineral (see Table 2 for compositions). (**E**) Energy spectrum for silicate inclusion in native Fe (Fig. 5D, point 9). (**F**) Backscatter electron image showing distribution of Si in silicate inclusion enclosed in native Fe. Large circle is the same view as in Fig. 5D.

Table 4. *Compositions of wüstite grains from the Luobusa chromitites*

Sample no.	38-10-2	38-10-3	48-3-4	48-3-5	48-3-6	3-2-2-6	3-2-2-7
SiO_2	1.95	2.54	0.00	0.00	0.00	0.00	0.00
TiO_2	0.00	0.02	0.00	0.00	0.00	2.32	1.70
Al_2O_3	1.16	1.33	0.00	0.00	0.00	0.00	0.00
Cr_2O_3	0.25	0.21	0.00	0.00	0.00	0.00	0.00
FeO	74.51	73.28	99.00	99.03	98.97	97.68	98.30
MnO	20.46	20.68	1.00	0.97	1.03	0.00	0.00
MgO	0.31	0.15	0.00	0.00	0.00	0.00	0.00
CaO	0.01	0.02	0.00	0.00	0.00	0.00	0.00
Total	98.65	98.23	100.00	100.00	100.00	100.00	100.00
			Number of cations based on 8O				
Si	0.1806	0.2342	0.0000	0.0000	0.0000	0.0000	0.0000
Ti	0.0000	0.0014	0.0000	0.0000	0.0000	0.1639	0.1207
Al	0.1267	0.1446	0.0000	0.0000	0.0000	0.0000	0.0000
Cr	0.0183	0.0153	0.0000	0.0000	0.0000	0.0000	0.0000
Fe	5.7722	5.6512	7.9190	7.9214	7.9166	7.6722	7.7586
Mn	1.6053	1.6152	0.0810	0.0786	0.0834	0.0000	0.0000
Mg	0.0428	0.0206	0.0000	0.0000	0.0000	0.0000	0.0000
Ca	0.0010	0.0020	0.0000	0.0000	0.0000	0.0000	0.0000
Sum	7.7469	7.6845	8.0000	8.0000	8.0000	7.8361	7.8793

Note: Samples 38-10-2 and 38-10-3 were determined with JXA 8800R electron microprobe, the others by scanning electron microprobe.

consisting mainly of light-grey material hosting large blade-like or acicular zones of dark material (Fig. 6C, D and E, respectively). Samples 1–18 also has many small irregular patches or blebs of dark material scattered through the grain (Fig. 6F). The light-grey areas and bands are relatively rich in both Fe and Ni and low in Cr (Table 7, Analyses 9 and 10), whereas the dark bands are characterized by lower Ni and higher Cr (Table 7, Analyses 11, 12 and 13). It is not clear whether these dark bands and patches are exsolution features or intergrowths of different composition.

The Ni–Fe–C alloys form greyish-white, granular grains, 200–300 μm across, with strong metallic lustre. These fall into two main compositional groups, Fe-rich and Ni-rich. Fe-rich varieties lack Ni and have atomic proportions close to $Fe_{0.5}C_{0.5}$ (Table 7, Analyses 5 and 6). The Ni-rich grains have variable amounts of Fe and C (Table 7, Analyses 7 and 8). Most grains of Fe–Ni–C are optically homogeneous and show little compositional variation. However, one grain has abundant lamellae that produce a pattern similar to Widmanstätten texture (Fig. 7A).

PGE alloys

Abundant PGE and base-metal alloys have been recovered from the Luobusa chromitites (Bai *et al.* 2000). Most of the base-metal alloys occur along cracks and fractures in the chromite and are considered to be secondary in origin,

associated with serpentinization (cf. Dick 1974). However, a number of Os–Ir–Ru, Pt–Fe and Ir–Fe–Ni alloys are completely enclosed in chromite grains and are considered primary.

The Os–Ir–Ru alloys are anhedral to subhedral, equidimensional grains up to 0.5 mm in diameter. They span a wide range of composition, but most contain less than 10% Ru and are classified as osmium and iridium (Bai *et al.* 2000). Many of the grains are compositionally zoned, typically with cores of osmium and rims of iridium (Fig. 7B). The Os–Ir ratios in these alloys range from approximately 2 : 1 to 1 : 3 (Bai *et al.* 2000). A few Os–Ir alloys contain small silicate inclusions ranging from about 5 to 50 microns across (Fig. 7C). These inclusions are perfectly spherical and have sharp boundaries with the host alloy. Energy-dispersive analyses indicate that they are Ca–Al–Fe–Mg silicates tentatively identifed as silicon spinel (Table 2, Analyses 11 and 12) (Fig. 7D).

The Pt–Fe alloys in these rocks occur as single grains or as intergrowths with Os–Ir–Ru alloys. The grains are typically 0.1–0.4 mm across and may be subrounded or tabular. They consist chiefly of Pt and Fe with up to 10% Rh and small amounts of Ni, Co, and Cu (Bai *et al.* 2000). Only one grain has a composition close to isoferroplatinum (Pt_3Fe). Small inclusions of Os–Ir–Ru alloy are present in some of the Pt–Fe grains.

The Ir–Ni–Fe alloys occur as single grains or as colliform intergrowths with Os–Ir–Ru alloys

Table 5. *X-ray diffraction data for wüstite and α-iron*

Sample 48-3			Standard (JCPDS 6-615)			Sample 38-10A–αFe			Standard (JCPDS 6-0693)		
d(A)	I/Io	hkl	d(A)	I/Io	hkl	d(A)	I/Io	hkl	d(A)	I/Io	hkl
2.490	80	111	2.490	80	111	2.026	100	110	2.027	100	110
2.160	80	200	2.153	100	200	1.426	20	200	1.433	20	200
1.528	70	220	1.523	60	220	1.170	20	211	1.170	30	211
1.305	40	311	1.299	25	311	1.014	20	220	1.013	10	220
1.249	20	222	1.243	15	222						

Table 6. *Chemical composition of native iron from chromitites of the Luobusa ophiolite, Tibet*

Sample no.	No. of analyses	Wt%					At%				Formula
		Fe	Si	Mn	Al	Total	Fe	Si	Mn	Al	
2-1	1	100.00	0.00	0.00	0.00	100.00	100.00	0.00	0.00	0.00	$Fe_{1.00}$
2-1B	1	98.08	0.03	0.54	1.35	100.00	96.65	0.06	0.54	2.75	$Fe_{0.96}Mn_{0.01}Al_{0.03}$
2-4	1	100.00	0.00	0.00	0.00	100.00	100.00	0.00	0.00	0.00	$Fe_{1.00}$
2-13	1	99.08	0.03	1.54	1.35	102.00	95.73	0.06	1.51	2.70	$Fe_{0.96}Mn_{0.01}Al_{0.03}$
16-16	1	97.73	0.00	0.65	0.00	98.38	99.33	0.00	0.67	0.00	$Fe_{0.99}Mn_{0.01}$
23-17	1	98.90	0.19	0.91	0.00	100.00	98.70	0.38	0.92	0.00	$Fe_{0.99}Mn_{0.01}$
38-10	2	97.86	0.53	1.62	0.00	100.00	97.33	1.03	1.64	0.00	$Fe_{0.98}Mn_{0.01}Si_{0.01}$
38-22	3	98.96	0.46	0.59	0.00	100.01	98.49	0.91	0.60	0.00	$Fe_{0.98}Mn_{0.01}Si_{0.01}$
48-3	3	100.00	0.00	0.00	0.00	100.00	100.00	0.00	0.00	0.00	$Fe_{1.00}$
50-2-3	3	100.00	0.00	0.00	0.00	100.00	100.00	0.00	0.00	0.00	$Fe_{1.00}$
50-15-1	1	92.45	1.50	6.05	0.00	100.00	91.01	2.94	6.05	0.00	$Fe_{0.91}Mn_{0.06}Si_{0.03}$

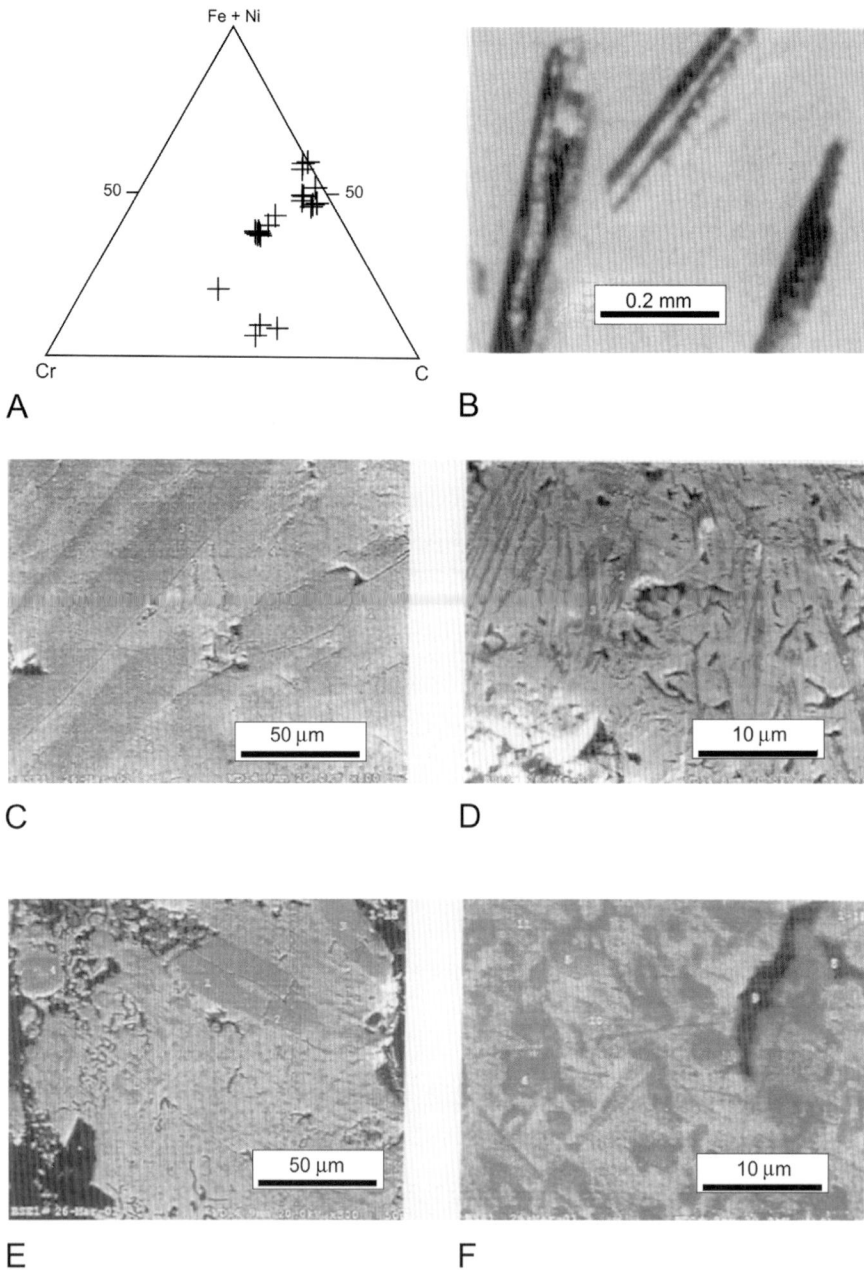

Fig. 6. Photographs and compositional plot of Cr–(Ni–Fe)–C alloys. (**A**) Ternary plot showing the three main compositional groups of Cr–(Ni–Fe)–C alloys in the Luobusa chromitites. (**B**) Photograph of CrC crystals. (**C**) SEM photograph of Fe–Ni–Cr–C alloy (Sample 1-17). The dark zones are rich in Fe and poor in Ni ($Fe_{43.4}Ni_{3.6}Cr_{5.5}C_{47.5}$), whereas the light zones are relatively low in Fe and high in Ni with the same C content ($Fe_{37.8}Ni_{13.3}Cr_{1.9}C_{47.0}$) (Table 7). (**D**) SEM photograph of Ni–(Fe–Cr)–C alloy (Sample 1-23) containing many needles of dark Cr–C with minor Fe and Ni ($Cr_{37.0}Ni_{7.0}Fe_{1.6}C_{54.4}$). (**E**) SEM photograph of Fe–Ni–Cr–C alloy (Sample 1-18). Light-coloured background material is mostly Ni–Fe–C with low Cr ($Ni_{35.3}Fe_{22.9}Cr_{1.8}C_{40.0}$), whereas the dark patches are higher in Fe and Cr and lower in Ni ($Fe_{29.8}Ni_{10.2}Cr_{22.0}C_{38.7}$) (Table 7). (**F**) Enlarged SEM photograph of Sample 1-18, showing distribution of dark, Fe–Cr-rich patches in the host grain. Compositions as in Fig. 6E. Dark patch (point 9) is graphite.

Table 7. *Chemical compositions of Cr–Ni(Fe)–C alloys from chromitites of the Luobusa ophiolite, Tibet*

Sample no	1	2	3	4	5	6	7	8	9	10	11	12	13
Grain no.	1	1	1	1-23	1-5	1-14	1-7	23-15	1-18a	1-17a	1-18b	1-17b	23-4
No. of analyses	1	1	1	3	3	3	3	3	2	2	6	3	1
Fe	1.92	1.66	1.57	2.91	83.88	85.88	2.02	24.55	32.56	61.13	42.55	69.41	5.53
Ni	1.59	1.40	1.40	13.23	0.00	0.00	81.26	72.28	52.79	19.70	15.60	6.01	70.22
Cr	79.69	80.02	76.71	62.54	0.00	0.00	0.00	0.00	2.40	3.12	29.76	8.23	9.35
Ti	0.16	0.13	0.00	0.00	0.00	0.00	0.00	0.00	0.00	0.00	0.00	0.00	0.00
C	16.64	16.79	20.32	21.32	16.12	14.12	16.73	3.17	12.26	16.06	12.11	16.36	13.90
Total	100.00	100.00	100.00	100.00	100.00	100.00	100.00	100.00	100.00	100.00	100.00	100.00	100.00
Atomic %													
Fe	1.15	0.99	0.87	1.60	52.87	56.66	1.28	22.73	22.85	38.69	29.20	43.35	3.76
Ni	0.91	0.80	0.74	6.91	0.00	0.00	49.19	63.65	35.29	11.90	10.18	3.57	45.43
Cr	51.42	51.44	45.81	37.01	0.00	0.00	0.00	0.00	1.81	2.12	21.95	5.53	6.83
Ti	0.00	0.00	0.00	0.00	0.00	0.00	0.00	0.00	0.00	0.00	0.00	0.00	0.00
C	46.52	46.77	52.58	54.48	47.13	43.34	49.53	13.63	40.04	47.28	38.67	47.54	43.98
Total	100.00	100.00	100.00	100.00	100.00	100.00	100.00	100.00	100.00	100.00	100.00	100.00	100.00

Note: Samples 1–3 Cr–C (Fig. 6B); Sample 4, Cr–C; Samples 5–6, Fe–C; Samples 7–8, Ni–C; Samples 9–10, Fe–Ni–Cr–C; Samples 11, 12, 13 Fe–Ni–Cr–C: light patches; Sample 11, 12, 13 Fe–Ni–Cr–C; dark patches.

A

B

C

D

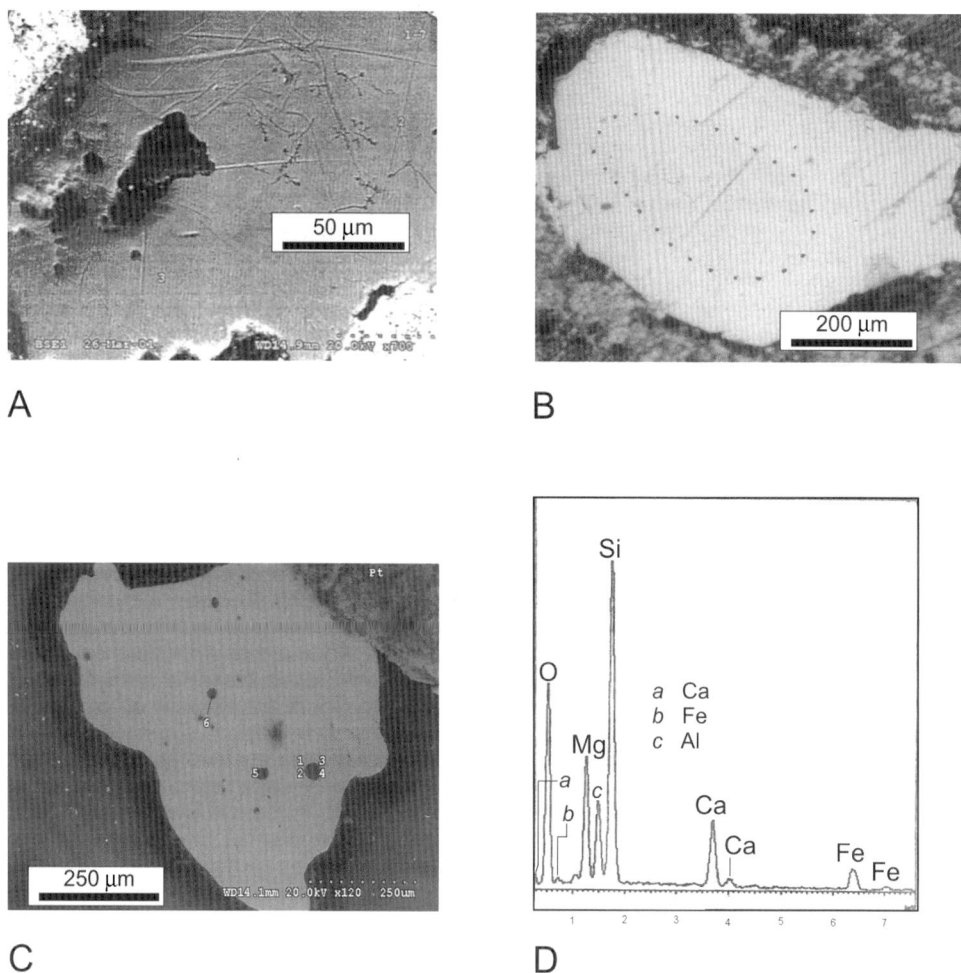

Fig. 7. SEM photographs and energy spectrum of Fe–Ni–C and Os–Ir. (**A**) SEM photograph of Sample 1–7, showing numerous lamellae and pseudo-Widmanstätten texture. (**B**) Zoned grain of Os–Ir with a core of Os and a rim of Ir (after Bai *et al.* 2000). (**C**) Grain of Os–Ir alloy containing small, dark, circular inclusions of a silicate minerals, probably silicon spinel (Table 2, Analyses 11 and 12). (**D**) Energy spectrum of the largest silicate inclusion in Fig. 7C.

(Bai *et al.* 2000). These alloys have relatively constant Fe (15–25 at%) but vary widely in Ir and Ni. Some grains also contain small amounts of Ru, Os and Cu.

Silicate minerals

The Luobusa chromitites contain a variety of silicate minerals, including olivine, orthopyroxene and Cr-diopside, minor phlogopite, serpentine and chlorite, some of which occur as inclusions in the minerals and alloys described above. Most interesting are Mg–Fe silicates with perfect octahedral morphology (Fig. 8A, B). These grains are 0.2 to 0.7 mm across and have ortho-

octahedral or modified octahedral forms. They are typically included in chromite grains or attached to them, leaving no doubt that they are natural in origin (Fig. 8A). Several hundred such grains have thus far been recovered, and most now consist of serpentine. However, some have cubic spinel structures – such as the grain in Figure 8 – indicating a high-pressure origin.

The structure of the grain was determined using a Rigaku AFC5R X-ray diffractometer with graphite monochromated MoKα radiation at Dalhousie University. The data were collected using the w-2Θ scan technique to a maximum 2Θ value of 60.1°. Weak reflections were rescanned (maximum of six scans) and

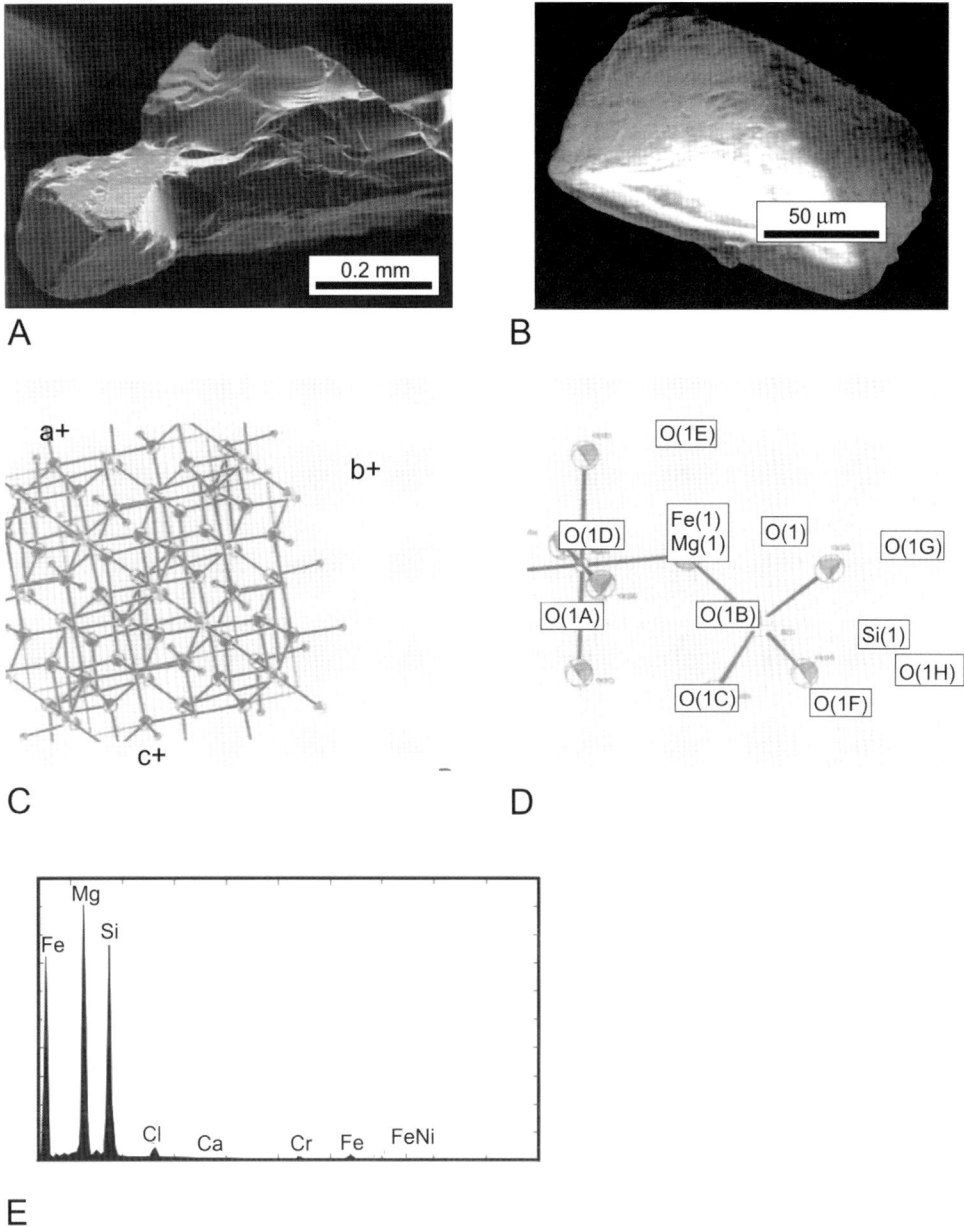

Fig. 8. SEM photos, crystal structure and composition of octahedral Mg–Si silicate minerals from the Luobusa chromitites. (**A**) Octahedral Mg–Fe silicate mineral attached to chromite. (**B**) A serpentinized octahedral Mg–Fe silicate grain. Grain is about 0.15 mm across. (**C**) Single-crystal X-ray structure of octahedral grain in Fig. 8A, showing cubic structure. (**D**) Bonds and bond angles of Si, O and Mg in the octahedral grain shown in Fig. 8A. (**E**) Energy-dispersive analysis of the octahedral grain in Fig. 8A, showing that it is a Mg silicate with small amounts of iron.

the counts accumulated to ensure good counting statistics. Stationary background counts were recorded on each side of the reflection and the ratio of peak to background counting time was 2 : 1. Of the 148 reflections obtained, 65 are unique ($R_{int} = 0.058$).

The structure was solved and expanded using Fourier techniques. The crystal has a spinel structure with a cubic, face-centred lattice (space group *Fd-3m*) with dimensions $a = 8.277(7)$ Å and $V = 567.1(2)$Å^3 (Fig. 8C). It is a Mg–Fe silicate with a formula of $(Mg,Fe)SiO_4$ (Fig. 8D). Refinement of the crystallographic data converges best when about 10% of the atoms at the Mg^{2+} position are allocated to Fe^{2+}. However, this refinement is very sensitive to the absorption correction used and the amount of Fe may vary somewhat. An energy-dispersive spectrum (Fig. 8E) confirms that the octahedral grain consists of nearly pure magnesium silicate with a very small amount of iron. Based on the crystallographic data, the mineral has an empirical formula of $(Mg_{1.80}Fe_{0.20})SiO_4$, and for $Z = 8$ and a formula weight of 147.00, the calculated density is $3.44\,g\,cm^{-3}$. The morphology, structure and composition of the octahedral grain strongly suggest that it is a high-pressure form of olivine, possibly ringwoodite.

Discussion

The unusual collection of minerals, including alloys and native elements recovered from the Luobusa chromitites, raises a number of important petrological and tectonic questions. Some of the minerals and alloys, particularly the diamonds, moissanite grains, octahedral Mg–Fe silicates and Os–Ir alloys are clearly UHP phases. Because the grains of Fe silicide, native Si, wüstite and native Fe are closely associated with the UHP phases, they are tentatively interpreted as having a similar origin. On the other hand, base-metal alloys and many other native elements may be secondary phases, although the degree of serpentinization is low in many of the rocks.

The native elements, PGM and alloys in the chromitites indicate formation under reducing conditions (Bai *et al.* 2000). Elsewhere, such minerals have been interpreted as primary (e.g. Stockman & Hlava 1984; Auge 1988; Garuti & Zaccarini 1997; Melcher *et al.* 1997; Bai *et al.* 2000) or secondary phases formed under reducing conditions during serpentinization (Jamieson 1905; Ramdohr 1950; Dick 1974). Dick (1974) has argued convincingly for a secondary origin for the Fe–Ni alloy (josephinite) in the Josephine ophiolite of Oregon, where the distribution of the alloy coincides closely with zones of serpentinization. However, serpentinization is very limited in the Luobusa ophiolite and many of the chromitites contain fresh silicate minerals. Thus, it seems unlikely that all of the native elements and alloys in the Luobusa chromitites could have been formed by secondary processes. However, it is equally unlikely

that they are primary minerals in the sense of having crystallized from the melt that formed the chromitites. The chromitites must have formed under hydrous, oxidizing conditions and the original magma would not have had sufficient concentrations of some elements such as Os and Ir to form primary minerals (Peach & Mathez 1996). Thus, we argue that the native elements, PGE alloys and UHP minerals, such as diamond, moissanite, Fe-silicides and octahedral silicates, are xenocrysts derived from deep-mantle sources that were incorporated into the chromitites during crystallization.

The depth of formation of the UHP minerals is uncertain, but diamond is generally considered to form at depths of 150–200 km, whereas the octahedral Mg–Fe silicates may have formed as deep as 560 km (Xu *et al.* 1997). Bird *et al.* (1999) have argued that some Os–Ir–Ru minerals may be derived from the core–mantle boundary region.

Tectonic implications

Traditionally, the ophiolites of the Indus–Yarlung–Zangbo suture zone have been thought of as remnant lithosphere from the Neo-Tethyan ocean, formed at the same age all along the belt (Nicolas *et al.* 1981; Allegre & Courtillot *et al.* 1984; Xiao 1988). It has recently been shown, however, that they are variable in age and more probably represent the remnants of a number of small, collapsed oceanic basins formed as a result of intra-oceanic subduction (Hsü *et al.* 1995; Malpas *et al.* 2003). All of the ophiolite massifs between Luobusa and Xigaze display geochemical signatures of suprasubduction zone magmatism. Available age dates, both radiometric and biostratigraphic, indicate that the suprasubduction zone magmatism occurred at approximately 126 Ma. Unlike the Xigaze and Dazhuqu ophiolites, the Luobusa Massif also contains older mid-ocean ridge (MOR) mantle, apparently trapped in the wedge above the subducting plate and modified by the subsequent boninitic magmatism. This older MOR lithosphere provides age dates of *c.* 175 Ma. Therefore, the Luobusa ophiolite appears to have formed in two stages, the first of which was generation of oceanic lithosphere at a mid-ocean ridge. The second stage involved modification of the MOR mantle in a mantle wedge above a subduction zone (Zhou, M.-F. *et al.* 1996; Malpas *et al.* 2003). Remelting of the depleted MOR peridotites in the mantle wedge produced tholeiitic and boninitic magmas that locally modified the mantle by melt–rock interaction. This suprasubduction zone magmatism produced an island-arc com-

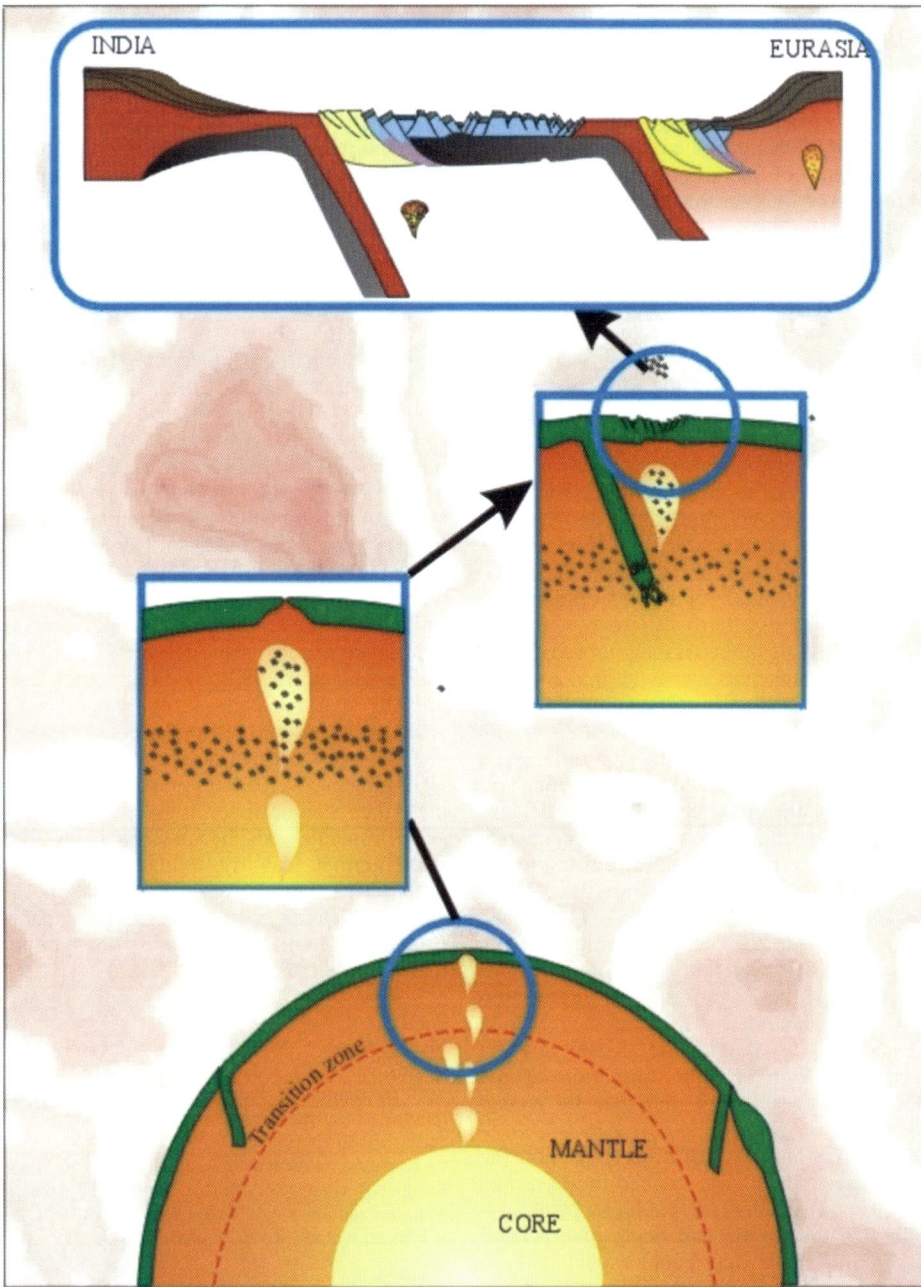

Fig. 9. Interpretative model providing a possible explanation for the presence of UHP minerals in the Luobusa chromitites. In stage 1, at the base of the diagram, subduction carries lithospheric material deep into the mantle, and convection brings deep-mantle material to shallow levels. Recent models suggest that whole-mantle convection is the norm, allowing UHP minerals and rocks to migrate upward. In stage 2, (second figure from base) sea-floor spreading produces the MORB-like mantle, which makes up the bulk of the Luobusa ophiolite. A subduction zone then penetrates this weakly depleted mantle (stage 3), third figure from base. Hydrous melting in the overlying mantle wedge produces boninitic melts, which rise upward, react with the weakly depleted mantle and undergo fractional crystallization to form the podiform chromitites (top figure). As these magmas migrate upward, they entrain blocks of mantle material containing UHP minerals, and these disaggregate during the final stages of chromitite crystallization.

plex, the Zhedong terrane, now found in close association with the ophiolite (McDermid *et al.* 2002; Malpas *et al.* 2003).

The UHP minerals found in the Luobusa mantle could only have been derived during the initial generation of MOR lithosphere, because the indicated pressures of formation require depths of 550 km or more for their formation. These depths are equivalent to the mantle transition zone, and clearly well below that of mantle trapped above a subducting plate. The UHP minerals must therefore have been already present in the mantle prior to development of the intra-oceanic subduction zone at 126 Ma.

Our preferred model for the formation of the Luobusa ophiolite and its UHP minerals is shown in Figure 9. In stage 1, crustal carbon and silicon are subducted into the lower mantle, where they are mixed with mantle carbon and converted to diamond, moissanite and CrC. A convection cell or plume rising from depths of at least 560 km, and perhaps from the core−mantle boundary, mixes with and entrains the diamond-bearing mantle rocks (stages 1 and 2). Sea-floor spreading produces the Luobusa mantle section, which is then incorporated into a mantle wedge above a subduction zone (stage 3). Boninitic melts formed by hydrous melting of the mantle wedge rise through the ophiolite, react with the host peridotites and form the podiform chromitites by fractional crystallization. Blocks of mantle rock containing the UHP minerals are picked up by the boninitic melts, transported to shallow levels and incorporated into the chromitites upon crystallization.

This model provides an explanation for the presence of UHP minerals in shallow chromitites, and for the high nitrogen aggregation state of the diamonds, which implies a long residence time in the mantle. Rifting of the Gondwana margin and the formation of the Luobusa mid-ocean ridge may also be related to the plume activity.

On the other hand, it is difficult to see how the UHP minerals could have retained their original structures and euhedral morphologies while rising through the mantle and being incorporated in a boninitic melt. One would expect them to have been deformed, dissolved or, at the least, severely etched and corroded. It also seems likely that the diamonds would have retrograded to graphite during cooling of the chromitite body. In fact, graphite is common in the Luobusa chromitites, and some of it may represent retrograded diamond. Most of the octahedral Mg−Fe silicates in the chromitites have been retrograded to serpentine, presumably at shallow levels. Preservation of the UHP minerals in the boninitic

melts would have been enhanced if they were contained in mantle blocks that disaggregated only in the final stages of emplacement.

One possible alternative is that the chromitites themselves formed at great depth and incorporated the UHP minerals during crystallization. UHP minerals enclosed in chromite would be more easily preserved during transport to shallow levels. However, the chromitites are believed to have formed at shallow levels by crystallization of a hydrous boninitic melt (Zhou, M.-F. *et al.* 1996). The chromitites show no evidence of high-temperature deformation (Fig. 2B), and most are enclosed in dunite envelopes that grade outward into the host peridotites, suggesting *in situ* melt−rock reaction (Zhou *et al.* 1996, 2003). Gabbro dykes in the peridotites, as well as gabbros, pyroxenites and lavas in the mélange zone, retain their original igneous textures and mineralogies. Eclogites have been reported from the Pakistan Himalaya, but, to our knowledge, no ultra-high pressure metamorphic rocks have been reported from the Indus−Yarlung Zangbo suture zone. Thus, there is no evidence that the ophiolite as a whole has ever experienced UHP conditions, and it seems unlikely that the chromitites could have been selectively transported from depth.

Whatever the exact mechanism by which the UHP minerals were incorporated into the chromitites, their demonstrated presence indicates that such minerals can be preserved far outside their normal stability fields. Preservation of the minerals in a high-temperature melt that was precipitating chromitite implies that they were hosted in mantle xenoliths, which protected them from the melt.

We thank Badengzhu of the Tibetan Geological Survey, Melanie Griselin, and Gareth Davies for assistance with sample collection and handling. S. Whattam assisted with preparation of the figures. Financial support for this project was provided by a Natural Sciences and Engineering Research Council of Canada (NSERC) grant to P. Robinson, a Research Grant Council of the Hong Kong SAR, China (HKU7086/01P) to J. Malpas, Ministry of Lands and Resources, China grant (2001010101) to W. J. Baj and by the University of Hong Kong.

References

AITCHISON, J. C., ABRAJEVITCH, A. *et al.* 2002. New insight into the evolution of the Yarlung−Tsangpo suture zone, Xizang (Tibet), China. *Episodes*, **25**, 90−94.

AITCHISON, J. C., BADENGZHU *et al.* 2000. Remnants of a Cretaceous intra-oceanic subduction system within the Yarlung−Zangbo suture (southern Tibet). *Earth and Planetary Science Letters*, **183**, 231−244.

ALLEGRÉ, C. J. & COURTILLOT, V. *et al.* 1984. Structure and evolution of the Himalaya–Tibet orogenic belt. *Nature*, **307**, 17–22.

ANTHONY, J. W, BIDEAUX, R. A., BLADH, K. W. & NICOLS, M. C. 1990. *Handbook of Mineralogy*. Mineral Data Publishing, Tucson, Arizona.

AUGE, T. 1988. Platinum-group minerals in the Tiebaghi and Vourinos ophiolitic complex: genetic implications. *Canadian Mineralogist*, **26**, 177–192.

BAI, W.-J., ZHOU, M.-F. & ROBINSON, P. T. 1993. Possibly diamond-bearing mantle peridotites and chromitites in the Luobusa and Dongiao ophiolites, Tibet. *Canadian Journal of Earth Sciences*, **30**, 1650–1659.

BAI, W.-J., ROBINSON, P. T. *et al.* 2000. The PGE and base-metal alloys in the podiform chromitites of the Luobusa ophiolite, Southern Tibet. *Canadian Mineralogist*, **38**, 585–598.

BAI, W.-J., TAO, S.-F. & SHI, R.-D. 2001. A new intergrowth consisting of FeO and SiO_2 phases from the lower mantle. *Continental Dynamics*, **6**, 1–7.

BIRD, J. M. & WEATHERS, M. S. 1975. Josephinite: specimens from the Earth's core? *Earth and Planetary Science Letters*, **28**, 51–64.

BIRD, J. M., MEIBOM, A., FREI, R. & NAGLER, TH. F. 1999. Osmium and lead isotopes of rare OsIrRu minerals: derivation from the core–mantle boundary region? *Earth and Planetary Science Letters*, **170**, 83–92.

CALON, T. J. 1979. *A study of the Alpine-type peridotites in the Seve–Koli nappe complex, Central Swedish Caledonides, with special reference to the Kittelfjall peridotite*. Unpublished PhD thesis, Leiden University 236 pp.

DAVIES, G. R., NIXON, P. G., PEARSON, D. G. & OBATA, M. 1992. Tectonic implications of graphitised diamonds in the Ronda peridotite, S. Spain. *Geology*, **21**, 471–474.

DICK, H. J. B. 1974. Terrestrial nickel–iron from the Josephine peridotite, its geologic occurrence, associations and origin. *Earth and Planetary Science Letters*, **24**, 291–298.

DI PIERRO, S., GNOS, E., GROBETY, B. H., ARMBRUSTER, T., BERNASCONI, S. M. & ULMER, P. 2003. Rock-forming moissanite (natural α-silicon carbide). *American Mineralogist*, **88**, 1817–1822.

DOBRZHINETSKAYA, L. F., EIDE, E. A. *et al.* 1995. Microdiamond in high-grade metamorphic rocks of the Western Gneiss region, Norway. *Geology*, **23**, 597–600.

FANG, Q.-S. & BAI, W.-J. 1981. The discovery of Alpine-type diamond-bearing ultrabasic intrusions in Tibet. *Geological Review*, **22**, 455–457 (in Chinese).

FANG, Q.-S. & BAI, W.-J. 1986. Characteristics of diamond and diamond-bearing ultramafic rocks in Qiaoxi and Hongqu, Xizang. *Institute of Geology Bulletin, Chinese Academy of Geological Sciences*, **14**, 62–125 (in Chinese).

GARUTI, G. & ZACCARINI, F. 1997. In situ alteration of platinum-group minerals at low temperature: evidence from serpentinized and weathered chromitite

of the Vourinos complex, Greece. *Canadian Mineralogist*, **35**, 611–626.

GHIRIBELLI, B, FREZZOTTI, M.-L., & PALMERI, R. 2002. Coesite in eclogites of the Lanterman Range (Antarctica): evidence from textural and Raman studies. *European Journal of Mineralogy*, **14**, 355–360.

HARRIS, D. C. & CABRI, L. J. 1991. Nomenclature of platinum-group element alloys: review and revision. *Canadian Mineralogist*, **29**, 231–237.

HE, G. 1987. Mantle xenoliths from kimberlites in China. *In*: NIXON, P. (ed.) *Mantle Xenoliths*. John Wiley & Sons, Ltd, New York, 181–185.

HIRAJIMA., T., ISHIWATARI, A., CONG, B., ZHANG, R. Y., BANNO, S. & NOZAKA, T. 1990. Coesite from Mengzhong eclogite at Donghai county, Northeastern Jiangsu Province, China. *Mineralogical Magazine*, **54**, 579–583.

HOUGH, R. M., GILMOUR, I. *et al.* 1995. Diamond and silicon carbide in impact melt rock from the Ries impact crater. *Nature*, **378**, 41–44.

HSÜ, K. J., PAN, G. *et al.* 1995. Tectonic evolution of the Tibetan Plateau: a working hypothesis based on the Archipelago model of orogenesis. *International Geology Review*, **37**, 473–508.

HU, X.-F. 1999. Origin of diamonds in chromitites of the Luobusa ophiolite, South Tibet, China. MSc thesis, Dalhousie University, Halifax, Nova Scotia.

JACOB, D., JAGOUTZ, E., LOWRY, D., MATTEY, D. & KUDRJAVTSEVA, G. 1994. Diamondiferous eclogites from Siberia: remnants of Archean oceanic crust. *Geochimica et Cosmochimica Acta*, **58**, 5191–5207.

JAMIESON, G. S. 1905. On the natural iron–nickel alloy awaruite. *American Journal of Science*, **19**, 413.

JAQUES, A. L. & HALL, A. E. *et al.* 1989. Composition of crystalline inclusions and C-isotopic composition of Argyle and Ellendale diamonds. *In*: ROSS, J. (ed.) *Kimberlites and Related Rocks*, Geological Society of Australia, Special Publications, **14(2)**, 966–989.

KAMINSKY, F. V. & VAGANOV, V. I. 1977. Petrological conditions for diamond occurrences in Alpine-type ultrabasic rocks. *International Geology Review*, **19**, 1151–1162.

LEUNG, I. S. 1990. Silicon carbide cluster entrapped in a diamond from Fuxian, China. *American Mineralogist*, **75**, 1110–1119.

LEUNG, I. S., GUO, W., FRIEDMAN, I. & GLEASON, J. 1990. Natural occurrence of silicon carbide in a diamondiferous kimberlite from Fuxian. *Nature*, **346**, 352–354.

LEUNG, I. S., TAYLOR, L. A., TSAO, C. S. & HAN, Z. 1996. SiC in diamond and kimberlites: implications for nucleation and growth of diamond. *International Geology Review*, **38**, 595–606.

LUKYANOVA, L. L., SMIRNOV, YU., ZILBERMAN, A. M. & CHERNYSHOVA, YE. M. 1980. A diamond find in picrites from the Urals. *International Geology Review*, **22**, 1189–1193.

MCDERMID, I. R. C., AITCHISON, J. C., DAVIS, A. M., HARRISON, T. M. & GROVE, M. 2002. The Zedong terrane: a Late Jurassic intra-oceanic magmatic arc within the Yarlung–Tsangpo suture zone, southern Tibet. *Chemical Geology*, **187**, 267–277.

MALPAS, J., ZHOU, M.-F., ROBINSON, P. T. & REYNOLDS, P. H. 2003. Geochemical and geochronological constraints on the origin and emplacement of the Yarlung–Zangbo ophiolites, Southern Tibet. *In*: DILEK, Y. & ROBINSON, P. T. (eds) *Ophiolites in Earth History*, Geological Society, London, Special Publications (in press).

MASSONNE, H.-J. 1999. A new occurrence of microdiamonds in quartzofeldspathic rocks of the Saxonian Erzgebirge, Germany, and their metamorphic evolution. *Proceedings of the 7th International Kimberlite Conference, Cape Town*, 2, 533–539.

MASSONNE, H. J. 2001. First find of coesite in the ultrahigh-pressure metamorphic area of the central Erzgebirge, Germany. *European Journal of Mineralogy*, 13, 565–570.

MATHEZ, E. A., FOGEL, F. A., HUTCHEON, I. D. & MARSHINTSEV, V. K. 1995. Carbon isotopic composition and origin of SiC from kimberlites of Yakutia, Russia. *Geochimica et Cosmochimica Acta*, 59, 781–791.

MELCHER, F., GRUM, W., SIMON, G., THALHAMMER, T. V. & STUMPFL, E. F. 1997. Petrogenesis of giant chromite deposits of Kempirsai, Kazakhstan: a study of solid and fluid inclusions in chromite. *Journal of Petrology*, 38, 1419–1458.

MILTON, C. & VITALIANO, D. B. 1985. Moissanite SiC, a geological aberration. *Geological Society of America, Abstracts with Programs*, 17, p. 665.

MOORE, R. O. & GURNEY, J. J. 1989. Mineral inclusions in diamond from the Monastery kimberlite, South Africa. *In*: ROSS, J. (ed.) *Kimberlites and Related Rocks*, Geological Society of Australia, Sydney, N. S. W., Australia, 1029–1041.

MOORE, R. O., OTTER, M. L., RICKARD, R. S., HARRIS, J. W. & GURNEY, J. J. 1986. The occurrence of moissanite and ferro-periclase as inclusion in diamond. *Fourth International Kimberlite Conference, Extended Abstracts, Geological Society of Australia*, 16, 409–411.

MPOSKOS, E. D. & KOSTOPOULOS, D. K. 2001. Diamond, former coesite and supersilicic garnet in metasedimentary rocks from the Greek Rhodope: a new ultrahigh-pressure metamorphic province established. *Earth and Planetary Science Letters*, 192, 497–506.

MUKHERJEE, B. K. & SACHAN, H. K. 2001. Discovery of coesite from Indian Himalayas: a record of ultrahigh pressure metamorphism in Indian continental crust. *Current Science*, 81, 1358–1361.

NICOLAS, A., GIRARDEAU, J. *et al.* 1981. The Xigaze ophiolite (Tibet): a peculiar oceanic lithosphere. *Nature*, 294, 414–417.

NICOLAS, A. & POIRIER, J. P. 1976. *Crystalline Plasticity and Solid State Flow in Metamorphic Rocks*. John Wiley & Sons, Ltd, London, 444 pp.

OKAY, A. I., XU, S.-T. & SENGOR, A. M. C. 1989. Coesite from the Dabie Shan eclogies, central China. *European Journal of Mineralogy*, 1, 595–598.

OTTER, M. L. & GURNEY, J. J. 1989. Mineral inclusions in diamonds from the Sloan diatreme, Colorado–Wyoming State Line kimberlite district, North America. *In*: ROSS, J. (ed.) *Kimberlites and*

Related Rocks, Geological Society of Australia, Sydney, N.S.W., Australia, 2, 1042–1053.

PEACH, C. L. & MATHEZ, E. A. 1996. Constraints in the formation of platinum-group element deposits in igneous rocks. *Economic Geology*, 91, 439–450.

PEARSON, D. G., DAVIES, G. R., NIXON, P. H. & MILLEDGE, H. J. 1989. Graphitised diamonds from a peridotite massif in Morocco and implications for anomalous diamond occurrences. *Nature*, 338, 60–62.

RAMDOHR, P. 1950. Uber Josephinite, Awaruit, Souesit, ihrer Eigenschaften, Entstehung und Paragenesis. *Mineralogical Magazine*, 12, p. 374.

RUDASHEVSKY, N. S., DMITRENKO, G. G., MOCHALOV, A. G. & MENSHIKOV, Yu. P. 1983. Native metals and carbides in alpine-type ultramafites of Koryak Highland. *Mineral. Zh.*, 9(4), 1983, 71–82 (in Russian).

SCHMAEDIEKE, E. 1991. Quartz pseudomorphs after coesite in eclogites from the Saxonian Erzgebirge. *European Journal of Mineralogy*, 1, 231–238.

SHATSKY, V. S., SOBOLEV, N. V. & VAVILOV, M. A. 1995. Diamond-bearing metamorphic rocks of the Kokchetav massif (northern Kazakistan) *In*: COLEMAN, R. & WANG, X. (eds) *Ultrahigh Pressure Metamorphism*. Cambridge University Press, 427–455.

SMITH, D. C. 1984. Coesite in clinopyroxene in the Caledonides and its implications for geodynamics. *Nature*, 310, 641–644.

SOBOLEV, N. V. & SHATSKY, V. S. 1990. Diamond inclusion in garnets from metamorphic rocks: a new environment for diamond formation. *Nature*, 343, 742–746.

SOBOLEV, N. V., KAMINSKY, F. V. *et al.* 1997. Mineral inclusions in diamonds from the Sputnik kimberlite pipe, Yakutia. *Lithos*, 39, 135–157.

STACHEL, T., HARRIS, J. W. & BREY, G. P. 1998. Rare and unusual mineral inclusions in diamonds from Mwadui, Tanzania. *Contributions to Mineralogy and Petrology*, 132, 34–47.

STOCKMAN, H. W. & HLAVA, P. F. 1984. Platinum-group minerals in Alpine chromitites from southwestern Oregon. *Economic Geology*, 79, 491–508.

TAYLOR, W. R., MILLEDGE, H. J., GRIFFIN, B. J., NIXON, P. H., KAMPERMAN, M. & MATTEY, D. P. 1995. Characteristics of microdiamonds from ultramafic massifs in Tibet: authentic ophiolitic diamonds or contamination. *Extended Abstracts, 6th International Kimberlite Conference*, Russian Academy of Science, Hovosibirsk 623–624.

XIAO, X. C. 1988. *General Review of the Tectonic Evolution of the Crustal–Mantle of the Qinghai–Xizhang (Tibet) Plateau* (in Chinese). Geological Publishing House, Beijing, 236 pp.

XU, S., OKAY, A. I., JI, S., SENGOR, A. M. C., SU, W., LIU, Y. & JIANG, L. 1992. Diamond from the Dabie Shan metamorphic rocks and its implication for tectonic setting. *Science*, 256, 80–82.

XU, Y., POE, B. T., SHANKLAND, T. J. & RUBIE, D. C. 1997. Electrical conductivity of olivine, wadsleyite, and ringwoodite under upper mantle conditions. *Science*, 277, 352–355.

YAN, B., LIANG, R., FANG, Q., YANG, F. & WAN, C. 1986. Characteristics of diamond and associated minerals in Qiaoxi and Hongqu, Xizang. *Bulletin* of the Institute of Geology, Chinese Academy of Geological Sciences, **14**, 61–125 (in Chinese).

YANG, J.-S., XU, Z.-Q, SONG, S.-G., ZHANG, X., WU, C., SHI, R., LI, H. & BRUNEL, M. 2001. Discovery of coesite in the North Qaidam Early Paleozoic ultra-high-pressure metamorphic belt, NW China. *Comptes Rendu Académe Science Paris (Earth and Planetary Science)*, **333**, 719–724.

YANG, J.-S., BAI, W.-J. *et al.*, 2003. Silicon rutile from podiform chromite of the Luobusa ophiolite, Tibet. *Progress in Natural Sciences*, **23**, 528–531.

ZHANG, H.-Y., BA, D.-Z. *et al.* 1996. *Study of Luobusa Typical Chromite Ore Deposit, Qusong County, Tibet (Xizang)*. Xizang Peoples Press, Lhasa, China, p. 181.

ZHOU, M.-F. 1995. *Petrogenesis of the Podiform Chromitites in the Luobusa ophiolites, Southern Tibet.* Unpublished PhD thesis, Dalhousie University.

ZHOU, M.-F., ROBINSON, P. T., MALPAS, J. & LI, Z. T. 1996. Podiform chromitites in the Luobusa ophiolite (Southern Tibet): implications for melt–rock interaction and chromite segregation in the upper mantle. *Journal of Petrology*, **37**, 3–21.

ZHOU, S., MO, X. X., MAHONEY, J. J., ZHANG, S. Q., GUO, T. J. & ZHAO, Z. D. 2002. Geochronology and Nd and Pb isotope characteristics of gabbro dikes in the Luobusha ophiolite, Tibet. *Chinese Science Bulletin*, **47**, 143–146.

Cretaceous palaeomagnetism of Indochina and surrounding regions: Cenozoic tectonic implications

CUNG THUONG CHI[1] & STEVEN L. DOROBEK[2]

[1]*Institute of Geological Sciences, National Center for Natural Science and Technology, Hoang Quoc Viet Str., Cau Giay, Hanoi, Vietnam (e-mail: chicung@netnam.vn)*
[2]*Department of Geology and Geophysics, Texas A&M University, College Station, TX 77843, USA*

Abstract: Results of a detailed palaeomagnetic study of Cretaceous-age volcanic, intrusive and sedimentary rock formations from southern Vietnam (24 sites, 163 core samples) are presented. The palaeomagnetic and supplementary rock magnetic studies indicate that magnetite and titanomagnetite are the predominant magnetic carriers in the volcanic and intrusive rock samples, whereas hematite is the principal carrier in the red-beds. The mean palaeomagnetic direction of twenty-one sites from southern Vietnam yields $D = 14.5°$, $I = 33.3°$, $\alpha_{95} = 6.3°$, $k_s/k_g = 1.04$, which corresponds with a VGP at $\lambda = 74.2°N$, $\phi = 171.1°E$, $A_{95} = 5.9°$. Comparison of the pole with the Eurasia mean Cretaceous palaeopole shows that relative to Eurasia southern Vietnam has experienced a southward displacement of $6.5° \pm 5.1°$, but with insignificant rotation since the Cretaceous.

Previously reported Cretaceous palaeomagnetic data, combined with new palaeomagnetic data from this study and analysis of regional structural trends, indicate that Sundaland can be divided into several fault-bounded tectonic domains (Shan–Thai, Indochina, offshore Sundaland), each with a different rotation and/or translation history. Such differential motion might explain, for example, Oligocene transtension and basin formation in the Gulf of Thailand and central onshore Thailand (between the Shan–Thai and Indochina blocks). Our data combined with previously acquired palaeomagnetic data across Southeast Asia, also suggest that, during the Cenozoic, Indochina and parts of Sundaland underwent complex internal deformation and did not behave as a rigid block.

There are two general schools of thought regarding the effects of the collision between India and Eurasia on the subsequent tectonic history of E–SE Asia. Proponents of extrusion tectonics suggest that convergence between the Indian subcontinent and the Eurasian plate was mainly accommodated by east–southeastward translation and rotation of large-scale, continental lithospheric blocks such as 'Sundaland' (i.e. Indochina, Shan–Thai, southwestern South China Sea, and southwestern Borneo), South China, and Tibet along major left-lateral strike-slip faults (Tapponnier *et al.* 1982, 1986; Peltzer & Tapponnier 1988). In contrast, other workers argue that crustal shortening and thickening in the Himalaya and Tibet is the principal mechanism for accommodating this collision (Dewey *et al.* 1989; England & Houseman 1989; England & Molnar 1990). One major consequence predicted by both models, however, is the large clockwise rotation of Sundaland, which behaved either as a rigid lithospheric block (a basic tenet of the extrusion model) or as a series of upper-crustal blocks that were translated southeastward along north–south-trending dextral megashear zones (as in crustal shortening models).

Limited palaeomagnetic data from previous studies of the Khorat Plateau, northeast Thailand, appear to support the extrusion model with clockwise rotation of Indochina (Yang & Besse 1993). However, more recent studies from the Shan–Thai Block (western Thailand) reveal differential clockwise rotations and insignificant latitudinal translation of the Shan–Thai Block with respect to Eurasia (Funahara *et al.* 1992, 1993; Huang & Opdyke 1992a, 1993; Yang *et al.* 1995; Richter & Fuller 1996). All of these studies, combined with palaeomagnetic results from Peninsular Malaysia and Borneo (large counter-clockwise rotations: McElhinny *et al.* 1974; Haile *et al.* 1983; Schmidtke *et al.* 1990), indicate that Sundaland did not behave as simply as that suggested by the extrusion model.

In this paper new palaeomagnetic data are presented from Cretaceous volcanic and intrusive

From: MALPAS, J., FLETCHER, C. J. N., ALI, J. R. & AITCHISON, J. C. (eds) 2004. *Aspects of the Tectonic Evolution of China*. Geological Society, London, Special Publications, **226**, 273–287.
0305-8719/04/$15 © The Geological Society of London 2004.

rocks and continental redbeds in Vietnam. The new results are integrated with other published palaeomagnetic data from South China and Sundaland. The larger data-set suggests that Sundaland experienced complex internal deformation during the Cenozoic.

Geological setting and sample sites

The structural features and stratigraphic units of Vietnam, which include the Sinovietnamian terrane, Indosinian terrane, Caledonian and Hercynian fold-belts, and the complex, smaller-scale, structural and magmatic features of Mesozoic and Cenozoic age that post-date and cross-cut these older elements, are probably southern extensions of more regional tectono-stratigraphic terranes that extend across much of mainland SE Asia (Tran et al. 1979; Hutchison 1989a, b; Nguyen et al. 1994; Metcalfe 1996).

Late Mesozoic to Early Cenozoic igneous rocks in Vietnam correspond with two periods of magmatic activity which occurred during the Late Jurassic–Early Cretaceous and Late Cretaceous–Palaeogene (Phan et al. 1991; Nguyen et al. 1994). Palaeomagnetic samples for this study were mostly collected from volcanic and intrusive rocks that are exposed across large parts of southern Vietnam (Fig 1). The volcanic–plutonic rocks are interpreted to be the southern extension of the 'Yenshanian' magmatic arc complex that formed in southeastern China during the Middle Jurassic to Late Cretaceous. Development of the Yenshanian Arc was related to westward subduction of the proto-Pacific plate beneath the Eurasian plate (Jahn et al. 1976; Taylor & Hayes 1983; Charvet et al. 1994). The apparent southeastward displacement of age-equivalent, arc-related igneous rocks in Vietnam relative to the Yenshanian magmatic arc complex in southeastern China has been cited as evidence for the Cenozoic extrusion of Indochina/Sundaland that was caused by the India/Eurasia collision (e.g. Leloup et al. 1995).

All of the palaeomagnetic samples used in this study were collected using a portable gasoline-powered drill. In Southern Vietnam, Late Mesozoic volcanic rocks crop out in the Da Lat depression, a structural low that extends from the NE–SW-trending Tuy Hoa–Tay Ninh fault in the northwest to the Vietnamese continental margin in the southeast (Fig. 1). Twenty-four sites with a total of 163 cored samples were collected from this area; 14 from the volcanic rocks, six from the granites, and four from continental red-bed siltstones. The volcanic rock samples are assigned to two formations: the Upper Jurassic–Lower Cretaceous Da Lat

Formation and the Upper Cretaceous Nha Trang (Don Duong) Formation. The Da Lat Formation unconformably overlies Lower to Middle Jurassic sedimentary rocks that are slightly tilted from horizontal (bedding attitude varies from 6 to 24°), which indicates that this region was not strongly folded during the Cenozoic (Rangin et al. 1995b). Basal beds of the Da Lat Formation contain Late Jurassic–Cretaceous fossils, such as Estheria, Zamites sp., Dicksonia sp., Ligodium sp., Picea sp., Schizosporites sp., Cedrus sp. and Pagiophyllum sp. (Phan et al. 1991), and pollen spores of Cretaceous age such as Lygodium sp., Classopollis sp., Osmundacidites sp. and Leiotriletes sp. The basal shale beds of the Nha Trang Formation contain pollen spores such as Selaginella sp., Taxodium sp., Lygodium sp., Seitotylus sp. The volcanic rocks sampled in this study are mainly from the Nha Trang Formation and consist mainly of small bodies of andesito-dacite, dacite, rhyolite, and their related tuffs, that are intercalated with numerous, larger-scale granitic bodies. Radiometric dates from plagioclase crystals in andesite and rhyolite were obtained by the K–Ar method, and give ages ranging from 80 to 110 Ma (Nguyen 2000). Radiometric ages obtained by different methods (K–Ar, Ar/Ar and Rb–Sr) from different minerals in granitic rocks (e.g. biotite, hornblende, amphibole, feldspar and plagioclase) and whole-rock samples indicate ages between 71 and 150 Ma; with the majority 80–120 Ma (Phan et al. 1991; Nguyen 2000). The Cretaceous continental redbeds of the Dak Rium Formation, are slightly folded with west–southwest trends and bedding varying from 5 to 20°. The basal beds of the Dak Rium Formation comprise conglomerate, breccias and purple coarse-grained sandstones which are overlain by purple siltstones that contain Late Jurassic–Cretaceous fossils such as Licopodites cf., Tenerrium Hear, Elatocladus sp., Pagiophyllum cf., Carssifolium, Nemestherria sp., and Cretaceous–Palaeogene pollen spores such as Schizeaceae, Euphorbiaceae, Rhizophoraceae, Albizia sp., Cenopodiaceae and Fagacea (Vu et al. 1989).

Palaeomagnetic study

Magnetic measurements were carried out at Texas A&M University using a CTF cryogenic magnetometer in a magnetic-shielded room and at the National Institute of Geology and Mining in Poland using a JR-5 spinner magnetometer (for sites 8703–8713). Stepwise alternating field (AF) demagnetization was carried out in 2.5 mT steps from 0 to 10 mT; 5 mT steps from 10 to 30 mT; and 10 mT steps from 30 to 100 mT,

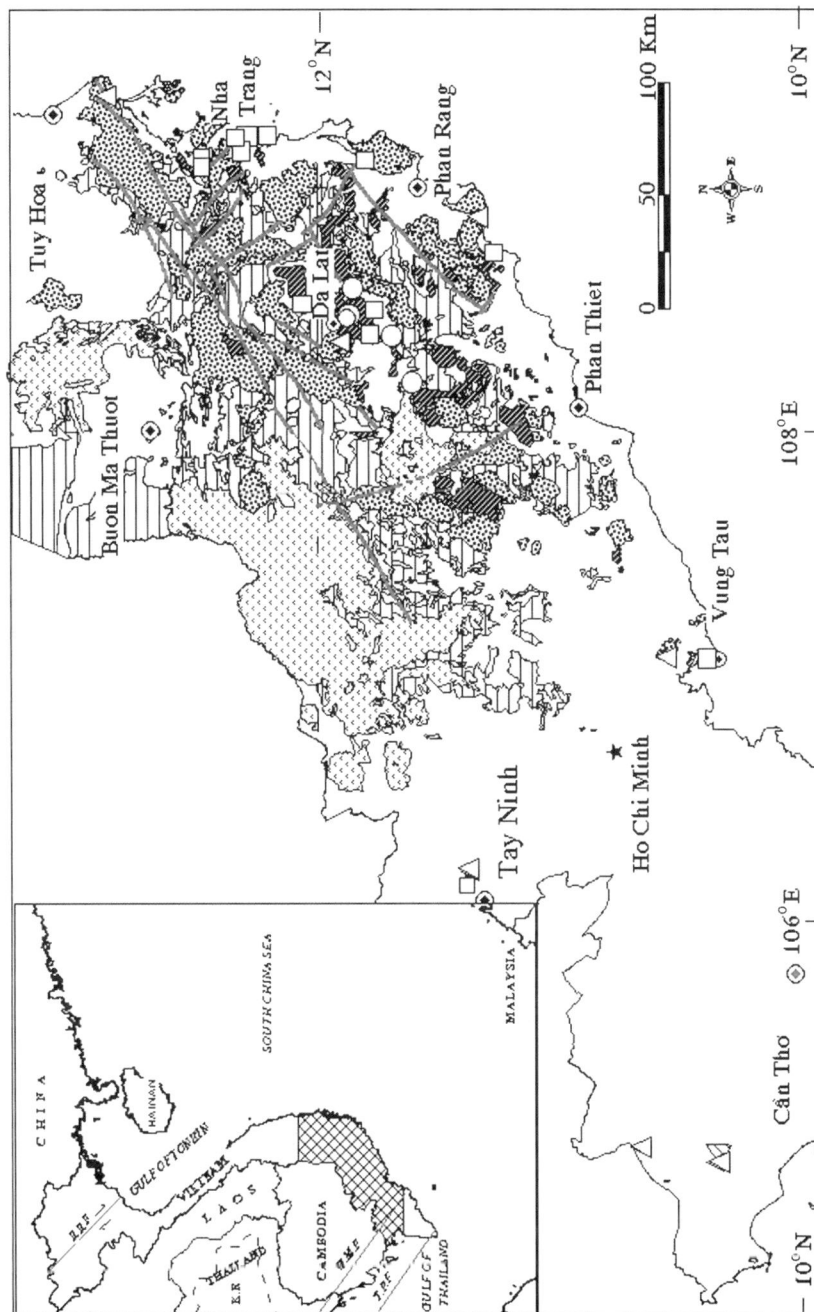

Fig. 1. Simplified geological map of the study area, with sample site locations. ⠿ Late Neogene basalt; ▤ Early to Middle Jurassic sediments; ⟋ Fault; ⠿ Late Jurassic–Cretaceous intrusive rocks; ▦ Late Jurassic–Cretaceous volcanic rocks. Sampling site on instrusive rocks △; Sampling site on volcanic rock □; Sampling site on red-bed sediments ○. Inset figure: KR, Khorat Plateau; RRF, Red River Fault; WMF, Wang Maeping Fault; TPF Three Pagoda Fault.

using a Schonstedt GSD1 single-axis demagne-
tizer. Stepwise thermal demagnetization was
performed using a Schonstedt thermal furnace in
50 °C steps from 100 to 600 °C and up to 700 °C
for red-bed samples and for some volcanic
samples that contain hematite as the magnetic
carrier. For each site, a pilot sample (we obtained
two specimens from each core sample) was
subjected to both thermal and AF demagnetiza-
tions, in order to choose the appropriate treatment
method for the remaining samples. Orthogonal
vector component projection (Zijderveld 1967)
and principal component analysis (Kirschvink
1980) were used for directional analysis. Site-
mean directions were calculated using Fisher
(1953) statistics, and followed the procedure of
McElhinny (1973). Isothermal remanent mag-
netizations (IRM) were imparted with an ASC-
IM10 pulse magnetizer with peak field strength
of 1.2 T.

The natural remanent magnetization (NRM) of
samples is dependent on rock type and varies
between 1×10^{-2} to 4.6 Am^{-1} for the volcanic
rocks, 1×10^{-3} to 1×10^{-2} Am^{-1} for the granites, and 1×10^{-4} to 1×10^{-4} Am^{-1} for the
sedimentary rocks. Thermal demagnetization
and IRM acquisition curves of representative
samples (Fig. 2) indicate that most volcanic and
intrusive samples contain a magnetic carrier
mineral that belongs to the titanomagnetite
family, with unblocking temperatures ranging
from 500 to 600 °C and IRM saturation
magnetization fields <300 mT. A few samples
also contained a high coercivity component,
probably hematite, with titanomagnetite as the
predominant magnetic carrier (e.g. NH6b, BH2b)
or hematite as the sole magnetic carrier (e.g.
NT15b). For red siltstone samples, hematite is the
principal magnetic carrier. These sedimentary
samples sometimes needed heating to 750 °C to
completely remove their NRM. This high
unblocking temperature suggests that the magnetic carrier is hematite with either defective
crystal lattices or impurities (Collinson 1983;
Tarling 1983). In order to better characterize the
magnetic carrier mineral in igneous rock
samples, the Lowrie–Fuller test (Lowrie &
Fuller 1971), and the technique described by
Cisowski (1981), were applied to representative
specimens. Application of these techniques
reveals that the predominant magnetic carriers
are single domain (SD) or pseudo-single domain
(PSD) magnetite and/or titanomagnetite, with
an average $H_{rc} = 24.3$ mT for volcanic rocks
(ten samples) and $H_{rc} = 15.7$ mT for intrusive
rocks (nine samples). A mean Wohlfarth's
ratio $R = 0.28$ was obtained for both rock
types (Cung 1996). A few samples also showed

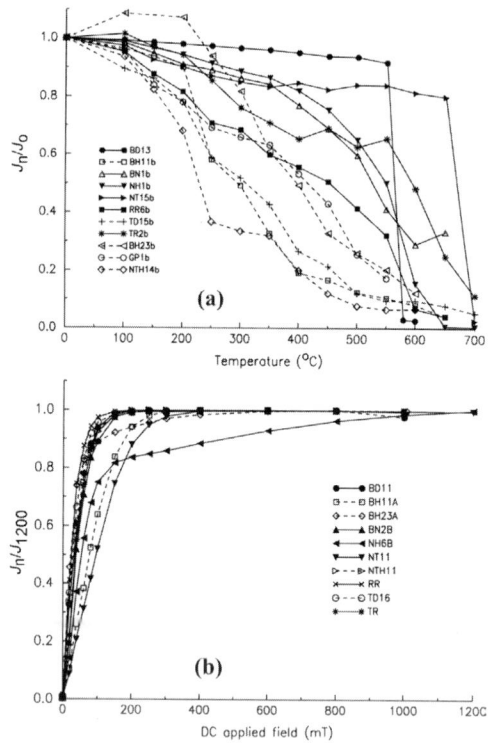

Fig. 2. Normalized magnetization intensities of
representative Late Jurassic–Cretaceous volcanic rocks
from southern Vietnam, during (**a**) thermal
demagnetization and (**b**) IRM acquisition.

a multiple-domain (MD)–SD transitional behaviour, which probably indicates the degree of
deuteric oxidation (Dunlop 1983).

In general, most samples showed a stable
behaviour both in AF and thermal demagnetization treatments, and revealed similar characteristic
directions (Fig. 3). At least two magnetic components were successfully isolated. The secondary
component was removed at fields <20 mT, or at
temperatures above 350 °C. At higher fields and/
or higher temperatures, the single characteristic
component underwent stable decay toward the
origin until 600 °C, or 100 mT.

The site-mean ChRM directions of 24 sites
obtained from this study are listed in Table 1 and
are plotted on an equal-area projection diagram
(Fig. 4). The mean ChRM direction of 21 sites
from southern Vietnam yields $D_s = 14.5°$,
$I_s = 33.3°$, $\alpha_{95} = 6.3$, $k_s = 26.7$. The corresponding VGP is located at $\lambda = 74.2$ °N, $\phi = 171.1$ °E
($A_{95} = 5.9°$). Three sites from massive outcrops
of intrusive igneous rocks were not included
in the calculation of the mean palaeomagnetic direction for the southern Vietnam region,

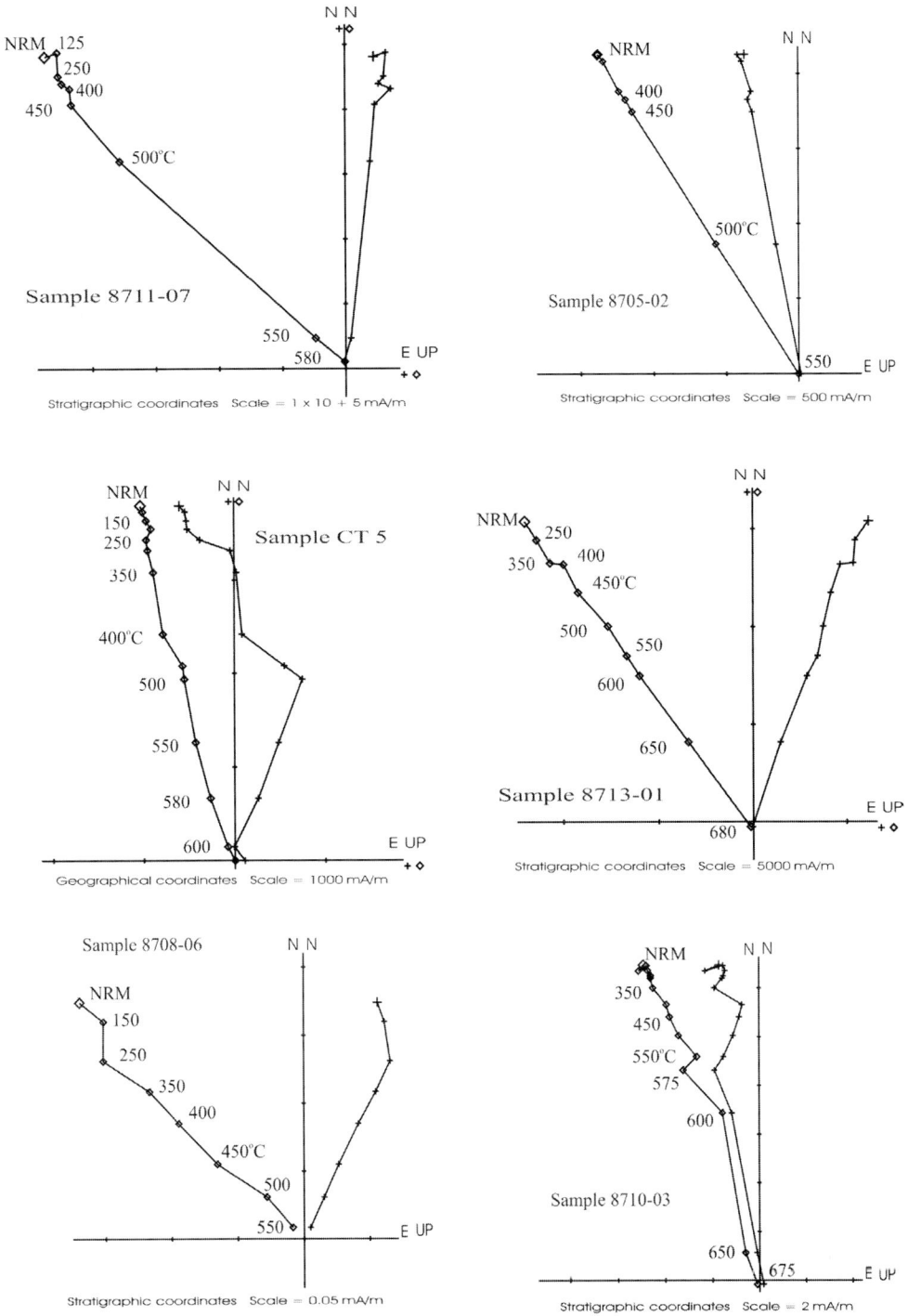

Fig. 3. Orthogonal vector component plots of representative samples. Diamond and cross symbols represent vector end-point projections in the horizontal (declination) and vertical (inclination) planes, respectively.

Table 1. *Palaeomagnetic results of Late Jurassic–Cretaceous rocks from southern Vietnam*

Site	Location Lat (λ°N)	Location Long (φ°E)	Rock type	St/Dp (degrees)	n/N	ChRM direction Dg (°)	Ig (°)	Ds (°)	Is (°)	α95 (°)	k	VGP φg (°E)	λg (°N)	φs (°E)	λs (°N)	A95 (°)	Palaeo-lat. (°N)
8703	12.47	109.13	Rhyolitic tuff	18/24	7/7	18.1	36.9	35.1	33.2	7.4	67.7	171.7	70.9	185.1	55.8	6.8	18.1
8705	12.29	109.21	Rhyolite	–	6/6	354.7	34.6	354.7	34.6	2.8	561.5	72.8	81.5	72.8	81.5	2.4	19.0
8706	12.20	109.21	Trachyriolite	34/18	5/5	2.5	37.3	16.9	44.8	3.0	661.1	124.4	81.0	155.0	68.7	3.0	26.4
8707*	12.06	108.53	Dacite	–	4/5	65.7	34.0	65.7	34.0	16.7	31.3	183.6	26.6	183.6	26.6	14.4	18.6
8708	11.85	108.58	Shalestone	265/04	4/6	28.0	44.4	26.1	41.0	5.7	261.8	166.2	60.0	169.9	62.6	5.4	23.5
8709	11.76	108.51	Andesitic tuff	254/06	6/10	24.5	40.6	21.2	35.0	13.2	26.8	169.1	64.0	175.6	68.3	11.5	19.3
8710	11.88	108.47	Siltstone	295/15	6/6	12.7	36.8	14.5	22.0	11.5	34.6	161.5	75.1	198.8	75.8	8.8	11.4
8711	11.78	108.42	Dacite	07/23	6/7	349.1	40.1	9.8	43.2	1.8	999.9	66.6	74.8	141.8	73.7	1.8	25.1
8713	11.69	108.38	Siltstone	281/06	7/9	21.3	41.9	20.4	35.7	7.9	58.9	163.8	66.3	173.5	68.8	7.0	19.8
PH	11.62	108.20	Siltstone	266/18	7/7	13.1	51.3	8.8	33.9	4.2	207.8	136.8	66.3	157.8	79.0	3.6	18.6
NH	12.47	109.13	Rhyolitic tuff	18/24	8/8	5.6	39.5	26.1	40.4	1.9	876.3	136.6	78.7	172.4	63.0	1.8	23.1
BD2*	11.39	106.19	Granodiorite	–	8/8	70.7	26.4	70.7	26.4	12.1	22.0	185.5	21.2	185.5	21.2	9.7	13.9
BD1	11.39	106.15	Andesite	–	6/6	26.4	22.5	26.4	22.5	7.2	99.8	192.8	64.1	192.8	64.1	5.6	11.7
DL	11.90	108.45	Felsite	–	5/7	15.9	33.9	15.9	33.9	10.9	49.9	173.0	73.3	173.0	73.3	9.4	18.6
BN	11.80	109.11	Dacite	–	7/7	27.8	34.4	27.8	34.4	3.2	359.7	180.8	62.3	180.8	62.3	2.8	18.9
RR	12.33	109.20	Andesite	–	8/8	6.3	54.3	6.3	54.3	4.0	193.4	122.4	66.8	122.4	66.8	4.7	34.8
TR	12.31	109.19	Andesite	–	8/8	4.5	23.6	4.5	23.6	4.0	191.6	198.8	85.6	198.8	85.6	3.1	12.3
NT	11.27	108.73	Rhyolite	–	6/6	13.0	37.7	13.0	37.7	6.0	123.9	158.8	74.1	158.8	74.1	5.4	21.1
VT	10.35	107.07	Rhyolite	–	3/3	354.5	15.6	354.5	15.6	17.6	50.0	353.7	84.1	353.7	84.1	13.0	7.9
DC	12.88	107.38	Granite	–	6/6	13.6	26.3	13.6	26.3	5.4	96.8	193.5	76.7	193.5	76.7	4.3	13.9
CN	11.36	108.87	Granite	–	6/6	34.5	30.9	34.5	30.9	7.1	59.5	185.8	56.2	185.8	56.2	5.9	16.7
NS	10.68	105.08	Granite	–	7/8	155.7	-15.6	155.7	-15.6	4.0	258.9	10.7	65.9	10.7	65.9	2.9	7.9
CT	10.37	105.02	Granodiorite	–	8/8	23.3	23.4	23.3	23.4	6.3	77.6	188.2	67.1	188.2	67.1	4.9	12.2
THI*	10.56	107.08	Granite	–	4/6	315.9	6.8	315.9	6.8	7.9	136.7	11.1	45.7	11.1	45.7	5.6	3.4
Mean of 21 sites					21/24	11.5	35.3	14.5	33.5	6.3	26.7	160.9	76.1	171.1	74.2	5.9	18.2

Note: St = Strike, Dp = deep. n = number of samples (sites) used in calculation of mean directions, N = total number of samples (sites), Dg (Ig) = geographical declination (inclination), Ds (Is) = stratigraphic declination (inclination), $α_{95}$ (A_{95}) = circle of 95% confidence, k = precision parameter, $λ_g$ ($φ_g$) = geographical latitude (longitude), $λ_s$ ($φ_s$) = stratigraphic latitude (longitude). (*) = indicates the sites which were not included in the mean calculation.

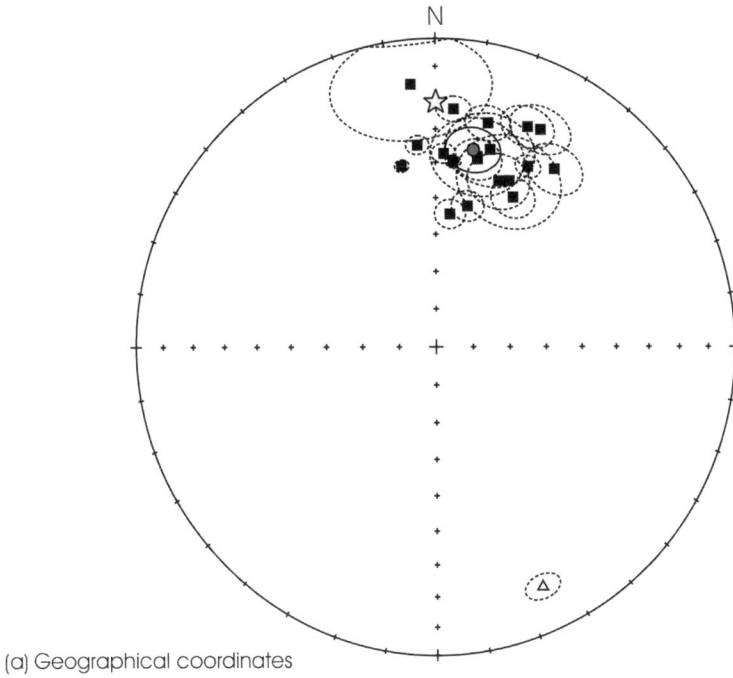

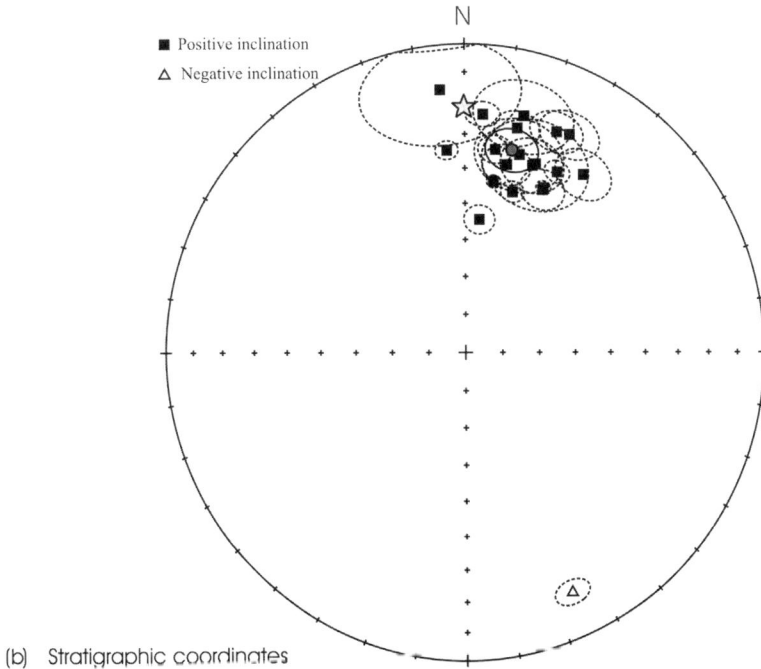

Fig. 4. Equal-area projection of the site-mean directions with their circles of confidence (α_{95}). Solid squares (open triangles) represent projections in the lower (upper) hemisphere. Shaded circle with solid ellipse represents the mean direction of 21 sites; shaded star represents the present magnetic field direction for the sampling area.

because of their anomalous directions. We attribute these anomalous directions to unrecognized tectonic tilting within the massive granites, which probably reflects localized deformation. The stability of the ChRM component carried by magnetite and titanomagnetite during demagnetization treatment, the isolation of secondary magnetic components, the similar characteristic directions obtained from a variety of sample lithologies, and the better cluster of site-mean palaeomagnetic directions obtained after applying tectonic correction, suggest that the ChRM directions resulting from this study are of primary origin.

Discussion

The mean palaeomagnetic direction of Cretaceous igneous and sedimentary rocks from southern Vietnam is distinctly different from that obtained from Late Neogene basalts in central and southern Vietnam (Cung *et al.* 1998), which further corroborates the primary origin of palaeomagnetic directions recorded from these rocks.

As mentioned above, the available radiometric ages of volcanic and granitic rocks in the study area and palaeontological constraints show that the palaeomagnetic samples of this study are of Cretaceous age. In order to determine the relative displacement of the southern Vietnam with respect to Eurasia, the average of stage palaeopoles from 140 to 60 Ma for Eurasia derived by Besse & Courtillot (1991) was computed ($\lambda = 75.9\,°$N, $\phi = 196.0\,°$E, $A_{95} = 2.5\,°$). Using this average pole, the expected direction for a reference point at the sampling area at $11.7\,°$N, $108.2\,°$E is $D_{ex} = 14.4\,°$, $I_{ex} = 22.8\,°$. Respectively, the relative rotation and latitudinal translation of southern Vietnam with respect to Eurasia can be determined from the expected and observed palaeomagnetic directions, and gives the values: $R = 0.4\,° \pm 5.4\,°$, $\lambda_{diff} = -6.5\,° \pm 5.1\,°$; (using the 100 Ma stage pole, $R = 1.2\,° \pm 6.4\,°$, $\lambda_{diff} = -6.8\,° \pm 6.2\,°$). These values indicate that there has been no significant rotation of southern Vietnam relative to Eurasia since the Cretaceous, but there has been southward translation. Also worth noting is that the southward displacement of eastern Indochina deduced from this study is smaller than that deduced by Yang & Besse (1993) from western Indochina.

In order to place the data obtained from this study in a regional tectonic framework, it was first necessary to compare them with the other published palaeomagnetic results from Indochina and the surrounding regions (compiled dataset in Table 2). The Late Jurassic–Cretaceous palaeopoles for the Indochina, Shan–Thai and South China blocks were plotted, along with the synthetic apparent polar wander path for Eurasia (Besse & Courtillot 1991), on a Lambert equal-area projection diagram (Fig. 5). It is important to note that the Cretaceous palaeopoles from the Indochina (Yang & Besse 1993; this study) and Shan–Thai blocks (Funahara *et al.* 1992; Huang & Opdyke 1993; Richter & Fuller 1996) lie more or less along a small circle centred on this region, which suggests that any tectonic rotation is related to complexly distributed faulting and/or folding across Indochina (MacDonald 1980). The degree of intraplate deformation across SE Asia that is related to the India–Eurasia collision, and other interactions between the Eurasia, Indo-Australia and Pacific plates, are further revealed from the palaeodeclination diagram shown in Figure 6. In general, a pattern of clockwise (CW) rotation is documented in terranes east of the eastern syntaxis of the Himalayas and Tibet, with larger CW rotations in the regions closer to the indentor, and less or no significant rotations in the far-field regions. These block movements are consistent with the stress field predicted by Huchon *et al.* (1994) as a result of the indentation of India into Eurasia. No significant movement of the South China block with respect to Eurasia has been observed (Figs 6 & 7). Chen *et al.* (1993) integrated Cretaceous palaeomagnetic data from South China, and showed that the block had experienced negligible motion with respect to Eurasia since Cretaceous times.

Differential rotational domains have been observed across the Sundaland region, including large CW rotation of the Shan–Thai block, minor CW rotation of the Indochina block, and large CCW rotation of the Malay Peninsula and Borneo. These patterns of block rotation suggest that not only has the India–Eurasia collision affected this region, but interactions between Australia, Eurasia and the CW motion of the Philippine Sea plate might also have caused some of the internal deformation across Sundaland, especially in its southeastern part (Hall *et al.* 1995; Hall 1996). Three different palaeomagnetic studies of Cretaceous red-bed rocks from the Shan–Thai block (Funahara *et al.* 1993; Huang & Opdyke 1993; Richter & Fuller 1996) revealed large CW rotations up to $65\,°$ and insignificant latitudinal translation of this block with respect to Eurasia (see Table 2 and Figs 6 & 7). The sample locations of these studies are, however, situated close to block boundaries and the strike-slip deformation zones, such as the Sagaing and Red River fault systems, which bound the blocks. Therefore, these palaeomagnetic

Table 2. *Parameters of rotation and latitudinal translation of cretaceous rocks from South China and Sundaland regions with respect to Eurasia*

No. Reference	Location λ (°N)	φ (°E)	Age (Ma)	Obs. VGP λ (°N)	φ (°E)	Ref. VGP λ (°N)	φ (°E)	R (°)	Δλ (°)	Significant?	
South China											
1	26.5	102.4	K1	81.5	220.9	76.7	197.1	−6.5 ± 7.6	2.5 ± 6.9	No/No	(1)
2	26.8	102.5	K1	69.0	204.6	73.3	206.5	4.5 ± 5.7	1.0 ± 5.3	No/No	(1)
3	29.7	98.5	K1	40.6	170.5	73.3	206.5	40.8 ± 13.0	−8.4 ± 11.2	Yes/No	(2)
4	29.7	98.7	K2	56.7	172.7	76.7	197.1	24.2 ± 11.1	−6.2 ± 9.4	Yes/No	(2)
5	32.0	119.0	K2	76.3	172.6	76.7	197.1	−1.4 ± 11.7	−5.5 ± 9.2	No/No	(4)
6	22.7	108.7	K1	86.5	26.4	76.7	197.1	−18.2 ± 7.3	−0.7 ± 6.8	Yes/No	(5)
7	26.0	117.3	K2	65.1	207.2	76.7	197.1	12.5 ± 6.3	4.1 ± 4.0	Yes/Yes	(5)
8	23.1	113.3	K2	56.2	211.5	76.7	197.1	20.2 ± 5.4	9.2 ± 5.1	Yes/Yes	(5)
9	30.1	103.0	K	76.3	274.5	76.7	197.1	−13.0 ± 10.3	11.8 ± 9.7	Yes/Yes	(6)
10	29.7	98.6	K1	48.5	175.9	73.3	206.5	30.9 ± 9.9	−6.4 ± 8.6	Yes/No	(6)
11	30.1	96.9	K1	71.8	289.2	73.3	206.5	−21.1 ± 13.9	11.1 ± 13.4	Yes/No	(6)
12	25.0	101.5	K	49.2	178.0	76.7	197.1	31.3 ± 11.1	−4.3 ± 10.0	Yes/No	(7)
13	26.0	117.3	K1	66.9	221.4	73.3	206.5	5.2 ± 6.4	6.6 ± 5.9	No/Yes	(8)
14	22.2	114.2	J3−K	78.2	171.9	74.8	210.9	−4.8 ± 10.2	−8.4 ± 9.1	No/No	(9)
15	30.0	102.9	K2	72.8	241.1	76.7	197.1	−3.2 ± 5.7	11.5 ± 5.2	No/Yes	(10)
16	30.0	103.0	J3−K	78.6	273.4	76.7	197.1	−13.1 ± 5.0	9.5 ± 4.4	Yes/Yes	(11)
17	29.7	120.3	K1	77.1	227.6	74.8	210.9	−3.7 ± 6.2	3.2 ± 5.5	No/No	(12)
18	26.6	102.4	K2	78.9	186.6	76.7	197.1	−2.2 ± 5.8	−2.4 ± 5.3	No/No	(13)
19	25.0	116.4	K2	67.9	186.2	76.7	197.1	9.5 ± 9.7	−4.2 ± 8.4	No/No	(14)
20	27.9	102.3	K1	77.4	196.2	73.3	206.5	−3.5 ± 11.2	−3.6 ± 10.1	No/No	(13)
21	35.9	128.5	K1	−	−	74.8	210.9	10.7 ± 10.5	−1.2 ± 6.7	Yes/No	(15)
Shan−Thai											
22	25.6	100.2	K2	83.6	152.7	76.7	197.1	−8.6 ± 10.2	−6.1 ± 8.9	No/No	(3)
23	23.4	100.9	K2	18.9	170.0	76.7	197.1	65.3 ± 9.0	−4.7 ± 8.2	Yes/No	(3)
24	21.6	101.4	K2	33.7	179.3	76.7	197.1	46.9 ± 8.2	−1.8 ± 7.7	Yes/No	(3)
25	25.5	99.5	K1	50.9	167.3	73.3	206.5	27.3 ± 20.4	−13.6 ± 17.0	Yes/No	(7)
26	20.7	96.6	J3−K	46.4	190.6	76.7	197.1	31.0 ± 6.5	5.6 ± 6.3	Yes/No	(22)
Indochina											
27	16.5	103.0	J3−K1	63.8	175.6	74.8	210.9	12.4 ± 3.7	−11.0 ± 3.6	Yes/Yes	(16)
28	21.7	104.2	J3−K	83.9	233.1	76.7	197.1	−9.2 ± 10.9	2.6 ± 9.1	No/No	(18)
29	11.7	108.2	K	74.2	171.1	75.9	196.0	0.4 ± 5.4	−6.5 ± 5.1	No/Yes	(23)
						76.7	197.1	1.2 ± 6.4	−6.8 ± 6.2	No/Yes	(23)
Borneo											
30	1.5	109.9	65−163	−2.3	11.8	78.5	178.7	−102.9 ± 6.6	13.8 ± 6.5	Yes/Yes	(17)
31	1.3	110.3	98−163	1.9	15.9	76.7	197.1	−101.3 ± 5.8	6.3 ± 5.8	Yes/Yes	(17)
32	−1.2	110.0	80−90	41.0	21.0	76.2	198.9	−62.8 ± 5.7	−0.9 ± 5.7	Yes/No	(21)
Malaya											
33	2.7	102.8	62−110	44.0	35.0	78.5	178.7	−55.6 ± 8.5	−12.3 ± 8.1	Yes/Yes	(19)
34	3.5	103.3	77−131	57.7	51.8	76.2	198.9	−40.6 ± 5.6	−20.4 ± 5.2	Yes/Yes	(20)

Note: K1 = Early Cretaceous, K2 = Late Cretaceous, K = Cretaceous, J3−K = Late Jurassic−Cretaceous. (1) = Huang & Opdyke (1992a), (2) = Huang & Opdyke (1992b), (3) = Huang & Opdyke (1993), (4) = Kent *et al.* (1986), (5) = Gilder *et al.* (1993), (6) = Otofuji *et al.* (1990), (7) = Funahara *et al.* (1992), (8) = Zhai *et al.* (1992), (9) = Chan *et al.* (1991), (10) = Enkin *et al.* (1992), (11) = Enkin *et al.* (1991), (12) = Lin (1984), (13) = Zhu *et al.* (1988), (14) = Hu *et al.* (1990), (15) = Lee *et al.* (1987), (16) = Yang & Besse (1993), (17) = Schmidtke *et al.* (1990), (18) = Cung T. C. (1996), (19) = McElhinny (1974), (20) = Haile *et al.* (1983), (21) = Haile (1977), (22) = Richter & Fuller (1996), (23) = this study. Rotation (R) and latitudinal translation (D$_l$) were calculated following Butler (1992) and Demarest (1983). Expected VGPs are from Besse and Courtillot (1991).

directions should probably reflect a combination of both local-scale and subregional-scale tectonic rotations rather than only simple rigid-block rotation. This interpretation is also revealed by the distribution of associated VGPs, which lie along a small circle (Fig. 5). Insignificant latitudinal translation of the Shan−Thai block relative to Eurasia indicates that the pole of

rotation must be located within the Shan−Thai Block (Packham 1996).

A number of palaeomagnetic studies have been carried out on Cretaceous red-beds in the Three Rivers region (Simao Terrane, Yunnan Province, China), situated just in front of the eastern syntaxis of the Himalayas where strong folding and faulting are observed (Otofuji *et al.*

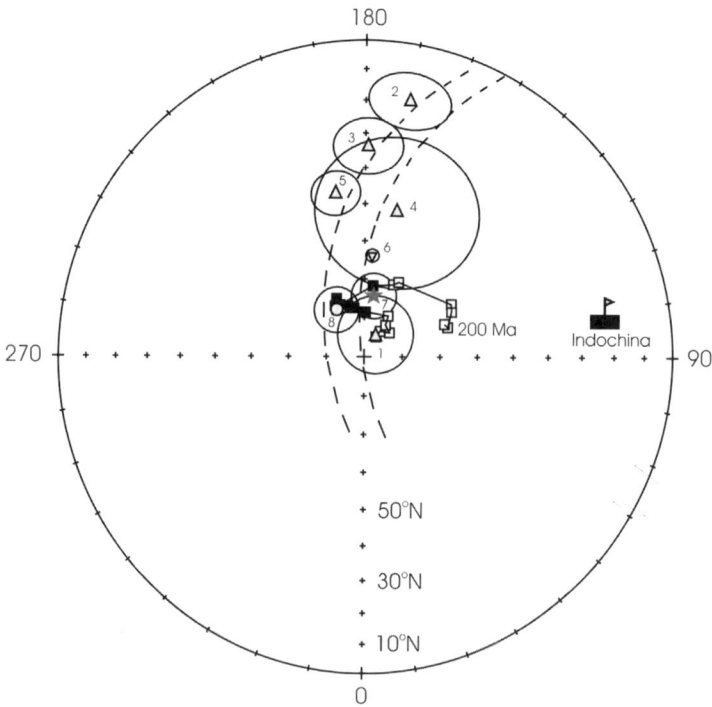

Fig. 5. Equal-area projection of the Cretaceous palaeopoles of Indochina, Shan–Thai and South China blocks and the Eurasian Apparent Polar Wander Path from 200 Ma to 10 Ma (from Besse & Courtillot 1991). The triangles (**1–5**) are Shan–Thai palaeopoles (Funahara *et al.* 1992; Huang & Opdyke 1993; Richter & Fuller 1996); reversed triangle (**6**) is the palaeopole of the Khorat Plateau–Thailand (Yang & Besse 1993); star (**7**) is the palaeopole of southern Vietnam (this study); circle (**8**) is the palaeopole of the South China block (Chen *et al.* 1993). Connected squares represent the Eurasian APWP, stage poles from 140 to 60 Ma that are used to compute the Cretaceous reference pole are highlighted in black. Dashed line shows the distribution of palaeopoles along a small circle approximately centred on the Indochina block. The ellipses represent the level of confidence A_{95}.

1990; Funahara *et al.* 1993; Huang & Opdyke 1993; Sato *et al.* 1999). Different degrees of CW rotation are reported by these studies, and the distribution of their observed palaeopoles along a small circle centred on the sampling area (cf. Sato *et al.* 1999) further corroborates the incoherent behaviour of the Indochina Block in response to the India–Eurasia collision. It is difficult, however, to attribute the palaeomagnetic data from Cretaceous rocks in the Three Rivers region to large CW rotation and southward displacement for the whole Indochina region. Instead, these results might be better explained by local, upper crustal ('thin-skinned') deformation (MacDonald 1980).

Yang & Besse (1993) presented a composite high-resolution data-set for Indochina. Comparison of the Late Jurassic–Early Cretaceous VGP for the Khorat Plateau (northeast Thailand) with the coeval pole for the South China block led Yang & Besse (1993) to conclude that Indochina

had rotated $14.2° ± 7.1°$ CW and moved $11.5° ± 6.7°$ southward relative to South China, inferring $1500 ± 800$ km of sinistral displacement of Indochina along the Red River fault in response to the India–Eurasia collision. It is important to note that their Late Jurassic–Early Cretaceous palaeomagnetic data have an inconclusive fold test (due to minor tilting of sedimentary bedding) and no reversal evidence; the primary origin of these data is assumed on the basis of the positive fold test of the underlying Upper Triassic–Lower Jurassic sedimentary rock samples.

In recent years (1998–2001), we have collected many samples of Lower–Middle Jurassic marine sedimentary formations from southern Vietnam and continental red-beds of the same age from northwestern Vietnam. Palaeomagnetic analyses of these samples revealed that they have been remagnetized (Cung 2001, 2002). Remagnetization of Mesozoic sedimentary rock

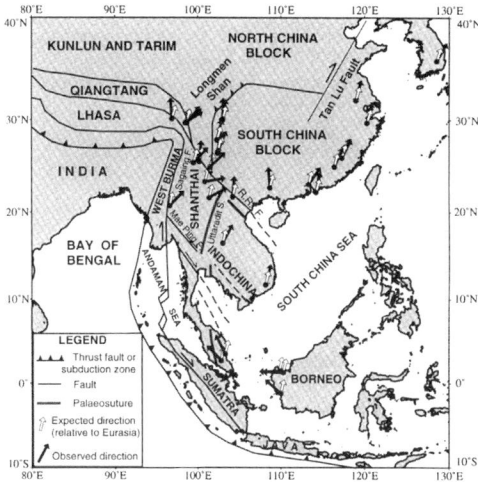

Fig. 6. Cretaceous observed (black arrows) and expected (white arrows) declination directions (with respect to Eurasia) for South China and Southeast Asian region.

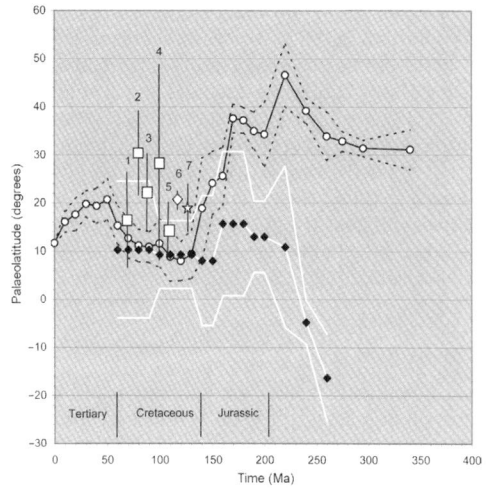

Fig. 7. Palaeolatitudes calculated at a reference point (RP, 11.7 °N, 108.2 °E) from the APWP for Eurasia (from Besse & Courtillot 1993; Van der Voo 1990) and for the South China block (from Enkin *et al.* 1992), and from the Cretaceous palaeopoles for Indochina, and Shan–Thai blocks. Solid line with open dots represents palaeolatitudes of the reference point, assuming it remains fixed with Eurasia; the dashed lines on either side of the solid line represent the error envelope. White line with black diamonds and its error envelope represent palaeolatitudes of RP, assuming it remains fixed with the South China block. Symbols associated with numbers represent palaeolatitudes of RP calculated with Cretaceous palaeopoles of: Shan–Thai (squares 1–5); Khorat Plateau (diamond – 6); and southern Vietnam (star – 7). Vertical error bars are shown on each palaeolatitude value. The symbols are situated arbitrarily and have no error bars on the horizontal axis within the Cretaceous period.

formations from the SE Asia region has been previously reported by a number of studies (Chen & Courtillot 1989; Dobson & Heller 1992; Yang & Besse 1993; Wang & Van Der Voo 1993; Kent *et al.* 1987). The poor age constraints on many of these stratigraphic units (which largely consist of continental red-beds with limited diagnostic fossils for age determinations) reduces the reliability of palaeomagnetic data obtained from these rocks.

Tectonic implications

Palaeomagnetic data from Cretaceous rocks that are distributed across Sundaland, together with our new data obtained from Vietnam, show that the Sundaland region has not behaved as a rigid lithospheric block in response to the India–Eurasia collision, but instead has been internally deformed. Regions close to India (the indentor) have been most strongly deformed by the collision, whereas regions further afield have been less affected. It thus appears that Sundaland comprises several tectonic blocks with different rotational histories. The Shan–Thai Block has apparently undergone large CW rotation and insignificant latitudinal change; the Indochina Block has undergone little or no rotation but has been subjected to a small amount of southward translation; Borneo and peninsula Malaysia experienced a large CCW rotation.

The different amounts of CW rotation for Indochina and the Shan–Thai block would have resulted in both extensional and compressional structures along the boundary between these two blocks. The Cenozoic transtensional basins of central Thailand and the Gulf of Thailand were created in areas of extension between these differentially rotated blocks. In contrast, the north–south-trending Eocene thrust belt to the north of the central Thailand intermontane basins probably accommodated the necessary crustal shortening where the Shan–Thai Block rotated CW relative to Indochina (Huchon *et al.* 1994). According to Polachan *et al.* (1991), the pull-apart basins in the Gulf of Thailand and central Thailand formed during the Oligocene, because of concomitant right-lateral strike-slip displacement along the Mae Ping and Three Pagoda faults. The Mae Ping and Three Pagoda faults, however, have experienced right-lateral displacement only during Neogene times, and are interpreted to have been left-lateral strike-slip faults during Oligocene times (Bunopas 1982; Maluski *et al.* 1993; Huchon *et al.* 1994). On the other hand, basin subsidence analyses by Pigott & Sattayarak (1993) show that extension and basin development in the Gulf of Thailand began during Early Oligocene times. These geological observations suggest that an earlier phase of crustal extension occurred when the Shan–Thai block rotated away from Indochina during Oligocene times with a later phase of extension that was related to right-lateral movement along the Mae Ping and Three Pagoda faults during Neogene times (cf. Polachan et al. 1991).

Conclusions

Palaeomagnetic data from Indochina and other parts of Sundaland provide critical information regarding the amount of large-scale tectonic rotation and lateral translation that has been previously suggested for this region. Results from this study provide important constraints on the kinematic history of large-scale tectonic deformation across SE Asia during the Cenozoic. Our data show that southern Vietnam has been subjected to a southward displacement of $6.5° \pm 5.1°$ and with an insignificantly small CW rotation since the Cretaceous. Our review of previously published palaeomagnetic data from Cretaceous rocks across the Sundaland region, together with new data obtained from this study, reveals that Sundaland has undergone complex internal deformation in response to the India–Eurasia collision, although interactions between the India–Australia, Eurasia, Philippine Sea, and Pacific plates also probably influenced deformation patterns across Sundaland. The combination of reliable palaeomagnetic data and regional structural information indicates that Sundaland was broken into several tectonic blocks during Early Cenozoic times. Each block apparently has a different rotational and translational history. Many of the sedimentary basins, regional fault systems, and other structural features that are attributed to rigid-block extrusion of Sundaland can also be explained by our model for more localized Cenozoic block movements across western SE Asia. For example, CW rotation of the Shan–Thai block with respect to Indochina probably caused extension and basin formation in the Gulf of Thailand and central Thailand during the Early Oligocene.

The sponsors of the Texas A&M Indochina Research Consortium (Agip, Amerada Hess, Amoco, Arco, British Gas, British Petroleum, Chevron, Dupont–Conoco, Elf Aquitaine, Enterprise, Mobil, Occidental, Petrofina, Phillips, Shell, Total and Unocal) are thanked for supporting this research. M. Flower and R. McCabe graciously provided technical insights and logistical support during the early stages of this research. The review comments and suggestions by M. Fuller and H. Leloup proved very helpful. The research was partly supported by a grant for a basic research project (No. 735701) from NSC of Vietnam to CTC.

References

BESSE, J. & COURTILLOT, V. 1991. Revised and synthetic apparent polar wander paths of the African, Eurasian, North American and Indian Plates, and true polar wander since 200 Ma. Journal of Geophysical Research, 96B, 4029–4050.

BUNOPAS, S. 1982. Paleogeographic History of Western Thailand and Adjacent Parts of Southeast Asia: A Plate Tectonic Interpretation. Geological Survey Special Paper, Issue 5, Department of Mineral Resources, Bangkok, Thailand.

BUTLER, R. 1992. Paleomagnetism. Blackwell Scientific Publications, Boston, USA.

CHAN, L. S. 1991. Paleomagnetism of late Mesozoic granitic intrusions in Hong Kong: implications for Upper Cretaceous reference pole of South China. Journal of Geophysical Research, 96B, 327–335.

CHARVET, J., LAPIERRE, H. & YU, Y. 1994. Geodynamic significance of the Mesozoic volcanism of southeastern China. Journal Southeast Asian Earth Sciences, 9, 387–396.

CHEN, Y. & COURTILLOT, V. 1989. Widespread Cenozoic(?) remagnetization in Thailand and its implications for the India-Asia collision. Earth and Planetary Science Letters, 93, 113–122.

CHEN, Y., COURTILLOT, V., COGNE, J.-P., BESSE, J., YANG, Z. & ENKIN, R. 1993. The configuration of Asia prior to the collision of India: Cretaceous paleomagnetic constraints. Journal of Geophysical Research, 98B, 21 927–21 941.

CISOWSKI, S. 1981. Interacting vs. non-interacting single domain behavior in natural and synthetic samples. Physics of the Earth and Planetary Interiors, 26, 56–62.

COLLINSON, D. W. 1983. Methods in Rock Magnetism and Palaeomagnetism, Chapman and Hall, London.

CUNG, T. C. 1996. Paleomagnetism of Mesozoic and Cenozoic rocks from Vietnam: implications for the Tertiary tectonic history of Indochina and a test of the extrusion model. PhD thesis, Texas A&M University, College Station, Texas.

CUNG, T. C. 2001. Results of paleomagnetic study on Jurassic sandstone and siltstone of the Ha Coi formation and their tectonic implications. Journal of Geology, Series B, 17–18/2001, 35–44.

CUNG, T. C. 2002. Paleomagnetic study on Early–Middle Jurassic continental redbeds from northwestern Vietnam and their tectonic implications. IGCP 430 Second Annual Workshop on 'Mantle Responses to Tethyan Closure', Program with Abstracts, Ha Long Bay, Vietnam, April 1–10 2002, 12.

CUNG, T. C., DOROBEK, S., RICHTER, C., FLOWER, M., KIKAWA, E., NGUYEN, Y. T. & MCCABE, R. 1998. Paleomagnetism of late Neogene basalts in Vietnam and Thailand: implications for the post-Miocene tectonic history of Indochina.

In: FLOWER, M. F. J., CHUNG, S.-L., LO, C.-H. & LEE, T.-Y. (eds) *Mantle Dynamics and Plate Interactions in East Asia.* American Geophysical Union Geodynamics Series, **27**, 289–299.

DEMAREST, H. H. 1983. Error analysis for the determination of tectonic rotation from paleomagnetic data. *Journal of Geophysical Research,* **88B**, 4321–4328.

DEWEY, J. F., CANDE, S. & PITMAN, W. C. III. 1989. Tectonic evolution of the India/Eurasia collision zone. *Eclogae Geologicae Helvetiae,* **82**, 717–734.

DOBSON, J. P. & HELLER, F. 1992. Remagnetization in southeast China and the collision and suturing of the Huanan and Yangtze blocks. *Earth and Planetary Science Letters,* **111**, 11–21.

DUNLOP, D. J. 1983. Determination of domain structure in igneous rocks by alternating field and other methods. *Earth and Planetary Science Letters,* **63**, 353–367.

ENGLAND, P. C. & HOUSEMAN, G. A. 1989. Extension during continental convergence, with application to the Tibetan Plateau. *Journal of Geophysical Research,* **94B**, 17 561–17 579.

ENGLAND, P. C. & MOLNAR, P. 1990. Right lateral shear and rotation as the explanation for strike-slip faulting in eastern Tibet. *Nature,* **344**, 140–142.

ENKIN, R. J., CHEN, Y., COURTILLOT, V. & BESSE, J. 1991. A Cretaceous pole from South China, and the Mesozoic hairpin turn of the Eurasian apparent polar wander path. *Journal Geophysical Research,* **96B**, 4007–4027.

ENKIN, R. J., YANG, Z., CHEN, Y. & COURTILLOT, V. 1992. Paleomagnetic constraints on the geodynamic history of the major blocks of China from the Permian to the present. *Journal of Geophysical Research,* **97B**, 13953–13989.

FISHER, R. A. 1953. Dispersion on a sphere. *Proceedings of the Royal Society of London,* **217A**, 295–305.

FUNAHARA, S., NISHIWAKI, N., MIKI, M., MURATA, F., OTOFUJI, Y.-I. & WANG, Y. Z. 1992. Paleomagnetic study of Cretaceous rocks from the Yangtze block, central Yunnan, China: Implications for the India–Asia collision. *Earth and Planetary Science Letters,* **113**, 77–91.

FUNAHARA, S., NISHIWAKI, N., MURATA, F., OTOFUJI, Y.-I. & WANG, Y. Z. 1993. Clockwise rotation of the Red River fault inferred from paleomagnetic study of Cretaceous rocks in the Shan–Thai–Malay block of western Yunnan, China. *Earth and Planetary Science Letters,* **117**, 29–42.

GILDER, S. A., COE, R. S., WU, H., KUANG, G., ZHAO, X., WU, Q. & TANG, X. 1993. Cretaceous and Tertiary paleomagnetic results from southeast China and their tectonic implications. *Earth and Planetary Science Letters,* **117**, 637–652.

HAILE, N. S., McELHINNY, M. W. & McDOUGALL, I. 1977. Paleomagnetic data and radiometric ages from the Cretaceous of west Kalimantan (Borneo) and their significance in interpreting regional structure. *Journal of the Geological Society of London,* **133**, 133–144.

HAILE, N. S., BECKINSALE, R. D., CHAKRABORTY, K. R., HUSSEIN, A. H. & HARDJONO, T. 1983.

Paleomagnetism, geochronology and petrology of the dolerite dikes and basaltic lavas from Kuantan, west Malaysia. *Geological Society of Malaysia Bulletin,* **16**, 71–85.

HALL, R. 1996. Reconstructing Cenozoic SE Asia. *In:* HALL, R. & BLUNDELL, D. (eds) *Tectonic Evolution of Southeast Asia.* Geological Society, London, Special Publications, **106**, 153–184.

HALL, R., ALI, J. R. & ANDERSON, C. D. 1995. Cenozoic motion of the Philippine Sea plate: palaeomagnetic evidence from eastern Indonesia. *Tectonics,* **14**, 1117–1132.

HU, L., LI, P. & MA, X. 1990. A magnetostratigraphic study of Cretaceous red beds from Shanghan, western Fujian, China (in Chinese). *Geology of Fujian,* **1**, 33–42.

HUANG, K. & OPDYKE, N. D. 1992a. Paleomagnetism of Cretaceous to lower Tertiary rocks from southwestern Sichuan: a revisit. *Earth and Planetary Science Letters,* **112**, 29–40.

HUANG, K. & OPDYKE, N. D. 1992b. Paleomagnetism of Cretaceous rocks from eastern Qiangtang terrane of Tibet. *Journal of Geophysical Research,* **97B**, 1789–1799.

HUANG, K. & OPDYKE, N. D. 1993. Paleomagnetic results from Cretaceous and Jurassic rocks of south and southwest Yunnan: evidence for large clockwise rotations in the Indochina and Shan–Thai–Malay terranes. *Earth and Planetary Science Letters,* **117**, 507–524.

HUCHON, P., LE PICHON, X. & RANGIN, C. 1994. Indochina Peninsula and the collision of India and Eurasia. *Geology,* **22**, 27–30.

HUTCHISON, C. S. 1989a. *Geological Evolution of Southeast Asia.* Clarendon Press, Oxford.

HUTCHISON, C. S. 1989b. The palaeo-Tethyan realm and Indosinian orogenic system of Southeast Asia. *In:* SENGÖR, A. M. C., YILMAZ, Y., OKAY, A. I. & GORUR, N. (eds) *Tectonic Evolution of the Tethyan Region.* NATO ASI Series, Series C, Mathematical and Physical Sciences, **259**, 585–643.

JAHN, B., CHEN, P. Y. & YEN, T. P. 1976. Rb–Sr ages of granitic rocks in southeastern China and their tectonic significance. *Geological Society of America Bulletin,* **87**, 763–776.

KENT, D. V., XU, G., HUANG, K., ZHANG, W. Y. & OPDYKE, N. D. 1986, Paleomagnetism of Upper Cretaceous rocks from South China. *Earth and Planetary Science Letters,* **79**, 179–184.

KENT, D. V., ZHENG, X. S., ZHANG, W. Y. & OPDYKE, N. D. 1987. Widespread Mesozoic to Recent remagnetization of Paleozoic and Triassic sedimentary rocks from South China. *Tectonophysics,* **139**, 133–143.

KIRSCHVINK, J. L. 1980. The least-square line and plane and the analysis of paleomagnetic data. *Geophysical Journal of the Royal Astronomical Society,* **62**, 699–718

LELOUP, P. H., HARRISON, T. M., RYERSON, F. J., CHEN, W., QI, L., TAPPONNIER, P. & LACASSIN, R. 1993. Structural, petrological and thermal evolution of a Tertiary ductile shear zone, Diancang Shan, Yunnan. *Journal of Geophysical Research,* **98B**, 6715–6744.

LELOUP, P. H., LACASSIN, R. *et al.* 1995. The Ailao Shan–Red River shear zone (Yunnan, China), Tertiary transform boundary of Indochina. *Tectonophysics*, **251**, 3–84.

LIN, J. 1984. *The apparent polar wander paths for the North and South China blocks.* PhD thesis, University of California, Santa Barbara.

LOWRIE, W. & FULLER, M. 1971. On the alternating field demagnetization characteristics of multidomain thermoremanent magnetization in magnetite. *Journal of Geophysical Research*, **76B**, 6339–6349.

MACDONALD, W. D. 1980. Net tectonic rotation, apparent tectonic rotation, and the structural tilt correction in paleomagnetic studies. *Journal of Geophysical Research*, **85B**, 3659–3669.

MCELHINNY, M. W. 1973. *Palaeomagnetism and Plate Tectonics.* Cambridge University Press, London.

MCELHINNY, M. W., HAILE, N. S. & CRAWFORD, A. R. 1974. Paleomagnetic evidence shows Malay Peninsula was not a part of Gondwanaland. *Nature*, **252**, 641–645.

MALUSKI, H., LACASSIN, R. *et al.* 1993. Mid-Oligocene left-lateral shear along the Wang Chao fault zone (NW Thailand). *European Union of Geologists, VII, Terra Abstracts*, p. 262.

METCALFE, I, 1996. Pre-Cretaceous evolution of SE Asian terranes. *In:* HALL, R. & BLUNDELL, D. (eds) *Tectonic Evolution of Southeast Asia.* Geological Society, London, Special Publications, **106**, 97–122.

NGUYEN, X. B. (ed.) 2000. *Project Report on: Tectonics and Mineral Genesis in southern Vietnam Region.* Geological Survey of Vietnam Press, Hanoi, Vietnam.

NGUYEN, X. B., TRAN, D. L. & HUYNH, T. 1994. *Explanatory Note to the Geological Map of Vietnam on 1 : 500,000 Scale.* Geological Survey of Vietnam Press, Hanoi, Vietnam.

OTOFUJI, Y., INOUE, Y., FUNAHARA, S., MURATA, F. & ZHENG, X. 1990. Paleomagnetic study of eastern Tibet-deformation of the Three Rivers region. *Geophysical Journal International*, **103**, 85–94.

PACKHAM, G. 1996. Cenozoic SE Asia: reconstructing its aggregation and reorganization. *In:* HALL, R. & BLUNDELL, D. (eds) *Tectonic Evolution of Southeast Asia.* Geological Society, London, Special Publications, **106**, 123–152.

PELTZER, G. & TAPPONNIER, P. 1988. Formation and evolution of strike-slip faults, rifts, and basins during the India–Asia collision: an experimental approach. *Journal of Geophysical Research*, **93B**, 15 085–15 117.

PHAN, C. T. *et al.* 1991. *Geology of Cambodia, Laos and Vietnam.* Geological Survey of Vietnam, Hanoi, Vietnam.

PIGOTT, J. D. & SATTAYARAK, N. 1993. Aspects of sedimentary basin evolution assessed through tectonic subsidence analysis. Example: Northern Gulf of Thailand. *Journal of Southeast Asian Earth Sciences*, **8**, 407–420.

POLACHAN, S., SURAWIT, P., CHALERMKIAT, T., SOMKIAT, J., KANOK, I. & CHUTAMAT, S. 1991. Development of Cenozoic basins in Thailand. *Marine and Petroleum Geology*, **8**, 84–97.

RANGIN, C., HUCHON, P. *et al.* 1995*b.* Cenozoic deformation of central and southern Vietnam. *Tectonophysics*, **251**, 179–196.

RICHTER, B. & FULLER, M. 1996. Palaeomagnetism of the Sibumasu and Indochina blocks: implications for the extrusion tectonic model. *In:* HALL, R. & BLUNDELL, D. (eds) *Tectonic Evolution of Southeast Asia.* Geological Society, London, Special Publications, **106**, 203–224.

SATO, K., LIU, Y., ZHU, Z., YANG, Z. & OTOFUJI, Y. 1999. Paleomagnetic study of middle Cretaceous rocks from Yunlong, western Yunnan, China: evidence of southward displacement of Indochina. *Earth and Planetary Science Letters*, **165**, 1–15.

SCHMIDTKE, E. A., FULLER, M. D. & HASTON, R. B. 1990. Paleomagnetic data from Sarawak, Malaysian Borneo, and the late Mesozoic and Cenozoic tectonics of Sundaland. *Tectonics*, **9**, 123–140.

TARLING, D. H. 1983. *Palaeomagnetism, Principles and Applications in Geology, Geophysics and Archaeology*, Chapman and Hall, London.

TAPPONNIER, P., PELTZER, G., LE DAIN, A. Y. & ARMIJO, R. 1982. Propagating extrusion tectonics in Asia: new insights from simple experiments with plasticine. *Geology*, **10**, 611–616.

TAPPONNIER, P., PELTZER, G. & ARMIJO, R. 1986. On the mechanics of the collision between India and Asia. *In:* COWARD, M. P. & RIES, A. C. (eds) *Collision Tectonics.* Geological Society, London, Special Publications, **19**, 115–157.

TAYLOR, B. & HAYES, D. E. 1983. Origin and history of the South China Sea Basin. *In:* HAYES, D. E. (ed.) *The Tectonic and Geologic Evolution of Southeast Asian Seas and Islands, Part 2.* American Geophysical Union, Geophysical Monograph Series, **27**, 23–56.

TRAN, V. T., TRAN, K. T. & TRUONG, C. B. (eds). 1979. *Geology of Vietnam (Northern Part)* (in Vietnamese). General Department of Geology, Research Institute of Geology and Mineral Resources, Hanoi, Vietnam.

VAN DER VOO, R. 1990. Phanerozoic paleomagnetic poles from Europe and North America and comparisons with continental reconstructions. *Research Geophysics*, **28(2)**, 167–199.

VU, K., TONG, D. T., PHAN, C. T. & NGUYEN, V. (eds) 1989. *Geology of Vietnam, Part I: Stratigraphy* (in Vietnamese). General Department of Mining and Geology, Hanoi, Vietnam.

WANG, Z. & VAN DER VOO, R. 1993. Pervasive remagnetization of Paleozoic rocks acquired at the time of Mesozoic folding in the South China block. *Journal of Geophysical Research*, **98B**, 1729–1741.

YANG, Z. & BESSE, J. 1993. Paleomagnetic study of Permian and Mesozoic sedimentary rocks from northern Thailand supports the extrusion model for Indochina. *Earth and Planetary Science Letters*, **117**, 525–552.

YANG, Z. Y., BESSE, J., SUTHEETORN, V., BASSOULLET, J. P., FONTAINE, H. & BUFFETAUT, E. 1995. Lower-Middle Jurassic paleomagnetic data from the Mae Sot area (Thailand): paleogeographic evolution and deformation history of Southeastern Asia. *Earth and Planetary Science Letters*, **136**, 325–341.

ZHAI, Y., SEGUIN, M. K., ZHOU, Y., DONG, J. & ZHENG, Y. 1992. New paleomagnetic data from the Huanan block, China, and Cretaceous tectonics in eastern China. *Physics of the Earth and Planetary Interiors*, **73**, 163–188.

ZHU, Z., HAO, T. & ZHAO, H. 1988. Paleomagnetic study on the tectonic motion of Pan–Xi block and adjacent area during Yin Zhi–Yan Shan period (in Chinese). *Acta Geophysica Sinica*, **31**, 420–431.

ZIJDERVELD, J. D. A. 1967. AC demagnetization of rocks: analysis of results. *In:* COLLINSON, D. W., CREER, K. M. & RUNCON, S. K. (eds) *Methods in Paleomagnetism.* Elsevier, New York, 254–286.

Geology of the Zamboanga Peninsula, Mindanao, Philippines: an enigmatic South China continental fragment?

GRACIANO P. YUMUL, JR[1,2], CARLA B. DIMALANTA[1], RODOLFO A. TAMAYO, JR[1], RENE C. MAURY[3], HERVE BELLON[3], MIREILLE POLVÉ[4], VICTOR B. MAGLAMBAYAN[1], CLIFF L. QUERUBIN[1,5] & JOSEPH COTTEN[3]

[1]*National Institute of Geological Sciences, College of Science, University of the Philippines, Diliman, Quezon City, Philippines (e-mail: rwg@i-next.net)*

[2]*Philippine Council for Industry and Energy Research & Development, Department of Science and Technology, Bicutan, Taguig, Metro Manila, Philippines,*

[3]*UMR 6538, Université de Bretagne Occidentale, 6, Avenue Le Gorgeu, B.P. 809, 29285 Brest Cedex, France*

[4]*UMR 5563, Université Paul Sabatier, 38, Rue des 36 Ponts, 31400 Toulouse, France*

[5]*Geosciences Division, Mines and Geosciences Bureau-IX, Zamboanga City, Mindanao, Philippines*

Abstract: Mindanao Island in the southern Philippines is made up of two blocks: the island-arc-related eastern-central Mindanao block and the continental Zamboanga Peninsula, which contains several ophiolitic bodies and mélanges. The Middle Miocene Siayan–Sindangan Suture Zone represents the tectonic boundary between the island-arc and continental blocks. A Middle Miocene age of collision is interpreted from the unconformity between the Late Miocene Motibot Formation and the underlying Middle Miocene Gunyan Mélange, which serves as basement to the suture zone. The Middle Miocene Siayan–Sindangan Suture Zone was formerly a subduction zone complex that was reactivated as a sinistral strike-slip fault following the collision of eastern-central Mindanao with the Zamboanga Peninsula. New ^{40}K–^{40}Ar whole-rock dating of lava flows from the Zamboanga Peninsula has revealed Middle to Late Miocene ages, which is consistent with the possible existence of an Early Miocene Sulu Trench. The possibility that the Zamboanga Peninsula could be part of the Palawan microcontinental block has been forwarded by previous workers, due to their similarity in stratigraphy, geological structure and metamorphic rock suites. The Palawan microcontinental block separated from southern China during the opening of the South China Sea in Oligo-Miocene times. If indeed the Zamboanga Peninsula was once part of Palawan, it represents the southernmost part of the rifted southeastern China continental margin.

The Philippine archipelago is an appropriate target to investigate a number of tectonic and magmatic processes, e.g. collision and subduction of bathymetric highs (Scarborough Seamount Chain in the South China Sea, Palawan microcontinental block), large-scale strike-slip faulting involving transtensional and transpressional features (Philippine Fault Zone), island-arc magmatism and oceanic lithosphere emplacement. However, not all of the country has already been mapped and studied in detail, including the Zamboanga Peninsula in Mindanao, southern Philippines. The last regional geological survey of the Zamboanga Peninsula was done in the 1950s to 1960s (e.g. Santos-Yñigo 1953) and has not

been updated systematically since that time. The Zamboanga Peninsula is reportedly made up of continental fragments, which could have been derived from either the Palawan microcontinental block, which in itself was rifted from southern China during the opening of the South China Sea, or the Borneo block (e.g. Faure *et al.* 1989; Rangin *et al.* 1990; Tamayo *et al.* 2000). The Zamboanga Peninsula is strategically important, as it hosts volcanogenic massive sulphides, epithermal gold, porphyry copper and chromitites (e.g. Querubin & Yumul 2001; Jimenez *et al.* 2002). Given the paucity of data in this part of Mindanao island, we conducted a four-year programme (1998–2001) to systematically map

From: MALPAS, J., FLETCHER, C. J. N., ALI, J. R. & AITCHISON, J. C. (eds) 2004. *Aspects of the Tectonic Evolution of China*. Geological Society, London, Special Publications, **226**, 289–312. 0305-8719/04/$15 © The Geological Society of London 2004.

and sample as much as possible of the Zamboanga Peninsula. Field mapping was complemented by palaeontological, geochemical and whole-rock K–Ar dating analyses. The present work reports the main results of this integrated study, which has led to further constraining of our understanding of the geological evolution of the Zamboanga Peninsula. Furthermore, our data provide additional evidence for the Zamboanga Peninsula being part of the Palawan microcontinental block. Such a correlation would have repercussions on the evolution of the rifting of the southeastern continental margin of China through space and time.

Geological outline

The Philippine archipelago

The Philippine archipelago is a composite terrane made up of two major blocks: the Philippine Mobile Belt and the Palawan microcontinental block (Fig. 1). The Philippine Mobile Belt is a seismically active zone with moving faults, numerous volcanoes and frequent earthquakes, contrasting with the aseismic character of the Palawan microcontinental block. The Philippines, as a whole, is characterized by a complex pattern of subduction zones, collision zones, island arcs and marginal basins which have and are, rapidly evolving (e.g. Aurelio 2000; Yumul *et al.* 2001). The Philippine archipelago is bounded on both flanks by oppositely dipping subduction systems (Fig. 1). In the east, the Philippine Sea Plate subducts west-northwestward along the incipient East Luzon Trough and the Philippine Trench. An approximately 150 km left-lateral transform fault connects the Philippine Trench with the East Luzon Trough. The Philippine Trench extends from eastern Luzon all the way to eastern Mindanao, where it is propagating southward (Rangin *et al.* 1996; Lallemand *et al.* 1998). The western margin is defined by the Manila–Negros–Sulu–Cotabato Trench system, along which the South China Sea, Sulu Sea and Celebes Sea basins are subducting. Through shear partitioning, any excess stress not accommodated by the two trench systems is taken up by the Philippine Fault Zone. All areas west of the Manila–Negros Trench system and the Sindangan–Cotabato–Daguma Lineament have Sundaland affinity, whereas those east of the East Luzon Trough–Philippine Trench system correspond with the Philippine Sea Plate; in between is the Philippine Mobile Belt containing rocks of either the Philippine Sea Plate or Sundaland–Eurasian origin (e.g. Rangin 1991; Hall 2002), (Fig. 1).

Mindanao Island

The Philippine Trench east of the Mindanao island marks the Philippine Sea Plate subduction zone. The Sulu Trench to the northwest accommodates the south-southeast-directed subduction of the SE Sulu Sea Basin, whereas the eastward consumption of the Celebes Basin occurs along the Cotabato Trench (Fig. 2). Ancient to present-day volcanism can be attributed to the different modern-day trenches or their ancient counterparts (e.g. the proto-Philippine Trench) (Sajona *et al.* 1997).

Two major sinistral fault zones, the north–south trending Philippine Fault Zone and the northwest–southeast Sindangan–Cotabato–Daguma Lineament (Pubellier *et al.* 1991), traverse Mindanao (Fig. 2). The Sindangan–Cotabato–Daguma Lineament is postulated to be a product of the 'soft' collision between Sundaland and the Philippine Mobile Belt, during the Late Miocene to Pliocene (Pubellier *et al.* 1996). The Sindangan–Cotabato–Daguma Lineament, just like the Philippine Fault Zone, is classified as an active fault (e.g. Quebral *et al.* 1996), and divides Mindanao into the Zamboanga Peninsula, with postulated continental affinity and eastern-central Mindanao with island-arc affinity (e.g. Sajona *et al.* 1997) (Fig. 2).

Methodology and results of analyses

Major- and trace-element, except for Rb, concentrations in the volcanic rocks of the ophiolites of the Zamboanga Peninsula (Table 1) were determined using a Jobin–Yvon (JY50 plus) inductively coupled plasma–atomic emission spectrometer (ICP–AES) at the Université de Bretagne Occidentale, Brest, France. Rubidium concentrations were determined using Atomic Absorption Spectrometry. The analytical techniques, operating conditions and detection limits are to be found in Cotten *et al.* (1995).

The multi-element patterns of basalts from the volcanic sequences of the ophiolites located in northeast Zamboanga (i.e. Polanco Ophiolite Complex) and Central Zamboanga (i.e. basalts associated with the ZNAC Ultramafics) are plotted in Fig. 7B. The geochemical values were normalized to the primitive mantle values of Sun and McDonough (1989). Samples collected from both blocks display two distinct patterns. First, there are generally flat spectra from the heavy rare-earth elements (HREE) towards the light (LREE) followed by decreasing normalized values from the LREE towards Th and, second, there are decreasing spectra from the HREE towards the LREE followed by an increase

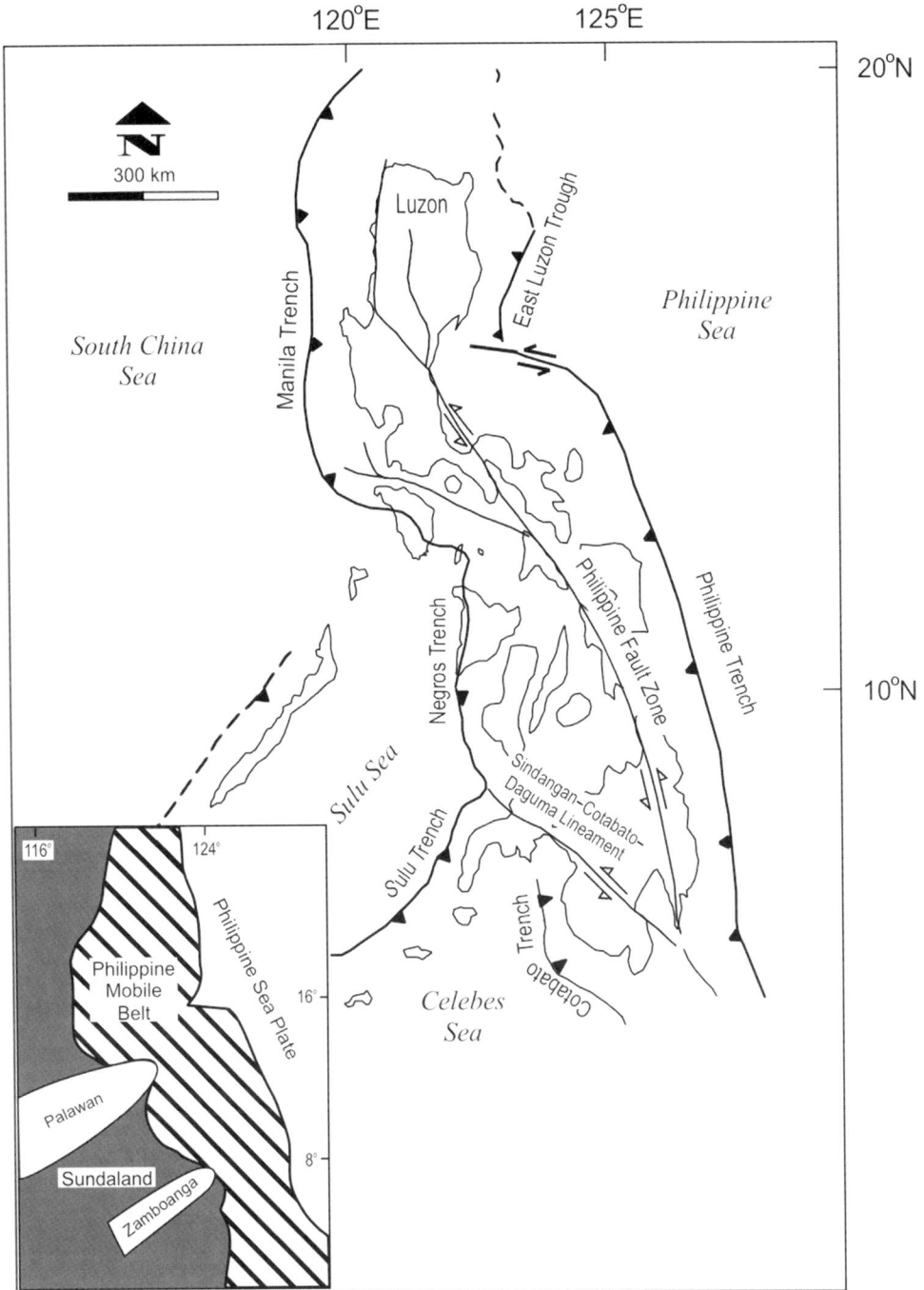

Fig. 1. Generalized tectonic map of the Philippines, showing the major tectonic features. Inset shows the Philippine Mobile Belt (diagonally lined area), bounded to the east by the Philippine Sea Plate, and Sundaland which includes Palawan and Zamboanga. See text for details.

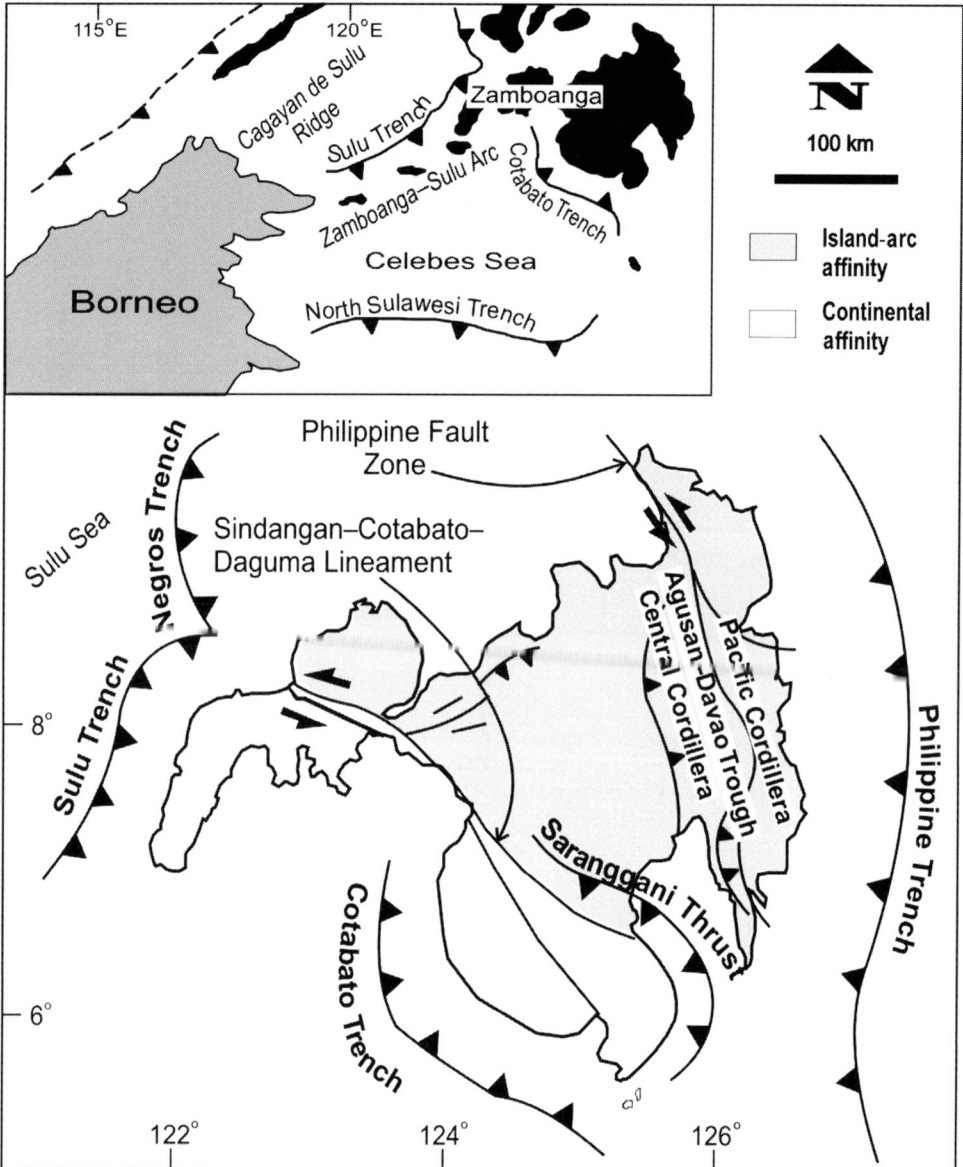

Fig. 2. Tectonic map of Mindanao, showing the trench systems surrounding the island, and the two major sinistral fault zones (Philippine Fault Zone, Sindangan–Cotabato–Daguma Lineament). Mindanao is divided into two blocks with different affinities – the eastern-central block has an island-arc affinity, while the Zamboanga–Cotabato–Daguma area exhibits continental affinity (modified from Pubellier *et al.* 1996). Earlier models divided Mindanao into the Pacific Cordillera, and Central Cordillera, which were juxtaposed along the Agusan–Davao Trough. The Sulu Sea basin, according to the model of Rangin (1989), opened through the separation of the Cagayan de Sulu Ridge and a proto-Zamboanga arc. Inset shows the regional setting of Mindanao.

in Th. The second pattern is spiked by weak to strong negative Ti and Nb anomalies. In addition, a sample collected from northeast Zamboanga shows relatively increasing multi-element patterns from the HREE towards Th, and is broken by strong negative spikes in Ti and Nb. Pattern 1 is similar to that of normal-mid-oceanic ridge or back-arc basin basalts, whereas patterns

Table 1. *Representative analyses of the major and trace elements of volcanic rocks from ophiolites in the Zamboanga Peninsula*

Wt%	PH98-31 B	PH98-37 B	PH98-38 D	PH98-50 D
SiO_2	51.6	49.7	50.2	52.5
TiO_2	1.39	1.07	0.40	1.33
Al_2O_3	14.85	15.65	14.70	14.35
$Fe_2O_3{}^*$	11.90	9.88	9.60	10.75
MnO	0.17	0.18	0.17	0.16
MgO	6.80	7.40	9.70	6.03
CaO	4.60	12.15	9.45	8.55
Na_2O	4.96	2.99	2.05	4.10
K_2O	0.06	0.05	0.22	0.10
P_2O_5	0.16	0.03	0.04	0.11
LOI	3.12	1.02	2.90	1.67
Total	99.61	100.12	99.43	99.65
ppm				
Sc	32	44	52	35
V	445	210	280	330
Cr	16	270	400	112
Co	16	40	45	35
Ni	11	64	80	45
Rb	0.4	0.1	2.8	1.1
Sr	72	156	81	190
Y	30.0	11.4	10.8	26.5
Zr	58	10	13	71
Nb	1.9	0.3	0.5	1.3
Ba	7	4	13	7
La	4.0	0.6	0.7	2.3
Ce	10.5	1.5	2.0	6.5
Nd	10.0	1.5	1.2	7.5
Sm	3.2	1.0	0.5	2.5
Eu	1.1	0.8	0.3	1.0
Gd	4.1	1.3	1.5	3.8
Dy	4.8	2.0	1.7	4.4
Er	3.0	1.1	1.1	2.7
Yb	3.2	1.0	1.2	2.6
Th	0.2	0.1	0.2	0.2

$Fe_2O_3{}^*$, Total iron as Fe_2O_3; B, Basalt; D, Diabase.

2 and 3 mimic those of basalts from subduction-related basins (e.g. Falloon *et al.* 1999; Caprarelli & Leitch 2002; Bortolotti *et al.* 2002).

Spinel compositions were determined using a CAMECA SX50 electron probe at the Université de Bretagne Occidentale, Brest, France. The measurements were made with the following analytical conditions: 15 kV, 10–11 nA and 10 seconds counting time. Sample preparation and analytical techniques followed Defant *et al.* (1991). In addition, some samples were analysed using a JEOL JCMA 733 mkII Electron Probe at the Geological Institute, University of Tokyo, Japan. Analytical conditions and procedures are discussed in Yumul (1989) (Table 2). Plots of the X_{Cr} (=Cr/[Cr + Al]) and X_{Mg} (= Mg/[Mg + Fe^{2+}]) of spinels from peridotites and chromitites sampled in the mantle sequences of the Zamboanga ophiolites are presented in Fig. 7A. Although most of the peridotites show X_{Cr} and X_{Mg} plots similar to those observed in modern abyssal peridotites, several samples from northeast Zamboanga display X_{Cr} values slightly higher than the upper limit of most modern abyssal peridotites. In addition, spinels from the chromitites of northeast Zamboanga exhibit relatively high X_{Cr} values as well (e.g. Arai 1992; Arai & Matsukage 1998).

Whole-rock $^{40}K-^{40}Ar$ analyses were performed at the Geochronology Laboratory of the University of Bretagne Occidentale, using the methodology described in Bellon & Rangin (1991) (Table 3). Ages were calculated using the constants of Steiger & Jäger (1977) with the

Table 2. *Representative analyses of spinel from harzburgites (Hz), chromitite (Chr) and dunite (Dun) from the Zamboanga ophiolites*

Wt%	NE Zamboanga							
	PH98-33 sp4moy	Hz 39sp1moy	PH98-34 sp1moy	Hz sp4moy	S-7 S7-SPC3[1]	Chr S7-SPC4[1]	S-18 S18-SPC1[1]	Dun S18-SPC4[1]
SiO_2	0.01	0.01	0.00	0.01	0.01	0.01	0.03	0.01
TiO_2	0.02	0.01	0.03	0.00	0.22	0.17	0.17	0.18
Al_2O_3	19.86	18.34	20.38	21.35	17.08	17.23	17.98	16.35
V_2O_3	–	–	–	–	0.04	0.08	0.17	0.16
Cr_2O_3	48.71	49.41	50.88	49.69	53.79	53.02	47.22	49.51
FeO^T	19.16	20.44	17.30	16.97	13.31	13.56	21.53	20.79
MnO	0.20	00.28	0.25	0.25	0.16	0.23	0.35	0.25
MgO	11.46	10.85	12.09	12.40	15.44	15.68	11.61	12.28
CaO	0.01	0.00	0.02	0.00	0.01	0.00	0.01	0.02
NiO	0.02	0.01	0.17	0.00	0.12	0.14	0.16	0.11
Total	99.45	99.36	101.11	100.66	100.19	100.13	99.22	99.65
X_{Cr}	0.622	0.644	0.626	0.610	0.679	0.674	0.638	0.670
X_{Mg}	0.541	0.519	0.563	0.573	0.711	0.723	0.554	0.583

Wt%	Central Zamboanga						SW Zamboanga	
	PH98-46 sp1moy	Hz 63sp	SPC4*	SPC5*	PH98-47 sp1moy	Hz sp3moy	ZC06 #88	Hz #92
SiO_2	0.06	0.09	0.04	0.03	0.02	0.00	0.00	0.00
TiO_2	0.05	0.05	0.04	0.04	0.21	0.28	0.04	0.06
Al_2O_3	30.54	26.37	30.77	26.99	31.24	29.79	26.11	28.69
V_2O_3	–	–	0.20	0.10	–	–	–	–
Cr_2O_3	37.74	43.42	38.15	43.18	35.29	34.49	41.44	38.74
FeO^T	18.20	15.51	18.99	14.67	18.78	22.13	18.44	18.81
MnO	0.19	0.23	0.26	0.22	0.22	0.15	0.21	0.06
MgO	13.17	14.60	12.28	14.91	13.87	12.65	12.74	12.44
CaO	0.04	0.04	0.03	0.02	0.00	0.02	0.02	0.00
NiO	0.05	0.09	0.10	0.17	0.15	0.18	0.09	0.00
Total	100.04	100.42	100.85	100.32	99.79	99.67	99.10	98.80
X_{Cr}	0.453	0.525	0.454	0.518	0.431	0.437	0.516	0.475
X_{Mg}	0.585	0.653	0.547	0.666	0.613	0.566	0.584	0.563

uncertainties computed following the method of Mahood & Drake (1982). Twelve new whole-rock $^{40}K-^{40}Ar$ ages were measured on samples collected from Sindangan, near the Siayan–Sindangan Suture Zone, and Zamboanga City (Fig. 5b). The two samples collected from Sindangan, belonging to the Sindangan Volcanics, gave ages of 10.9 Ma and 13.6 Ma, respectively (Table 3). The ten Zamboanga City samples, mostly basaltic to andesitic flows from the Anungan and Curuan Formations, gave a range of 8.95 Ma to 18.2 Ma.

Palaeontological analyses were undertaken at the University of the Philippines – National Institute of Geological Sciences and Philippine Mines and Geosciences Bureau – Central Office (Table 4).

Geology of the Zamboanga Peninsula

The Zamboanga Peninsula can be subdivided into southwest and northeast areas with the Siayan–Sindangan Suture Zone as the boundary. Pre-Cenozoic granite, quartz–sericite–albite schist, gneiss and amphibolite are the reported basement rocks of southwest Zamboanga, and are believed to have continental affinity (Santos-Yñigo 1953; Faure *et al.* 1989; Tamayo *et al.* 2000). In contrast, northeast Zamboanga is underlain by ophiolitic slivers with metamorphosed volcanic and sedimentary rocks (Antonio 1972; Pubellier *et al.* 1991). These terranes are overlain by thick Miocene clastic and carbonate rock formations, which are capped by Plio-Pleistocene volcanic rocks. The geology of the three areas where detailed mapping was carried

Table 3. *Whole-rock $^{40}K - ^{40}Ar$ dating of samples from the Zamboanga Peninsula*

Sample no.	Location	Formation	Age (Ma)	K_2O (wt%)	$^{40}Ar^*$ (%)	$^{40}Ar^*/g$ (10^{-7} cm^3)
ZA5-417-09	Mayalal River, Mayalal	Anungan Formation	15.9 ± 0.4	1.495	56.7	7.682
ZAR-2	San Roque, Zamboanga City	Anungan Formation	18.2 ± 0.6	0.65	37.0	3.828
ZA5-411-02	Capisan, Zamboanga City	Anungan Formation	14.1 ± 0.4	0.75	37.6	3.435
ZA1-411-02	Bandera, Zamboanga City	Anungan Formation	17.1 ± 0.5	1.02	42.2	5.665
ZA1-415-01	La Paz, Patalon	Anungan Formation	12.7 ± 0.3	1.364	45.1	5.607
ZA5-410-03	San Roque, Zamboanga City	Anungan Formation	15.4 ± 0.4	2.1	44.4	10.47
Mt Maria	Marangan	Curuan Formation	10.6 ± 0.4	1.64	27.1	5.611
ZAF-422-04	Marangan	Curuan Formation	9.0 ± 0.35	1.71	25.1	4.948
ZA6-415-02	Ayala	Anungan Formation	14.8 ± 0.4	0.72	38.6	3.457
Mt Nancy	Mt Nancy	Anungan Formation	14.3 ± 0.4	1.59	53.7	7.382
ZN 06-103	Sto. Niño, Sindangan	Sindangan Volcanics	10.9 ± 0.3	1.96	56.4	6.897
ZN 06-502	Bonbon, Sindangan	Sindangan Volcanics	13.6 ± 0.3	2.81	64.3	12.34

out is presented here. For simplicity (Figs 3a–c & 4) the areas are designated Northeast (Dapitan–Dipolog–Siayan–Sindangan area), Central (Liloy–Titay–Ipil area) and Southwest Blocks (Zamboanga City and its immediate vicinity).

Northeast Block

The Northeast Block encompasses northeastern Zamboanga (Dipolog–Dapitan area), the northwest portion of the Sindangan–Cotabato–Daguma Lineament which, in the mapped area, is called the Siayan–Sindangan Suture Zone, and a small portion of SW Zamboanga, as represented by Sindangan (Figs 3a & 4). Some portions of the geology of the block have been reported by Jimenez *et al.* (2002). Unpublished reports resulting from the 1999 UP–NIGS field mapping provided additional information.

Northeast of the Siayan–Sindangan Suture Zone, the block has the Polanco Ophiolite Complex for its basement. This NW–SE-trending ophiolite suite, which is recognized for the first time to be a complete crust–mantle sequence, is overlain by thick Miocene sedimentary formations and Plio-Pleistocene volcanic rocks (Fig. 5a). The ultramafic and volcanic rocks generally occur in the northwest and southeast, respectively. The harzburgites that dominate the peridotite sequence are variably serpentinized. It is cut by 5 to 15 cm wide anorthosite, aplite and diabase dykes. Podiform chromitites, as observed in Gunyan where some of the ophiolitic members outcrop as part of the mélange, are associated with the harzburgites and dunites. Wehrlite and gabbro dominate, with troctolite occurring as minor intrusions. These rocks, together with the dyke and volcanic rock complex, are exposed in the Sergio Osmeña–Dapitan road. The sheeted dyke complex is made up of microgabbro, diabase and basalt. The dykes, which vary in width from a few centimetres to half a metre, are subparallel to one another. The volcanic rocks are made up of massive flows and pillow basalts. All of the volcanic and hypabyssal rocks have undergone ocean-floor greenschist metamorphism. Unconformably overlying the Polanco Ophiolite Complex is the Siari Breccia, as exposed in Siari which is situated northeast of Sindangan (Fig. 3a). Based on its stratigraphic relationship with the Late Miocene Motibot Formation, a late Middle Miocene age is assigned to the Siari Breccia. This is an epiclastic (reworked) andesitic breccia containing angular to subangular, pebble to cobble-sized andesite, basalt and minor chert clasts.

The Siayan–Sindangan Suture Zone is a northwest–southeast trending suture zone that

Table 4. *Results of palaeontological analyses of samples from the Zamboanga Peninsula*

Sample no.	Rock Name	Formation	Locality	Analyst	Faunal List	Palaeoenvironment	Age
ZN-08-204	Siltstone	Motibot	Madalum, Sindangan	APA	*Ceratolithus acutus* *Triquetrorhabdulus rugosus*	Marine shelfal	Late Miocene
ZN-11-501	Siltstone	Motibot	North of Sergio Osmeña	APA	*Triquetrorhabdulus serratus* *Triquetrorhabdulus rugosus*	Marine shelfal	Late Miocene
ZN-07-X7B	Siltstone	Motibot	Madalag, Sergio Osmeña	APA	*Triquetrorhabdulus serratus* *Triquetrorhabdulus rugosus* *Amaurolithus amplificus* *Sphenolithus moriformis* *Sphenolithus* sp.	Marine shelfal	Late Miocene
ZN-13-602	Mudstone	Motibot	Sitog, Katipunan	APA	*Amaurolithus primus* *Triquetrorhabdulus serratus*	Marine shelfal	Late Miocene
ZN-09-606	Limestone clast	Gunyan Mélange	Gunyan, Siayan	MMDL	*Cyclicargolithus floridanus* *Sphenolithus moriformis*	Marine shelfal	Late Oligocene to Early Middle Miocene
ZN-09-101	Tuffaceous sandstone	Motibot	Disacan, Manucan	APA	*Discoaster quinqueramus* *Discoaster berggrenii* *Helicosphaera carteri* *Reticulofenestra* sp. *Reticulofenestra pseudoumbilicus* *Sphenolithus abies* *Calcidiscus leptoporus* *Discoaster pentaradiatus* *Ceratolithus* sp.	Marine shelfal	Late Miocene
ZN-09-503	Limestone	Ipil Volcanics	Bualbual, Titay	MGB	*Lepidocyclina sumatrensis* *Miogypsina borhensis* *Cycloclypeus lidae*	Marine shelfal	Early to Middle Miocene
ZS4052K-103	Siltstone	Anungan	Taloptaptap River	AGSF	*Discoaster* spp. *Sphenolithus* spp. *S. heteromorphus* Planktonic foraminifera		Early to Middle Miocene
ZS4132K-230-A1	Sandstone	Anungan	Nanka–Lamisahan	AGSF	*Cyclicargolithus abisectus* (?) *Discoaster* spp. *Sphenolithus* spp. *Coccolithus pelagicus*		Probably Early to Middle Miocene

ZS4132K-230-A2	Siltstone	Anungan (Pico Clastic member)	Nanka–Lamisahan	AGSF	*Helicosphaera euphratis* *Sphenolithus* spp. *Discoaster* sp. *Sphenolithus heteromorphus* *Braarudosphaera* sp. *Pontosphaera* sp. *Helicosphaera carteri* *H. euphratis*	Shelfal	Early to Middle Miocene
ZS4072K-207B	Limestone	Anungan (Manicahan Limestone member)	Quinipot	LPDS	*Miogypsina* sp. *Lepidocyclina* (*Nephrolepedina*) *Amphistegina* sp. *Operculina* sp. (?) Coral fragments Planktonic foraminifera	Inner shelf to lagoonal	Early to Middle Miocene
ZS4212K-PasonancaE	Limestone	Anungan (Manicahan Limestone member)	Pasonanca	LPDS	*Miogypsina* sp. *Lepidocyclina* (*Nephrolepedina*) *Amphistegina* sp. *Operculina* sp.	Inner shelf to lagoonal	Early to Middle Miocene
ZS4122K-229B1	Limestone	Anungan (Manicahan Limestone member)	Manicahan River	LPDS	*Miogypsina* sp. *Lepidocyclina* sp.	Inner shelf to lagoonal	Early to Middle Miocene
ZS4122K-229B2	Limestone	Anungan (Manicahan Limestone member)	Manicahan River	LPDS	*Cycloclypeus* (?) sp. *Amphistegina* sp. *Operculina* sp. (?)		Oligocene to Recent
ZS4152K-302C1	Limestone	Anungan (Manicahan Limestone member)	Campo Dos	LPDS	*Miogypsina* sp. *Lepidocyclina* (*Nephrolepedina*) (?) *Amphistegina* sp. *Heterostegina*(?) sp.	Inner shelf to lagoonal	Early to Middle Miocene
ZA4-411-04	Sandstone-siltstone	Curuan (Pasonanca Clastic member)	Bandera	LPDS	*Miogypsina* sp. *Cycloclypeus* sp. *Lepidocyclina* sp. *Amphistegina radiata* (?) *Operculina* sp. *Lithothamnium* sp.	Shallow marine	Early to Middle Miocene

(continued)

Table 4. *continued*

Sample no.	Rock Name	Formation	Locality	Analyst	Faunal List	Palaeoenvironment	Age
ZA2/3-419-08	Limestone clast	Limpapa Mélange	Mayalal–Limpapa	LPDS	*Cycloclypeus* sp. *Operculina* sp. *Lepidocyclina* sp.	Shallow marine	Not younger than Late Miocene
ZA2/3-419-09	Sandstone clast	Limpapa Mélange	Mayalal–Limpapa	LPDS	*Globoquadrina altispira globosa* (?) *Sphaeroidinellopsis disjuncta Globorotalia mayeri Globigerinoides ruber Hastigerina praesiphonera Globorotalia foshi foshi*		Early Middle Miocene
ZA3-410-5	Limestone	Anungan (Manicahan Limestone member)	Pasonanca watershed	MGB	*Spiroclypeus* sp. *Operculina* sp. Algae	Littoral–neritic	*Spiroclypeus* sp. ranges from Oligocene to Early Miocene
ZA6-411-13A	Bentonitic siltstone	Curuan (Pasonanca Clastic member)	Maasin River	MGB	*Globorotalia acostaensis Orbulina universa Globoquadrina altispira globosa Globigerina venezuelana Amphistegina quoyii Liscopulvilina barthelot Nodosaria radiculai Bolivina* sp. *Siphonina bradyana Sigmoilopsis schlumbergeri*	Littoral–neritic	Late Miocene to probably Early Pliocene
ZA4-422-9	Sandstone	Anungan (Pico Clastic member)	Marangan	MGB	*Orbulina universa* (Mid-Miocene to Recent)	Probably littoral	Undiagnostic

APA, A. P. Alampay; MMDL, M. M. de Leon; LPDS, L. P. de Silva, Jr; AGSF, A. G. S. Fernando; MGB, Palaeontological Unit, Lands Geology Division, Mines and Geosciences Bureau – Central Office.

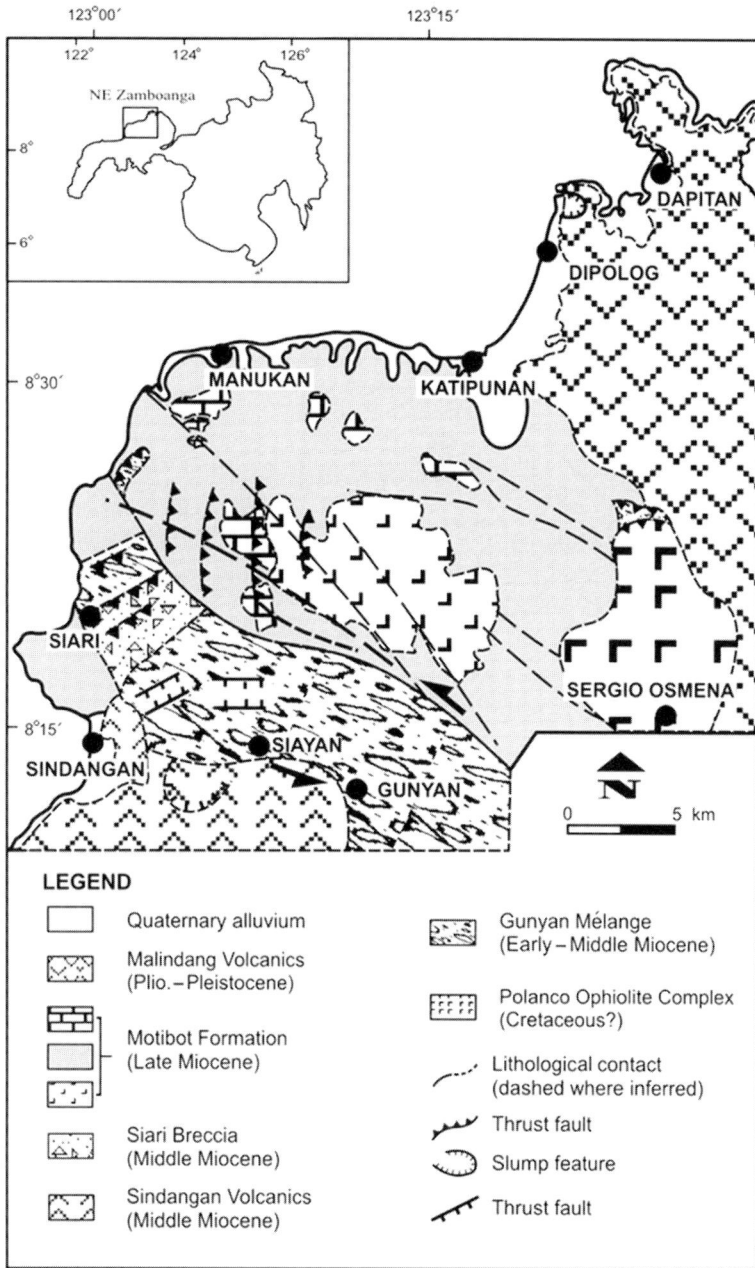

Fig. 3. (a) Geological map of the northeastern Zamboanga block. The northeast block encompasses northeastern Zamboanga (Dipolog–Dapitan area). It is floored by the Polanco Ophiolite Complex, which is overlain by Miocene sedimentary formations and Plio-Pleistocene volcanic rocks. Note the thrust and extensional fault within the Siayan–Sindangan Suture Zone bounded by left-lateral faults and floored by the Gunyan Mélange. (b) Geological map of the Central Zamboanga block. The central block includes the Liloy–Titay–Ipil areas. This block has an ultramafic rock suite associated with metamorphic rocks and a mélange as basement. These are overlain by younger sediments and volcanic rocks. Details are in the text. (c) Geological map of the southwestern Zamboanga block. Zamboanga City and its immediate vicinity make up the southwest block. The Tungauan Schist comprises the basement of this block. The overlying units consist of mélanges, sedimentary sequences and volcanic rocks. See text for details.

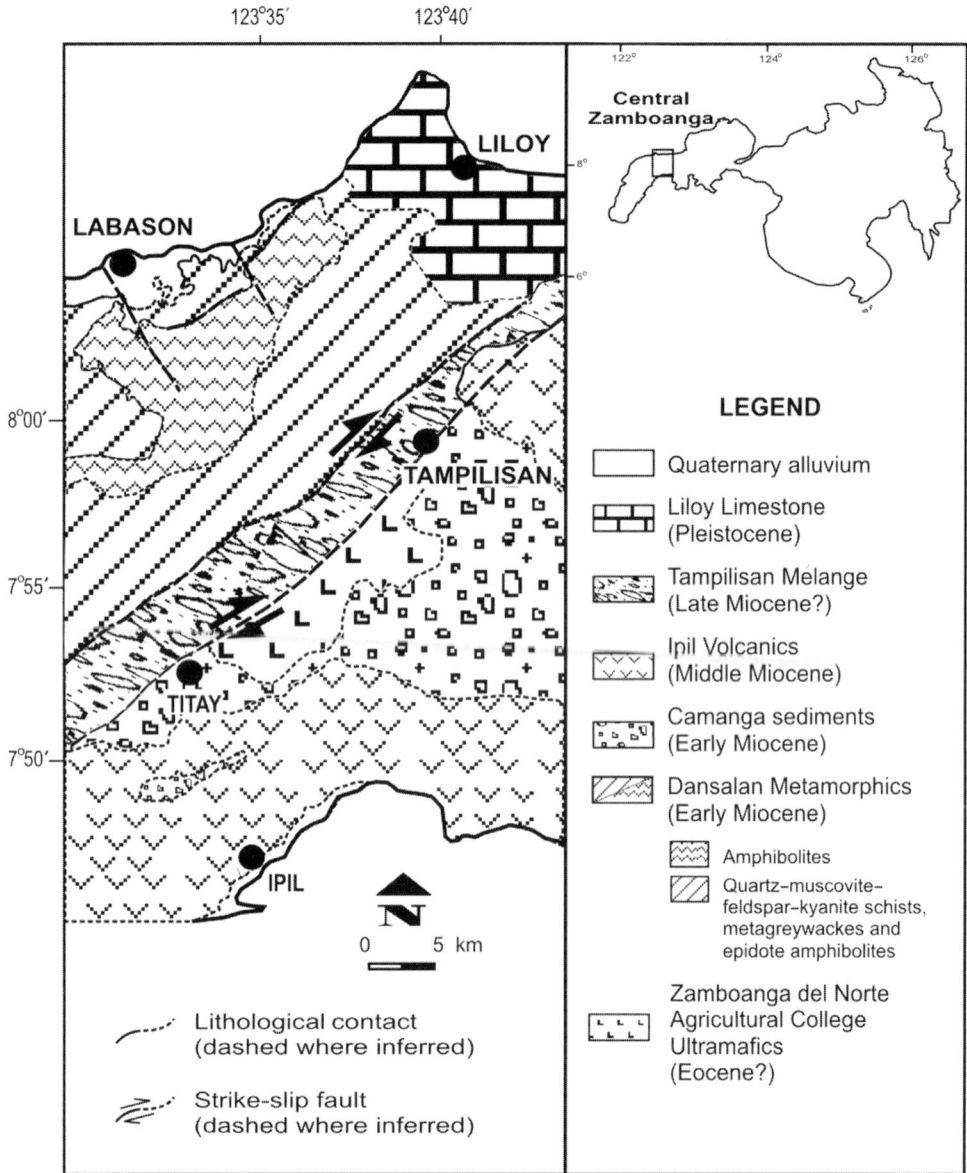

(b)

Fig. 3. (*Continued*)

is bounded by two sinistral faults. This almost 5 kilometre-wide zone is underlain by the Gunyan Mélange, Siari Breccia and the Sindangan Volcanics. The Gunyan Mélange contains both serpentinite- and shale-matrix mélanges. The serpentinite-matrix mélange occurs in a linear belt roughly near the centre of the Siayan–Sindangan Suture Zone, as exposed in Gunyan in the municipality of Siayan. This serpentinite mélange is made up of ophiolite-derived harzburgites, gabbros, basalts and cherts. Chromitite is being mined among the harzburgite clasts, tens to hundreds of metres in width, which are sheared and parallel the northwest-verging thrust faults within the suture zone. The shale-matrix mélange, on the other hand, is found along the periphery of the Siayan–Sindangan Suture Zone. The clasts within the shale-matrix mélange include

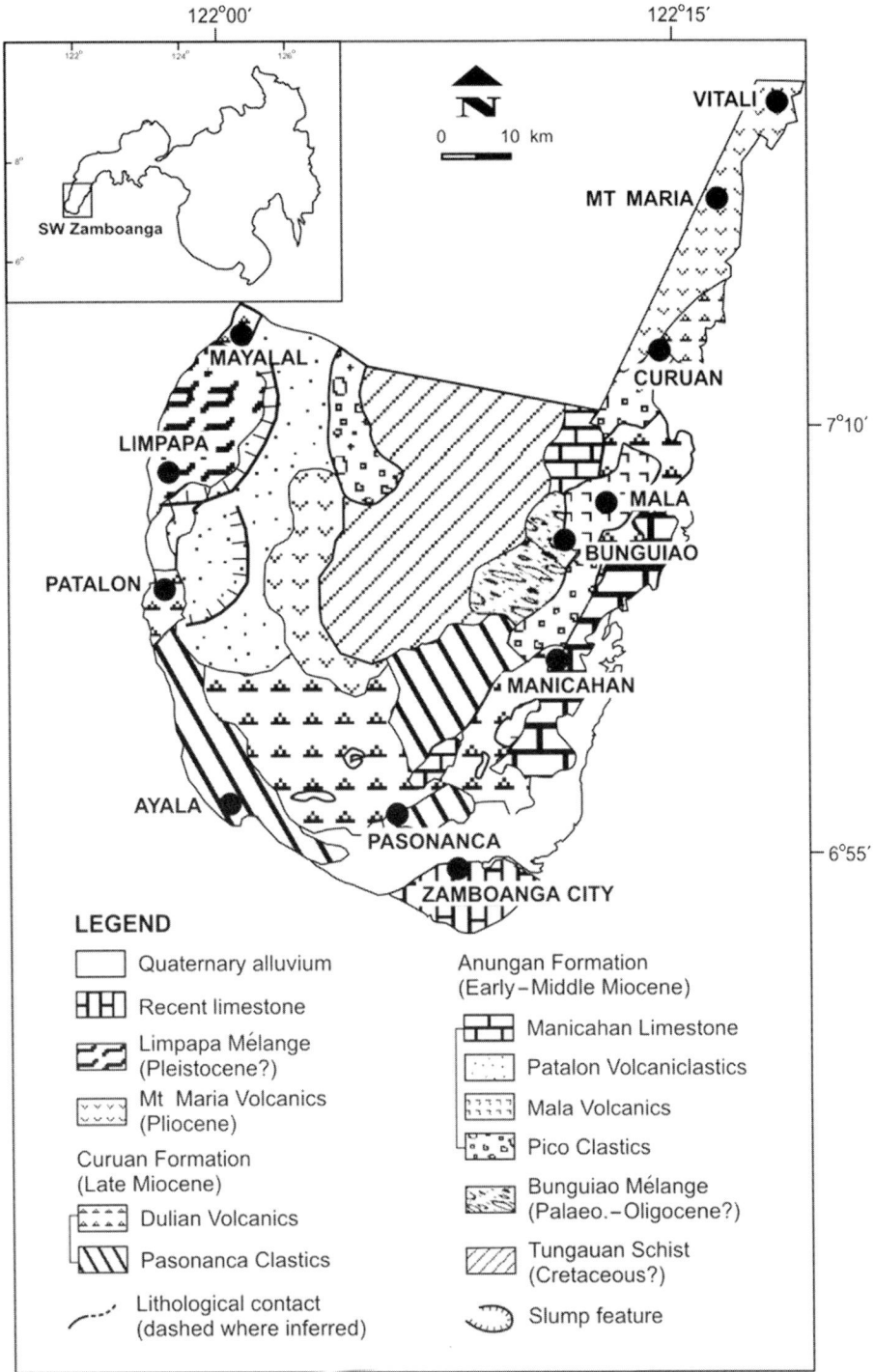

(c)

Fig. 3. *(Continued)*

AGE		East-Central Zamboanga (BMG 1982)	SW Zamboanga (BMG 1982)	SW Zamboanga (this study)	Central Zamboanga (Querubin *et al.* 1999)	NE Zamboanga (this study)
HOLOCENE		Quaternary Deposits				
PLEISTOCENE		Labangan Fm	Sta. Maria Volc. Fm	Limpapa Mélange	Liloy Limestone	Malindang Volcanics
		Zamboanga Volcanics	Tigpalay Conglomerate		Mt Maria Volcanics	
PLIOCENE		Aurora Fm	Pangamuran Volcanic Fm	Mt Maria Volcanics		
		Timonan Fm				
MIOCENE	Late	Nato Fm	Pictoran Fm	Curuan Fm — Vitali Diorite	Curuan Fm — Pasonanca Clastics / Dulian Volc.	Tampilisan Mélange	Motibot Formation
				Soleplep Volcanics		
	Middle	Sibuguey Diorite			Manicahan Limestone	Ipil Volcanics	Siari Breccia / Sindangan Volcanics
		Zamboanga Formation	Anungan Clastic Fm	Vitali Diorite — Anungan Fm — Mala Volc.	Dansalan Metamorphics / Camanga Sediments	Cunyan Mélange
	Early			Patalon Volcaniclastics		
				Pico Clastics		
OLIGOCENE		Sibuguey Fm			Bungiao Mélange	ZNAC Ultramafics
EOCENE		Mangabel Fm				
PALAEOCENE						
CRETACEOUS		Sindangan Volcanics	Tungauan Schist / Ultramafic Rocks	Tungauan Schist		

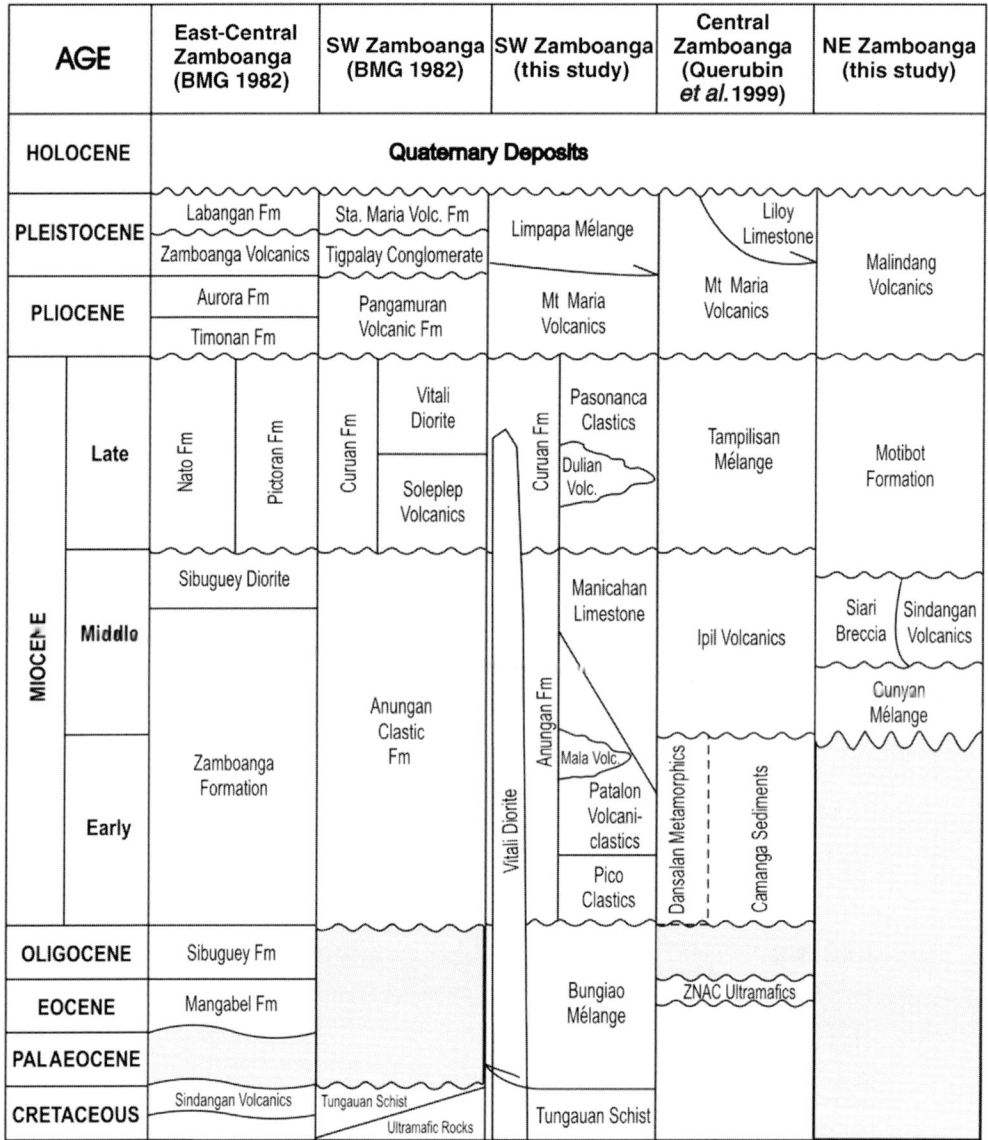

Fig. 4. Stratigraphy of the northeastern, central and southwestern Zamboanga blocks based on the results of the four-year (1998–2001) field mapping in the Zamboanga Peninsula. This is compared with the stratigraphy proposed by the BMG (1982) for the east-central and southwest Zamboanga areas. The southwest Zamboanga, which is believed to have continental affinity, is made up of a basement of Pre-Cenozoic granitic rocks, quartz–sericite–albite schists, gneiss and amphibolites. In contrast, northeast Zamboanga, which consists of oceanic and arc-related lithologies, is underlain by ophiolitic slivers with metamorphosed volcanic and sedimentary rocks.

sandstone, andesite and metamorphic rocks. Some of the serpentinite-matrix mélange are also found as blocks within the shale-matrix mélange. A limestone clast (Sample ZN-09-606; Table 4) gave a Late Oligocene to early Middle Miocene age, thus, constraining the age of the Gunyan Mélange to post-early Middle

Miocene. This is consistent with the Middle Miocene dating of the mélange matrix by Pubellier *et al.* (1991). Two contemporaneous late Middle Miocene formations overlie the Gunyan Mélange: the Siari Breccia, which also overlies the Polanco Ophiolite Complex, and the Sindangan Volcanics. Thick basaltic and andesitic flows

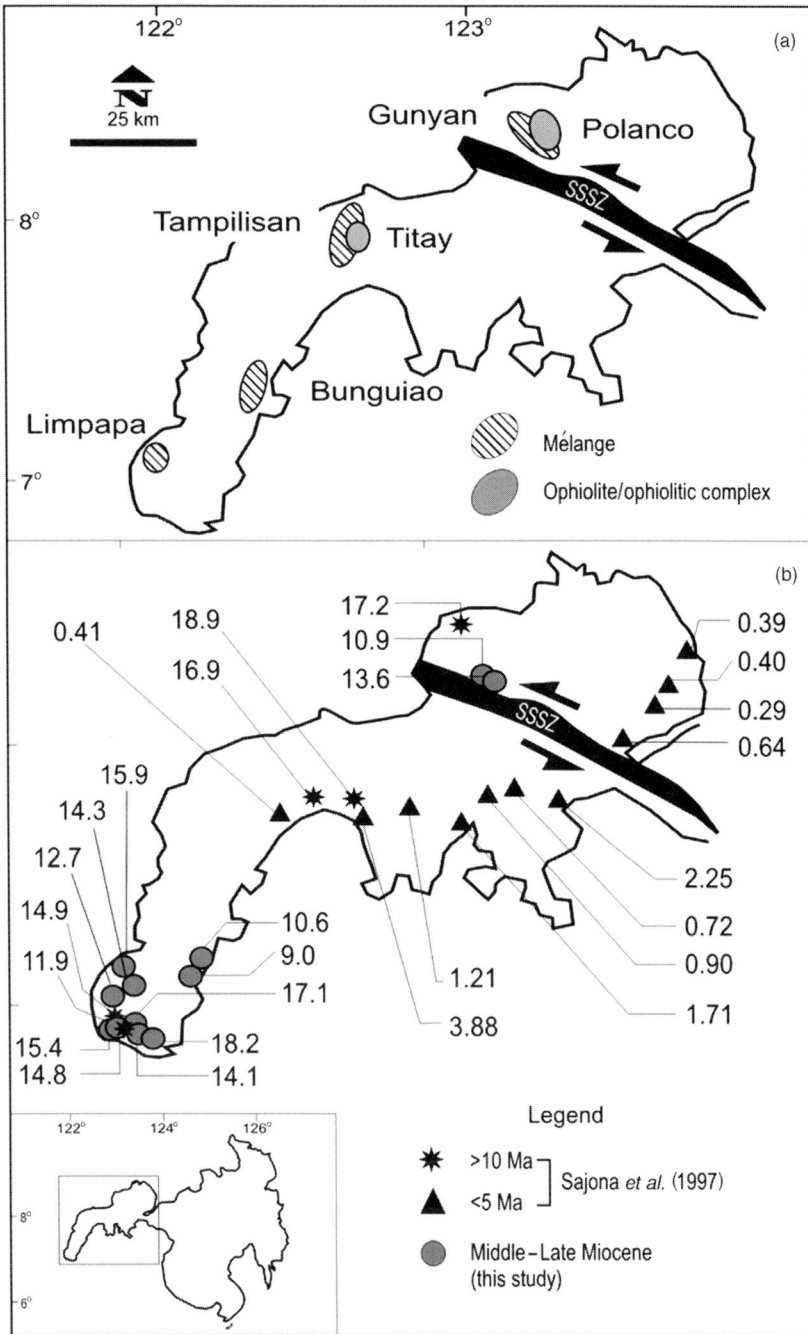

Fig. 5. (**a**) The four-year field investigations conducted in the Zamboanga Peninsula led to the recognition of complete ophiolite suites and several mélange units. SSSZ, Siayan–Sindangan Suture Zone. See text for discussion. (**b**) Map showing the $^{40}K-^{40}Ar$ ages of samples collected from the Zamboanga Peninsula adopted from Sajona *et al.* 1997 The map also shows the new whole-rock ages determined in this study using basalt and andesite samples collected from the Sindangan Volcanics, Anungan and Curuan Formations. The ages range from Middle to Late Miocene, and suggest the existence of an Early Miocene(?) Sulu Trench. The new dates also confirm the presence of a Miocene arc in the Zamboanga Peninsula. SSSZ, Siayan–Sindangan Suture Zone.

that grade upsequence into alternating tuffaceous sediments, porphyritic basalt and andesite flows characterize the Sindangan Volcanics. The basalt–andesite flows of the Sindangan Volcanics extensively cover a wide area southwest of the Siayan–Sindangan Suture Zone. The Motibot Formation conformably overlies the Siari Breccia. This formation, which is dated as Late Miocene and was deposited on a marine shelf, covers part of the Siayan–Sindangan Suture Zone, and is also exposed on both sides of the suture zone (Fig. 3a ZN-08-204; Table 4). The Motibot Formation, which marks the onset of a common history for the northeastern and southwestern Zamboanga blocks, has three members: a lower limestone member, a middle basaltic flow to tuffaceous sedimentary unit and a clastic sequence (conglomerate, limestone and tuffaceous sediments) (Fig. 3a & 4). Chert interbeds are also noted. The whole sequence generally dips to the northeast. Unconformably overlying the Motibot Formation are the Plio-Pleistocene Malindang Volcanics, which are made up of thick, massive lava flows, volcaniclastic sediments and tuff deposits. Good exposures of the tuff deposits can be observed along the coast from Dipolog to Dapitan. Near Dapitan, the tuff deposits are particularly coarse and grade into massive volcaniclastic deposits with poorly sorted, boulder-sized andesite and basalt blocks. Andesitic to basaltic lava flows with monomictic agglomerates belonging to this formation are hydrothermally altered and associated with gold mineralization (Jimenez *et al.* 2002). In terms of geological structures, the two northwest–southeast-trending sinistral faults that bound the Siayan–Sindangan Suture Zone represent the northwestern extensions of the Sindangan–Cotabato–Daguma Lineament. A number of lineaments parallels the two northwest–southeast sinistral faults and appears to be related to fault splays that branch northwards. Aside from the strike-slip faults identified within the Siayan–Sindangan Suture Zone, a number of parallel thrust faults cut across the suture zone and are best observed within the Gunyan Mélange area.

Central Block

This block encompasses the area from Liloy in the north through Titay all the way to Ipil to the south. Some portions of the geology of the Central Block have been reported by Tamayo *et al.* (2000) and Querubin & Yumul (2001). The geology has been described in Querubin *et al.* (1999). An ultramafic rock suite associated with metamorphic rocks and a mélange serve as basement and are covered by younger sediments and volcanic rocks (Figs 3 & 4). Quartz–sericite to quartz–chlorite–sericite schists, mapped and reported as part of the regionally metamorphosed Cretaceous Tungauan Schist (Santos-Yñigo 1953), are exposed in the area. These metamorphic rocks are grouped as the Dansalan Metamorphics in this study. Tamayo *et al.* (2000) also reported the presence of metagreywackes, epidote-bearing amphibolites and quartz–mica–feldspar–kyanite schists. The $^{40}K–^{40}Ar$ analysis of amphibole separates from amphibolites exposed near Mount Dansalan in Titay gave an Early Miocene (24.6 ± 1.4 Ma to 21.2 ± 1.2 Ma) age of metamorphism. The precursors of the amphibolites, based on relict textures, are isotropic to layered gabbros. The amphibolites are interpreted to represent a pile of arc-related gabbroic cumulate rocks (Tamayo *et al.* 2000). Fractures and joints commonly observed in the metamorphic rocks are generally oriented along a N060° trend. Harzburgites, dunites and chromitites are found in the vicinity of the Zamboanga del Norte Agricultural College (ZNAC) in Titay. These are considered part of the ZNAC Ultramafics, and are correlated with Antonio's (1972) Mindanao Ultramafics. Dismembered parts of an ophiolite (gabbros, dyke swarms, basalt flows and pillows with capping red and green chert) are also exposed along a stretch of the Liloy–Ipil Road. Field relationships show that the ZNAC Ultramafics are thrust over the Dansalan Metamorphics. The Tampilisan Mélange, which contains both clay-matrix and serpentinite-matrix mélange units, is included in the ZNAC Ultramafics. The Tampilisan Mélange outcrops in a NE–SW-trending zone characterized by intensely sheared, rounded to subrounded boulder to cobble-sized quartz, silicified limonite and schist clasts embedded in a reddish brown, clay-rich matrix (Fig. 5). In other parts of the mélange, intensely sheared to crushed harzburgites, dunites and pyroxenites contained in a serpentinite matrix outcrop as thrusted elongate bodies or in erosional windows in younger formations. An alternating sandstone and siltstone sequence belonging to the Early Miocene Camanga Sediments unconformably overlies the ZNAC Ultramafics (Fig. 4). The indurated sandstones are well sorted and grade upward to fine-grained siltstones. This clastic unit, in turn, is unconformably overlain by massive, grey, fine to coarsely crystalline limestone found within the Ipil Volcanics. Limestone outcrops of this unit occur as northeast–southwest-trending erosional remnants fringing the south to southeastern margin of the Liloy–Ipil area. A sample from this limestone deposited in a marine shelf setting was dated as Early to Middle Miocene (Sample ZN-09-503; Table 4).

The Ipil Volcanics were previously designated as either part of the Mio-Pliocene Andesite–Basalt Series (Santos-Yñigo 1953) or Pleistocene Zamboanga Volcanics (Antonio 1972). These volcanic rocks are light to dark grey, massive, fine-grained to porphyritic andesitic flows and breccias associated with medium to coarse-grained tuffaceous sandstones and tuffs. Their outcrops fringe the southern margin of the area extending from Ipil to Titay, as well as areas in between the boundary of Liloy and Tampilisan. The Ipil Volcanics partly cap the Camanga Sediments. The Plio-Pleistocene Mount Maria Volcanics, which include northeasterly-trending andesitic to basaltic plugs and pyroclastic flow deposits, are overlain by the Liloy Limestone – the youngest formation in central Zamboanga. Within the Central Block, the most prominent structure is the ten to fifteen kilometre wide, northeast–southwest-trending Titay Shear Zone (Fig. 3b). This zone extends from the southeast limit of the Mount Dansalan gabbros and amphibolites in the NW, to the ZNAC Ultramafics in the SE. The rocks within the Titay Shear Zone are intensely fractured to crushed, with shear directions usually trending N060°. Shear-sense indicators (S-surfaces and slickensides) suggest right-lateral motion. A consequence of the movement along the Titay Shear Zone is the formation of the Tampilisan Mélange (Fig. 5a). Another important feature found in the area is that most of the thrust faults are north-northwestward directed, and the ophiolitic materials are almost always thrusted on top of the Dansalan Metamorphics and Tungauan Schist, at least in the central to southern portions of the Zamboanga Peninsula, respectively.

Southwest Block

This block outcrops in and around Zamboanga City (Fig. 3c). Our mapping (2000–2001) identified seven formations: the Pre-Cenozoic Tungauan Schist, Pre-Miocene Bungiao Mélange, Early to Middle Miocene Anungan Formation, Late Miocene Curuan Formation, Late Miocene Vitali Diorite, Pliocene to Pleistocene Mount Maria Volcanics and Pleistocene Limpapa Mélange, which are all capped by Quaternary alluvium (Fig. 4). The Tungauan Schist, as recognized before (Santos-Yñigo 1953), represents the basement of southwestern Zamboanga and includes mica schists, phyllites, talc–chlorite mica schists and marbles. Granule to pebble-sized quartz sweats are common in the schists. Hydrothermally altered schists, in which gold panning is reported, and reddish to brown ferruginous schists are also present.

The Bungiao Mélange, which is best exposed in Bungiao, is thrust on to the Tungauan Schist (Fig. 5a). The Bungiao Mélange is a tectonic mélange characterized by cobble to hill-sized (around 100 m) clasts of slate, phyllite, low-grade schists, metasedimentary rocks, metavolcanics, marble, andesites and harzburgites set in a serpentinite matrix. The exposures are commonly cut by thrust faults, resulting in intense shearing. Harzburgite phacoids are also common. Unconformably overlying the Bungiao Mélange is the Anungan Formation. It is divided into four members: the Pico Clastics, Mala Volcanics, Patalon Volcaniclastics and Manicahan Limestone. The Pico Clastic member, dated as Early to Middle Miocene based on calcareous nannofossils, grades upward from breccia and conglomerate to quartz-rich sandstone, siltstone and shale (Fig. 4; Sample ZS4132K-230-A2; Table 4). This member of the Anungan Formation unconformably overlies the Bungiao Mélange as well as the Tungauan Schist. The breccias and conglomerates are poorly sorted and clast-supported. Clasts include pebble- to cobble-sized schists, quartz and andesites, with minor ultramafic rocks. Quartz-rich sandstones are interbedded with limestones, siltstones and claystones. Bentonite and zeolite deposits belong to this member of the Anungan Formation. These are interbedded with tuffaceous to arkosic sandstones, siltstones and, rarely, limestones. The Mala Volcanic member, on the other hand, is made up of andesitic pyroclastic flow and airfall deposits. In Mala, dacitic and andesitic pyroclastic flow deposits are found. The dacitic pyroclastic flow deposit contains clasts of chlorite schist, gabbro, aphanitic to porphyritic andesite and siltstone. The andesitic pyroclastic flows occur as clast- to matrix-supported with poorly to moderately well-sorted, granule to boulder-sized andesite, quartz and amphibole clasts. Unconformably overlying the Mala Volcanic member is the Patalon Volcaniclastic member. This member comprises interbedded lava and pyroclastic flow deposits, along with minor sandstones. The latter represents normal river deposits, lahar deposits or airfall tephra. In its type locality, each unit is <5 m thick. The andesitic composition of the units of the Patalon Volcaniclastic member distinguishes the formation from the underlying Mala Volcanic member.

The Manicahan Limestone member is interfingered with the Pico Clastic and Mala Volcanic members. These limestones, which are crystalline and fossiliferous, were dated as Early to Middle Miocene (Sample ZS4122K-229B1; Table 4). The limestones are generally grey, bedded, and, in some outcrops, crystalline to coralline. Some limestones are interbedded with arenites.

The Late Miocene Curuan Formation unconformably overlies the Anungan Formation

(Samples ZA4-411-04; ZA6-411-13A; Table 4). The Curuan Formation has two interfingered members, the Pasonanca Clastics and the Dulian Volcanics. The Pasonanca Clastic member comprises sandstones, siltstones, claystones and conglomerates. The sandstones range from volcanic lithic through tuffaceous and arkosic to calcareous. Unconformably overlying the conglomerate unit of this member is the Pico Clastic member sedimentary breccia. In other localities, fine-grained sandstone and siltstone with conglomerate lenses of the Pasonanca Clastic member unconformably overlie the Manicahan Limestone member of the Anungan Formation. The Dulian Volcanic member is composed of tuff, andesitic lava and limestone clast-bearing pyroclastic flows which differentiate this unit from the older Mala Volcanic member. Debris-flow deposits with pebble- to cobble-sized andesite to limestone clasts and monomictic hyperconcentrated pyroclastic flow deposits with porphyritic andesite blocks are common in the Dulian Volcanic member. The Plio-Pleistocene Mount Maria Volcanics cap the older sequence. As noted in the area, this unit includes andesitic to dacitic flows, plugs and tuffs similar to those exposed in the Central Block. Finally, the Limpapa Mélange, which is found in the northwest of the block, has a hummocky surface and is cut by a number of NE-trending normal and reverse faults (Figs 3c and 5a). The clasts include andesites, pyroclastic flows, limestones, sandstones–siltstones and gravel deposits, embedded in an intensely fractured limestone and calcareous sandstone matrix. The clasts sampled from the mélange (samples which include limestones and sandstones) gave Early Middle Miocene to Late Miocene ages (Samples ZA2/3-419-08, ZA2/3-419-09; Table 4). This suggests an uppermost Miocene to Pliocene age for the mélange unit. In addition, the Vitali Diorite is observed to be intruded into the Pico Clastic member and the Bungiao Mélange. This plutonic body also includes andesite porphyry, monzonite and microdiorite. This unit has been assigned a Late Miocene age (BMG 1982).

Discussion

Siayan–Sindangan Suture Zone: from subduction to strike-slip collisional boundary setting

Mindanao was previously divided into the eastern Mindanao–Halmahera Block (also referred to as the Pacific Cordillera) and the western Mindanao–Sangihe Block (also known as the Central Cordillera) (e.g. Quebral *et al.* 1996).

These two blocks were modelled to have been juxtaposed along a collision zone marked by the Agusan–Davao Trough (e.g. Hamilton 1979; Hawkins *et al.* 1985) (Fig. 2). Reactivation of this suture zone is believed to have formed the Philippine Fault Zone in this part of the Philippine archipelago. This model was drawn from focal mechanism solutions showing the presence of a doubly plunging subduction zone in the Molucca Sea (Cardwell *et al.* 1980), with the collision zone closing in a scissor-type manner. However, other workers pointed out that the basements of the Pacific and Central Cordilleras in Mindanao are the same (e.g. Mitchell *et al.* 1986; Pubellier *et al.* 1993; Quebral *et al.* 1996). They also questioned the presence of a collision zone in the Agusan–Davao Trough. In addition, structural and geomorphological studies support the existence of a collision zone along the Sindangan–Cotabato–Daguma area (e.g. Pubellier *et al.* 1991) (Fig. 2).

The Sindangan–Cotabato–Daguma Lineament, which is interpreted as a soft-collision boundary, is a NW–SE trending strike-slip fault that separates the island-arc-related eastern-central Mindanao and the Zamboanga Peninsula with continental affinity (Fig. 2). We mapped the northwestern portion of the Sindangan–Cotabato–Daguma Lineament, which for simplicity we have called the Siayan–Sindangan Suture Zone. The Gunyan Mélange floors the suture zone, and contains mainly ophiolitic clasts. This feature led us to conclude that the collisional boundary, as exemplified by the Siayan–Sindangan Suture Zone, could have started as a subduction boundary. After collision of the eastern-central Mindanao and the Zamboanga Peninsula, the convergent boundary was converted into today's strike-slip fault system. Available GPS information shows that the Sindangan–Cotabato–Daguma Lineament is still active. On the basis of the Late Oligocene–early Middle Miocene limestone clast in the Siayan–Sindangan Suture Zone basement Gunyan Mélange, the age of the Siayan–Sindangan Suture Zone is likely to be Middle Miocene, and thus older than that previously reported by Pubellier *et al.* (1991). An additional argument for this age of the Siayan–Sindangan Suture Zone is the Late Miocene Motibot Formation, which extensively covers the northwestern part of the Siayan–Sindangan Suture Zone and its surrounding areas.

Southwest Zamboanga volcanic rocks: evidence for an Early Miocene Sulu Trench?

Rangin (1989) modelled the opening of the Sulu Sea Basin through the separation of the Early to

Middle Miocene Cagayan de Sulu Ridge and the proto-Zamboanga arc (Fig. 2). The opening was attributed to back-arc spreading related to the subduction along a north-northwest-dipping proto-Cotabato Trench. This model, however, was discarded for two reasons. First, Rangin & Silver (1991), based on geophysical evidence, related to ODP Leg 124, found no evidence for the existence of a trench in the supposed location of the proto-Cotabato Trench. Second, no well-developed Early to Middle Miocene magmatic arc, that might be considered as the counterpart of the Cagayan de Sulu Ridge remnant arc, was recognized in the Zamboanga Peninsula. This led to the conclusion that the opening of the Sulu Sea basin was due to the southward subduc-tion of the proto-South China Sea beneath the Cagayan de Sulu Ridge. Sajona et al. (1997) pub-lished whole-rock ^{40}K-^{40}Ar ages for several vol-canic centres in the Zamboanga Peninsula, and came up with two age population sets. Five of their arc samples gave an age range of 11.9 Ma to 18.9 Ma, whereas ten samples gave an age range of 0.29 Ma to 3.88 Ma (Fig. 5b). These authors attributed the older volcanic arcs to an Early Miocene Sulu Trench. The younger set of dated arc rocks was attributed to the present-day Sulu Trench, which was reactivated during the Pliocene. The termination of the Early Miocene Sulu Trench, which was attributed to the collision of Palawan–Mindoro with the Phi-lippine Mobile Belt (e.g. Bellon & Rangin 1991).

The 12 samples (two from the Sindangan Vol-canics, eight from the Anungan Formation and two from the Curuan Formation samples) dated in this study define a whole-rock ^{40}K–^{40}Ar age range of Middle to Late Miocene. The Middle to Late Miocene arc may be related to an Early Miocene(?) Sulu Trench along which the Oligocene–Miocene Sulu Sea crust may have subducted. Our new isotopic dates also confirm the presence of a Miocene arc in the Zamboanga Peninsula, which was earlier postulated by Ran-gin (1989). Nonetheless, the Middle to Late Miocene volcanic arc in the Zamboanga Penin-sula may not necessarily be related to an Early Miocene(?) Sulu Trench. Hall (2002), in his reconstruction, favoured the existence of a Middle Miocene SE-directed subduction of the Sulu Sea, which was followed by a Late Miocene, NW-directed subduction of the Celebes Sea. He noted that this is consistent with the geology of Sabah and the deformations observed in the Makassar Strait. Although the age pro-posed by Hall (2002) is not consistent with an Early Miocene(?) Sulu Trench, this study drew conclusions in favour of the existence of a north-west-dipping subduction zone consuming the

Celebes Sea. More work needs to be done to further define and refine the possible relationship between the Middle to Late Miocene Zamboanga Peninsula volcanic arc and the Early Miocene(?) Sulu Trench.

Geological history of the Zamboanga Peninsula: a snapshot of the evolution of Sundaland

The Early Miocene times saw several important changes in the Philippines similar to what has been observed elsewhere in SE Asia (e.g. Polve et al. 1997; Maury et al. 1998). Arc polarity reversal in northern Luzon from the west-dipping proto-Philippine Trench to the Manila Trench occurred during this period (Bellon & Yumul 2000), as a result of the Early Miocene collision of the Palawan microcontinental block with the Philippine Mobile Belt (Yumul et al. 2001). Closer to Mindanao, the proto-Sulu Trench is believed to have been active during the Early Miocene. The subduction zone that was initially responsible for the Middle Miocene soft-collisional boundary between the Zamboanga Peninsula and the east-central Mindanao could have also been initiated during the Early Miocene. The Zamboanga Peninsula, together with the Cotabato–Daguma area, is considered to be part of Sundaland (Fig. 1) (e.g. Rangin et al. 1999a, b; Yumul et al. 2001). The basement of the Peninsula is made up of the regionally metamorphosed Tungauan Schist. Tamayo et al. (2000) gave geochemical analyses of the meta-morphic rocks, confirming the continental affinity of the Zamboanga Peninsula. The conti-nental affinity of the Peninsula was recognized on the basis of regional correlations with Palawan and Mindoro (Faure et al. 1989). Contrary to pre-vious reports, the prominent thrusting directions recognized in this area are north directed. Ophio-litic materials are thrusted on to the metamorphic rocks and not the other way around (Fig. 5a). Vol-canic activities also characterize the whole his-tory of the Zamboanga Peninsula. Turbiditic deposits with interbedded cherts, as observed in the Late Miocene Motibot Formation, are also found in the NE. Calcareous (e.g. the Pasonanca Clastic member of the Late Miocene Curuan For-mation) to debris flow-hyperconcentrated lahar deposits (e.g. the Dulian Volcanic member of the Curuan Formation) arc mostly found in the south. At a first order of approximation, there appears to be a shallowing of the depositional environment from the northeast (open-marine to shelf in the Dipolog–Manukan area) to the south-west (neritic to lagoonal in Zamboanga City) during the Late Miocene (Fig. 6). The southwest block of the Peninsula evolved in a shallow-marine environment, as shown by the extensive bentonite and zeolite deposits of the Early to Middle Miocene Anungan Formation. The Middle to Late Miocene and Plio-Pleistocene periods also saw the formation of base and precious-metal mineralizations in the area (e.g. Jimenez *et al.* 2002).

The Polanco Ophiolite Complex, a supra-sub-duction zone ophiolite, floors the area northeast of the Siayan–Sindangan Suture Zone. The tectonized harzburgites from the Polanco ophiolite contains olivines and spinels with X_{Mg} 0.909 to 0.918 and X_{Cr} 0.61 to 0.644, respectively (Tamayo 2001). These values are slightly higher than those observed in mantle peridotites recovered from modern mid-oceanic ridge settings (Dick 1989; Niu & Hékinian 1997). Instead, they are similar to those exhibited by peridotites originating from modern suprasubduction zone environments (Ishii *et al.* 1992; Parkinson & Pearce 1998). Results of calculations by Tamayo (2001) on the Polanco Ophiolite Complex peridotites suggest that they underwent a >20% degree of partial melting, similar to the upper limit of modern mid-ocean ridge peridotites and within the range of suprasubduction zone mantle rocks (Fig. 7A). Volcanic rocks collected

from the sheeted dyke complex of the Polanco Ophiolite Complex display multi-element spectra similar to transitional mid-ocean ridge basalt (MORB)–island arc tholeiites (IAT) (samples PH98-38, PH98-39, PH98-40 and PH98-42; Fig. 7B). A sample (PH98-37) from massive lava flow deposits overlying the sheeted dyke complex also shows a transitional MORB–IAT-like multi-element pattern (Fig. 7B). This feature suggests a subduction-related environment of formation for the sheeted dyke and lava sequences of the Polanco Ophiolite Complex, and is consistent with the geochemical signatures recorded by the spatially associated mantle rocks.

The evolution of the Polanco Ophiolite Complex is intimately related to the evolution of the Siayan–Sindangan Suture Zone. The Siayan–Sindangan Suture Zone is floored by the Gunyan Mélange, which is characterized by ophiolitic clasts that were derived from the Polanco Ophiolite Complex. The presence of the ophiolitic materials and the associated mélange presumably indicate a closed marginal basin. The Zamboanga Peninsula and eastern-central Mindanao, which had collided by the Middle Miocene, had by then followed a common history by the Late Miocene.

One remaining question is whether there is now enough field evidence to conclude that the

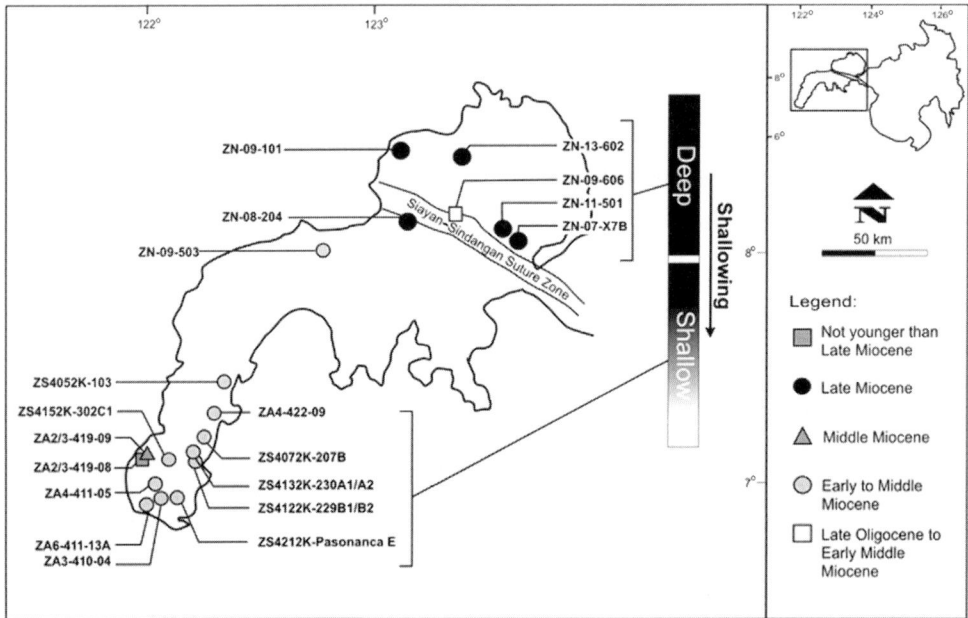

Fig. 6. Map showing the location of samples submitted for palaeontological analyses. The results suggest a general shallowing of the depositional environment from the northeast (open marine to shelf) to the southwest (neritic to lagoonal setting). See text for discussion.

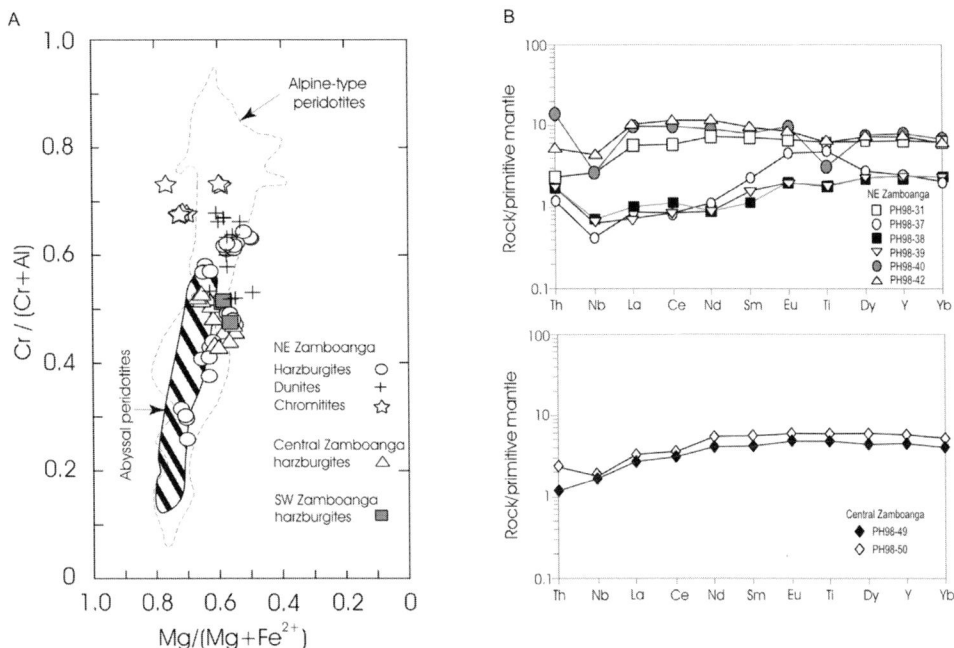

Fig. 7. (**a**) Plot of the X_{Cr} and X_{Mg} of spinels from peridotites and chromitites sampled in the mantle sequences of the Zamboanga ophiolites. Also plotted for comparison are the fields of alpine-type peridotites (Dick & Bullen 1984) and abyssal peridotites from modern mid-oceanic ridge systems (Dick 1989; Niu & Hékinian 1997). Although most of the peridotites show X_{Cr} and X_{Mg} plots similar to those observed in modern abyssal peridotites, several samples from northeast Zamboanga display X_{Cr} values slightly higher than the upper limit of most modern abyssal peridotites. In addition, spinels from the chromitites of northeast Zamboanga exhibit relatively high X_{Cr} values as well. (**b**) Multi-element patterns normalized to the primitive mantle values of Sun & McDonough (1989) of volcanic rocks from ophiolites from the Zamboanga Peninsula.

Zamboanga Peninsula was part of the Palawan microcontinental block and, by extension, if it also originated from China. The geochemical work of Tamayo *et al.* (2000) showed the continental affinity of the Zamboanga Peninsula. This is further supported by the presence of quartz-rich to arenaceous sandstones in the Pico Clastic member of the Anungan Formation. The basement complexes of both Palawan and Zamboanga are characterized by metamorphic and ophiolitic suites. Palawan is dominated by limestones, while the Zamboanga Peninsula is mostly an assemblage of clastic, non-clastic and volcaniclastic sedimentary rocks. A possible explanation for this difference is that Palawan and Zamboanga Peninsula represents an oceanward evolution of a passive margin, with the limestones occupying the shelf and the clastic rocks deposited in deeper waters. However, it is also possible that, during the Miocene, Palawan and Zamboanga followed completely different geological evolutionary paths. In summary, the available data do not negate the possibility that the

Zamboanga Peninsula is the southernmost part of a rifted segment of the southeastern continental margin of China. Such a conclusion would constrain the evolution of the rifted continental margin of southeastern China through space and time.

Conclusions

The island of Mindanao in the southern Philippines is made up of two major blocks: eastern-central Mindanao, which has island arc affinity, and the Zamboanga Peninsula which displays continental affinity. Field geological mapping has shown that the boundary of these two blocks is characterized by a collision zone: the Sindangan–Cotabato–Daguma Lineament, that evolved from a subduction zone during the Early to Middle Miocene to an active strike-slip fault. The northwestern part of this soft collision zone, the Siayan–Sindangan Suture Zone, is underlain by the Gunyan Mélange associated with the Polanco Ophiolite Complex. Collision occurred during the Middle Miocene, with the two blocks following a common geological history by the Late Miocene with the deposition of the capping Late Miocene Motibot Formation. Our set of whole-rock $^{40}K-^{40}Ar$ ages for volcanic rocks in the Zamboanga Peninsula is consistent with the existence of an Early Miocene(?) proto-Sulu Trench. Recent available geochemical data also confirm the continental affinity of the Zamboanga Peninsula. The Zamboanga Peninsula could possibly be a part of the Palawan microcontinental block, thus representing the southernmost part of the rifted southeastern continental margin of China.

This paper was written in fond memory of the late Dr Luis Santos-Yñigo, an excellent field geologist who did pioneering work in the Zamboanga Peninsula. Financial and logistic support came from the Department of Science and Technology Grants-In-Aid, University of the Philippines (UP), DOST-Philippine Council for Industry and Energy Research and Development, DOST Regional Office 9 and UP-National Institute of Geological Sciences (Research Grant on Philippine Ophiolites). Our appreciation goes to the UP–NIGS students (1999, 2000, 2001 Geology 170 and 215), who helped us to map and understand the geology of the Peninsula. Palaeontological analyses by M. Moraleda-De Leon, A. Peleo-Alampay, L. De Silva, Jr, A. Fernando and the palaeontologists of the Mines and Geosciences Bureau – Petrolab are acknowledged with gratitude. Our thanks go to: A. Cabantog, B. Nazareth, F. Huot, D. Faustino, J. De Jesus, E. Marquez, E. Andal, A. Digdigan, F. Jimenez, F. Olaguera, J. Hernandez, J. Lucero and to all of our other NIGS and MGB-9 colleagues for their field and laboratory assistance. Support from the French Embassy in the Philippines, University of Bretagne Occidentale and the University of Paul Sabatier for scholarship, travel and laboratory analyses under the RP-France Program on the Geosciences: Ophiolite Component and Adakite–Gold Mineralization Program are also acknowledged. GPY thanks J. Malpas and M. F. Zhou for the invitation to participate in the University of Hong Kong DES Croucher Advanced Institute Workshop Meeting. Constructive comments and reviews by S. Suzuki, M. Pubellier and J. Ali greatly improved the paper, and are acknowledged with thanks.

References

ANTONIO, L. R. 1972. *Geology and Mineral Resources of East-central Zamboanga Peninsula, Mindanao, Philippines*. Philippine Bureau of Mines Technical Report, 87 pp.

ARAI, S. 1992. Chemistry of chromian spinel in volcanic rocks as a potential guide to magma chemistry. *Mineralogical Magazine*, **56**, 173–184.

ARAI, S. & MATSUKAGE, K. 1998. Petrology of a chromitite micropod from Hess Deep, equatorial Pacific: a comparison between abyssal and alpine-type podiform chromitites. *Lithos*, **43**, 1–14.

AURELIO, M. A. 2000. Shear partitioning in an island arc setting: constraints from Philippine Fault and recent GPS data. *The Island Arc*, **9**, 585–598.

BELLON, H. & RANGIN, C. 1991. Geochemistry and isotopic dating of Cenozoic volcanic arc sequences around the Celebes and Sulu Seas. *In*: SILVER, E., RANGIN, C. *et al.* (eds) *Proceedings of the Ocean Drilling Program, Scientific Results*, **124**, 321–338.

BELLON, H. & YUMUL, G. P. JR. 2000. Mio-Pliocene magmatism in the Baguio Mining District (Philippines): clues to its mineralization and geodynamic setting. *Comptes Rendus de l' Académie des Sciences de Paris, Science de la Terre et des Planets*, **331**, 1–8.

BORTOLOTTI, V., MARRONI, M., NICOLAE, I., PANDOLFI, L., PRINCIPI, G. & SACCANI, E. 2002. Geodynamic implications of Jurassic ophiolites associated with island-arc volcanics, south Apuseni Mountains, western Romania. *International Geology Review*, **44**, 938–955.

BUREAU OF MINES AND GEOSCIENCES (BMG) 1982. *Geology and Mineral Resources of the Philippines. Volume 1*. Philippine Ministry of Natural Resources, 406 pp.

CAPRARELLI, G. & LEITCH, E. C. 2002. MORB-like rocks in a Paleozoic convergent margin setting, northeast New South Wales. *Australian Journal of Earth Sciences*, **49**, 367–374.

CARDWELL, R. K., ISACKS, B. L. & KARIG, D. E. 1980. The spatial distribution of earthquakes, focal mechanism solutions and subducted lithosphere in the Philippines and northeast Indonesian islands. *In*: HAYES, D. E. (ed.) *The Tectonic and Geologic Evolution of Southeast Asian Seas and Islands*. American Geophysical Union, Geophysical Monograph Series, **23**, 1–35.

COTTEN, J., LE DEZ, A. *et al.* 1995. Origin of anomalous rare-earth element and yttrium enrichments in subaereally exposed basalts: evidence from French Polynesia. *Chemical Geology*, **119**, 115–138.

DEFANT, M. J., RICHERSON, P. M. *et al.* 1991. Dacite genesis via both slab melting and differentiation: petrogenesis of La Yeguada volcanic complex, Panama. *Journal of Petrology*, **32**, 1101–1142.

DICK, H. J. B. 1989. Abyssal peridotites, very slow spreading ridges and ocean ridge magmatism. *In*: SAUNDERS, A. D. & NORRY, M. J. (eds) *Magmatism in the Ocean Basins*. Geological Society, London, Special Publications, **42**, 71–105.

DICK, H. J. B. & BULLEN, T. 1984. Chromian spinel as a petrogenetic indicator in abyssal and alpine-type peridotites and spatially associated lavas. *Contributions to Mineralogy and Petrology*, **86**, 54–76.

FALLOON, T. J., GREEN, D. H., JACQUES, A. L. & HAWKINS, J. W. 1999. Refractory magmas in back-arc basin settings: experimental constraints on the petrogenesis of a Lau basin example. *Journal of Petrology*, **40**, 255–277.

FAURE, M., MARCHADIER, Y. & RANGIN, C. 1989. Pre-Eocene synmetamorphic structure in the Mindoro–Romblon–Palawan area, West Philippines and implications for the history of Southeast Asia. *Tectonics*, **8**, 963–979.

HALL, R. 2002. Cenozoic geological and plate tectonic evolution of SE Asia and the SW Pacific: computer-based reconstructions, models and animations. *Journal of Asian Earth Sciences*, **20**, 353–431.

HAMILTON, W. 1979. *Tectonics of the Indonesian Region*. US Geological Survey Professional Paper, **1078**, 345 pp.

HAWKINS, J. W., MOORE, G. F., VILLAMOR, R., EVANS, C. & WRIGHT, E. 1985. Geology of the composite terranes of east and central Mindanao. *In*: HOWELL, D. (ed) *Tectonostratigraphic Terranes of the Circum-Pacific Region*, Circum-Pacific Council on Energy and Mineral Resources, Earth Series, **1**, 437–463.

ISHII T., ROBINSON, P. T., MAEKAWA, H. & FISKE, R. 1992. Petrological studies of peridotites from diapiric serpentinities seamounts in the Izu–Ogasawara–Mariana forearc, Leg 125. *In*: FRYER, P., PEARCE, J. A., STOKKING I. B. *et al.* (eds) *Proceedings of the Ocean Drilling Program, Scientific Results*, **125**, 445–486.

JIMENEZ, F. A JR, YUMUL, G. P. JR, MAGLAMBAYAN, V. B. & TAMAYO, R. A. JR 2002. Shallow to near-surface, vein-type epithermal gold mineralization at Lalab in the Sibutad gold deposit, Zamboanga del Norte, Mindanao, Philippines. *Journal of Asian Earth Sciences*, **21**, 119–133.

LALLEMAND, S. E., POPOFF, M., CADET, J.-P., BADER, A.-G., PUBELLIER, M., RANGIN, C. & DEFFONTAINES, B. 1998. Genetic relations between the central and southern Philippine Trench and Sangihe Trench. *Journal of Geophysical Research*, **B103**, 933–950.

MAHOOD, G. A. & DRAKE, R. E. 1982. K–Ar dating young rhyolitic rocks: a case study of the Sierra La Primavera, Mexico. *Geological Society of American Bulletin*, **93**, 1232–1241.

MAURY, R. C., DEFANT, M. J., BELLON, H., JACQUES, D., JORON, J.-L., MCDERMOTT, F. & VIDAL, P. 1998. Temporal geochemical trends in northern Luzon arc lavas (Philippines): implications on metasomatic processes in the island are mantle. *Bulletin de la Société Géologique de France*, **169**, 69–80.

MITCHELL, A. H. G., HERNANDEZ, F. & DELA CRUZ, A. P. 1986. Cenozoic evolution of the Philippine archipelago. *Journal of Southeast Asian Earth Sciences*, **1**, 3–22.

NIU, Y. & HÉKINIAN, R. 1997. Spreading-rate dependence of the extent of mantle melting beneath ocean ridges. *Nature*, **385**, 326–329.

PARKINSON, I. J. & PEARCE, J. A. 1998. Peridotites from the Izu–Bonin–Mariana forearc (ODP Leg 125): evidence for mantle melting and melt–mantle interaction in a supra-subduction zone setting. *Journal of Petrology*, **39**, 1577–1618.

POLVÉ, M., MAURY, R. C. *et al.* 1997. Magmatic evolution of Sulawesi (Indonesia): constraints on the cenozoic geodynamic history of the Sundaland active margin. *Tectonophysics*, **272**, 69–92.

PUBELLIER, M., QUEBRAL, R., RANGIN, C., DEFFONTAINES, B., MULLER, C., BUTTERLIN, J. & MANZANO, J. 1991. The Mindanao Collision Zone: a soft collision event with a continuous Neogene strike-slip setting. *Journal of Southeast Asian Earth Sciences*, **6**, 239–248.

PUBELLIER, M., QUEBRAL, R., DEFFONTAINES, B. & RANGIN, C. 1993. *Neotectonic Map of Mindanao* with a 23 pp. explanatory note. Asia Geodyne Corporation, Quezon City, Philippines.

PUBELLIER, M., QUEBRAL, R., AURELIO, M. & RANGIN, C. 1996. Docking and post-docking escape tectonics in the Southern Philippines. *In*: HALL, R. B. & BLUNDEL, D. J. (eds) *Tectonic Evolution of Southeast Asia*. Geological Society, London, Special Publications, **106**, 511–523.

QUEBRAL, R., PUBELLIER, M., RANGIN, C. & DEFFONTAINES, B. 1996. Eastern Mindanao, Philippines: a transition zone from a collision to a strike-slip environment. *Tectonics*, **15**, 713–726.

QUERUBIN, C. L. & YUMUL, G. P. JR 2001. Stratigraphic correlation of the Malusok volcanogenic massive sulfide deposits, southern Mindanao, Philippines. *In*: YUMUL, G. P. Jr & IMAIL, A. (eds) *Mineralization and Hydrothermal Systems in the Philippines*. Resource Geology, **51**, 135–143.

QUERUBIN, C. L., YUMUL, G. P. JR, CABANTOG, A.V. & LUCERO, J. N. 1999. *The Central Zamboanga Rift Margin, Mindanao, Philippines: Implications on the Tectonic Evolution of Western Mindanao*. Philippine Mines and Geosciences Bureau – Region IX Technical Report.

RANGIN, C. 1989. The Sulu Sea: a back-arc basin setting within a collision zone. *Tectonophysics*, **161**, 119–141.

RANGIN, C. 1991. The Philippine Mobile Belt: a complex plate boundary. *Journal of Southeast Asian Earth Sciences*, **6**, 209–220.

RANGIN, C. & SILVER, E. A. 1991. Neogene tectonic evolution of the Celebes–Sulu basins: new insights from Leg 124 drilling. *In*: SILVER, E. A., RANGIN, C. *et al.* (eds) *Proceedings of the Ocean Drilling Program, Scientific Results*, **124**, 51–63.

RANGIN, C., DAHRIN, D., QUEBRAL, R. & the MODEC SCIENTIFIC PARTY. 1996. Collision and strike-slip faulting in the northern Molucca Sea (Philippines and Indonesia): preliminary results of a morphotectonic study. *In*: HALL, R. & BLUNDELL, D. (eds) *Tectonic Evolution of Southeast Asia*. Geological Society, London, Special Publications, **106**, 29–46.

RANGIN, C., JOLIVET, L. & PUBELLIER, M. 1990. A simple model for the tectonic evolution of southeast Asia and Indonesia region for the past 43 m.y. *Bulletin de la Société Géologique de France*, **8**, 889–905.

RANGIN, C., LE PICHON, X. *et al.* 1999a. Plate convergence measured by GPS across the Sundaland/ Philippine Sea Plate deformed boundary: the Philippines and eastern Indonesia. *Geophysical Journal International*, **139**, 296–316.

RANGIN, C., SPAKMAN, W., PUBELLIER, M. & BIJWAARD, H. 1999b. Tomographic and geological constraints on subduction along the eastern Sundaland continental margin (South-East Asia). *Bulletin de la Société Géologique de France*, **170**, 775–788.

SAJONA, F. G., BELLON, H. *et al.* 1997. Tertiary and Quaternary magmatism in Mindanao and Leyte

(Philippines): geochronology, geochemistry and tectonic setting. *Journal of Asian Earth Sciences*, **15**, 121–153.

SANTOS-YÑIGO, L. M. 1953. Geology of Southern Zamboanga Province. *Philippine Geologist*, **7**, 45–64.

STEIGER, R. H. & JÄGER, E. 1977. Subcommission on geochronology: convention on the use of decay constants in geo- and cosmochronology. *Earth and Planetary Science Letters*, **36**, 359–362.

SUN, S. S. & MCDONOUGH, W. F. 1989. Chemical and isotopic systematics of oceanic basalts: implications for mantle composition and process. *In*: SAUNDERS, A. D. & NORRY, M. J. (eds) *Magmatism in the Ocean Basins*. Geological Society, London, Special Publications, **42**, 313–345.

TAMAYO, RODOLFO, JR 2001. *Caractérisation pétrologique et géochimique, origines et évolutions géodynamiques des ophiolites des Philippines*. Thèse de Doctorat de L'Université de Bretagne Occidentale, 318 pp.

TAMAYO, R. A. JR., YUMUL, G. P. JR *ET AL.* 2000. A complex origin for the SW Zamboanga metamorphic basement complex, Western Mindanao, Philippines. *The Island Arc*, **9**, 639–653.

YUMUL, G. P. JR 1989. Petrological characterization of the residual–cumulate sequences of the Zambales Ophiolite Complex, Luzon, Philippines. *Ofioliti*, **14**, 253–291.

YUMUL, G. P. JR, DE JESUS, J. V. & JIMENEZ, F. A. JR 2001. Collision boundaries along the Western Philippine archipelago. *Gondwana Research*, **4**, 837–838.

Cenozoic tectonics of the China continental margin: insights from Taiwan

LOUIS S. TENG[1] & ANDREW T. LIN[2]

[1]*Institute of Geosciences, National Taiwan University, 1 Roosevelt Road, Sections 4, Taipei, Taiwan, ROC (e-mail: tengls@ntu.edu.tw)*

[2]*Institute of Geophysics, National Central University, 300 Jongda Road, Jongli, Taiwan, ROC (e-mail: lin@earth.ncu.edu.tw)*

Abstract: The continental margin to the east and south of China comprises an active margin in the East China Sea, a collision mountain belt in Taiwan, and a passive margin in the South China Sea. These three segments were generally regarded as separate tectonic entities and their interrelations have long been the subject of debate. Here we synthesize available information to outline the tectonic and geological background of the China margin, examine the link between Taiwan and the neighbouring China margins, and thereby establish a Cenozoic evolutionary model.

The China margin is floored with a pre-Cenozoic continental basement covered with an up to 10-km-thick pile of Cenozoic sedimentary strata. The continental basement has been invariably stretched and moulded into a series of northeast-trending horsts and grabens. Except in the Okinawa Trough of the East China Sea, the Cenozoic sedimentary cover typically exhibits a two-tier tectonostratigraphic structure, with narrow Palaeogene rift basins draped by a blanket-like Neogene–Quaternary sequence. The two-tier structure prevails in the entire inner part of the China margin, including the Taiwan Strait off western Taiwan. In the outer China margin, however, the two-tier structure persists only in the South China Sea, and is in stark contrast with the collisional orogen of Taiwan and the Ryukyu arc of the East China Sea.

By untangling the contractional deformation of the northern Taiwan mountain belt, it has been possible to reconstruct a precollisional tectonostratigraphic section with a distinctive two-tier structure shown by a Palaeogene half-graben covered with a Miocene drape sequence. When put together with Palaeogene rift basins of the Taiwan Strait, it becomes clear that the precollisional continental margin of Taiwan resembles that of the South China Sea, characterized by two lines of Palaeogene rift basins. Hence before the collision started in Late Miocene times, Taiwan was part of the passive South China margin that extended northward to the southern Ryukyu area.

Ever since the end of the Cretaceous, the China continental margin has been dominated by extensional tectonics, regardless of the presence or absence of subduction zones. In the Early Cenozoic, extensive crustal attenuation resulted in region-wide subsidence and formation of rift basins. Extension in the South China Sea culminated in Late Oligocene times, when part of the outer margin was drifted away by the opening ocean basin. In the East China Sea, the margin remained intact and became separated from the South China Sea margin by a transform fault. From the Miocene onwards, the South China Sea margin has been passively subsiding, sporadically punctuated with basaltic volcanism. In the East China Sea margin, the Okinawa Trough has opened and the Ryukyu volcanic arc thrived. The NE edge of the South China Sea margin was deformed as the Taiwan orogen.

China consists of a mosaic of continental blocks and accretionary complexes that had undergone a prolonged history of subduction, collision, and terrain accretion since the Proterozoic (Sengor & Natal'in 1996; Zhao *et al.* 1996). By Late Mesozoic times, the process of terrain amalgamation was completed, and China has since been an integral part of the Eurasian continent (Hsü *et al.* 1990; Enkin *et al.* 1992; Li 1998). In the Cenozoic era, continental China has not remained stable, but rather has been strongly influenced by plate interactions around the southern and eastern edges of the Eurasian plate (Fig. 1). In western China, indentation of the Indian subcontinent at the SW Eurasian margin has caused extensive contractional defor-

From: MALPAS, J., FLETCHER, C. J. N., ALI, J. R. & AITCHISON, J. C. (eds) 2004. *Aspects of the Tectonic Evolution of China.* Geological Society, London, Special Publications, **226**, 313–332.
0305-8719/04/$15 © The Geological Society of London 2004.

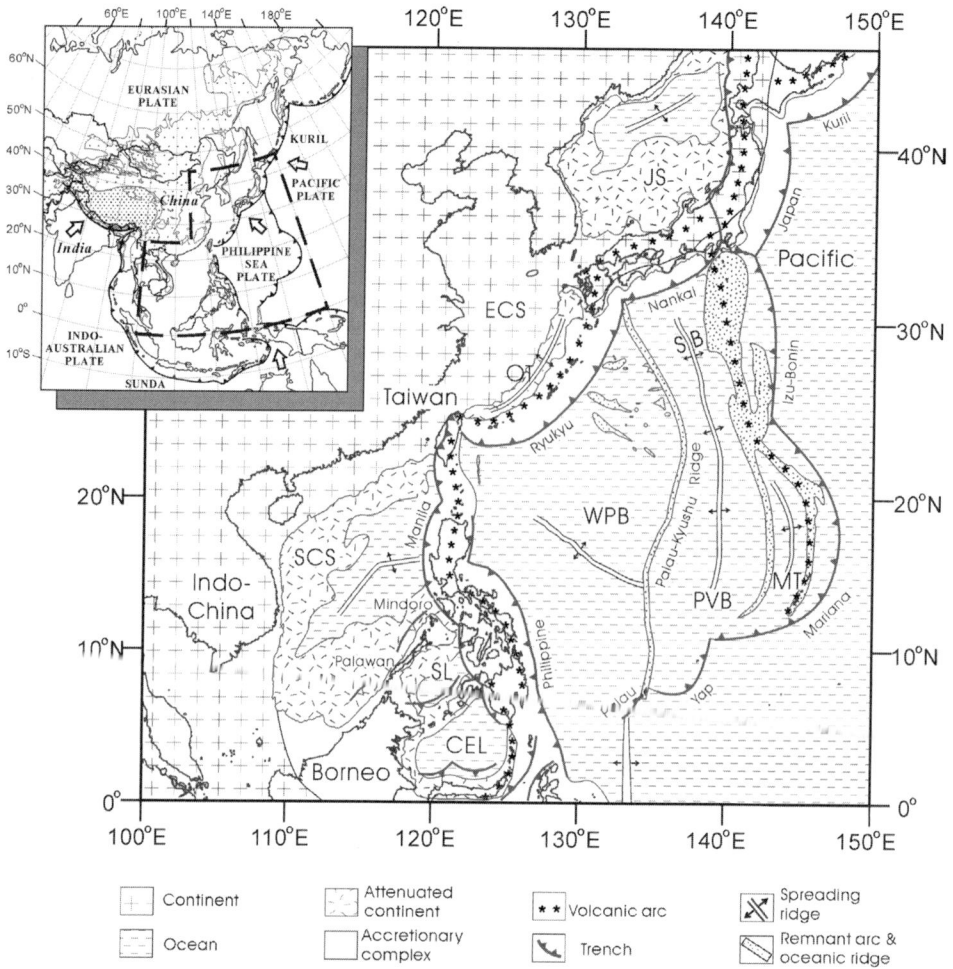

Fig. 1. Plate-tectonic setting of the continental margin of China. Marginal basins: CEL, Celebes Sea; ECS, East China Sea; HB, Huatong Basin; JS, Japan Sea; MT, Mariana Trough; OT, Okinawa Trough; PVB, Parece Vela Basin; SB, Shikoku Basin; SCS, South China Sea; SL, Sulu Sea; WPB, West Philippine Basin.

mation in the continental interior, not only raised the Himalayan orogen and Tibetan plateau but also rejuvenated a series of inland mountain chains (Tapponier & Molnar 1979; Tapponier *et al.* 1982). The eastern China continent, in contrast, has been dominated by extensional tectonism shown by region-wide subsidence and formation of intracontinental rift basins (Chen & Dickinson 1986; Ren *et al.* 2002).

As the offshore extension of the eastern China continent, the China continental margin is tectonically more complicated than the continental interior. From Japan to Taiwan, the margin of the East China Sea is fringed with an east-facing Ryukyu arc–trench system and underlain by

a west-dipping subduction zone. From Taiwan to Indochina, the margin of the South China Sea is passively coupled with an extinct oceanic basin. In between, the margin has been deformed as a rising mountain belt by arc–continent collision in Taiwan. The varying tectonic styles along the margin have prompted previous workers to treat the East China Sea, South China Sea, and Taiwan as separate tectonic entities (e.g. Wang 1987; Liu 1989; Zhou *et al.* 1989; Teng 1990; Zhou *et al.* 1995; Sibuet and Hsü 1997). However, as noted by some researchers (e.g. Li 1984; Yu 1994; Ren *et al.* 2002), the Cenozoic stratigraphy and structural features are quite comparable throughout the China margin, regardless of the differences at the outer rim. This suggests that the margin might have a common tectonic history in the Cenozoic. But how the margin has evolved as a whole remains little understood.

Located at the junction between the East and South China Sea margins, Taiwan is the critical place to explore the Cenozoic tectonic history of the China margin. In the Taiwan Island, vast tracts of rock strata of the outer China margin have been exhumed and exposed in the mountain belt, which provides a rare opportunity for field observations. In offshore western Taiwan, the inner part of the China margin remains unscathed and can be readily compared with other parts of the margin. Geological and geophysical surveys in the past 100 years, both onshore and offshore, has produced a wealth of basic data unparalleled in the neighbouring areas. These data allow previous workers to reconstruct the Cenozoic history of Taiwan and tie it in with other parts of the China margin (Suppe 1981; Teng 1992; Hsü & Sibuet 1995; Huang et al. 1997; Sibuet & Hsü 1997). Nevertheless, whether Taiwan is affiliated with the South China Sea margin, or with the East China Sea margin is still controversial (e.g. Huang et al. 1997 and Sibuet & Hsü 1997).

Here we integrate tectonic and geological information of Taiwan and adjacent areas to investigate the Cenozoic tectonic history of the China continental margin. We restore the precollision continental margin of Taiwan and find it closely affiliated with the South China Sea margin. However, the entire China margin might have had a similar tectonic history in the early Cenozoic, characterized by continual crustal extension that propagated from inland toward the outer margin. In the East China Sea margin, continental rifting continued in the late Cenozoic, and is still active in the Okinawa Trough. In the South China Sea margin, extension abated in the Late Oligocene when the outer continental margin was drifted away by the opening South China Sea Basin. The margin has since been smoothly subsiding, and the northeastern edge of the margin has later been deformed as the mountain belt of Taiwan.

Tectonic setting

China is presently a part of the Eurasian plate that is bordered by the Philippine Sea and Pacific plates in the east and the Indo-Australian plate in the south (Fig. 1). Throughout the Cenozoic, the Eurasian margin has been subducted by neighbouring oceanic plates, leading to festoons of arc–trench systems from Kuril in the north to Sunda in the south. Situated in the middle of

the eastern Eurasian plate, the China continent faces mainly the Philippine Sea plate.

The Philippine Sea plate is a composite ocean that can largely be separated into two parts by the north-trending Palau–Kyushu Ridge (Fig. 1). West of the Ridge, the West Philippine Basin is an extinct Palaeogene ocean with a segmented east-southeast-trending spreading ridge (Hilde & Lee 1984; Deschamps et al. 1999). The eastern Philippine Sea plate is generally younger, consisting of a series of north-trending volcanic arcs, remnant arcs and back-arc basins developed in Neogene–Quaternary times (Karig 1971). Currently, the Philippine Sea plate is rotating clockwise about a pole northeast of Japan (Seno et al. 1993). It is subducting beneath the Eurasian plate at the Ryukyu Trench but overriding the Eurasian plate at the Manila Trench. In the southern Philippines, the Philippine Sea plate is not in contact with the Eurasian plate, but is underthrusting the Philippine archipelagos – a tectonic collage sandwiched between opposite-facing subduction zones (Rangin et al. 1991).

Taiwan is a key point at the China continental margin where two arc–trench systems of opposite polarity meet. To the north, the southeast-facing Ryukyu arc–trench system at the eastern edge of the Eurasian plate stretches from Kyushu into northeast Taiwan (Kao et al. 1998; Sibuet et al. 1998). In the south, the northwest-facing Luzon Arc–Manila Trench system at the western edge of the Philippine Sea plate extends from Luzon into Taiwan (Huang et al. 1997; Kao et al. 2000). The boundary between the two arc systems lies at the western edge of the northwest-subducting Philippine Sea plate beneath northern Taiwan (WEP, Fig. 2).

At a first glance, the tectonic configuration of the Eurasian continental margin appears simple and well related to neighbouring plates. However, the apparent simplicity holds only for the Late Cenozoic and may easily break down for early Cenozoic times, owing to the ever-changing plate motion. Particularly noteworthy is the Philippine Sea plate, which originated in the Southern Hemisphere in Early Tertiary times (Seno & Maruyama 1984; Haston & Fuller 1991; Hall et al. 1995). The plate did not enter its present location in the Northern Hemisphere until the West Philippine and Parece Vela basins successively opened up through time. Before the Philippine Sea plate progressively moved in and interacted with the Eurasian margin, the vast area east of the Eurasian margin was probably occupied by the Pacific (Engebretson et al. 1985) and/or some other small ocean basins (Hall et al. 1995; Sibuet et al. 2002). The motion of these ocean plates has important bearings on

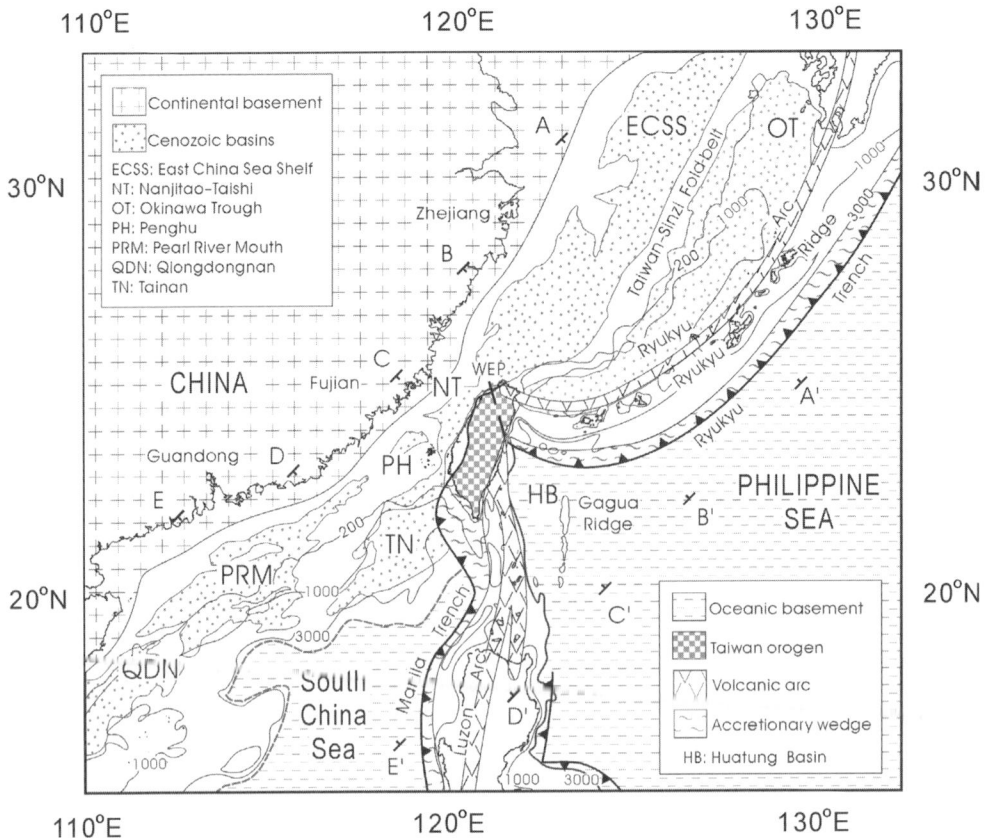

Fig. 2. Cenozoic tectonics and geology of the continental margin of China. Sections A–A' to E–E' shown in Figure 3.

continental margin tectonics, and it should be taken into account in reconstructing the tectonic history of the China margin.

Regional background

The China margin is floored with an attenuated continental basement, upon which Cenozoic rift basins have developed. The intensity of basement attenuation increases from onshore China to the outer margin, as shown by outward thinning of the underlying continental crust (Fig. 3). In coastal China, the crust is about 30 km thick (Li & Mooney 1998), progressively decreasing to 22–24 km at the shelf break and to 16–18 km in the Okinawa Trough (Liu 1989; Hirata et al. 1991). In the South China Sea, basement attenuation is even more severe in the slope area, where crustal thickness rapidly falls from 22 km to <12 km (Nissen et al. 1995; Zhou et al. 1995; Yan et al. 2001). In Taiwan, however, the continental crust has been thickened to more

than 40 km by the collision (Lin 1996; Shih et al. 1998; Yen et al. 1998).

The basement rocks of the China margin are similar to those exposed in neighbouring areas, including China, Korea, Japan, Taiwan and the Philippines (Figs 1 and 2). On the inner continental shelf, the basement is composed of Proterozoic–Palaeozoic metamorphic complexes, Palaeozoic to Early Mesozoic sedimentary sequences, and Late Mesozoic igneous intrusions and extrusions, comparable with rocks exposed in coastal China and southern Korea (Wageman et al. 1970; Guong et al. 1989; Liu 1989; Zhou et al. 1989). In the outer shelf and slope area, the basement rocks are younger, composed of Palaeozoic–Mesozoic metamorphic and igneous complexes that crop out extensively in Japan, Taiwan and Philippine islands (Faure et al. 1989; Taira et al. 1989; Zhou et al. 1989; Taira 2001).

The continental basement has been invariably stretched and moulded into a series of northeast-trending horst-and-graben structures that are

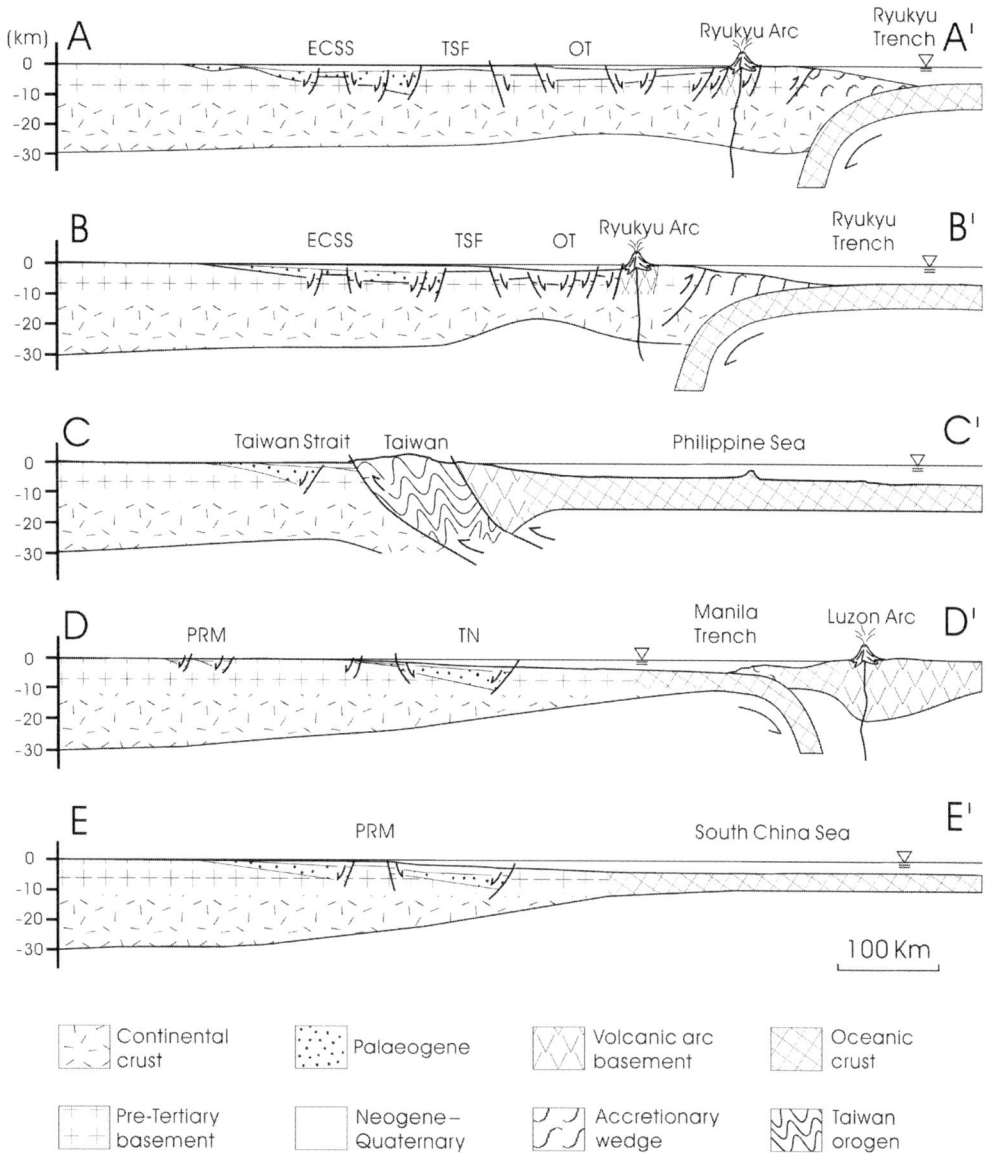

Fig. 3. Schematic cross-sections of the continental margin of China. TSF, Taiwan–Sinzi Fold-belt. Surface structures are slightly exaggerated for clarity. Locations and basic terms shown in Figure 2.

buried by 2–10 km of Late Mesozoic and Cenozoic sedimentary strata (Figs 2 & 3). The depositional basins, which largely follow the structural grains of the basement, typically exhibit a two-tier tectonostratigraphic structure (Li 1984; Yu 1994). The lower part of the sedimentary fill, composed of Palaeogene sequences, is usually ponded in narrow half-graben troughs separated by intervening basement highs. The upper part,

mainly Neogene–Quaternary in age, forms a sheet-like sequence draping the horst-and-graben structures and filling in wide and shallow depressions. The only exception is the Okinawa Trough, which is a rift basin filled with Neogene–Quaternary sediments (Kimura 1985; Letouzey & Kimura 1986).

Aside from the above common features, there are prominent disparities in the outer part of the China margin. Along the East China Sea margin, there is the Ryukyu arc–trench system and associated Okinawa Trough, which is separated from the East China Sea Shelf Basin by a subsurface basement high, termed the Taiwan–Sinzi folded zone (Wageman *et al.* 1970). Another basement high, the Ryukyu Ridge (Wageman *et al.* 1970), that includes most of the Ryukyu islands, separates the Okinawa Trough from the Ryukyu fore-arc basin. The Ryukyu volcanic arc runs along the southern flank of the Okinawa Trough (Sibuet *et al.* 1998).

The South China Sea margin is characterized by a shelf–slope–rise configuration typical of passive continental margins. The continental basement is bordered by a magnetic quiet zone interpreted as the continent–ocean boundary (Taylor & Hayes 1983; Xia *et al.* 1994; Yan *et al.* 2001) and coupled with the Late Oligocene–Early Miocene ocean basin to the south (Taylor & Hayes 1983; Briais *et al.* 1993).

In Taiwan, the outer China margin has been deformed with the impinging Luzon Arc, whereas the inner margin remains intact in the offshore Taiwan Strait. The collision was first initiated in southern Ryukyu in Late Miocene times, and has been propagating from northeast to southwest (Teng 1996). Currently, the collision is under way in south-central Taiwan, where the mountain belt is being pushed westward over the continental shelf (Yu *et al.* 1997). In northeastern Taiwan, however, the subduction polarity has flipped, and the mountain belt has been transferred to the Ryukyu Arc system (Teng 1996).

In summary, the Cenozoic China margin, excluding the Taiwan collision zone, appears to have been dominated by extensional tectonics demonstrated by the omnipresent rift basins. Except the Okinawa Trough, rift basins are of Palaeogene age and are draped by Neogene–Quaternary deposits in a typical two-tier structure. In the South China Sea, this structure is believed to reflect continual rifting of the continental margin since the Late Cretaceous, that has resulted in breakaway of the outer margin and spreading of the South China Sea Basin (Halloway 1982; Taylor & Hayes 1983; Ru & Pigott 1986). The boundary between the two tiers, often referred to as the breakup unconformity, is thought to be indicative of the onset of oceanic spreading (Halloway 1982; Taylor & Hayes 1983). The breakup model, however, does not quite apply to the East China Sea margin, because no oceanic spreading has ever taken place. Hence, while the inner margin looks similar in the East and South China Seas,

the outer margin varies considerably and the link between the two lies in the Taiwan area.

Taiwan Strait

Lying west of the Taiwan mountain belt, the Taiwan Strait is part of the inner China margin that has not yet been involved in the collisional orogeny. It is floored with an attenuated continental basement overlain by four northeast-trending Cenozoic rift basins, namely Penghu, Nanjihtao, Tainan and Taishi (Hsiao *et al.* 1991a, b; Chow *et al.* 1991; Lee *et al.* 1996; Lin 2001). The eastern part of the strait has been warped down as the foreland basin of the Taiwan mountain belt (Lin & Watts 2002).

The continental basement of the Taiwan Strait is exposed in coastal China, where Precambrian metamorphic complexes, Palaeozoic sedimentary strata, and Jurassic–Cretaceous granitoids and rhyolites are widely distributed (Hsü *et al.* 1990; Jahn *et al.* 1990; Zhou & Li 2000). The basement slants eastward and is deeply buried by the Cenozoic sedimentary strata of the Taiwan Strait (Fig. 4). Drilling into horst-like basement highs yielded deformed Jurassic and Early Cretaceous sedimentary rocks and minor Early Cretaceous microdiorite (Lee *et al.* 1996; Chen *et al.* 1997; Lin 2001).

The Cenozoic sedimentary cover can be divided into three tectonostratigraphic units, namely the Palaeogene synrift, Miocene post-rift, and Pliocene–Quaternary foreland basin sequences, according to their roles in the context of continental rifting and arc–continent collision (Lin 2001). The synrift sequence, which forms southeast-thickening sedimentary wedges filling in half-graben troughs, overlies the Mesozoic basement with a prominent unconformity (ROU) indicative of the onset of rifting. The post-rift sequence covers both rift basins and intervening basement highs like a blanket and is based on an unconformity (BU) that marks the cessation of rifting and the onset of thermal subsidence. The foreland sequence is a southeast-thickening sedimentary wedge accumulated in front of the collisional orogen. The boundary between the post-rift and foreland sequences is a disconformity (BFU).

The synrift sequence consists mainly of siliciclastic sedimentary strata, with subordinate amounts of volcanic rocks and carbonate. In the Penghu Basin, the synrift sequence can be further divided into two parts by an intra-rift unconformity (IRU, Fig. 5). Several boreholes penetrated the upper synrift sequence into the upper part of the lower synrift sequence, and encountered a thick sequence of marine shales intercalated with a few sandstones and volcanic layers

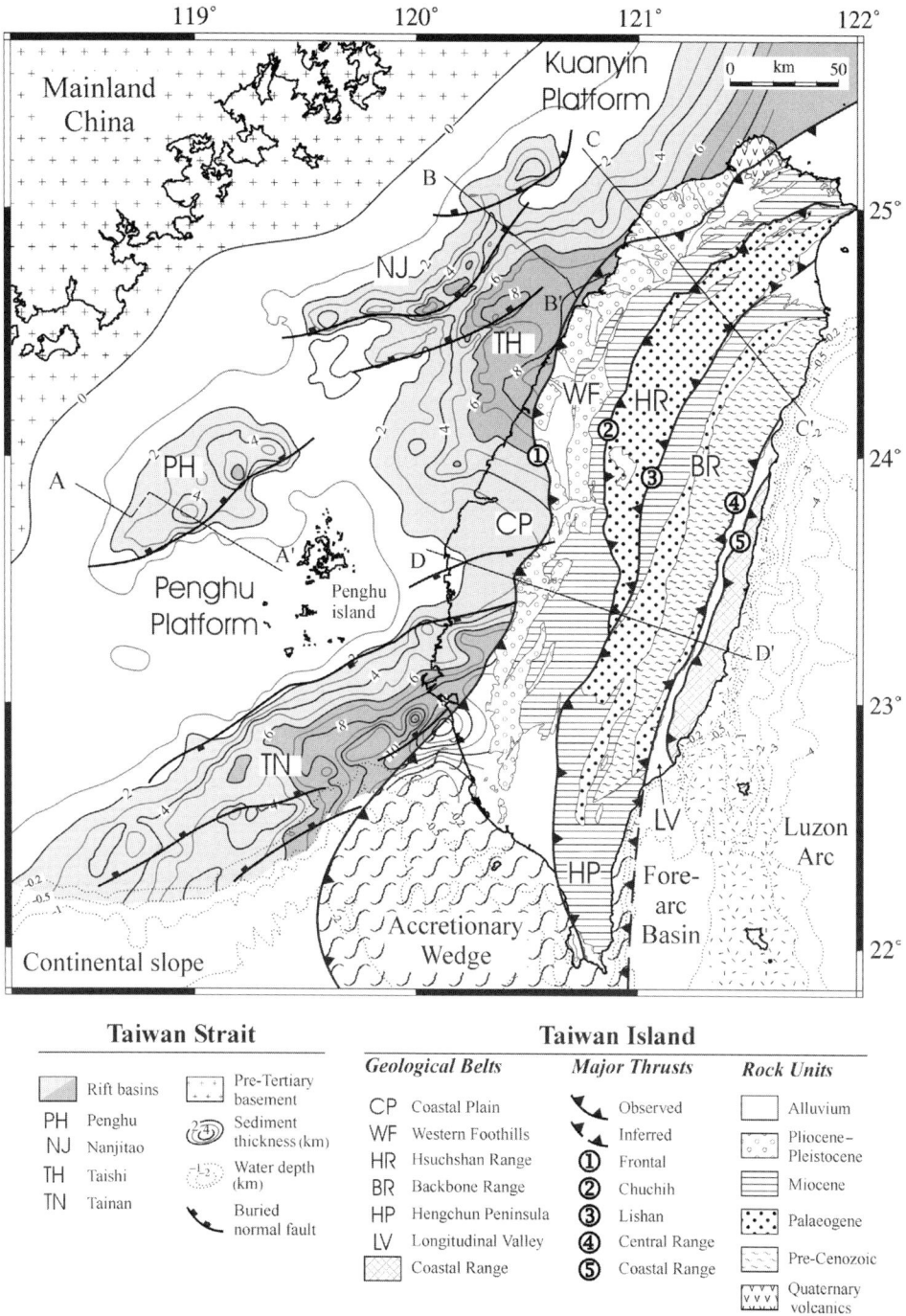

Fig. 4. Cenozoic geology of Taiwan and the Taiwan Strait. Summarized from Ho (1988), Teng (1992), Liu *et al.* (1997) and Lin (2001). Sections A–A′ and B B′ shown in Figure 5; C–C′ and D–D′ in Figure 6.

(Hsiao *et al.* 1991a; Lin 2001). Nannofossils recovered from borehole rock samples show that the upper synrift sequence accumulated in Late Eocene times, whereas the lower synrift sequence was deposited in the Early to Middle Eocene (Hsiao *et al.* 1991a). In the Nanjihtao and Taishi Basins, the synrift sequence is lithologically and chronologically similar to the lower synrift sequence of the Penghu Basin (Chow *et al.* 1991; Lin 2001). However, the oldest age of the synrift sequence is still unknown, because the thickest synrift deposits in the deep basins have not yet been cored. Regional geological analyses and seismic interpretations suggest that Palaeocene and upper Cretaceous deposits may lie in the deep parts of the basins (Chow *et al.* 1991; Hsiao *et al.* 1991a, b; Lee *et al.* 1996). Thin Palaeocene marine shales and carbonates encountered in boreholes on the neighbouring basement highs (Hsiao *et al.* 1991b; Lee *et al.* 1996; Lin 2001) lend support to the presence of marine incursion in the Late Palaeocene.

In contrast with the synrift sequence, the post-rift and foreland basin sequences have been extensively penetrated with boreholes, and their stratigraphies are much better understood (Hsiao *et al.* 1991a, b; Lin 2001). The post-rift sequence is the most widespread, consisting of flat-lying, subparallel, and laterally persistent layers of paralic to shallow-marine sandstones/ mudstones with minor volcanics and carbonates. The foreland basin sequence, distributed mainly in the eastern Taiwan Strait, comprises a west-onlapping package of continental to shallow-marine sandstones and mudstones. Because the post-rift sequence thickens toward the Taiwan Island (Fig. 6a), the time gaps associated with the top (BFU) and basal (BU) unconformities vary from west to east (Fig. 7). In the western and central Taiwan Strait, the Early to Middle Miocene post-rift sequence directly overlies the Middle Eocene synrift sequence, with the entire Upper Eocene and Oligocene missing. The post-rift sequence is, in turn, onlapped at the top by the foreland basin sequence with a time gap spanning the late Middle Miocene to Late Pliocene. In the eastern Taiwan Strait, more and more Late Oligocene strata appear at the base and more Late Miocene at the top of the post-rift sequence. The time gaps of BFU

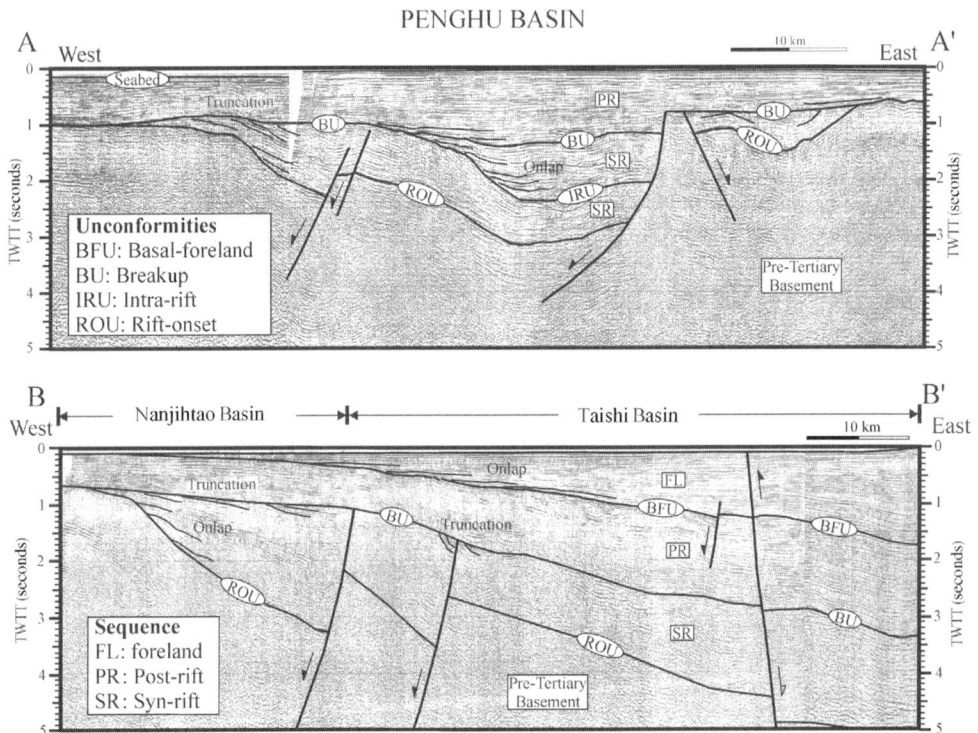

Fig. 5. Seismic sections of the Taiwan Strait. Note the truncation of a synrift sequence (SR) at the breakup unconformity (BU), and the structural inversion near Taiwan in B–B'. Locations shown in Figure 4; stratigraphy detailed in Figure 7. Modified from Lin (2001).

and BU consequently shrink as they extend eastward into a more continuous stratigraphic section in Taiwan.

Comparing the Taiwan Strait with the adjacent China margin, it is easy to observe that the Strait is different because it contains a thick foreland basin sequence. If that sequence were removed, the Strait would display a distinct two-tier structure characteristic of the inner continental shelf. The structural grain of the northeast-trending Palaeogene basins is compatible with that of other Palaeogene basins in the China margin. There is no tectonic break of any sort that can be invoked to separate the Strait from either the East China Sea margin or the South China Sea margin. The Taiwan Strait was, thus, a coherent link between the East and South China Sea margins.

Taiwan Island

Except for the Coastal Range of eastern Taiwan, the rest of Taiwan, including the mountain ranges and the coastal plain, pertains to the China continental margin (Fig. 4). The Coastal Plain is the onshore extension of the Taiwan Strait and part of the unscathed inner margin, whereas the mountain ranges are the deformed outer margin. Rock strata of the mountain ranges have been deformed into imbricate fold and thrust sheets trending north–northeast. There is an obvious, albeit progressive, eastward increase in the intensity of structural deformation, the metamorphic grade and the stratigraphic age (Fig. 6), which warrants division of the mountain ranges into three lithotectonic belts – namely the Western Foothills, Hsuehshan Range and Backbone Range (Ho 1988).

The Coastal Plain of western Taiwan is floored with a subsurface pre-Cenozoic basement that includes tilted Lower Cretaceous sedimentary sequences and Permian crystalline limestones (Yuan *et al.* 1985; Jahn *et al.* 1992). Unconformably overlying the basement are flat-bedded Cenozoic sedimentary strata comparable with those in the eastern Taiwan Strait. Palaeogene strata are generally thin and may be partly or totally absent at different places (Yuan *et al.* 1985). The Neogene–Quaternary sequence is thick and continuous, covering almost all of the coastal plain. Basaltic rocks are occasionally intercalated within the Eocene and Miocene strata. These Cenozoic sedimentary sequences can be followed into the Western Foothills, where extensive Neogene–Quaternary strata crop out. Oligocene and older sequences are sporadically exposed, but have been widely sampled by drilling (Chiu 1975).

The Hsuehshan Range and neighbouring western Backbone Range constitute a slate terrane composed of Middle Eocene to Early Miocene metasedimentary rocks (Huang 1980; Chou 1990). The eastern Backbone Range is a pre-Cenozoic metamorphic complex that includes various kinds of crystalline limestone, gneiss and schist, transformed from Permian and Lower Cretaceous strata (Wang Lee & Wang 1987; Yui & Lan 1991). In the Backbone Range, the contacts between the Eocene–Miocene slates and between the Eocene slate and the pre-Cenozoic metamorphic complex are both unconformable. In places the entire Eocene is missing, and the Miocene slate may directly overlie the metamorphic complex (Suppe *et al.* 1976).

In the southern tip of Taiwan (Fig. 4), the Backbone Range plunges south toward the Hengchung Peninsula, and the Eocene–Miocene slate is conformably overlain by Middle and Upper Miocene sedimentary strata (Pelletier & Stephan 1986; Sung & Wang 1986). Around the western and southern rim of the peninsula, the Miocene strata are unconformably overlain by Pliocene–Quaternary sandstone–shale and reef limestone.

In spite of the structural complexities, the stratigraphic architecture of Taiwan is similar to that of East and South China Sea margins, aside from the Pliocene–Quaternary foreland deposits. It basically consists of a pre-Cenozoic continental basement and a thick Cenozoic sedimentary cover (Fig. 7). The basement is partly shown by the Permian and Cretaceous rocks beneath the Coastal Plain and the coeval metamorphic complex in the Backbone Range. Although not exposed in the Western Foothills and Hsuehshan Range, the basement is believed to underlie the whole mountain belt as part of the thickened continental crust. The Cenozoic sedimentary cover is widely distributed and has a stratigraphy comparable with that of the neighbouring areas. Nevertheless, the original depositional basin was destroyed and the tectonostratigraphic relations disrupted. It requires the pre-collisional Taiwan margin to be reconstructed before *vis-à-vis* comparison with other parts of the China margin can be made.

Reconstructing the pre-collisional margin

As noted by Suppe (1981), the Taiwan mountain belt is the onshore extension of the accretionary wedge associated with subduction and accretion of the China continental margin at the Manila Trench off southern Taiwan (Fig. 2). Within the wedge, rock strata of the continental margin are compressively deformed and stacked as imbricate thrust sheets (Fig. 6). The wedge expands from south to north and may reach a steady state in north-central Taiwan with constant

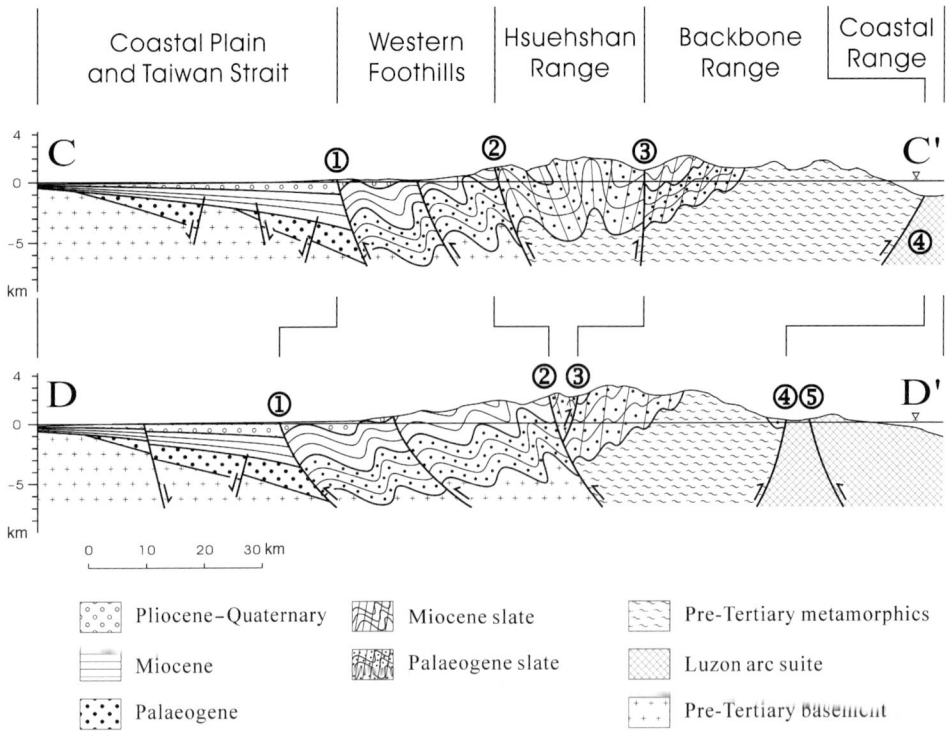

Fig. 6. Geological framework and lithotectonic belts of Taiwan. Locations and major thrusts 1–5 shown in Figure 4. Summarized from Ho (1988), Teng *et al.* (1991) and Lin (2001).

width and height. In the steady-state mountain belt, the geological characteristics also seem to have attained an equilibrium state exhibited by persistence of structural and stratigraphic features along strike (Fig. 4).

To reconstruct the pre-collisional continental margin in Taiwan, it is necessary to first undo the contractional deformation incurred during the accretionary process (Fig. 8A). Using the theory of fault-bend folding, Suppe (1980) constructed a retrodeformable section, near C-C' in Figure 4, across the northern Taiwan mountain belt (Fig. 8B), in which each thrust sheet of the Western Foothills and Hsuehshan Range can be geometrically delineated. By untangling the imbricate fold-and-thrust structures, Suppe was able to relocate each thrust sheet to its predeformational position and obtain a 120-km shortening for the Western Foothills and Hsueshan Range (Fig. 8C). This reconstruction, although involving some assumptions and uncertainties, provides a good first approximation of the structural deformation caused by collisional orogeny.

Following Suppe's retrodeformable section, Teng *et al.* (1991) established a stratigraphic section across northern Taiwan, based on six Tertiary stratigraphic columns (Fig. 8D). Each column in the Western Foothills and Hsuehshan Range can be restored to its precollisional position according to Suppe's reconstruction. When restored to a key Middle Miocene horizon, the stratigraphic section shows a distinct two-tier structure with a Miocene drape-like sequence covering a Palaeogene rift basin. Although established only for northern Taiwan, this stratigraphic section can probably apply to the entire mountain belt. The reconstructed Palaeogene half-graben basin, named the Hsuehshan Trough by Teng *et al.* (1991), clearly demonstrates that the Taiwan mountain belt was indeed part of the rifting China margin before collision.

Put together with the rift basins of the Taiwan Strait, it is easy to see that the western Hsuehshan Trough overlaps the Taishi Basin in the eastern Taiwan Strait (Fig. 4), and the Palaeogene sequence of the Taishi Basin (Fig. 5A) may well be part of the synrift deposits in the Hsuehshan Trough. The precollisional margin of Taiwan seems to have comprised two lines of Palaeogene rift basins (Fig. 9B). The inner margin basins, including Penghu and Nanjihtao, formed in Palaeocene to Middle Eocene times.

Fig. 7 chart labels (Taiwan Strait / Taiwan Island stratigraphic columns):

BELTS — EPOCH — TAIWAN STRAIT: NJ/PH, KYP/PHP, TH/TN — TAIWAN ISLAND: CP/WF, HR, BR — Sequence

Epochs: PLEISTOCENE, PLIOCENE (EAR./LATE), MIOCENE (LATE, MIDDLE, EARLY), OLIGOCENE (LATE, EARLY), EOCENE (LATE, MIDDLE, EARLY), PALAEOCENE (LATE, EARLY), CRETACEOUS, PERMIAN

Chart annotations:
(BFU)
Collision Orogeny
Opening of South China Sea (BU)
inferred
Rifting of China Continental Margin (ROU)
Yenshan Orogeny

Sequence column: Fore-land, Post-rift, Syn-rift, Pre-rift

Legend:
- Terrestrial
- Coastal
- Nearshore
- Offshore
- Deep marine
- Limestone
- Volcanics
- Major events
- Unconformity

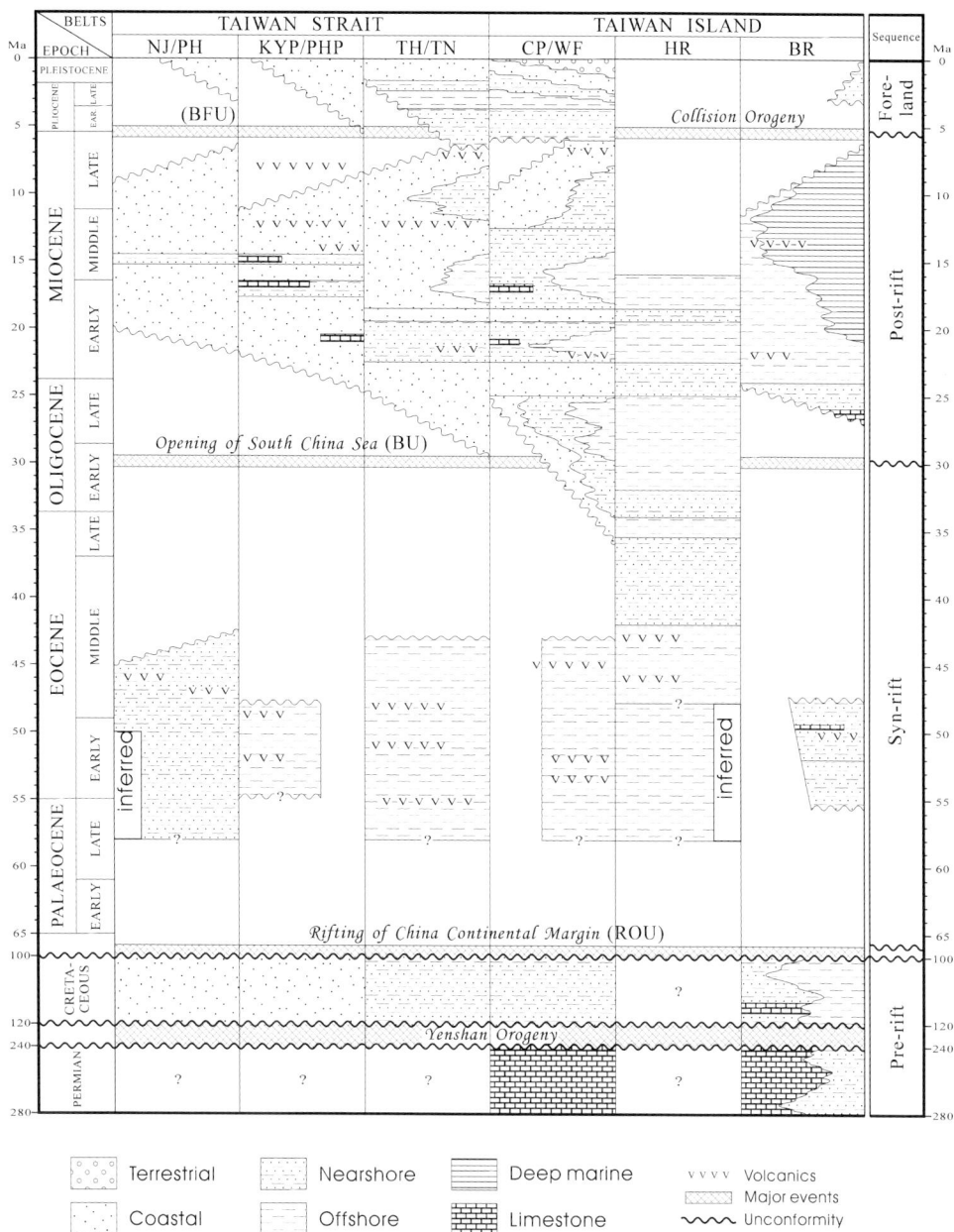

Fig. 7. Reconstructing the pre-collisional rift basin of Taiwan. The imbricate fold-and-thrust structures of Taiwan (A) can be modeled as stacked fault-bend folds (B) and retrodeformed accordingly (C). The stratigraphies of the relocated thrust sheets allow reconstruction of a half-graben basin (D), namely the Hsuehshan Trough, which is filled with Eocene-Oligocene deposits and covered by Miocene strata. The Trough is underlain by the pre-Tertiary rocks of the Kuany Platform in the west and bordered with the metamorphic basement of the Backbone Range in the east. (A) modified from section C-C′ in Fig. 6; (B) and (C) from Suppe (1980); (D) from Teng *et al* (1991).

The outer margin basins, like the Hsuehshan Trough and Tainan Basin, may continue to develop into the Oligocene. Because the Palaeogene basin-fill and Miocene cover sequences are both composed of coastal to shallow-marine deposits (Chou 1973, 1990; Teng *et al.* 1991; Lin 2001), these rift basins must have been positioned in the continental shelf. Compared with

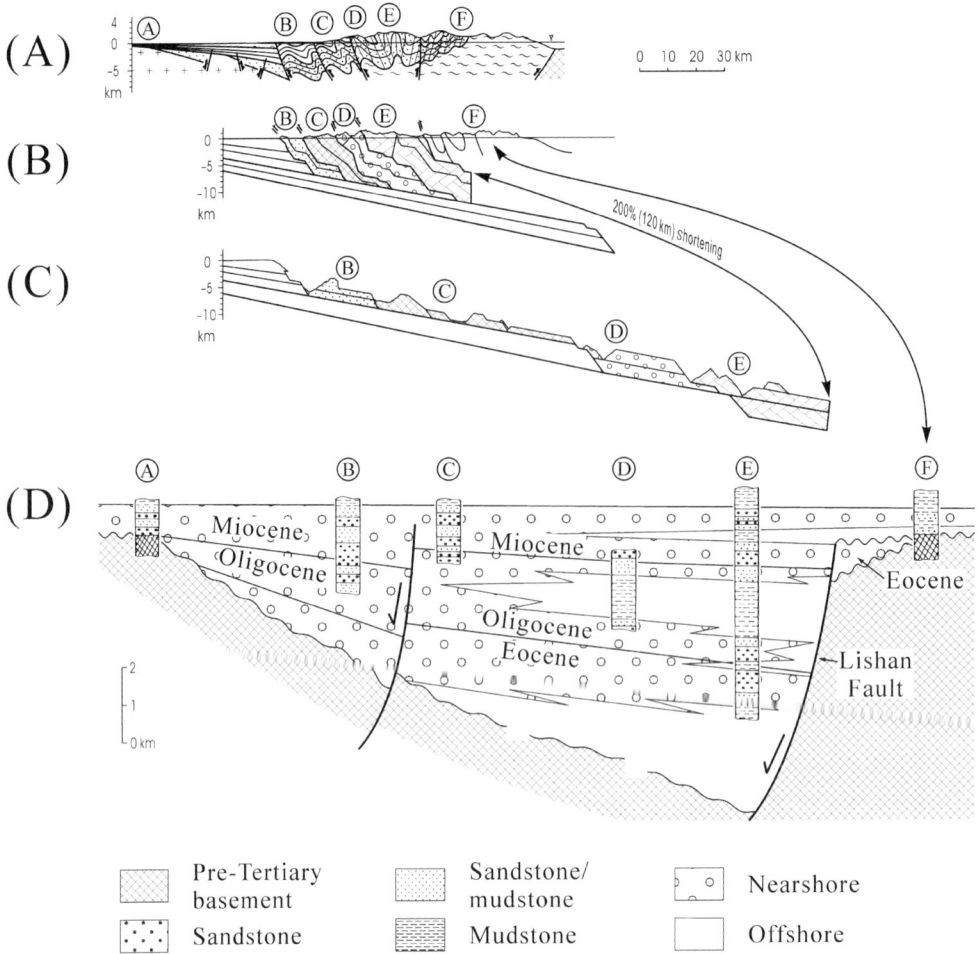

Fig. 8. Reconstructing of pre-collisional rift basin of Taiwan. The imbricate fold-and-thrust structures of Taiwan (A) can be modelled as stacked fault-bend folds (B) and retrodeformed accordingly (C). The stratigraphies of the relocated thrust sheets allow reconstruction of a half-graben basin (D), namely the Hsuehshan Trough, which is filled with Eocene-Oligocene deposits and covered by Miocene strata. The trough is underlain by the pre-Tertiary rocks of the Kuanyin Platform in the west and bordered with the metamorphic basement of the Backbone Range in the east. (A) modified from section C–C′ in Fig. 6; (B) and (C) from Suppe (1980); (D) from Teng et al. (1991).

the East and South China Seas, the precollisional continental shelf of Taiwan clearly resembles that of the South China Sea characterized by two lines of Palaeogene rift basins rather than the East China Sea marked by the Late Neogene Okinawa Trough (Fig. 3). This suggests that Taiwan was part of the South China Sea margin before collision.

In accordance with the morphotectonic configuration of the passive South China Sea margin, a more complete precollisional margin of Taiwan can be reconstructed by disentangling the collisional orogen. The margin is featured with a

rifted continental shelf coupled with a piece of attenuated continental basement in the continental slope and a deep ocean basin (Fig. 9B). The Coastal Range of eastern Taiwan can be easily moved to a position >500 km to the southeast (Teng 1990) by backtracking the 5–10 cm/year motion of the Luzon Arc and underlying Philippine Sea plate in the past 10 Ma (Seno & Maruyama 1984; Hall et al. 1995). The reconstructed tectonic scenario looks very similar to the one just in front of the collision zone off southwestern Taiwan (Fig. 3D). During the course of collision, the attenuated continental

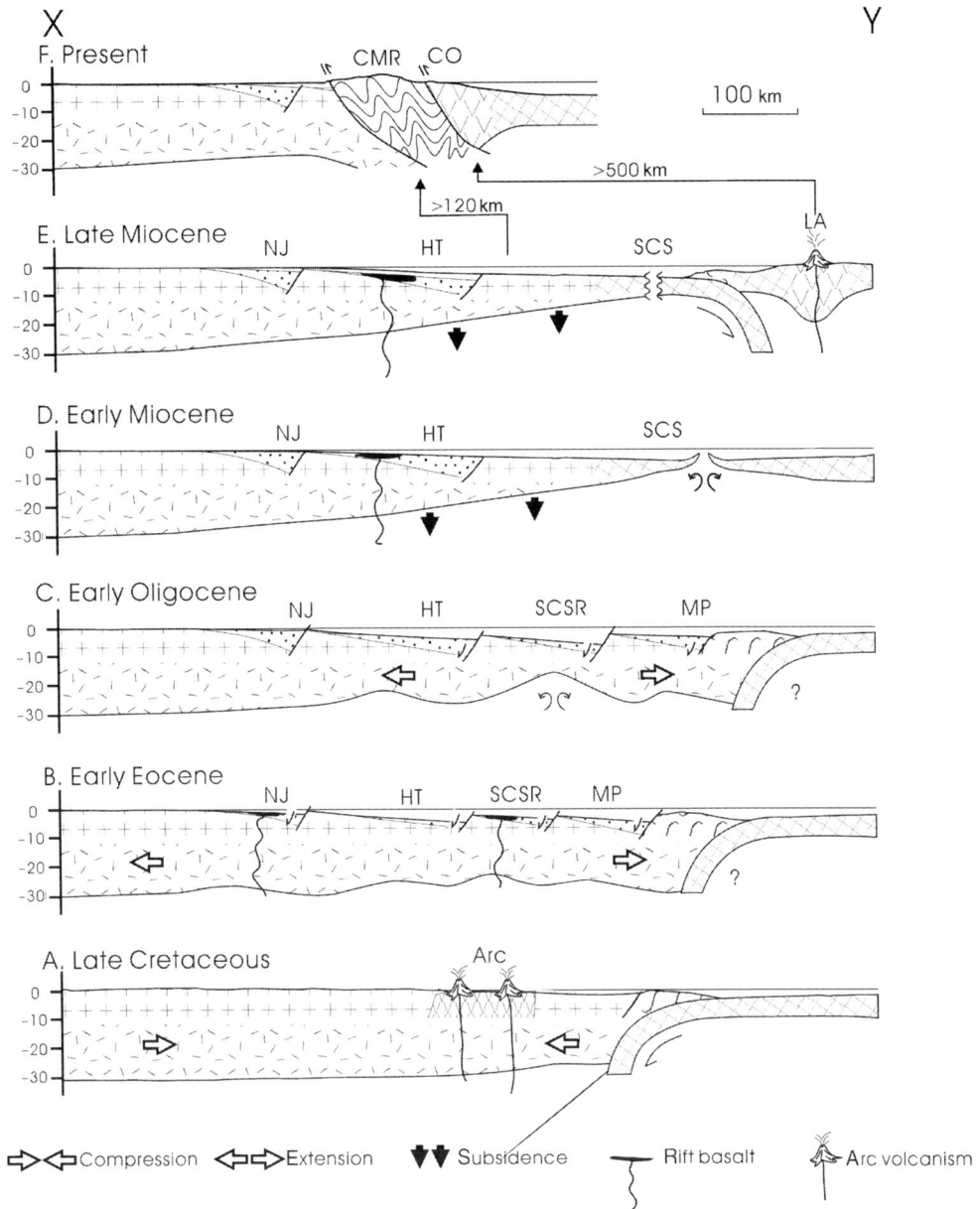

Fig. 9. Reconstructed Cenozoic continental margin of Taiwan. CMR, Central mountain ranges; CO, Coastal Range; HT, Hsuehshan Trough; LA, Luzon Arc; MP, Mindoro–Palawan block; NJ, Nanjihtao Basin; SCS, South China Sea Basin; SCSR, South China Sea rift. Symbols are the same as Figure 3.

basement and associated ocean basin would have been successively pulled into the subduction zone and would have become invisible on the surface. However, ophiolitic blocks akin to the oceanic South China Sea Basin have been found in the mélange deposit of the Coastal

Range (Suppe 1981), and continental slope deposits that overlay the attenuated continental basement are widely distributed in the Heng-chung Peninsula (Pelletier & Stephan 1986; Sung & Wang 1986). These lines of evidence suggest that the South China Sea Basin and the attenuated continental basement have indeed extended to the southeast of Taiwan before being consumed in the subduction zone (Suppe 1981; Wang 1987). This argument can be supported by geophysical investigations in the offshore southern Taiwan area, which indicate that the attenuated continental basement and associated ocean basin are being subducted beneath the Hengchung Peninsula and its southern offshore extension (Nakamura *et al.* 1998; Kao *et al.* 2000).

Having established the similarities between Taiwan and the South China Sea margin, it is worth examining how Taiwan is related to the East China Sea margin. From the inner margin point of view, there is no obvious difference or discontinuity between the Taiwan Strait and the East China Sea shelf. But, for the outer margin, the difference is clear. There is no arc–trench system in the precollisional Taiwan margin equivalent to the Ryukyu in the East China Sea margin (cf. Figs 3B and 9E). Admittedly, the present Ryukyu volcanic arc and associated Okinawa Trough can be traced into northeastern Taiwan (Teng 1996; Sibuet *et al.* 1998), but this connection was not established until the Late Pliocene (Teng 1996). Although it has been suggested that the Ryukyu arc–trench system could extend into Taiwan in Miocene and Early Tertiary times (Jahn 1972; Hsü & Sibuet 1995; Juan 1975; Sibuet & Hsü 1997), no evidence for such a system, like the presence of Miocene arc volcanics and/or coeval subduction complex, has been presented. In contrast, there is a significant break in the continuity of basement rocks and the Miocene volcanic arc between the southern and north-central Ryukyu Islands (Kizaki 1986; Letouzey & Kimura 1986; Shinjo 1999). Marine reflection seismic data also show that the southern Ryukyu Islands were involved in the collisional orogeny of Taiwan (Letouzey & Kimura 1986; Hsiao *et al.* 1998), and may have been part of the passive Taiwan margin before collision (Teng 1996).

Tectonic evolution

Given the reconstructed pre-collisional continental margin of Taiwan and relevant tectonostratigraphic data (Figs 7 and 9E), it is possible to establish the Cenozoic evolutionary history of the margin by successively stripping off the sedimentary strata and restoring the extensional deformation (Fig. 9 A–E). Since Taiwan was part of the South China Sea margin, the interpretations will be made in the context of South China Sea tectonics (Fig. 10). The East China Sea margin will be discussed separately.

During much of the Mesozoic, southeast China was rimmed with an Andean-type continental margin subducted by west-moving oceanic plates (Faure & Natal'in 1992; Zhou & Li 2000). This subduction system is believed to last into the Late Cretaceous (Figures 9A and 10A), as shown by widespread subduction-related granitic intrusions in Japan (Kinoshita 1995; Taira 2001), Taiwan (Lan *et al.* 1996), and the South China Sea margin (Guong *et al.* 1989). The margin was compressed with the northwest-moving Pacific plate, such that Early Cretaceous and older rocks were folded and eroded.

The stress regime dramatically changed in latest Cretaceous through Early Palaeocene times, when the margin began to extend. It is not clear whether the west-dipping subduction system still existed, but crustal extension clearly prevailed, and resulted in submergence of the margin and formation of rift basins (Figs 9B and 10B). Meanwhile, widespread volcanism took place and spewed basaltic rocks in the rift basins (Chen 1991; Yui *et al.* 1994; Chen *et al.* 1997). Marine incursion almost reached the present shoreline of coastal China, as evidenced by the marine synrift deposits in the Taiwan Strait (Figs 5 and 7).

Crustal extension continued into the Early Oligocene, but the centre of extension seemed to have shifted toward the outer margin (Figs 9C and 10C). In the inner margin, rifting ceased and the early rift basin deposits may have been uplifted and partially eroded (Fig. 5). In the outer margin, rift basins continued to deepen and accumulate synrift deposits.

In the Late Oligocene, extension seemed to have shifted outward and resulted in the breakup of the outer part of the continental margin. The oceanic South China Sea Basin began to spread (Figs 9D and 10D), pushing the splintered piece of continental margin to drift southward to the Mindoro–Palawan area. Rifting dwindled but continued in the remnant South China Sea margin (Figs 9E and 10E), which has been smoothly subsiding and occasionally punctuated with basaltic volcanism (Chung *et al.* 1994, 1995). This passive margin presently persists in offshore Guangdong (Fig. 3E), but, in Taiwan, it was tectonized by the impinging Luzon Arc (Figs 9F and 10F).

The Late Cretaceous to Early Tertiary history of the East China Sea margin was probably similar (Maruyama *et al.* 1997; Taira 2001), characterized by westward subduction of the Pacific plate, followed by regional extension (Fig. 10A, B). However, as the South China Sea began to open in Oligocene times (Fig. 10C), the East China Sea margin did not

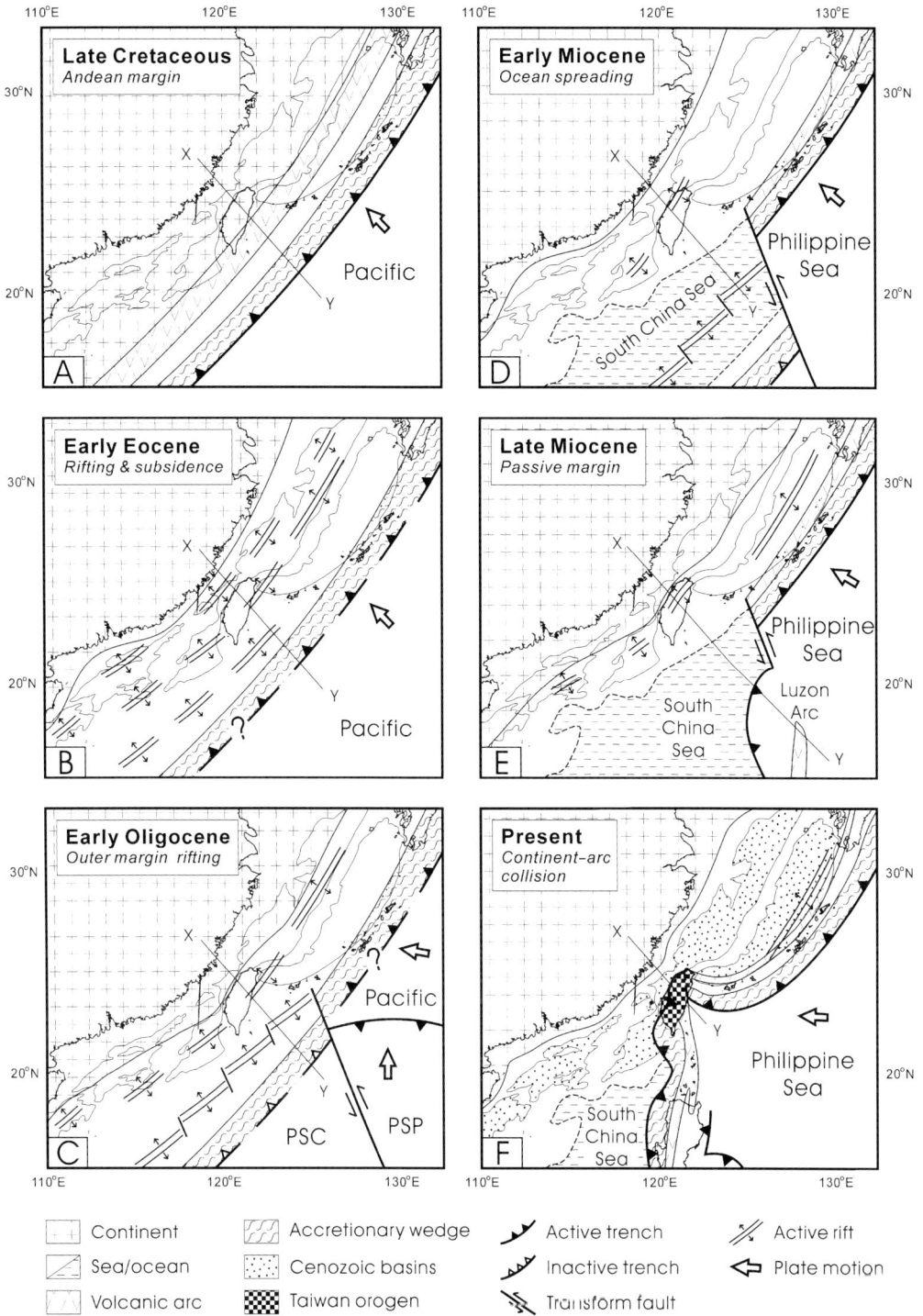

Fig. 10. Cenozoic tectonic evolution of China continental margin. PSC, Proto-South China Sea; PSP, Philippine Sea plate. Modern coastline and Cenozoic basins are drawn for reference. Sections X–Y shown in Figure 9.

seem to rift correspondingly. As the South China Sea opened up, the East China Sea margin must have been separated from the South China Sea margin by a transform fault in southern Ryukyu (Fig. 10D). In the meantime, the Philippine Sea plate moved in, and might have begun to subduct beneath the East China Sea margin. By the Late Miocene, the Ryukyu arc–trench system was well established, and the Okinawa Trough started rifting (Fig. 10E). Following the westward migration of arc–continent collision, the Ryukyu volcanic arc and the Okinawa Trough extended into northeastern Taiwan.

Conclusions and discussion

After summarizing the tectonic and geological information of Taiwan and neighbouring areas, we find that the Cenozoic China continental margin has been dominated by extensional tectonics, regardless of the presence or absence of subduction zones. In the Early Tertiary, the margin was subjected to extensive crustal attenuation that resulted in region-wide subsidence and formation of rift basins. Extension culminated in the Late Oligocene, when the outer part of the South China Sea margin was drifted away by the opening ocean basin. The East China Sea margin, however, did not rift correspondingly, and became separated from the South China Sea margin. From the Miocene on, the South China Sea margin has been smoothly subsiding as a passive margin, whereas the East China Sea margin has been subjected to subduction and back-arc spreading. As the NE edge of the South China Sea margin was tectonized into the Taiwan orogen, the Ryukyu arc–trench system of the East China Sea margin followed the west-migrating arc–continent collision into northeast Taiwan.

The tectonic analysis presented herein results from a straightforward synthesis of available data and may be the simplest of interpretations. More complicated scenarios that involve multiple stages of subduction, obduction and collision (e.g. Jahn 1972; Juan 1975; Pelletier & Stephan 1986; Lu & Hsü 1992; Hsü & Sibuet 1995; Sibuet et al. 2002) have been proposed and should be reserved as alternative models. In spite of the extensive survey and intensive studies in the past, a number of fundamental questions relating to regional tectonics have not been properly answered. For instance, was there a Cenozoic volcanic arc in western Taiwan (cf. Yui et al. 1994, 1995; Chung 1995; Chung et al. 1995)? Is there a piece of Cretaceous ocean off southern and eastern Taiwan (cf. Hilde & Lee 1984; Deschamps et al. 2000; Sibuet et al. 2002)? When and how did the Philippine Sea plate

move in and interact with the China margin (cf. Maruyama et al. 1997; Hall 2002)? Until we have clear constraints on these issues, the Cenozoic tectonics of China margin will continue to be debated.

In a broader tectonic framework, the extensional tectonics that have dominated the Cenozoic China margin are probably not a local phenomenon. Coeval crustal extension has been widely reported from NE Russia (Worrall et al. 1996), Japan (Jolivet et al. 1994), SE Asia (Morley 2002), and inland China (Ren et al. 2002). How were the different types of crustal extension in East Asia related to one another (e.g. Worrall et al. 1996; Ren et al. 2002)? Was crustal extension driven by India–Asia collision (Tapponier et al. 1982; Leloup et al. 2001) or by changes in subducting ocean plates (Northrop et al. 1995; Morley 2002)? Was it a reflection of the upper-mantle flow (Flower et al. 2001)? These are just a few among the numerous exciting topics waiting to be explored.

We are grateful to M. C. Blake, S. L. Chung, and T. Y. Lee for critical reading and constructive comments that greatly improved the manuscript. Thanks are due to C. S. Liu, W. R. Chi, S. L. Chung and T. Y. Lee for kindly providing topographic data and valuable geological information, to Y. T. Huang, Y. L. Tsai and P. Y. Lin for assistance in typing and drafting, and to the National Science Council, ROC, for financial support. L. Teng wishes to pay special thanks to J. Malpas for generous invitation to the Croucher Workshop and to C. Fletcher for his encouragements and patience that have made this publication possible.

References

BRIAIS, A., PATRIAT, P. & TAPPONNIER, P. 1993. Updated interpretation of magnetic anomalies and seafloor spreading stages in the South China Sea: implications for the Tertiary tectonics of Southeast Asia. *Journal of Geophysical Research*, **98**, 6299–6328.

CHEN, C. H., LEE, C. Y., HUANG, T. C. & TING, J. S. 1997. Radiometric ages and petrological and geochemical aspects of some late Cretaceous and Paleogene volcanic rocks beneath the northern offshore of Taiwan. *Petroleum Geology of Taiwan*, **31**, 61–88.

CHEN, P. Y. 1991. Basaltic–andesitic volcanic rocks from the areas of Changshihchiao and Hsiangyang, southern E–W cross-island highway, Taiwan. *Special Publication of the Central Geological Survey*, **5**, 127–159.

CHEN, Q. & DICKINSON, W. R. 1986. Contrasting nature of petroliferous Mesozoic–Cenozoic basins in eastern and western China. *American Association of Petroleum Geologists Bulletin*, **70**, 263–275.

CHIU, H. T. 1975. Miocene stratigraphy and its relation to the Paleogene rocks in west-central Taiwan. *Petroleum Geology of Taiwan*, **12**, 51–80.

CHOU, J. T. 1973. Sedimentology and paleogeography of the Upper Cenozoic System of Western Taiwan. *Proceedings of the Geological Society of China*, **16**, 111–143.

CHOU, J. T. 1990. Paleogene formations of the central and Hsuehshan ranges in Taiwan. *Special Publication of the Central Geological Survey*, **4**, 177–192.

CHOW, J., CHEN, H. M., CHANG, T. Y., KUO, C. L. & TSAI, S. F. 1991. Preliminary study on hydrocarbon plays around Nanjihtao Basin, Taiwan Strait. *Petroleum Geology of Taiwan*, **26**, 45–56.

CHUNG, S. L. 1995. Geochemical characteristics of metabasites from the slate formations of Taiwan: discussion. *Journal of the Geological Society of China*, **38**, 173–177.

CHUNG, S. L., JAHN, B. M., CHEN, S. J., LEE, T. & CHEN, C. H. 1995. Miocene basalts in Northwestern Taiwan: evidence for EM-type mantle sources in the continental lithosphere. *Geochimica et Cosmochimica Acta*, **59**, 549–555.

CHUNG, S. L., SUN, S. S., TU, K., Chen, C-H. & LEE, C. Y. 1994. Late Cenozoic basaltic volcanism around the Taiwan Strait, SE China: product of lithosphere–asthenosphere interaction during continental extension. *Chemical Geology*, **112**, 1–20.

DESCHAMPS, A., LALLEMAND, S. & DOMINGUEZ, S. 1999. The last spreading episode of the West Philippine Basin revisited. *Geophysical Research Letters*, **26**, 2073–2076.

DESCHAMPS, A., MONIE, P., LALLEMAND, S., HSÜ, S. K. & YEH, K. Y. 2000. Evidence for Early Cretaceous oceanic crust trapped in the Philippine Sea Plate. *Earth and Planetary Science Letters*, **179**, 503–516.

ENGEBRETSON, D. C., COX, A. & GORDON, R. G. 1985. Relative motions between oceanic and continental plates in the Pacific basin. *Geological Society of America, Special Paper*, **206**, p. 59.

ENKIN, R., YANG, Z., CHEN, Y. & COURTILLOT, V. 1992. Paleomagnetic constraints on the geodynamic history of the major blocks of China from the Permian to the present. *Journal of Geophysical Research*, **97**, 13 953–13 989.

FAURE, M. & NATAL'IN, B. 1992. The geodynamic evolution of the eastern Eurasian margin in Mesozoic times. *Tectonophysics*, **208**, 397–411.

FAURE, M., MARCHADIER, Y. & RANGIN, C. 1989. Pre-Eocene synmetamorphic structure in the Mindoro–Romblon–Palawan Area, West Philippines, and implications for the history of Southeast Asia. *Tectonics*, **8**, 963–979.

FLOWER, M. F. J., RUSSO, R. M., TAMAKI, K. & HOANG, N. 2001. Mantle contamination and the Izu–Bonin–Mariana (IBM) 'high-tide mark': evidence for mantle extrusion caused by Tethyan closure. *Tectonophysics*, **333**, 9–34.

GUONG, Z., JIN, Q., QIU, Z., WANG, S. & MENG, J. 1989. Geology, tectonics and evolution of the Pearl River Mouth Basin. *In*: ZHU, X. (ed.) *Chinese Sedimentary Basins*. Elsevier, Amsterdam, 181–196.

HALL, R. 2002. Cenozoic geological and plate tectonic evolution of SE Asia and the SW Pacific: computer-based reconstructions, model and animations. *Journal of Asian Earth Sciences*, **20**, 353–431.

HALL, R., ALI, J. R. & ANDERSON, C. D. 1995. Cenozoic motion of the Philippine Sea Plate: paleomagnetic evidence from Eastern Indonesia. *Tectonics*, **14**, 1117–1132.

HALLOWAY, N. H. 1982. North Palawan Block, Philippines – its relation to Asian Mainland and role in evolution of South China Sea. *American Association of Petroleum Geologists Bulletin*, **63**, 1355–1383.

HASTON, R. B. & FULLER, M. 1991. Paleomagnetic data from the Philippine Sea plate and their tectonic significance. *Journal of Geophysical Research*, **96**, 6073–6098.

HILDE, T. W. C. & LEE, C. S. 1984. Origin and evolution of West Philippine Basin: a new interpretation. *Tectonophysics*, **102**, 85–104.

HIRATA, N., KINOSHITA, H. *et al.* 1991. Report on DELP 1988 cruises in the Okinawa Trough, part 3. Crustal structure of the southern Okinawa Trough. *Earthquake Research Institute, University of Tokyo, Bulletin*, **66**, 37–70.

HO, C. S. 1988. An introduction to the geology of Taiwan: explanatory text of the geologic map of Taiwan. *Central Geological Survey, The Ministry of Economic Affairs, Taipei, Taiwan, ROC*, second edition, p. 192.

HSIAO, L. Y., HUANG, S. T., TENG, L. S. & LIN, K. A. 1998. Structural characteristics of the Southern Taiwan–Sinzi Folded Zone. *Petroleum Geology of Taiwan*, **32**, 133–153.

HSIAO, P. T., HU, C. C. *et al.* 1991*a*. Hydrocarbon potential evaluation of the Penghu Basin. *Petroleum Geology of Taiwan*, **26**, 215–230.

HSIAO, P. T., LIN, K. A. *et al.* 1991*b*. Petroleum appraisal on Tungyintao Basin. *Petroleum Geology of Taiwan*, **26**, 183–213.

HSÜ, S. K. & SIBUET, J.-C. 1995. Is Taiwan the result of arc–continent or arc–arc collision? *Earth and Planetary Science Letters*, **136**, 315–324.

HSÜ, K. J., LI, J., CHEN, H., WANG, Q., SUN, S. & SENGÖR, A. M. C. 1990. Tectonics of South China: key to understanding West Pacific geology. *Tectonophysics*, **183**, 9–39.

HUANG, C. Y., WU, W. Y. *et al.* 1997. Tectonic evolution of accretionary prism in the arc–continent collision terrane of Taiwan. *Tectonophysics*, **281**, 31–51.

HUANG, T. C. 1980. Calcareous nannofossils from the slate terrane West of Yakou, Southern Cross-Island Highway, Taiwan. *Petroleum Geology of Taiwan*, **17**, 59–74.

HUANG, T. C. 1982. Tertiary calcareous nannofossil stratigraphy and sedimentation cycles in Taiwan. *Proceedings 2nd ASCOPE Conference and Exhibition, Manila, Philippines, 1981*, 873–886.

JAHN, B. M. 1972. Reinterpretation of geologic evolution of the Coastal Range, eastern Taiwan. *Geological Society of America Bulletin*, **83**, 241–247.

JAHN, B. M., CHI, W. R. & YUI, T. F. 1992. A late Permian formation of Taiwan (marbles from Chia-Li well No.1): Pb–Pb isochron and Sr isotopic evidence, and its regional geological significance. *Journal of the Geological Society of China*, **35**, 193–218.

JAHN, B. M., ZHOU, X. H. & LI, J. L. 1990. Formation and tectonic evolution of Southeastern China and Taiwan: isotopic and geochemical constraints. *Tectonophysics*, **183**, 145–160.

JOLIVET, L., TAMAKI, K. & FOURNIER, M. 1994. Japan Sea, opening history and mechanism: a synthesis. *Journal of Geophysical Research*, **99**, 22 237–22 259.

JUAN, V. C. 1975. Tectonic evolution of Taiwan. *Tectonophysics*, **26 (3–4)**, 197–212.

KAO, H., HUANG, G. C. & LIU, C. S. 2000. Transition from oblique subduction to collision in the Northern Luzon Arc–Taiwan region: constraints from bathymetry and seismic observation. *Journal of Geophysical Research*, **105**, 3059–3079.

KAO, H., SHEN, S. J. & MA, K. F. 1998. Transition from oblique subduction to collision: earthquakes in the southernmost Ryukyu Arc–Taiwan region. *Journal of Geophysical Research – Solid Earth*, **103**, 7211–7229.

KARIG, D. E. 1971. Origin and development of marginal basins in the Western Pacific. *Journal of Geophysical Research*, **76**, 2542–2561.

KIMURA, M. 1985. Back-arc rifting in the Okinawa trough. *Marine and Petroleum Geology*, **2**, 222–240.

KINOSHITA, O. 1995. Migration of igneous activities related to ridge subduction in Southwest Japan and the East Asian continental margin from the Mesozoic to the Paleogene. *Tectonophysics*, **245**, 25–35.

KIZAKI, K. 1986. Geology and tectonics of the Ryukyu islands. *Tectonophysics*, **125**, 193–207.

LAN, C. Y., JAHN, B. M., MERTZMAN, S. A. & WU, T. W. 1996. Subduction-related granitic rocks of Taiwan. *Journal of Asian Earth Sciences*, **14**, 11–28.

LEE, T. Y., HSÜ, Y. Y. & TANG, C. H. 1996. Sequence stratigraphy and depositional cycles in the Tungyintao Basin, offshore Northern Taiwan. *Petroleum Geology of Taiwan*, **30**, 1–30.

LELOUP, P. H., ARNAUD, N. 2001. New constraints on the structure, thermochronology, and timing of the Ailao Shan–Red River shear zone, SE Asia. *Journal of Geophysical Research*, **106**, 6683–6732.

LETOUZEY, J. & KIMURA, M. 1986. The Okinawa Trough: genesis of a back-arc basin developing along a continental margin. *Tectonophysics*, **125**, 209–230.

LI, D. 1984. Geologic evolution of petroliferous basins on continental shelf of China. *American Association of Petroleum Geologists Bulletin*, **68**, 993–1003.

LI, S. & MOONEY, W. D. 1998. Crustal structure of China from deep seismic sounding profiles. *Tectonophysics*, **288**, 105–113.

LI, Z. X. 1998. Tectonic history of the major East Asia lithospheric blocks since the Mid-Proterozoic – a synthesis. *In*: FLOWER, M. F. J., CHUNG, S. L., LO, C. H. & LEE, T. Y. (eds), *Mantle Dynamics and Plate Interactions in East Asia, American Geophysical Union, Washington, Geodynamic Series*, **27**, 221–243.

LIN, A. T. 2001. *Cenozoic stratigraphy and tectonic development of the West Taiwan Basin*. PhD thesis, Oxford University.

LIN, A. T. & WATTS, A. B. 2002. Origin of the West Taiwan basin by orogenic loading and flexure of a rifted continental margin. *Journal of Geophysical Research*, **107**, ETG2-1-2-19.

LIN, C. H. 1996. Crustal structures estimated from arrival differences of the first P-waves in Taiwan. *Journal of the Geological Society of China*, **39**, 1–12.

LIU, C. S., HUANG, I. L. & TENG, L. S. 1997. Structural features off Southwestern Taiwan. *Marine Geology*, **137**, 305–319.

LIU, G. 1989. Geophysical and geological exploration and hydrocarbon prospects of the east China Sea. *China Earth Sciences*, **1**, 43–58.

LU, C. Y. & HSU, K. J. 1992. Tectonic evolution of the Taiwan mountain belt. *Petroleum Geology of Taiwan*, **27**, 21–46.

MARUYAMA, S., ISOZAKI, Y., KIMURA, G. & TERABAYASHI, M. 1997. Paleogeographic maps of the Japanese Islands: plate tectonic synthesis from 750 Ma to the present. *The Island Arc*, **6**, 121–142.

MORLEY, C. K. 2002. A tectonic model for the Tertiary evolution of strike-slip faults and rift basins in SE Asia. *Tectonophysics*, **347**, 189–215.

NAKAMURA, Y., MCINTOSH, K. & CHEN, A. T. 1998. Preliminary results of a large offset seismic survey west of Hengchun Peninsula, Southern Taiwan. *Terrestrial, Atmospheric and Oceanic Sciences*, **9**, 395–408.

NISSEN, S. S., HAYES, D. E., BOCHU, Y., WEIJUN, Z., YONGQIN, C. & XIAUPIN, N. 1995. Gravity, heat flow, and seismic constraints on the processes of crustal extension: northern margin of the South China Sea. *Journal of Geophysical Research*, **100**, 22 447–22 483.

NORTHRUP, C. J., ROYDEN, L. H. & BURCHFIEL, B. C. 1995. Motion of the Pacific Plate relative to Eurasia and its potential relation to Cenozoic extension along the eastern margin of Eurasia. *Geology*, **23**, 719–722.

PELLETIER, B. & STEPHAN, J. F. 1986. Middle Miocene obduction and Late Miocene beginning of collision registered in the Hengchun Peninsula: geodynamic implications for the evolution of Taiwan. *Tectonophysics*, **125**, 133–160.

RANGIN, C., JOLIVET, L. & PUBELLIER, M. 1990. A simple model for the tectonic evolution of Southeast Asia and Indonesia region for the past 43 m.y. *Bullétin de la Société Géologique de France*, **6**, 889–905.

REN, J. Y., TAMAKI, K., LI, S. T. & JUNXIA, Z. 2002. Late Mesozoic and Cenozoic rifting and its dynamic setting in Eastern China and adjacent areas. *Tectonophysics*, **344**, 175–205.

RU, K. & PIGOTT, J. D. 1986. Episodic rifting and subsidence in the South China Sea. *American Association of Petroleum Geologists Bulletin*, **70**, 1136–1155.

SENGOR, A. M. C. & NATAL'IN, B. A. 1996. Paleotectonics of Asia: fragments of a synthesis. *In*: YIN, A. & HARRISON, T. M. (eds) *The Tectonic Evolution of Asia. Cambridge University Press*, 486–640.

SENO, T. & MARUYAMA, S. 1984. Paleogeographic reconstruction and origin of the Philippine Sea. *Tectonophysics*, **102**, 53–84.

SENO, T., STEIN, S. & GRIPP, A. E. 1993. A model for the motion of the Philippine Sea Plate consistent with NUVEL-1 and geological data. *Journal of Geophysical Research – Solid Earth*, **98**, 17 941–17 948.

SHIH, R. C., LIN, C. H., LAI, H. L., YEH, Y. H., HUANG, B. B. & YEN, H. Y. 1998. Preliminary crustal structures across central Taiwan from modeling of the onshore–offshore wide-angle seismic data. *Terrestrial, Atmospheric and Oceanic Sciences*, **9**, 317–328.

SHINJO, R. 1999. Geochemistry of high Mg andesites and the tectonics evolution of the Okinawa Trough–Ryukyu arc system. *Chemical Geology*, **157**, 69–88.

SIBUET, J. C. & HSÜ, S. K. 1997. Geodynamics of the Taiwan arc–arc collision. *Tectonophysics*, **274**, 221–251.

SIBUET, J. C., DEFFONTAINES, B., HSÜ, S. K., THAREAU, N., LE FORMAL, J. P., LIU, C. S. & ACT PARTY 1998. Okinawa trough backarc basin: early tectonic and magmatic evolution. *Journal of Geophysical Research*, **103**, 30 245–30 267.

SIBUET, J. C., HSÜ, S. K., LE PICHON, X., LE FORMAL, J. P., REED, D., MOORE, G. & LIU, C. S. 2002. East Asia plate tectonics since 15 Ma: constraints from the Taiwan region. *Tectonophysics*, **344**, 103–134.

SUNG, Q. & WANG, Y. 1986. Sedimentary environments of the Miocene sediments in the Hengchun Peninsula and their tectonic implication. *Memoir of the Geological Society of China*, **7**, 325–340.

SUPPE, J. 1980. A retrodeformable cross section of Northern Taiwan. *Proceedings of the Geological Society of China*, **23**, 46–55.

SUPPE, J. 1981. Mechanics of mountain-building and metamorphism in Taiwan. *Memoir of the Geological Society of China*, **4**, 67–90.

SUPPE, J., WANG, Y., LIOU, J. G. & ERNST, W. G. 1976. Observation of some contacts between basement and Cenozoic cover in the Central Mountains, Taiwan. *Proceedings of the Geological Society of China*, **19**, 59–70.

TAIRA, A. 2001. Tectonic evolution of the Japanese island arc system. *Annual Review of Earth and Planetary Sciences*, **29**, 109–134.

TAIRA, A., TOKUYAMA, H. & SOH, W. 1989. Accretion tectonics and evolution of Japan. *In*: BEN-AVRAHAM, Z. (ed.) *The Evolution of the Pacific Ocean Margins*, Oxford University Press, Oxford, 100–123.

TAPPONNIER, P. & MOLNAR, P. 1979. Active faulting and Cenozoic tectonics of the Tie shan, Mongolia, and Baikal regions. *Journal of Geophysical Research*, **84**, 3425–3459.

TAPPONIER, P., PELTZER, G., LE DAIN, A. Y., ARMIJO, R. & COBBOLD, P. 1982. Propagating extrusion tectonics in Asia: new insights from simple experiments with plasticine. *Geology*, **7**, 611–616.

TAYLOR, B. & HAYES, D. E. 1983. Origin and history of the South China Sea basin. *In*: HAYES, D. E. (ed.) *The Tectonic and Geologic evolution of Southeast Asian Seas and Islands, Part 2, Geophysical Monograph, American Geophysical Union*, **27**, 23–56.

TENG, L. S. 1990. Geotectonic evolution of late Cenozoic arc–continental collision in Taiwan. *Tectonophysics*, **183**, 67–76.

TENG, L. S. 1992. Geotectonic evolution of Tertiary continental margin basins of Taiwan. *Petroleum Geology of Taiwan*, **27**, 1–19.

TENG, L. S. 1996. Extensional collapse of the Northern Taiwan mountain belt. *Geology*, **24**, 949–952.

TENG, L. S., WANG, Y., TANG, C. H., HUANG, C. Y., HUANG, T. C., YU, M. S. & KE, A. 1991. Tectonic aspects of the Paleogene depositional basin of Northern Taiwan. *Proceedings of the Geological Society of China*, **34**, 313–336.

WAGEMAN, J. M., HILDE, T. W. C. & EMERY, K. O. 1970. Structural framework of East China Sea and Yellow Sea. *American Association of Petroleum Geologists Bulletin*, **54**, 1611–1643.

WANG LEE, C. & WANG, Y. 1987. Tananao terrane of Taiwan – its relation to the late Mesozoic collision and accretion of the southeast China margin. *Acta Geologica Taiwanica*, **25**, 225–239.

WANG, Y. 1987. Continental margin rifiting and Cenozic tectonics around Taiwan. *Memoir of the Geological Society of China*, **9**, 227–240.

WORRALL, D. M., KRUGLYAK, V., KUNST, F. & KUZNETSOV, V. 1996. Tertiary tectonics of the Sea of Okhotsk, Russia: far-field effects of the India–Eurasia collision. *Tectonics*, **15**, 813–826.

XIA, K., HUANG, C., JIANG, S., ZHANG, Y., SU, D., XIA, S. & CHEN, Z. 1994. Comparison of the tectonics and geophysics of the major structural belts between the northern and southern continental margins of the South China Sea. *Tectonophysics*, **235**, 99–116.

YAN, P., ZHOU, D. & LIU, Z. S. 2001. A crustal structure profile across the northern continental margin of the South China Sea. *Tectonophysics*, **338**, 1–21.

YEN, H. Y., YEN, Y. H. & WU, F. T. 1998. Two-dimensional crustal structures of Taiwan from gravity data. *Tectonics*, **17**, 104–111.

YU, H. S. 1994. Structure, stratigraphy and basin subsidence of Tertiary basins along the Chinese southeastern continental margin. *Tectonophysics*, **235**, 63–76.

YU, S. B., CHEN, H. Y. & KUO, L. C. 1997. Velocity field of GPS stations in the Taiwan area. *Tectonophysics*, **274**, 41–59.

YUAN, J., LIN, S. J., HUANG, S. T. & SHAW, C. L. 1985. Stratigraphic study on the pre-Miocene under the Peikang area, Taiwan. *Petroleum Geology of Taiwan*, **21**, 115–128.

YUI, T. F. & LAN, C. Y. 1991. Isotopic compositions of Tananao marble in the Tungao area, northeastern Taiwan: a chronological consideration. *Special Publication of the Central Geological Survey*, **5**, 161–172.

YUI, T. F., WU, T. W. & LU, J. Y. 1994. Geochemical characteristics of metabasites from the slate formations of Taiwan. *Journal of the Geological Society of China*, **37**, 53–67.

YUI, T. F., WU, T. W. & LU, C. Y. 1995. Geochemical characteristics of metabasites from the slate formations of Taiwan: reply. *Journal of the Geological Society of China*, **38**, 179–182.

ZHAO, X., COE, R. S., GILDER, S. A. & FROST, G. M. 1996. Palaeomagnetic constraints on the palaeogeography of China: implications for Gondwanaland. *Australian Journal of Earth Sciences*, **43**, 643–672.

ZHOU, D., RU, K., & CHEN, H. 1995. Kinematics of Cenozoic extension on the South China Sea continental margin and its implications for the tectonic evolution of the region. *Tectonophysics*, **251**, 161–177.

ZHOU, X. M. & LI, W. X. 2000. Origin of late Mesozoic igneous rocks in Southeastern China: implications for lithosphere subduction and underplating of mafic magmas. *Tectonophysics*, **326**, 269–287.

ZHOU, Z., ZHAO, J. & YIN, P. 1989. Characteristics and tectonic evolution of the East China Sea. *In*: ZHU, X. (ed.) *Chinese Sedimentary Basins*, Elsevier, Amsterdam, 165–179.

Precisely relocated hypocentres, focal mechanisms and active orogeny in Central Taiwan

F. T. WU[1], C. S. CHANG[2] AND Y. M. WU[2]

[1]*Department of Geological Sciences, State University of New York, Binghamton,
NY 13902 USA (e-mail: wu@binghamton.edu)*
[2]*Central Weather Bureau, Taipei, Taiwan*

Abstract: The 1999 Chi-Chi earthquake series occurred in Central Taiwan, where ongoing mountain building is most active. The pre- and post-Chi-Chi seismicity helps to clarify the internal orogenic activity. The 27 000 earthquakes from the 1993–2002 catalogues have been relocated with greater precision. By associating the seismicity with focal mechanisms, many structures inside the orogen have been mapped. Among them are a steeply dipping thrust fault in the deep crust; a 50-km-long left-lateral strike-slip fault in the south; and an Eastern Central Range NNE-striking normal fault. While the deep crustal thrust appears to contribute to the root-building, the southern strike-slip slip fault accommodates the main-shock fault motion, and the Eastern Central Range normal faulting probably occurs mainly after a major western Taiwan thrust type earthquake. Much of the Backbone Range and the Eastern Central Range were seismically quiescent before and after the Chi-Chi earthquake. The contrast in the seismicity of the Central Range and the surrounding regions implies different material behaviour in these different regimes of the orogen.

Taiwan is an active orogen created from the convergence between the Eurasian and the Philippine Sea plate (Fig. 1). The rate of convergence is 70 mm/year, based on NUVEL-1 of De Mets *et al.* (1990), and recent estimates from GPS measurement are about 8 cm/year (Yu *et al.* 1999). This convergence is nearly totally absorbed in the shortening across the Taiwan, as the rate decreases from over 50 mm/year on the east coast to essentially zero in the Coastal Plain. The total rate of uplift is estimated to be in excess of 5 mm/year in the last million years, with a 1 cm/year rock uplift rate in the last 30 years (Liu, C. C., pers. comm., 2002). The seismicity in the orogen is coupled with these high rates of crustal movement. Since seismicity is a response to the tectonic stresses, the location and the focal mechanisms of earthquakes will provide information concerning the stresses, as well as giving the kinematics of faults. In this paper we utilize a relatively new and effective method to relocate more precisely the hypocentres of earthquakes (Waldhauser and Ellsworth 2000). The better-resolved hypocentre distribution forms the basis from which zones of active deformation can be mapped. Certainly, the presence of seismicity is an unmistakable sign of brittle deformation, but clearly defined zones of seismically quiescence in the orogen, especially in between regions of high

activity, call for understanding of possible rheological behaviour of rocks – brittle or ductile – in the orogen, and the thermal conditions there. In these studies, the precision of the hypocentral location is the key; only with precise locations can the correlation of the clear patterns of seismicity with known geological entities, such as major faults, crustal roots and so on, be explored.

A detailed seismicity study in Taiwan is made feasible because of a number of factors. Foremost among them is the timely upgrade of the seismic network. The Central Weather Bureau (CWB) seismic network evolved in stages from a historical analogue network in the late 1890s to the present digital network (Yeh *et al.* 2000). For the most recent major overhaul, completed in 1991, the number of stations was doubled, and the network was switched to digital technology (Rau *et al.* 1996). In an area of 36 000 km^2, there are 72 stations, with interstation spacing varying from a few kilometres to about 30 km (Fig. 2). On the average, more than 15 000 events are located every year, but most of them were located offshore of eastern Taiwan. In central Taiwan, where the mountain ranges are relatively high (reaching nearly 4000 m at the highest) and the orogeny is known to be most active the seismicity had been relatively low, historically as well as in the recent years. From 1991 through 19 September 1999, there were few noticeable

From: MALPAS, J., FLETCHER, C. J. N., ALI, J. R. & AITCHISON, J. C. (eds) 2004. *Aspects of the Tectonic Evolution of China*. Geological Society, London, Special Publications, **226**, 333–354.
0305-8719/04/$15 © The Geological Society of London 2004.

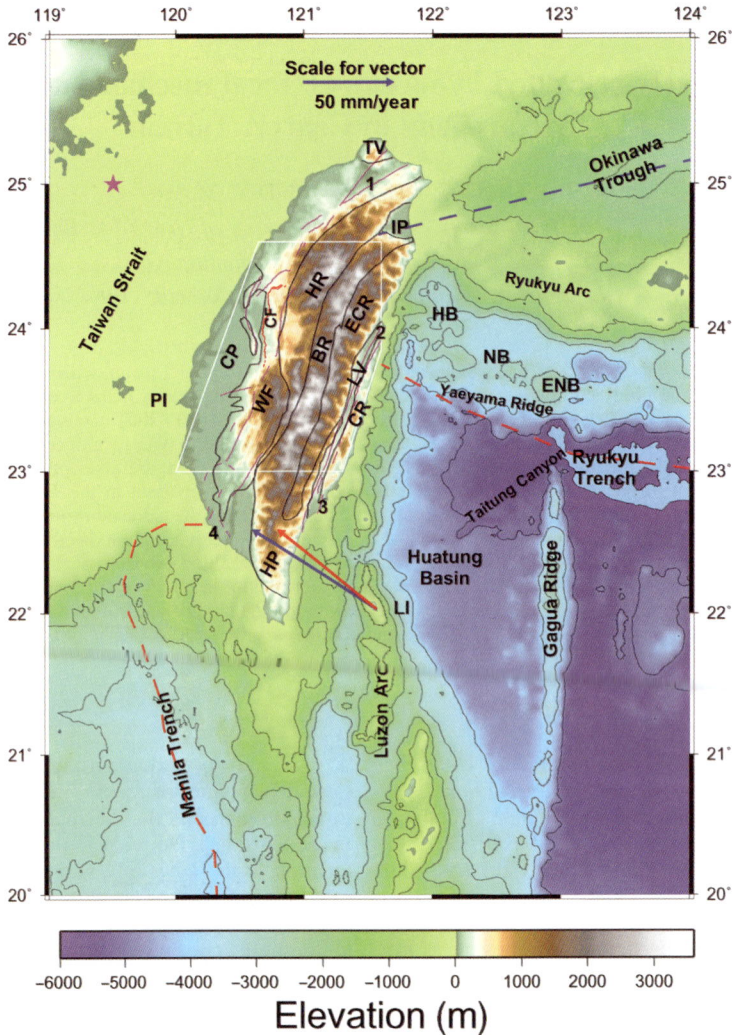

Fig. 1. Plate tectonics in the vicinity of Taiwan. The two-letter codes marking the physiological/geological regions are: CR, Coastal Range; ECR, East Central Range; BR, Backbone Range; HR, Hsueshan Range; WF, Western Foothills; CP, Coastal Plain; HP, Hengchun Peninsula; HB, Hoping Basin; NB, Nanao Basin, and ENB, East Nanao Basin. Place names are: 1, Taipei; 2, Hualian; 3, Taitung; 4, Kaohsiung; PI, Penghu Islands and LI, Lanhsu Island. The boundary between CP and WF is often viewed as the 'deformation front', i.e. the western limit of collision-related deformation, although the Taiwan Strait is actually active, as shown by the occurrence of a large earthquake (marked by a star; *Mc*.8). The arrows centred on the Lanhsu island south-east of Taiwan are the plate motion vector predicted by NUVEL-1 (the red vector) and the motion vector measured by GPS (red vector; Yu *et al.* 1997). The dashed red lines mark the approximate positions of the plate boundaries. The thin red line marks the Chelungpu Fault (CF) and thin magenta lines the other 'active faults'; the fault west of CF is the Changhua Fault. Our area of interest is indicated by the outlined box, extending from the Coastal Plain in the west to the Coastal Range.

events in the study area (between 23°N and 24°N in central Taiwan). Then, on 20 September 1999, the Chi-Chi earthquake ($M_w = 7.6$) occurred; more than 20 $M_w > 6$ (NEIC/USGS) and over 20 000 M > 2 aftershocks followed. The seismic voids in this region were filled, and new seismicity patterns developed. Some of the earthquake zones more clearly distinguishable before the Chi-Chi event in fact disappeared after the main-shock, and the aftershocks occurred not only in the main rupture zone, but even skipped the high Backbone Range and occurred in the

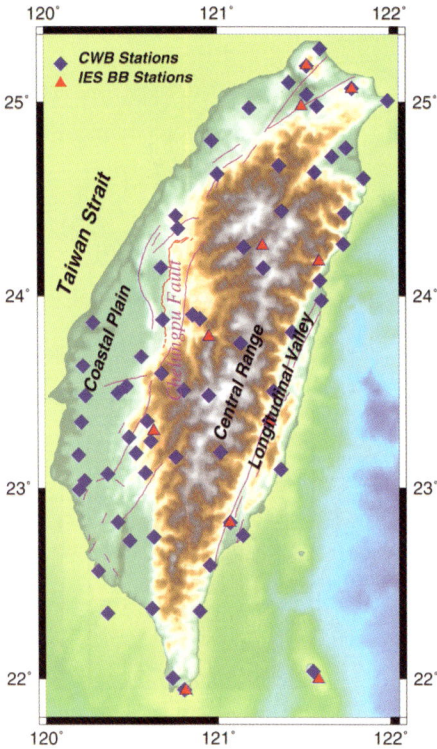

Fig. 2. Locations of the Central Weather Bureau (CWB) and Institute of Earth Sciences (IES), Academia Sinica, seismic stations. The diamonds indicate the CWB narrowband stations, and the squares the IES broadband stations; phase arrival times are from the CWB network, and most of the focal mechanisms used in this paper are from IES stations.

Eastern Central Range, a very low seismic region before the main-shock.

In addition to the short-period CWB network, appropriate for mapping detailed seismicity, the Broadband Array for Taiwan Seismology (BATS) network data are routinely used in the determination of the focal mechanisms of $ML > 3.5$ events. The focal mechanisms derived from waveform inversion are generally more robust than the short-period first-motion solutions (Kao *et al.* 2002). For the large earthquakes ($M > 5.5$), Harvard and USGS focal mechanism solutions are also available. Together with the seismicity, the motions along the fault internal to the orogen can be assessed using these mechanisms.

Seismicity has previously been used in studying the Taiwan orogeny. The most recent work by Carena *et al.* (2002) proposed the presence of a detachment fault by clustering seismic data on to perceived planar structures. Previously,

Tsai *et al.* (1977) mapped plate tectonics around Taiwan on the basis of seismicity, and Wu *et al.* (1989; 1997) included seismicity in their overall studies of Taiwan tectonics. They identified zones of seismic quiescence and activity, and associated them with crustal rheology. However, the relocation of hypocentres and the aftershocks of the Chi-Chi earthquake provide better data on known and unknown structures. Although the time window of our good seismicity data is quite short (actually from July 1993 through 2001) a number of significant points regarding the orogenic processes can be addressed.

Tectonics and geology of Taiwan

The plate-tectonic framework and tectonic units are shown in Figure 1. The Philippine Sea plate subducts toward the north along the Ryukyu Trench, and the Eurasian plate subducts toward the east along the Manila Trench. However, the Ryukyu Trench as a bathymetric low disappears west of 123°E, offshore of east Taiwan, and the Manila Trench loses its definition offshore of SW Taiwan north of 21.5°N. As a result, the plate boundaries shown in Figure 1 are only approximate in the immediate vicinity of Taiwan. The section of Taiwan in between the two ends of the boundaries depicted in Figure 1 is the most active section, and the Chi-Chi earthquake occurred in this area. As mentioned earlier, in between the subduction zones the convergence between the Philippine Sea plate and the Eurasian plate appears to be totally absorbed by shortening. The relative motion between Lanhsu, an extinct volcanic island offshore of southeastern Taiwan, and the Penghu Islands in the Taiwan Strait, is about 8 cm/year according to GPS data, slightly larger than the NUVEL-1 prediction (Yu *et al.* 1999; Fig. 1). This situation is in contrast to a typical cross-section across the middle part of the Andes in South America, where shortening takes up only 10–15 mm/year of the total convergence of 70 mm/year between the Nazca and South American plate, while subduction eventually consumes the rest of it (Norabuena *et al.* 1998).

The geology of Taiwan is often represented in a two-dimensional section, and for this paper such a method is adequate. Thus, starting from the east, and moving onshore from the Philippine Sea basin, the Coastal Range is encountered first. It is a telescoped ensemble including all the materials between the former Luzon volcanic arc and the trench (*Terrestrial, Atmospheric and Oceanic Sciences* 1987). Separating the Coastal Range from the Central Range to the west is the Longitudinal Valley (LV), for some time a depositional trough of continental sediments between two topographical highs. In the part of the Central Range just west of the LV, Mesozoic metamorphic rocks are exposed (ECR in Fig. 1); this is overlain by a suite of Eocene–Miocene sediments to the west; these rocks have been metamorphosed to slates, in what is usually called the Backbone Range (BR in Fig. 1). Further west is the Hsuehshan Range, which is well developed in the north but tapers out toward the south (HR in Fig. 1); the strata are Eocene–Oligocene in age, but are folded and less metamorphosed than the BR rocks. Neogene rocks underlie the Western Foothills. The boundaries between geological units are mostly faults. The major ones are the Longitudinal Valley fault(s), a left-lateral oblique thrust fault zone, and the Lishan fault – the boundary between the Backbone Range and the Hsuehshan Range. Although the Lishan fault is recognized by some (Ho 1988) as a major fault, based on the differences in lithology, age of strata, grades of metamorphism and styles of deformation (Lee *et al.* 1997), it is not shown on the most recent geological map of Taiwan (Central Geological Survey, Taiwan 2000). The Lishan Fault is actually an important boundary for seismicity, as we shall show later. Between the Foothills and the Coastal Plain is the Chelungpu Fault, and further west is the Changhua Fault (Fig. 1). Incidentally, the Changhua Fault is often depicted as the 'deformation front', in other words, the western limit of the orogenic deformation. But the Taiwan Strait is also seismically active, even though the tectonics may be dominated by N–S tension (Kao and Wu 1996).

The subsurface structures of Taiwan are complex, and recent tomography studies (e.g. Rau and Wu 1995; Ma *et al.* 1996) have only just begun to provide some key details. Figure 3 shows two cross-sections through central Taiwan in our region of interest, using results of Rau and Wu (1995). The crustal root under the Central Range reaches a depth of about 50 km in the north and somewhat less toward the south. It is also clear that the rocks under the Central Range is in the range have velocities of 4.5–5.5 km sec^{-1}. The low-velocity sediments are confined mainly to the Coastal Plain. Under the Coastal Plain the crustal thickness is about 25–30 km.

The current orogeny of Taiwan is geologically quite young, with an estimated age of 4–6 Ma. The present Taiwan began to emerge above sea-level at that time (e.g. Liu *et al.* 2001). The rapid uplift rates of *c.*0.6–0.9 cm/year in the last 0.6 million years, based on fission-track dating (Liu *et al.* 1982) and >1 cm/year in the

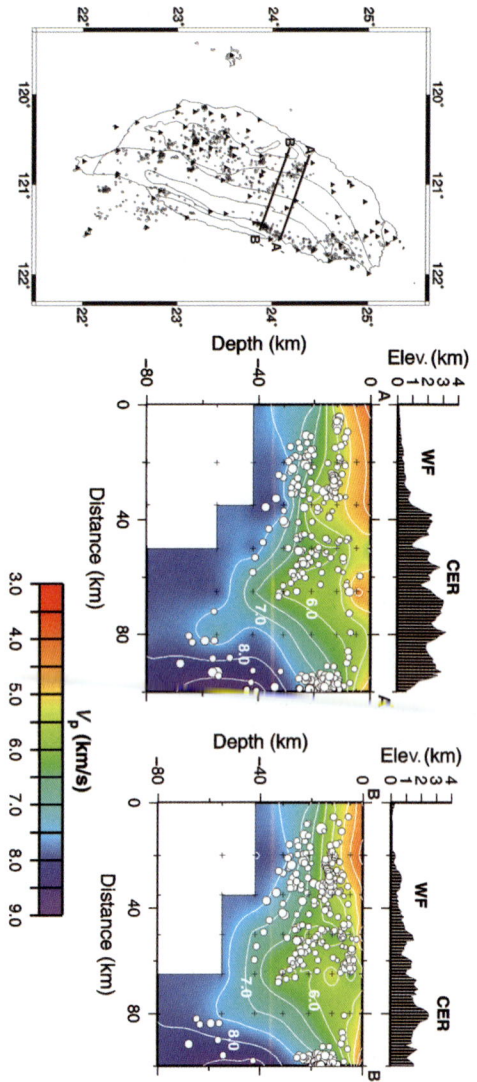

Fig. 3. Tomographic sections from Rau and Wu (1995). The locations of the profiles are shown in the figure, on the left. In both sections A–A′ and B–B′ the deepest part of the crust is nearly 50 km (as marked by the 7.5 km sec^{-1} contour). Materials with 5.5–6.0 km sec^{-1} rise under the Central Range (CER) to within a few kilometres of the surface. Under the Central Range the 5.5 km sec^{-1} materials reach to within a few kilometres of the surface, but in western Taiwan, under the Foothills and the Coastal Plain, the relatively low-velocity materials are thick. Section B–B′ crosses the area where the fault displacements of Chi-Chi are at their maximum.

last 30 years based on repeated levelling data (Liu, C. C., Institute of Earth Sciences, Academia Sinica, pers. comm., 2002); these rates imply that vigorous orogenic processes are continuing.

While it is the general view that the Taiwan orogen was created as a result of the convergence, or collision, of the Philippine Sea and Eurasian plates, its exact geometry and the mechanics of mountain-building are debatable. Models proposed for the Taiwan orogeny include those of Suppe (1981), Lallemand et al. (2001), Carena et al. (2002) and Wu et al. (1997). The first three authors proposed models that involve eastward subduction of the Eurasian plate and deformation of the Tertiary sedimentary wedge to create the Central Range. However, Wu et al. argued that no subduction has been identified and that collision involves the shortening of the whole lithosphere. The seismicity data that we present here should provide further information for constraining these models.

Earthquake data in Taiwan

The two seismic networks mentioned earlier have supplied the seismic data pertinent to the current study. The 72 stations of the Central Weather Bureau (CWB) narrowband (centred around 1 Hz) network are located on Taiwan and its neighbouring islands (Fig. 2). All stations are equipped with three-component seismometers, and the signals are digitized on-site at the rate of 100 samples per second. The 12-bit data are then transmitted to the recording centre in Taipei. Although most of the stations are surface installations, at sites near cities a switch to borehole sensors at the depth of tens to over a hundred metres was made in July 1993; we only use data from after the switch. No significant differential time delays between stations were introduced in the transmission, and thus the relative arrival times, which double difference relocation relies on, are not affected by the transmission. The arrival times are read and used by the CWB for routine earthquake location and magnitude determination. Times are read with a precision of 1/100th of a second, determined by the sampling interval. The arrival times and the locations are archived at the CWB, and are used as the initial locations for our relocation program. For this study, all $M > 2$ events in the period July 1993 to the end of 2002, with a minimum of eight observations in a polygon (Fig. 1) are candidates for relocation. The other network of concern here is the BATS (Broadband Array in Taiwan for Seismology) network of the Institute of Earth Sciences (IES), Academia Sinica. Installation began in 1994, and by mid-1995 there were enough stations for waveform moment tensor inversion, from which the double-couple focal mechanisms can be derived. The station locations

are shown in Figure 2. At each site, Streckeisen (STS-1 or STS-2) seismometers and digital data loggers are generally used. The BATS focal mechanism solutions are obtained for $M > 3.5$ events and published regularly (Kao et al. 2002). These results are used in this paper.

Double-difference relocation

Traditional earthquake location involves the adjustment of x, y, z and t_0, namely the three spatial coordinates and the origin time, to minimize the residuals between observed and calculated arrival times. The predicted arrival time is calculated on the basis of the Earth being a stack of layers with different velocities – a poor approximation to reality. The process is begun by assuming an initial trial location, and successive iterations are programmed to minimize the sum of the square of the residuals. In double-difference methods (Waldhauser 2001) the residuals in arrival times between events i and j (the double difference) are used. They are defined as:

$$\mathrm{d}r_k^{ij} = (t_k^i - t_k^j)^{\mathrm{obs}} - (t_k^i - t_k^j)^{\mathrm{cal}}$$

where 'obs' refers to observed, and 'cal' refers to calculated arrival times. For two events i and j we may write:

$$\frac{\partial t_k^i}{\partial m}\Delta m^i - \frac{\partial t_k^j}{\partial m}\Delta m^j = \mathrm{d}r_k^{ij}$$

where Δm^i is the adjustment of the hypocentral parameters ($\mathrm{d}x$, $\mathrm{d}y$, $\mathrm{d}z$ and $\mathrm{d}t$) for event i; if the events are very close, then the same slowness can be used:

$$\frac{\partial t_k^{ij}}{\partial m}\Delta m^{ij} = \mathrm{d}r_k^{ij}$$

The resulting normal equation obtained from both the neighboring and the more distant events can be solved, if very large, by LQSR (Paige & Saunders 1982) and, for a small cluster of events, by using Singular Value Decomposition (SVD; e.g. Press et al. 1986).

To start off the iterated solution, the catalogue locations are used. In contrast to the routine earthquake location, in which each event location is determined independently, the double-difference relocation utilizes the fact that, if two events are very close, then the relative locations of the two events can be determined more efficaciously by the differences in arrival times at stations that record both events. By so doing, the effect of the assumed velocity structures for calculating travel times is minimized. The formulation above also implies that the locations of the whole set of events are linked and that their relative locations with respect to an average location are determined by minimizing the squares of the double-difference residuals. The relative location should be quite good – within a few hundred metres (Waldhauser 2001).

A comparison of the CWB catalogue and the relocated seismicity

To demonstrate the effects of relocation, we present maps of epicentres, and five cross-sections each of both the CWB catalogue and the double-difference results for more than 6000 events in 2000 (Fig. 4). In the maps (Fig. 4a and 4c) the tightening of the epicentral clusters can clearly be seen after relocation. The five cross-sections in Figure 4b and 4d demonstrate how clouds of foci segregate into tighter groups. Compare, for example, the depth distributions (Fig. 4a and b, profiles 3) for a cluster of events just west of the Chelungpu fault at about 24°N, the relocated foci indicate a gently inclined east-dipping zone, while the catalogue results are quite scattered. Also, the relocated events under the Eastern Central Range in profile appear as a line of hypocentres, mapping a possible steep east-dipping fault. Although there is no a priori reason that tightly clustered foci are correct or better, the decrease in overall residuals and the improvement of formal location errors from a few kilometres to a few hundred metres, as well as the extensive demonstrations of Waldhauser *et al.* (1999) and Waldhauser & Ellsworth (2000, 2002), indicate that the tightly clustered results are meaningful.

Seismicity before and after the 1999 Chi-Chi earthquake

Pre-Chi-Chi seismicity

Seismicity. The Chi-Chi main rupture zone (e.g. Ji *et al.* 2001) had not been very active seismically before the Chi-Chi earthquake. In fact, only one moderately damaging event is known to have occurred there in the 1650–1999 period – the 1917 Puli event (Wu 1978). For the activity a few years before the Chi-Chi main-shock, we show the epicentral map and the cross-sections for the 1 July 1993–20 September 1999, seismicity in Figure 5a and Figure 6, respectively. One of the main features that can be recognized in Figure 5a was known since the first telemetered network was established

in 1973 (Wang *et al.* 1983). This is the linear belt of seismicity that starts near the east-trending section at the northern end of the Chelungpu Fault, and continues south eastward for about seventy kilometres (Fig. 5a). It was often considered as a potentially hazardous seismogenic structure, and designated the Sanyi–Puli Belt (Lee *et al.* 1997; Wu & Rau 1998). A series of cross-sections parallel and perpendicular to this feature show that there are actually two zones: the upper one in the depth range of 5–15 km, and the lower one in the depth range 20 to 40 km; the top zone is flat or dipping to the south-west, and the deeper zone has a more complex structure and dips to the NE in places (Fig. 7). This zone essentially underlies the SW edge of the Hsuehshan Range and it is probably intimately related to the creation of the Range (see later discussion).

In the middle part of the region (Fig. 6, profiles 3–11) seismicity is relatively high in the Foothills, but noticeably low under the higher ranges; this gap under the Central Range has been recognized, although not as clearly, in earlier studies (Wu *et al.* 1989, 1997). As shown in Figure 5a the gap is essentially bounded by the western limit of the Backbone Range, or the Lishan fault as mentioned earlier. It is also defined clearly in profiles 4 through 12 in Figure 6. Seismicity does affect the region at higher elevation in profiles 1 through 3 and profiles 13 through 15. Interestingly, the events under the higher ranges are mostly above 10 km, explained elsewhere as a consequence of the rheological behaviour of rocks under the Central Range (Wu *et al.* 1997). In the southern part of the Central Range in our study area, we find a curious arch-shaped seismic zone under the Range, without much of a gap and with the events shallower under the high Central Range than on its flanks (Fig. 6, profiles 13, 14 and especially 15). In plan view (Fig. 5a) a series of relatively short (10–20 km in length) linear seismic zones can be seen under the high Ranges; they appear as narrow steep zones in cross-sections (Fig. 6, profiles 13–14). Along the coast of eastern Taiwan, between 24°N and 24.3°N the seismic zones are near vertical or dipping to the west, but in the Coastal Range – a region that has seen a few large earthquakes – the seismic zones are quite complex (Fig. 6, profiles 4–14). In some sections east-dipping zones dominate, but in others clearly west-dipping structures are displayed. In profiles 12 and 13 (Fig. 6), narrow east-dipping zones are quite clear.

Focal mechanisms. There are relatively few earthquakes, between July 1993 and the time of the Chi-Chi main-shock, in our study area that

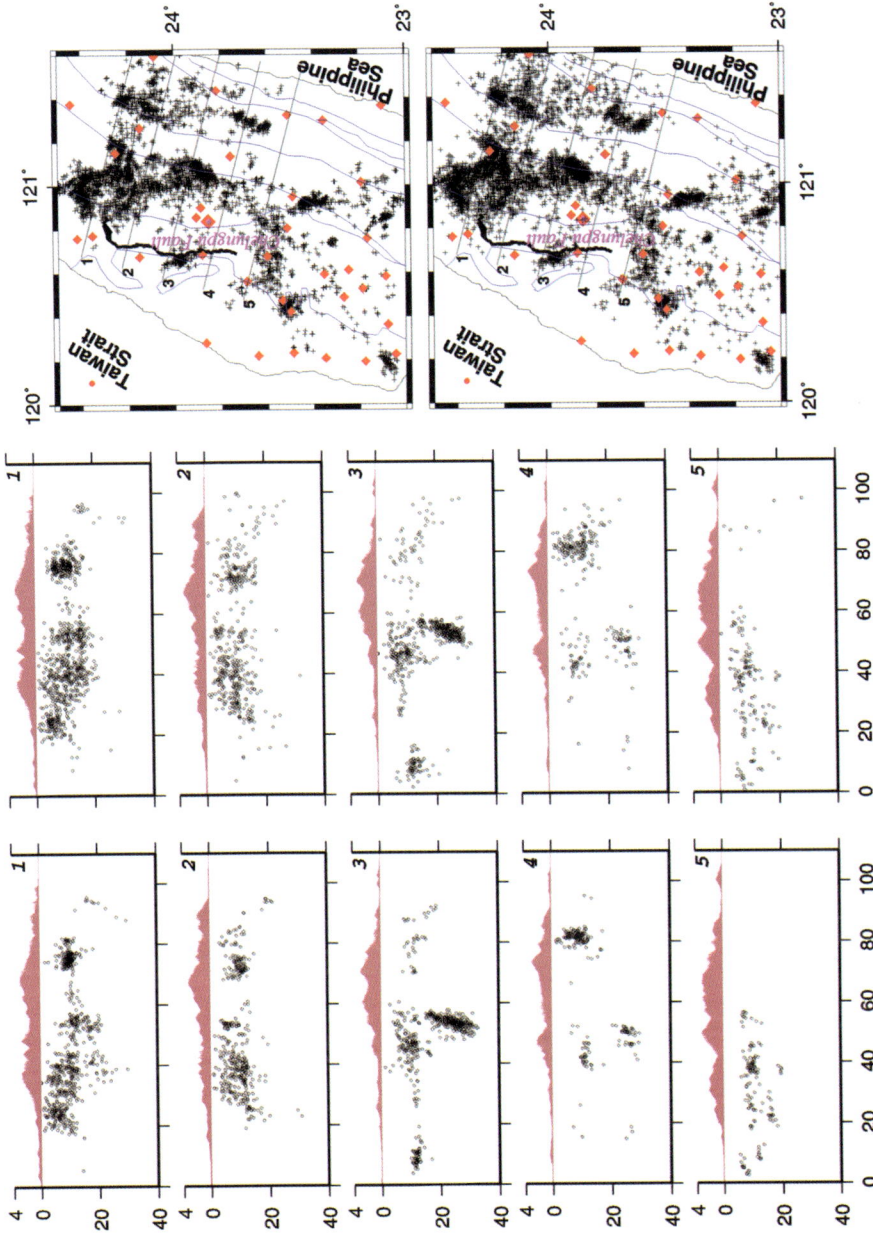

Fig. 4. Comparisons of the CWB catalogue and relocated event locations in plan view and in cross-sections. Aftershocks in 2000 are used. The five left-hand panels show the relocated events, and to their right the corresponding catalogue locations are shown; the locations of the profiles are shown in the maps on the right; the top figure shows the relocated and the bottom the catalogue epicentres. Note that the relocated events are more tightly clustered. In the two maps on the right, the top one shows the relocated and the bottom one the catalogue epicentres. The diamonds in this and later figures indicate the locations of the CWB network.

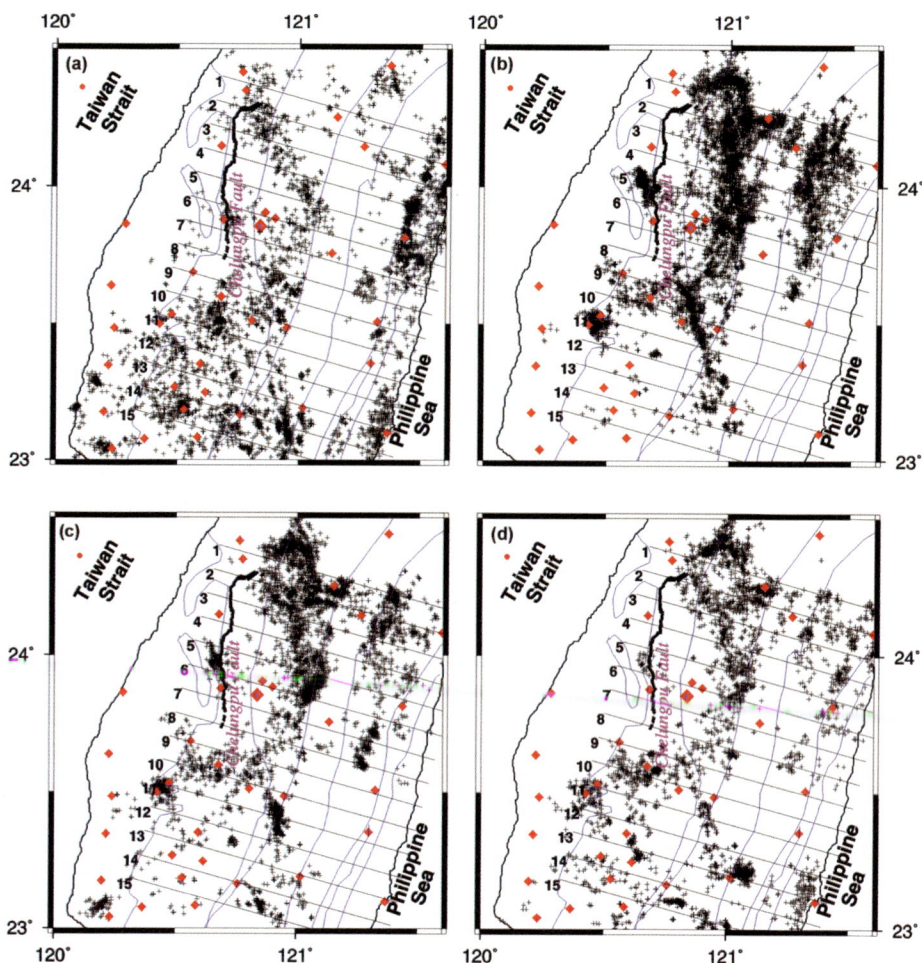

Fig. 5. Epicentral map of events in (**a**) 1 July 1993 to 20 September 1999, just before the Chi-Chi main-shock; (**b**) 20 September 1999, to 31 December 1999; (**c**) 1 January 2000, to 31 December 2000; (**d**) 1 January 2001, to 31 December 2002. Diagonal lines and numbers refer to cross-sections in Figures 6, 11, 12 & 13; for (**c**) there are no corresponding sections in other figures, but they are plotted on the map for reference.

are large enough ($M > 3.5$) for determination of focal mechanisms using BATS data (Kao *et al.* 2002). The mechanisms shown in Figure 8a are the BATS solutions, but they are placed at relocated epicentres; this is justified on the grounds that the moment tensor inversion results are not sensitive to a shift of a few kilometres in the location (H. Kao, pers. comm., 2003). We note that there was an *M*3.7 earthquake with nearly the same epicentre as the main-shock (Fig. 8a), but the BATS and the relocated depths of the event are both about 25 km, and thus much deeper than the Chi-Chi main-shock. It therefore cannot be considered a foreshock of the Chi-Chi earthquake. The largest earthquake in this period is the so-called 'Rueyli event' (mechanism

1998.07.04.51 in Fig. 9), about 30 km to the south of the southern end of the Chelungpu fault – the surface trace of the Chi-Chi earthquake fault. The fault plane solution and the hypocentral distribution favour the presence of an east-dipping plane. The solutions for events in the northern Longitudinal Valley are mainly high-angle thrust type, consistent with the focal distributions shown in Figure 6 (profiles 1–6 and 12–13).

Post-Chi-Chi seismicity

Since the main-shock, over 20 000 $M > 2.0$ aftershocks have been located by the CWB up to the end of 2002. The aftershocks immediately following

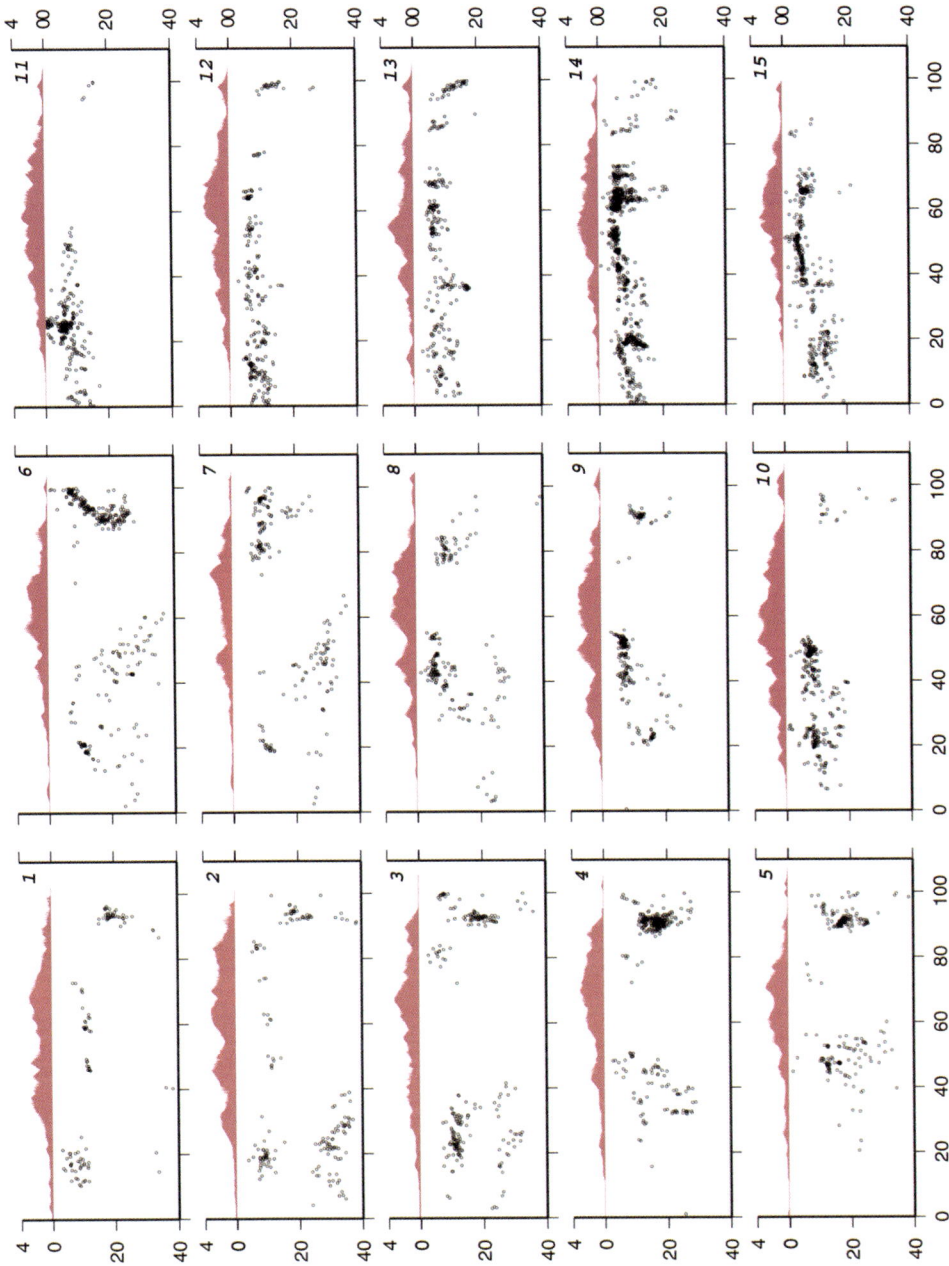

Fig. 6. Cross-sections of pre-Chi-Chi seismicity in Central Taiwan corresponding to Figure 5a. The locations of the profiles are also shown in Figure 5a. For each profile in this and later figures, events within 5 km on both sides of the profile are plotted. The profiles are 10 km apart. As in these and all the following sections, the horizontal axis is marked in kilometres from the end of the profile lines and the vertical axis is depth in kilometres.

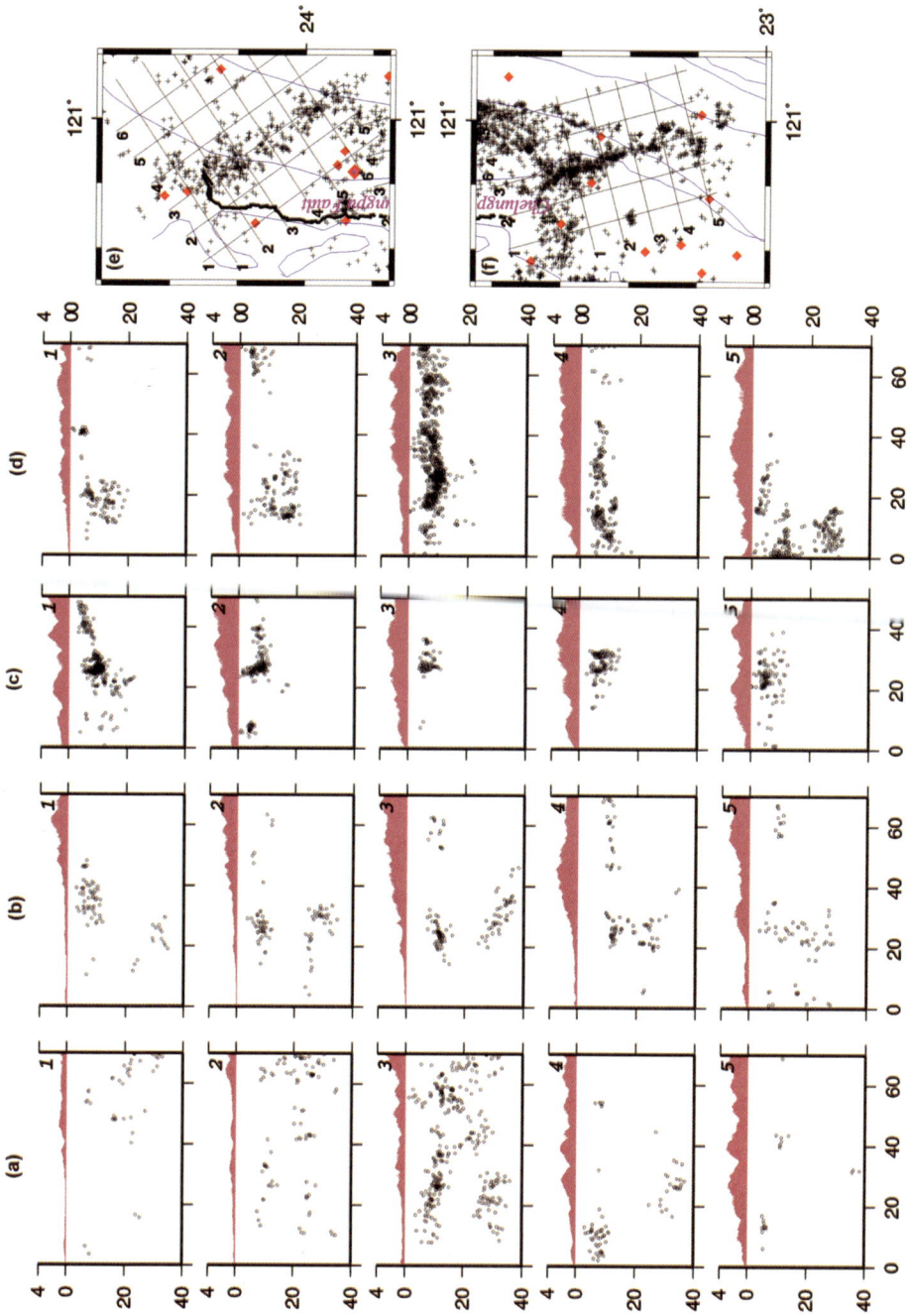

Fig. 7. Locations of profiles in two areas to show the details of two seismic zones. (**a**) Along strike of the Sanyi–Puli zone in northern Taiwan before the Chi-Chi main-shock; (**b**) Perpendicular to strike of the Sanyi-Puli zone before the Chi-Chi mair-shock; (**c**) Along the strike of the Liliao Fault, 20 September 1999 to 31 December 1999; (**d**) Perpendicular to strike of the Liliao Fault, 20 September 1999 to 31 December 1999; (**e**) Epicentral map and location of profiles for the Sanyi-Puli zone; and (**f**) Epicentral map and location of profiles for the Luliao Fault.

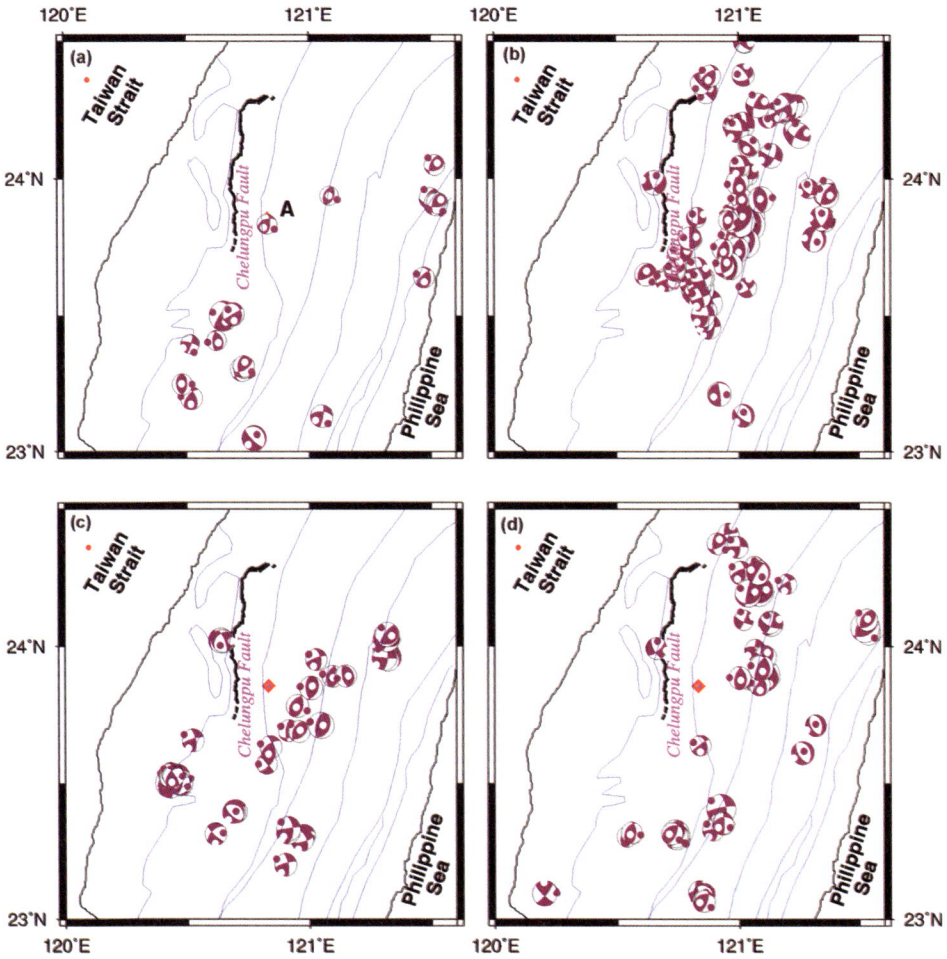

Fig. 8. Focal mechanisms of $M > 3.5$ earthquakes in the Chi-Chi area. (**a**) from 1996 to 1999, before the main-shock; event A occurred near the epicentre of the main-shock, but is at much greater depth; (**b**) from 20 September 1999, after the main-shock for 10 days; (**c**) 1 October to December 1999; (**d**) from 1 January 2000 to June 30, 2001.

the main-shock are so closely spaced in time that a portion of the events within the first ten days remains unlocated. But, with the majority of events already processed, future additions are not expected to alter the pattern of seismicity defined by the available events. From the time of the main-shock on 20 September 1999, to the end of 2000, 18 $M_w > 5.5$ events had occurred. Eight of them are listed in Table 1 with the USGS National Earthquake Information Center and our relocated hypocentres. Focal mechanisms for these events are available and, together with their own aftershock seismicity, seven of them can be associated with fault planes and particular types of faulting.

The post-Chi-Chi seismicity is shown in plan view in Figs 5b (20 September to December 1999), 5c (2000) and 5d (2001 and 2002). To see the initial development of the seismicity patterns, we have plotted the seismicity maps of September, October, November and December 1999, separately in maps (Fig. 10) with all post-Chi-Chi foci in 1999 shown together in cross-sections (Fig. 11).

Seismicity. In the first two and one half hours, the aftershock seismicity was limited to the east of the Chelungpu Fault, west of the Lishan Fault (Fig. 1), and above 13 km or so. Events in this zone continued to be the dominant seismicity in this area for the next two years. However, in the main rupture area, as indicated by the results of dynamic fault modelling (Ji *et al.* 2001;

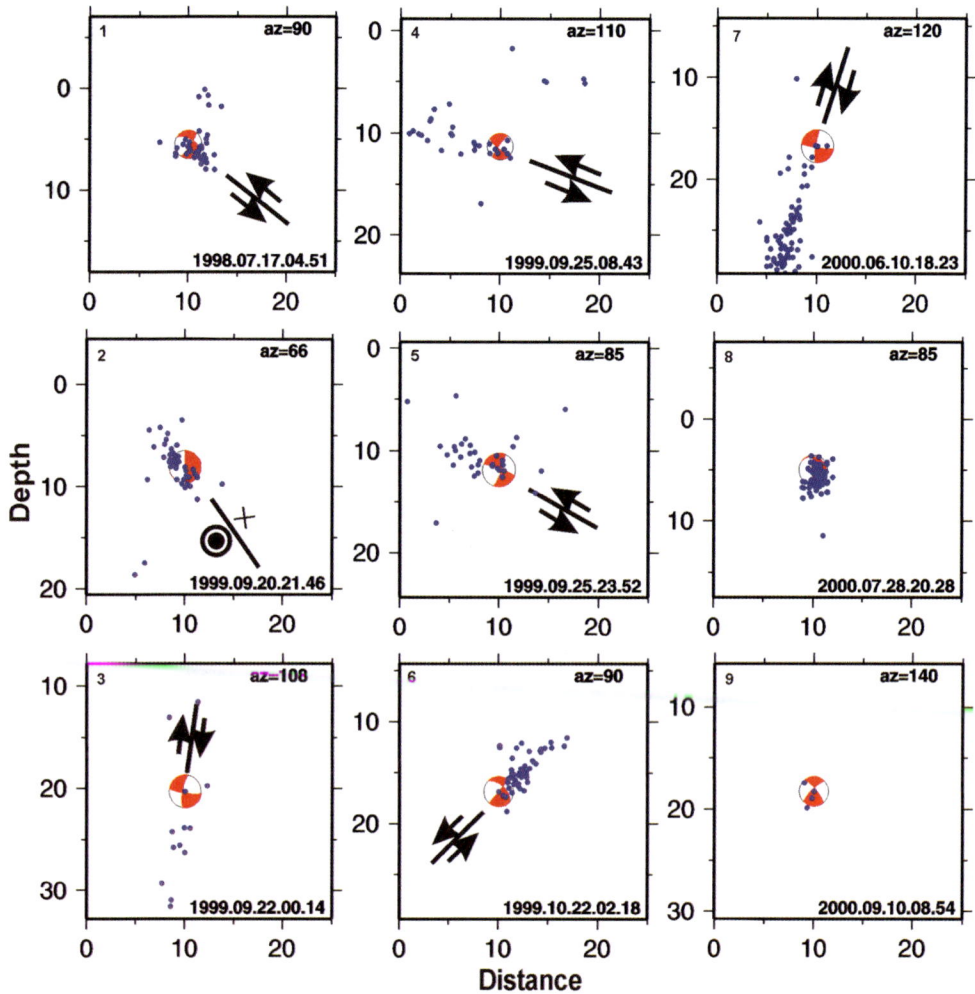

Fig. 9. Harvard CMT (all but 1999.09.25.08.43, which is a BATS solution) focal mechanisms for $M_w > 5.5$ events. The first 2–24 hours of aftershocks after each earthquake are plotted in the same frame. The date, hour and minute for each event are shown at the lower right-hand corner; the azimuth of the profile in the upper right-hand corner and the preferred fault orientation and movements are shown by a line and also by arrows. For events 8 and 9, too little information is available to decipher the associated fault.

Ma *et al.* 2001), directly east of the Chelungpu Fault, the aftershock activities were noticeably lower. The high activity concentrates mainly beyond the eastern edge of the rupture zone. We shall refer to this zone as the Chi-Chi zone in further discussion.

By viewing the time-lapse seismicity display in maps and in sections, we have followed the evolution of the seismicity in more detail than we can show in this paper. Three remarkable developments began approximately two and a half hours after the main-shock. One is the appearance of an Eastern Central Range seismi-

city belt, skipping the intervening higher mountain ranges; the first shock (of $M > 2$) of this NNE-trending zone was located toward the south, but, within two days, the zone grew toward the north, and, within one month, the Eastern Central Range zone attained nearly the same length as the Chelungpu Fault (Fig. 10b). The southern part of the eastern zone is relatively narrow and dipping steeply toward the east (Fig. 11, profile 8). The northern part of the zone is more complex. In fact, a few days after the main-shock shallow seismicity developed across the Backbone Range as well (Fig. 10b and Fig. 11, profiles 1, 2 and 3).

Table 1. $M_w > 5.5$ events from USGS/NEIS*

Number	Year	Month	Day	Hour	Minute	Second	Lat	Long	Depth	M_w
1	1998	7	17	4	51	15	23.5	120.65	5.49	5.7
2	1999	9	20	21	46	37.3	23.59	120.83	7.89	6.5
3	1999	9	22	0	14	40.8	23.83	121.05	23.22	6.4
4	1999	9	25	8	43	29.2	23.68	120.97	11.32	5.6
5	1999	9	25	23	52	49.5	23.85	121.01	11.47	6.5
6	1999	10	22	2	18	56.9	23.5	120.43	16.85	5.9
7	2000	6	10	18	23	29.5	23.89	121.11	16.77	6.4
8	2000	7	28	20	28	7.7	23.4	120.93	4.93	5.7
9	2000	9	10	8	54	46.5	24.07	121.52	18.27	5.8

* The locations are the hypoDD results.

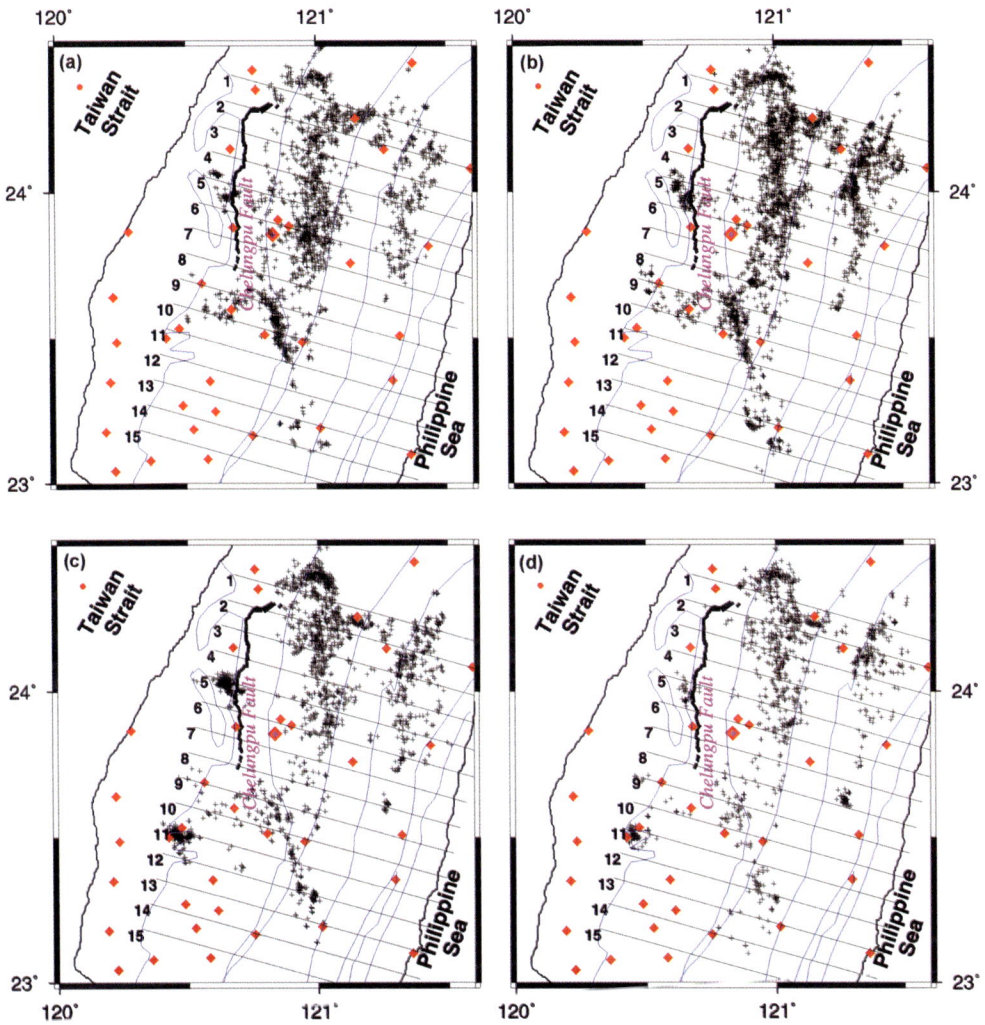

Fig. 10. Development of seismicity within the first hundred days of the Chi-Chi main-shock, as shown in epicentral maps in the following periods: (**a**) Within the first 11 days (from 20 September to 30 September 1999); (**b**) October 1999; (**c**) November 1999; (**d**) December 1999.

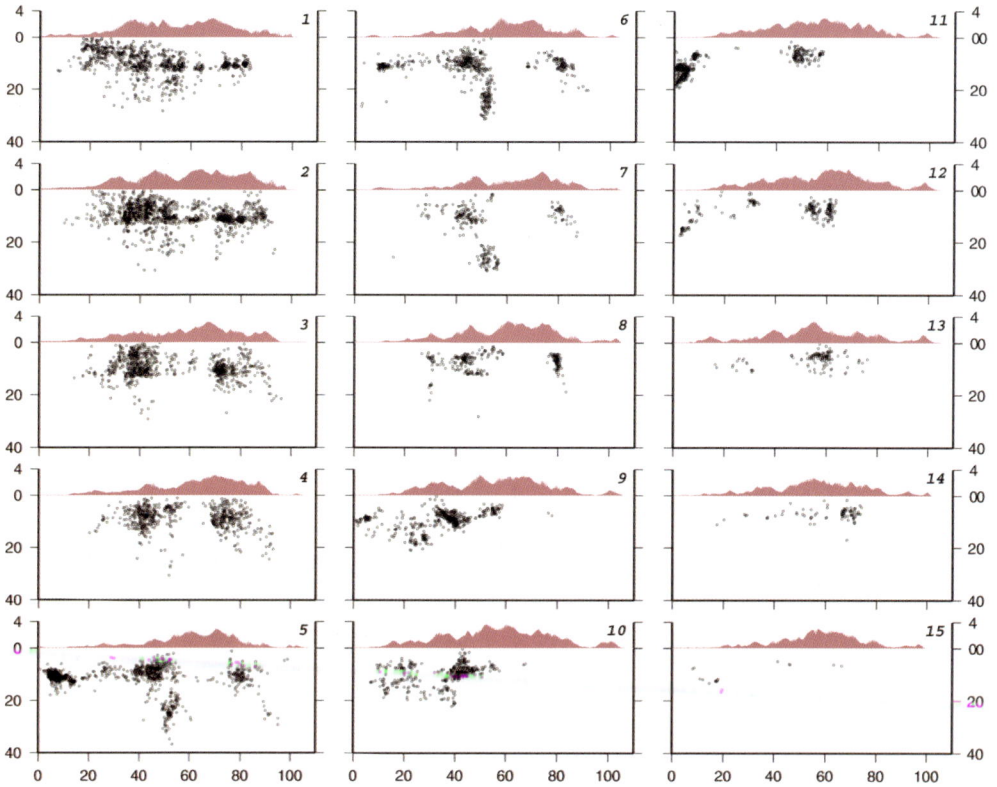

Fig. 11. Cross-sections of post-Chi-Chi seismicity from 20 September to 31 December 1999.

The pre-existing gap was bridged in this area, although the Backbone Range between latitudes of 23.3°N to 24°N was still quiescent, as shown in Fig. 5b. The seismicity in the eastern zone is evidently 'triggered' by the Chi-Chi main-shock, as earthquakes there began to appear within a few hours of the main-shock in an area that had been relatively quiet for many years.

The second remarkable development was the initiation of the lower crustal seismicity, between depths of 15 to 35 km immediately to the southeast of the main-shock zone (Fig. 4, profile 3; Figs 11 and 13, profiles 5 through 7). This zone, about 30 km long, dips steeply toward the west and strikes N–S; in the same profiles one can find other less-extensive zones to the west. Bounding the quiescent zone to the east and lying essentially under the Lishan Fault, this zone may play an important role in the orogen. Before the Chi-Chi earthquake, deeper earthquakes in this area did occur, but they were more scattered (Fig. 6, profiles 5, 6 and 7). At shallower depths (less than approximately 15 km) no corresponding zones of concentration under the Lishan Fault can be discerned; the western

boundary of the quiescent zone is somewhat irregular (Figs 11 and 13, profiles 4–12).

The third development was the well-delineated, long and narrow belt of NNW-trending seismicity to the south of the Chi-Chi zone (Figs 5b, c and d). By taking sections along and across the zone, it can be seen that the zone dips steeply to the east and is contained mainly within the depth range of about 12 km (Fig. 7c, profile 2 and Fig. 7d, profile 3). This zone began to develop at its northern end, again within about two and a half hours of the main-shock, with its northern end connecting to the fairly complex zone of deeper crustal events (Fig. 7d, profile 5). It propagated southward for a distance of more than 50 km, and the seismicity is confined above *c.*13 km. This feature is somewhat unique, in that it cuts across geological boundaries, including the Lishan Fault. This zone has been linked to en echelon cracks on the ground (Li Yuanhsi, Central Geological Survey, Taiwan, pers. comm., 2003) and is identified as a part of the pre-existing Liliao fault.

The seismicity near the Chelungpu fault is very low, except for a group of events between the Chelungpu and the Changhua Faults (Fig. 1 and Figs 5b, c and d). Activities began shortly after the main-shock, and continued through 2001. This cluster of aftershocks forms a shallow dipping zone (Fig. 11, profile 5 and Fig. 4, profile 3). When extrapolated to reach the surface, its trace would be about 20 km west of the Changhua Fault. However, the geometry of this zone may not be related to a thrust fault plane, as the focal mechanisms of the larger events show high-angle normal faulting (see below).

In the pre-Chi-Chi seismicity we have noted the presence of double-layered seismicity (e.g. Fig. 6, profiles 1, 2 and 3 in particular). But, in the corresponding post-Chi-Chi sections (Fig. 11, profiles 1, 2 and 3), the gaps are filled with events. After 2001 however, the gaps become visible again (Fig. 12, profiles 1, 2 and 3). However, it is interesting to note that even in the most active post-Chi-Chi period some of the basic features of seismicity remain. The seismically quiescent zone under the Backbone Range east of the Lishan Fault was clearly outlined during the first three months after the main-shock (Fig. 5b and Fig. 13, profiles 3–10). Later, in 2001–2002,

more events were found there, but the shallow events dominated (Fig. 12, profiles 4 and 5). However, under the northern Longitudinal Valley and the Coastal Range, as well as in the Sanyi–Puli zone, seismicity was largely absent from the time of the main-shock until 2001.

Focal mechanisms and seismicity of $M > 5.5$ events. Of the $M > 5.5$ earthquakes after the Chi-Chi main-shock that are listed in the USGS/NEIS catalogue, many occurred within a few hours of the main event, or followed closely behind a previous large event, so that focal mechanism solutions for these events from either global or BATS data cannot be determined. For those with focal mechanisms, the choice of which of the two planes is the fault can be made with the relocated seismicity following the events. We have singled out nine events, including one in 1998, before the Chi-Chi earthquake, in the $M > 5.5$ catalogue obtained from the USGS/NEIS website for our study (Table 1). Except for one event (No. 4 in Table 1) Harvard moment tensor solutions are available; there are differences between some of the BATS and

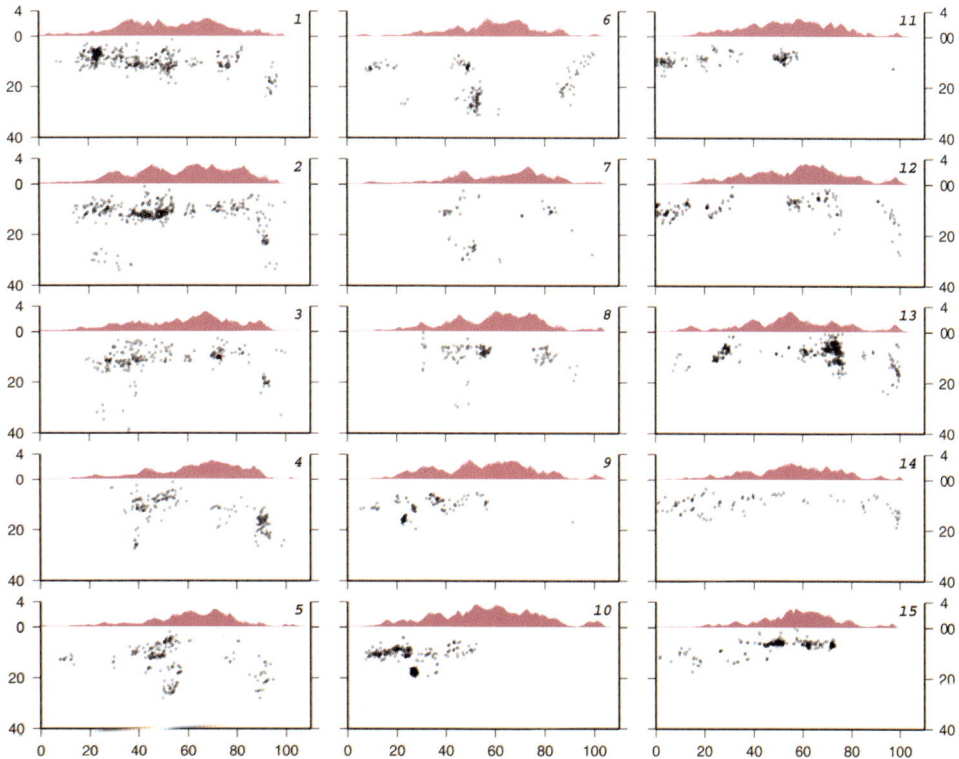

Fig. 12. Cross-sections of post-Chi-Chi seismicity for 2001 and 2002. For locations of the profiles, refer to Figure 5d. In profile 2 note the absence of earthquakes at depths of around 20 km.

Harvard solutions, but most of them are quite similar. We isolate the aftershocks within 48 hours after these events and show both the focal mechanisms and the relocated aftershocks in cross-sections (Fig. 9). The shocks are plotted at the locations determined in our study, in order to match the main-shock with the aftershocks; their location parameters and magnitudes are listed also in Table 1. We should note here that the teleseismic mechanism solutions are not sensitive to the relatively small adjustments in hypocentre locations in any case.

In Figure 9 event No. 2 is located near the northern end of the southern linear seismicity zone, or the Liliao Fault zone; the $M_w = 6.5$ event took place within four hours of the Chi-Chi main-shock. The seismicity lined up quite well in cross-section and on the map with the NNW-oriented left-lateral fault (Figs 7c, d and f; Fig. 8b); the Harvard CMT solution shows an essentially vertical plane, but the seismic zone is inclined (Fig. 9, no. 2). Events No. 3 and No. 7 are two of the deeper events, and the aftershocks indicate that the steeply west-dipping planes were the likely fault planes. Thus these two represent high-angle thrust fault, agreeing with conclusions made by Chen *et al.* (2002). The shallower events, No. 4 and No. 5, are associated with relatively shallow-dipping thrust faults, based on aftershock distributions. Event No. 6 was the main-shock that occurred to the west of the pre-Chi-Chi Rueyli earthquake (No. 1), quite distant and isolated from the main Chi-Chi seismic zone. It was preceded by a few foreshocks two days before the earthquake on 22 October. The aftershocks in the 24 hours after the main-shock clearly delineate a west-dipping zone, as shown in Figure 9 and also in Figure 11, profiles 11 and 12. The aftershock zone for event 8 is at the southern end of the Liliao seismic zone, and is evidently associated with a N–S-trending left-lateral strike-slip fault, but, as for event No. 9, too few aftershocks are present in the figures to define the fault plane dip.

Relation between BATS focal mechanism for 3.5 < M < 5.5 events and seismicity. For smaller events, there is an abundance of BATS solutions in the few months after the main-shock. Figures 8b, c, and d show the results for September, October, and November plus December, respectively. The aggregate of thrust mechanisms in the area east of the Chelungpu Fault in Figure 8b shows that nearly horizontal E–W compressional stress (Kao and Angelier 2001) controls the tectonics in this region. The strike-slip faulting in the Liliao seismic zone is consistent with this stress field. However, there are several exceptions to this rule for events in the surrounding areas. Some of the clear exceptions are the normal faulting events with a E–W tensile stress axis in the Eastern Central Range, shown in Figures 8b and d. Furthermore, events west of the Chelungpu fault around 24°N also show similar mechanisms (Figs 8b, c, d). Although the events in these two regions have similar mechanisms they probably arise from totally different reasons, as we shall argue later.

Discussion: seismicity and brittle/ductile deformation of the Taiwan Orogen

While surface geology provides the boundary conditions for understanding a young mountain range, the earthquakes inside it can track a part of its internal deformation. The recognition of the deformation pattern is made easier when the catalogue locations of such events can be improved, as shown in Figure 4. Relocated hypocentres in Central Taiwan show a number of well-delineated zones. Some of these zones are nearly 'planar', conforming to our concept of fault zones. Certainly, each earthquake, large or small, has its own source zone and sense of motion. It is natural to assume that the distribution of the aftershocks in the first few tens of hours following a large event is related to the causative fault. Our results show that, in some cases, the alignment of earthquake foci within one or two days of the large event with one of the planes of the focal mechanism solution lends strong support to the choice of one of the planes of the mechanism solution as the fault plane. Experiences elsewhere (Waldhauser and Ellsworth 2000, 2002) show that the epicentres aligned very closely with the surface trace of faults. Figure 13 is the same as Figure 11, except that the interpreted faults based on seismicity and focal mechanisms are marked.

One of the clear features determined from the post-Chi-Chi seismicity is the deep crustal seismic zone. The epicentres of this zone lie very close to the Lishan Fault, and the steeply west-dipping zone lines up very closely with the steeply dipping plane of the focal mechanism solutions (both thrusts) of two large Chi-Chi aftershocks in this zone (Fig. 9, events 3 and 7) leaving little doubt that the rather intense seismicity in this zone is associated with a significant reverse fault at a depth of 20–35 km. Chen *et al.* (2002) also found this zone using their temporary network data, and call it a conjugate fault to the main Chi-Chi rupture. Since the compressive stress in this region is nearly horizontal (Kao and Angelier 2001), the angles between the main-shock rupture and the steep zone are quite

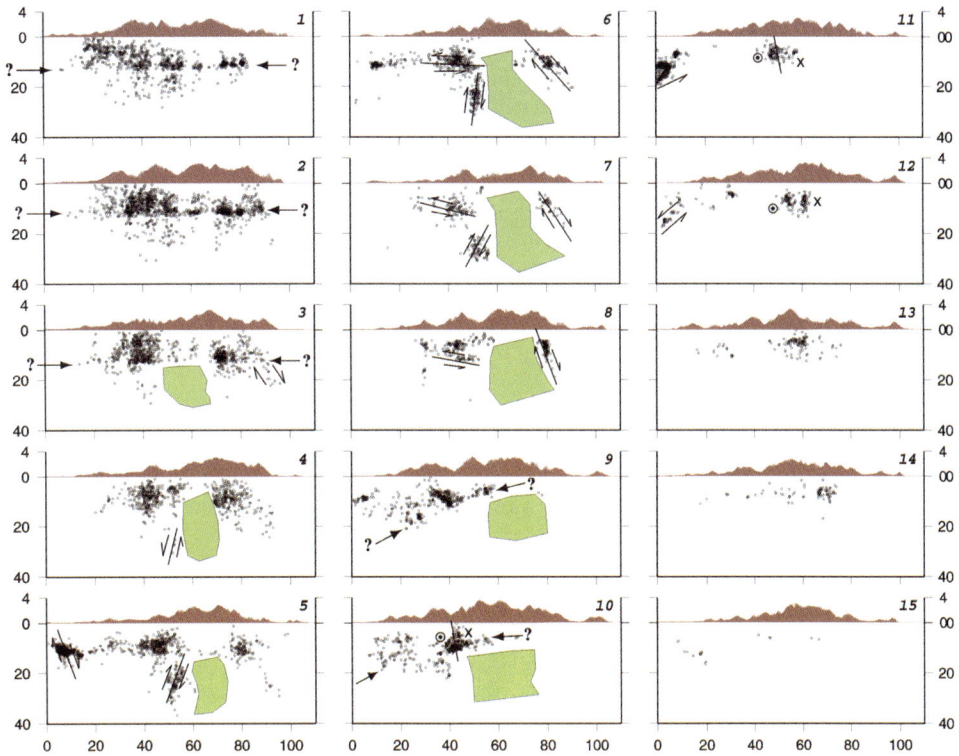

Fig. 13. Cross-sections of post-Chi-Chi seismicity from 20 September to 31 December; the data are the same as for Figure 11, except that an interpretation has been added. The arrow and question marks note the position of the discontinuity in the density of foci – much denser above the discontinuity than below it. Faulted areas are drawn with the sense of motion indicated. For a strike-slip fault: × indicates motion away from the reader and ⊙ motion toward the reader. The hatching indicates areas of seismic quiescence.

different, and they are not conjugate faults in the traditional sense. Conjugate or not, it is agreed that along this fault the Central Range is on the footwall side. It therefore does not help in the building of the Central Range. On the other hand, this fault does contribute to the creation of the root under the Central Range. It is known that Taiwan already has a substantial root (45–50 km thick) under the high ranges (Fig. 3), and motion along this fault could be a mechanism to bring the mid-crustal rock down to deeper level, as shown in Figure 13. Is the whole Range built in this way? So far, this presumed fault zone only extends for about 30 km in a north–south direction. However, these deep crustal earthquakes, occurring under the high temperatures that normally would inhibit seismicity, clearly show that significant deformation in that part of the crust in Taiwan is taking place. As far as the orogeny is concerned, the deepened root would lead to isostatic rebound, although, as has been argued elsewhere (Wu *et al.* 1997), pure shear deformation in the upper and the lower crust may ultimately be responsible for the morphogenesis; the multiple thrust faults in the upper crust under the eastern Foothills (Fig. 8b) may contribute to the rising of the Central Range.

The seismicity in the shallow part (<10 km) of the Central Range presents a different situation. Before the main-shock, the seismicity in the southern part of our study area was apparently greater than further north, and it was nearly continuous across the southern Central Range. If thermal conditions control the seismicity further north, does a change in thermal conditions bring about the seismicity in this part of the Range? Two factors could be taken into consideration in this regard. First, this part of the Central Range is generally younger than further north-based on the geometry of the collision (Suppe 1981) or on the interpretation of palaeomagnetic data (Lee 1989). Therefore one may conclude that the deformation of this part of the Central Range may not have progressed as far as in the north, and therefore that the temperature is relatively low at the top level. Secondly, the generally con-cave-upward shape of the zone of concentrated seismicity in this part of the Range, as shown in Figure 6, profiles 14 and 15, with the deeper foci under the Coastal Plain than under the high Central Range, is also consistent with the uplift of the Central Range and the elevated geotherm there, even though the top part is generally at a low enough temperature for the rocks to remain brittle. The variation in seismicity along the trend of the Central Range can thus be hypoth-esized as the result of a southward-propagating orogeny. A related question is: what does the Southern Central Range seismicity represent? Is the zone of concentrated seismicity related to the presence of a fault, following the interpretation of Carena *et al.* (2002)? But then the west-dipping plane would not fit Carena *et al.*'s prediction.

Among the differences between pre-Chi-Chi and post-Chi-Chi foci distributions is the mid-crustal (10–20 km) seismicity in some areas. The pre-Chi-Chi focal gap between the two zones of foci (Fig. 7, profiles 2, 3 and 4 in a and b) was filled with foci in the post-Chi-Chi period. There is a rather clear discontinuity at about 13 km (Fig. 11 and 13, profiles 1–3) in northern Taiwan; above this level the event density is evidently higher, but in the gap defined by pre-Chi-Chi seismicity there are a significant number of events after the main-shock. Then the two-layered seismicity became recognizable again in 2001 and 2002. Since the filling and emptying of the gap occurred over a period of more than a year, a cause related to the possible changes during this period must be sought. The most obvious process is that of a change in the stresses in the region as a function of time after the main-shock. Could the filling of this gap be an effect of the strain rate? In other words, just after the main-shock a sudden readjustment of stresses would have occurred, and during this period the rate of loading would be higher than the rate of tec-tonic loading in the pre-main-shock time; and thus the normally ductile materials became brittle (the 'silly putty' effect). Perhaps after 2001 the stresses had relaxed sufficiently and the materials in the gap became ductile. In contrast, the seismi-cally quiescent Central Range did not seem to be affected by the assumed strain rate change; the zone remained essentially aseismic throughout the post-Chi-Chi period (Figs 11 and 13, profiles 4–10). With regard to the vertical discontinuity in seismicity itself, profiles 1–3 in Figure 11 show that it actually continues eastward to the Eastern Central Range, and is rather flat. It crosses regions with different geological characteristics at the surface, and most probably at depth, as indi-cated by the velocity changes across the Central

Range in the tomographic velocity images (Fig. 3). The simplest explanation of this layering is that it is related to the level of groundwater cir-culation – controlled by gravity and porosity. It is conceivable that electrical conductivity from mag-netotelluric studies may help to constrain such an interpretation, but at present it is difficult to be cer-tain. The other possible explanation was that of Carena *et al.* (2002), i.e. it indicates the presence of a décollement, albeit one with a much limited spatial extent (only 30 km or so), in the N–S direction and one that is very flat all across the western Coastal Plain and the Eastern Central Range. This discontinuity in seismicity deserves further investigation.

Seismicity and its relation to brittle faulting in the orogen are obvious, but the easily recognized quiescent regions next to areas of high seismicity under parts of the Central Range east of the Lishan Fault require careful examination. Before the Chi-Chi main-shock, the presence of seismicity in the southern part of the Central Range in our study area is quite well defined (Figs 5a and 6, pro-files 12–15); however, in the north (Figs 5a and 6, profiles 1 and 2) relatively few *M* > 2 earthquakes occurred under the Central Range, and they con-centrated mainly above 10 km or so. In between these areas (Fig. 6, profiles 3–11) the Central Range is nearly aseismic, while the seismicity in the Foothills and the Coastal Range was compara-tively high. This seismic quiescence is rather cur-ious in view of the relatively high rate of uplift in the high Central Range. Wu *et al.* (1997) hypoth-esized that rocks under the high ranges are at a higher temperature, as a result of lower, hotter crustal materials having been elevated during the orogeny and thus leading to ductility of the rocks. Wu *et al.* also invoke the thermally induced ductility in quartzo-feldsparthic rocks (Kohlstedt *et al.* 1995) to explain the double-layered seis-micity under parts of the Foothills; the double seismic layers can be seen in profiles 2, 3, 7 and 8 in Fig. 6. The presence of shallow seismicity in parts of the Central Range can be explained as a result of cooling of rocks, or perhaps due to the higher fluid pressure.

Tectonically, the double seismic zone whose details are shown in Figures 7a and 7b could be quite significant. Named the Sanyi–Puli Zone, it sits at the southwest edge of the Hsuehshan Range. For events in this zone, the focal mechan-isms, as determined by Wu and Rau (1998), show both shallow and deep thrusts, striking generally in the NNE direction, but strike-slip and normal events are also present. So far, the events in this zone tend to be in the M2–3 range. If the deeper northwest-dipping zones seen in some profiles in Figure 7 do represent fault planes, and thrust motion occurs on these faults then the Hsuehshan Range could ramp up on the lower thrust – similar to a mechanism proposed by Clark *et al.* (1993).

In terms of seismic zones activated after the Chi-Chi earthquake, the zone in the East Central Range is enigmatic. The zone is spatially distinct from the Chi-Chi zone, with the Central Range intervening, and most of mechanisms obtained for events in this zone are normal faults, with E–W tensile axes (Figs 8 & 13). Incidentally, these events are located in the same area where Crespi *et al.* (1996) found a number of normal faults in the outcrops of metamorphic rocks. Thus, this type of normal faulting had occurred during the lifetime of the Taiwan Orogeny – probably repeatedly. Normal faulting mechanisms for $M < 3.5$ events in recent times were not unusual (Rau *et al.* 1996), although there were not enough to define a major structure. Does large-scale normal faulting mainly occur after a large earth-

quake on the west side of the island? From the pattern of rapid after-slip following the main-shock (Hsü *et al.* 2002), the Backbone Range was moving faster westward than the Coastal Range (*c.*8 cm v. *c.*1–2 cm from September 1999 to December 1999), creating a post-earth-quake stress field conducive to such faulting. The E–W tension also raises questions regarding how the compressive stresses in the Chi-Chi area are generated from the convergence of the Philippine Sea and Eurasian plates.

Yet another enigmatic zone is the cluster of events just west of the Chelungpu fault (Figs. 5b, 6b and 11, profile 5). In the profile, a shallowly west-dipping zone can be seen. However, the mechanisms as shown in Fig. 8b, c and d are all normal faulting mechanisms with E–W tensile axis and 45° planes. In the seismicity profiles there are steeply dipping features extending below the concentrated seismic zone. These are perhaps related to the 45°-dip normal faulting activities. The cur-

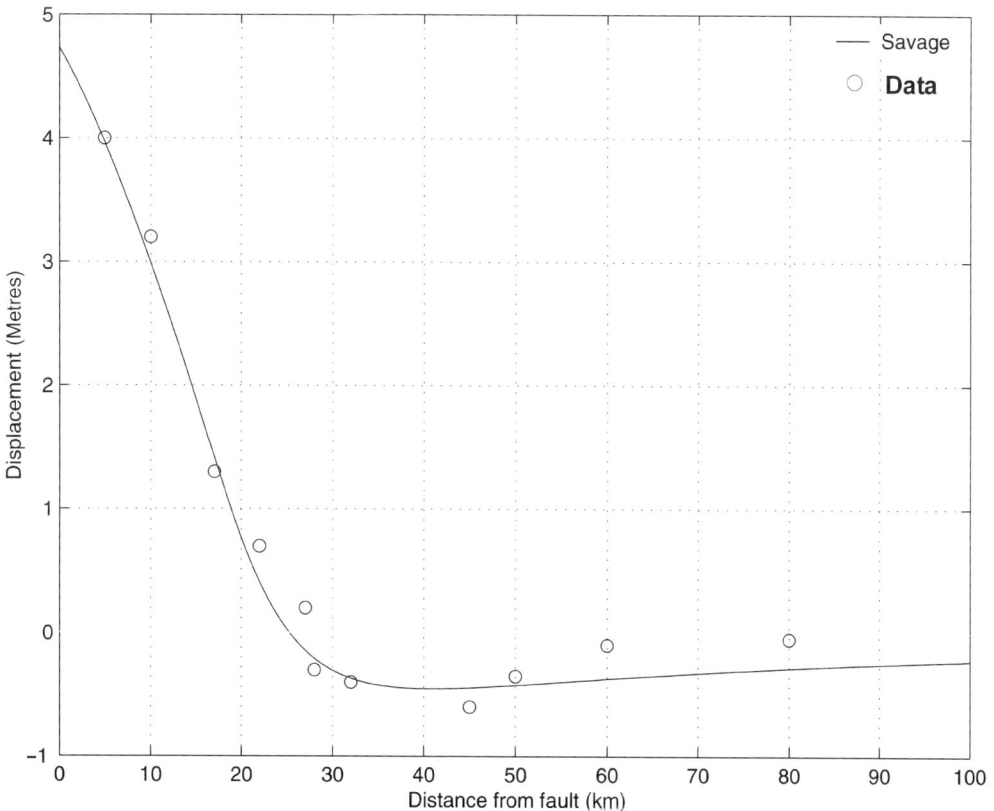

Fig. 14. Circles indicate vertical displacements of the hanging-wall side of the Chelungpu fault, along an E–W-profile at a latitude of 24.1°N with 0 on the fault. The data are sampled from Yang *et al.* (2000), and the line shows the theoretical displacement from a dislocation with a width of 20 km, dipping eastward at an angle of 25°, and the amplitude of displacement is 13 m.

ious juxtaposition of tensile and compressional faults calls for a mechanism that can switch the stresses within a fairly short distance. One consequence of the westward thrust along the Chelungpu fault by over 10 metres is that a wedge of mass was shifted in that direction. The addition of mass on the western crust may lead to flexure and therefore east–west tensile conditions in the upper crust.

Finally, judging from the deformation field accompanying the Chi-Chi faulting, one could say that the Chi-Chi earthquake itself did not contribute directly to the building of the Central Range. Yang et al. (2000) show that a part of the Backbone Range actually subsided about 1 m after the earthquake. A small part of this subsidence can be ascribed to the typical displacement field around a thrust fault; Figure 14 shows a fitting of the vertical displacement across an east–west section at 24.1°N across the Chelungpu Fault, showing the amount of subsidence expected from elastic rebound; the dislocation model (Savage 1983) used for the theoretical curve consists of a thrust fault 20 km wide, dipping at 25° and with a thrust displacement of 13 metres. While the theoretical model is a two-dimensional approximation of the fault, the parameters obtained are close to those obtained from dynamic modelling of the Chi-Chi fault (Ji et al. 2001; Ma et al. 2001). It illustrates the fact that the Chi-Chi earthquake itself led to subsidence beyond the buried tip of the Chelungpu Fault. The fault that the Central Range keeps on rising indicates the complexity of deformation in the orogen. The long-term uplift may very well be related to the ductile response of the crust.

Conclusion

The 1999 Chi-Chi earthquake in Taiwan activated a series of seismic zones, not only in the fringe area of the main rupture but also across the Central Range, to the south of the main rupture and to the lower crust east of the main rupture. Through the correlation of focal mechanisms of large and moderate earthquakes in these zones with their own detailed aftershock seismicity, the corresponding fault zones can be mapped. These faults and their senses of motion help us understand some of the deformation in the Taiwan orogen. A possible model for the creation of the root under the Central Range is thus proposed. The four-dimensional (space plus time) changes in the patterns of seismicity demonstrate the complex deformation in the orogen. Combining observations of presence and absence of seismicity and GPS (Yu and Chen 1994), it is clear that brittle deformation

under the Foothills and ductile deformation under the Central Range are both important in the creation of the mountains of Taiwan (Hsu et al. 2003).

References

CARENA, S., SUPPE, J. & KAO, H. 2002. Active detachment of Taiwan illuminated by small earthquakes and its control of first order topography. Geology, **30**, 935–938.

CENTRAL GEOLOGICAL SURVEY OF ROC. 2000. Geological Map of Taiwan, 1:500,000, Central Geological Survey, Taipei, Taiwan.

CHEN, K. C., HUANG, B. S. & WANG, J. H. 2002. Conjugate thrust faulting associated with the 1999 Chi-Chi, Taiwan earthquake sequence, Geophysical Research Letters, **29**, (8), doi: 10.029/2001GL014250.

CLARK, M. B., FISHER, D. M., LU, C. Y. & CHEN, C. H. 1993. Kinematic analyses of the Hsuehshan Range, Taiwan; a large-scale pop-up structure. Tectonics, **12**, 205–218.

CRESPI, J. M., CHAN, Y. C. & SWAIM, M. S. 1996. Synorogenic extension and exhumation of the Taiwan hinterland. Geology, **24**(3), 247–250.

DE METS, C., GORDON, R. G., ARGUS, D. F., & STEIN, S. 1990. Current plate motions. Geophysical Journal International, **101**, 425–478.

HO, C. S. 1988. An Introduction to the Geology of Taiwan: Explanatory Text of the Geologic Map of Taiwan. Central Geological Survey, the Ministry of Economic Affairs, Taipei, Taiwan.

HSU, Y. J., BECHOR, N., SEGALL, P., YU, S. B. & KUO, L. C. 2002. Rapid afterslip following the 1999 Chi-Chi Taiwan earthquake. Geophysical Research Letters, **29**, (16), 1754, doi: 10.1029/2002GL014967.

HSU, Y. J., SIMONS, M., YU, S. B., KUO, L. C. & CHEN, H. Y. 2003. A two-dimensional dislocation model for interseismic deformation of the Taiwan mountain belt. Earth and Planetary Science Letters, **211**, 287–294.

JI, C., HELMBERGER, D. V., SONG, T. R. A., MA, K. F. & WALD, D. J. 2001. Slip distribution and tectonic implication of the 1999 Chi-Chi, Taiwan, earthquake. Geophysical Research Letters, **28**, 4379–4382.

KAO, H. & ANGELIER, J. 2001. Stress tensor inversion for the Chi-Chi earthquake sequence and its implications on regional collision. Bulletin, Seismological Society of America, **91**, 1028–1040.

KAO, H. & CHEN, W. P. 2000. The Chi-Chi earthquake sequence; active, out-of-sequence thrust faulting in Taiwan. Science, **288**, 2346–2349.

KAO, H. & WU, F. T. 1996. The 16 September 1994 earthquake (mb = 6.5) in the Taiwan Strait and its tectonic implications. Terrestrial Atmosphere Ocean, **7**, 13–29.

KAO, H., LIU, Y. H., LIANG W. T. & CHEN, W. P. 2002. Source parameters of regional earthquakes in Taiwan: 1999–2000 including the Chi-Chi earthquake sequence. *Terrestrial Atmosphere Ocean*, **13**, 279–298.

KOHLSTEDT, D. L., EVANS, B. & MACKWELL, S. J. 1995. Strength of the lithosphere: constraints imposed by laboratory experiments. *Journal of Geophysical Research*, **100**, 17 585–17 602.

LALLEMAND, S., FONT, Y., BIJWAARD, H. & KAO, H. 2001. New insights on 3-D plates interaction near Taiwan from tomography and tectonic implication, *Tectonophysics*, **335**, 229–253.

LEE, J. C., ANGELIER, J. & CHU, H. T. 1997. Polyphase history and kinematics of a complex major fault zone in the northern Taiwan mountain belt: the Lishan Fault. *Tectonophysics*, **274**, 97–115.

LEE, T. Q. 1989. *Evolution Tectonique et Geodynamique Neogene et Quaternaire de la Chaine Cotiere de Taiwan: Apport du Paleomagnetisme.* Thèse de Doctoral de l'Université Pierre et Marie Curie (Paris VI), Paris, France, 328 pp (in French).

LIU, T. K. 1982. Tectonic implication of fission track ages from the Central Range, Taiwan, *Proceedings Geological Society of China*, **25**, 22–37.

LIU, T. K., HSIEH, S., CHEN, Y. G. & CHEN, W. S. 2001. Thermo-kinematic evolution of the Taiwan oblique-collision mountain belt as revealed by zircon fission track dating. *Earth and Planetary Science Letters*, **186**, 45–56.

MA, K. F., MORI, J., LEE, S. J. & YU, S. B. 2001. Spatial and temporal distribution of slip for the 1999 Chi-Chi, Taiwan, earthquake, *Bulletin, Seismological Society of America*, **91**, 1069–1087.

MA, K. F., WANG, J. H. & ZHAO, D. 1996. Three-dimensional seismic velocity structure of the crust and uppermost mantle beneath Taiwan. *Journal of Physics of the Earth*, **44**, 85–105.

NORABUENA, E., LEFFLER-GRIFFIN *et al.* 1998. Space geodetic observations of Nazca-South America convergence along the Central Andes. *Science*, **279**, 358–362.

PAIGE, C. C. & Saunders, M. A. 1982. LSQR: sparse linear equations and least-squares problems. *ACM Transactions on Mathematical and Software*, **8/2**, 195–209.

PRESS, W. H., FLANNERY, B. P., TEUKOLSKI, S. A. & VETTERLING, W. T. 1986. *Numerical Recipes*, Cambridge Univ. Press, 818 pp.

RAU, R. J. & WU, F. T. 1995, Tomographic imaging of lithospheric structures under Taiwan. *Earth and Planetary Science Letters*, **133**, 517–532.

RAU, R. J., WU, F. T. & Shin,T. C. 1996. Regional network focal mechanism determination using 3D velocity model and SH/P amplitude ratio. *Bulletin Seismological Society of America*, **86**, 1270–1283.

RAU, R. J., LIANG, W. T., KAO, H. & HUANG, B. S. 2000. Shear wave anisotropy beneath the Taiwan orogen. *Earth and Planetary Science Letters*, **177**, 177–192.

SAVAGE, J. C. 1983. A dislocation model of strain accumulation and release at a subduction zone. *Journal of Geophysical Research*, **88**, 4984–4996.

SUPPE, J. 1981. Mechanics of mountain building and metamorphism in Taiwan. *Memoir of the Geological Society of China*, **4**, 67–89.

TENG, L. S. 1987. Tectostratigraphic facies and geologic evolution of the Coastal Range, eastern Taiwan. *Memoir of the Geological Society of China*, **8**, 229–250.

TSAI, Y. B., TENG, T., CHIU, J. M. & LIU, H. L. 1977. Tectonic implications of the seismicity in the Taiwan region. *Memoir of the Geological Society of China*, **2**, 13–41.

WALDHAUSER, F. 2001. HypoDD: a computer Program to compute double-difference hypocenter locations, US Geological Survey Open-File Report **01–113**, Menlo Park, California.

WALDHAUSER, F. & ELLSWORTH, W. L. 2000. A double-difference earthquake location algorithm: method and application to the northern Hayward fault. *Bulletin Seismological Society of America*, **90**, 1353–1368.

WALDHAUSER, F. & ELLSWORTH, W. L. 2002. Fault structure and mechanics of the Hayward Fault, California, from double difference earthquake locations. *Journal of Geophysical Research*, **107**, ESE 3-1–3-15.

WALDHAUSER, F., ELLSWORTH, W. L. & COLE, A. 1999. Slip-parallel seismic lineations on the Northern Hayward Fault, California. *Geophysical Research Letters*, **26**, 3525–3528.

WANG, J. H., TSAI, Y. B. & CHEN, K. C. 1983. Some aspects of seismicity in Taiwan region. *Bulletin Institute of Earth Sciences, Academia Sinica*, **3**, 87–104.

WU, F. T. 1978. Recent tectonics of Taiwan. *Journal of Physics of the Earth*, **26**, S265–S299.

WU, F. T. & RAU, R. J. 1998. Seismotectonics and identification of potential seismic source zones in Taiwan. *Terrestrial, Atmospheric and Oceanic Sciences*, **9**, 739–754.

WU, F. T., CHEN, K. C., WANG, J. H., McCAFFERY, R. & SALZBERG, D. 1989. Focal mechanisms of recent large earthquakes and the nature of faulting in the longitudinal valley of eastern Taiwan. *Proceedings, Geological Society of China*, **32**, 157–177.

WU, F. T., RAU, R. J. & SALZBERG, D. 1997. Taiwan Orogeny; thin-skinned or lithospheric collision? An introduction to active collision in Taiwan. *Tectonophysics*, **274**, 191–220.

YANG, M., RAU, R. J., YU, J. Y. & YU, T. T. 2000. Geodetically observed surface displacements of the 1999 Chi-Chi, Taiwan, earthquake. *Earth Planets Space*, **52**, 403–413.

YEH, Y. T., LIU, C. C. & WANG, J. H. 2000. Seismic networks in Taiwan 1989. *Proceedings of the National Science Council, Republic of China, Part A: Physical Science and Engineering*, **13**, 23–31.

YU, S. B. & CHEN, H. Y. 1994. Global Positioning System measurements of crustal deformation in the Taiwan arc–continent collision zone. *Terrestrial, Atmospheric and Oceanic Sciences*, **5**, 477–498.

YU, S. B., KUO, L. C., PUNONGBAYAN, R. S. & RAMOS, E. G. 1999. GPS observation of crustal deformation in the Taiwan–Luzon region. *Geophysical Research Letters*, **26**, 923–926.

Index

Page numbers in *italics* refer to Figures and page numbers in **bold** refer to Tables.